GAMES OF STRATEGY

THIRD EDITION

GAMES OF STRATEGY

THIRD EDITION

Avinash Dixit
Princeton University

Susan Skeath
Wellesley College

David Reiley
University of Arizona and Yahoo! Research

W. W. Norton & Company

New York • London

W. W. Norton & Company has been independent since its founding in 1923, when William Warder Norton and Mary D. Herter Norton first published lectures delivered at the People's Institute, the adult education division of New York City's Cooper Union. The firm soon expanded its program beyond the Institute, publishing books by celebrated academics from America and abroad. By mid-century, the two major pillars of Norton's publishing program—trade books and college texts—were firmly established. In the 1950s, the Norton family transferred control of the company to its employees, and today—with a staff of four hundred and a comparable number of trade, college, and professional titles published each year—W. W. Norton & Company stands as the largest and oldest publishing house owned wholly by its employees.

PRINTED IN THE UNITED STATES OF AMERICA.

Editor: Jack Repcheck
Managing Editor, College: Marian Johnson
Copy Editor: Barbara Curialle
Production Manager: Christine D'Antonio
Project Editor: Melissa Atkin
Book Design: Jack Meserole
Editorial Assistant: Jason Spears

Composition by A-R Editions
Manufacturing by Quebecor World—Taunton Division

Library of Congress Cataloging-in-Publication Data
Dixit, Avinash K.
Games of strategy / Avinash Dixit, Susan Skeath, David Reiley.—3rd ed.
p. cm.
Includes bibliographical references and index.

ISBN 978-0-393-93112-9 (hardcover)
ISBN 978-0-393-11751-6 (softcover)

1. Game theory. 2. Policy sciences. 3. Decision making. I. Skeath, Susan.
II. Reiley, David. III. Title.
HB144.D59 2009
519.3—dc22 2009004484

W. W. Norton & Company, Inc., 500 Fifth Avenue, New York, N.Y. 10110
www.wwnorton.com

W. W. Norton & Company Ltd., Castle House, 75/76 Wells Street, London W1T 3QT

2 3 4 5 6 7 8 9 0 Hardcover
1 2 3 4 5 6 7 8 9 0 Softcover

■

To the memory of my father,

Kamalakar Ramachandra Dixit

— Avinash Dixit

To the memory of my father,

James Edward Skeath

— Susan Skeath

To my mother,

Ronie Reiley

— David Reiley

Contents

2 How to Think About Strategic Games 17

PART TWO
Concepts and Techniques

Games with Sequential Moves 47

PART FOUR
Applications to Specific Strategic Situations

Preface to the Third Edition

We wrote this textbook to make possible the teaching of game theory to first- or second-year college students at an introductory or "principles" level, without requiring any prior knowledge of the field where game theory is used—economics, political science, evolutionary biology, etc.—and requiring only minimal high-school mathematics. Our aim has succeeded beyond our expectations. Many such courses now exist where none did ten years ago; indeed, some of these courses have been inspired by our textbook. Feedback from teachers and students in these courses as well as our own experiences of using this textbook led us to make many improvements of substance and exposition in the second edition. The process has continued; and new suggestions and ideas have accumulated to the point that a third edition is needed.

The most frequent comment on the earlier editions was that there should be more exercises and some solutions available to students. Improved exercises were made possible by the addition of David Reiley to our team. Ably helped by his student, Mike Urbancic, he has revised the exercises in all chapters. These revisions include a near doubling of the number of exercises in the early chapters where students solve many games to master the chapters' concepts. Most of the later chapters also now include at least a few additional problems, including some discussion-type exercises in the less analytical chapters. The exercises in each chapter are split in two sets—solved and unsolved. In most cases, these sets run in parallel; for each solved exercise, there is a corresponding unsolved one that presents variation and gives students further practice. The solutions to the solved set are available on a free and open-access Web site: wwnorton.com/books/games_of_strategy. The solutions to the unsolved set will be reserved for instructors who have adopted the textbook. Instructors should contact the publisher about getting access to the instructors' Web site.

The next major change is in Chapters 7 and 8 on mixed strategies. Originally, we had divided the treatment into zero-sum and nonzero-sum games. But this is an artificial distinction. Most users preferred a distinction between simple and complex topics. Their suggested separation was as follows: the simple topics would include the solution and interpretation of mixed-strategy equilibria in two-by-two games; the main complex topic would be the general theory of mixing in games with more than two pure strategies, when some of them may go unused in equilibrium. We found this division more usable and have therefore changed the presentation. We now treat two-by-two games along with empirical and experimental evidence on mixing in Chapter 7 and the more general and more complex theory in Chapter 8. The latter can be omitted without loss of continuity, but it forms an important part of the subject and is useful to teachers of somewhat more mathematical courses.

The third substantial innovation concerns the treatment of information in games. This topic continues to grow in importance within game theory and beyond, in part because of its applicability to many real-world situations. An especially important subtopic is mechanism design—how to find optimal incentives for workers and managers to exert effort, for insurance policy holders to take care and mitigate their risks, for suppliers not to pad costs, and so on. Indeed, this was the research for which the 2007 Nobel Prize in Economics was awarded. We exposit it here in a simple but rigorous way probably for the first time. In fact, in recognition of the growing importance of mechanism design theory, we have created a new chapter in which to present it (Chapter 14). The basic chapter on information (Chapter 9) now treats only the topics of external uncertainty and signaling games. We describe strategies for coping with uncertainty in the chapter itself rather than in an appendix and include a newer and simpler example of the various types of equilibria that can arise in signaling games. In Chapter 11, we have included a new section showing how asymmetric information can help sustain cooperation in a finitely repeated prisoner's dilemma.

Updates were also made to several other chapters, including those on the prisoners' dilemma (Chapter 11), collective action (Chapter 12), and auctions (Chapter 17). In addition to the new section on asymmetric information and the prisoners' dilemma, we include a new and timely example explaining the Kyoto Protocol as an example of a prisoners' dilemma. We have substantially revised the chapter on collective action to enhance readability and to improve connections between our examples and the general theory of collective action games. The chapter on auctions needed updating to recognize the fast-changing reality of this subject; we now include references to the relationship between auction theory and mechanism design as well as new examples of online auctions. Many other chapters have also been revised to improve the exposition and to update the information presented.

We thank numerous readers of previous edition who provided comments and suggestions. Feedback from Vincent Crawford (University of California, San Diego), Thomas Prusa (Rutgers University), and Greg Trandel (University of Georgia) was expecially useful in the second edition. Their influence endures in the third. For this edition, we thank Andrew Tokarczyk for ideas on new examples. We have also had the added benefit of extensive comments from Katherine Bawn (University of California, Los Angeles), Randall Calvert (Washington University, St. Louis), Lisa Carlson (University of Idaho), Mary Gade (Oklahoma State University, Stillwater), David McAdams (Massachusetts Institute of Technology), John F. Schnell (University of Alabama, Huntsville), Theodore Turocy (Texas A&M University), and Dennis Patterson (Texas Tech University). Thank you all.

Avinash Dixit
Susan Skeath
David Reiley

PART ONE

Introduction and General Principles

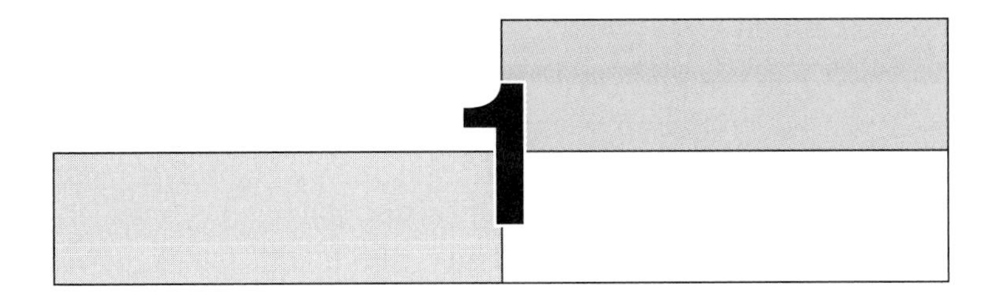

Basic Ideas and Examples

ALL INTRODUCTORY TEXTBOOKS begin by attempting to convince the student readers that the subject is of great importance in the world and therefore merits their attention. The physical sciences and engineering claim to be the basis of modern technology and therefore of modern life; the social sciences discuss big issues of governance—for example, democracy and taxation; the humanities claim that they revive your soul after it has been deadened by exposure to the physical and social sciences and to engineering. Where does the subject games of strategy, often called game theory, fit into this picture, and why should you study it?

We offer a practical motivation much more individual and closer to your personal concerns than most other subjects. You play games of strategy all the time: with your parents, siblings, friends, and enemies, and even with your professors. You have probably acquired a lot of instinctive expertise, and we hope you will recognize in what follows some of the lessons that you have already learned. We will build on this experience, systematize it, and develop it to the point where you will be able to improve your strategic skills and use them more methodically. Opportunities for such uses will appear throughout the rest of your life; you will go on playing such games with your employers, employees, spouses, children, and even strangers.

Not that the subject lacks wider importance. Similar games are played in business, politics, diplomacy, and wars—in fact, whenever people interact to strike mutually agreeable deals or to resolve conflicts. Being able to recognize such games will enrich your understanding of the world around you and will make you a better participant in all its affairs.

It will also have a more immediate payoff in your study of many other subjects. Economics and business courses already use a great deal of game-theoretic thinking. Political science is rapidly catching up. Biology has been importantly influenced by the concepts of evolutionary games and has in turn exported these ideas to economics. Psychology and philosophy also interact with the study of games of strategy. Game theory has become a provider of concepts and techniques of analysis for many disciplines, one might say all disciplines except those dealing with completely inanimate objects.

1 WHAT IS A GAME OF STRATEGY?

The word *game* may convey an impression that the subject is frivolous or unimportant in the larger scheme of things—that it deals with trivial pursuits such as gambling and sports when the world is full of weightier matters such as war and business and your education, career, and relationships. Actually, games of strategy are not "just a game"; all of these weighty matters are instances of games, and game theory helps us understand them all. But it will not hurt to start with gambling or sports.

Most games include chance, skill, and strategy in varying proportions. Playing double or nothing on the toss of a coin is a game of pure chance, unless you have exceptional skill in doctoring or tossing coins. A hundred-yard dash is a game of pure skill, although some chance elements can creep in; for example, a runner may simply have a slightly off day for no clear reason.

Strategy is a skill of a different kind. In the context of sports, it is a part of the mental skill needed to play well; it is the calculation of how best to use your physical skill. For example, in tennis, you develop physical skill by practicing your serves (first serves hard and flat, second serves with spin or kick) and passing shots (hard, low, and accurate). The strategic skill is knowing where to put your serve (wide, or on the T) or passing shot (crosscourt, or down the line). In football, you develop such physical skills as blocking and tackling, running and catching, and throwing. Then the coach, knowing the physical skills of his own team and those of the opposing team, calls the plays that best exploit his team's skills and the other team's weaknesses. The coach's calculation constitutes the strategy. The physical game of football is played on the gridiron by jocks; the strategic game is played in the offices and on the sidelines by coaches and by nerdy assistants.

A hundred-yard dash is a matter of exercising your physical skill as best you can; it offers no opportunities to observe and react to what other runners in the race are doing and therefore no scope for strategy. Longer races do entail strategy—whether you should lead to set the pace, how soon before the finish you should try to break away, and so on.

Strategic thinking is essentially about your interactions with others: someone else is also doing similar thinking at the same time and about the same situation. Your opponents in a marathon may try to frustrate or facilitate your attempts to lead, as they think best suits their interests. Your opponent in tennis tries to guess where you will put your serve or passing shot; the opposing coach in football calls the play that will best counter what he thinks you will call. Of course, just as you must take into account what the other player is thinking, he is taking into account what you are thinking. Game theory is the analysis, or science, if you like, of such interactive decision making.

When you think carefully before you act—when you are aware of your objectives or preferences and of any limitations or constraints on your actions and choose your actions in a calculated way to do the best according to your own criteria—you are said to be behaving rationally. Game theory adds another dimension to rational behavior—namely, interaction with other equally rational decision makers. In other words, game theory is the science of rational behavior in interactive situations.

We do not claim that game theory will teach you the secrets of perfect play or ensure that you will never lose. For one thing, your opponent can read the same book, and both of you cannot win all the time. More importantly, many games are complex and subtle enough, and most actual situations include enough idiosyncratic or chance elements, that game theory cannot hope to offer surefire recipes for action. What it does is to provide some general principles for thinking about strategic interactions. You have to supplement these ideas and some methods of calculation with many details specific to your situation before you can devise a successful strategy for it. Good strategists mix the science of game theory with their own experience; one might say that game playing is as much art as science. We will develop the general ideas of the science but will also point out its limitations and tell you when the art is more important.

You may think that you have already acquired the art from your experience or instinct, but you will find the study of the science useful nonetheless. The science systematizes many general principles that are common to several contexts or applications. Without general principles, you would have to figure out from scratch each new situation that requires strategic thinking. That would be especially difficult to do in new areas of application—for example, if you learned your art by playing games against parents and siblings and must now practice strategy against business competitors. The general principles of game theory provide you with a ready reference point. With this foundation in place, you can proceed much more quickly and confidently to acquire and add the situation-specific features or elements of the art to your thinking and action.

2 SOME EXAMPLES AND STORIES OF STRATEGIC GAMES

With the aims announced in Section 1, we will begin by offering you some simple examples, many of them taken from situations that you have probably encountered in your own lives, where strategy is of the essence. In each case we will point out the crucial strategic principle. Each of these principles will be discussed more fully in a later chapter, and after each example we will tell you where the details can be found. But don't jump to them right away; for a while, just read all the examples to get a preliminary idea of the whole scope of strategy and of strategic games.

A. Which Passing Shot?

Tennis at its best consists of memorable duels between top players: John McEnroe versus Ivan Lendl, Pete Sampras versus Andre Agassi, and Martina Navratilova versus Chris Evert. Picture the 1983 U.S. Open final between Evert and Navratilova.[1] Navratilova at the net has just volleyed to Evert on the baseline. Evert is about to hit a passing shot. Should she go down the line or crosscourt? And should Navratilova expect a down-the-line shot and lean slightly that way or expect a crosscourt shot and lean the other way?

Conventional wisdom favors the down-the-line shot. The ball has a shorter distance to travel to the net, so the other player has less time to react. But this does not mean that Evert should use that shot all of the time. If she did, Navratilova would confidently come to expect it and prepare for it, and the shot would not be so successful. To improve the success of the down-the-line passing shot, Evert has to use the crosscourt shot often enough to keep Navratilova guessing on any single instance.

Similarly in football, with a yard to go on third down, a run up the middle is the percentage play—that is, the one used most often—but the offense must throw a pass occasionally in such situations "to keep the defense honest."

Thus the most important general principle of such situations is not what Evert *should* do but what she *should not* do: she should not do the same thing all the time or systematically. If she did, then Navratilova would learn to cover that, and Evert's chances of success would fall.

Not doing any one thing systematically means more than not playing the same shot in every situation of this kind. Evert should not even mechanically switch back and forth between the two shots. Navratilova would spot and exploit

[1]Chris Evert won her first title at the U.S. Open in 1975. Navratilova claimed her first title in the 1983 final.

this *pattern* or indeed any other detectable system. Evert must make the choice on each particular occasion *at random* to prevent this guessing.

This general idea of "mixing one's plays" is well known, even to sports commentators on television. But there is more to the idea, and these further aspects require analysis in greater depth. Why is down-the-line the percentage shot? Should one play it 80% of the time or 90% or 99%? Does it make any difference if the occasion is particularly big; for example, does one throw that pass on third down in the regular season but not in the Super Bowl? In actual practice, just how does one mix one's plays? What happens when a third possibility (the lob) is introduced? We will examine and answer such questions in Chapters 7 and 8.

The movie *The Princess Bride* (1987) illustrates the same idea in the "battle of wits" between the hero (Westley) and a villain (Vizzini). Westley is to poison one of two wineglasses out of Vizzini's sight, and Vizzini is to decide who will drink from which glass. Vizzini goes through a number of convoluted arguments as to why Westley should poison one glass. But all of the arguments are innately contradictory, because Westley can anticipate Vizzini's logic and choose to put the poison in the other glass. Conversely, if Westley uses any specific logic or system to choose one glass, Vizzini can anticipate that and drink from the other glass, leaving Westley to drink from the poisoned one. Thus, Westley's strategy has to be random or unsystematic.

The scene illustrates something else as well. In the film, Vizzini loses the game and with it his life. But it turns out that Westley had poisoned both glasses; over the last several years, he had built up immunity to the poison. So Vizzini was actually playing the game under a fatal information disadvantage. Players can sometimes cope with such asymmetries of information; Chapter 9 examines when and how they can do so.

B. The GPA Rat Race

You are enrolled in a course that is graded on a curve. No matter how well you do in absolute terms, only 40% of the students will get As, and only 40% will get Bs. Therefore you must work hard, not just in absolute terms, but relative to how hard your classmates (actually, "class enemies" seems a more fitting term in this context) work. All of you recognize this, and after the first lecture you hold an impromptu meeting in which all students agree not to work too hard. As weeks pass by, the temptation to get an edge on the rest of the class by working just that little bit harder becomes overwhelming. After all, the others are not able to observe your work in any detail; nor do they have any real hold over you. And the benefits of an improvement in your grade point average are substantial. So you hit the library more often and stay up a little longer.

The trouble is, everyone else is doing the same. Therefore your grade is no better than it would have been if you and everyone else had abided by the agreement. The only difference is that all of you have spent more time working than you would have liked.

This is an example of the prisoners' dilemma.[2] In the original story, two suspects are being separately interrogated and invited to confess. One of them, say A, is told, "If the other suspect, B, does not confess, then you can cut a very good deal for yourself by confessing. But if B does confess, then you would do well to confess, too; otherwise the court will be especially tough on you. So you should confess no matter what the other does." B is told to confess, with the use of similar reasoning. Faced with this choice, both A and B confess. But it would have been better for both if neither had confessed, because the police had no really compelling evidence against them.

Your situation is similar. If the others slack off, then you can get a much better grade by working hard; if the others work hard, then you had better do the same or else you will get a very bad grade. You may even think that the label "prisoner" is very fitting for a group of students trapped in a required course.

There is a prisoners' dilemma for professors and schools, too. Each professor can make his course look good or attractive by grading it slightly more liberally, and each school can place its students in better jobs or attract better applicants by grading all of its courses a little more liberally. Of course, when all do this, none has any advantage over the others; the only result is rampant grade inflation, which compresses the spectrum of grades and therefore makes it difficult to distinguish abilities.

People often think that in every game there must be a winner and a loser. The prisoners' dilemma is different—both or all players can come out losers. People play (and lose) such games every day, and the losses can range from minor inconveniences to potential disasters. Spectators at a sports event stand up to get a better view but, when all stand, no one has a better view than when they were all sitting. Superpowers acquire more weapons to get an edge over their rivals but, when both do so, the balance of power is unchanged; all that has happened is that both have spent economic resources that they could have used for better purposes, and the risk of accidental war has escalated. The magnitude of the potential cost of such games to all players makes it important to understand the ways in which mutually beneficial cooperation can be achieved and sustained. All of Chapter 11 deals with the study of this game.

Just as the prisoners' dilemma is potentially a lose-lose game, there are win-win games, too. International trade is an example; when each country produces more of what it can do relatively best, all share in the fruits of this international

[2]There is some disagreement regarding the appropriate grammatical placement of the apostrophe in the term *prisoners' dilemma.* Our placement acknowledges the facts that there must be at least two prisoners in order for there to be any dilemma at all and that the (at least two) prisoners therefore jointly possess the dilemma.

division of labor. But successful bargaining about the division of the pie is needed if the full potential of trade is to be realized. The same applies to many other bargaining situations. We will study these in Chapter 18.

C. "We Can't Take the Exam, Because We Had a Flat Tire"

Here is a story, probably apocryphal, that circulates on the undergraduate e-mail networks; each of us independently received it from our students:

> There were two friends taking chemistry at Duke. Both had done pretty well on all of the quizzes, the labs, and the midterm, so that going into the final they each had a solid A. They were so confident the weekend before the final that they decided to go to a party at the University of Virginia. The party was so good that they overslept all day Sunday, and got back too late to study for the chemistry final that was scheduled for Monday morning. Rather than take the final unprepared, they went to the professor with a sob story. They said they each had gone up to UVA and had planned to come back in good time to study for the final but had had a flat tire on the way back. Because they didn't have a spare, they had spent most of the night looking for help. Now they were really too tired, so could they please have a makeup final the next day? The professor thought it over and agreed.
>
> The two studied all of Monday evening and came well prepared on Tuesday morning. The professor placed them in separate rooms and handed the test to each. The first question on the first page, worth 10 points, was very easy. Each of them wrote a good answer, and greatly relieved, turned the page. It had just one question, worth 90 points. It was: "Which tire?"

The story has two important strategic lessons for future party goers. The first is to recognize that the professor may be an intelligent game player. He may suspect some trickery on the part of the students and may use some device to catch them. Given their excuse, the question was the likeliest such device. They should have foreseen it and prepared their answer in advance. This idea that one should look ahead to future moves in the game and then reason backward to calculate one's best current action is a very general principle of strategy, which we will elaborate on in Chapter 3. We will also use it, most notably, in Chapter 10.

But it may not be possible to foresee all such professorial countertricks; after all, professors have much more experience of seeing through students' excuses than students have of making up such excuses. If the pair are unprepared, can they independently produce a mutually consistent lie? If each picks a tire at random, the chances are only 25% that the two will pick the same one. (Why?) Can they do better?

You may think that the front tire on the passenger side is the one most likely to suffer a flat, because a nail or a shard of glass is more likely to lie closer to the side of the road than the middle and the front tire on that side will encounter it first. You may think this is good logic, but that is not enough to make it a good choice. What matters is not the logic of the choice but making the same choice as your friend does. Therefore you have to think about whether your friend would use the same logic and would consider that choice equally obvious. But even that is not the end of the chain of reasoning. Would your friend think that the choice would be equally obvious to you? And so on. The point is not whether a choice is obvious or logical, but whether it is obvious to the other that it is obvious to you that it is obvious to the other. . . . In other words, what is needed is a convergence of expectations about what should be chosen in such circumstances. Such a commonly expected strategy on which the players can successfully coordinate is called a focal point.

There is nothing general or intrinsic to the structure of all such games that creates such convergence. In some games, a focal point may exist because of chance circumstances about the labeling of strategies or because of some experience or knowledge shared by the players. For example, if the passenger's front side of a car were for some reason called the Duke's side, then two Duke students would be very likely to choose it without any need for explicit prior understanding. Or, if the driver's front side of all cars were painted orange (for safety, to be easily visible to oncoming cars), then two Princeton students would be very likely to choose that tire, because orange is the Princeton color. But without some such clue, tacit coordination might not be possible at all.

We will study focal points in more detail in Chapter 4. Here in closing we merely point out that when asked in classrooms, more than 50% of students choose the driver's front side. They are generally unable to explain why, except to say that it seems the obvious choice.

D. Why Are Professors So Mean?

Many professors have inflexible rules not to give makeup exams and never to accept late submission of problem sets or term papers. Students think the professors must be really hardhearted to behave in this way. The true strategic reason is often exactly the opposite. Most professors are kindhearted and would like to give their students every reasonable break and accept any reasonable excuse. The trouble lies in judging what is reasonable. It is hard to distinguish between similar excuses and almost impossible to verify their truth. The professor knows that on each occasion he will end up by giving the student the benefit of the doubt. But the professor also knows that this is a slippery slope. As the students come to know that the professor is a soft touch, they will procrastinate more and produce ever-flimsier excuses. Deadlines will cease to mean anything, and examinations will become a chaotic mix of postponements and makeup tests.

Often the only way to avoid this slippery slope is to refuse to take even the first step down it. Refusal to accept any excuses at all is the only realistic alternative to accepting them all. By making an advance commitment to the "no excuses" strategy, the professor avoids the temptation to give in to all.

But how can a softhearted professor maintain such a hardhearted commitment? He must find some way to make a refusal firm and credible. The simplest way is to hide behind an administrative procedure or university-wide policy. "I wish I could accept your excuse, but the university won't let me" not only puts the professor in a nicer light, but removes the temptation by genuinely leaving him no choice in the matter. Of course, the rules may be made by the same collectivity of professors as hides behind them but, once they are made, no individual professor can unmake the rules in any particular instance.

If the university does not provide such a general shield, then the professor can try to make up commitment devices of his own. For example, he can make a clear and firm announcement of the policy at the beginning of the course. Any time an individual student asks for an exception, he can invoke a fairness principle, saying, "If I do this for you, I would have to do it for everyone." Or the professor can acquire a reputation for toughness by acting tough a few times. This may be an unpleasant thing for him to do and it may run against his true inclination, but it helps in the long run over his whole career. If a professor is believed to be tough, few students will try excuses on him, so he will actually suffer less pain in denying them.

We will study commitments, and related strategies, such as threats and promises, in considerable detail in Chapter 10.

E. Roommates and Families on the Brink

You are sharing an apartment with one or more other students. You notice that the apartment is nearly out of dishwasher detergent, paper towels, cereal, beer, and other items. You have an agreement to share the actual expenses, but the trip to the store takes time. Do you spend your time or do you hope that someone else will spend his, leaving you more time to study or relax? Do you go and buy the soap or stay in and watch TV to catch up on the soap operas?[3]

In many situations of this kind, the waiting game goes on for quite a while before someone who is really impatient for one of the items (usually beer) gives in and spends the time for the shopping trip. Things may deteriorate to the point of serious quarrels or even breakups among the roommates.

This game of strategy can be viewed from two perspectives. In one, each of the roommates is regarded as having a simple binary choice—to do

[3]This example comes from Michael Grunwald's "At Home" column, "A Game of Chicken," in the *Boston Globe Magazine*, April 28, 1996.

the shopping or not. The best outcome for you is where someone else does the shopping and you stay at home; the worst is where you do the shopping while the others get to use their time better. If both do the shopping (unknown to each other, on the way home from school or work), there is unnecessary duplication and perhaps some waste of perishables; if neither does, there can be serious inconvenience or even disaster if the toilet paper runs out at a crucial time.

This is analogous to the game of chicken that used to be played by American teenagers. Two of them drove their cars toward each other. The first to swerve to avoid a collision was the loser (chicken); the one who kept driving straight was the winner. We will analyze the game of chicken further in Chapter 4 and in Chapters 7 and 13.

A more interesting dynamic perspective on the same situation regards it as a "war of attrition," where each roommate tries to wait out the others, hoping that someone else's patience will run out first. In the meantime, the risk escalates that the apartment will run out of something critical, leading to serious inconvenience or a blowup. Each player lets the risk escalate to the point of his own tolerance; the one revealed to have the least tolerance loses. Each sees how close to the brink of disaster the others will let the situation go. Hence the name "brinkmanship" for this strategy and this game. It is a dynamic version of chicken, offering richer and more interesting possibilities.

One of us (Dixit) was privileged to observe a brilliant example of brinkmanship at a dinner party one Saturday evening. Before dinner, the company was sitting in the living room when the host's fifteen-year-old daughter appeared at the door and said, "Bye, Dad." The father asked, "Where are you going?" and the daughter replied, "Out." After a pause that was only a couple of seconds but seemed much longer, the host said, "All right, bye."

Your strategic observer of this scene was left thinking how it might have gone differently. The host might have asked, "With whom?" and the daughter might have replied, "Friends." The father could have refused permission unless the daughter told him exactly where and with whom she would be. One or the other might have capitulated at some such later stage of this exchange or it could have led to a blowup.

This was a risky game for both the father and the daughter to play. The daughter might have been punished or humiliated in front of strangers; an argument could have ruined the father's evening with his friends. Each had to judge how far to push the process, without being fully sure whether and when the other might give in or whether there would be an unpleasant scene. The risk of an explosion would increase as the father tried harder to force the daughter to answer and as she defied each successive demand.

In this respect the game played by the father and the daughter was just like that between a union and a company's management who are negotiating a labor contract or between two superpowers who are encroaching on each other's sphere of influence in the world. Neither side can be fully sure of the other's intentions, so each side explores them through a succession of small incremental steps, each of which escalates the risk of mutual disaster. The daughter in our story was exploring previously untested limits of her freedom; the father was exploring previously untested—and perhaps unclear even to himself—limits of his authority.

This was an example of brinkmanship, a game of escalating mutual risk, par excellence. Such games can end in one of two ways. In the first way, one of the players reaches the limit of his own tolerance for risk and concedes. (The father in our story conceded quickly, at the very first step. Other fathers might be more successful strict disciplinarians, and their daughters might not even initiate a game like this.) In the second way, before either has conceded, the risk that they both fear comes about, and the blowup (the strike or the war) occurs. The feud in our host's family ended "happily"; although the father conceded and the daughter won, a blowup would have been much worse for both.

We will analyze the strategy of brinkmanship more fully in Chapter 10; in Chapter 15, we will examine a particularly important instance of it—namely, the Cuban missile crisis of 1962.

F. The Dating Game

When you are dating, you want to show off the best attributes of your personality to your date and to conceal the worst ones. Of course, you cannot hope to conceal them forever if the relationship progresses; but you are resolved to improve or hope that by that stage the other person will accept the bad things about you with the good ones. And you know that the relationship will not progress at all unless you make a good first impression; you won't get a second chance to do so.

Of course, you want to find out everything, good and bad, about the other person. But you know that, if the other is as good at the dating game as you are, he or she will similarly try to show the best side and hide the worst. You will think through the situation more carefully and try to figure out which signs of good qualities are real and which ones can easily be put on for the sake of making a good impression. Even the worst slob can easily appear well groomed for a big date; ingrained habits of courtesy and manners that are revealed in a hundred minor details may be harder to simulate for a whole evening. Flowers are relatively cheap; more expensive gifts may have value, not for intrinsic reasons, but as credible evidence of how much the other person is willing to sacrifice for you. And the "currency" in which the gift is given may have different significance, depending on the context; from a millionaire, a diamond may be worth less in this regard than the act of giving up valuable time for your company or time spent on some activity at your request.

You should also recognize that your date will similarly scrutinize your actions for their information content. Therefore you should take actions that are credible signals of your true good qualities, and not just the ones that anyone can imitate.

This is important not just on a first date; revealing, concealing, and eliciting information about the other person's deepest intentions remain important throughout a relationship. Here is a story to illustrate that.

Once upon a time in New York City there lived a man and a woman who had separate rent-controlled apartments, but their relationship had reached the point at which they were using only one of them. The woman suggested to the man that they give up the other apartment. The man, an economist, explained to her a fundamental principle of his subject: it is always better to have more choice available. The probability of their splitting up might be small but, given even a small risk, it would be useful to retain the second low-rent apartment. The woman took this very badly and promptly ended the relationship!

Economists who hear this story say that it just confirms their principle that greater choice is better. But strategic thinking offers a very different and more compelling explanation. The woman was not sure of the man's commitment to the relationship, and her suggestion was a brilliant strategic device to elicit the truth. Words are cheap; anyone can say, "I love you." If the man had put his property where his mouth was and had given up his rent-controlled apartment, that would have been concrete evidence of his love. The fact that he refused to do so constituted hard evidence of the opposite, and the woman did right to end the relationship.

These are examples, designed to appeal to your immediate experience, of a very important class of games—namely, those where the real strategic issue is manipulation of information. Strategies that convey good information about yourself are called signals; strategies that induce others to act in ways that will credibly reveal their private information, good or bad, are called screening devices. Thus the woman's suggestion of giving up one of the apartments was a screening device, which put the man in the situation of offering to give up his apartment or else revealing his lack of commitment. We will study games of information, as well as signaling and screening, in Chapters 9 and 14.

3 OUR STRATEGY FOR STUDYING GAMES OF STRATEGY

We have chosen several examples that relate to your experiences as amateur strategists in real life to illustrate some basic concepts of strategic thinking and strategic games. We could continue, building a whole stock of dozens of similar stories. The hope would be that, when you face an actual strategic situation, you might recognize a parallel with one of these stories, which would help you decide the appropriate strategy for your own situation. This is the *case study* approach taken by most business schools. It offers a concrete and memorable vehicle for the underlying concepts. However, each new strategic situation typically consists of a unique combination of so many variables that an intolerably large stock of cases is needed to cover all of them.

An alternative approach focuses on the general principles behind the examples and so constructs a *theory* of strategic action—namely, formal game theory. The hope here is that, facing an actual strategic situation, you might recognize which principle or principles apply to it. This is the route taken by the more academic disciplines, such as economics and political science. A drawback to this approach is that the theory is presented in a very abstract and mathematical manner, without enough cases or examples. This makes it difficult for most beginners to understand or remember the theory and to connect the theory with reality afterward.

But knowing some general theory has an overwhelming compensating advantage. It gives you a deeper understanding of games and of *why* they have the outcomes they do. This helps you play better than you would if you merely read some cases and knew the recipes for *how* to play some specific games. With the knowledge of why, you can think through new and unexpected situations where a mechanical follower of a "how" recipe would be lost. A world champion of checkers, Tom Wiswell, has expressed this beautifully: "The player who knows how will usually draw, the player who knows why will usually win."[4] This is not to be taken literally for all games; some games may be hopeless situations for one of the players no matter how knowledgable he may be. But the statement contains the germ of an important general truth—knowing why gives you an advantage beyond what you can get if you merely know how. For example, knowing the why of a game can help you foresee a hopeless situation and avoid getting into such a game in the first place.

Therefore we will take an intermediate route that combines some of the advantages of both approaches—case studies (how) and theory (why). We will organize the subject around its general principles, generally one in each of the chapters to follow. Therefore you don't have to figure them out on your own from the cases. But we will develop the general principles through illustrative cases rather than abstractly, so the context and scope of each idea will be clear and evident. In other words, we will focus on theory but build it up through cases, not abstractly.

Of course, such an approach requires some compromises of its own. Most important, you should remember that each of our examples serves the purpose of conveying some general idea or principle of game theory. Therefore we will

[4]Quoted in Victor Niederhoffer, *The Education of a Speculator* (New York: Wiley, 1997), p. 169. We thank Austin Jaffe of Pennsylvania State University for bringing this aphorism to our attention.

leave out many details of each case that are incidental to the principle at stake. If some examples seem somewhat artificial, please bear with us; we have generally considered the omitted details and left them out for good reasons.

A word of reassurance. Although the examples that motivate the development of our conceptual or theoretical frameworks are deliberately selected for that purpose (even at the cost of leaving out some other features of reality), once the theory has been constructed, we pay a lot of attention to its connection with reality. Throughout the book, we examine factual and experimental evidence in regard to how well the theory explains reality. The frequent answer—very well in some respects and less well in others—should give you cautious confidence in using the theory and should be a spur to contributing to the formulation of better theories. In appropriate places, we examine in great detail how institutions evolve in practice to solve some problems pointed out by the theories; note in particular the discussion in Chapter 11 of how prisoners' dilemmas arise and are solved in reality and a similar discussion of more general collective-action problems in Chapter 12. Finally, in Chapter 15 we will examine the use of brinkmanship in the Cuban missile crisis. Such theory-based case studies, which take rich factual details of a situation and subject them to an equally detailed theoretical analysis, are becoming common in such diverse fields as business studies, political science, and economic history; we hope our original study of an important episode in the diplomatic and military areas will give you an interesting introduction to this genre.

To pursue our approach, in which examples lead to general theories that are then tested against reality and used to interpret reality, we must first identify the general principles that serve to organize the discussion. We will do so in Chapter 2 by classifying or dichotomizing games along several key dimensions of different strategic matters or concepts. Along each dimension, we will identify two extreme pure types. For example, one such dimension concerns the order of moves, and the two pure types are those in which the players take turns making moves (sequential games) and those in which all players act at once (simultaneous games). Actual games rarely correspond to exactly one of these conceptual categories; most partake of some features of each extreme type. But each game can be located in our classification by considering which concepts or dimensions bear on it and how it mixes the two pure types in each dimension. To decide how to act in a specific situation, one then combines in appropriate ways the lessons learned for the pure types.

Once this general framework has been constructed in Chapter 2, the chapters that follow will build on it, developing several general ideas and principles for each player's strategic choice and the interaction of all players' strategies in games.

How to Think About Strategic Games

CHAPTER 1 GAVE SOME simple examples of strategic games and strategic thinking. In this chapter, we begin a more systematic and analytical approach to the subject. We choose some crucial conceptual categories or dimensions, in each of which there is a dichotomy of types of strategic interactions. For example, one such dimension concerns the timing of the players' actions, and the two pure types are games where the players act in strict turns (sequential moves) and where they act at the same time (simultaneous moves). We consider some matters that arise in thinking about each pure type in this dichotomy, as well as in similar dichotomies, with respect to other matters, such as whether the game is played only once or repeatedly and what the players know about each other.

In the chapters that follow, we will examine each of these categories or dimensions in more detail and show how the analysis can be used in several specific applications. Of course, most actual applications are not of a pure type but rather a mixture. Moreover, in each application, two or more of the categories have some relevance. The lessons learned from the study of the pure types must therefore be combined in appropriate ways. We will show how to do this by using the context of our applications.

In this chapter, we state some basic concepts and terminology—such as strategies, payoffs, and equilibrium—that are used in the analysis and briefly describe solution methods. We also provide a brief discussion of the uses of game theory and an overview of the structure of the remainder of the book.

1 DECISIONS VERSUS GAMES

When a person (or team or firm or government) decides how to act in dealings with other people (or teams or firms or governments), there must be some cross-effect of their actions; what one does must affect the outcome for the other. When George Pickett (of Pickett's Charge at the battle of Gettysburg) was asked to explain the Confederacy's defeat in the Civil War, he responded, "I think the Yankees had something to do with it."[1]

For the interaction to become a strategic game, however, we need something more—namely, the participants' mutual awareness of this cross-effect. What the other person does affects you; if you know this, you can react to his actions, or take advance actions to forestall the bad effects his future actions may have on you and to facilitate any good effects, or even take advance actions so as to alter his future reactions to your advantage. If you know that the other person knows that what you do affects him, you know that he will be taking similar actions. And so on. It is this mutual awareness of the cross-effects of actions and the actions taken as a result of this awareness that constitute the most interesting aspects of strategy.

This distinction is captured by reserving the label **strategic games** (or sometimes just **games,** because we are not concerned with other types of games, such as those of pure chance or pure skill) for interactions between mutually aware players and **decisions** for action situations where each person can choose without concern for reaction or response from others. If Robert E. Lee (who ordered Pickett to lead the ill-fated Pickett's Charge) had thought that the Yankees had been weakened by his earlier artillery barrage to the point that they no longer had any ability to resist, his choice to attack would have been a decision; if he was aware that the Yankees were prepared and waiting for his attack, then the choice became a part of a (deadly) game. The simple rule is that unless there are two or more players, each of whom responds to what others do (or what each thinks the others might do), it is not a game.

Strategic games arise most prominently in head-to-head confrontations of two participants: the arms race between the United States and the Soviet Union from the 1950s through the 1980s; wage negotiations between General Motors and the United Auto Workers; or a Super Bowl matchup between two "pirates," the Tampa Bay Buccaneers and the Oakland Raiders. In contrast, interactions among a large number of participants seem less susceptible to the issues raised by mutual awareness. Because each farmer's output is an insignificant part of

[1]James M. McPherson, "American Victory, American Defeat," in *Why the Confederacy Lost,* ed. Gabor S. Boritt (New York: Oxford University Press, 1993), p. 19.

the whole nation's or the world's output, the decision of one farmer to grow more or less corn has almost no effect on the market price, and not much appears to hinge on thinking of agriculture as a strategic game. This was indeed the view prevalent in economics for many years. A few confrontations between large companies—as in the U.S. auto market, which was once dominated by GM, Ford, and Chrysler—were usefully thought of as strategic games, but most economic interactions were supposed to be governed by the impersonal forces of supply and demand.

In fact, game theory has a much greater scope. Many situations that start out as impersonal markets with thousands of participants turn into strategic interactions of two or just a few. This happens for one of two broad classes of reasons—mutual commitments or private information.

Consider commitment first. When you are contemplating building a house, you can choose one of several dozen contractors in your area; the contractor can similarly choose from several potential customers. There appears to be an impersonal market. Once each side has made a choice, however, the customer pays an initial installment, and the builder buys some materials for the plan of this particular house. The two become tied to each other, separately from the market. Their relationship becomes *bilateral.* The builder can try to get away with a somewhat sloppy job or can procrastinate, and the client can try to delay payment of the next installment. Strategy enters the picture. Their initial contract has to anticipate their individual incentives and specify a schedule of installments of payments that are tied to successive steps in the completion of the project. Even then, some adjustments have to be made after the fact, and these adjustments bring in new elements of strategy.

Next, consider private information. Thousands of farmers seek to borrow money for their initial expenditures on machinery, seed, fertilizer, and so forth, and hundreds of banks exist to lend to them. Yet the market for such loans is not impersonal. A borrower with good farming skills who puts in a lot of effort will be more likely to be successful and will repay the loan; a less-skilled or lazy borrower may fail at farming and default on the loan. The risk of default is highly personalized. It is not a vague entity called "the market" that defaults, but individual borrowers who do so. Therefore each bank will have to view its lending relation with each individual borrower as a separate game. It will seek collateral from each borrower or will investigate each borrower's creditworthiness. The farmer will look for ways to convince the bank of his quality as a borrower; the bank will look for effective ways to ascertain the truth of the farmer's claims.

Similarly, an insurance company will make some efforts to determine the health of individual applicants and will check for any evidence of arson when a claim for a fire is made; an employer will inquire into the qualifications of individual employees and monitor their performance. More generally, when participants in a transaction possess some private information bearing on the outcome, each bilateral deal becomes a game of strategy, even though the larger picture may have thousands of very similar deals going on.

To sum up, when each participant is significant in the interaction, either because each is a large player to start with or because commitments or private information narrow the scope of the relationship to a point where each is an important player *within* the relationship, we must think of the interaction as a strategic game. Such situations are the rule rather than the exception in business, in politics, and even in social interactions. Therefore the study of strategic games forms an important part of all fields that analyze these matters.

2 CLASSIFYING GAMES

Games of strategy arise in many different contexts and accordingly have many different features that require study. This task can be simplified by grouping these features into a few categories or dimensions, along each of which we can identify two pure types of games and then recognize any actual game as a mixture of the pure types. We develop this classification by asking a few questions that will be pertinent for thinking about the actual game that you are playing or studying.

A. Are the Moves in the Game Sequential or Simultaneous?

Moves in chess are sequential: White moves first, then Black, then White again, and so on. In contrast, participants in an auction for an oil-drilling lease or a part of the airwave spectrum make their bids simultaneously, in ignorance of competitors' bids. Most actual games combine aspects of both. In a race to research and develop a new product, the firms act simultaneously, but each competitor has partial information about the others' progress and can respond. During one play in football, the opposing offensive and defensive coaches simultaneously send out teams with the expectation of carrying out certain plays but, after seeing how the defense has set up, the quarterback can change the play at the line of scrimmage or call a time-out so that the coach can change the play.

The distinction between **sequential** and **simultaneous moves** is important because the two types of games require different types of interactive thinking. In a sequential-move game, each player must think: if I do this, how will my opponent react? Your current move is governed by your calculation of its *future* consequences. With simultaneous moves, you have the trickier task of trying to figure out what your opponent is going to do *right now*. But you must recognize that, in making his own calculation, the opponent is also trying to figure out your current move, while at the same time recognizing that you are doing the same with him. . . . Both of you have to think your way out of this circle.

In the next three chapters, we will study the two pure cases. In Chapter 3, we examine sequential-move games, where you must look ahead to act now; in Chapters 4 and 5, the subject is simultaneous-move games, where you must square the circle of "He thinks that I think that he thinks . . ." In each case, we will devise some simple tools for such thinking—trees and payoff tables—and obtain some simple rules to guide actions.

The study of sequential games also tells us when it is an advantage to move first and when second. Roughly speaking, this depends on the relative importance of commitment and flexibility in the game in question. For example, the game of economic competition among rival firms in a market has a first-mover advantage if one firm, by making a firm commitment to compete aggressively, can get its rivals to back off. But, in political competition, a candidate who has taken a firm stand on an issue may give his rivals a clear focus for their attack ads, and the game has a second-mover advantage.

Knowledge of the balance of these considerations can also help you devise ways to manipulate the order of moves to your own advantage. That in turn leads to the study of strategic moves, such as threats and promises, which we will take up in Chapter 10.

B. Are the Players' Interests in Total Conflict, or Is There Some Commonality?

In simple games such as chess or football, there is a winner and a loser. One player's gain is the other's loss. Similarly, in gambling games, one player's winnings are the others' losses, so the total is zero. This is why such situations are called *zero-sum games*. More generally, the idea is that the players' interests are in complete conflict. Such conflict arises when players are dividing up any fixed amount of possible gain, whether it be measured in yards, dollars, acres, or scoops of ice cream. Because the available gain need not always be exactly zero, the term *constant-sum game* is often substituted for zero-sum; we will use the two interchangeably.

Most economic and social games are not zero-sum. Trade, or economic activity more generally, offers scope for deals that benefit everyone. Joint ventures can combine the participants' different skills and generate synergy to produce more than the sum of what they could have produced separately. But the interests are not completely aligned either; the partners can cooperate to create a larger total pie, but they will clash when it comes to deciding how to split this pie among them.

Even wars and strikes are not zero-sum games. A nuclear war is the most striking example of a situation where there can be only losers, but the concept is far older. Pyrrhus, the king of Epirus, defeated the Romans at Heraclea in 280 B.C. but at such great cost to his own army that he exclaimed, "Another such victory and we are lost!" Hence the phrase "Pyrrhic victory." In the 1980s, at the height of the frenzy of business takeovers, the battles among rival bidders led to such costly escalation that the successful bidder's victory was often similarly Pyrrhic.

Most games in reality have this tension between conflict and cooperation, and many of the most interesting analyses in game theory come from the need to handle it. The players' attempts to resolve their conflict—distribution of territory or profit—are influenced by the knowledge that, if they fail to agree, the outcome will be bad for all of them. One side's threat of a war or a strike is its attempt to frighten the other side into conceding its demands.

Even when a game is constant-sum for all players, when there are three (or more) players, we have the possibility that two of them will cooperate at the expense of the third; this leads to the study of alliances and coalitions. We will examine and illustrate these ideas later, especially in Chapter 18 on bargaining.

C. Is the Game Played Once or Repeatedly, and with the Same or Changing Opponents?

A game played just once is in some respects simpler and in others more complicated than one with a longer interaction. You can think about a one-shot game without worrying about its repercussions on other games you might play in the future against the same person or against others who might hear of your actions in this one. Therefore actions in one-shot games are more likely to be unscrupulous or ruthless. For example, an automobile repair shop is much more likely to overcharge a passing motorist than a regular customer.

In one-shot encounters, each player doesn't know much about the others; for example, what their capabilities and priorities are, whether they are good at calculating their best strategies or have any weaknesses that can be exploited, and so on. Therefore in one-shot games, secrecy or surprise is likely to be an important component of good strategy.

Games with ongoing relationships require the opposite considerations. You have an opportunity to build a reputation (for toughness, fairness, honesty, reliability, and so forth, depending on the circumstances) and to find out more about your opponent. The players together can better exploit mutually beneficial prospects by arranging to divide the spoils over time (taking turns to "win") or to punish a cheater in future plays (an eye for an eye or tit-for-tat). We will consider these possibilities in Chapter 11 on the prisoners' dilemma.

More generally, a game may be zero-sum in the short run but have scope for mutual benefit in the long run. For example, each football team likes to win, but they all recognize that close competition generates more spectator interest, which benefits all teams in the long run. That is why they agree to a drafting scheme where teams get to pick players in reverse order of their current standing, thereby reducing the inequality of talent. In long-distance races, the runners or cyclists often develop a lot of cooperation; two or more of them can help one another by taking turns following in one another's slipstream. Near the end of the race, the cooperation collapses as all of them dash for the finish line.

Here is a useful rule of thumb for your own strategic actions in life. In a game that has some conflict and some scope for cooperation, you will often think up a great strategy for winning big and grinding a rival into dust but have a nagging worry at the back of your mind that you are behaving like the worst 1980s yuppie. In such a situation, the chances are that the game has a repeated or ongoing aspect that you have overlooked. Your aggressive strategy may gain you a short-run advantage, but its long-run side effects will cost you even more. Therefore you should dig deeper and recognize the cooperative element and then alter your strategy accordingly. You will be surprised how often niceness, integrity, and the golden rule of doing to others as you would have them do to you turn out to be not just old nostrums, but good strategies as well, when you consider the whole complex of games that you will be playing in the course of your life.

D. Do the Players Have Full or Equal Information?

In chess, each player knows exactly the current situation and all the moves that led to it, and each knows that the other aims to win. This situation is exceptional; in most other games, the players face some limitation of information. Such limitations come in two kinds. First, a player may not know all the information that is pertinent for the choice that he has to make at every point in the game. This type of information problem arises because of the player's uncertainty about relevant variables, both internal and external to the game. For example, he may be uncertain about external circumstances, such as the weekend weather or the quality of a product he wishes to purchase; we call this situation one of **external uncertainty.** Or he may be uncertain about exactly what moves his opponent has made in the past or is making at the same time he makes his own move; we call this **strategic uncertainty.** If a game has neither external nor strategic uncertainty, we say that the game is one of **perfect information;** otherwise the game has **imperfect information.** We will give a more precise technical definition of perfect information in Chapter 6, Section 3.A, after we have introduced the concept of an information set. We will develop the theory of games with imperfect information (uncertainty) in three future chapters. In Chapter 4, we discuss games with contemporaneous (simultaneous) actions, which entail strategic uncertainty, and we analyze methods for making choices under external uncertainty in the Appendix to Chapter 7 and in Chapter 9.

Trickier strategic situations arise when one player knows more than another does; they are called situations of **incomplete** or, better, **asymmetric information.** In such situations, the players' attempts to infer, conceal, or sometimes convey their private information become an important part of the game and the strategies. In bridge or poker, each player has only partial knowledge of the cards held by the others. Their actions (bidding and play in bridge, the number of cards taken and the betting behavior in poker) give information to opponents. Each player tries to manipulate his actions to mislead the opponents (and, in bridge, to inform his partner truthfully), but in doing so each must be aware that the opponents know this and that they will use strategic thinking to interpret that player's actions.

You may think that if you have superior information, you should always conceal it from other players. But that is not true. For example, suppose you are the CEO of a pharmaceutical firm that is engaged in an R&D competition to develop a new drug. If your scientists make a discovery that is a big step forward, you may want to let your competitors know, in the hope that they will give up their own searches. In war, each side wants to keep its tactics and troop deployments secret; but, in diplomacy, if your intentions are peaceful, then you desperately want other countries to know and believe this fact.

The general principle here is that you want to release your information selectively. You want to reveal the good information (the kind that will draw responses from the other players that work to your advantage) and conceal the bad (the kind that may work to your disadvantage).

This raises a problem. Your opponents in a strategic game are purposive, rational players and know that you are one, too. They will recognize your incentive to exaggerate or even to lie. Therefore they are not going to accept your unsupported declarations about your progress or capabilities. They can be convinced only by objective evidence or by actions that are credible proof of your information. Such actions on the part of the more-informed player are called **signals,** and strategies that use them are called **signaling.** Conversely, the less-informed party can create situations in which the more-informed player will have to take some action that credibly reveals his information; such strategies are called **screening,** and the methods they use are called **screening devices.** The word *screening* is used here in the sense of testing in order to sift or separate, not in the sense of concealing. Recall that in the dating game in Section 2.F of Chapter 1, the woman was screening the man to test his commitment to their relationship, and her suggestion that the pair give up one of their two rent-controlled apartments was the screening device. If the man had been committed to the relationship, he might have acted first and volunteered to give up his apartment; this action would have been a signal of his commitment.

Now we see how, when different players have different information, the manipulation of information itself becomes a game, perhaps more important than the game that will be played after the information stage. Such information games are ubiquitous, and playing them well is essential for success in life. We will study more games of this kind in greater detail in Chapter 9 and also in Chapter 14.

E. Are the Rules of the Game Fixed or Manipulable?

The rules of chess, card games, or sports are given, and every player must follow them, no matter how arbitrary or strange they seem. But in games of business, politics, and ordinary life, the players can make their own rules to a greater or lesser extent. For example, in the home, parents constantly try to make the rules, and children constantly look for ways to manipulate or circumvent those rules. In legislatures, rules for the progress of a bill (including the order in which amendments and main motions are voted on) are fixed, but the game that sets the agenda—which amendments are brought to a vote first—can be manipulated; that is where political skill and power have the most scope, and we will address these matters in detail in Chapter 16.

In such situations, the real game is the "pregame" where rules are made, and your strategic skill must be deployed at that point. The actual playing out of the subsequent game can be more mechanical; you could even delegate it to someone else. However, if you "sleep" through the pregame, you might find that you have lost the game before it ever began. For many years, American firms ignored the rise of foreign competition in just this way and ultimately paid the price. Others, such as oil magnate John D. Rockefeller, Sr., adopted the strategy of limiting their participation to games in which they could also participate in making the rules.[2]

The distinction between changing rules and acting within the chosen rules will be most important for us in our study of strategic moves, such as threats and promises. Questions of how you can make your own threats and promises credible or how you can reduce the credibility of your opponent's threats basically have to do with a pregame that entails manipulating the rules of the subsequent game in which the promises or threats may have to be carried out. More generally, the strategic moves that we will study in Chapter 10 are essentially ploys for such manipulation of rules.

But if the pregame of rule manipulation is the real game, what fixes the rules of the pregame? Usually these pregame rules depend on some hard facts related to the players' innate abilities. In business competition, one firm can take preemptive actions that alter subsequent games between it and its rivals; for example, it can expand its factory or advertise in a way that twists the results of subsequent price competition more favorably to itself. Which firm can do this best or most easily depends on which one has the managerial or organizational resources to make the investments or to launch the advertising campaigns.

Players may also be unsure of their rivals' abilities. This often makes the pregame one of unequal information, requiring more subtle strategies and occasionally resulting in some big surprises. We will comment on all these matters in the appropriate places in the chapters that follow.

[2]For more on the methods used in Rockefeller's rise to power, see Ron Chernow, *Titan* (New York: Random House, 1998).

F. Are Agreements to Cooperate Enforceable?

We saw that most strategic interactions consist of a mixture of conflict and common interest. Then there is a case to be made that all participants should get together and reach an agreement about what everyone should do, balancing their mutual interest in maximizing the total benefit and their conflicting interests in the division of gains. Such negotiations can take several rounds in which agreements are made on a tentative basis, better alternatives are explored, and the deal is finalized only when no group of players can find anything better. The concept of the core in Chapter 19 embodies such a process and its outcome. However, even after the completion of such a process, additional difficulties often arise in putting the final agreement into practice. For instance, all the players must perform, in the end, the actions that were stipulated for them in the agreement. When all others do what they are supposed to do, any one participant can typically get a better outcome for himself by doing something different. And, if each one suspects that the others may cheat in this way, he would be foolish to adhere to his stipulated cooperative action.

Agreements to cooperate can succeed if all players act immediately and in the presence of the whole group, but agreements with such immediate implementation are quite rare. More often the participants disperse after the agreement has been reached and then take their actions in private. Still, if these actions are observable to the others and a third party—for example, a court of law—can enforce compliance, then the agreement of joint action can prevail.

However, in many other instances individual actions are neither directly observable nor enforceable by external forces. Without enforceability, agreements will stand only if it is in all participants' individual interests to abide by them. Games among sovereign countries are of this kind, as are many games with private information or games where the actions are either outside the law or too trivial or too costly to enforce in a court of law. In fact, games where agreements for joint action are not enforceable constitute a vast majority of strategic interactions.

Game theory uses a special terminology to capture the distinction between situations in which agreements are enforceable and those in which they are not. Games in which joint-action agreements are enforceable are called **cooperative;** those in which such enforcement is not possible, and individual participants must be allowed to act in their own interests, are called **noncooperative.** This has become standard terminology, but it is somewhat unfortunate because it gives the impression that the former will produce cooperative outcomes and the latter will not. In fact, individual action can be compatible with the achievement of a lot of mutual gain, especially in repeated interactions. The important distinction is that in so-called noncooperative games, cooperation will emerge only if it is in the participants' separate and individual interests to continue to take the prescribed actions. This emergence of cooperative outcomes from noncooperative behavior is one of the most interesting findings of game theory, and we will develop the idea in Chapters 11, 12, and 13.

We will adhere to the standard usage, but emphasize that the terms *cooperative* and *noncooperative* refer to the way in which actions are implemented or enforced—collectively in the former mode and individually in the latter—and not to the nature of the outcomes.

As we said earlier, most games in practice do not have adequate mechanisms for external enforcement of joint-action agreements. Therefore most of our analytical development will proceed in the noncooperative mode. The few exceptions include the discussion of bargaining in Chapter 18 and a brief treatment of markets and competition in Chapter 19.

3 SOME TERMINOLOGY AND BACKGROUND ASSUMPTIONS

When one thinks about a strategic game, the logical place to begin is by specifying its structure. This includes all the strategies available to all the players, their information, and their objectives. The first two aspects will differ from one game to another along the dimensions discussed in the preceding section, and one must locate one's particular game within that framework. The objectives raise some new and interesting considerations. Here we consider aspects of all these matters.

A. Strategies

Strategies are simply the choices available to the players, but even this basic notion requires some further study and elaboration. If a game has purely simultaneous moves made only once, then each player's strategy is just the action taken on that single occasion. But if a game has sequential moves, then the actions of a player who moves later in the game can respond to what other players have done (or what he himself has done) at earlier points. Therefore each such player must make a complete plan of action, for example: "If the other does A, then I will do X but, if the other does B, then I will do Y." This complete plan of action constitutes the strategy in such a game.

There is a very simple test to determine whether your strategy is complete. It should specify how you would play the game in such full detail—describing your action in every contingency—that, if you were to write it all down, hand it to someone else, and go on vacation, this other person acting as your representative could play the game just as you would have played it. He would know what to do on each occasion that could conceivably arise in the course of play, without ever needing to disturb your vacation for instructions on how to deal with some situation that you had not foreseen.

This test will become clearer in Chapter 3, when we develop and apply it in some specific contexts. For now, you should simply remember that a strategy is a complete plan of action.

This notion is similar to the common usage of the word *strategy* to denote a longer-term or larger-scale plan of action, as distinct from tactics that pertain to a shorter term or a smaller scale. For example, the military makes strategic plans for a war or a large-scale battle, while tactics for a smaller skirmish or a particular theater of battle are often left to be devised by lower-level officers to suit the local conditions. But game theory does not use the term *tactics* at all. The term *strategy* covers all the situations, meaning a complete plan of action when necessary and meaning a single move if that is all that is needed in the particular game being studied.

The word *strategy* is also commonly used to describe a person's decisions over a fairly long time span and a sequence of choices, even though there is no game in our sense of purposive and aware interaction with other people. Thus you have probably already chosen a career strategy. When you start earning an income, you will make saving and investment strategies and eventually plan a retirement strategy. This usage of the term *strategy* has the same sense as ours—a plan for a succession of actions in response to evolving circumstances. The only difference is that we are reserving it for a situation—namely, a game—where the circumstances evolve because of actions taken by other purposive players.

B. Payoffs

When asked what a player's objective in a game is, most newcomers to strategic thinking respond that it is "to win"; but matters are not always so simple. Sometimes the margin of victory matters; for example, in R&D competition, if your product is only slightly better than the nearest rival's, your patent may be more open to challenge. Sometimes there may be smaller prizes for several participants, so winning isn't everything. Most important, very few games of strategy are purely zero-sum or win-lose; they combine some common interest and some conflict among the players. Thinking about such mixed-motive games requires more refined calculations than the simple dichotomy of winning and losing—for example, comparisons of the gains from cooperating versus cheating.

We will give each player a complete numerical scale with which to compare all logically conceivable outcomes of the game, corresponding to each available combination of choices of strategies by all the players. The number associated with each possible outcome will be called that player's **payoff** for that outcome. Higher payoff numbers attach to outcomes that are better in this player's rating system.

Sometimes the payoffs will be simple numerical ratings of the outcomes, the worst labeled 1, the next worst 2, and so on, all the way to the best. In other games, there may be more natural numerical scales—for example, money income or profit for firms, viewer-share ratings for television networks, and so on. In many situations, the payoff numbers are only educated guesses; then we should do some sensitivity tests by checking that the results of our analysis do not change significantly if we vary these guesses within some reasonable margin of error.

Two important points about the payoffs need to be understood clearly. First, the payoffs for one player capture everything in the outcomes of the game that he cares about. In particular, the player need not be selfish, but his concern about others should be already included in his numerical payoff scale. Second, we will suppose that, if the player faces a random prospect of outcomes, then the number associated with this prospect is the average of the payoffs associated with each component outcome, each weighted by its probability. Thus, if in one player's ranking, outcome A has payoff 0 and outcome B has payoff 100, then the prospect of a 75% probability of A and a 25% probability of B should have the payoff $0.75 \times 0 + 0.25 \times 100 = 25$. This is often called the **expected payoff** from the random prospect. The word *expected* has a special connotation in the jargon of probability theory. It does not mean what you think you will get or expect to get; it is the mathematical or probabilistic or statistical expectation, meaning an average of all possible outcomes, where each is given a weight proportional to its probability.

The second point creates a potential difficulty. Consider a game where players get or lose money and payoffs are measured simply in money amounts. In reference to the preceding example, if a player has a 75% chance of getting nothing and a 25% chance of getting $100, then the expected payoff as calculated in that example is $25. That is also the payoff that the player would get from a simple nonrandom outcome of $25. In other words, in this way of calculating payoffs, a person should be indifferent to whether he receives $25 for sure or faces a risky prospect of which the average is $25. One would think that most people would be averse to risk, preferring a sure $25 to a gamble that yields only $25 on the average.

A very simple modification of our payoff calculation gets around this difficulty. We measure payoffs not in money sums but by using a nonlinear rescaling of the dollar amounts. This is called the expected utility approach, and we will present it in detail in the Appendix to Chapter 7. For now, please take our word that incorporating differing attitudes toward risk into our framework is a manageable task. Almost all of game theory is based on the expected utility approach, and it is indeed very useful, although not without flaws. We will adopt it in this book, but we also indicate some of the difficulties that it leaves unresolved, with the use of a simple example in Chapter 8, Section 6.

C. Rationality

Each player's aim in the game will be to achieve as high a payoff for himself as possible. But how good is each player at pursuing this aim? This question is not about whether and how other players pursuing their own interests will impede him; that is in the very nature of a game of strategic interaction. We mean how good each player is at calculating the strategy that is in his own best interests and at following this strategy in the actual course of play.

Much of game theory assumes that players are perfect calculators and flawless followers of their best strategies. This is the assumption of **rational behavior.** Observe the precise sense in which the term *rational* is being used. It means that each has a consistent set of rankings (values or payoffs) over all the logically possible outcomes and calculates the strategy that best serves these interests. Thus rationality has two essential ingredients: complete knowledge of one's own interests, and flawless calculation of what actions will best serve those interests.

It is equally important to understand what is *not* included in this concept of rational behavior. It does not mean that players are selfish; a player may rate highly the well-being of some other and incorporate this high rating into his payoffs. It does not mean that players are short-run thinkers; in fact, calculation of future consequences is an important part of strategic thinking, and actions that seem irrational from the immediate perspective may have valuable long-term strategic roles. Most important, being rational does not mean having the same value system as other players, or sensible people, or ethical or moral people would use; it means merely pursuing one's own value system consistently. Therefore, when one player carries out an analysis of how other players will respond (in a game with sequential moves) or of the successive rounds of thinking about thinking (in a game with simultaneous moves), he must recognize that the other players calculate the consequences of their choices by using their own value or rating system. You must not impute your own value systems or standards of rationality to others and assume that they would act as you would in that situation. Thus many "experts" commenting on the Persian Gulf conflict in late 1990 predicted that Saddam Hussein would back down "because he is rational"; they failed to recognize that Saddam's value system was different from the one held by most Western governments and by the Western experts.

In general, each player does not really know the other players' value systems; this is part of the reason that in reality many games have incomplete and asymmetric information. In such games, trying to find out the values of others and trying to conceal or convey one's own become important components of strategy.

Game theory assumes that all players are rational. How good is this assumption, and therefore how good is the theory that employs it? At one level, it is obvious that the assumption cannot be literally true. People often don't even have full advance knowledge of their own value systems; they don't think ahead about how they would rank hypothetical alternatives and then remember these rankings until they are actually confronted with a concrete choice. Therefore they find it very difficult to perform the logical feat of tracing all possible consequences of their and other players' conceivable strategic choices and ranking the outcomes in advance in order to choose which strategy to follow. Even if they knew their preferences, the calculation would remain far from easy. Most games in real life are very complex, and most real players are limited in their thinking and computational abilities. In games such as chess, it is known that the calculation for the best strategy can be performed in a finite number of steps, but no one has succeeded in performing it, and good play remains largely an art.

The assumption of rationality may be closer to reality when the players are regulars who play the game quite often. Then they benefit from having experienced the different possible outcomes. They understand how the strategic choices of various players lead to the outcomes and how well or badly they themselves fare. Then we as analysts of the game can hope that their choices, even if not made with full and conscious computations, closely approximate the results of such computations. We can think of the players as implicitly choosing the optimal strategy or behaving as if they were perfect calculators. We will offer some experimental evidence in Chapter 5 that the experience of playing the game generates more rational behavior.

The advantage of making a complete calculation of your best strategy, taking into account the corresponding calculations of a similar strategically calculating rival, is that then you are not making mistakes that the rival can exploit. In many actual situations, you may have specific knowledge of the way in which the other players fall short of this standard of rationality, and you can exploit this in devising your own strategy. We will say something about such calculations, but very often this is a part of the "art" of game playing, not easily codifiable in rules to be followed. You must always beware of the danger that the others are merely pretending to have poor skills or strategy, losing small sums through bad play and hoping that you will then raise the stakes, when they can raise the level of their play and exploit your gullibility. When this risk is real, the safer advice to a player may be to assume that the rivals are perfect and rational calculators and to choose his own best response to them. In other words, one should play to the opponents' capabilities instead of their limitations.

D. Common Knowledge of Rules

We suppose that, at some level, the players have a common understanding of the rules of the game. In a *Peanuts* cartoon, Lucy thought that body checking was allowed in golf and decked Charlie Brown just as he was about to take his swing. We do not allow this.

The qualification "at some level" is important. We saw how the rules of the immediate game could be manipulated. But this merely admits that there is another game being played at a deeper level—namely, where the players choose the rules of the superficial game. Then the question is whether the rules of this deeper game are fixed. For example, in the legislative context, what are the rules of the agenda-setting game? They may be that the committee chairs have the power. Then how are the committees and their chairs elected? And so on. At some basic level, the rules are fixed by the constitution, by the technology of campaigning, or by general social norms of behavior. We ask that all players recognize the given rules of this basic game, and that is the focus of the analysis. Of course, that is an ideal; in practice, we may not be able to proceed to a deep enough level of analysis.

Strictly speaking, the rules of the game consist of (1) the list of players, (2) the strategies available to each player, (3) the payoffs of each player for all possible combinations of strategies pursued by all the players, and (4) the assumption that each player is a rational maximizer.

Game theory cannot properly analyze a situation where one player does not know whether another player is participating in the game, what the entire sets of actions available to the other players are from which they can choose, what their value systems are, or whether they are conscious maximizers of their own payoffs. But in actual strategic interactions, some of the biggest gains are to be made by taking advantage of the element of surprise and doing something that your rivals never thought you capable of. Several vivid examples can be found in historic military conflicts. For example, in 1967 Israel launched a preemptive attack that destroyed the Egyptian air force on the ground; in 1973 it was Egypt's turn to spring a surprise by launching a tank attack across the Suez Canal.

It would seem, then, that the strict definition of game theory leaves out a very important aspect of strategic behavior, but in fact matters are not that bad. The theory can be reformulated so that each player attaches some small probability to the situation where such dramatically different strategies are available to the other players. Of course, each player knows his own set of available strategies. Therefore the game becomes one of asymmetric information and can be handled by using the methods developed in Chapter 9.

The concept of common knowledge itself requires some explanation. For some fact or situation X to be common knowledge between two people, A and B, it is not enough for each of them separately to know X. Each should also know that the other knows X; otherwise, for example, A might think that B does not know X and might act under this misapprehension in the midst of a game. But then A should also know that B knows that A knows X, and the other way around, otherwise A might mistakenly try to exploit B's supposed ignorance of A's knowledge. Of course, it doesn't even stop there. A should know that B knows that A knows that B knows, and so on ad infinitum. Philosophers have a lot of fun exploring the fine points of this infinite regress and the intellectual paradoxes that it can generate. For us, the general notion that the players have a common understanding of the rules of their game will suffice.

E. Equilibrium

Finally, what happens when rational players' strategies interact? Our answer will generally be in the framework of **equilibrium.** This simply means that each player is using the strategy that is the best response to the strategies of the other players. We will develop game-theoretic concepts of equilibrium in Chapters 3 through 8 and then use them in subsequent chapters.

Equilibrium does not mean that things don't change; in sequential-move games the players' strategies are the complete plans of action and reaction, and the position evolves all the time as the successive moves are made and responded to. Nor does equilibrium mean that everything is for the best; the interaction of rational strategic choices by all players can lead to bad outcomes for all, as in the prisoners' dilemma. But we will generally find that the idea of equilibrium is a useful descriptive tool and organizing concept for our analysis. We will consider this idea in greater detail later, in connection with specific equilibrium concepts. We will also see how the concept of equilibrium can be augmented or modified to remove some of its flaws and to incorporate behavior that falls short of full calculating rationality.

Just as the rational behavior of individual players can be the result of experience in playing the game, the fitting of their choices into an overall equilibrium can come about after some plays that involve trial and error and nonequilibrium outcomes. We will look at this matter in Chapter 5.

Defining an equilibrium is not hard; actually finding an equilibrium in a particular game—that is, solving the game—can be a lot harder. Throughout this book we will solve many simple games in which there are two or three players, each of them having two or three strategies or one move each in turn. Many people believe this to be the limit of the reach of game theory and therefore believe that the theory is useless for the more complex games that take place in reality. That is not true.

Humans are severely limited in their speed of calculation and in their patience for performing long calculations. Therefore humans can easily solve only the simple games with two or three players and strategies. But computers are very good at speedy and lengthy calculations. Many games that are far beyond the power of human calculators are easy for computers. The level of complexity in many games in business and politics is already within the powers of computers. Even in games such as chess that are far too complex to solve completely, computers have reached a level of ability comparable to that of the best humans; we consider chess in more detail in Chapter 3.

Computer programs for solving quite complex games exist, and more are appearing rapidly. Mathematica and similar program packages contain routines for finding mixed-strategy equilibria in simultaneous-move games. Gambit, a

National Science Foundation project led by Professors Richard D. McKelvey of the California Institute of Technology and Andrew McLennan of the University of Minnesota, is producing a comprehensive set of routines for finding equilibria in sequential- and simultaneous-move games, in pure and mixed strategies, and with varying degrees of uncertainty and incomplete information. We will refer to this project again in several places in the next several chapters. The biggest advantage of the project is that its programs are open source and can easily be obtained from its Web site with the URL http://gambit.sourceforge.net.

Why then do we set up and solve several simple games in detail in this book? The reason is that understanding the concepts is an important prerequisite for making good use of the mechanical solutions that computers can deliver, and understanding comes from doing simple cases yourself. This is exactly how you learned and now use arithmetic. You came to understand the ideas of addition, subtraction, multiplication, and division by doing many simple problems mentally or using paper and pencil. With this grasp of basic concepts, you can now use calculators and computers to do far more complicated sums than you would ever have the time or patience to do manually. If you did not understand the concepts, you would make errors in using calculators; for example, you might solve $3 + 4 \times 5$ by grouping additions and multiplications incorrectly as $(3 + 4) \times 5 = 35$ instead of correctly as $3 + (4 \times 5) = 23$.

Thus the first step of understanding the concepts and tools is essential. Without it, you would never learn to set up correctly the games that you ask the computer to solve. You would not be able to inspect the solution with any feeling for whether it was reasonable and, if it was not, would not be able to go back to your original specification, improve it, and solve it again until the specification and the calculation correctly capture the strategic situation that you want to study. Therefore please pay serious attention to the simple examples that we solve and the drill exercises that we ask you to solve, especially in Chapters 3 through 8.

F. Dynamics and Evolutionary Games

The theory of games based on assumptions of rationality and equilibrium has proved very useful, but it would be a mistake to rely on it totally. When games are played by novices who do not have the necessary experience to perform the calculations to choose their best strategies, explicitly or implicitly, their choices, and therefore the outcome of the game, can differ significantly from the predictions of analysis based on the concept of equilibrium.

However, we should not abandon all notions of good choice; we should recognize the fact that even poor calculators are motivated to do better for their own sakes and will learn from experience and by observing others. We should allow for a dynamic process in which strategies that proved to be better in previous plays of the game are more likely to be chosen in later plays.

The **evolutionary** approach to games does just this. It is derived from the idea of evolution in biology. Any individual animal's genes strongly influence its behavior. Some behaviors succeed better in the prevailing environment, in the sense that the animals exhibiting those behaviors are more likely to reproduce successfully and pass their genes to their progeny. An evolutionary stable state, relative to a given environment, is the ultimate outcome of this process over several generations.

The analogy in games would be to suppose that strategies are not chosen by conscious rational maximizers, but instead that each player comes to the game with a particular strategy "hardwired" or "programmed" in. The players then confront other players who may be programmed to apply the same or different strategies. The payoffs to all the players in such games are then obtained. The strategies that fare better—in the sense that the players programmed to play them get higher payoffs in the games—multiply faster, whereas the strategies that fare worse decline. In biology, the mechanism of this growth or decay is purely genetic transmission through reproduction. In the context of strategic games in business and society, the mechanism is much more likely to be social or cultural—observation and imitation, teaching and learning, greater availability of capital for the more successful ventures, and so on.

The object of study is the dynamics of this process. Does it converge to an evolutionary stable state? Does just one strategy prevail over all others in the end, or can a few strategies coexist? Interestingly, in many games the evolutionary stable limit is the same as the equilibrium that would result if the players were consciously rational calculators. Therefore the evolutionary approach gives us a backdoor justification for equilibrium analysis.

The concept of evolutionary games has thus imported biological ideas into game theory; there has been an influence in the opposite direction, too. Biologists have recognized that significant parts of animal behavior consist of strategic interactions with other animals. Members of a given species compete with one another for space or mates; members of different species relate to one another as predators and prey along a food chain. The payoff in such games in turn contributes to reproductive success and therefore to biological evolution. Just as game theory has benefited by importing ideas from biological evolution for its analysis of choice and dynamics, biology has benefited by importing game-theoretic ideas of strategies and payoffs for its characterization of basic interactions between animals. We have a true instance of synergy or symbiosis. We provide an introduction to the study of evolutionary games in Chapter 13.

G. Observation and Experiment

All of Section 3 to this point has concerned how to think about games or how to analyze strategic interactions. This constitutes theory. This book will cover an extremely simple level of theory, developed through cases and illustrations instead of formal mathematics or theorems, but it will be theory just the same. All theory should relate to reality in two ways. Reality should help structure the theory, and reality should provide a check on the results of the theory.

We can find out the reality of strategic interactions in two ways: (1) by observing them as they occur naturally and (2) by conducting special experiments that help us pin down the effects of particular conditions. Both methods have been used, and we will mention several examples of each in the proper contexts.

Many people have studied strategic interactions—the participants' behavior and the outcomes—under experimental conditions, in classrooms among "captive" players, or in special laboratories with volunteers. Auctions, bargaining, prisoners' dilemmas, and several other games have been studied in this way. The results are a mixture. Some conclusions of the theoretical analysis are borne out; for example, in games of buying and selling, the participants generally settle quickly on the economic equilibrium. In other contexts, the outcomes differ significantly from the theoretical predictions; for example, prisoners' dilemmas and bargaining games show more cooperation than theory based on the assumption of selfish, maximizing behavior would lead us to expect, whereas auctions show some gross overbidding.

At several points in the chapters that follow, we will review the knowledge that has been gained by observation and experiments, discuss how it relates to the theory, and consider what reinterpretations, extensions, and modifications of the theory have been made or should be made in the light of this knowledge.

4 THE USES OF GAME THEORY

We began Chapter 1 by saying that games of strategy are everywhere—in your personal and working life; in the functioning of the economy, society, and polity around you; in sports and other serious pursuits; in war; and in peace. This should be motivation enough to study such games systematically, and that is what game theory is about, but your study can be better directed if you have a clearer idea of just how you can put game theory to use. We suggest a threefold method.

The first use is in *explanation.* Many events and outcomes prompt us to ask: Why did that happen? When the situation requires the interaction of decision makers with different aims, game theory often supplies the key to understanding the situation. For example, cutthroat competition in business is the result of the rivals being trapped in a prisoners' dilemma. At several points in the book we will mention actual cases where game theory helps us to understand how and why the events unfolded as they did. This includes the detailed case study of the Cuban missile crisis from the perspective of game theory.

The other two uses evolve naturally from the first. The second is in *prediction*. When looking ahead to situations where multiple decision makers will interact strategically, we can use game theory to foresee what actions they will take and what outcomes will result. Of course, prediction for a particular context depends on its details, but we will prepare you to use prediction by analyzing several broad classes of games that arise in many applications.

The third use is in *advice* or *prescription*: we can act in the service of one participant in the future interaction and tell him which strategies are likely to yield good results and which ones are likely to lead to disaster. Once again such work is context specific, and we equip you with several general principles and techniques and show you how to apply them to some general types of contexts. For example, in Chapters 7 and 8 we will show how to mix moves, in Chapter 10 we will examine how to make your commitments, threats, and promises credible, and in Chapter 11 we will examine alternative ways of overcoming prisoners' dilemmas.

The theory is far from perfect in performing any of the three functions. To explain an outcome, one must first have a correct understanding of the motives and behavior of the participants. As we saw earlier, most of game theory takes a specific approach to these matters—namely, the framework of rational choice of individual players and the equilibrium of their interaction. Actual players and interactions in a game might not conform to this framework. But the proof of the pudding is in the eating. Game-theoretic analysis has greatly improved our understanding of many phenomena, as reading this book should convince you. The theory continues to evolve and improve as the result of ongoing research. This book will equip you with the basics so that you can more easily learn and profit from the new advances as they appear.

When explaining a past event, we can often use historical records to get a good idea of the motives and the behavior of the players in the game. When attempting prediction or advice, there is the additional problem of determining what motives will drive the players' actions, what informational or other limitations they will face, and sometimes even who the players will be. Most important, if game-theoretic analysis assumes that the other player is a rational maximizer of his own objectives when in fact he is unable to do the calculations or is a clueless person acting at random, the advice based on that analysis may prove wrong. This risk is reduced as more and more players recognize the importance of strategic interaction and think through their strategic choices or get expert advice on the matter, but some risk remains. Even then, the systematic thinking made possible by the framework of game theory helps keep the errors down to this irreducible minimum, by eliminating the errors that arise from faulty logical thinking about the strategic interaction. Also, game theory can take into account many kinds of uncertainty and incomplete information, including that about the strategic possibilities and rationality of the opponent. We will consider a few examples in the chapters to come.

5 THE STRUCTURE OF THE CHAPTERS TO FOLLOW

In this chapter we introduced several considerations that arise in almost every game in reality. To understand or predict the outcome of any game, we must know in greater detail all of these ideas. We also introduced some basic concepts that will prove useful in such analysis. However, trying to cope with all of the concepts at once merely leads to confusion and a failure to grasp any of them. Therefore we will build up the theory one concept at a time. We will develop the appropriate technique for analyzing that concept and illustrate it.

In the first group of chapters, from Chapters 3 to 8, we will construct and illustrate the most important of these concepts and techniques. We will examine purely sequential-move games in Chapter 3 and introduce the techniques—game trees and rollback reasoning—that are used to analyze and solve such games. In Chapters 4 and 5, we will turn to games with simultaneous moves and develop for them another set of concepts—payoff tables, dominance, and Nash equilibrium. Both chapters will focus on games where players use pure strategies; in Chapter 4, we will restrict players to a finite set of pure strategies and, in Chapter 5, we will allow strategies that are continuous variables. Chapter 5 will also examine some mixed empirical evidence and conceptual criticisms and counterarguments on Nash equilibrium, and a prominent alternative to Nash equilibrium—namely, rationalizability. In Chapter 6, we will show how games that have some sequential moves and some simultaneous moves can be studied by combining the techniques developed in Chapters 3 through 5. In Chapters 7 and 8, we will turn to simultaneous-move games that require the use of randomization or mixed strategies. In Chapter 7, we introduce the basic ideas about mixing in two-by-two games, develop the simplest techniques for finding mixed-strategy Nash equilibria, and consider empirical evidence on mixing. Chapter 8 will then develop a little general theory of mixed strategies.

The ideas and techniques developed in Chapters 3 through 8 are the most basic ones: (1) correct forward-looking reasoning for sequential-move games and (2) equilibrium strategies—pure and mixed—for simultaneous-move games. Equipped with these concepts and tools, we can apply them to study some broad classes of games and strategies in Chapters 9 through 13.

Chapter 9 studies what happens in games when players are subject to uncertainty or when they have asymmetric information. We will examine strategies for coping with risk and even for using risk strategically. We will also study the important strategies of signaling and screening that are used for manipulating and eliciting information. We develop the appropriate generalization of Nash equilibrium in the context of uncertainty, namely Bayesian Nash equilibrium, and show the different kinds of equilibria that can arise. In Chapter 10, we will continue to examine the role of player manipulation in games as we consider strategies that players use to manipulate the rules of a game, by seizing a first-mover advantage and making a strategic move. Such moves are of three kinds—commitments, threats, and promises. In each case, credibility is essential to the success of the move, and we will outline some ways of making such moves credible.

In Chapter 11, we will move on to study the best-known game of them all—the prisoners' dilemma. We will study whether and how cooperation can be sustained, most importantly in a repeated or ongoing relationship. Then, in Chapter 12, we will turn to situations where large populations, rather than pairs or small groups of players, interact strategically, games that concern problems of collective action. Each person's actions have an effect—in some instances beneficial, in others, harmful—on the others. The outcomes are generally not the best from the aggregate perspective of the society as a whole. We will clarify the nature of these outcomes and describe some simple policies that can lead to better outcomes.

All these theories and applications are based on the supposition that the players in a game fully understand the nature of the game and deploy calculated strategies that best serve their objectives in the game. Such rationally optimal behavior is sometimes too demanding of information and calculating power to be believable as a good description of how people really act. Therefore Chapter 13 will look at games from a very different perspective. Here, the players are not calculating and do not pursue optimal strategies. Instead, each player is tied, as if genetically preordained, to a particular strategy. The population is diverse, and different players have different predetermined strategies. When players from such a population meet and act out their strategies, which strategies perform better? And if the more successful strategies proliferate better in the population, whether through inheritance or imitation, then what will the eventual structure of the population look like? It turns out that such evolutionary dynamics often favor exactly those strategies that would be used by rational optimizing players. Thus our study of evolutionary games lends useful indirect support to the theories of optimal strategic choice and equilibrium that we will have studied in the previous chapters.

In the final group, Chapters 14 through 19, we will take up specific applications to situations of strategic interactions. Here, we will use as needed the ideas and methods from all the earlier chapters. Chapter 14 uses the methods developed in Chapter 9 to analyze the strategies that people and firms have to use when dealing with others who have some private information. We will illustrate the screening mechanisms that are used for eliciting information, for example, the multiple fares with different restrictions that airlines use for separating the business travelers who are willing to pay more from the tourists who are more price sensitive. We will also develop the methods for designing incentive payments to elicit effort from workers when direct monitoring is difficult or too costly. Chapter 15 then applies the ideas from Chapter 10 to examine a particularly interesting dynamic version of a threat, known as the strategy of brinkmanship. We will elucidate its nature and apply the idea to study the Cuban missile crisis of 1962. Chapter 16 is about voting in committees and elections. We will look at the variety of voting rules available and some paradoxical results that can arise. In addition, we will address the potential for strategic behavior not only by voters but also by candidates in a variety of election types.

Chapters 17 through 19 will look at mechanisms for the allocation of valuable economic resources: Chapter 17 will treat auctions, Chapter 18 will consider bargaining processes, and Chapter 19 will look at markets. In our discussion of auctions, we will emphasize the roles of information and attitudes toward risk in the formulation of optimal strategies for both bidders and sellers. We will also take the opportunity to apply the theory to the newest type of auctions, those that take place online. Chapter 18 will present bargaining in both cooperative and noncooperative settings. Finally, Chapter 19 will consider games of market exchange, building on some of the concepts used in bargaining theory and including some theory of the core.

All of these chapters together provide a lot of material; how might readers or teachers with more specialized interests choose from it? Chapters 3 through 7 constitute the core theoretical ideas that are needed throughout the rest of the book. Chapters 10 and 11 are likewise important for the general classes of games and strategies considered therein. Beyond that, there is a lot from which to pick and choose. Section 1 of Chapter 5 and all of Chapter 8 consider some more advanced topics and go somewhat deeper into theory and mathematics. These chapters will appeal to those with more scientific and quantitative backgrounds and interests, but those who come from the social sciences or humanities and have less quantitative background can omit them without loss of continuity. Chapter 9 deals with an important topic in that most games in practice have incomplete and asymmetric information, and the players' attempts to manipulate information is a critical aspect of many strategic interactions. However, the concepts and techniques for analyzing information games are inherently somewhat more complex. Therefore some readers and teachers may choose to study just the examples that convey the basic ideas of signaling and screening and leave out the rest. We have placed this chapter early in Part Three, however, in view of the importance of the subject. Chapters 10 and 11 are key to understanding many phenomena in the real world, and most teachers will want to include them in their courses, but Section 5 of Chapter 11 is mathematically a little more advanced and can be omitted. Chapters 12 and 13 both look at games with large numbers of players. In Chapter 12, the focus is on social interactions; in Chapter 13, the focus is on evolutionary biology. The topics in Chapter 13 will be of greatest interest to those with interests in biology, but similar themes are emerging in the social sciences, and students from that background should aim to get the gist of the ideas even if they skip the details. Chapter 14 is most important for students of business and organization theories. Chapters 15 and 16 present topics from political science—international diplomacy and elections, respectively—and Chapters 17 through 19 cover topics from economics—auctions, bargaining, and markets. Those teaching courses with more specialized audiences may choose a subset from Chapters 12 through 19, and indeed expand on the ideas considered therein.

Whether you come from mathematics, biology, economics, politics, other sciences, or from history or sociology, the theory and examples of strategic games will stimulate and challenge your intellect. We urge you to enjoy the subject even as you are studying or teaching it.

SUMMARY

Strategic *games* situations are distinguished from individual decision-making situations by the presence of significant interactions among the players. Games can be classified according to a variety of categories including the timing of play, the common or conflicting interests of players, the number of times an interaction occurs, the amount of information available to the players, the type of rules, and the feasibility of coordinated action.

Learning the terminology for a game's structure is crucial for analysis. Players have *strategies* that lead to different *outcomes* with different associated *payoffs*. Payoffs incorporate everything that is important to a player about a game and are calculated by using probabilistic averages or *expectations* if outcomes are random or include some risk. *Rationality,* or consistent behavior, is assumed of all players, who must also be aware of all of the relevant rules of conduct. *Equilibrium* arises when all players use strategies that are best responses to others' strategies; some classes of games allow learning from experience and the study of dynamic movements toward equilibrium. The study of behavior in actual game situations provides additional information about the performance of the theory.

Game theory may be used for explanation, prediction, or prescription in various circumstances. Although not perfect in any of these roles, the theory continues to evolve; the importance of strategic interaction and strategic thinking has also become more widely understood and accepted.

KEY TERMS[3]

asymmetric information (23)
cooperative game (26)
decision (18)
equilibrium (33)
evolutionary game (35)
expected payoff (29)
external uncertainty (23)
game (18)
imperfect information (23)
incomplete information (23)
noncooperative game (26)
payoff (28)
perfect information (23)
rational behavior (30)
screening (24)
screening device (24)
sequential moves (20)
signal (24)
signaling (24)
simultaneous moves (20)
strategic game (18)
strategic uncertainty (23)
strategies (27)

SOLVED EXERCISES[4]

S1. Determine which of the following situations describe games and which describe decisions. In each case, indicate what specific features of the situation caused you to classify it as you did.

(a) A group of grocery shoppers in the dairy section, with each shopper choosing a flavor of yogurt to purchase

(b) A pair of teenage girls choosing dresses for their prom

(c) A college student considering what type of postgraduate education to pursue

(d) The *New York Times* and the *Wall Street Journal* choosing the prices for their online subscriptions this year

(e) A presidential candidate picking a running mate

S2. Consider the strategic games described below. In each case, state how you would classify the game according to the six dimensions outlined in the text. (i) Are moves sequential or simultaneous? (ii) Is the game zero-sum or not? (iii) Is the game repeated? (iv) Is there imperfect information, and if so, is there incomplete (asymmetric) information? (v) Are the rules fixed or not? (vi) Are cooperative agreements possible or not? If you do not have enough information to classify a game in a particular dimension, explain why not.

(a) *Rock-Paper-Scissors:* On the count of three, each player makes the shape of one of the three items with his hand. Rock beats Scissors, Scissors beats Paper, and Paper beats Rock.

[3]The number in parentheses after each key term is the page on which that term is defined or discussed.

[4]**Note to Students:** The solutions to the **Solved Exercises** are found on the following Web site, which is free and open to all: **wwnorton.com/books/games_of_strategy**.

(b) *Roll-call voting:* Voters cast their votes orally as their names are called. The choice with the most votes wins.

(c) *Sealed-bid auction:* Bidders on a bottle of wine seal their bids in envelopes. The highest bidder wins the item and pays the amount of his bid.

S3. "A game player would never prefer an outcome in which every player gets a little profit to an outcome in which he gets all the available profit." Is this statement true or false? Explain why in one or two sentences.

S4. You and a rival are engaged in a game in which there are three possible outcomes: you win, your rival wins (you lose), or the two of you tie. You get a payoff of 50 if you win, a payoff of 20 if you tie, and a payoff of zero if you lose. What is your expected payoff in each of the following situations?

(a) There is a 50% chance that the game ends in a tie, but only a 10% chance that you win. (There is thus a 40% chance that you lose.)

(b) There is a 50-50 chance that you win or lose. There are no ties.

(c) There is an 80% chance that you lose, a 10% chance that you win, and a 10% chance that you tie.

S5. Explain the difference between game theory's use as a predictive tool and its use as a prescriptive tool. In what types of real-world settings might these two uses be most important?

UNSOLVED EXERCISES

U1. Determine which of the following situations describe games and which describe decisions. In each case, indicate what specific features of the situation caused you to classify it as you did.

(a) A party nominee for president of the United States must choose whether to use private financing or public financing for her campaign.

(b) Frugal Fred receives a $20 gift card for downloadable music and must choose whether to purchase individual songs or whole albums.

(c) Beautiful Belle receives 100 replies to her online dating profile and must choose whether to reply to each of them.

(d) NBC chooses how to distribute its television shows online this season. They consider Amazon.com, iTunes, and/or NBC.com. The fee they might pay to Amazon or to iTunes is open to negotiation.

(e) China chooses a level of tariffs to apply to American imports.

U2. Consider the strategic games described below. In each case, state how you would classify the game according to the six dimensions outlined in the text. (i) Are moves sequential or simultaneous? (ii) Is the game zero-sum or not? (iii) Is the game repeated? (iv) Is there imperfect information, and if so, is there incomplete (asymmetric) information? (v) Are the rules fixed or not? (vi) Are cooperative agreements possible or not? If you do not have enough information to classify a game in a particular dimension, explain why not.

(a) Garry and Ross are sales representatives for the same company. Their manager informs them that of the two of them, whoever sells more this year wins a Cadillac.

(b) On the game show *The Price is Right,* four contestants are asked to guess the price of a television set. Play starts with the leftmost player, and each player's guess must be different from the guesses of the previous players. The person who comes closest to the real price, without going over it, wins the television set.

(c) Six thousand players each pay $10,000 to enter the World Series of Poker. Each starts the tournament with $10,000 in chips, and they play No-Limit Texas Hold 'Em (a type of poker) until someone wins all the chips. The top six hundred players each receive prize money according to the order of finish, with the winner receiving more than $8,000,000.

(d) Passengers on Desert Airlines are not assigned seats; passengers choose seats once they board. The airline assigns the order of boarding according to the time the passenger checks in, either on the Web site up to 24 hours before takeoff, or in person at the airport.

U3. "Any gain by the winner must harm the loser." Is this statement true or false? Explain your reasoning in one or two sentences.

U4. Alice, Bob, and Confucius are bored during recess, so they decide to play a new game. Each of them puts a dollar in the pot, and each tosses a quarter. Alice wins if the coins land all heads or all tails. Bob wins if two heads and one tail land, and Confucius wins if one head and two tails land. The quarters are fair, and the winner receives a net payment of $2 ($3 − $1 = $2), and the losers lose their $1.

(a) What is the probability that Alice will win and the probability that she will lose?

(b) What is Alice's expected payoff?

(c) What is the probability that Confucius will win and the probability that he will lose?

(d) What is Confucius' expected payoff?

(e) Is this a zero-sum game? Please explain your answer.

U5. "When one player surprises another, this indicates that the players did not have common knowledge of the rules." Give an example that illustrates this statement, and give a counterexample that shows that the statement is not always true.

PART TWO

Concepts and Techniques

Games with Sequential Moves

Sequential-move games entail strategic situations in which there is a strict order of play. Players take turns making their moves, and they know what players who have gone before them have done. To play well in such a game, participants must use a particular type of interactive thinking. Each player must consider: If I make this move, how will my opponent respond? Whenever actions are taken, players need to think about how their current actions will influence future actions, both for their rivals and for themselves. Players thus decide their current moves on the basis of calculations of future consequences.

Most actual games combine aspects of both sequential- and simultaneous-move situations. But the concepts and methods of analysis are more easily understood if they are first developed separately for the two pure cases. Therefore in this chapter we study purely sequential games. Chapters 4 and 5 deal with purely simultaneous games, and Chapter 6 and parts of Chapters 7 and 8 show how to combine the two types of analysis in more realistic mixed situations. The analysis presented here can be used whenever a game includes sequential decision making. Analysis of sequential games also provides information about when it is to a player's advantage to move first and when it is better to move second. Players can then devise ways, called *strategic moves*, to manipulate the order of play to their advantage. The analysis of such moves is the focus of Chapter 10.

1 GAME TREES

We begin by developing a graphical technique for displaying and analyzing sequential-move games, called a **game tree.** This tree is referred to as the **extensive form** of a game. It shows all the component parts of the game that we introduced in Chapter 2: players, actions, and payoffs.

You have probably come across **decision trees** in other contexts. Such trees show all the successive decision points, or nodes, for a single decision maker in a neutral environment. Decision trees also include branches corresponding to the available choices emerging from each node. Game trees are just joint decision trees for all of the players in a game. The trees illustrate all of the possible actions that can be taken by all of the players and indicate all of the possible outcomes of the game.

A. Nodes, Branches, and Paths of Play

Figure 3.1 shows the tree for a particular sequential game. We do not supply a story for this game, because we want to omit circumstantial details and to help you focus on general concepts. Our game has four players: Ann, Bob, Chris, and Deb. The rules of the game give the first move to Ann; this is shown at the leftmost point, or **node,** which is called the **initial node** or **root** of the game tree. At this node, which may also be called an **action** or **decision node,** Ann has two choices available to her. Ann's possible choices are labeled "Stop" and "Go" (remember that these labels are abstract and have no necessary significance) and are shown as **branches** emerging from the initial node.

If Ann chooses "Stop," then it will be Bob's turn to move. At his action node, he has three available choices labeled 1, 2, and 3. If Ann chooses "Go," then Chris gets the next move, with choices "Risky" and "Safe." Other nodes and branches follow successively and, rather than list them all in words, we draw your attention to a few prominent features.

If Ann chooses "Stop" and then Bob chooses 1, Ann gets another turn, with new choices, "Up" and "Down." It is quite common in actual sequential-move games for a player to get to move several times and to have her available moves differ at different turns. In chess, for example, two players make alternate moves; each move changes the board and therefore the available moves at subsequent turns.

B. Uncertainty and "Nature's Moves"

If Ann chooses "Go" and then Chris chooses "Risky," something happens at random—a fair coin is tossed and the outcome of the game is determined by whether that coin comes up "heads" or "tails." This aspect of the game is an

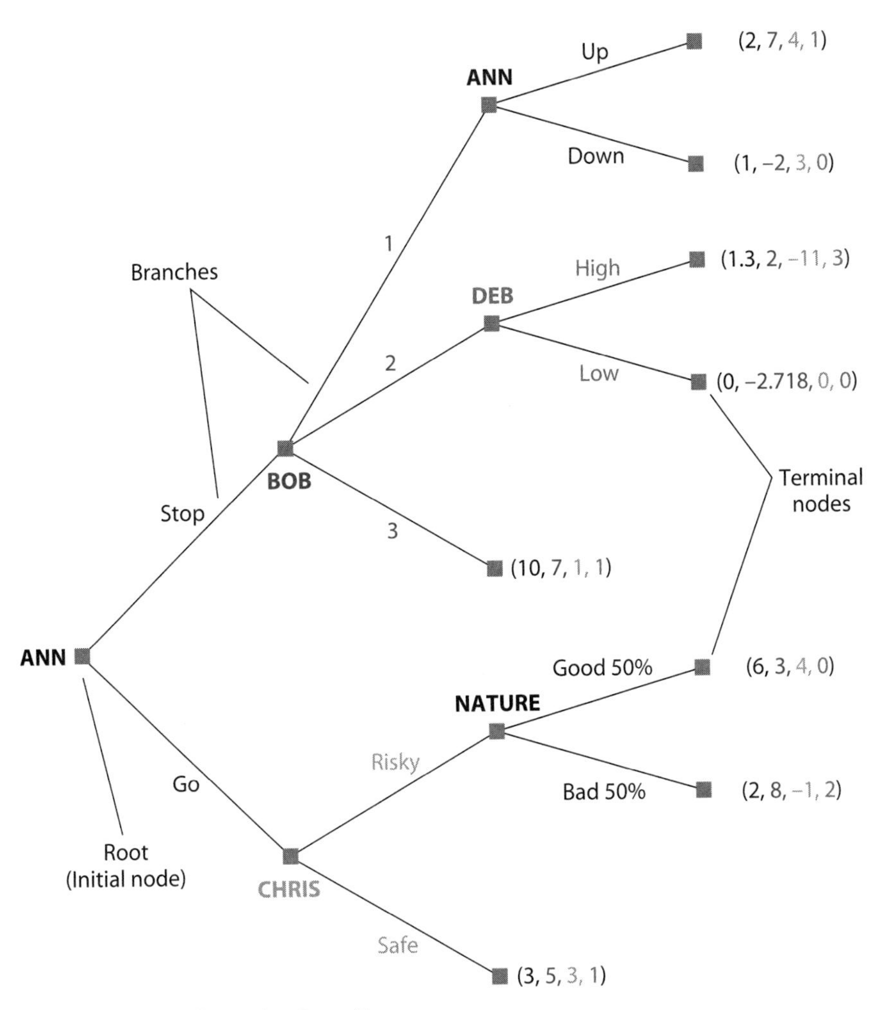

FIGURE 3.1 An Illustrative Game Tree

example of external uncertainty and is handled in the tree by introducing an outside player called "Nature." Control over the random event is ceded to the player known as Nature, who chooses, as it were, one of two branches, each with 50% probability. The probabilities here are fixed by the type of random event, a coin toss, but could vary in other circumstances; for example, with the throw of a die, Nature could specify six possible outcomes, each with $16\frac{2}{3}$% probability. Use of the player Nature allows us to introduce external uncertainty in a game and gives us a mechanism to allow things to happen that are outside the control of any of the actual players.

You can trace a number of different paths through the game tree by following successive branches. In Figure 3.1, each path leads you to an end point of the game after a finite number of moves. An end point is not a necessary feature of all games; some may in principle go on forever. But most applications that we will consider are finite games.

C. Outcomes and Payoffs

At the last node along each path, called a **terminal node,** no player has another move. (Note that terminal nodes are thus distinguished from *action* nodes.) Instead, we show the outcome of that particular sequence of actions, as measured by the payoffs for the players. For our four players, we list the payoffs in order (Ann, Bob, Chris, Deb). It is important to specify which payoff belongs to which player. The usual convention is to list payoffs in the order in which the players make the moves. But this method may sometimes be ambiguous; in our example, it is not clear whether Bob or Chris should be said to have the second move. Thus we have used alphabetical order. Further, we have color-coded everything so that Ann's name, choices, and payoffs are all in black; Bob's in dark green; Chris's in grey; and Deb's in light green. When drawing trees for any games that you analyze, you can choose any specific convention you like, but you should state and explain it clearly for the reader.

The payoffs are numerical, and generally for each player a higher number means a better outcome. Thus, for Ann, the outcome of the bottommost path (payoff 3) is better than that of the topmost path (payoff 2) in Figure 3.1. But there is no necessary comparability across players. Thus there is no necessary sense in which, at the end of the topmost path, Bob (payoff 7) does better than Ann (payoff 2). Sometimes, if payoffs are dollar amounts, for example, such interpersonal comparisons may be meaningful.

Players use information about payoffs when deciding among the various actions available to them. The inclusion of a random event (a choice made by Nature) means that players need to determine what they get on average when Nature moves. For example, if Ann chooses "Go" at the game's first move, Chris may then choose "Risky," giving rise to the coin toss and Nature's "choice" of "Good" or "Bad." In this situation, Ann could anticipate a payoff of 6 half the time and a payoff of 2 half the time, or a statistical average or *expected payoff* of $4 = (0.5 \times 6) + (0.5 \times 2)$.

D. Strategies

Finally, we use the tree in Figure 3.1 to explain the concept of a strategy. A single action taken by a player at a node is called a **move.** But players can, do, and should make plans for the succession of moves that they expect to make in all of the various eventualities that might arise in the course of a game. Such a plan of action is called a strategy.

In this tree, Bob, Chris, and Deb each get to move at most once; Chris, for example, gets a move only if Ann chooses "Go" on her first move. For them, there is no distinction between a move and a strategy. We can qualify the move by specifying the contingency in which it gets made; thus, a strategy for Bob might be,

"Choose 1 if Ann has chosen Stop." But Ann has two opportunities to move, so her strategy needs a fuller specification. One strategy for her is, "Choose Stop, and then if Bob chooses 1, choose Down."

In more complex games such as chess, where there are long sequences of moves with many choices available at each, descriptions of strategies get very complicated; we consider this aspect in more detail later in this chapter. But the general principle for constructing strategies is simple, except for one peculiarity. If Ann chooses "Go" on her first move, she never gets to make a second move. Should a strategy in which she chooses "Go" also specify what she would do in the hypothetical case in which she somehow found herself at the node of her second move? Your first instinct may be to say *no,* but formal game theory says *yes,* and for two reasons.

First, Ann's choice of "Go" at the first move may be influenced by her consideration of what she would have to do at her second move if she were to choose "Stop" originally instead. For example, if she chooses "Stop," Bob may then choose 1; then Ann gets a second move and her best choice would be "Up," giving her a payoff of 2. If she chooses "Go" on her first move, Chris would choose "Safe" (because his payoff of 3 from "Safe" is better than his expected payoff of 1.5 from "Risky"), and that outcome would yield Ann a payoff of 3. To make this thought process clearer, we state Ann's strategy as, "Choose Go at the first move, and choose Up if the next move arises."

The second reason for this seemingly pedantic specification of strategies has to do with the stability of equilibrium. When considering stability, we ask what would happen if players' choices were subjected to small disturbances. One such disturbance is that players make small mistakes. If choices are made by pressing a key, for example, Ann may intend to press the "Go" key, but there is a small probability that her hand may tremble and she may press the "Stop" key instead. In such a setting, it is important to specify how Ann will follow up when she discovers her error because Bob chooses 1 and it is Ann's turn to move again. More advanced levels of game theory require such stability analyses, and we want to prepare you for that by insisting on your specifying strategies as such complete plans of action right from the beginning.

E. Tree Construction

Now we sum up the general concepts illustrated by the tree of Figure 3.1. Game trees consist of nodes and branches. Nodes are connected to one another by the branches and come in two types. The first node type is called a decision node. Each decision node is associated with the player who chooses an action at that node; every tree has one decision node that is the game's initial node, the starting point of the game. The second type of node is called a terminal node. Each terminal node has associated with it a set of outcomes for the players taking part in the game; these outcomes are the payoffs received by each player if the game has followed the branches that lead to this particular terminal node.

The branches of a game tree represent the possible actions that can be taken from any decision node. Each branch leads from a decision node on the tree either to another decision node, generally for a different player, or to a terminal node. The tree must account for all of the possible choices that could be made by a player at each node; so some game trees include branches associated with the choice "Do nothing." There must be at least one branch leading from each decision node, but there is no maximum. Every decision node can have only one branch leading to it, however.

Game trees are often drawn from left to right across a page. However, game trees can be drawn in any orientation that best suits the game at hand: bottom up, sideways, top down, or even radially outward from a center. The tree is a metaphor, and the important feature is the idea of successive branching, as decisions are made at the tree nodes.

2 SOLVING GAMES BY USING TREES

We illustrate the use of trees in finding equilibrium outcomes of sequential-move games in a very simple context that many of you have probably confronted—whether to smoke. This situation and many other similar one-player strategic situations can be described as games if we recognize that future choices are actually made by a different player. That player is one's future self who will be subject to different influences and will have different views about the ideal outcome of the game.

Take, for example, a teenager named Carmen who is deciding whether to smoke. First, she has to decide whether to try smoking at all. If she does try it, she has the further decision of whether to continue. We illustrate this example as a simple decision in the tree of Figure 3.2.

The nodes and the branches are labeled with Carmen's available choices, but we need to explain the payoffs. Choose the outcome of never smoking at all as the standard of reference, and call its payoff 0. There is no special significance to the number zero in this context; all that matters for comparing outcomes, and thus for Carmen's decision, is whether this payoff is bigger or smaller than the others. Suppose Carmen best likes the outcome in which she tries smoking for a while but does not continue. The reason may be that she just likes to have experienced many things first-hand or so that she can more convincingly be able to say "I have been there and know it to be a bad situation" when she tries in the future to dissuade her children from smoking. Give this outcome the payoff 1. The outcome in which she tries smoking and then continues is the worst. Leav-

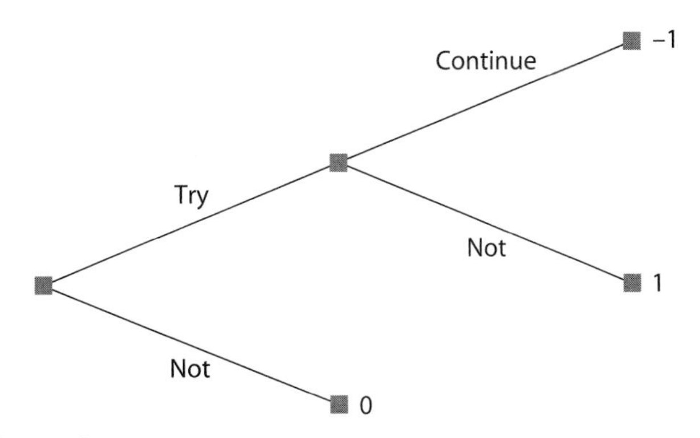

FIGURE 3.2 The Smoking Decision

ing aside the long-term health hazards, there are immediate problems—her hair and clothes will smell bad, and her friends will avoid her. Give this outcome the payoff -1. Carmen's best choice then seems clear—she should try smoking but she should not continue.

However, this analysis ignores the problem of addiction. Once Carmen has tried smoking for a while, she becomes a different person with different tastes, as well as different payoffs. The decision of whether to continue will be made not by "Today's Carmen" with today's assessment of outcomes as shown in Figure 3.2, but by a different "Future Carmen" with a different ranking of the alternatives then available. When she makes her choice today, she has to look ahead to this consequence and factor it into her current decision, which she should make on the basis of her current preferences. In other words, the choice problem concerning smoking is not really a decision in the sense explained in Chapter 2—a choice made by a single person in a neutral environment—but a game in the technical sense also explained in Chapter 2, where the other player is Carmen's future self with her own distinct preferences. When Today's Carmen makes her decision, she has to play against her future self.

We convert the decision tree of Figure 3.2 into a game tree in Figure 3.3, by distinguishing between the two players who make the choices at the two nodes. At the initial node, "Today's Carmen" decides whether to try smoking. If her decision is to try, then the addicted "Future Carmen" comes into being and chooses whether to continue. We show the healthy, non-polluting Today's Carmen, her actions, and her payoffs in green, and the addicted Future Carmen, her actions, and her payoffs in black, the color that her lungs have become. The payoffs of Today's Carmen are as before. But Future Carmen will enjoy continuation and will suffer terrible withdrawal symptoms if she does not continue. Let Future Carmen's payoff from "continue" be $+1$ and that from "not" be -1.

Given the preferences of the addicted Future Carmen, she will choose "continue" at her decision node. Today's Carmen should look ahead to this prospect and fold it into her current decision, recognizing that the choice to

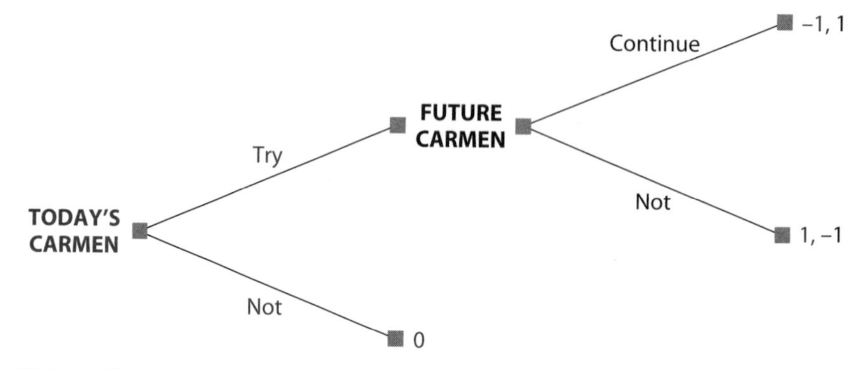

FIGURE 3.3 The Smoking Game

try smoking will inevitably lead to continuation. Even though Today's Carmen does not want continuation to happen given her preferences today, she will not be able to implement her currently preferred choice at the future time, because a different Carmen with different preferences will make that choice. So Today's Carmen should foresee that the choice "Try" will lead to "Continue" and get her the payoff −1 as judged by her today, whereas the choice "Don't Try" will get her the payoff 0. So she should choose the latter.

This argument is shown more formally and with greater visual effect in Figure 3.4. In Figure 3.4a, we cut off, or **prune,** the branch "Not" emerging from the second node. This pruning corresponds to the fact that Future Carmen, who makes the choice at that node, will not choose the action associated with that branch, given her preferences as shown in black.

The tree that remains has two branches emerging from the first node where Today's Carmen makes her choice; each of these branches now leads directly to a terminal node. The pruning allows Today's Carmen to forecast completely the eventual consequence of each of her choices. "Try" will be followed by "Continue" and yield a payoff −1, as measured in the preferences of Today's Carmen, while "Not" will yield 0. Carmen's choice today should then be "Not" rather than "Try." Therefore we can prune the "Try" branch emerging from the first node (along with its foreseeable continuation). This pruning is done in Figure 3.4b. The tree shown there is now "fully pruned," leaving only one branch emerging from the initial node and leading to a terminal node. Following the only remaining path through the tree shows what will happen in the game when all players make their best choices with correct forecasting of all future consequences.

In pruning the tree in Figure 3.4, we crossed out the branches not chosen. Another equivalent but alternative way of showing player choices is to "highlight" the branches that *are* chosen. To do so, you can place check marks or arrowheads on these branches or show them as thicker lines. Any one method will do; Figure 3.5 shows them all. You can choose whether to prune or to highlight,

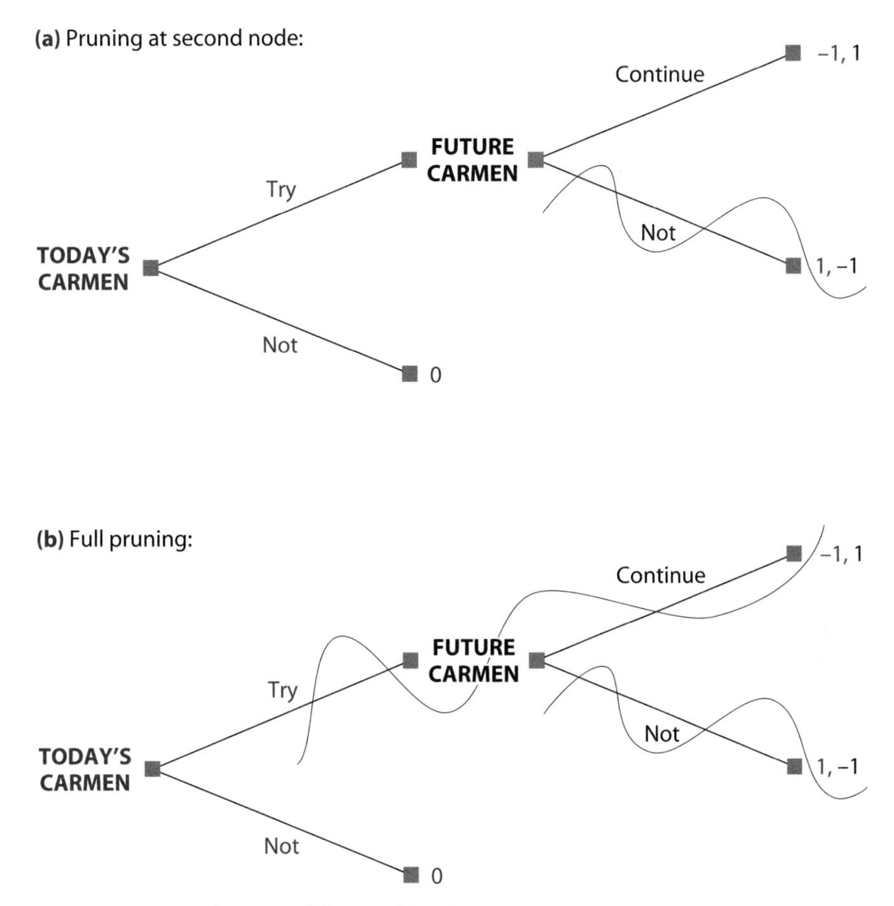

FIGURE 3.4 Pruning the Tree of the Smoking Game

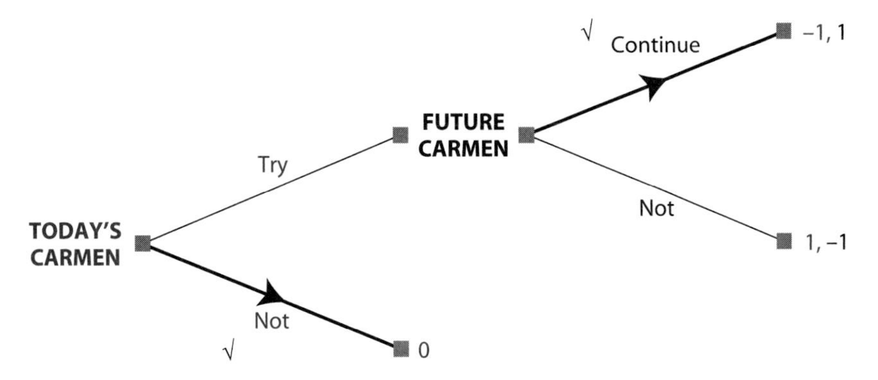

FIGURE 3.5 Showing Branch Selection on the Tree of the Smoking Game

but the latter, especially in its arrowhead form, has some advantages. First, it produces a cleaner picture. Second, the mess of the pruning picture sometimes does not clearly show the order in which various branches were cut. For example, in Figure 3.4b, a reader may get confused and incorrectly think that the "Continue" branch at the second node was cut first and that the "Try" branch at the first node followed by the "Not continue" branch at the second node were cut next. Finally, and most important, the arrowheads show the outcome of the sequence of optimal choices most visibly as a continuous link of arrows from the initial node to a terminal node. Therefore, in subsequent diagrams of this type, we generally use arrows instead of pruning. When you draw game trees, you should practice showing both methods for a while; when you are comfortable with trees, you can choose either to suit your taste.

No matter how you display your thinking in a game tree, the logic of the analysis is the same and important. You must start your analysis by considering those action nodes that lead directly to terminal nodes. The optimal choices for a player moving at such a node can be found immediately by comparing her payoffs at the relevant terminal nodes. With the use of these end-of-game choices to forecast consequences of earlier actions, the choices at nodes just preceding the final decision nodes can be determined. Then the same can be done for the nodes before them, and so on. By working backward along the tree in this way, you can solve the whole game.

This method of looking ahead and reasoning back to determine behavior in sequential-move games is known as **rollback.** As the name suggests, using rollback requires starting to think about what will happen at all the terminal nodes and literally "rolling back" through the tree to the initial node as you do your analysis. Because this reasoning requires working backward one step at a time, the method is also called **backward induction.** We use the term rollback because it is simpler and becoming more widely used, but other sources on game theory will use the older term backward induction. Just remember that the two are equivalent.

When all players choose their optimal strategies found by doing rollback analysis, we call this set of strategies the **rollback equilibrium** of the game; the outcome that arises from playing these strategies is the *rollback equilibrium outcome.* Game theory predicts this outcome as the equilibrium of a sequential game when all players are rational calculators in pursuit of their respective best payoffs. Later in this chapter, we address how well this prediction is borne out in practice. For now, you should know that all finite sequential-move games presented in this book have at least one rollback equilibrium. In fact, most have exactly one. Only in exceptional cases where a player gets equal payoffs from two or more different sets of moves, and is therefore indifferent between them, will games have more than one rollback equilibrium.

In the smoking game, the rollback equilibrium is where Today's Carmen chooses the strategy "Not" and Future Carmen chooses the strategy "Continue."

When Today's Carmen takes her optimal action, the addicted Future Carmen does not come into being at all and therefore gets no actual opportunity to move. But Future Carmen's shadowy presence and the strategy that she would choose if Today's Carmen chose "Try" and gave her an opportunity to move are important parts of the game. In fact, they are instrumental in determining the optimal move for Today's Carmen.

We introduced the ideas of the game tree and rollback analysis in a very simple example, where the solution was obvious from verbal argument. Now we proceed to use the ideas in successively more complex situations, where verbal analysis becomes harder to conduct and the visual analysis with the use of the tree becomes more important.

3 ADDING MORE PLAYERS

The techniques developed in Section 2 in the simplest setting of two players and two moves can be readily extended. The trees get more complex, with more branches, nodes, and levels, but the basic concepts and the method of rollback remain unchanged. In this section, we consider a game with three players, each of whom has two choices; with slight variations, this game reappears in many subsequent chapters.

The three players, Emily, Nina, and Talia, all live on the same small street. Each has been asked to contribute toward the creation of a flower garden at the intersection of their small street with the main highway. The ultimate size and splendor of the garden depends on how many of them contribute. Furthermore, although each player is happy to have the garden—and happier as its size and splendor increase—each is reluctant to contribute because of the cost that she must incur to do so.

Suppose that, if two or all three contribute, there will be sufficient resources for the initial planting and subsequent maintenance of the garden; it will then be quite attractive and pleasant. However, if one or none contribute, it will be too sparse and poorly maintained to be pleasant. From each player's perspective, there are thus four distinguishable outcomes:

- She does not contribute, both of the others do (pleasant garden, saves cost of own contribution)
- She contributes, and one or both of the others do (pleasant garden, incurs cost of contribution)
- She does not contribute, only one or neither of the others does (sparse garden, saves cost of own contribution)
- She contributes, but neither of the others does (sparse garden, incurs cost of own contribution)

Of these outcomes, the one listed at the top is clearly the best and the one listed at the bottom is clearly the worst. We want higher payoff numbers to indicate outcomes that are more highly regarded; so we give the top outcome the payoff 4 and the bottom one the payoff 1. (Sometimes payoffs are associated with an outcome's rank order; so, with four outcomes, 1 would be best and 4 worst, and smaller numbers would denote more preferred outcomes. When reading, you should carefully note which convention the author is using; when writing, you should carefully state which convention you are using.)

There is some ambiguity about the two middle outcomes. Let us suppose that each player regards a pleasant garden more highly than her own contribution. Then the outcome listed second gets payoff 3, and the outcome listed third gets payoff 2.

Suppose the players move sequentially. Emily has the first move, and chooses whether to contribute. Then, after observing what Emily has chosen, Nina makes her choice between contributing and not contributing. Finally, having observed what Emily and Nina have chosen, Talia makes a similar choice.[1]

Figure 3.6 shows the tree for this game. We have labeled the action nodes for easy reference. Emily moves at the initial node, *a*, and the branches corresponding to

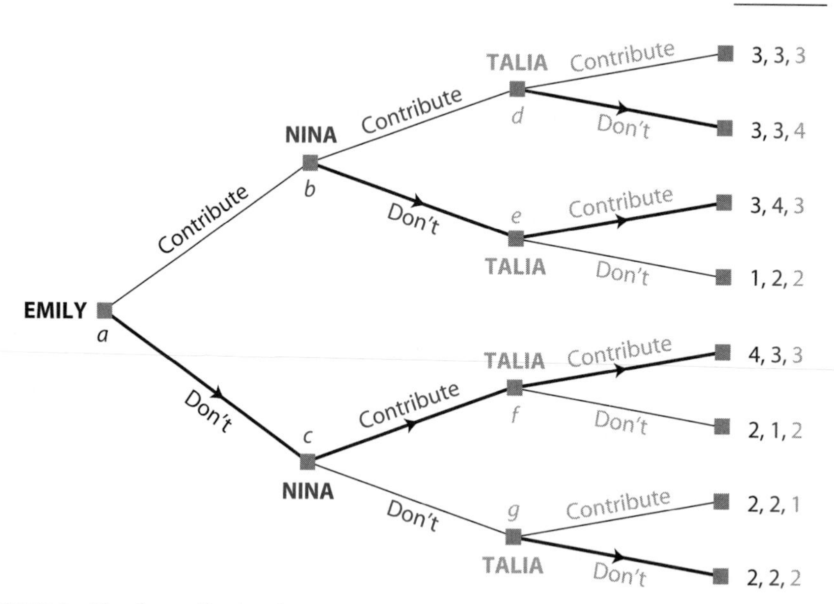

FIGURE 3.6 The Street Garden Game

[1]In later chapters, we vary the rules of this game—the order of moves and payoffs—and examine how such variation changes the outcomes.

her two choices, Contribute and Don't, respectively, lead to nodes *b* and *c*. At each of these nodes, Nina gets to move and to choose between Contribute and Don't. Her choices lead to nodes *d, e, f,* and *g,* at each of which Talia gets to move. Her choices lead to eight terminal nodes, where we show the payoffs in order (Emily, Nina, Talia).[2] For example, if Emily contributes, then Nina does not, and finally Talia does, then the garden is pleasant, and the two contributors get payoffs 3 each, while the noncontributor gets her top outcome with payoff 4; in this case, the payoff list is (3, 4, 3).

To apply rollback analysis to this game, we begin with the action nodes that come immediately before the terminal nodes—namely, *d, e, f,* and *g*. Talia moves at each of these nodes. At *d,* she faces the situation where both Emily and Nina have contributed. The garden is already assured to be pleasant; so, if Talia chooses Don't, she gets her best outcome, 4, whereas, if she chooses Contribute, she gets the next best, 3. Her preferred choice at this node is Don't. We show this preference both by thickening the branch for Don't and by adding an arrowhead; either one would suffice to illustrate Talia's choice. At node *e,* Emily has contributed and Nina has not; so Talia's contribution is crucial for a pleasant garden. Talia gets the payoff 3 if she chooses Contribute and 2 if she chooses Don't. Her preferred choice at *e* is Contribute. You can check Talia's choices at the other two nodes similarly.

Now we roll back the analysis to the preceding stage—namely, nodes *b* and *c,* where it is Nina's turn to choose. At *b,* Emily has contributed. Nina's reasoning now goes as follows: "If I choose Contribute, that will take the game to node *d,* where I know that Talia will choose Don't, and my payoff will be 3. (The garden will be pleasant, but I will have incurred the cost of my contribution.) If I choose Don't, the game will go to node *e,* where I know that Talia will choose Contribute, and I will get a payoff of 4. (The garden will be pleasant, and I will have saved the cost of my contribution.) Therefore I should choose Don't." Similar reasoning shows that at *c,* Nina will choose Contribute.

Finally, consider Emily's choice at the initial node, *a.* She can foresee the subsequent choices of both Nina and Talia. Emily knows that, if she chooses Contribute, these later choices will be Don't for Nina and Contribute for Talia. With two contributors, the garden will be pleasant but Emily will have incurred a cost; so her payoff will be 3. If Emily chooses Don't, then the subsequent choices will both be Contribute, and, with a pleasant garden and no cost of her own contribution, Emily's payoff will be 4. So her preferred choice at *a* is Don't.

The result of rollback analysis for this street garden game is now easily summarized. Emily will choose Don't, then Nina will choose Contribute, and finally

[2]Recall from the discussion of the general tree in Section 1 that the usual convention for sequential-move games is to list payoffs in the order in which the players move; however, in case of ambiguity or simply for clarity, it is good practice to specify the order explicitly.

Talia will choose Contribute. These choices trace a particular **path of play** through the tree—along the lower branch from the initial node, *a,* and then along the upper branches at each of the two subsequent nodes reached, *c* and *f.* In Figure 3.6, the path of play is easily seen as the continuous sequence of arrowheads joined tail to tip from the initial node to the terminal node fifth from the top of the tree. The payoffs that accrue to the players are shown at this terminal node.

Rollback analysis is simple and appealing. Here, we emphasize some features that emerge from it. First, notice that the **equilibrium path of play** of a sequential-move game misses most of the branches and nodes. Calculating the best actions that would be taken if these other nodes were reached, however, is an important part of determining the ultimate equilibrium. Choices made early in the game are affected by players' expectations of what would happen if they chose to do something other than their best actions and by what would happen if any opposing player chose to do something other than what was best for her. These expectations, based on predicted actions at out-of-equilibrium nodes (nodes associated with branches pruned in the process of rollback), keep players choosing optimal actions at each node. For instance, Emily's optimal choice of Don't at the first move is governed by the knowledge that, if she chose Contribute, then Nina would choose Don't, followed by Talia choosing Contribute; this sequence would give Emily the payoff 3, instead of the 4 that she can get by choosing Don't at the first move.

The rollback equilibrium gives a complete statement of all this analysis by specifying the optimal *strategy* for each player. Recall that a strategy is a complete plan of action. Emily moves first and has just two choices, so her strategy is quite simple and is effectively the same thing as her move. But Nina, moving second, acts at one of two nodes, at one if Emily has chosen Contribute and at the other if Emily has chosen Don't. Nina's complete plan of action has to specify what she would do in either case. One such plan, or strategy, might be "choose Contribute if Emily has chosen Contribute, choose Don't if Emily has chosen Don't." We know from our rollback analysis that Nina will not choose this strategy, but our interest at this point is in describing all the available strategies from which Nina can choose within the rules of the game. We can abbreviate and write C for Continue and D for Don't; then this strategy can be written as "C if Emily chooses C so that the game is at node *b,* D if Emily chooses D so that the game is at node *c,*" or, more simply, "C at *b,* D at *c,*" or even "CD" if the circumstances in which each of the stated actions is taken are evident or previously explained. Now it is easy to see that, because Nina has two choices available at each of the two nodes where she might be acting, she has available to her four plans, or strategies—"C at *b,* C at *c,*" "C at *b,* D at *c,*" "D at *b,* C at *c,*" and "D at *b,* D at *c,*" or "CC," "CD," "DC," and "DD." Of these strategies, the rollback analysis and the arrows at nodes *b* and *c* of Figure 3.6 show that her optimal strategy is "DC."

Matters are even more complicated for Talia. When her turn comes, the history of play can, according to the rules of the game, be any one of four possibilities. Talia's turn to act comes at one of four nodes in the tree, one after Emily has chosen C and Nina has chosen C (node *d*), the second after Emily's C and Nina's D (node *e*), the third after Emily's D and Nina's C (node *f*), and the fourth after both Emily and Nina choose D (node *g*). Each of Talia's strategies, or complete plans of action, must specify one of her two actions for each of these four scenarios, or one of her two actions at each of her four possible action nodes. With four nodes at which to specify an action and with two actions from which to choose at each node, there are 2 times 2 times 2 times 2, or 16 possible combinations of actions. So Talia has available to her 16 possible strategies. One of them could be written as

"C at *d*, D at *e*, D at *f*, C at *g*" or "CDDC" for short,

where we have fixed the order of the four scenarios (the histories of moves by Emily and Nina) in the order of nodes *d*, *e*, *f*, and *g*. Then, with the use of the same abbreviation, the full list of 16 strategies available to Talia is

CCCC, CCCD, CCDC, CCDD, CDCC, CDCD, CDDC, CDDD,
DCCC, DCCD, DCDC, DCDD, DDCC, DDCD, DDDC, DDDD.

Of these strategies, the rollback analysis of Figure 3.6 and the arrows at nodes *d*, *e*, *f*, and *g* show that Talia's optimal strategy is DCCD.

Now we can express the findings of our rollback analysis by stating the strategy choices of each player—Emily chooses D from the two strategies available to her, Nina chooses DC from the four strategies available to her, and Talia chooses DCCD from the sixteen strategies available to her. When each player looks ahead in the tree to forecast the eventual outcomes of her current choices, she is calculating the optimal strategies of the other players. This configuration of strategies, D for Emily, DC for Nina, and DCCD for Talia, then constitutes the rollback equilibrium of the game.

We can put together the optimal strategies of the players to find the actual path of play that will result in the rollback equilibrium. Emily will begin by choosing D. Nina, following her strategy DC, chooses the action C in response to Emily's D. (Remember that Nina's DC means "choose D if Emily has played C, and choose C if Emily has played D.") According to the convention that we have adopted, Talia's actual action after Emily's D and then Nina's C—from node *f*—is the third letter in the four-letter specification of her strategies. Because Talia's optimal strategy is DCCD, her action along the path of play is C. Thus the actual path of play consists of Emily playing D, followed successively by Nina and Talia playing C.

To sum up, we have three distinct concepts:

1. The lists of available strategies for each player. The list, especially for later players, may be very long, because their actions in situations corresponding to all conceivable preceding moves by other players must be specified.
2. The optimal strategy, or complete plan of action for each player. This strategy must specify the player's best choices at each node where the rules of the game specify that she moves, even though many of these nodes will never be reached in the actual path of play. This specification is in effect the preceding movers' forecasting of what would happen if they took different actions and is therefore an important part of their calculation of their own best actions at the earlier nodes. The optimal strategies of all players together yield the rollback equilibrium.
3. The actual path of play in the rollback equilibrium, found by putting together the optimal strategies for all the players.

4 ORDER ADVANTAGES

In the rollback equilibrium of the street-garden game, Emily gets her best outcome (payoff 4), because she can take advantage of the opportunity to make the first move. When she chooses not to contribute, she puts the onus on the other two players—each can get her next-best outcome if and only if both of them choose to contribute. Most casual thinkers about strategic games have the preconception that such **first-mover advantage** should exist in all games. However, that is not the case. It is easy to think of games in which an opportunity to move second is an advantage. Consider the strategic interaction between two firms that sell similar merchandise from catalogs—say, Land's End and L.L. Bean. If one firm had to release its catalog first, and then the second firm could see what prices the first had set before printing its own catalog, then the second mover could just undercut its rival on all items and gain a tremendous competitive edge.

First-mover advantage comes from the ability to commit oneself to an advantageous position and to force the other players to adapt to it; **second-mover advantage** comes from the flexibility to adapt oneself to the others' choices. Whether commitment or flexibility is more important in a specific game depends on its particular configuration of strategies and payoffs; no generally valid rule can be laid down. We will come across examples of both kinds of advantages throughout this book. The general point that there need not be first-mover advantage, a point that runs against much common perception, is so important that we felt it necessary to emphasize at the outset.

When a game has a first- or second-mover advantage, each player may try to manipulate the order of play so as to secure for herself the advantageous position. Tactics for such manipulation are strategic moves, which we consider in Chapter 10.

5 ADDING MORE MOVES

We saw in Section 3 that adding more players increases the complexity of the analysis of sequential-play games. In this section, we consider another type of complexity that arises from adding additional moves to the game. We can do so most simply in a two-person game by allowing players to alternate moves more than once. Then the tree is enlarged in the same fashion as a multiple-player game tree would be, but later moves in the tree are made by the players who have made decisions earlier in the same game.

Many common games, such as tic-tac-toe, checkers, and chess, are two-person strategic games with such alternating sequential moves. The use of game trees and rollback should allow us to "solve" such games—to determine the rollback equilibrium outcome and the equilibrium strategies leading to that outcome. Unfortunately, as the complexity of the game grows and as strategies become more and more intricate, the search for an optimal solution becomes more and more difficult as well. In such cases, when manual solution is no longer really feasible, computer routines such as Gambit, mentioned in Chapter 2, become useful.

A. Tic-Tac-Toe

Start with the most simple of the three examples mentioned in the preceding paragraph, tic-tac-toe, and consider an easier-than-usual version in which two players (X and O) each try to be the first to get two of their symbols to fill any row, column, or diagonal of a two-by-two game board. The first player has four possible positions in which to put her X. The second player then has three possible actions at each of four decision nodes. When the first player gets to her second turn, she has two possible actions at each of 12 (4 times 3) decision nodes. As Figure 3.7 shows, even this mini-game of tic-tac-toe has a very complex game tree. This tree is actually not too complex, because the game is guaranteed to end after the first player moves a second time; but there are still 24 terminal nodes to consider.

We show this tree merely as an illustration of how complex game trees can become in even simple (or simplified) games. As it turns out, using rollback on the mini-game of tic-tac-toe leads us quickly to an equilibrium. Rollback shows

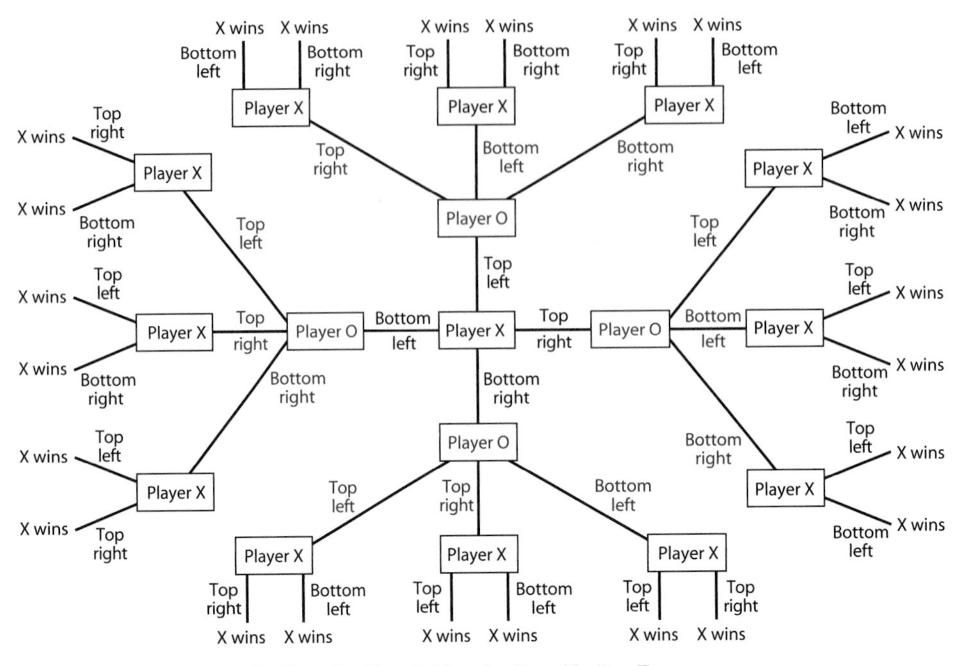

FIGURE 3.7 The Complex Tree for Simple Two-by-Two Tic-Tac-Toe

that all of the choices for the first player at her second move lead to the same outcome. There is no optimal action; any move is as good as any other move. Thus when the second player makes her first move, she also sees that each possible move yields the same outcome and she, too, is indifferent among her three choices at each of her four decision nodes. Finally, the same is true for the first player on her first move; any choice is as good as any other, so she is guaranteed to win the game.

Although this version of tic-tac-toe has an interesting tree, its solution is not as interesting. The first player always wins; so choices made by either player cannot affect the ultimate outcome. Most of us are more familiar with the three-by-three version of tic-tac-toe. To illustrate that version with a game tree, we would have to show that the first player has nine possible actions at the initial node, the second player has eight possible actions at each of nine decision nodes, and then the first player, on her second turn, has seven possible actions at each of 8 times 9 = 72 nodes, while the second player, on her second turn, has six possible actions at each of 7 times 8 times 9 = 504 nodes. This pattern continues until eventually the tree stops branching so rapidly because certain combinations of moves lead to a win for one player and the game ends. But no win is possible until at least the fifth move. Drawing the complete tree for this game requires a very large piece of paper or very tiny handwriting.

Most of you know, however, how to achieve at worst a tie when you play three-by-three tic-tac-toe. So there is a simple solution to this game that can be found by rollback, and a learned strategic thinker can reduce the complexity of the game considerably in the quest for such a solution. It turns out that, as in the two-by-two version, many of the possible paths through the game tree are strategically identical. Of the nine possible initial moves, there are only three types; you put your X in either a corner position (of which there are four possibilities), a side position (of which there are also four possibilities), or the (one) middle position. Using this method to simplify the tree can help reduce the complexity of the problem and lead you to a description of an optimal rollback equilibrium strategy. Specifically, we could show that the player who moves second can always guarantee at least a tie with an appropriate first move and then by continually blocking the first player's attempts to get three symbols in a row.[3]

B. Chess

Although relatively small games, such as tic-tac-toe, can be solved using rollback, we showed above how rapidly the complexity of game trees can increase even in two-player games. Thus when we consider more complicated games, such as chess, finding a complete solution becomes much more difficult.

In chess, the players, White and Black, have a collection of 16 pieces in six distinct shapes, each of which is bound by specified rules of movement on the eight-by-eight game board shown in Figure 3.8.[4] White opens with a move, Black responds with one, and so on, in turns. All the moves are visible to the other player, and nothing is left to chance, as it would be in card games that include shuffling and dealing. Moreover, a chess game must end in a finite number of moves. The rules declare that a game is drawn if a given position on the board is repeated three times in the course of play. Because there are a finite number of ways to place the 32 (or fewer after captures) pieces on 64 squares, a game could not go on infinitely long without running up against this rule. Therefore in principle chess is amenable to full rollback analysis.

That rollback analysis has not been carried out, however. Chess has not been "solved" as tic-tac-toe has been. And the reason is that, for all its simplicity of rules, chess is a bewilderingly complex game. From the initial set position of

[3]If the first player puts her first symbol in the middle position, the second player must put her first symbol in a corner position. Then the second player can guarantee a tie by taking the third position in any row, column, or diagonal that the first player tries to fill. If the first player goes to a corner or a side position first, the second player can guarantee a tie by going to the middle first and then following the same blocking technique. Note that, if the first player picks a corner, the second player picks the middle, and the first player then picks the corner opposite from her original play, then the second player must not pick one of the remaining corners if she is to ensure at least a tie.

[4]An easily accessible statement of the rules of chess and much more is at Wikipedia, at http://en.wikipedia.org/wiki/Chess.

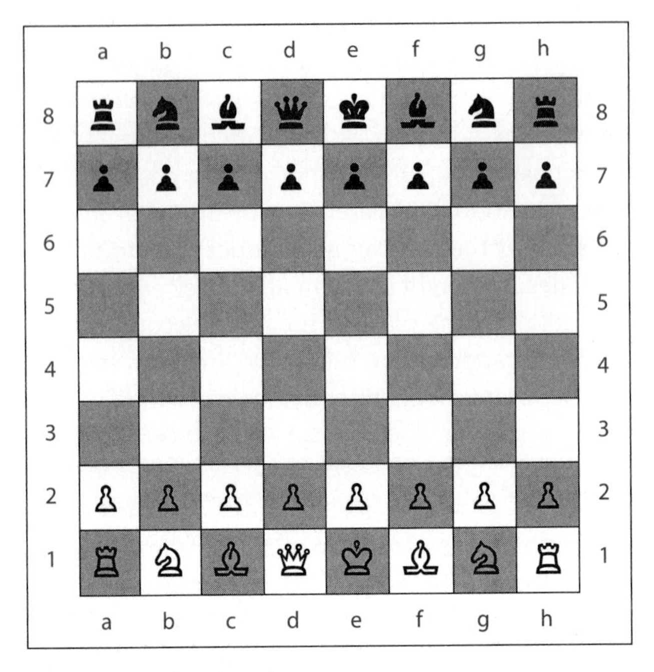

FIGURE 3.8 Chessboard

the pieces illustrated in Figure 3.8, White can open with any one of 20 moves,[5] and Black can respond with any of 20. Therefore 20 branches emerge from the first node of the game tree, each leading to a second node from each of which 20 more branches emerge. After only two moves, there are already 400 branches, each leading to a node from which many more branches emerge. And the total number of possible moves in chess has been estimated to be 10^{120}, or a "one" with 120 zeros after it. A supercomputer a thousand times as fast as your PC, making a trillion calculations a second, would need more than 10^{100} years to check out all these moves.[6] Astronomers offer us less than 10^{10} years before the sun turns into a red giant and swallows the earth.

The general point is that, although a game may be amenable in principle to a complete solution by rollback, its complete tree may be too complex to permit such solution in practice. Faced with such a situation, what is a player to do? We can learn a lot about this by reviewing the history of attempts to program computers to play chess.

[5]He can move one of eight pawns forward either one square or two or he can move one of the two knights in one of two ways (to squares a3, c3, f3, or h3).

[6]This would have to be done only once because, after the game has been solved, anyone can use the solution and no one will actually need to play. Everyone will know whether White has a win or whether Black can force a draw. Players will toss to decide who gets which color. They will then know the outcome, shake hands, and go home.

When computers first started to prove their usefulness for complex calculations in science and business, many mathematicians and computer scientists thought that a chess-playing computer program would soon beat the world champion. It took a lot longer, even though computer technology improved dramatically while human thought progressed much more slowly. Finally, in December 1992, a German chess program called Fritz2 beat world champion Gary Kasparov in some blitz (high-speed) games. Under regular rules, where each player gets $2\frac{1}{2}$ hours to make 40 moves, humans retained greater superiority for longer. A team sponsored by IBM put a lot of effort and resources into the development of a specialized chess-playing computer and its associated software. In February 1996, this package, called Deep Blue, was pitted against Gary Kasparov in a best-of-six series. Deep Blue caused a sensation by winning the first game, but Kasparov quickly figured out its weaknesses, improved his counterstrategies, and won the series handily. In the next 15 months, the IBM team improved Deep Blue's hardware and software, and the resulting Deeper Blue beat Kasparov in another best-of-six series in May 1997.

To sum up, computers have progressed in a combination of slow patches and some rapid spurts, while humans have held some superiority but have not been able to improve sufficiently fast to keep ahead. Closer examination reveals that the two use quite different approaches to thinking through the very complex game tree of chess.

When contemplating a move in chess, looking ahead to the end of the whole game may be too hard (for humans and computers both). How about looking part of the way—say, 5 or 10 moves ahead—and working back from there? The game need not end within this limited horizon; that is, the nodes that you reach after 5 or 10 moves will not generally be terminal nodes. Only terminal nodes have payoffs specified by the rules of the game. Therefore you need some indirect way of assigning plausible payoffs to nonterminal nodes, because you are not able to explicitly roll back from a full look-ahead. A rule that assigns such payoffs is called an **intermediate valuation function.**

In chess, humans and computer programs both use such partial look-ahead in conjunction with an intermediate valuation function. The typical method assigns values to each piece and to positional and combinational advantages that can arise during play. Quantification of values for different positions are made on the basis of the whole chess-playing community's experience of play in past games starting from such positions or patterns; this is called "knowledge." The sum of all the numerical values attached to pieces and their combinations in a position is the intermediate value of that position. A move is judged by the value of the position to which it is expected to lead after an explicit forward-looking calculation for a certain number—say, five or six—of moves.

The evaluation of intermediate positions has progressed furthest with respect to chess openings—that is, the first dozen or so moves of a game. Each opening can lead to any one of a vast multitude of further moves and positions, but experience enables players to sum up certain openings as being more or less likely to favor one player or the other. This knowledge has been written down in massive books of openings, and all top players and computer programs remember and use this information.

At the end stages of a game, when only a few pieces are left on the board, backward reasoning on its own is often simple enough to be doable and complete enough to give the full answer. The midgame, when positions have evolved into a level of complexity that will not simplify within a few moves, is the hardest to analyze. To find a good move from a midgame position, a well-built intermediate valuation function is likely to be more valuable than the ability to calculate another few moves further ahead.

This is where the art of chess playing comes into its own. The best human players develop an intuition or instinct that enables them to sniff out good opportunities and avoid subtle traps in a way that computer programs find hard to match. Computer scientists have found it generally very difficult to teach their machines the skills of pattern recognition that humans acquire and use instinctively—for example, recognizing faces and associating them with names. The art of the midgame in chess also is an exercise in recognizing and evaluating patterns in the same, still mysterious way. This is where Kasparov has his greatest advantage over Fritz2 or Deep Blue. It also explains why computer programs do better against humans at blitz or limited-time games: a human does not have the time to marshal his art of the midgame.

In other words, the best human players have subtle "chess knowledge," based on experience or the ability to recognize patterns, which endows them with a better intermediate valuation function. Computers have the advantage when it comes to raw or brute-force calculation. Thus although both human and computer players now use a mixture of look-ahead and intermediate valuation, they use them in different proportions: humans do not look so many moves ahead but have better intermediate valuations based on knowledge; computers have less sophisticated valuation functions but look ahead further by using their superior computational powers.

Recently, chess computers have begun to acquire more knowledge. When modifying Deep Blue in 1996 and 1997, IBM enlisted the help of human experts to improve the intermediate valuation function in its software. These consultants played repeatedly against the machine, noted its weaknesses, and suggested how the valuation function should be modified to correct the flaws. Deep Blue benefited from the contributions of the experts and their subtle kind of thinking, which results from long experience and an awareness of complex interconnections among the pieces on the board.

If humans can gradually make explicit their subtle knowledge and transmit it to computers, what hope is there for human players who do not get reciprocal help from computers? At times in their 1997 encounter, Kasparov was amazed by the human or even superhuman quality of Deep Blue's play. He even attributed one of the computer's moves to "the hand of God." And matters can only get worse: the brute-force calculating power of computers is increasing rapidly while they are simultaneously, but more slowly, gaining some of the subtlety that constitutes the advantage of humans.

The abstract theory of chess says that it is a finite game that can be solved by rollback. The practice of chess requires a lot of "art" based on experience, intuition, and subtle judgment. Is this bad news for the use of rollback in sequential-move games? We think not. It is true that theory does not take us all the way to an answer for chess. But it does take us a long way. Looking ahead a few moves constitutes an important part of the approach that mixes brute-force calculation of moves with a knowledge-based assessment of intermediate positions. And, as computational power increases, the role played by brute-force calculation, and therefore the scope of the rollback theory, will also increase.

Evidence from the study of the game of checkers, as we describe below, suggests that a solution to chess may yet be feasible.

C. Checkers

An astonishing number of computer and person hours have been devoted to the search for a solution to chess. Less famously, but just as doggedly, researchers have been working on solving the somewhat less complex game of checkers. Recently, this work was rewarded; the game of checkers was declared "solved" in July 2007.[7]

Checkers is another two-player game played on an eight-by-eight board. Each player has 12 round game pieces of different colors, as shown in Figure 3.9, and players take turns moving their pieces diagonally on the board, jumping (and capturing) the opponent's pieces when possible. As in chess, the game ends and Player A wins when Player B is either out of pieces or unable to move; the game can also end in a draw if both players agree that neither can win.

Although the complexity of checkers pales somewhat in comparison to that of chess—the number of possible positions in checkers is approximately the square root of the number in chess—there are still 5×10^{20} possible positions, so drawing a game tree is out of the question. Conventional wisdom and evidence from world championships for years suggested that good play should lead to a draw, but there was no proof. Now a computer scientist in Canada has the proof—a computer program named Chinook that can play to a guaranteed tie.

[7]Our account is based on two reports in the journal *Science*. See "Program Proves That Checkers, Perfectly Played, Is a No-Win Situation," Adrian Cho, *Science* (317), 20 July 2007, pp. 308–309, and "Checkers Is Solved," Jonathan Schaeffer et al., *Science* (317), 14 September 2007, pp. 1518–1522.

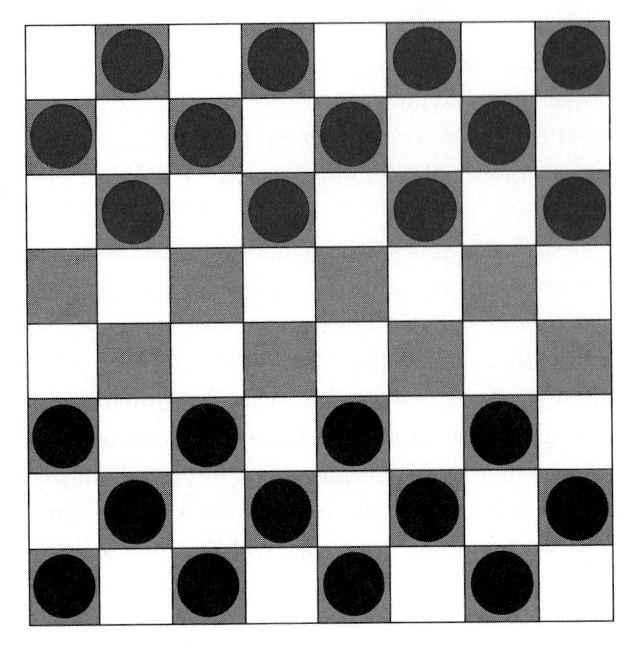

FIGURE 3.9 Checkers

Chinook was first created in 1989; played the world champion, Marion Tinsley, in 1992 (losing four to two with 33 draws) and again in 1994 (when Tinsley's health failed during a series of draws); was put on hold between 1997 and 2001 while its creators waited for computer technology to improve; and finally exhibited a loss-proof algorithm in the spring of 2007. That algorithm uses a combination of endgame rollback analysis and starting position forward analysis along with the equivalent of an intermediate valuation function to trace out the best moves within a database including all possible positions on the board.

The creators of Chinook describe the full game of checkers as "weakly solved"; they know that they can generate a tie and they have a strategy for reaching that tie from the start of the game. For all 39×10^{12} possible positions that include 10 or fewer pieces on the board, they describe checkers as "strongly solved"; not only do they know they can play to a tie, they can reach that tie from any of the possible positions that can arise once only 10 pieces remain. Their algorithm first solved the 10-piece endgames, then went back to the start to search out paths of play in which both players make optimal choices. The search mechanism, involving a complex system of evaluating the value of each intermediate position, invariably led to those 10-piece positions that generate a draw.

Thus our hope for the future of rollback analysis may not be misplaced. We know that for really simple games, we can find the rollback equilibrium by verbal reasoning without having to draw the game tree explicitly. For games having an intermediate range of complexity, verbal reasoning is too hard, but a complete tree can be drawn and used for rollback. Sometimes we may enlist the aid of a computer to draw and analyze a moderately complicated game tree. For the most complex games, such as checkers and chess, we can draw only a small part of the game tree, and we must use a combination of two methods: (1) calculation based on the logic of rollback; and (2) rules of thumb for valuing intermediate positions on the basis of experience. The computational power of current algorithms has shown that even some games in this category are amenable to solution, provided one has the time and resources to devote to the problem.

Thankfully, most of the strategic games that we encounter in economics, politics, sports, business, and daily life are far less complex than chess or even checkers. The games may have a number of players who move a number of times; they may even have a large number of players or a large number of moves. But we have a chance at being able to draw a reasonable-looking tree for those games that are sequential in nature. The logic of rollback remains valid, and it is also often the case that, once you understand the idea of rollback, you can carry out the necessary logical thinking and solve the game without explicitly drawing a tree. Moreover, it is precisely at this intermediate level of difficulty, between the simple examples that we solved explicitly in this chapter and the insoluble cases such as chess, that computer software such as Gambit is most likely to be useful; this is indeed fortunate for the prospect of applying the theory to solve many games in practice.

6 EVIDENCE CONCERNING ROLLBACK

How well do actual participants in sequential-move games perform the calculations of rollback reasoning? There is very little systematic evidence, but classroom and research experiments with some games have yielded outcomes that appear to counter the predictions of the theory. Some of these experiments and their outcomes have interesting implications for the strategic analysis of sequential-move games.

For instance, many experimenters have had subjects play a single-round bargaining game in which two players, designated A and B, are chosen from a class or a group of volunteers. The experimenter provides a dollar (or some known total), which can be divided between them according to the following procedure. Player A proposes a split—for example, “75 to me, 25 to B.” If player B accepts this proposal, the dollar is divided as proposed by A. If B rejects the proposal, neither player gets anything.

Rollback in this case predicts that B should accept any sum, no matter how small, because the alternative is even worse—namely, zero—and, foreseeing this, A should propose “99 to me, 1 to B.” This particular outcome almost never happens.

Most players assigned the A role propose a much more equal split. In fact, 50–50 is the single most common proposal. Furthermore, most players assigned the B role turn down proposals that leave them 25% or less of the total and walk away with nothing; some reject proposals that would give them 40% of the pie.[8]

Many game theorists remain unpersuaded that these findings undermine the theory. They counter with some variant of the following argument: "The sums are so small as to make the whole thing trivial in the players' minds. The B players lose 25 or 40 cents, which is almost nothing, and perhaps gain some private satisfaction that they walked away from a humiliatingly small award. If the total were a thousand dollars, so that 25% of it amounted to real money, the B players would accept." But this argument does not seem to be valid. Experiments with much larger stakes show similar results. The findings from experiments conducted in Indonesia, with sums that were small in dollars but amounted to as much as three months' earnings for the participants, showed no clear tendency on the part of the A players to make less equal offers, although the B players tended to accept somewhat smaller shares as the total increased; similar experiments conducted in the Slovak Republic found the behavior of inexperienced players unaffected by large changes in payoffs.[9]

The participants in these experiments typically have no prior knowledge of game theory and no special computational abilities. But the game is extremely simple; surely even the most naive player can see through the reasoning, and answers to direct questions after the experiment generally show that most participants do. The results show not so much the failure of rollback as the theorist's error in supposing that each player cares only about her own money earnings. Most societies instill in their members a strong sense of fairness, which then causes most A players to offer 50–50 or something close and the B players to reject anything that is grossly unfair. This argument is supported by the observation that even in a most drastic variant called the *dictator game*, where the A player decides on the split and the B player has no choice at all, many As give significant shares to the Bs.[10]

[8]Read Richard H. Thaler, "Anomalies: The Ultimate Game," *Journal of Economic Perspectives*, vol. 2, no. 4 (fall 1988), pp. 195–206, and Douglas D. Davis and Charles A. Holt, *Experimental Economics* (Princeton: Princeton University Press, 1993), pp. 263–269, for a detailed account of this game and related ones.

[9]The results of the Indonesian experiment are reported in Lisa Cameron, "Raising the Stakes in the Ultimatum Game: Experimental Evidence from Indonesia," *Economic Inquiry*, vol. 37, no. 1 (January 1999), pp. 47–59. Slonim and Roth report results similar to Cameron's, but also found that offers (in all rounds of play) were rejected less often as the payoffs were raised. See Robert Slonim and Alvin Roth, "Learning in High Stakes Ultimatum Games: An Experiment in the Slovak Republic," *Econometrica*, vol. 66, no. 3 (May 1998), pp. 569–596.

[10]One could argue that this social norm of fairness may actually have value in the ongoing evolutionary game being played by the whole of society. Players who are concerned with fairness reduce transaction costs and the costs of fights that can be beneficial to society in the long run. These matters will be discussed in Chapters 11 and 12.

Some experiments have been conducted to determine the extent to which fairness plays into behavior in ultimatum and dictator games. The findings show that changes in the information available to players and to the experimenter have a significant effect on the final outcome. In particular, when the experimental design is changed so that not even the experimenter can identify who proposed (or accepted) the split, the extent of sharing drops noticeably. Evidence from the new field of "neuroeconomics" includes similar results. Alan Sanfey and colleagues conducted experiments of the ultimatum game in which they took MRI readings of the subjects' brains as they made their choices. They found "that activity in a region [of the brain] well known for its involvement in negative emotion" was stimulated in the responders (B players) when they rejected "unfair" (less than 50:50) offers. Thus deep instincts or emotions of anger and disgust seem to be implicated in these rejections. They also found that "unfair" (less than 50:50) offers were rejected less often when responders knew that the offerer was a computer than when they knew that the offerer was human.[11]

Another experimental game with similarly paradoxical outcomes goes as follows. Two players are chosen and designated A and B. The experimenter puts a dime on the table. Player A can take it or pass. If A takes the dime, the game is over, with A getting the 10 cents and B getting nothing. If A passes, the experimenter adds a dime, and now B has the choice of taking the 20 cents or passing. The turns alternate, and the pile of money grows until reaching some limit—say, a dollar—that is known in advance by both players.

We show the tree for this game in Figure 3.10. Because of the appearance of the tree, this type of game is often called the *centipede game*. You may not even need the tree to use rollback on this game. Player B is sure to take the dollar at the last stage; so A should take the 90 cents at the penultimate stage, and so on. Thus A should take the very first dime and end the game.

However, in most classroom or experimental settings, such games go on for at least a few rounds. Remarkably, by behaving "irrationally," the players as a

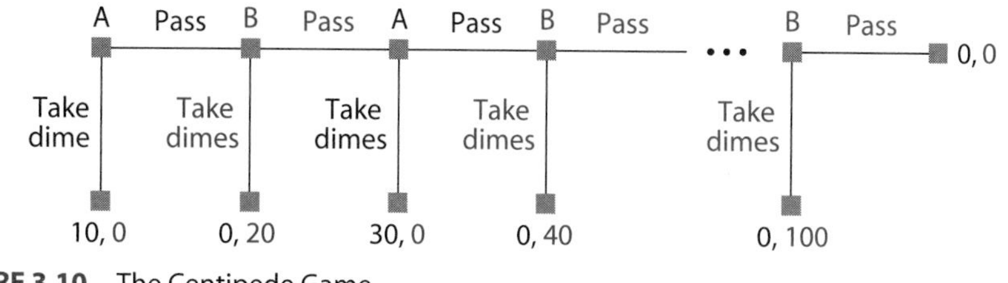

FIGURE 3.10 The Centipede Game

[11]See Alan Sanfey, James Rilling, Jessica Aronson, Leigh Nystrom, and Jonathan Cohen, "The Neural Basis of Economic Decision-Making in the Ultimatum Game," *Science*, vol. 300 (June 13, 2003), pp. 1755–1758.

group make more money than they would if they followed the logic of backward reasoning. Sometimes A does better and sometimes B, but sometimes they even solve this conflict or bargaining problem. In a classroom experiment that one of us (Dixit) conducted, one such game went all the way to the end. Player B collected the dollar, and quite voluntarily gave 50 cents to player A. Dixit asked A, "Did you two conspire? Is B a friend of yours?" and A replied, "No, we didn't even know each other before. But he is a friend now."

Once again, what is revealed is not that players cannot calculate and use game-theoretic logic but that their value systems and payoffs are different from those attributed to them by the theorist who predicted that the game should end with A taking the dime on the first step. We will come across some similar evidence of cooperation that seems to contradict rollback reasoning when we look at finitely repeated prisoners' dilemma games in Chapter 11.[12]

The examples discussed here seem to indicate that apparent violations of strategic logic can be explained by recognizing that people do not care merely about their own money payoffs, but internalize concepts such as fairness. But not all observed plays, contrary to the precepts of rollback, have some such explanation. People do fail to look ahead far enough, and they do fail to draw the appropriate conclusions from attempts to look ahead. For example, when issuers of credit cards offer favorable initial interest rates or no fees for the first year, many people fall for them without realizing that they may have to pay much more later. Therefore the game-theoretic analysis of rollback and rollback equilibria serves an advisory or prescriptive role as much as it does a descriptive role. People equipped with the theory of rollback are in a position to make better strategic decisions and get higher payoffs, no matter what is included in their payoff calculations. And game theorists can use their expertise to give valuable advice to those who are placed in complex strategic situations but lack the skill to determine their own best strategies.

7 STRATEGIES IN THE *SURVIVOR* GAME

The examples in the preceding sections were deliberately constructed to illustrate and elucidate basic concepts such as nodes, branches, moves, and strategies, as well as the technique of rollback. Now we show how all of them can be applied, by considering a real-life (or at least "reality-TV-life") situation.

In the summer of 2000, CBS television broadcast the first of the series of *Survivor* shows, which became an instant hit and helped launch the whole

[12]Once again, one wonders what would happen if the sum added at each step were a thousand dollars instead of a dime.

new genre of "reality TV." Leaving aside many complex details and some earlier stages not relevant for our purpose, the concept was as follows. A group of contestants, called a "tribe," was put on an uninhabited island and left largely to fend for themselves for food and shelter. Every 3 days they had to vote one of themselves out of the tribe. The person who had the most votes cast against him or her at a meeting of the remaining players (called the "tribal council") was the victim of the day. However, before each meeting of the tribal council, the survivors up to that point competed in a game of physical or mental skill that was devised by the producers of the game for that occasion, and the winner of this competition, called a "challenge," was immune to being voted off at the following meeting. Also, one could not vote against oneself. Finally, when two people were left, the seven who had been voted off most recently returned as a "jury" to pick one of the two remaining survivors as the million-dollar winner of the whole game.

The strategic problems facing all contestants were: (1) to be generally regarded as a productive contributor to the tribe's search for food and other tasks of survival, but to do so without being regarded as too strong a competitor and therefore a target for elimination, (2) to form alliances to secure blocks of votes to protect oneself from being voted off, (3) to betray these alliances when the numbers got too small and one had to vote against someone, but (4) to do so without seriously losing popularity with the other players, who would ultimately have the power of the vote on the jury.

We pick up the story when just three contestants were left: Rudy, Kelly, and Rich. Of them, Rudy was the oldest contestant, an honest and blunt person who was very popular with the contestants who had been previously voted off. It was generally agreed that, if he was one of the last two, then he would be voted the million-dollar winner. So it was in the interests of both Kelly and Rich that they should face each other, rather than face Rudy, in the final vote. But neither wanted to be seen as instrumental in voting off Rudy. With just three contestants left, the winner of the immunity challenge is effectively decisive in the cast-off vote, because the other two must vote against each other. Thus the jury would know who was responsible for voting off Rudy and, given his popularity, would regard the act of voting him off with disfavor. The person doing so would harm his or her chances in the final vote. This was especially a problem for Rich, because he was known to have an alliance with Rudy.

The immunity challenge was one of stamina; each contestant had to stand on an awkward support and lean to hold one hand in contact with a totem on a central pole, called the "immunity idol." Anyone whose hand lost contact with the idol, even for an instant, lost the challenge; the one to hold on longest was the winner.

An hour and a half into the challenge, Rich figured out that his best strategy was to deliberately lose this immunity challenge. Then, if Rudy won immunity, he would maintain his alliance and keep Rich—Rudy was known to be a man who always kept his word. Rich would lose the final vote to Rudy in this case, but that would make him no worse off than if he won the challenge and kept Rudy. If Kelly won immunity, the much more likely outcome, then it would be in her interest to vote off Rudy—she would have at least some chance against Rich but zero against Rudy. Then Rich's chances of winning were quite good. Whereas, if Rich himself held on, won immunity, and then voted off Rudy, his chances against Kelly would be decreased by the fact that he voted off Rudy.

So Rich deliberately stepped off, and later explained his reasons quite clearly to the camera. His calculation was borne out. Kelly won that challenge and voted off Rudy. And, in the final jury vote between Rich and Kelly, Rich won by one vote.

Rich's thinking was essentially a rollback analysis along a game tree. He did this analysis instinctively, without drawing the tree, while standing awkwardly and holding on to the immunity idol, but it took him an hour and a half to come to his conclusion. With all due credit to Rich, we show the tree explicitly, and can reach the answer faster.

Figure 3.11 shows the tree. You can see that it is much more complex than the trees encountered in earlier sections. It has more branches and moves; in addition, there are uncertain outcomes, and the chances of winning or losing in various alternative situations have to be estimated instead of being known precisely. But you will see how we can make reasonable assumptions about these chances and proceed with the analysis.

At the initial node, Rich decides whether to continue or to give up in the immunity challenge. In either case, the winner of the challenge cannot be forecast with certainty; this is indicated in the tree by letting "Nature" make the choice, as we did with the coin-toss situation in Figure 3.1. If Rich continues, Nature chooses the winner from the three contestants. We don't know the actual probabilities, but we will assume particular values for exposition and point out the crucial assumptions. The supposition is that Kelly has a lot of stamina and that Rudy, being the oldest, is not likely to win. So we posit the following probabilities of a win when Rich chooses to continue: 0.5 (50%) for Kelly, 0.45 for Rich, and only 0.05 for Rudy. If Rich gives up on the challenge, Nature picks the winner of the immunity challenge randomly between the two who remain; in this case, we assume that Kelly wins with probability 0.9 and Rudy with probability 0.1.

The rest of the tree follows from each of the three possible winners of the challenge. If Rudy wins, he keeps Rich as he promised, and the jury votes Rudy the winner.[13] If Rich wins immunity, he has to decide whether to keep Kelly or

[13]Technically, Rudy faces a choice between keeping Rich or Kelly at the action node after he wins the immunity challenge. Because everyone placed zero probability on his choosing Kelly (owing to the Rich–Rudy alliance), we illustrate only Rudy's choice of Rich. The jury, similarly, has a choice between Rich and Rudy at the last action node along this branch of play. Again, the foregone conclusion is that Rudy wins in this case.

Rudy. If he keeps Rudy, the jury votes for Rudy. If he keeps Kelly, it is not certain whom the jury chooses. We assume that Rich alienates some jurors by turning on Rudy and that, despite being better liked than Kelly, he gets the jury's vote in this situation only with probability 0.4. Similarly, if Kelly wins immunity, she can either keep Rudy and lose the jury's vote, or keep Rich. If she keeps Rich, his probability of winning the jury's vote is higher, at 0.6, because in this case he is both better liked by the jury and hasn't voted off Rudy.

What about the players' actual payoffs? We can safely assume that both Rich and Kelly want to maximize the probability of his or her emerging as the ultimate winner of the $1 million. Rudy similarly wants to get the prize, but keeping

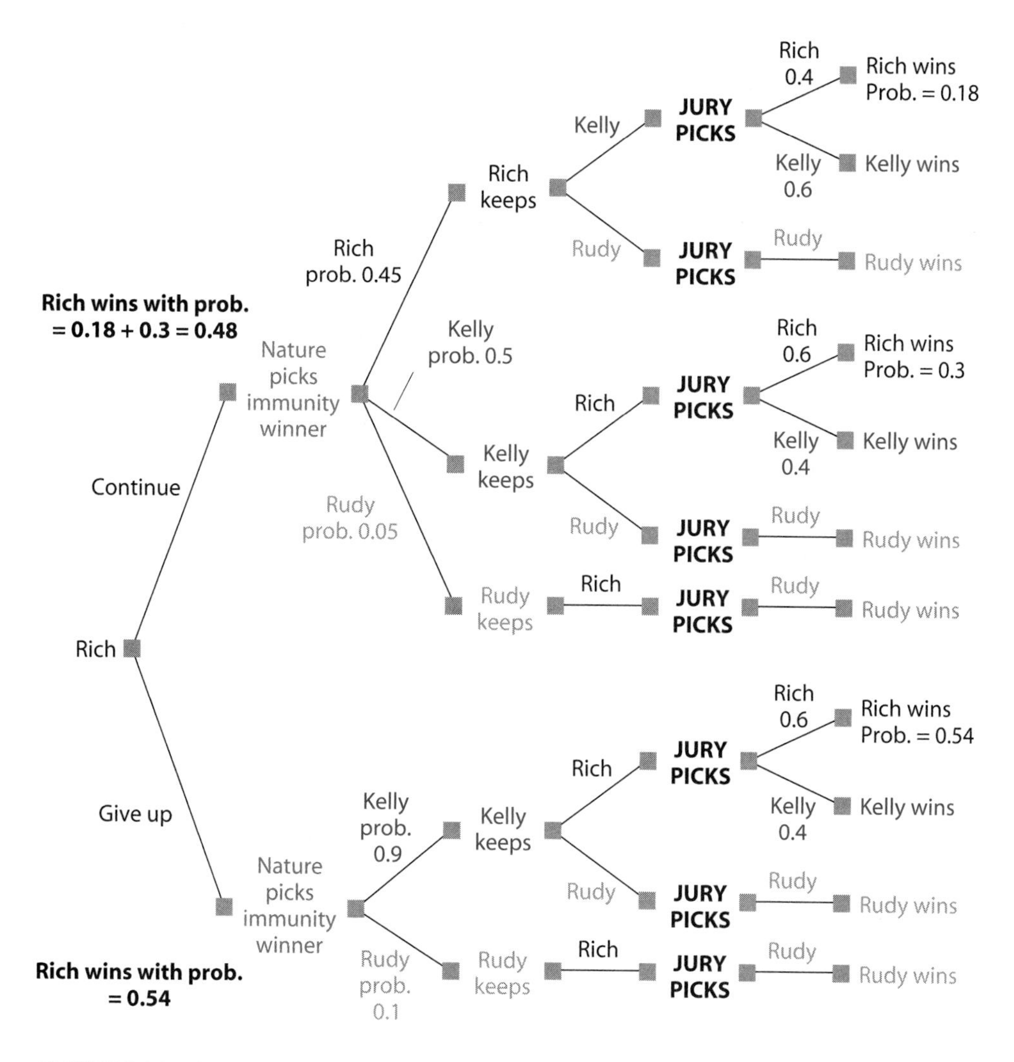

FIGURE 3.11 *Survivor* Immunity Game Tree

his word to Rich is paramount. With these preferences of the various players in mind, Rich can now do rollback analysis along the tree to determine his own initial choice.

Rich knows that, if he wins the immunity challenge (the uppermost path after his own first move and Nature's move), he will have to keep Kelly to have a 40% chance of eventual victory; keeping Rudy at this stage would mean a zero probability of eventual victory. Rich can also calculate that, if Kelly wins the immunity challenge (which occurs once in each of the upper and lower halves of the tree), she will choose to keep him for similar reasons, and then the probability of his eventual victory will be 0.6.

What are Rich's chances as he calculates them at the initial node? If Rich chooses Give Up at the initial node, then there is only one way for him to emerge as the eventual winner—if Kelly wins immunity (probability 0.9), if she then keeps Rich (probability 1), and if the jury votes for Rich (probability 0.6). Because all three things need to happen for Rich to win, his overall probability of victory is the product of the three probabilities—namely, $0.9 \times 1 \times 0.6 = 0.54$.[14] If Rich chooses Continue at the initial node, then there are two ways in which he can win. First, he wins the game if he wins the immunity challenge (probability 0.45), if he then eliminates Rudy (probability 1), and if he still wins the jury's vote against Kelly (probability 0.4); the total probability of winning in this way is $0.45 \times 0.4 = 0.18$. Second, he wins the game if Kelly wins the challenge (probability 0.5), if she eliminates Rudy (probability 1), and if Rich gets the jury's vote (probability 0.6); total probability here is $0.5 \times 0.6 = 0.3$. Rich's overall probability of eventual victory if he chooses Continue is the sum of the probabilities of these two paths to victory—namely, $0.18 + 0.3 = 0.48$.

Rich can now compare his probability of winning the million dollars when he chooses Give Up (0.54) with his probability of winning when he chooses Continue (0.48). Given the assumed values of the various probabilities in the tree, Rich has a better chance of victory if he gives up. Thus, Give Up is his optimal strategy. Although this result is based on assuming specific numbers for the probabilities, Give Up remains Rich's optimal strategy as long as (1) Kelly is very likely to win the immunity challenge once Rich gives up and (2) Rich wins the jury's final vote more often when Kelly has voted out Rudy than when Rich has done so.[15]

This example serves several purposes. Most important, it shows how a complex tree, with much external uncertainty and missing information about precise probabilities, can still be solved by using rollback analysis. We

[14]Readers who need instruction or a refresher course in the rules for combining probabilities will find a quick tutorial in the Appendix to Chapter 7.

[15]Readers who can handle the algebra of probabilities can solve this game by using more general symbols instead of specific numbers for the probabilities, as in Exercise U9 of this chapter.

hope this gives you some confidence in using the method and some training in converting a somewhat loose verbal account into a more precise logical argument. You might counter that Rich did this reasoning without drawing any trees. But knowing the system or general framework greatly simplifies the task even in new and unfamiliar circumstances. Therefore it is definitely worth the effort to acquire the systematic skill.

A second purpose is to illustrate the seemingly paradoxical strategy of "losing to win." Another instance of this strategy can be found in some sporting competitions that are held in two rounds, such as the soccer World Cup. The first round is played on a league basis in several groups of four teams each. The top two teams from each group then go to the second round, where they play others chosen according to a prespecified pattern; for example, the top-ranked team in group A meets the second-ranked team in group B, and so on. In such a situation, it may be good strategy for a team to lose one of its first-round matches if this loss causes it to be ranked second in its group; that ranking might earn it a subsequent match against a team that, for some particular reason, it is more likely to beat than the team that it would meet if it had placed first in its group in the first round.

SUMMARY

Sequential-move games require players to consider the future consequences of their current moves before choosing their actions. Analysis of pure sequential-move games generally requires the creation of a *game tree*. The tree is made up of *nodes* and *branches* that show all of the possible actions available to each player at each of her opportunities to move, as well as the payoffs associated with all possible outcomes of the game. Strategies for each player are complete plans that describe actions at each of the player's decision nodes contingent on all possible combinations of actions made by players who acted at earlier nodes. The equilibrium concept employed in sequential-move games is that of *rollback equilibrium*, in which players' equilibrium strategies are found by looking ahead to subsequent nodes and the actions that would be taken there and by using these forecasts to calculate one's current best action. This process is known as *rollback*, or *backward induction*.

Different types of games entail advantages for different players, such as *first-mover advantages*. The inclusion of many players or many moves enlarges the game tree of a sequential-move game but does not change the solution process. In some cases, drawing the full tree for a particular game may require more space or time than is feasible. Such games can often be solved by identifying strategic similarities between actions that reduces the size of the tree or by simple logical thinking.

When solving larger games, verbal reasoning can lead to the rollback equilibrium if the game is simple enough or a complete tree may be drawn and analyzed. If the game is sufficiently complex that verbal reasoning is too difficult and a complete tree is too large to draw, we may enlist the help of a computer program. Checkers has been "solved" with the use of such a program, although full solution of chess will remain beyond the powers of computers for a long time. In actual play of these truly complex games, elements of both art (identification of patterns and of opportunities versus peril) and science (forward-looking calculations of the possible outcomes arising from certain moves) have a role in determining player moves.

Tests of the theory of sequential-move games seem to suggest that actual play shows the irrationality of the players or the failure of the theory to predict behavior adequately. The counterargument points out the complexity of actual preferences for different possible outcomes and the usefulness of strategic theory for identifying optimal actions when actual preferences are known.

KEY TERMS

action node (48)
backward induction (56)
branch (48)
decision node (48)
decision tree (48)
equilibrium path of play (60)
extensive form (48)
first-mover advantage (62)
game tree (48)
initial node (48)
intermediate valuation function (67)
move (50)
node (48)
path of play (60)
prune (54)
rollback (56)
rollback equilibrium (56)
root (48)
second-mover advantage (62)
terminal node (50)

SOLVED EXERCISES

S1. Suppose two players, Hansel and Gretel, take part in a sequential-move game. Hansel moves first, Gretel moves second, and each player moves only once.

(a) Draw a game tree for a game in which Hansel has two possible actions (Up or Down) at each node and Gretel has three possible actions (Top, Middle, or Bottom) at each node. How many of each node type—decision and terminal—are there?

(b) Draw a game tree for a game in which Hansel and Gretel each have three possible actions (Sit, Stand, or Jump) at each node. How many of the two node types are there?

(c) Draw a game tree for a game in which Hansel has four possible actions (North, South, East, or West) at each node and Gretel has two possible actions (Stay or Go) at each node. How many of the two node types are there?

S2. In each of the following games, how many pure strategies (complete plans of action) are available to each player? List out all of the pure strategies for each player.

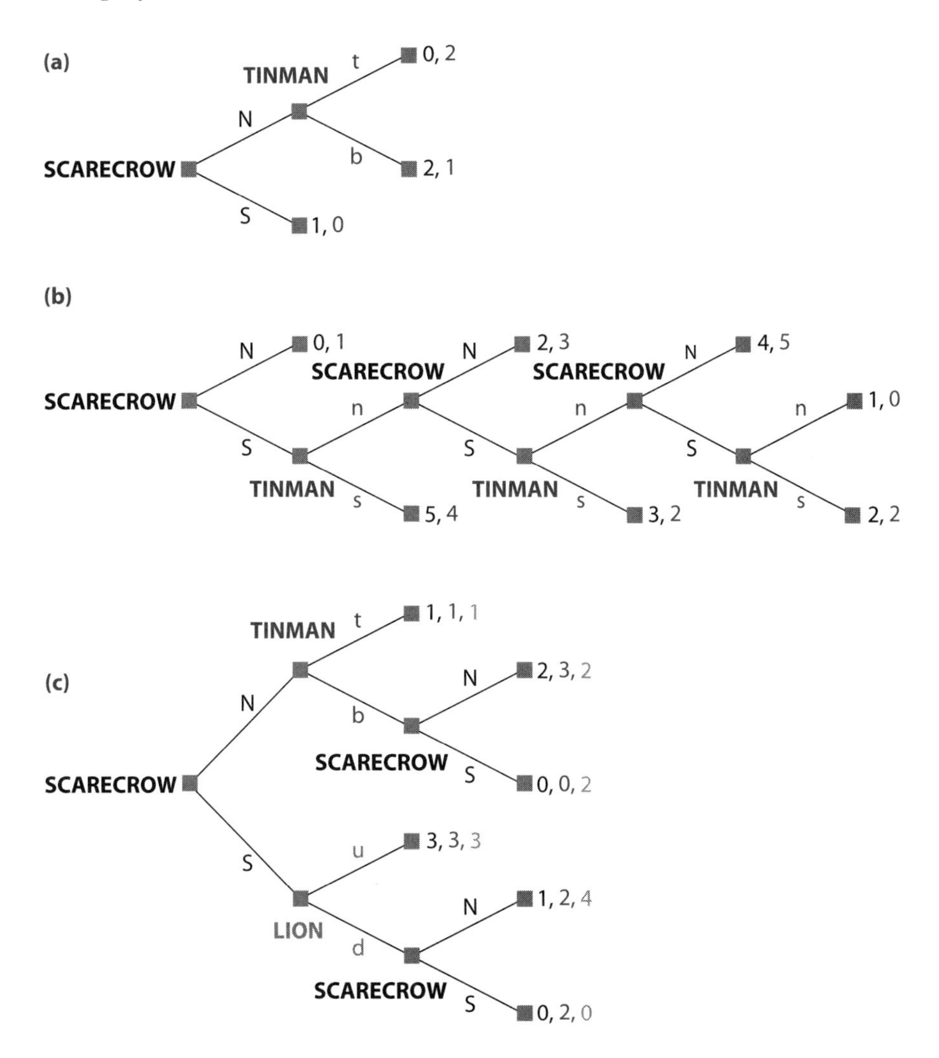

S3. For each of the games illustrated in Exercise S2, identify the rollback equilibrium outcome and the complete equilibrium strategy for each player.

S4. Consider the rivalry between Airbus and Boeing to develop a new commercial jet aircraft. Suppose Boeing is ahead in the development process and

Airbus is considering whether to enter the competition. If Airbus stays out, it earns zero profit, whereas Boeing enjoys a monopoly and earns a profit of $1 billion. If Airbus decides to enter and develop the rival airplane, then Boeing has to decide whether to accommodate Airbus peaceably or to wage a price war. In the event of peaceful competition, each firm will make a profit of $300 million. If there is a price war, each will lose $100 million because the prices of airplanes will fall so low that neither firm will be able to recoup its development costs.

Draw the tree for this game. Find the rollback equilibrium and describe the firms' equilibrium strategies.

S5. Consider a game in which two players, Fred and Barney, take turns removing matchsticks from a pile. They start with 21 matchsticks, and Fred goes first. On each turn, each player may remove either one, two, three, or four matchsticks. The player to remove the last matchstick wins the game.

(a) Suppose there are only six matchsticks left, and it is Barney's turn. What move should Barney make to guarantee himself victory? Explain your reasoning.

(b) Suppose there are 12 matchsticks left, and it is Barney's turn. What move should Barney make to guarantee himself victory? (Hint: Use your answer to part (a) and roll back.)

(c) Now start from the beginning of the game. If both players play optimally, who will win?

(d) What are the optimal strategies (complete plans of action) for each player?

S6. Consider the game in the previous exercise. Suppose the players have reached a point where it is Fred's move and there are just five matchsticks left.

(a) Draw the game tree for the game starting with five matchsticks.

(b) Find the rollback equilibria for this game starting with five matchsticks.

(c) Would you say this five-matchstick game has a first-mover advantage or a second-mover advantage?

(d) Explain why you found more than one rollback equilibrium. How is your answer related to the optimal strategies you found in part (c) of the previous exercise?

S7. A slave has just been thrown to the lions in the Roman Colosseum. Three lions are chained down in a line, with Lion 1 closest to the slave. Each lion's chain is short enough that he can only reach the two players immediately adjacent to him.

The game proceeds as follows. First, Lion 1 decides whether or not to eat the slave.

If Lion 1 has eaten the slave, then Lion 2 decides whether or not to eat Lion 1 (who is then too heavy to defend himself). If Lion 1 has not eaten the slave, then Lion 2 has no choice: he cannot try to eat Lion 1, because a fight would kill both lions.

Similarly, if Lion 2 has eaten Lion 1, then Lion 3 decides whether or not to eat Lion 2.

Each lion's preferences are fairly natural: best (4) is to eat and stay alive, next best (3) is to stay alive but go hungry, next (2) is to eat and be eaten, and worst (1) is to go hungry and be eaten.

(a) Draw the game tree, with payoffs, for this three-player game.
(b) What is the rollback equilibrium to this game? Make sure to describe the strategies, not just the payoffs.
(c) Is there a first-mover advantage to this game? Explain why or why not.
(d) How many complete strategies does each lion have? List them.

S8. Consider three major department stores—Big Giant, Titan, and Frieda's—contemplating opening a branch in one of two new Boston-area shopping malls. Urban Mall is located close to the large and rich population center of the area; it is relatively small and can accommodate at most two department stores as "anchors" for the mall. Rural Mall is farther out in a rural and relatively poorer area; it can accommodate as many as three anchor stores. None of the three stores wants to have branches in both malls because there is sufficient overlap of customers between the malls that locating in both would just mean competing with itself. Each store prefers to be in a mall with one or more other department stores than to be alone in the same mall, because a mall with multiple department stores will attract sufficiently many more total customers that each store's profit will be higher. Further, each store prefers Urban Mall to Rural Mall because of the richer customer base. Each store must choose between trying to get a space in Urban Mall (knowing that if the attempt fails, it will try for a space in Rural Mall) and trying to get a space in Rural Mall right away (without even attempting to get into Urban Mall).

In this case, the stores rank the five possible outcomes as follows: 5 (best), in Urban Mall with one other department store; 4, in Rural Mall with one or two other department stores; 3, alone in Urban Mall; 2, alone in Rural Mall; and 1 (worst), alone in Rural Mall after having attempted to get into Urban Mall and failed, by which time other nondepartment stores have signed up the best anchor locations in Rural Mall.

The three stores are sufficiently different in their managerial structures that they experience different lags in doing the paperwork required to request an expansion space in a new mall. Frieda's moves quickly, followed by Big Giant, and finally by Titan, which is the least efficient in readying a location plan. When all three have made their requests, the malls decide which stores to let in. Because of the name recognition that both Big Giant and Titan have with the potential customers, a mall would take either (or both) of those stores before it took Frieda's. Thus, Frieda's does not get one of the two spaces in Urban Mall if all three stores request those spaces; this is true even though Frieda's moves first.

(a) Draw the game tree for this mall location game.

(b) Illustrate the rollback pruning process on your game tree and use the pruned tree to find the rollback equilibrium. Describe the equilibrium by using the (complete) strategies employed by each department store. What are the payoffs to each store at the rollback equilibrium outcome?

S9. **(Optional)** Consider the following ultimatum bargaining game, which has been studied in laboratory experiments. The Proposer moves first, and proposes a split of $10 between himself and the Responder. Any whole-dollar split may be proposed. For example, the Proposer may offer to keep the whole $10 for himself, he may propose to keep $9 for himself and give $1 to the Responder, $8 to himself and $2 to the Responder, and so on. (Note that the Proposer therefore has eleven possible choices.) After seeing the split, the Responder can choose to accept the split or reject it. If the Responder accepts, both players get the proposed amounts. If she rejects, both players get $0.

(a) Write out the game tree for this game.

(b) How many complete strategies does each player have?

(c) What is the rollback equilibrium to this game, assuming the players care only about their cash payoffs?

(d) Suppose Rachel the Responder would accept any offer of $3 or more, and reject any offer of $2 or less. Suppose Pete the Proposer knows Rachel's strategy, and he wants to maximize his cash payoff. What strategy should he use?

(e) Rachel's true payoff (her "utility") might not be the same as her cash payoff. What other aspects of the game might she care about? Given your answer, propose a set of payoffs for Rachel that would make her strategy optimal.

(f) In laboratory experiments, players typically do not play the rollback equilibrium. Proposers typically offer an amount between $2 and $5 to the Responder. Responders often reject offers of $3, $2, and especially $1. Explain why you think this might occur.

UNSOLVED EXERCISES

U1. "In a sequential-move game, the player who moves first is sure to win." Is this statement true or false? State the reason for your answer in a few brief sentences, and give an example of a game that illustrates your answer.

U2. In each of the following games, how many pure strategies (complete plans of action) are available to each player? List all of the pure strategies for each player.

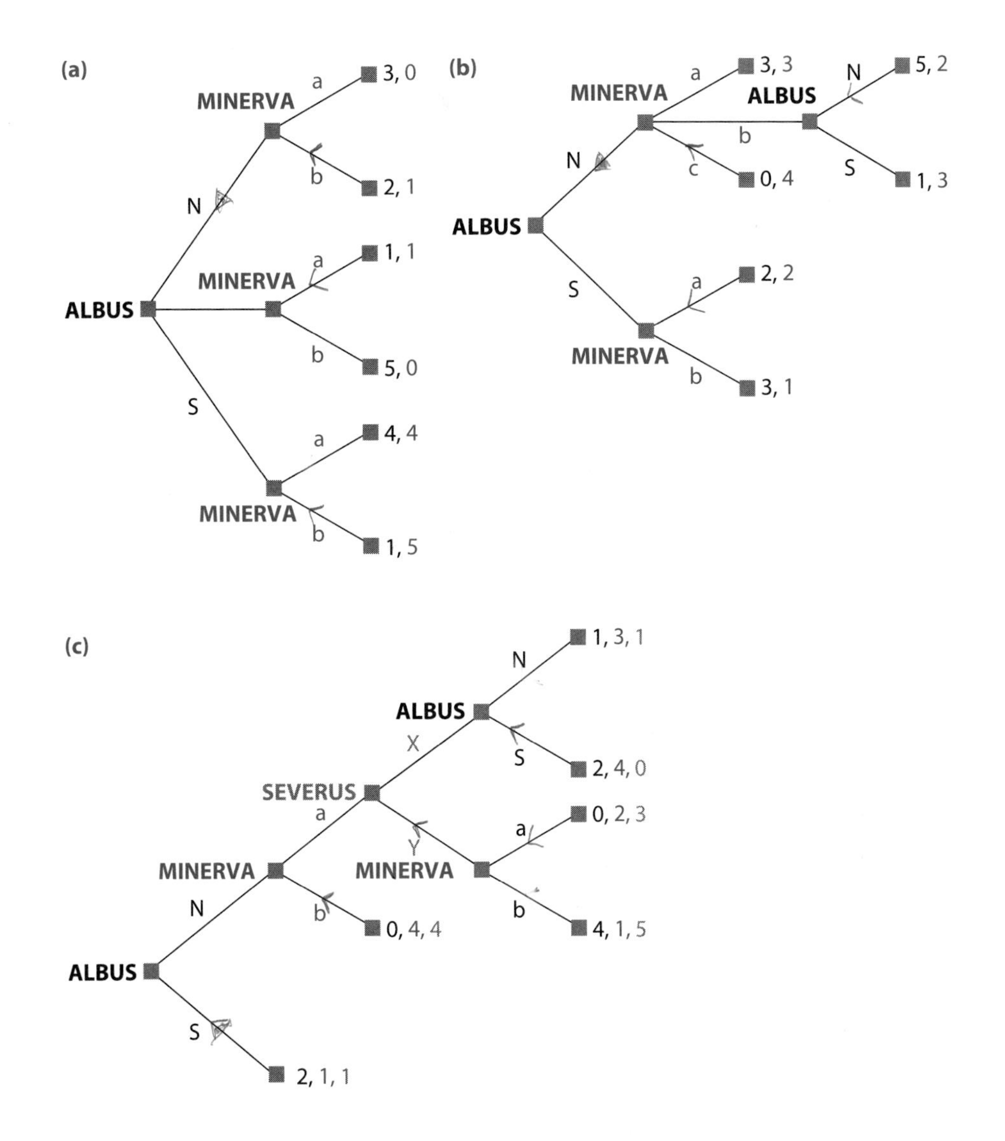

U3. For each of the games illustrated in Exercise U2, identify the rollback equilibrium outcome and the complete equilibrium strategy for each player.

U4. Two distinct proposals, A and B, are being debated in Washington. The Congress likes proposal A, and the president likes proposal B. The proposals are not mutually exclusive; either or both or neither may become law. Thus there are four possible outcomes, and the rankings of the two sides are as follows, where a larger number represents a more favored outcome.

Outcome	Congress	President
A becomes law	4	1
B becomes law	1	4
Both A and B become law	3	3
Neither (status quo prevails)	2	2

(a) The moves in the game are as follows. First, the Congress decides whether to pass a bill and whether it is to contain A or B or both. Then the president decides whether to sign or veto the bill. Congress does not have enough votes to override a veto. Draw a tree for this game and find the rollback equilibrium.

(b) Now suppose the rules of the game are changed in only one respect: the president is given the extra power of a line-item veto. Thus, if the Congress passes a bill containing both A and B, the president may choose not only to sign or veto the bill as a whole, but also to veto just one of the two items. Show the new tree and find the rollback equilibrium.

(c) Explain intuitively why the difference between the two equilibria arises.

U5. Two players, Amy and Beth, play the following game with a jar containing 100 pennies. The players take turns; Amy goes first. Each time it is a player's turn, she takes between one and 10 pennies out of the jar. The player whose move empties the jar wins.

(a) If both players play optimally, who will win the game? Does this game have a first-mover advantage? Explain your reasoning.

(b) What are the optimal strategies (complete plans of action) for each player?

(c) Now suppose we change the rules so that the player whose move empties the jar loses. Does this game have a first-mover advantage? Explain your reasoning.

(d) In this second variant, what are the optimal strategies for each player?

U6. Now Amy and Beth play a game with two jars, each containing 100 pennies. The players take turns; Amy goes first. Each time it is a player's turn, she chooses one of the jars and removes anywhere from one to 10 pennies from it. The player whose move leaves both jars empty wins. (Note that when a player empties the second jar, the first jar must already have been emptied in some previous move by one of the players.)

(a) Does this game have a first-mover advantage or a second-mover advantage? Explain which player can guarantee victory, and how she can do it.

(Hint: simplify the game by starting with a smaller number of pennies in each jar, and see if you can generalize your finding to the actual game.)

(b) What are the optimal strategies (complete plans of action) for each player? (Hint: First think of a starting situation in which both jars have equal numbers of pennies. Then consider starting positions in which the two jars differ by one to 10 pennies. Finally, consider starting positions in which the jars differ by more than 10 pennies.)

U7. Modify Exercise S7 so that there are now four lions.

(a) Draw the game tree, with payoffs, for this four-player game.

(b) What is the rollback equilibrium to this game? Make sure to describe the strategies, not just the payoffs.

(c) Is the additional lion good or bad for the slave? Explain.

U8. To give Mom a day of rest, Dad plans to take his two children, Bart and Cassie, on an outing on Sunday. Bart prefers to go to the amusement park (A), whereas Cassie prefers to go to the science museum (S). Each child gets 3 units of utility from his/her more preferred activity, and only 2 units of utility from his/her less preferred activity. Dad gets 2 units of utility for either of the two activities.

To choose their activity, Dad plans first to ask Bart for his preference, then to ask Cassie after she hears Bart's choice. Each child can choose either the amusement park (A) or the science museum (S). If both children choose the same activity, then that is what they will all do. If the children choose different activities, Dad will make a tie-breaking decision. As the parent, Dad has an additional option: he can choose the amusement park, the science museum, or his personal favorite, the mountain hike (M). Bart and Cassie each get 1 unit of utility from the mountain hike, and Dad gets 3 units of utility from the mountain hike.

Because Dad wants his children to cooperate with one another, he gets 2 extra units of utility if the children choose the same activity (no matter which one of the two it is).

(a) Draw the game tree, with payoffs, for this three-person game.

(b) What is the rollback equilibrium to this game? Make sure to describe the strategies, not just the payoffs.

(c) How many different complete strategies does Bart have? Explain.

(d) How many complete strategies does Cassie have? Explain.

U9. (Optional—more difficult) Consider the *Survivor* game tree illustrated in Figure 3.11. We might not have guessed exactly the values Rich estimated for the various probabilities, so let's generalize this tree by considering other possible values. In particular, suppose that the probability of winning the immunity challenge when Rich chooses Continue is x for Rich, y for Kelly, and $1 - x - y$ for Rudy; similarly, the probability of winning when Rich gives up is z for Kelly and $1 - z$ for Rudy. Further, suppose that Rich's chance of being picked by the jury is p if he has won immunity and has voted Rudy off the island; his chance of being picked is q if Kelly has won immunity and has voted Rudy off. Continue to assume that if Rudy wins immunity, he keeps Rich with probability 1, and that Rudy wins the game with probability 1 if he ends up in the final two. Note that in the example of Figure 3.11, we had $x = 0.45$, $y = 0.5$, $z = 0.9$, $p = 0.4$, and $q = 0.6$. (In general, the variables p and q need not sum to 1, though this happened to be true in Figure 3.11.)

(a) Find an algebraic formula, in terms of x, y, z, p, and q, for the probability that Rich wins the million dollars if he chooses Continue. (Note: Your formula might not contain all of these variables.)

(b) Find a similar algebraic formula for the probability that Rich wins the million dollars if he chooses Give Up. (Again, your formula might not contain all of the variables.)

(c) Use these results to find an algebraic inequality telling us under what circumstances Rich should choose Give Up.

(d) Suppose all the values are the same as in Figure 3.11 except for z. How high or low could z be so that Rich would still prefer to Give Up? Explain intuitively why there are some values of z for which Rich is better off choosing Continue.

(e) Suppose all the values are the same as in Figure 3.11 except for p and q. Assume that since the jury is more likely to choose a "nice" person who doesn't vote Rudy off, we should have $p > 0.5 > q$. For what values of the ratio (p/q) should Rich choose Give Up? Explain intuitively why there are some values of p and q for which Rich is better off choosing Continue.

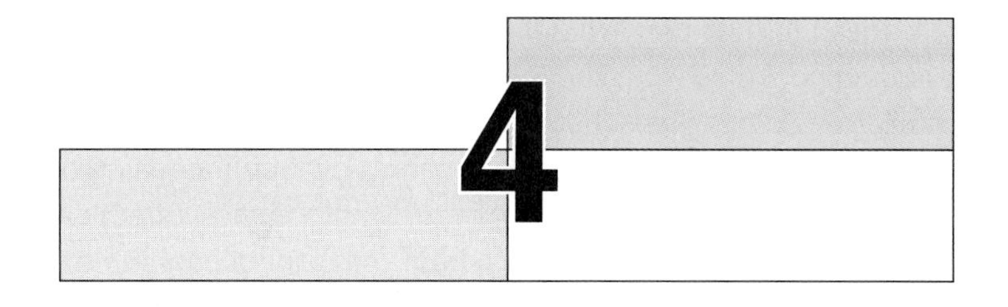

Simultaneous-Move Games with Pure Strategies I: Discrete Strategies

RECALL FROM CHAPTER 2 that games are said to have simultaneous moves if players must move without knowledge of what their rivals have chosen to do. It is obviously so if players choose their actions at exactly the same time. A game is also simultaneous when players choose their actions in isolation, with no information about what other players have done or will do, even if the choices are made at different hours of the clock. (For this reason, simultaneous-move games have *imperfect information* in the sense we defined in Chapter 2, Section 2.D.) This chapter focuses on games that have such purely simultaneous interactions among players. We consider a variety of types of simultaneous games, introduce a solution concept called Nash equilibrium for these games, and study games with one equilibrium, many equilibria, or no equilibrium at all.

Many familiar strategic situations can be described as simultaneous-move games. The various producers of television sets, stereos, or automobiles make decisions about product design and features without knowing what rival firms are doing about their own products. Voters in U.S. elections simultaneously cast their individual votes; no voter knows what the others have done when she makes her own decision. The interaction between a soccer goalie and an opposing striker during a penalty kick requires both players to make their decisions simultaneously—the goalie cannot afford to wait until the ball has actually been kicked to decide which way to go, because then it would be far too late.

When a player in a simultaneous-move game chooses her action, she obviously does so without any knowledge of the choices made by other players. She also cannot look ahead to how they will react to her choice, because they act simultaneously and do not know what she is choosing. Rather, each player must figure out what others are doing when what the others are doing is figuring out what this player is doing. This circularity makes the analysis of simultaneous-move games somewhat more intricate than that of sequential-move games, but the analysis is not difficult. In this chapter, we develop a simple concept of equilibrium for such games that has considerable explanatory and predictive power.

1 DEPICTING SIMULTANEOUS-MOVE GAMES WITH DISCRETE STRATEGIES

In Chapters 2 and 3, we emphasized that a strategy is a complete plan of action. But in a purely simultaneous-move game, each player can have at most one opportunity to act (although that action may have many component parts); if a player had multiple opportunities to act, that would be an element of sequentiality. Therefore there is no real distinction between strategy and action in simultaneous-move games, and the terms are often used as synonyms in this context. There is only one complication. A strategy can be a probabilistic choice from the basic actions initially specified. For example, in sports, a player or team may deliberately randomize its choice of action to keep the opponent guessing. Such probabilistic strategies are called **mixed strategies,** and we consider them in Chapters 7 and 8. In this chapter, we confine our attention to the basic initially specified actions, which are called **pure strategies.**

In many games, each player has available to her a finite number of discrete pure strategies—for example, Dribble, Pass, or Shoot in basketball. In other games, each player's pure strategy can be any number from a continuous range—for example, the price charged by a firm.[1] This distinction makes no difference to the general concept of equilibrium in simultaneous-move games, but the ideas are more easily conveyed with discrete strategies; solution of games with continuous strategies needs slightly more advanced tools. Therefore, in this chapter, we restrict the analysis to the simpler case of discrete pure strategies and take up continuously variable strategies in Chapter 5.

Simultaneous-move games with discrete strategies are most often depicted with the use of a **game table** (also called a **game matrix** or **payoff table**). The table is called the **normal form** or the **strategic form** of the game. Games with any number of players can be illustrated by using a game table, but its dimen-

[1]In fact, prices must be denominated in the minimum unit of coinage—for example, whole cents—and can therefore take on only a finite number of discrete values. But this unit is usually so small that it makes more sense to think of the price as a continuous variable.

sion must equal the number of players. For a two-player game, the table is two dimensional and appears similar to a spreadsheet. The row and column headings of the table are the strategies available to the first and second players, respectively. The size of the table, then, is determined by the numbers of strategies available to the players.[2] Each cell within the table lists the payoffs to all players that arise under the configuration of strategies that placed players into that cell. Games with three players require three-dimensional tables; we consider them later in this chapter.

We illustrate the concept of a payoff table for a simple game in Figure 4.1. The game here has no special interpretation; so we can develop the concepts without the distraction of a "story." The players are named Row and Column. Row has four choices (strategies or actions) labeled Top, High, Low, and Bottom; Column has three choices labeled Left, Middle, and Right. Each selection of Row and Column generates a potential outcome of the game. Payoffs associated with each outcome are shown in the cell corresponding to that row and that column. By convention, of the two payoff numbers, the first is Row's payoff and the second is Column's. For example, if Row chooses High and Column chooses Right, the payoffs are 6 to Row and 4 to Column. For additional convenience, we show everything pertaining to Row—player name, strategies, and payoffs—in black, and everything pertaining to Column in green.

Remember that, in some games, most notably in sports contexts, the interests of the two sides are exactly the opposite of each other. Then, for each combination of the players' choices, the payoffs of one can be obtained by reversing the sign of the payoffs to the other. As noted in Chapter 2, we call these **zero-sum** (or **constant-sum**) games.

		COLUMN		
		Left	Middle	Right
ROW	Top	3, 1	2, 3	10, 2
	High	4, 5	3, 0	6, 4
	Low	2, 2	5, 4	12, 3
	Bottom	5, 6	4, 5	9, 7

FIGURE 4.1 Representing a Simultaneous-Move Game in a Table

[2]If each firm can choose its price at any number of cents in a range that extends over a dollar, each has 100 distinct discrete strategies, and the table becomes 100 by 100. That is surely too unwieldy to analyze. Algebraic formulas with prices as continuous variables provide a simpler approach, not a more complicated one as some readers might fear. We develop this "Algebra is our friend" method in Chapter 5.

		DEFENSE		
		Run	Pass	Blitz
OFFENSE	Run	2	5	13
	Short Pass	6	5.6	10.5
	Medium Pass	6	4.5	1
	Long Pass	10	3	-2

FIGURE 4.2 Representing a Zero-Sum Simultaneous-Move Game in a Table

In zero-sum games, we can simplify the game table by showing the payoffs of just one player, generally the Row player. Those of the Column player are left implicit. Figure 4.2 shows an example of this shorthand notation for a very simplified version of a single play in (American) football. The team on the offense is attempting to move the ball forward to improve its chances of kicking a field goal. It has four possible strategies: a run and one of three different-length passes (short, medium, and long). The defense can adopt one of three strategies to try to keep the offense at bay: a run defense, a pass defense, or a blitz of the quarterback. The game is zero-sum; the offense tries to gain yardage while the defense tries to prevent it from doing so. Suppose we have enough information about the underlying strengths of the two teams to work out the probabilities of completing different plays and to determine the average gain in yardage that could be expected under each combination of strategies. For example, when Offense chooses the Medium Pass and Defense counters with its Pass defense, we estimate Offense's payoff to be 4.5 (yards).[3] Defense's "payoff" is the loss of 4.5 yards, or negative 4.5 yards (–4.5), but this number is not explicitly shown in the table. The other cells similarly show Offense's payoff, with Defense's payoff implicit and equal to the negative of whatever Offense receives.

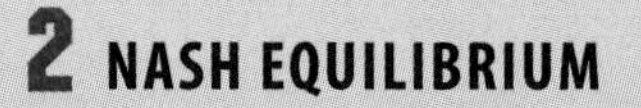

2 NASH EQUILIBRIUM

To analyze simultaneous games, we need to consider how players choose their actions. Return to the game table in Figure 4.1. Focus on one specific outcome—

[3]Here is how the payoffs for this case were constructed. When Offense chooses the Medium Pass and Defense counters with its Pass defense, our estimate is that with probability 50% the pass will be completed for a gain of 15 yards, with probability 40% the pass will fall incomplete (0 yards), and with probability 10% the pass will be intercepted with a loss of 30 yards; this makes an average of $0.5 \times 15 + 0.4 \times 0 + 0.1 \times (-30) = 4.5$ yards. The numbers in the table were constructed by a small panel of expert neighbors and friends convened by Dixit on one fall Sunday afternoon. They received a liquid consultancy fee.

namely, the one where Row chooses Low and Column chooses Middle; payoffs there are 5 to Row and 4 to Column. Each player wants to pick an action that yields her the highest payoff, and in this outcome each indeed makes such a choice, given what her opponent chooses. Given that Row is choosing Low, can Column do any better by choosing something other than Middle? No, because Left would give her the payoff 2, and Right would give her 3, neither of which is better than the 4 she gets from Middle. Thus Middle is Column's **best response** to Row's choice of Low. Conversely, given that Column is choosing Middle, can Row do better by choosing something other than Low? Again no, because the payoffs from switching to Top (2), High (3), or Bottom (4) would all be no better than what Row gets with Low (5). Thus Low is Row's best response to Column's choice of Middle.

The two choices, Low for Row and Middle for Column, have the property that each is the chooser's best response to the other's action. If they were making these choices, neither would want to switch to anything different *on her own*. By the definition of a noncooperative game, the players are making their choices independently; therefore such unilateral changes are all that each player can contemplate. Because neither wants to make such a change, it is natural to call this state of affairs an equilibrium. This is exactly the concept of Nash equilibrium.

To state it a little more formally, a **Nash equilibrium**[4] in a game is a list of strategies, one for each player, such that no player can get a better payoff by switching to some other strategy that is available to her while all the other players adhere to the strategies specified for them in the list.

A. Some Further Explanation of the Concept of Nash Equilibrium

To understand the concept of Nash equilibrium better, we take another look at the game in Figure 4.1. Consider now a cell other than (Low, Middle)—say, the one where Row chooses High and Column chooses Left. Can this be a Nash equilibrium? No, because, if Column is choosing Left, Row does better to choose Bottom and get the payoff 5 rather than to choose High, which gives her only 4. Similarly, (Bottom, Left) is not a Nash equilibrium, because Column can do better by switching to Right, thereby improving her payoff from 6 to 7.

[4]This concept is named for the mathematician and economist John Nash, who developed it in his doctoral dissertation at Princeton in 1949. Nash also proposed a solution to cooperative games, which we consider in Chapter 18. He shared the 1994 Nobel Prize in economics with two other game theorists, Reinhard Selten and John Harsanyi; we will treat some aspects of their work in Chapters 9, 10, and 14. Sylvia Nasar's biography of Nash, *A Beautiful Mind* (New York: Simon & Schuster, 1998), was the (loose) basis for a movie starring Russell Crowe. Unfortunately, the movie's attempt to explain the concept of Nash equilibrium fails. We explain this failure in Exercise S12 of this chapter and in Exercise S14 of Chapter 8.

		COLUMN		
		Left	Middle	Right
Row	Top	3, 1	2, 3	10, 2
	High	4, 5	3, 0	6, 4
	Low	2, 2	5, 4	12, 3
	Bottom	5, 6	5, 5	9, 7

FIGURE 4.3 Variation on Game of Figure 4.1 with a Tie in Payoffs

The definition of Nash equilibrium does not require equilibrium choices to be strictly better than other available choices. Figure 4.3 is the same as Figure 4.1 except that Row's payoff from (Bottom, Middle) is changed to 5, the same as that from (Low, Middle). It is still true that, given Column's choice of Middle, Row *could not do any better* than she does when choosing Low. So neither player has a reason to change her action when the outcome is (Low, Middle), and that qualifies it for a Nash equilibrium.[5]

More important, a Nash equilibrium does not have to be jointly best for the players. In Figure 4.1, the strategy pair (Bottom, Right) gives payoffs (9, 7), which are better for both players than the (5, 4) of the Nash equilibrium. However, playing independently, they cannot sustain (Bottom, Right). Given that Column plays Right, Row would want to deviate from Bottom to Low and get 12 instead of 9. Getting the jointly better payoffs of (9, 7) would require cooperative action that made such "cheating" impossible. We examine this type of behavior later in this chapter and in more detail in Chapter 11. For now, we merely point out the fact that a Nash equilibrium may not be in the joint interests of the players.

To reinforce the concept of Nash equilibrium, look at the football game of Figure 4.2. If the Defense is choosing the Pass defense, then the best choice for the Offense is Short Pass (payoff of 5.6 versus 5, 4.5, or 3). Conversely, if the Offense is choosing the Short Pass, then the Defense's best choice is the Pass defense—it holds the Offense down to 5.6 yards, whereas the Run defense and the Blitz would be expected to concede 6 and 10.5 yards, respectively. (Remember that the entries in each cell of a zero-sum game are the Row player's payoffs; therefore the best choice for the Column player is the one that yields the smallest number, not the largest.) In this game, the strategy combination (Short Pass, Pass defense) is a Nash equilibrium, and the resulting payoff to the Offense is 5.6 yards.

[5]But note that (Bottom, Middle) with the payoffs of (5, 5) is not itself a Nash equilibrium. If Row was choosing Bottom, Column's own best choice would not be Middle; she could do better by choosing Right. In fact, you can check all the other cells in the table to verify that none of them can be a Nash equilibrium.

How does one find Nash equilibria in games? One can always check every cell to see if the strategies that generate it satisfy the definition of a Nash equilibrium. Such **cell-by-cell-inspection,** or **enumeration,** is foolproof but tedious and unmanageable except in simple games or unless one is using a good computer program for finding equilibria. Luckily, there are many other methods, applicable to special types of games, that not only find Nash equilibria more quickly when they apply, but also give us a better understanding of the process of thinking by which beliefs and then choices are formed. We develop several such methods in later sections.

B. Nash Equilibrium As a System of Beliefs and Choices

Before we proceed with further study and use of the Nash equilibrium concept, we should try to clarify something that may have bothered some of you. We said that in a Nash equilibrium each player chooses her "best response" to the other's choice. But the two choices are made simultaneously. How can one *respond* to something that has not yet happened, at least when one does not *know* what has happened?

People play simultaneous-move games all the time and do make choices. To do so, they must find a substitute for actual knowledge or observation of the others' actions. Players could make blind guesses and hope that they turn out to be inspired ones, but luckily there are more systematic ways to try to figure out what the others are doing. One method is experience and observation—if the players play this game or similar games with similar players all the time, they may develop a pretty good idea of what the others do. Then choices that are not best will be unlikely to persist for long. Another method is the logical process of thinking through the others' thinking. You put yourself in the position of other players and think what they are thinking, which of course includes their putting themselves in your position and thinking what you are thinking. The logic seems circular, but there are several ways of breaking into the circle, and we demonstrate these ways by using specific examples in the sections that follow. Nash equilibrium can be thought of as a culmination of this process of thinking about thinking, where each player has correctly figured out the others' choice.

Whether by observation or logical deduction or some other method, you, the game player, acquire some notion of what the others are choosing in simultaneous-move games. It is not easy to find a word to describe the process or its outcome. It is not anticipation, nor is it forecasting, because the others' actions do not lie in the future but occur simultaneously with your own. The word most frequently used by game theorists is **belief.** This word is not perfect either, because it seems to connote more confidence or certainty than is intended; in fact, in Chapters 7 and 8, we allow for the possibility that beliefs are held with some uncertainty. But for lack of a better word, it will have to suffice.

This concept of belief also relates to our discussion of uncertainty in Chapter 2, Section 2.D. There we introduced the concept of strategic uncertainty. Even when all the rules of a game—the strategies available to all players and the payoffs for each as functions of the strategies of all—are known without any uncertainty external to the game, such as weather, each player may be uncertain about what actions the others are taking at the same time. Similarly, if past actions are not observable, each player may be uncertain about what actions the others took in the past. How can players choose in the face of this strategic uncertainty? They must form some subjective views or estimates about the others' actions. That is exactly what the notion of belief captures.

Now think of Nash equilibrium in this light. We defined it as a configuration of strategies such that each player's strategy is her best response to that of the others. If she does not know the actual choices of the others but has beliefs about them, in Nash equilibrium those beliefs would have to be correct—the others' actual actions should be just what you believe them to be. Thus we can define Nash equilibrium in an alternative and equivalent way: it is a set of strategies, one for each player, such that (1) each player has correct beliefs about the strategies of the others and (2) the strategy of each is the best for herself, given her beliefs about the strategies of the others.[6]

This way of thinking about Nash equilibrium has two advantages. First, the concept of "best response" is no longer logically flawed. Each player is choosing her best response, not to the as yet unobserved actions of the others, but only to her own already formed beliefs about their actions. Second, in Chapters 7 and 8, where we allow mixed strategies, the randomness in one player's strategy may be better interpreted as uncertainty in the other players' beliefs about this player's action. For now, we proceed by using both interpretations of Nash equilibrium in parallel.

You might think that formation of correct beliefs and calculation of best responses is too daunting a task for mere humans. We discuss some criticisms of this kind, as well as empirical and experimental evidence concerning Nash equilibrium, in Chapter 5 for pure strategies and Chapters 7 and 8 for mixed strategies. For now, we simply say that the proof of the pudding is in the eating. We develop and illustrate the Nash equilibrium concept by applying it. We hope that seeing it in use will prove a better way to understand its strengths and drawbacks than would an abstract discussion at this point.

[6]In this chapter we consider only Nash equilibria in pure strategies, namely the ones initially listed in the specification of the game, and not mixtures of two or more of them. Therefore in such an equilibrium, each player has certainty about the actions of the others; strategic uncertainty is removed. When we consider mixed strategy equilibria in Chapters 7 and 8, the strategic uncertainty for each player will consist of the probabilities with which the various strategies are played in the other players' equilibrium mixtures.

3 DOMINANCE

Some games have a special property that one strategy is uniformly better than or worse than another. When this is the case, it provides one way in which the search for Nash equilibrium and its interpretation can be simplified.

The well-known game of the **prisoners' dilemma** illustrates this concept well. Consider a story line of the type that appears regularly in the television program *Law and Order.* Suppose that a husband and wife have been arrested under the suspicion that they were conspirators in the murder of a young woman. Detectives Green and Lupo place the suspects in separate detention rooms and interrogate them one at a time. There is little concrete evidence linking the pair to the murder, although there is some evidence that they were involved in kidnapping the victim. The detectives explain to each suspect that they are both looking at jail time for the kidnapping charge, probably 3 years, even if there is no confession from either of them. In addition, the husband and wife are told individually that the detectives "know" what happened and "know" how one had been coerced by the other to participate in the crime; it is implied that jail time for a solitary confessor will be significantly reduced if the whole story is committed to paper. (In a scene common to many similar programs, a yellow legal pad and a pencil are produced and placed on the table at this point.) Finally, they are told that, if both confess, jail terms could be negotiated down but not as much as they would be if there were one confession and one denial.

Both husband and wife are then players in a two-person, simultaneous-move game in which each has to choose between confessing and not confessing to the crime of murder. They both know that no confession leaves them each with a 3-year jail sentence for involvement with the kidnapping. They also know that, if one of them confesses, he or she will get a short sentence of 1 year for cooperating with the police, while the other will go to jail for a minimum of 25 years. If both confess, they figure that they can negotiate for jail terms of 10 years each.

The choices and outcomes for this game are summarized by the game table in Figure 4.4. The strategies Confess and Deny can also be called Defect and Cooperate to capture their roles in the relationship between the *two players;* thus Defect

		WIFE	
		Confess (Defect)	Deny (Cooperate)
HUSBAND	Confess (Defect)	10 yr, 10 yr	1 yr, 25 yr
	Deny (Cooperate)	25 yr, 1 yr	3 yr, 3 yr

FIGURE 4.4 Prisoners' Dilemma

means to defect from any tacit arrangement with the spouse, and Cooperate means to take the action that helps the spouse (not cooperate with the cops).

Payoffs here are the lengths of the jail sentences associated with each outcome, so low numbers are better for each player. In that sense, this example differs from those of most of the games that we analyze, in which large payoffs are good rather than bad. We take this opportunity to alert you that "large is good" is not always true. When payoff numbers indicate players' rankings of outcomes, people often use 1 for the best alternative and successively higher numbers for successively worse ones. Also, in the table for a zero-sum game that shows only one player's bigger-is-better payoffs, smaller numbers are better for the other. In the prisoners' dilemma here, smaller numbers are better for both. Thus, if you ever write a payoff table where large numbers are bad, you should alert the reader by pointing it out clearly. And when reading someone else's example, be aware of the possibility.

Now consider the prisoners' dilemma game in Figure 4.4 from the husband's perspective. He has to think about what the wife will choose. Suppose he believes that she will confess. Then his best choice is to confess; he gets a sentence of only 10 years, when denial would have meant 25 years. What if he believes the wife will deny? Again, his own best choice is to confess; he gets only 1 year instead of the 3 that his own denial would bring in this case. Thus, in this special game, Confess is better than Deny for the husband *regardless of his belief about the wife's choice.* We say that, for the husband, the strategy Confess is a **dominant strategy** or that the strategy Deny is a **dominated strategy.** Equivalently, we could say that the strategy Confess *dominates* the strategy Deny or that the strategy Deny is *dominated* by the strategy Confess.

If an action is clearly best for a player, no matter what the others might be doing, then there is compelling reason to think that a rational player would choose it. And if an action is clearly bad for a player, no matter what the others might be doing, then there is equally compelling reason to think that a rational player would avoid it. Therefore dominance, when it exists, provides a compelling basis for the theory of solutions to simultaneous-move games.

A. Both Players Have Dominant Strategies

In the preceding prisoners' dilemma, dominance should lead the husband to choose Confess. Exactly the same logic applies to the wife's choice. Her own strategy Confess dominates her own strategy Deny; so she also should choose Confess. Therefore (Confess, Confess) is the outcome predicted for this game. Note that it is a Nash equilibrium. (In fact it is the only Nash equilibrium.) Each player is choosing his or her own best strategy.

In this special game, the best choice for each is independent of whether their beliefs about the other are correct—that is the meaning of dominance—but each, attributing to the other the same rationality as he or she practices, should be able to form correct beliefs. And the actual action of each is the best response to the actual action of the other. Note that the fact that Confess dominates Deny for both players is completely independent of whether they are actually guilty, as in many episodes of *Law and Order,* or are being framed, as happened in the movie *LA Confidential.* It only depends on the pattern of payoffs dictated by the various sentence lengths.

Any game with the same general payoff pattern as that illustrated in Figure 4.4 is given the generic label "prisoners' dilemma." More specifically, a prisoners' dilemma has three essential features. First, each player has two strategies: to cooperate with one's rival (deny any involvement in the crime, in our example) or to defect from cooperation (confess to the crime, here). Second, each player also has a dominant strategy (to confess or to defect from cooperation). Finally, the dominance solution equilibrium is worse for both players than the nonequilibrium situation in which each plays the dominated strategy (to cooperate with rivals).

Games of this type are particularly important in the study of game theory for two reasons. The first is that the payoff structure associated with the prisoners' dilemma arises in many quite varied strategic situations in economic, social, political, and even biological competitions. This wide-ranging applicability makes it an important game to study and to understand from a strategic standpoint. The whole of Chapter 11 and sections in several other chapters deal with its study.

The second reason that prisoners' dilemma games are integral to any discussion of games of strategy is the somewhat curious nature of the equilibrium outcome achieved in such games. Both players follow conventional wisdom in choosing their dominant strategies, but the resulting equilibrium outcome yields them payoffs that are lower than they could have achieved if they had each chosen their dominated strategies. Thus the equilibrium outcome in the prisoners' dilemma is actually a bad outcome for the players. They could find another outcome that they both prefer to the equilibrium outcome; the problem is how to guarantee that someone will not cheat. This particular feature of the prisoners' dilemma has received considerable attention from game theorists who have asked an obvious question: What can players in a prisoners' dilemma do to achieve the better outcome? We leave this question to the reader momentarily, as we continue the discussion of simultaneous games, but return to it in detail in Chapter 11.

B. One Player Has a Dominant Strategy

When a rational player has a dominant strategy, she will use it, and the other player can safely believe this. In the prisoners' dilemma, it applied to both players. In some other games, it applies only to one of them. If you are playing in a game in which you do not have a dominant strategy but your opponent does, you can assume that she will use her dominant strategy and so you can choose your equilibrium action (your best response) accordingly.

We illustrate this case by using a game frequently played between the Congress, which is responsible for fiscal policy (taxes and government expenditures), and the Federal Reserve (Fed), which is in charge of monetary policy (primarily, interest rates).[7] In a version that simplifies the game to its essential features, the Congress's fiscal policy can have either a balanced budget or a deficit, and the Fed can set interest rates either high or low. In reality, the game is not clearly simultaneous; nor is who has the first move obvious if choices are sequential. We consider the simultaneous-move version here, and in Chapter 6 study how the outcomes differ for different rules of the game.

Almost everyone wants lower taxes. But there is no shortage of good claims on government funds: defense, education, health care, and so on. There are also various politically powerful special interest groups—including farmers and industries hurt by foreign competition—who want government subsidies. Therefore the Congress is under constant pressure both to lower taxes and to increase spending. But such behavior runs the budget into deficit, which can lead to higher inflation. The Fed's primary task is to prevent inflation. However, it also faces political pressure for lower interest rates from many important groups, especially homeowners who benefit from lower mortgage rates. Lower interest rates lead to higher demand for automobiles, housing, and capital investment by firms, and all this demand can cause higher inflation. The Fed is generally happy to lower interest rates but only so long as inflation is not a threat. And there is less threat of inflation when the government's budget is in balance. With all this in mind, we construct the payoff matrix for this game in Figure 4.5.

Congress likes best (payoff 4) the outcome with a budget deficit and low interest rates. This pleases all the immediate political constituents. It may entail trouble for the future, but political time horizons are short. For the same reason,

		FEDERAL RESERVE	
		Low interest rates	High interest rates
CONGRESS	Budget balance	3, 4	1, 3
	Budget deficit	4, 1	2, 2

FIGURE 4.5 Game of Fiscal and Monetary Policies

[7]Similar games are played in many other countries with central banks that have operational independence in the choice of monetary policy. Fiscal policies may be chosen by different political entities—the executive or the legislature—in different countries.

Congress likes worst (payoff 1) the outcome with a balanced budget and high interest rates. Of the other two outcomes, it prefers (payoff 3) the outcome with a balanced budget and low interest rates; this outcome pleases the important home-owning middle classes and, with low interest rates, less expenditure is needed to service the government debt, so the balanced budget still has room for many other items of expenditure or for tax cuts.

The Fed likes worst (payoff 1) the outcome with a budget deficit and low interest rates, because this combination is the most inflationary. It likes best (payoff 4) the outcome with a balanced budget and low interest rates, because this combination can sustain a high level of economic activity without much risk of inflation. Comparing the other two outcomes with high interest rates, the Fed prefers the one with budget balance because it reduces the risk of inflation.

We look now for dominant strategies in this game. The Fed does better by choosing low interest rates if it believes that the Congress is opting for budget balance (Fed's payoff 4 rather than 3), but it does better choosing high interest rates if it believes that the Congress is choosing to run a budget deficit (Fed's payoff 2 rather than 1). The Fed, then, does not have a dominant strategy. But the Congress does. If the Congress believes that the Fed is choosing low interest rates, it does better for itself by choosing a budget deficit rather than budget balance (Congress's payoff 4 instead of 3). If the Congress believes that the Fed is choosing high interest rates, again it does better for itself by choosing a budget deficit rather than budget balance (Congress's payoff 2 instead of 1). Choosing to run a budget deficit is then Congress's dominant strategy.

The choice for the Congress in the game is now clear. No matter what it believes the Fed is doing, the Congress will choose to run a budget deficit. The Fed can now take this choice into account when making its own decision. The Fed should believe that the Congress will choose its dominant strategy (budget deficit) and choose the best strategy for itself, given this belief. That means that the Fed should choose high interest rates.

In this outcome, each side gets payoff 2. But an inspection of Figure 4.5 shows that, just as in the prisoners' dilemma, there is another outcome—namely, a balanced budget and low interest rates—that can give both players higher payoffs—namely, 3 for the Congress and 4 for the Fed. Why is that outcome not achievable as an equilibrium? The problem is that Congress would be tempted to deviate from its stated strategy and sneakily run a budget deficit. The Fed, knowing this temptation and that it would then get its worst outcome (payoff 1), deviates also to its high interest rate strategy. In Chapters 6 and 10, we consider how the two sides can get around this difficulty to achieve their mutually preferred outcome. But we should note that, in most countries and at many times, the two policy authorities are indeed stuck in the bad outcome; the fiscal policy is too loose, and the monetary policy has to be tightened to keep inflation down.

C. Successive Elimination of Dominated Strategies

The games considered so far have had only two pure strategies available to each player. In such games, if one strategy is dominant, the other is dominated; so choosing the dominant strategy is equivalent to eliminating the dominated one. In larger games, some of a player's strategies may be dominated even though no single strategy dominates all of the others. If players find themselves in a game of this type, they may be able to reach an equilibrium by removing dominated strategies from consideration as possible choices. Removing dominated strategies reduces the size of the game, and then the "new" game may have another dominated strategy, for the same player or for her opponent, that can also be removed. Or the "new" game may even have a dominant strategy for one of the players. **Successive** or **iterated elimination of dominated strategies** uses this process of removal of dominated strategies and reduction in the size of a game until no further reductions can be made. If this process ends in a unique outcome, then the game is said to be **dominance solvable;** that outcome is the Nash equilibrium of the game, and the strategies that yield it are the equilibrium strategies for each player.

We can use the game of Figure 4.1 to provide an example of this process. Consider first Row's strategies. If any one of Row's strategies always provides worse payoffs for Row than another of her strategies, then that strategy is dominated and can be eliminated from consideration for Row's equilibrium choice. Here, the only dominated strategy for Row is High, which is dominated by Bottom; if Column plays Left, Row gets 5 from Bottom and only 4 from High; if Column plays Middle, Row gets 4 from Bottom and only 3 from High; and, if Column plays Right, Row gets 9 from Bottom and only 6 from High. So we can eliminate High. We now turn to Column's choices to see if any of them can be eliminated. We find that Column's Left is now dominated by Right (with similar reasoning, $1 < 2$, $2 < 3$, and $6 < 7$). Note that we could not say this before Row's High was eliminated; against Row's High, Column would get 5 from Left but only 4 from Right. Thus the first step of eliminating Row's High makes possible the second step of eliminating Column's Left. Then, within the remaining set of strategies (Top, Low, and Bottom for Row, and Middle and Right for Column), Row's Top and Bottom are both dominated by his Low. When Row is left with only Low, Column chooses his best response—namely, Middle.

The game is thus dominance solvable, and the outcome is (Low, Middle) with payoffs (5, 4). We identified this outcome as a Nash equilibrium when we first illustrated that concept by using this game. Now we see in better detail the thought process of the players that leads to the formation of correct beliefs. A rational Row will not choose High. A rational Column will recognize this, and thinking about how her various strategies perform for her against Row's remaining strategies, will not choose Left. In turn, Row will recognize this, and therefore will not choose either Top or Bottom. Finally, Column will see through all this, and choose Middle.

Other games may not be dominance solvable, or successive elimination of dominated strategies may not yield a unique outcome. Even in such cases, some elimination may reduce the size of the game and make it easier to solve by using one or more of the techniques described in the following sections. Thus eliminating dominated strategies can be a useful step toward solving a large simultaneous-play game, even when their elimination does not completely solve the game.

Thus far in our consideration of iterated elimination of dominated strategies, all the payoff comparisons have been unambiguous. What if there are some ties? Consider the variation on the preceding game that is shown in Figure 4.3. In that version of the game, High (for Row) and Left (for Column) also are eliminated. And, at the next step, Low still dominates Top. But the dominance of Low over Bottom is now less clear-cut. The two strategies give Row equal payoffs when played against Column's Middle, although Low does give Row a higher payoff than Bottom when played against Column's Right. We say that, from Row's perspective at this point, Low *weakly* dominates Bottom. In contrast, Low *strictly* dominates Top, because it gives strictly higher payoffs than does Top when played against both of Column's strategies, Middle and Right, under consideration at this point.

We give a more precise definition of the distinction between strict and weak dominance in the Appendix to this chapter. Here, though, we provide a word of warning. Successive elimination of weakly dominated strategies can get rid of some Nash equilibria.

Consider the game illustrated in Figure 4.6. For Row, Up is weakly dominated by Down; if Column plays Left, then Row gets a better payoff by playing Down than by playing Up, and, if Column plays Right, then Row gets the same payoff from her two strategies. Similarly, for Column, Right weakly dominates Left. Dominance solvability then tells us that (Down, Right) is a Nash equilibrium.

		COLUMN	
		Left	Right
ROW	Up	0, 0	1, 1
	Down	1, 1	1, 1

FIGURE 4.6 Elimination of Weakly Dominated Strategies

That is true, but (Down, Left) and (Up, Right) also are Nash equilibria. Consider (Down, Left). When Row is playing Down, Column cannot improve her payoff by switching to Right, and, when Column is playing Left, Row's best response is clearly to play Down. A similar reasoning verifies that (Up, Right) also is a Nash equilibrium.

Therefore, if you use weak dominance to eliminate some strategies, it is a good idea to make a quick cell-by-cell check to see if you have missed any other equilibria. The iterated dominance solution seems to be a reasonable outcome to predict as the likely Nash equilibrium of this simultaneous-play game, but it is also important to consider the significance of multiple equilibria as well as of the other equilibria themselves. We address these issues in later chapters, taking up a discussion of multiple equilibria in Chapter 5 and the interconnections between sequential- and simultaneous-move games in Chapter 6.

4 BEST-RESPONSE ANALYSIS

Many simultaneous-move games have no dominant strategies and no dominated strategies. Others may have one or several dominated strategies, but iterated elimination of dominated strategies will not yield a unique outcome. In such cases, we need a next step in the process of finding a solution to the game. We are still looking for a Nash equilibrium in which every player does the best she can, given the actions of the other player(s), but we must now rely on subtler strategic thinking than the simple elimination of dominated strategies requires.

Here, we develop another systematic method for finding Nash equilibria that will prove very useful in later analysis. We begin without imposing a requirement of correctness of beliefs. We take each player's perspective in turn and ask the following question: For each of the choices that the other player(s) might be making, what is the best choice for this player? Thus we find the best responses

		COLUMN		
		Left	Middle	Right
ROW	Top	3, 1	2, (3)	10, 2
	High	4, (5)	3, 0	6, 4
	Low	2, 2	(5), (4)	(12), 3
	Bottom	(5), 6	4, 5	9, (7)

FIGURE 4.7 Best Response Analysis

of each player to all available strategies of the others. In mathematical terms, we find each player's best-response strategy, depending on, or as a function of, the other players' available strategies.

Return to the game of Figure 4.1, reproduced as Figure 4.7, and consider Row first. If Column chooses Left, Row's best response is Bottom, yielding 5. We show this best response by circling that payoff in the game table. If Column chooses Middle, Row's best response is Low (also yielding 5). And, if Column chooses Right, Row's best choice is again Low (now yielding 12). As before, we show Row's best choices by circling the appropriate payoffs. Similarly, Column's best responses are shown by circling her payoffs 3 (Middle as best response to Row's Top), 5 (Left to High), 4 (Middle to Low), and 7 (Right to Bottom).[8] We see that one cell—namely, (Low, Middle)—has both its payoffs circled. Therefore the strategies Low for Row and Middle for Column are simultaneously best responses to each other. We have found the Nash equilibrium of this game. (Again.)

Best-response analysis is a comprehensive way of locating all possible Nash equilibria of a game. You should improve your understanding of it by trying it out on the other games that have been used in this chapter. The cases of dominance are of particular interest. If Row has a dominant strategy, that same strategy is her best response to all of Column's strategies; therefore her best responses are all lined up horizontally in the same row. Similarly, if Column has a dominant strategy, her best responses are all lined up vertically in the same column. You should see for yourself how the Nash equilibria of the preceding prisoners' dilemma and Congress–Fed games emerge from such a drawing.

There will be some games for which best-response analysis does not find a Nash equilibrium, just as dominance solvability sometimes fails. But in this case we can say something more specific than can be said when dominance fails. When best-response analysis of a discrete strategy game does not find a Nash equilibrium, then the game has no equilibrium in pure strategies. We address games of this type in Section 8 of this chapter. In Chapter 5, we extend best-response analysis to games where the players' strategies are continuous variables—for example, prices or advertising expenditures. There, we construct best-response *curves* to help us find Nash equilibria, and we see that such games are less likely—by virtue of the continuity of strategy choices—to have no equilibrium.

[8]Alternatively and equivalently, one could mark in some way the choices that are *not* made. For example, in Figure 4.3, Row will not choose Top, High, or Bottom as responses to Column's Right; one could show this by drawing slashes through Row's payoffs in these cases, respectively, 10, 6, and 9. When this is done for all strategies of both players, (Low, Middle) has both of its payoffs unslashed; it is then the Nash equilibrium of the game. The alternatives of circling choices that are made and slashing choices that are not made stand in a conceptually similar relation to each other, as do the alternatives of showing chosen branches by arrows and pruning unchosen branches for sequential-move games. We prefer the first alternative in each case, because the resulting picture looks cleaner and tells the story better.

5 THE MINIMAX METHOD FOR ZERO-SUM GAMES

For zero-sum games, an alternative to best-response analysis works by using the special logic of strict conflict that exists in such games. This approach, the **minimax method,** works only for zero-sum games and relies on a thought process that accounts for the fact that outcomes that are good for one player are, by definition, bad for the other. In this method, each player is assumed to choose her strategy by thinking: "Would this be the best choice for me, even if the other player found out that I was playing it?" She must then consider her opponent's best response to her chosen strategy. But in a zero-sum game, that best response is the worst one for her. In other words, each player believes that her opponent will choose an action that yields her the worst possible consequences of each of her own actions. Then acting on those beliefs she should choose the action that leads to the least-bad outcome.

This logic may seem extremely pessimistic, but it still relies on a type of best-response calculation and it is appropriate for finding the equilibrium of a zero-sum game. In equilibrium, each player is choosing her own best response, given her beliefs about what the other will do. In anticipating such best responses, each player will expect to receive the worst payoff associated with each action and will choose her own action accordingly. She is thus choosing her best payoff from among the set of worst payoffs.

Suppose the payoff table shows the row player's payoffs, and Row wants the outcome to be a cell with as high a number as possible. Then Column wants the outcome to be a cell with as low a number as possible. Using the pessimistic logic just described, Row figures that, for each of her rows, Column will choose the column with the lowest number in that row. Therefore Row should choose the row that gives her the highest among these lowest numbers, or the maximum among the minima—the **maximin** for short. Similarly, Column reckons that, for each of her columns, Row will choose the row with the largest number in that column. Then Column should choose the column with the smallest number among these largest ones, or the minimum among the maxima—the **minimax.** If Row's maximin value and Column's minimax value are in the same cell of the game table, then that outcome is a Nash equilibrium of the zero-sum game. This method of finding equilibria in zero-sum games should be called the maximin-minimax method, but it is called simply the *minimax method* for short. It will lead you to a Nash equilibrium in pure strategies if one exists.

To illustrate the minimax method, we use the football example of Figure 4.2. We already know the Nash equilibrium for that game, but now we obtain it by using the minimax method. We reproduce the game table in Figure 4.8, adding information that pertains to the minimax argument.

		DEFENSE			
		Run	Pass	Blitz	
OFFENSE	Run	2	5	13	min = 2
	Short Pass	6	5.6	10.5	min = 5.6
	Medium Pass	6	4.5	1	min = 1
	Long Pass	10	3	–2	min = –2
		max = 10	max = 5.6	max = 13	

FIGURE 4.8 The Minimax Method

Begin by finding the lowest number in each row (the offense's worst payoff from each strategy) and the highest number in each column (the defense's worst payoff from each strategy). The offense's worst payoff from Run is 2; its worst payoff from Short Pass, 5.6; its worst payoff from Medium Pass, 1; and its worst payoff from Long Pass, −2. We write the minimum for each row at the far right of that row. The defense's worst payoff from Run is 10; its worst payoff from Pass, 5.6; and its worst payoff from Blitz, 13. We write the maximum for each column at the bottom of that column.

The next step is to find the best of each player's worst possible outcomes, the largest row minimum and the smallest column maximum. The largest of the row minima is 5.6; so the offense can ensure itself a gain of 5.6 yards by playing the Short Pass; this is its maximin. The lowest of the column maxima is 5.6; so the defense can be sure of holding the offense down to a gain of 5.6 yards by deploying its Pass defense. This is the defense's minimax.

Looking at these two strategy choices, we see that the maximin and minimax values are found in the same cell of the game table. Thus the offense's maximin strategy is its best response to the defense's minimax and vice versa; we have found the Nash equilibrium of this game. That equilibrium entails the offense attempting a Short Pass while the defense defends against a Pass. A total of 5.6 yards will be gained by the offense (and given up by the defense).

The minimax method may fail to find an equilibrium in some zero-sum games. If so, then our conclusion is similar to that when best-response analysis fails: the game has no Nash equilibrium in pure strategies. We address this matter later in this chapter and examine mixed strategy equilibria in Chapters 7 and 8. And, to repeat, the minimax method cannot be applied to non-zero-sum games. In such games, your opponent's best is not necessarily your worst. Therefore the pessimistic assumption that leads you to choose the strategy that makes your minimum payoff as large as possible may not be your best strategy.

6 THREE PLAYERS

So far, we have analyzed only games between two players. All of the methods of analysis that have been discussed, however, can be used to find the pure-strategy Nash equilibria of any simultaneous-play game among any number of players. When a game is played by more than two players, each of whom has a relatively small number of pure strategies, the analysis can be done with a game table, as we did in the first five sections of this chapter.

In Chapter 3, we described a game among three players, each of whom had two pure strategies. The three players, Emily, Nina, and Talia, had to choose whether to contribute toward the creation of a flower garden for their small street. We assumed there that the garden when all three contributed was no better than when only two contributed and that a garden with just one contributor was so sparse that it was as bad as no garden at all. Now, let us suppose instead that the three players make their choices simultaneously and that there is a somewhat richer variety of possible outcomes and payoffs. In particular, the size and splendor of the garden will now differ according to the exact number of contributors; three contributors will produce the largest and best garden, two contributors will produce a medium garden, and one contributor will produce a small garden.

Suppose Emily is contemplating the possible outcomes of the street-garden game. There are six possibilities to consider. Emily can choose either to contribute or not to contribute when both Nina and Talia contribute or when neither of them contributes or when just one of them contributes. From her perspective, the best possible outcome, with a rating of 6, would be to take advantage of her good-hearted neighbors and to have both Nina and Talia contribute while she does not. Emily could then enjoy a medium-sized garden without putting up her own hard-earned cash. If both of the others contribute and Emily also contributes, she gets to enjoy a large, very splendid garden but at the cost of her own contribution; she rates this outcome second-best, or 5.

At the other end of the spectrum are the outcomes that arise when neither Nina nor Talia contributes to the garden. If that is the case, Emily would again prefer not to contribute, because she would foot the bill for a public garden that everyone could enjoy; she would rather have the flowers in her own yard. Thus, when neither other player is contributing, Emily ranks the outcome in which she contributes as a 1 and the outcome in which she does not as a 2.

In between these cases are the situations in which either Nina or Talia contributes to the flower garden but not both. When one of them contributes, Emily knows that she can enjoy a small garden without contributing; she also feels that the cost of her contribution outweighs the increase in benefit that she gets

TALIA chooses:

Contribute

		NINA	
		Contribute	Don't
EMILY	Contribute	5, 5, 5	3, 6, 3
	Don't	6, 3, 3	4, 4, 1

Don't Contribute

		NINA	
		Contribute	Don't
EMILY	Contribute	3, 3, 6	1, 4, 4
	Don't	4, 1, 4	2, 2, 2

FIGURE 4.9 Street-Garden Game

from being able to increase the size of the garden. Thus she ranks the outcome in which she does not contribute, but still enjoys the small garden, as a 4 and the outcome in which she does contribute, to provide a medium garden, as a 3. Because Nina and Talia have the same views as Emily on the costs and benefits of contributions and garden size, each of them orders the different outcomes in the same way—the worst outcome being the one in which each contributes and the other two do not, and so on.

If all three women decide whether to contribute to the garden without knowing what their neighbors will do, we have a three-person simultaneous-move game. To find the Nash equilibrium of the game, we then need a game table. For a three-player game, the table must be three-dimensional and the third player's strategies correspond to the new dimension. The easiest way to add a third dimension to a two-dimensional game table is to add pages. The first page of the table shows payoffs for the third player's first strategy, the second page shows payoffs for the third player's second strategy, and so on.

We show the three-dimensional table for the street-garden game in Figure 4.9. It has two rows for Emily's two strategies, two columns for Nina's two strategies, and two pages for Talia's two strategies. We show the pages side by side so that you can see everything at the same time. In each cell, payoffs are listed for the row player first, the column player second, and the page player third; in this case, the order is Emily, Nina, Talia.

Our first test should be to determine whether there are dominant strategies for any of the players. In one-page game tables, we found this test to be simple; we just compared the outcomes associated with one of a player's strategies with the outcomes associated with another of her strategies. In practice this comparison required, for the row player, a simple check within columns of the single page of the table and vice versa for the column player. Here, we must check in both pages of the table to determine whether any player has a dominant strategy.

For Emily, we compare the two rows of both pages of the table and note that, when Talia contributes, Emily has a dominant strategy not to contribute, and, when Talia does not contribute, Emily also has a dominant strategy not to contribute. Thus the best thing for Emily to do, regardless of what either of the other players does, is not to contribute. Similarly, we see that Nina's dominant strategy—in both pages of the table—is not to contribute. When we check for a dominant strategy for Talia, we have to be a bit more careful. We must compare outcomes that keep Emily's and Nina's behavior constant, checking Talia's payoffs from choosing Contribute versus Don't. That is, we compare cells across pages of the table—the top-left cell in the first page (on the left) with the top-left cell in the second page (on the right), and so on. As for the first two players, this process indicates that Talia also has a dominant strategy not to contribute.

Each player in this game has a dominant strategy, which must therefore be her equilibrium pure strategy. The Nash equilibrium of the street-garden game entails all three players choosing not to contribute to the street garden and getting their second-worst payoffs; the garden is not planted, but no one has to contribute either.

Notice that this game is yet another example of a prisoners' dilemma. There is a unique Nash equilibrium in which all players receive a payoff of 2. Yet there is another outcome in the game—in which all three neighbors contribute to the garden—that for all three players yields higher payoffs of 5. Even though it would be beneficial to each of them for all to pitch in to build the garden, no one has the individual incentive to do so. As a result, gardens of this type are either not planted at all or paid for through tax dollars—because the town government can require its citizens to pay such taxes. In Chapter 12, we will encounter more such dilemmas of collective action and study some methods for resolving them.

The Nash equilibrium of the game can also be found using the cell-by-cell inspection method. For example, consider another cell in Figure 4.9—say, the one where Emily and Nina contribute but Talia does not, with the payoffs (3, 3, 6). When Emily considers changing her strategy, as the row player she can change only the row position of the game's outcome. Emily can move the outcome only from a given cell in a given row, column, and page to another cell in a different row but the same column and same page of the table. If she does that in this instance, she improves her payoff from 3 to 4. Similarly, Nina can change only the column position of the outcome, moving it to a cell in another column but in the same row and same page of the table. Doing so improves Nina's payoff from 3 to 4. Finally, Talia can change only the page position of the game's outcome. She can move the outcome to a different page, but the row and column positions must remain the same. Doing so would worsen Talia's payoff from 6 to 5. Because at least one player can do better by unilaterally changing her strategy, the cell that we examined cannot be the outcome of a Nash equilibrium.

TALIA chooses:

Contribute

		NINA	
		Contribute	Don't
EMILY	Contribute	5, 5, 5	3, 6, 3
	Don't	6, 3, 3	4, 4, 1

Don't Contribute

		NINA	
		Contribute	Don't
EMILY	Contribute	3, 3, 6	1, 4, 4
	Don't	4, 1, 4	2, 2, 2

FIGURE 4.10 Best-Response Analysis in the Street-Garden Game

We can also use the best-response method, as shown in Figure 4.10, by drawing circles around the best responses, as in Figure 4.7. Because each player has Don't as her dominant strategy, all of Emily's best responses are on her Don't rows, all of Nina's on her Don't columns, and all of Talia's on her Don't page. The cell at the bottom right has all three best responses; therefore it gives us the Nash equilibrium.

7 MULTIPLE EQUILIBRIA IN PURE STRATEGIES

Each of the games considered in preceding sections has had a unique pure-strategy Nash equilibrium. In general, however, games need not have unique Nash equilibria. We illustrate this result by using a class of games that have many applications. As a group, they may be labeled **coordination games.** The players in such games have some (but not always completely) common interests. But, because they act independently (by virtue of the nature of noncooperative games), the coordination of actions needed to achieve a jointly preferred outcome is problematic.

A. Will Harry Meet Sally? Pure Coordination

To illustrate this idea, picture two undergraduates, Harry and Sally, who meet in their college library. They are attracted to each other and would like to continue the conversation but have to go off to their separate classes. They arrange to meet for coffee after the classes are over at 4:30. Sitting separately in class, each realizes that in the excitement they forgot to fix the place to meet. There are two possible choices, Starbucks and Local Latte. Unfortunately, these locations are on opposite sides of the large campus; so it is not possible to try both. And Harry and Sally have not exchanged cell-phone numbers, so they can't send messages. What should each do?

		SALLY	
		Starbucks	Local Latte
HARRY	Starbucks	1, 1	0, 0
	Local Latte	0, 0	1, 1

FIGURE 4.11 Pure Coordination

Figure 4.11 illustrates this situation as a game and shows the payoff matrix. Each player has two choices—Starbucks and Local Latte. The payoffs for each are 1 if they meet and 0 if they do not. Cell-by-cell inspection shows at once that the game has two Nash equilibria, one where both choose Starbucks and the other where both choose Local Latte. It is important for both that they achieve one of the equilibria, but which one is immaterial because the two yield equal payoffs. All that matters is that they coordinate on the same action; it does not matter which action. That is why the game is said to be one of **pure coordination.**

But will they coordinate successfully? Or will they end up in different cafés, each thinking that the other has let him or her down? Alas, that risk exists. Harry might think that Sally will go to Starbucks because she said something about the class to which she was going and that class is on the Starbucks side of the campus. But Sally may have the opposite belief about what Harry will do. When there are multiple Nash equilibria, if the players are to select one successfully, they need some way to coordinate their beliefs or expectations about each other's actions.

The situation is similar to that of the heroes of the "Which tire?" game in Chapter 1, where we labeled the coordination device a **focal point.** In the present context, one of the two cafés may be generally known as the student hangout. But it is not enough that Harry knows this to be the case. He must know that Sally knows, and that she knows that he knows, and so on. In other words, their expectations must *converge* on the focal point. Otherwise Harry might be doubtful about where Sally will go because he does not know what she is thinking about where he will go; and similar doubts may arise at the third or fourth or higher level of thinking about thinking.[9]

[9]Thomas Schelling presented the classic treatment of coordination games and developed the concept of a focal point in his book *The Strategy of Conflict* (Cambridge: Harvard University Press, 1960); see pp. 54–58, 89–118. His explanation of focal points included the results garnered when he posed several questions to his students and colleagues. The best-remembered of these is "Suppose you have arranged to meet someone in New York City on a particular day, but have failed to arrange a specific place or time, and have no way of communicating with the other person. Where will you go and at what time?" Fifty years ago when the question was first posed, the clock at Grand Central Station was the usual focal place; now it might be the observation platform atop the Empire State Building or Times Square. The focal time remains twelve noon.

When one of us (Dixit) posed this question to students in his class, the freshmen generally chose Starbucks and the juniors and seniors generally chose the local café in the campus student center. These responses are understandable—freshmen, who have not been on campus long, focus their expectations on a nationwide chain that is known to everyone, whereas juniors and seniors have acquired the local habits, which they now regard as superior, and expect their peers to believe likewise.

If one café had an orange decor and the other a crimson decor, then in Princeton the former may serve as a focal point because orange is the Princeton color, whereas at Harvard crimson may be focal for the same reason. If one person is a Princeton student and the other a Harvard student, they may fail to meet at all, either because each thinks that his or her color "should" get priority or because each thinks that the other will be inflexible and so tries to accommodate him or her. More generally, whether players in coordination games can find a focal point depends on their having some commonly known point of contact, whether historical, cultural, or linguistic.

B. Will Harry Meet Sally? And Where? Assurance

Now change the game payoffs a little. The behavior of juniors and seniors suggests that our pair may not be quite indifferent about which café they both choose. The coffee may be better at one or the ambiance better at one. Or they may want to choose the one that is not the general student hangout, to avoid the risk of running into former boyfriends or girlfriends. Suppose they both prefer Local Latte; so the payoff of each is 2 when they meet there versus 1 when they meet at Starbucks. The new payoff matrix is shown in Figure 4.12.

Again, there are two Nash equilibria. But in this version of the game, each prefers the equilibrium where both choose Local Latte. Unfortunately, their mere liking of that outcome is not guaranteed to bring it about. First of all (and as always in our analysis), the payoffs have to be common knowledge—both have to know the entire payoff matrix, both have to know that both know, and so on. Such detailed knowledge about the game can arise if the two discussed and

		SALLY	
		Starbucks	Local Latte
HARRY	Starbucks	1, 1	0, 0
	Local Latte	0, 0	2, 2

FIGURE 4.12 Assurance

agreed on the relative merits of the two cafés but simply forgot to decide definitely to meet at Local Latte. Even then, Harry might think that Sally has some other reason for choosing Starbucks, or think that she thinks that he does, and so on. Without genuine **convergence of expectations** about actions, they may choose the worse equilibrium or, worse still, they may fail to coordinate actions and get 0 each.

To repeat, players in the game illustrated in Figure 4.12 can get the preferred equilibrium outcome only if each has enough certainty or assurance that the other is choosing the appropriate action. For this reason, such games are called **assurance games.**[10]

In many real-life situations of this kind, such assurance is easily obtained, given even a small amount of communication between the players. Their interests are perfectly aligned; if one of them says to the other, "I am going to Local Latte," the other has no reason to doubt the truth of this statement and will follow to get the mutually preferred outcome. That is why we had to construct the story with the two students isolated in different classes with no means of communication. If the players' interests conflict, truthful communication becomes more problematic. We examine this problem further when we consider strategic manipulation of information in games in Chapter 9.

In larger groups, communication can be achieved by scheduling meetings or by making announcements. These devices work only if everyone knows that everyone else is paying attention to them, because successful coordination requires the desired equilibrium to be a focal point. The players' expectations must converge on it; everyone should know that everyone knows that . . . everyone is choosing it. Many social institutions and arrangements play this role. Meetings where the participants sit in a circle facing inward ensure that everyone sees everyone else paying attention. Advertisements during the Super Bowl, especially when they are proclaimed in advance as major attractions, ensure each viewer that many others are viewing them also. That makes such ads especially attractive to companies making products that are more desirable for any one buyer when many others are buying them, too; such products include those produced by the computer, telecommunication, and Internet industries.[11]

[10]The classic example of an assurance game usually offered is the stag hunt described by the 18th-century French philosopher Jean-Jacques Rousseau. Several people can successfully hunt a stag, thereby getting a large quantity of meat if they collaborate. If any one of them is sure that all of the others will collaborate, he also stands to benefit by joining the group. But if he is unsure whether the group will be large enough, he will do better to hunt for a smaller animal, a hare, on his own. However, it can be argued that Rousseau believed that each hunter would prefer to go after a hare regardless of what the others were doing, which would make the stag hunt a multiperson prisoners' dilemma, not an assurance game. We discuss this example in the context of collective action in Chapter 12.

[11]Michael Chwe develops this theme in *Rational Ritual: Culture, Coordination, and Common Knowledge* (Princeton: Princeton University Press, 2001).

		SALLY	
		Starbucks	Local Latte
HARRY	Starbucks	2, 1	0, 0
	Local Latte	0, 0	1, 2

FIGURE 4.13 Battle of the Sexes

C. Will Harry Meet Sally? And Where? Battle of the Sexes

Now introduce another complication to the café-choice game. Both players want to meet but prefer different cafés. So Harry might get a payoff of 2 and Sally a payoff of 1 from meeting at Starbucks, and the other way around from meeting at Local Latte. This payoff matrix is shown in Figure 4.13.

This game is called the **battle of the sexes.** The name derives from the story concocted for this payoff structure by game theorists in the sexist 1950s. A husband and wife were supposed to choose between going to a boxing match and a ballet, and (presumably for evolutionary genetic reasons) the husband was supposed to prefer the boxing match and the wife the ballet. The name has stuck and we will keep it, but our example—where either could easily have some non-gender-based reason to prefer either of the two cafés—should make it clear that it has no necessarily sexist connotations.

What will happen in this game? There are still two Nash equilibria. If Harry believes that Sally will choose Starbucks, it is best for him to do likewise, and the other way around. For similar reasons, Local Latte also is a Nash equilibrium. To achieve either of these equilibria and avoid the outcomes where the two go to different cafés, the players need a focal point, or convergence of expectations, exactly as in the pure-coordination and assurance games. But the risk of coordination failure is greater in the battle of the sexes. The players are initially in quite symmetric situations, but each of the two Nash equilibria gives them asymmetric payoffs; and their preferences between the two outcomes are in conflict. Harry prefers the outcome where they meet in Starbucks, and Sally prefers to meet in Local Latte. They must find some way of breaking the symmetry.

In an attempt to achieve his or her preferred equilibrium, each player may try to act tough and follow the strategy leading to the better equilibrium. In Chapter 10, we consider in detail such advance devices, called strategic moves, that players in such games can adopt to try to achieve their preferred outcomes. Or each may try to be nice, leading to the unfortunate situation where Harry goes to Local Latte because he wants to please Sally, only to find that she has chosen to please him and gone to Starbucks, like the couple choosing Christmas presents for each other in O. Henry's short story titled "The Gift of the Magi." Alternatively, if the game is repeated, successful coordination may be negotiated and maintained as an equilibrium. For example, the two can arrange to alternate between the cafés. In Chapter 11, we examine such tacit cooperation in repeated games in the context of a prisoners' dilemma.

D. Will James Meet Dean? Chicken

Our final example in this section is a slightly different kind of coordination game. In this game, the players want to avoid, not choose, actions with the same labels. Further, the consequences of one kind of coordination failure are far more drastic than those of the other kind.

The story comes from a game that was supposedly played by American teenagers in the 1950s. Two teenagers take their cars to opposite ends of Main Street, Middle-of-Nowhere, USA, at midnight and start to drive toward each other. The one who swerves to prevent a collision is the "chicken," and the one who keeps going straight is the winner. If both maintain a straight course, there is a collision in which both cars are damaged and both players injured.[12]

The payoffs for **chicken** depend on how negatively one rates the "bad" outcome—being hurt and damaging your car in this case—against being labeled chicken. As long as words hurt less than crunching metal, a reasonable payoff table for the 1950s version of chicken is found in Figure 4.14. Each player most prefers to win, having the other be chicken, and each least prefers the crash of the two cars. In between these two extremes, it is better to have your rival be chicken with you (to save face) than to be chicken by yourself.

This story has four essential features that define any game of chicken. First, each player has one strategy that is the "tough" strategy and one that is the "weak" strategy. Second, there are two pure-strategy Nash equilibria. These are the outcomes in which exactly one of the players is chicken, or weak. Third, each player strictly prefers that equilibrium in which the other player chooses

[12]A slight variant was made famous by the 1955 James Dean movie *Rebel Without a Cause.* There, two players drove their cars in parallel, very fast, toward a cliff. The first to jump out of his car before it went over the cliff was the chicken. The other, if he left too late, risked going over the cliff in his car to his death. The characters in the film referred to this as a "chicky game." In the mid-1960s, the British philosopher Bertrand Russell and other peace activists used this game as an analogy for the nuclear arms race between the United States and the USSR, and the game theorist Anatole Rapoport gave a formal game-theoretic statement. Other game theorists have chosen to interpret the arms race as a prisoners' dilemma or as an assurance game. For a review and interesting discussion, see Barry O'Neill, "Game Theory Models of Peace and War," in *The Handbook of Game Theory,* vol. 2, ed. Robert J. Aumann and Sergiu Hart (Amsterdam: North Holland, 1994), pp. 995–1053.

		DEAN	
		Swerve (Chicken)	Straight (Tough)
JAMES	Swerve (Chicken)	0, 0	−1, 1
	Straight (Tough)	1, −1	−2, −2

FIGURE 4.14 Chicken

chicken, or weak. Fourth, the payoffs when both players are tough are very bad for both players. In games such as this one, the real game becomes a test of how to achieve one's preferred equilibrium.

We are now back in a situation similar to that discussed for the battle-of-the-sexes game. One expects most real-life chicken games to be even worse as battles than most battles of the sexes—the benefit of winning is larger, as is the cost of the crash, and so all the problems of conflict of interest and asymmetry between the players are aggravated. Each player will want to try to influence the outcome. It may be the case that one player will try to create an aura of toughness that everyone recognizes so as to intimidate all rivals.[13] Another possibility is to come up with some other way to convince your rival that you will not be chicken, by making a visible and irreversible commitment to going straight. (In Chapter 10, we consider just how to make such commitment moves.) In addition, both players also want to try to prevent the bad (crash) outcome if at all possible.

As with the battle of the sexes, if the game is repeated, tacit coordination is a better route to a solution. That is, if the teenagers played the game every Saturday night at midnight, they would have the benefit of knowing that the game had both a history and a future when deciding their equilibrium strategies. In such a situation, they might logically choose to alternate between the two equilibria, taking turns being the winner every other week. (But if the others found out about this deal, both players would lose face.)

There is one final point, arising from these coordination games, that must be addressed. The concept of Nash equilibrium requires each player to have the correct belief about the other's choice of strategy. When we look for Nash equilibria in pure strategies, the concept requires each to be confident about the other's choice. But our analysis of coordination games shows that thinking about the other's choice in such games is fraught with strategic uncertainty. How can

[13]Why would a potential rival play chicken against someone with a reputation for never giving in? The problem is that participation in chicken, as in lawsuits, is not really voluntary. Put another way, choosing whether to play chicken is itself a game of chicken. As Thomas Schelling says, "If you are publicly invited to play chicken and say you would rather not, then you have just played [and lost]" (*Arms and Influence*, New Haven: Yale University Press, 1965, p. 118).

we incorporate such uncertainty in our analysis? In Chapter 7, we introduce the concept of a mixed strategy, where actual choices are made randomly among the available actions. This approach generalizes the concept of Nash equilibrium to situations where the players may be unsure about each other's actions.

8 NO EQUILIBRIUM IN PURE STRATEGIES

Each of the games considered so far has had at least one Nash equilibrium in pure strategies. Some of these games, such as those in Section 7, had more than one equilibrium, whereas games in earlier sections had exactly one. Unfortunately, not all games that we come across in the study of strategy and game theory will have such easily definable outcomes in which players always choose one particular action as an equilibrium strategy. In this section, we look at games in which there is not even one pure-strategy Nash equilibrium—games in which none of the players would consistently choose one strategy as that player's equilibrium action.

A simple example of a game with no equilibrium in pure strategies is that of a single point in a tennis match. Imagine a match between the two all-time best women players—Martina Navratilova and Chris Evert.[14] Navratilova at the net has just volleyed a ball to Evert on the baseline, and Evert is about to attempt a passing shot. She can try to send the ball either down the line (DL; a hard, straight shot) or crosscourt (CC; a softer, diagonal shot). Navratilova must likewise prepare to cover one side or the other. Each player is aware that she must not give any indication of her planned action to her opponent, knowing that such information will be used against her. Navratilova would move to cover the side to which Evert is planning to hit or Evert would hit to the side that Navratilova is not planning to cover. Both must act in a fraction of a second, and both are equally good at concealing their intentions until the last possible moment; therefore their actions are effectively simultaneous, and we can analyze the point as a two-player simultaneous-move game.

Payoffs in this tennis-point game are given by the fraction of times a player wins the point in any particular combination of passing shot and covering play.

[14]For those among you who remember only the latest phenom who shines for a couple of years and then burns out, here are some amazing facts about these two, who were at the top levels of the game for almost two decades and ran a memorable rivalry all that time. Navratilova was a left-handed serve-and-volley player. In grand-slam tournaments, she won 18 singles titles, 31 doubles, and 7 mixed doubles. In all tournaments, she won 167, a record. Evert, a right-handed baseliner, had a record win–loss percentage (90% wins) in her career and 150 titles, of which 18 were for singles in grand slam tournaments. She probably invented (and certainly popularized) the two-handed backhand that is now so common. From 1973 to 1988, the two played each other 80 times, and Navratilova ended up with a slight edge, 43–37.

		NAVRATILOVA	
		DL	CC
EVERT	DL	50	80
	CC	90	20

FIGURE 4.15 No Equilibrium in Pure Strategies

Given that a down-the-line passing shot is stronger than a crosscourt shot and that Evert is more likely to win the point when Navratilova moves to cover the wrong side of the court, we can work out a reasonable set of payoffs. Suppose Evert is successful with a down-the-line passing shot 80% of the time if Navratilova covers crosscourt; she is successful with the down-the-line shot only 50% of the time if Navratilova covers down the line. Similarly, Evert is successful with her crosscourt passing shot 90% of the time if Navratilova covers down the line. This success rate is higher than when Navratilova covers crosscourt, in which case Evert wins only 20% of the time.

Clearly, the fraction of times that Navratilova wins this tennis point is just the difference between 100% and the fraction of time that Evert wins. Thus the game is zero-sum (more precisely, constant-sum, because the two payoffs sum to 100), and we can represent all the necessary information in the payoff table with just the payoff to Evert in each cell. Figure 4.15 shows the payoff table and the fraction of time that Evert wins the point against Navratilova in each of the four possible combinations of their strategy choices.

The rules for solving simultaneous-move games tell us to look first for dominant or dominated strategies and then to try minimax (in that this is a zero-sum game) or use cell-by-cell inspection to find a Nash equilibrium. It is a useful exercise to verify that no dominant strategies exist here. Going on to cell-by-cell inspection, we start with the choice of DL for both players. From that outcome, Evert can improve her success from 50% to 90% by choosing CC instead. But then Navratilova can hold Evert down to 20% by choosing CC. After this, Evert can raise her success again to 80% by making her shot DL, and Navratilova in turn can do better with DL. In every cell, one player always wants to change her play, and we cycle through the table endlessly without finding an equilibrium.

An important message is contained in the absence of a Nash equilibrium in this game and similar ones. What is important in games of this type is not what players should do, but what players should *not* do. In particular, each player should neither always nor systematically pick the same shot when faced with this situation. If either player engages in any determinate behavior of that type, the other can take advantage of it. (So if Evert consistently went crosscourt with her passing shot, Navratilova would learn to cover crosscourt every time and would thereby reduce Evert's chances of success with her crosscourt shot.) The most reasonable thing for players to do here is to act somewhat unsystematically, hoping for the element of surprise in defeating their opponents. An unsystematic approach entails choosing each strategy part of the time. (Evert should be using her weaker shot with enough frequency to guarantee that Navratilova cannot predict which shot will come her way. She should not, however, use the two shots in any set pattern, because that, too, would cause her to lose the element of surprise.) This approach, in which players randomize their actions, is known as mixing strategies and is the focus of Chapters 7 and 8. The game illustrated in Figure 4.15 may not have an equilibrium in pure strategies, but it can still be solved by looking for an equilibrium in mixed strategies, as we do in Chapter 7, Section 1.

SUMMARY

In simultaneous-move games, players make their strategy choices without knowledge of the choices being made by other players. Such games are illustrated by *game tables,* where cells show payoffs to each player and the dimensionality of the table equals the number of players. Two-person *zero-sum games* may be illustrated in shorthand with only one player's payoff in each cell of the game table.

Nash equilibrium is the solution concept used to solve simultaneous-move games; such an equilibrium consists of a set of strategies, one for each player, such that each player has chosen her best response to the other's choice. Nash equilibrium can also be defined as a set of strategies such that each player has correct *beliefs* about the others' strategies and strategies are best for each player given beliefs about the other's strategies. Nash equilibria can be found by using *cell-by-cell inspection,* through a search for *dominant strategies,* by *successive elimination of dominated strategies,* or with *best-response analysis.* Zero-sum games can also be solved by using the *minimax method.*

There are many classes of simultaneous games. *Prisoners' dilemma* games appear in many contexts. Coordination games, such as *assurance, chicken,* and *battle of the sexes,* have multiple equilibria, and the solution of such games requires players to achieve coordination by some means. If a game has no equilibrium in *pure strategies,* we must look for an equilibrium in *mixed strategies,* the analysis of which is presented in Chapters 7 and 8.

KEY TERMS

assurance game (114)
battle of the sexes (115)
belief (95)
best response (93)
best-response analysis (105)
cell-by-cell inspection (95)
chicken (116)
constant-sum game (91)
convergence of expectations (114)
coordination game (111)
dominance solvable (102)
dominant strategy (98)
dominated strategy (98)
enumeration (95)
focal point (112)
game matrix (90)
game table (90)
iterated elimination of dominated strategies (102)
maximin (106)
minimax (106)
minimax method (106)
mixed strategy (90)
Nash equilibrium (93)
normal form (90)
payoff table (90)
prisoners' dilemma (97)
pure coordination game (112)
pure strategy (90)
strategic form (90)
successive elimination of dominated strategies (102)
zero-sum game (91)

SOLVED EXERCISES

S1. "If a player has a dominant strategy in a simultaneous-move game, then she is sure to get her best possible outcome." True or false? Explain and give an example of a game that illustrates your answer.

S2. Find all Nash equilibria in pure strategies for the following zero-sum games. First check for dominant strategies. If neither player has a dominant strategy, use iterated elimination of dominated strategies to find the Nash equilibrium.

(a)

		COLUMN	
		Left	Right
ROW	Up	4	3
	Down	2	1

(b)

		COLUMN	
		Left	Right
ROW	Up	3	2
	Down	4	1

(c)

		COLUMN		
		Left	Middle	Right
ROW	Up	1	2	5
	Straight	2	4	3
	Down	1	3	3

(d)

		COLUMN			
		North	South	East	West
ROW	Up	6	7	5	6
	High	7	3	4	5
	Low	8	6	3	2
	Down	3	5	4	5

S3. Use the minimax method to find the Nash equilibria for the games in Exercise S2.

S4. Find all Nash equilibria in pure strategies in the following non-zero-sum games. Describe the steps that you used in finding the equilibria.

(a)

		COLUMN	
		Left	Right
ROW	Up	2, 4	1, 0
	Down	6, 5	4, 2

(b)

		COLUMN	
		Left	Right
ROW	Up	1, 1	0, 1
	Down	1, 0	1, 1

(c)

		COLUMN		
		Left	Middle	Right
ROW	Up	0, 1	9, 0	2, 3
	Straight	5, 9	7, 3	1, 7
	Down	7, 5	10, 10	3, 5

(d)

		COLUMN		
		West	Center	East
ROW	North	2, 3	8, 2	7, 4
	Up	3, 0	4, 5	6, 4
	Down	10, 4	6, 1	3, 9
	South	4, 5	2, 3	5, 2

S5. Consider the following table:

		COLUMN			
		North	South	East	West
ROW	Earth	1, 3	3, 1	0, 2	1, 1
	Water	1, 2	1, 2	2, 3	1, 1
	Wind	3, 2	2, 1	1, 3	0, 3
	Fire	2, 0	3, 0	1, 1	2, 2

(a) Does either Row or Column have a dominant strategy? Explain why or why not.

(b) Use iterated elimination of dominated strategies to reduce the game as much as possible. Give the order in which the eliminations occur and give the reduced form of the game.

(c) Is this game dominance solvable? Explain why or why not.

(d) State the Nash equilibrium (or equilibria) of this game.

S6. An old lady is looking for help crossing the street. Only one person is needed to help her; more are okay but no better than one. You and I are the two people in the vicinity who can help; we have to choose simultaneously whether to do so. Each of us will get pleasure worth a 3 from her success (no matter who helps her). But each one who goes to help will bear a cost of 1, this being the value of our time taken up in helping. If neither player helps, the payoff for each player is zero. Set this up as a game. Write the payoff table, and find all pure-strategy Nash equilibria.

S7. A university is contemplating whether to build a new lab or a new theater on campus. The science faculty would rather see a new lab built, and the humanities faculty would prefer a new theater. However, the funding for the project (whichever it may turn out to be) is contingent on unanimous support from the faculty. If there is disagreement, neither project will go forward, leaving each group with no new building and their worst payoff. The meetings of the two separate faculty groups on which proposal to support occur simultaneously, with payoffs given in the following table:

		HUMANITIES FACULTY	
		Lab	Theater
SCIENCE FACULTY	Lab	4, 2	0, 0
	Theater	0, 0	1, 5

(a) What are the pure-strategy Nash equilibria of this game?

(b) Which game described in this chapter is most similar to this game? Explain your reasoning.

S8. Suppose two game-show contestants, Alex and Bob, each separately select one of three doors numbered 1, 2, and 3. Both players get dollar prizes if their choices match, as indicated in the following table.

		B		
		1	2	3
A	1	10, 10	0, 0	0, 0
	2	0, 0	15, 15	0, 0
	3	0, 0	0, 0	15, 15

(a) What are the Nash equilibria of this game? Which, if any, is likely to emerge as the (focal) outcome? Explain.

(b) Consider a slightly changed game in which the choices are again just numbers, but the two cells with (15, 15) in the table become (25, 25). What is the expected (average) payoff to each player if each flips a coin to decide whether to play 2 or 3? Is this better than focusing on both choosing 1 as a focal equilibrium? How should you account for the risk that Alex might do one thing while Bob does the other?

S9. Marta has three sons: Arturo, Bernardo, and Carlos. She discovers a broken lamp in her living room and knows that one of her sons must have broken it at play. In reality, Carlos was the culprit, but Marta doesn't know this. She cares more about finding out the truth than she does about punishing the child who broke the lamp, so Marta announces that her sons are to play the following game.

Each child will write down his name on a piece of paper and write down either "Yes, I broke the lamp," or "No, I didn't break the lamp." If at least one child claims to have broken the lamp, she will give the normal allowance of $2 to each child who claims to have broken the lamp, and $5 to each child who claims not to have broken the lamp. If all three children claim not to have broken the lamp, none of them receives any allowance (each receives $0).

(a) Write down the game table. Make Arturo the row player, Bernardo the column player, and Carlos the page player.

(b) Find all the Nash equilibria of this game.

(c) There are multiple Nash equilibria of this game. Which one would you consider to be a focal point?

S10. Consider a game in which there is a prize worth $30. There are three contestants, Larry, Curly, and Moe. Each can buy a ticket worth $15 or $30 or not buy a ticket at all. They make these choices simultaneously and independently. Then, knowing the ticket-purchase decisions, the game organizer awards the prize. If no one has bought a ticket, the prize is not awarded. Otherwise, the prize is awarded to the buyer of the highest-cost ticket if there is only one such player or is split equally between two or three if there are ties among the highest-cost ticket buyers. Show this game in strategic form, using Larry as the Row player, Curly as the Column player, and Moe as the Page player. Find all pure-strategy Nash equilibria.

S11. Anne and Bruce would like to rent a movie, but they can't decide what kind of movie to get: Anne wants to rent a comedy, and Bruce wants to watch a drama. They decide to choose randomly by playing "Evens or Odds." On the count of three, each of them shows one or two fingers. If the sum is even, Anne wins and they rent the comedy; if the sum is odd, Bruce wins and they rent the drama. Each of them earns a payoff of 1 for winning and 0 for losing "Evens or Odds."

(a) Draw the game table for "Evens or Odds."

(b) Demonstrate that this game has no Nash equilibrium in pure strategies.

S12. In the film *A Beautiful Mind,* John Nash and three of his graduate-school colleagues find themselves faced with a dilemma while at a bar. There are four brunettes and a single blonde available for them to approach. Each young man wants to approach and win the attention of one of the young women. The payoff to each of winning the blonde is 10; the payoff of winning a brunette is 5; the payoff from ending up with no girl is zero. The catch is that if two or more young men go for the blonde, she rejects all of them, and then the brunettes also reject the men because they don't want to be second choice. Thus, each player gets a payoff of 10 only if he is the sole suitor for the blonde.

(a) First consider a simpler situation in which there are only two young men instead of four. (There are two brunettes and one blonde, but these women merely respond in the manner just described and are not active players in the game.) Show the playoff table for the game, and find all of the pure-strategy Nash equilibria of the game.

(b) Now show the (three-dimensional) table for the case in which there are three young men (and three brunettes and one blonde who are not active players). Again, find all of the Nash equilibria of the game.

(c) Without the use of a table, give all of the Nash equilibria for the case in which there are four young men (as well as four brunettes and a blonde).

(d) **(Optional)** Use your results to parts (a), (b), and (c) to generalize your analysis to the case in which there are n young men. Do not attempt to write down an n-dimensional payoff table; merely find the payoff to one player when k of the others choose Blonde and $(n - k - 1)$ choose Brunette, for $k = 0, 1, \ldots (n-1)$. Can the outcome specified in the movie as the Nash equilibrium of the game—that all of the young men choose to go for brunettes—ever really be a Nash equilibrium of the game?

UNSOLVED EXERCISES

U1. Find all Nash equilibria in pure strategies for the zero-sum games in the following tables by checking for dominant strategies and using iterated dominance.

(a)

		COLUMN	
		Left	Right
ROW	Up	1	4
	Down	2	3

(b)

		COLUMN	
		Left	Right
ROW	Up	1	2
	Down	4	3

(c)

		COLUMN		
		Left	Middle	Right
ROW	Up	5	3	2
	Straight	6	4	3
	Down	1	6	2

(d)

		COLUMN		
		Left	Middle	Right
ROW	Up	5	1	3
	Straight	6	1	2
	Down	1	0	0

U2. Use the minimax method to find the Nash equilibria for the games in Exercise U1.

U3. Find all Nash equilibria in pure strategies in the following non-zero-sum games. Describe the steps that you used in finding the equilibria.

(a)

		COLUMN	
		Left	Right
ROW	Up	3, 1	4, 2
	Down	5, 2	2, 3

(b)

		COLUMN	
		Left	Right
ROW	Up	0, 0	0, 0
	Down	0, 0	1, 1

(c)

		COLUMN		
		Left	Middle	Right
ROW	Up	2, 9	5, 5	6, 2
	Straight	6, 4	9, 2	5, 3
	Down	4, 3	2, 7	7, 1

(d)

		COLUMN		
		Left	Middle	Right
ROW	Up	5, 3	7, 2	2, 1
	Straight	1, 2	6, 3	1, 4
	Down	4, 2	6, 4	3, 5

U4. Use successive elimination of dominated strategies to solve the following game. Explain the steps you followed. Show that your solution is a Nash equilibrium.

		COLUMN		
		Left	Middle	Right
ROW	Up	4, 3	2, 7	0, 4
	Down	5, 0	5, −1	−4, −2

U5. Find all of the pure-strategy Nash equilibria for the following game. Describe the process that you used to find the equilibria. Use this game to explain why it is important to describe an equilibrium by using the strategies employed by the players, not merely by the payoffs received in equilibrium.

		COLUMN		
		Left	Center	Right
ROW	Up	1, 2	2, 1	1, 0
	Level	0, 5	1, 2	7, 4
	Down	−1, 1	3, 0	5, 2

U6. Consider the following game table:

		COLUMN		
		Left	Center	Right
ROW	Top	4, __	__, 2	3, 1
	Middle	3, 5	2, __	2, 3
	Bottom	__, 3	3, 4	4, 2

(a) Complete the payoffs of the game table above so that Column has a dominant strategy. State which strategy is dominant and explain why. (Note: there are many equally correct answers.)

(b) Complete the payoffs of the game table above so that neither player has a dominant strategy, but also so that each player does have a dominated strategy. State which strategies are dominated and explain why. (Again, there are many equally correct answers.)

U7. The game known as the *Battle of the Bismarck Sea* (named for that part of the southwestern Pacific Ocean separating the Bismarck Archipelago from Papua–New Guinea) summarizes a well-known game actually played in a naval engagement between the United States and Japan during World War II. In 1943, a Japanese admiral was ordered to move a convoy of ships to New Guinea; he had to choose between a rainy northern route and a sunnier southern route, both of which required three days' sailing time. The Americans knew that the convoy would sail and wanted to send bombers after it, but they did not know which route it would take. The Americans had to send reconnaissance planes to scout for the convoy, but they had only enough reconnaissance planes to explore one route at a time. Both the Japanese and the Americans had to make their decisions with no knowledge of the plans being made by the other side.

If the convoy was on the route that the Americans explored first, they could send bombers right away; if not, they lost a day of bombing. Poor weather on the northern route would also hamper bombing. If the Americans explored the northern route and found the Japanese right away, they could expect only two (of three) good bombing days; if they explored the northern route and found that the Japanese had gone south, they could also expect two days of bombing. If the Americans chose to explore the southern route first, they could expect three full days of bombing if they found the Japanese right away but only one day of bombing if they found that the Japanese had gone north.

(a) Illustrate this game in a game table.

(b) Identify any dominant strategies in the game and solve for the Nash equilibrium.

U8. Two players, Jack and Jill, are put in separate rooms. Then each is told the rules of the game. Each is to pick one of six letters: G, K, L, Q, R, or W. If the two happen to choose the same letter, both get prizes as follows:

Letter	G	K	L	Q	R	W
Jack's Prize	3	2	6	3	4	5
Jill's Prize	6	5	4	3	2	1

If they choose different letters, each gets zero. This whole schedule is revealed to both players, and both are told that both know the schedules, and so on.

(a) Draw the table for this game. What are the Nash equilibria in pure strategies?

(b) Can one of the equilibria be a focal point? Which one? Why?

U9. Three friends (Julie, Kristin, and Larissa) independently go shopping for dresses for their high-school prom. On reaching the store, each girl sees only three dresses worth considering: one black, one lavender, and one yellow. Each girl furthermore can tell that her two friends would consider the same set of three dresses, because all three have somewhat similar tastes.

Each girl would prefer to have a unique dress, so a girl's utility is zero if she ends up purchasing the same dress as at least one of her friends. All three know that Julie strongly prefers black to both lavender and yellow, so she would get a utility of 3 if she were the only one wearing the black dress, and a utility of 1 if she were either the only one wearing the lavender dress or the only one wearing or the yellow dress. Similarly, all know that Kristin prefers lavender and secondarily prefers yellow, so her utility would be 3 for uniquely wearing lavender, 2 for uniquely wearing yellow, and 1 for uniquely wearing black. Finally, all know that Larissa prefers yellow and secondarily prefers black, so she would get 3 for uniquely wearing yellow, 2 for uniquely wearing black, and 1 for uniquely wearing lavender.

(a) Provide the game table for this three-player game. Make Julie the Row player, Kristin the Column player, and Larissa the Page player.

(b) Identify any dominated strategies in this game, or explain why there are none.

(c) What are the pure-strategy Nash equilibria in this game?

U10. Bruce, Colleen, and David are all getting together at Bruce's house on Friday evening to play their favorite game, Monopoly. They all love to eat sushi while they play. They all know from previous experience that two orders of sushi are just the right amount to satisfy their hunger. If they wind up with less than two orders, they all end up going hungry and don't enjoy the evening. More than two orders would be a waste, because they can't manage to eat a third order and the extra sushi just goes bad. Their favorite restaurant, Fishes in the Raw, packages its sushi in such large containers that each individual person can feasibly purchase at most one order of sushi. Fishes in the Raw offers takeout, but unfortunately doesn't deliver.

Suppose that each player enjoys \$20 worth of utility from having enough sushi to eat on Friday evening, and \$0 from not having enough to eat. The cost to each player of picking up an order of sushi is \$10.

Unfortunately, the players have forgotten to communicate about who should be buying sushi this Friday, and none of the players has a cell phone, so they must each make independent decisions of whether to buy (B) or not buy (N) an order of sushi.

(a) Write down this game in strategic form.

(b) Find all the Nash equilibria in pure strategies.

(c) Which equilibrium would you consider to be a focal point? Explain your reasoning.

U11. Roxanne, Sara, and Ted all love to eat cookies, but there's only one left in the package. No one wants to split the cookie, so Sara proposes the following extension of "Evens or Odds" (see Exercise **S11**) to determine who gets to eat it. On the count of three, each of them will show one or two fingers, they'll add them up, and then divide the sum by 3. If the remainder is zero Roxanne gets the cookie, if the remainder is 1 Sara gets it, and if it is 2 Ted gets it. Each of them receives a payoff of 1 for winning (and eating the cookie) and zero otherwise.

(a) Represent this three-player game in normal form, with Roxanne as the Row player, Sara as the Column player, and Ted as the Page player.

(b) Find all the pure-strategy Nash equilibria of this game. Is this game a fair mechanism for allocating cookies? Explain why or why not.

U12. **(Optional)** Construct the payoff matrix for your own two-player game that satisfies the following requirements. First, each player should have three strategies. Second, the game should not have any dominant strategies. Third, the game should not be solvable using minimax. Fourth, the game should have exactly two pure-strategy Nash equilibria. Provide your game matrix, and then demonstrate that all of the above conditions are true.

Simultaneous-Move Games with Pure Strategies II: Continuous Strategies and III: Discussion and Evidence

THE DISCUSSION OF SIMULTANEOUS-MOVE GAMES in Chapter 4 focused on games in which each player had a discrete set of actions from which to choose. Discrete strategy games of this type include sporting contests in which a small number of well-defined plays can be used in a given situation—soccer penalty kicks, in which the kicker can choose to go high or low, to a corner or the center, for example. Other examples include coordination and prisoners' dilemma games in which players have only two or three available strategies. Such games are amenable to analysis with the use of a game table, at least for situations with a reasonable number of players and available actions.

Many simultaneous-move games differ from those considered so far; they entail players choosing strategies from a wide range of possibilities. Games in which manufacturers choose prices for their products, philanthropists choose charitable contribution amounts, or contractors choose project bid levels are examples in which players have a virtually infinite set of choices. Technically, prices and other dollar amounts do have a minimum unit, such as a cent, and so there is actually only a finite and discrete set of price strategies. But in practice the unit is very small, and allowing the discreteness would require us to give each player too many distinct strategies and make the game table too large; therefore it is simpler and better to regard such choices as continuously variable real numbers. When players have such a large range of actions available, game tables become virtually useless as analytical tools; they become too unwieldy to be of practical use. For these games we need a different solution technique. We present the analytical tools for handling such **continuous strategy** games in the first part of this chapter.

This chapter also takes up some broader matters relevant to behavior in simultaneous-move games and to the concept of Nash equilibrium. We review the empirical evidence on Nash equilibrium play that has been collected both from the laboratory and from real-life situations. We also present some theoretical criticisms of the Nash equilibrium concept and rebuttals of these criticisms. You will see that game-theoretic predictions are often a reasonable starting point for understanding actual behavior, with some caveats, such as the level of player.

1 PURE STRATEGIES THAT ARE CONTINUOUS VARIABLES

In Chapter 4 we developed the method of best-response analysis for finding all pure-strategy Nash equilibria of simultaneous-move games. Now we extend that method to games in which each player has available a continuous range of choices—for example, firms setting prices of their products. To calculate best responses in this type of game, we find, for each possible value of one firm's price, the value of the other firm's price that is best for it (maximizes its payoff). The continuity of the sets of strategies allows us to use algebraic formulas to show how strategies generate payoffs and to show the best responses as curves in a graph, with each player's price (or any other continuous strategy) on one of the axes. In such an illustration, the Nash equilibrium of the game occurs where the two curves meet. We develop this idea and technique by using two stories.

A. Price Competition

Our first story is set in a small town, Yuppie Haven, that has two restaurants, Xavier's Tapas Bar and Yvonne's Bistro. To keep the story simple, we suppose that each place has a set menu. Xavier and Yvonne have to set the prices of their respective menus. Prices are their strategic choices in the game of competing with each other; each bistro's goal is to set price to maximize profit, the payoff in this game. We suppose that they must get their menus printed separately without knowing the other's price, so the game has simultaneous moves.[1] Because prices can take any value within an (almost) infinite range, we start with general or algebraic symbols for them. We then find **best-response rules** that we use to solve

[1] In reality, the competition extends over time, so each can observe the other's past choices. This repetition of the game introduces new considerations, which we cover in Chapter 11.

the game and to determine equilibrium prices. Let us call Xavier's price P_x and Yvonne's price P_y.

In setting its price, each restaurant has to calculate the consequences for its profit. To keep things relatively simple, we put the two restaurants in a very symmetric relationship, but readers with a little more mathematical skill can do a similar analysis by using much more general numbers or even algebraic symbols. Suppose the cost of serving each customer is \$8 for each restaurateur. Suppose further that experience or market surveys have shown that, when Xavier's price is P_x and Yvonne's price is P_y, the number of their respective customers, respectively Q_x and Q_y (measured in hundreds per month), are given by the equations[2]

$$Q_x = 44 - 2P_x + P_y,$$
$$Q_y = 44 - 2P_y + P_x.$$

The key idea in these equations is that, if one restaurant raises its price by \$1 (say, Yvonne increases P_y by \$1), its sales will go down by 200 per month (Q_y changes by -2) and those of the other restaurant will go up by 100 per month (Q_x changes by 1). Presumably, 100 of Yvonne's customers switch to Xavier's and another 100 stay at home.

Xavier's profit per week (in hundreds of dollars per week), call it Π_x—the Greek letter Π (pi) is the traditional economic symbol for profit—is given by the product of the net revenue per customer (price less cost or $P_x - 8$) and the number of customers served:

$$\Pi_x = (P_x - 8)\, Q_x = (P_x - 8)\,(44 - 2\,P_x + P_y).$$

By multiplying out and rearranging the terms on the right-hand side of the preceding expression, we can write profit as a function of increasing powers of P_x:

$$\Pi_x = -8(44 + P_y) + (16 + 44 + P_y)P_x - 2(P_x)^2$$
$$= -8(44 + P_y) + (60 + P_y)P_x - 2(P_x)^2.$$

Xavier sets his price P_x to maximize this payoff. Doing so for each possible level of Yvonne's price P_y gives us Xavier's best-response rule; we can then graph it.

Many simple illustrative examples where one real number (such as the price) is chosen to maximize another real number that depends on it (such as the profit or the payoff) have a similar form. (In mathematical jargon, we would describe the second number as a function of the first.) In the appendix to this

[2]Readers who know some economics will recognize that the equations linking quantities to prices are demand functions for the two products *X* and *Y*. The quantity demanded of each product is decreasing in its own price (demands are downward sloping) and increasing in the price of the other product (the two are substitutes).

chapter we develop a simple general technique for performing such maximization; you will find many occasions to use it. Here we just state the formula.

The function we want to maximize takes the general form

$$Y = A + BX - CX^2$$

where we have used the descriptor Y for the number we want to maximize and X for the number we want to choose to maximize that Y. In our specific example, profit, Π_x, would be represented by Y, and the price, P_x, by X. Similarly, although in any specific problem, the terms A, B, and C in the equation above would be known numbers, we have denoted them by general algebraic symbols so that our formula can be applied across a wide variety of similar problems. (The technical term for the terms A, B, and C is *parameters*, or *algebraic constants*.) Because most of our applications involve nonnegative X entities, such as prices, and the maximization of the Y entity, we require $B > 0$ and $C > 0$. Then the formula giving the choice of X to maximize Y in terms of the known parameters A, B, and C is simply $X = B/(2C)$. Observe that A does not appear in the formula, although it will of course affect the value of Y that results.

Comparing the general function in the equation above and the specific example of the profit function in the pricing game on the previous page, we have[3]

$$B = 60 + P_y \text{ and } C = 2.$$

Therefore Xavier's choice of price to maximize his profit will satisfy the formula $B/(2C)$ and will be

$$P_x = 15 + 0.25P_y.$$

This equation determines the value of P_x that maximizes Xavier's profit, given a particular value of Yvonne's price, P_y. In other words, it is exactly what we want, the rule for Xavier's best response.

Yvonne's best-response rule can be found similarly. Because the costs and sales of the two restaurants are entirely symmetric, the equation is obviously going to be

$$P_y = 15 + 0.25P_x.$$

Both rules are used in the same way to develop best-response graphs. If Xavier sets a price of 16, for example, then Yvonne plugs this value into her best-response rule to find $P_y = 15 + 0.25(16) = 19$; similarly, Xavier's best response to Yvonne's $P_y = 16$ is $P_x = 19$, and each restaurant's best response to the other's price of 4 is 16, that to 8 is 17, and so on.

[3]Although P_y, chosen by Yvonne, is a variable in the full game, here we are considering only a part of the game, namely Xavier's best response, where he regards Yvonne's choice as outside his control and therefore like a constant.

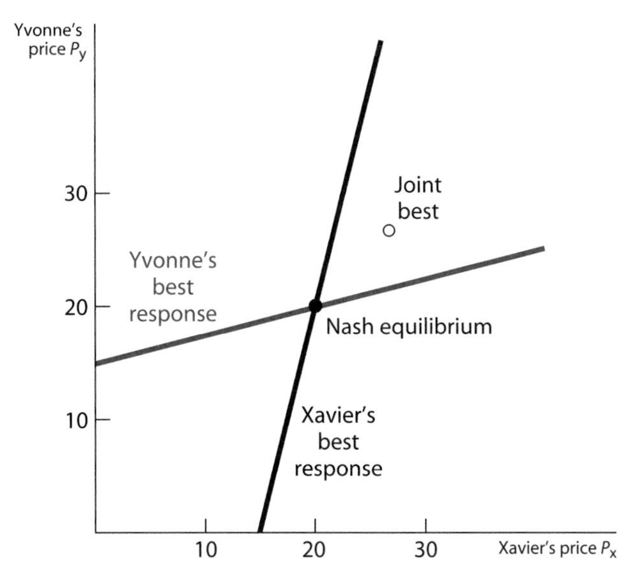

FIGURE 5.1 Best-Response Curves and Equilibrium in the Restaurant Pricing Game

Figure 5.1 shows the graphs of these two best-response relations. Owing to the special features of our example—namely, the linear relation between quantity sold and prices charged, and the constant cost of producing each meal—each of the two **best-response curves** is a straight line. For other specifications of demands and costs, the curves can be other than straight, but the method of obtaining them is the same: namely, first holding one restaurant's price (say, P_y) fixed and finding the value of the other's price (say, P_x) that maximizes the second restaurant's profit, and then the other way around.

The point of intersection of the two best-response curves is the Nash equilibrium of the pricing game between the two restaurants. That point represents the pair of prices, one for each firm, that are best responses to each other. The specific values for each restaurant's pricing strategy in equilibrium can be found algebraically by solving the two best-response rules jointly for P_x and P_y. We deliberately chose our example to make the equations linear, and the solution is easy. In this case, we simply substitute the expression for P_x into the expression for P_y to find

$$P_y = 15 + 0.25P_x = 15 + 0.25(15 + 0.25P_y) = 18.75 + 0.0625\,P_y.$$

This last equation simplifies to $P_y = 20$. Given the symmetry of the problem, it is simple to determine that $P_x = 20$ also.[4] Thus, in equilibrium, each restaurant charges \$20 for its menu and makes a profit of \$12 on each of the 2,400 customers

[4]Without this symmetry, the two best-response equations will be different, but given our other specifications, still linear. So it is not much harder to solve the nonsymmetric case. You will have a chance to do so in Exercise S2 at the end of this chapter.

[2,400 = (44−2 × 20 + 20) hundred] that it serves each month, for a total profit of $28,800 per month.

B. Some Economics of Oligopoly

Our main purpose in presenting the restaurant pricing example was to illustrate how the Nash equilibrium can be found in a game where the strategies are continuous variables, such as prices. But it is interesting to take a further look into this situation and to explain some of the economics behind pricing strategies and profits when a small number of firms (here just two) compete. In the jargon of economics, such competition is referred to as oligopoly, from the Greek words for "a small number of sellers."

Begin by observing that each of the two firm best-response curve slopes upward. Specifically, when one restaurant raises its price by $1, the other's best response is to raise its own price by 0.25, or 25 cents. When one restaurant raises its price, some of its customers switch to the other restaurant, and its rival can then best profit from them by raising its price part of the way. Thus a restaurant that raises its price is helping to increase the other's profit. In Nash equilibrium, where each restaurant chooses its price independently and out of concern for its own profit, it does not take into account this benefit that it conveys to the other. Could they get together and cooperatively agree to raise their prices, thereby raising both profits? Yes. Suppose the two restaurant charged $24 each. Then each would make a profit of $16 on each of the 2,000 customers [2,000 = (44 − 2 × 24 + 24) hundred] that it would serve each month, for a total profit of $32,000.

This pricing game is exactly like the prisoners' dilemma game presented in Chapter 4, but now the strategies are continuous variables. In the story in Chapter 4, the husband and wife were each tempted to cheat the other and confess to the police; but, when they both did so, both ended up with longer prison sentences (worse outcomes). In the same way, the more profitable price of $24 is not a Nash equilibrium. The separate calculations of the two restaurants will lead them to undercut such a price. Suppose that Yvonne somehow starts by charging $24. Using the best-response formula, we see that Xavier will then charge 15 + 0.25 × 24 = 21. Then Yvonne will come back with her best response to that: 15 + 0.25 × 21 = 20.25. Continuing this process, the prices of both will converge toward the Nash equilibrium price of $20.

But what price is jointly best for the two restaurants? Given the symmetry, suppose both charge the same price P. Then the profit of each will be

$$\Pi_x = \Pi_y = (P - 8)(44 - 2P + P) = (P - 8)(44 - P) = -352 + 52P - P^2.$$

The two can choose P to maximize this expression. Using the formula provided in Section 1.A, we see that the solution is $P = 52/2 = 26$. The resulting profit for each restaurant is $32,400 per month.

In the jargon of economics, such collusion to raise prices to the jointly optimal level is called a *cartel*. The high prices hurt consumers, and regulatory agencies of the U.S. government often try to prevent the formation of cartels and to make firms compete with one another. Explicit collusion over price is illegal, but it may be possible to maintain tacit collusion in a repeated prisoners' dilemma; we examine such repeated games in Chapter 11.[5]

Collusion need not always lead to higher prices. In the preceding example, if one restaurant lowers its price, its sales increase, in part because it draws some customers away from its rival because the products (meals) of the two restaurants are *substitutes* for each other. In other contexts, two firms may be selling products that are *complements* to each other—for example, hardware and software. In that case, if one firm lowers its price, the sales of both firms increase. In a Nash equilibrium, where the firms act independently, they do not take into account the benefit that a lower price of each would convey on the other. Therefore they keep prices higher than they would if they were able to coordinate their actions. Allowing them to cooperate would lead to lower prices and thus be beneficial to the consumers as well.

Competition need not always involve the use of prices as the strategic variables. For example, fishing fleets may compete to bring a larger catch to market; this is quantity competition as opposed to the price competition considered in this section. We consider quantity competition later in this chapter and in several of the end-of-chapter exercises.

C. Political Campaign Advertising

Our second example is one drawn from politics. It requires just a little more mathematics than we normally use, but we explain the intuition behind the calculations in words and with a graph.

Consider an election contested by two parties or candidates. Each is trying to win votes away from the other by advertising—either positive ads that highlight the good things about oneself or negative, attack ads that emphasize the bad things about the opponent. To keep matters simple, suppose the voters start entirely ignorant and unconcerned and are moved solely by the ads. (Many people would claim that this is a pretty accurate description of U.S. politics, but more advanced analyses in political science do recognize that there are informed and strategic voters. We address the behavior of such voters in detail in Chapter 16.) Even more simply, suppose the vote share of a party equals its share of the total campaign advertising that is done. Call the

[5]Firms do try to achieve explicit collusion when they think they can get away with it. An entertaining and instructive story of one such episode is in *The Informant*, by Kurt Eichenwald (New York: Broadway Books, 2000).

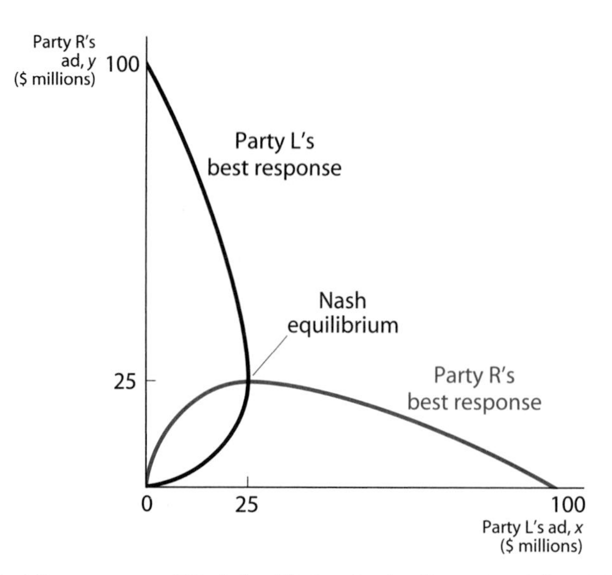

FIGURE 5.2 Best Responses and Nash Equilibrium in the Campaign Advertising Game

parties or candidates L and R; when L spends $\$x$ million on advertising and R spends $\$y$ million, L will get a share $x/(x + y)$ of the votes and R will get $y/(x + y)$. Once again, readers who get interested in this application can find more general treatments in specialized political science writings.

Raising money to pay for these ads includes a cost: money to send letters and make phone calls; time and effort of the candidates, party leaders, and activists; the future political payoff to large contributors; and possible future political costs if these payoffs are exposed and lead to scandals. For simplicity of analysis, let us suppose all these costs are proportional to the direct campaign expenditures x and y. Specifically, let us suppose that party L's payoff is measured by its vote percentage minus its advertising expenditure, $100x/(x + y) - x$. Similarly party R's payoff is $100y/(x + y) - y$.

Now we can find the best responses. Because we cannot do so without calculus, we derive the formula mathematically and then explain its general meaning intuitively, in words. For a given strategy x of party L, party R chooses y to maximize its payoff. The calculus first-order condition is found by holding x fixed and setting the derivative of $100y/(x + y) - y$ with respect to y equal to zero. It is $100x/(x + y)^2 - 1 = 0$, or $y = 10\sqrt{x} - x$. Figure 5.2 shows its graph and that of the analogous best-response function of party L—namely, $x = 10\sqrt{y} - y$.

Look at the best-response curve of party R. As the value of party L's x increases, party R's y increases for a while and then decreases. If the other party is advertising very little, then one's own ads have a high reward in the form of votes, and it pays to respond to a small increase in the other's expenditures by spending more oneself to compete harder. But if the other party already has a massive expenditure, then one's own ads get only a small return in relation to their cost, so it is better to respond to the other's increase by scaling back.

As it happens, the two parties' best-response curves intersect at their peak points. Again, some algebraic manipulation of the equations for the two curves yields us exact values for the equilibrium values of x and y. You should verify that here x and y are each equal to 25, or $25 million. (This is presumably a congressional election; Senate and presidential elections cost much more these days.)

As in the pricing game, we have a prisoners' dilemma. If both parties cut back on their ads in equal proportions, their vote shares would be entirely unaffected, but both would save on their expenditures and so both would have a larger payoff. Unlike a producers' cartel for substitute products that keeps prices high and hurts consumers, but like a producers' cartel for complements that leads to lower prices, a politicians' cartel to advertise less would probably benefit voters and society. We could all benefit from finding ways to resolve this particular prisoners' dilemma. In fact, Congress has been trying to do just that for several years and has imposed some partial curbs, but political competition seems too fierce to permit a full or lasting resolution.

What if the parties are not symmetrically situated? Two kinds of asymmetries can arise. One party (say, R) may be able to advertise at a lower cost, because it has favored access to the media. Or R's advertising dollars may be more effective than L's; for example, L's vote share may be $x/(x + 2y)$, while R's is $2y/(x + 2y)$.

In the first of these cases, R exploits its cheaper access to advertising by choosing a higher level of expenditures y for any given x for party L; that is, R's best-response curve in Figure 5.2 shifts upward. The Nash equilibrium shifts to the northwest along L's unchanged best-response curve. Thus R ends up advertising more, and L less, than before. It is as if the advantaged party uses its muscle and the disadvantaged party gives up to some extent in the face of this adversity.

In the second case, both parties' best-response curves shift in more complex ways. The outcome is that both spend equal amounts, but less than the 25 that they spent in the symmetric case. In our example where R's dollars are twice as effective as L's, it turns out that their common expenditure level is $200/9 = 22.2 < 25$. (Thus the symmetric case is the one of most intense competition.) When R's spending is more effective, it is also true that the best-response curves are asymmetric in such a way that the new Nash equilibrium, rather than being at the peak points of the two best-response curves, is on the downward part of L's best-response curve and on the upward part of R's best-response curve. That is to say, although both parties spend the same dollar amount, the favored party, R, spends more than the amount that would bring forth the maximum response from party L, and the underdog party, L, spends less than the amount that would bring forth the maximum response from party R. We include an optional exercise (Exercise U12) in this chapter that lets the mathematically advanced students derive these results.

D. General Method for Finding Nash Equilibria

Although the strategies (prices or campaign expenditures) and payoffs (profits or vote shares) in the two previous examples are specific to the context of competition between firms or political parties, the method for finding the Nash equilibrium of a game with continuous strategies is perfectly general. Here we state its steps so you can use it as a recipe for solving other games of this kind.

Suppose the players are numbered 1, 2, 3, . . . Label their strategies $x, y, z, \ldots$ in that order, and their payoffs by the corresponding upper-case letters $X, Y, Z, \ldots$ The payoff of each is in general a function of the choices of all; label the respective functions $F, G, H, \ldots$ Construct payoffs from the information about the game, and write them as

$$X = F(x, y, z, \ldots),\ Y = G(x, y, z, \ldots),\ Z = H(x, y, z, \ldots).$$

Using this general format to describe our example of price competition between two players (firms) makes the strategies x and y become the prices P_x and P_y. The payoffs X and Y are the profits Π_x and Π_y. The functions F and G are the quadratic formulas,

$$\Pi_x = -8(44 + P_y) + (16 + 44 + P_y)P_x - 2(P_x)$$

and similarly for Π_y.

In the general approach, player 1 regards the strategies of players 2, 3, . . . as outside his control, and chooses his own strategy to maximize his own payoff. Therefore for each given set of values of $y, z, \ldots$, player 1's choice of x maximizes $X = F(x, y, z, \ldots)$. If you use calculus, the condition for this maximization is that the derivative of X with respect to x holding $y, z, \ldots$ constant (the partial derivative) equals zero. For special functions, simple formulas are available, such as the one we stated and used above for the quadratic. And even if an algebra or calculus formulation is too difficult, computer programs can tabulate or graph best-response functions for you. Whatever method you use, you find an equation for player 1's optimal choice of x for given $y, z, \ldots$ that is player 1's best-response function. Similarly, you can find the best-response functions for each of the other players.

The best-response functions are equal in number to the number of the strategies in the game and can be solved simultaneously while regarding the strategy variables as the unknowns. The solution is the Nash equilibrium we seek. Some games may have multiple solutions, yielding multiple Nash equilibria. Other games may have no solution, requiring further analysis, such as inclusion of mixed strategies.

2 EMPIRICAL EVIDENCE CONCERNING NASH EQUILIBRIUM

In Chapter 3, when we considered empirical evidence on sequential-move games and rollback, we said that the evidence came from observations on games actually played in the world, as well as games deliberately constructed for testing the theory in the laboratory or the classroom. We pointed out the different merits and drawbacks of the two methods. Similar issues arise in securing and interpreting the evidence on simultaneous-move games.

Real-world games are played for substantial stakes by experienced players, who therefore have the knowledge and the incentives to employ good strategies. But these situations include many factors beyond those considered in the theory. Therefore, if the data do not bear out the predictions of the theory, we cannot tell whether the theory is wrong or whether some other factor is having an effect that overwhelms the strategic considerations.

Laboratory experiments can control for the other factors and therefore provide cleaner tests. But they bring in inexperienced players and provide them meager time and incentives to learn the game and play it well. Confronted with a new game, most of us would flounder and try things out at random. Thus the first several plays of the game in an experimental setting may represent this learning phase and not the equilibrium that experienced players would learn to play. Some experiments do control for inexperience and learning by discarding several initial plays from their data. But the learning phase may last longer than the one morning or one afternoon that is the typical limit of laboratory sessions.

A. Laboratory and Classroom Experiments

Researchers have conducted numerous laboratory experiments in the past three decades to test how people act when placed in certain interactive strategic situations. In particular, do they play Nash equilibrium strategies? Reviewing this work, Douglas Davis and Charles Holt conclude that, in relatively simple single-move games with a unique Nash equilibrium, that outcome "has considerable drawing power . . . after some repetitions with different partners."[6] But in more complex or repeated situations or when coordination is required because there are multiple Nash equilibria or when the calculations required for finding a Nash equilibrium are more complex, the theory's success is more mixed. We briefly consider the performance of Nash equilibrium in such circumstances.

[6]Douglas D. Davis and Charles A. Holt, *Experimental Economics* (Princeton: Princeton University Press, 1993), Ch. 2.

I. CHOOSING AMONG MULTIPLE EQUILIBRIA When there are multiple equilibria, players generally fail to coordinate unless they have some common cultural background (and this fact is common knowledge among them) that is needed for locating focal points. Thomas Schelling and David Kreps report on several experiments of coordination games.[7] Kreps played the following game between pairs of his students. One student was assigned Boston, and the other was assigned San Francisco. Each was given a list of nine other U.S. cities—Atlanta, Chicago, Dallas, Denver, Houston, Los Angeles, New York, Philadelphia, and Seattle—and asked to choose a subset of these cities. The two chose simultaneously and independently. If their choices divided up the nine cities completely and without any overlap between them, both got a prize. Otherwise, neither got anything. This game has numerous (512) Nash equilibria in pure strategies. But, when both players were Americans or long-time U.S. residents, more than 80% of the time they chose the division geographically; the student assigned Boston chose all the cities east of the Mississippi, and the student assigned San Francisco chose those west of the Mississippi. Such coordination was much less likely when one or both students were non-U.S. residents. In such pairs the choices were sometimes made alphabetically, but even then there was no clear dividing point.

II. REVELATION OF INNATE ALTRUISM OR PUBLIC-SPIRITEDNESS IN EXPERIMENTS One respect in which the behavior of players in some experimental situations does not often conform to the experimenter's predicted Nash equilibrium is that people seem to "err" on the side of niceness or fairness. Thus in prisoners' dilemma games, we observe "too much" cooperation and, in bargaining games, people concede "too much" to opponents. But the reason may not be any failure to calculate or to learn to play Nash equilibrium. It may instead be that the players' payoffs are different from those assumed by the experimenter.

As with observations of naturally occurring games, participants in experimental situations also may know some complexities of the situation better than the experimenter knows them. For example, the possibility of repetition or a separate ongoing relationship with the other player may affect their choices in this game. Or the players' value systems may have internalized some social norms of niceness and fairness that have proved useful in the larger social context and that therefore carry over to their behavior in the experimental game.[8]

[7]Thomas Schelling, *The Strategy of Conflict* (New York: Oxford University Press, 1960), pp. 54–58; David Kreps, *A Course in Microeconomic Theory* (Princeton: Princeton University Press, 1990), pp. 302–303, 414–415.

[8]The distinguished game theorist Jörgen Weibull argues this position in detail in "Testing Game Theory," in *Advances in Understanding Strategic Behaviour: Game Theory, Experiments and Bounded Rationality: Essays in Honour of Werner Güth,* ed. Steffen Huck (Basingstoke: Palgrave MacMillan, 2004).

These observations do not show any deficiency of the Nash equilibrium concept itself. However, they do warn us against using the concept under naive or mistaken assumptions about people's payoffs; it might be a mistake, for example, to assume that players are always driven by the selfish pursuit of money.

III. LEARNING FROM EXPERIENCE One game, often used in classrooms or laboratories, asks each participant to choose a number between 0 and 100. Typically, the players are handed cards on which to write their names and a choice, so this game is a simultaneous-move game. When the cards are collected, the average of the numbers is calculated. The person whose choice is closest to a specified fraction—say half—of the average is the winner. The rules of the game (this whole procedure) are announced in advance.

The Nash equilibrium of this game is for everyone to choose 0. In fact the game is dominance solvable. Even if everyone chooses 100, half of the average can never exceed 50; so, for each player, any choice above 50 is dominated by 50.[9] But all players should rationally figure this out, so the average can never exceed 50 and half of it can never exceed 25, and so any choice above 25 is dominated by 25. The iteration goes on until only 0 is left.

However, when a group actually plays this game for the first time, the winner is typically a player who has chosen a number just a little less than 25. This outcome seems to suggest that the winner assumes that everyone else will choose randomly (so their average is 50) and then chooses her own best response to that. The outcome is quite far from the Nash equilibrium.

What happens if the game is repeated? Our experience in classroom trials has been that the winning choice falls rapidly in successive plays. In Skeath's class, half the class played the game first while the other half watched, then the other half played, and finally the whole class played. In Dixit's class, the game was played by different groups of 10 students at a time. By the third round, the winner's choice was usually as low as 2 or 3.

How should one interpret this result? Critics would say that, unless the exact Nash equilibrium is reached, the theory is refuted. Indeed, they would argue, if you have good reason to believe that other players will not play their Nash equilibrium strategies, then your best choice is not your Nash equilibrium strategy either. If you can figure out how others will deviate from their Nash equilibrium strategies, then you should play your best response to what you believe they are choosing. Others would argue that theories in social science can never hope for the kind

[9]If you factor in your own choice, the calculation is strengthened. Suppose there are N players. In the "worst-case scenario" where all the other $(N - 1)$ players choose 100 and you choose x, the average is $[x + (N - 1)100]/N$. Then your best choice is half of this; so $x = [x + (N - 1)100]/(2N)$, or $x = 100(N - 1)/(2N - 1)$. If $N = 10$, then $x = 900/19 = 47$ (approximately). So any choice above 47 is dominated by 47. The same reasoning applies to the successive rounds.

of precise prediction that we expect in sciences such as physics and chemistry. If the observed outcomes are close to the Nash equilibrium, that is a vindication of the theory. In this case, the experiment not only produces such a vindication, but illustrates the process by which people gather experience and learn to play strategies close to Nash equilibrium. We sympathize with this latter viewpoint.

Interestingly, we have found that people learn somewhat faster by observing others play a game than while they play it themselves. This may be because, as observers, they are free to focus on the game as a whole and think about it analytically. Players' brains are occupied with the task of making their own choices and they are less able to take the broader perspective.

We should clarify the concept of gaining experience by playing the game. The quotation from Davis and Holt at the start of this section spoke of "repetitions with different partners." In other words, experience should be gained by playing the game frequently, but with different opponents each time. However, for any learning process to generate outcomes increasingly closer to the Nash equilibrium, the whole *population* of learners needs to be stable. If novices keep appearing on the scene and trying new experimental strategies, then the original group may unlearn what they had learned by playing against one another.

If a game is played repeatedly between two players or even among the same small group of known players, then any pair is likely to play each other repeatedly. In such a situation, the whole repeated game becomes a game in its own right. It can have very different Nash equilibria from those that simply repeat the Nash equilibrium of a single play. For example, tacit cooperation may emerge in repeated prisoners' dilemmas, owing to the expectation that any temporary gain from cheating will be more than offset by the subsequent loss of trust. If games are repeated in this way, then learning about them must come from playing whole sets of the repetitions frequently, against different partners each time.

B. Real-World Evidence

The predictions of game theory have been subjected to real-world empirical evidence in two distinct ways. One is to see if the theory can explain some observed phenomena in general terms. The other is to test statistically some implications of the theory against data. We briefly consider each in turn.

I. EXPLANATORY POWER The first approach uses game-theoretic reasoning to explain phenomena that are observed in reality, of which there are numerous successful examples. One of the earliest was in the area of international relations. Thomas Schelling pioneered the use of game theory to explain phenomena such as the escalation of arms races, even between countries that have no intention of attacking each other, and the credibility of deterrent threats. Subsequent applications in this area have included the questions of when and how a country can credibly signal its resolve in diplomatic negotiation or in the face of a potential war. Game theory began to be used systematically in economics and business in the mid-1970s, and such applications continue to proliferate. We have space for only a couple of prominent examples.

The theory has helped us to understand when and how the established firms in an industry can make credible commitments to deter new competition—for example, to wage a destructive price war against any new entrant. The prisoners' dilemma game, in its one-time and repeated forms, has helped us to understand what kinds of industries will see fierce competition and exhibit low prices and what kinds will sustain tacit agreements to keep prices and profits high. More recently, game theory has become the tool of choice for the study of political systems and institutions within a country as well as for cross-country comparisons. For example, game theory has shown how voting and agenda setting in committees and elections can be strategically manipulated in pursuit of one's ultimate objectives. In this introductory book, we can develop only a few elementary examples of this kind. We already saw an example (price competition) in this chapter. More examples appear later, including a case study of the Cuban missile crisis and analyses of auctions, voting, and bargaining.[10]

Some critics remain unpersuaded by these successful applications of the theory. They claim that the same understanding of these phenomena can be obtained without using game theory, by basing one's analysis on previously known general principles of economics, political science, and so on. In one sense they are right. A few of these analyses existed before game theory came along. For example, the equilibrium of the interaction between two price-setting firms, which we developed in Section 1 of this chapter, was known in economics for more than a hundred years; one can think of Nash equilibrium as just a general formulation of that equilibrium concept for all games. Some theories of strategic voting date to the 18th century, and some notions of credibility can be found in history as far back as Thucydides' *Peloponnesian War.* However, what game theory does is to unify all these applications and thereby facilitate the development of new ones.

In the past 30 years, several new ideas and applications have been identified. For example, we now understand how different forms of auctions (English

[10]For those who would like to see more applications, here are some suggested sources. Thomas Schelling's *Strategy of Conflict* (New York: Oxford University Press, 1960) and *Arms and Influence* (New Haven: Yale University Press, 1966) are still required reading for all students of game theory. The classic textbook on game-theoretic treatment of industries is Jean Tirole, *Industrial Organization* (Cambridge: MIT Press, 1988). In political science, an early classic is William Riker, *Liberalism Against Populism* (San Francisco: W. H. Freeman, 1982). For surveys of more recent work, see several articles in *The Handbook of Game Theory,* ed. Robert J. Aumann and Sergiu Hart (Amsterdam: North-Holland, 1992, 1994, 2002), particularly Barry O'Neill, "Game Theory Models of Peace and War," in volume 2, and Kyle Bagwell and Asher Wolinsky, "Game Theory and Industrial Organization," and Jeffrey Banks, "Strategic Aspects of Political Systems," in volume 3.

and Dutch, sealed bid and open outcry) lead to differences in bidding strategies and in the seller's revenues. We understand how the existence of a second-strike capability reduces the fear of surprise attack. And we understand how governments can successfully manipulate fiscal and monetary policies to improve their chances of reelection even when the voters are sophisticated and aware of such attempts. If these examples were all amenable to previously known approaches, they would have been discovered long ago.

II. STATISTICAL TESTING The second approach to examining empirical evidence is quantitative and statistical. The general procedure in this work is to assume that Nash equilibrium prevails and to derive the implications of this assumption in the form of equations linking various magnitudes—the players' choices and outcomes—that may be observable in the situation being studied. These equations can then be estimated by using real data. In industrial economics, firms compete by choosing their quantities and prices as illustrated in the examples in this chapter; they also have other strategic choices at their disposal, including product quality, investment, R & D, and so on. While the choice of quantities or prices may be studied in a static context (at a given time), games of strategic competition in investment or R & D are dynamic. Numerous studies of both kinds of interactions have been carried out.[11] This work has produced encouraging results. Game-theoretic models, based on the Nash equilibrium concept and its dynamic generalizations, fit the data for many major industries, such as automobile manufacturers, reasonably well and give us a better understanding of the determinants of competition than the older analysis, which assumed perfect competition and estimated supply-and-demand curves.

In politics, the votes on various issues within legislatures are the outcome of the legislators' strategic interaction. The equilibrium of this game depends on the legislators' underlying preferences. Detailed voting records in the U.S. Congress are public information. On the basis of the relation between preferences and voting in a Nash equilibrium, these data can be used to infer the legislators' preferences. This method has been used with remarkable success by Keith Poole and Howard Rosenthal.[12] They find that U.S. politics can be

[11]A survey of static studies of prices and quantity competition is "Empirical Studies of Industries with Market Power," by Timothy F. Bresnahan, in *Handbook of Industrial Organization,* ed. Richard L. Schmalansee and Robert D. Willig, vol. 2 (Amsterdam: North-Holland, 1989). A general method for dynamic studies is developed in "A Framework for Applied Dynamic Analysis in Industrial Organization," by Ulrich Doraszelski and Ariel Pakes, in *Handbook of Industrial Economics,* ed. Mark Armstrong and Robert Porter, vol. 3 (Amsterdam: North-Holland, 2007).

[12]Keith Poole and Howard Rosenthal, "Patterns of Congressional Voting," *American Journal of Political Science,* vol. 35, no. 1 (February 1991), pp. 228–278, and *Congress: A Political-Economic History of Roll Call Voting* (New York: Oxford University Press, 1996).

adequately summarized by conflicts over issues in a two-dimensional space, one representing economic inequality and the other racial inequality.

Pankaj Ghemawat, a professor at the Harvard Business School, has developed a mixed mode of quantitative analysis, using case studies of individual firms or industries and statistical analysis of larger data samples.[13] His game-theoretic models are remarkably successful in improving our understanding of several initially puzzling business decisions on pricing, capacity, innovation, and so on. However, his work also brings out the need to construct models that have sufficiently rich detail to do justice to the circumstances of the firms or industries being analyzed. In a general and introductory textbook such as this one, we lack the space and eschew the more advanced techniques that are needed to construct such models. But we will set you on the way to further study that will bring these methods within your grasp. And in Chapter 15 we develop one such theory-based case study from the field of international politics to illustrate the method.

III. A REAL-WORLD EXAMPLE OF LEARNING We conclude by offering an interesting illustration of equilibrium and the learning process in a real-world game. The setting is outside the laboratory or classroom, where people play the game frequently and the stakes are high, creating strong motivation and good opportunities to learn. Stephen Jay Gould discovered this beautiful example.[14] Through most of the 20th century, the best batting averages recorded in a baseball season have been declining. In particular, the number of instances of a player averaging .400 or better used to be much more frequent than they are now. Devotees of baseball history often explain this decline by invoking nostalgia: "There were giants in those days." A moment's thought should make one wonder why there were no corresponding pitching giants who would keep batting averages low. But Gould demolishes such arguments in a more systematic way. He points out that we should look at all batting averages, not just the top ones. The worst batting averages are not as bad as they used to be; there are also many fewer .150 hitters in the major leagues than there used to be. He argues that this overall decrease in *variation* is a standardization or stabilization effect:

> When baseball was very young, styles of play had not become sufficiently regular to foil the antics of the very best. Wee Willie Keeler could "hit 'em where they ain't" (and compile an average of .432 in 1897) because fielders didn't yet know where they should be. Slowly, players moved toward *optimal* methods of positioning, fielding, pitching, and batting—and variation inevitably declined. The best [players] now met an opposition too finely honed to its own perfection to permit the extremes of achievement that characterized a more casual age. [emphasis added]

[13] Pankaj Ghemawat, *Games Businesses Play: Cases and Models* (Cambridge: MIT Press, 1997).

[14] "Losing the Edge," in *The Flamingo's Smile* (New York: Norton, 1985), pp. 215–229.

In other words, through a succession of adjustments of strategies to counter one another, the system settled down into its (Nash) equilibrium.

Gould marshals decades of hitting statistics to demonstrate that such a decrease in variation did indeed occur, except for occasional "blips." And indeed the blips confirm his thesis, because they occur soon after an equilibrium is disturbed by an externally imposed change. Whenever the rules of the game are altered (the strike zone is enlarged or reduced, the pitching mound is lowered, or new teams and many new players enter when an expansion takes place) or the technology changes (a livelier ball is used or perhaps, in the future, aluminum bats are allowed), the preceding system of mutual best responses is thrown out of equilibrium. Variation increases for a while as players experiment, and some succeed while others fail. Finally a new equilibrium is attained, and variation goes down again. That is exactly what we should expect in the framework of learning and adjustment to a Nash equilibrium.

We take up the evidence concerning mixed strategies in Chapter 8 and the evidence for some specific games or types of games—for example, the prisoners' dilemma, bargaining, and auctions—at appropriate points in later chapters. For now, the experimental and empirical evidence that we have presented should make you cautiously optimistic about using Nash equilibrium as a first approach or as the point of departure for your analysis. On the whole, we believe you should have considerable confidence in using the Nash equilibrium concept when the game in question is played frequently by players from a reasonably stable population and under relatively unchanging rules and conditions. When the game is new or is played just once and the players are inexperienced, you should use the equilibrium concept more cautiously and should not be surprised if the outcome that you observe is not the equilibrium that you calculate. But even then, your first step in the analysis should be to look for a Nash equilibrium; then you can judge whether it seems a plausible outcome and, if not, proceed to the further step of asking why not. Often the reason will be your misunderstanding of the players' objectives, not the players' failure to play the game correctly, given their true objectives.

3 CRITICAL DISCUSSION OF THE NASH EQUILIBRIUM CONCEPT

In addition to the critiques lodged against the Nash equilibrium concept by those who have examined the empirical evidence, there have also been theoretical criticisms of the concept. In this section, we briefly review some such criticisms and some rebuttals, in each case by using an example.[15] Some of the

[15]David M. Kreps, *Game Theory and Economic Modelling* (Oxford: Clarendon Press, 1990) gives an excellent in-depth discussion.

criticisms are mutually contradictory, and some can be countered by thinking of the games themselves in a better way. Others tell us that the Nash equilibrium concept by itself is not enough and suggest some augmentations or relaxations of it that have better properties. We develop one such alternative here and point to some others that appear in later chapters. We believe our presentation will leave you with renewed but cautious confidence in using the Nash equilibrium concept. But some serious doubts remain unresolved, indicating that game theory is not yet a settled science. Even this should give encouragement, not the opposite, to budding game theorists, because it shows that there is a lot of room for new thinking and new research in the subject. A totally settled science would be a dead science.

We begin by considering the basic appeal of the Nash equilibrium concept. Most of the games in this book are noncooperative, in the sense that every player takes her action independently. Therefore it seems natural to suppose that, if her action is not the best according to her own value system (payoff scale), given what everyone else does, then she will change it. In other words, it is appealing to suppose that every player's action will be the best response to the actions of all the others. Nash equilibrium has just this property of "simultaneous best responses"; indeed, that is its very definition. In any purported final outcome that is not a Nash equilibrium, at least one player could have done better by switching to a different action.

This consideration leads eminent game theorist Roger Myerson to rebut those criticisms of the Nash equilibrium that are based on the intuitive appeal of playing a different strategy. His rebuttal simply shifts the burden of proof onto the critic. "When asked why players in a game should behave as in some Nash equilibrium," he says, "my favorite response is to ask 'Why not?' and to let the challenger specify what he thinks the players should do. If this specification is not a Nash equilibrium, then . . . we can show that it would destroy its own validity if the players believed it to be an accurate description of each other's behavior."[16]

A. The Treatment of Risk in Nash Equilibrium

Some critics argue that the Nash equilibrium concept does not pay due attention to risk. In some games, people might find strategies different from their Nash equilibrium strategies to be safer and may therefore choose those strategies. We offer two examples of this kind. The first is due to John Morgan, an economics professor at the University of California, Berkeley; Figure 5.3 shows the game table.

Cell-by-cell inspection quickly reveals that this game has a unique Nash equilibrium—namely, (A, A), yielding the payoffs (2, 2). But you may think, as

[16]Roger Myerson, *Game Theory* (Cambridge: Harvard University Press, 1991), p. 106.

		COLUMN		
		A	B	C
ROW	A	2, 2	3, 1	0, 2
	B	1, 3	2, 2	3, 2
	C	2, 0	2, 3	2, 2

FIGURE 5.3 A Game with a Questionable Nash Equilibrium

did several participants in an experiment conducted by Morgan, that playing C has a lot of appeal, for the following reasons. It *guarantees* you the same payoff as you would get in the Nash equilibrium—namely, 2; whereas if you play your Nash equilibrium strategy A, you will get a 2 only so long as the other player also plays A. Why take that chance? What is more, if you think the other player might use this rationale for playing C, then you would be making a serious mistake by playing A; you would get only a 0 when you could have gotten your guaranteed 2 by playing C.

Myerson would respond, "Not so fast. If you really believe that the other player would think thus and would play C, then you should play B to get the payoff 3. And if you think the other person would think thus and so would play B, then your best response to B is A. And if you think the other would figure this out too, you should be playing your best response to A, namely A. Back to the Nash equilibrium!" As you can see, criticizing Nash equilibrium and rebutting the criticisms is itself something of an intellectual game, and quite a fascinating one.

The second example, due to David Kreps, is even more dramatic. The payoff matrix is in Figure 5.4. Before doing any theoretical analysis of this game, you should pretend that you are actually playing the game and that you are player A. Which of your two actions would you choose?

Keep in mind your answer to the preceding question and let us proceed to analyze the game. If we start by looking for dominant strategies, we see that player A has no dominant strategy but player B does. Playing Left guarantees B a payoff of 10, no matter what A does, versus the 9.9 that is gained by playing Right (also no matter what A does). Thus, player B should play Left. Given that player B is going to go Left, player A does better to go Down. The unique pure-strategy Nash equilibrium of this game is (Down, Left); each player achieves a payoff of 10 at this outcome.

The problem that arises here is that many people (but not all) would not, as player A, choose to play Down. (What did you choose?) This is true for those who have been students of game theory for years as well as for those who have

		B	
		Left	Right
A	Up	9, 10	8, 9.9
	Down	10, 10	−1000, 9.9

FIGURE 5.4 Disastrous Nash Equilibrium?

never heard of the subject. If A has *any* doubts about *either* B's payoffs *or* B's rationality, then it is a lot safer for A to play Up than to play her Nash equilibrium strategy of Down. What if A thought the payoff table was as illustrated in Figure 5.4 but in reality B's payoffs were the reverse—the 9.9s went with Left and the 10s went with Right? What if the 9.9s were only an approximation and the exact payoffs were actually 10.1? What if B was a player with a substantially different value system or was not a truly rational player who might choose the "wrong" action just for fun? Obviously, our assumptions of perfect information and rationality can really be crucial to the analysis that we use in the study of strategy. Doubts of players can alter equilibria from those that we would normally predict and can call the reasonableness of the Nash equilibrium concept into question.

However, the real problem with many such examples is not that the Nash equilibrium concept is inappropriate but that the examples choose to use it in an inappropriately simplistic way. In this example, if there are any doubts about B's payoffs, then this fact should be made an integral part of the analysis. If A does not know B's payoffs, the game is one of asymmetric information, and we do not develop the general techniques for studying such games until Chapter 9. But this particular example is a relatively simple game of that kind, and we can figure out its equilibrium very easily.

Suppose A thinks there is a probability p that B's payoffs from Left and Right are the reverse of those shown in Figure 5.4; so $(1 - p)$ is the probability that B's payoffs are as stated in that figure. Because A must take her action without knowing which is the case, she must choose her strategy to be "best on the average." In this game the calculation is simple, because in each case B has a dominant strategy; the only problem for A is that in the two different cases different strategies are dominant for B. With probability $(1 - p)$, B's dominant strategy is Left (the case shown in the figure) and, with probability p, it is Right (the opposite case). Therefore if A chooses Up, then with probability $(1 - p)$ he will meet B playing Left and so get a payoff of 9; with probability p, he will meet B playing Right and so get a payoff of 8. Thus A's statistical or probability-weighted average payoff from playing Up is $9(1 - p) + 8p$. Similarly, A's statistical average payoff from playing Down is $10(1 - p) - 1000p$. Therefore it is better for A to choose Up if

$$9(1 - p) + 8p > 10(1 - p) - 1000p, \quad \text{or} \quad p > 1/1009.$$

Thus, even if there is only a very slight chance that B's payoffs are the opposite of those in Figure 5.4, it is optimal for A to play Up. In this case, analysis based on rational behavior, when done correctly, contradicts neither the intuitive suspicion nor the experimental evidence after all.

In the preceding calculation, we supposed that, facing an uncertain prospect of payoffs, player A would calculate the statistical average payoffs from her different actions and would choose that action which yields her the highest statistical average payoff. This implicit assumption, though it serves the purpose in this example, is not without its own problems. For example, it implies that a person faced with two situations, one having a 50-50 chance of winning or losing \$10 and the other having a 50-50 chance of winning \$10,001 and losing \$10,000, should choose the second situation, because it yields a statistical average winning of 50 cents ($\frac{1}{2} \times 10{,}001 - \frac{1}{2} \times 10{,}000$), whereas the first yields 0 ($\frac{1}{2} \times 10 - \frac{1}{2} \times 10$). But most people would think that the second situation carries a much bigger risk and would therefore prefer the first situation. This difficulty is quite easy to resolve. In the Appendix to Chapter 7, we show how the construction of a scale of payoffs that is suitably nonlinear in money amounts enables the decision maker to allow for risk as well as return Then in Chapter 9, we show how the concept can be used for understanding how people respond to the presence of risk in their lives, for example, by arranging the sharing of risk with others, or through the provision of insurance.

B. Multiplicity of Nash Equilibria

Another criticism of the Nash equilibrium concept is based on the observation that many games have multiple Nash equilibria. Thus, the argument goes, the concept fails to pin down outcomes of games sufficiently precisely to give unique predictions. This argument does not automatically require us to abandon the Nash equilibrium concept. Rather, it suggests that if we want a unique prediction from our theory, we must add some criterion for deciding which one of the multiple Nash equilibria we want to select.

In Chapter 4, we studied many games of coordination with multiple equilibria. From among these equilibria, the players may be able to select one as a focal point if they have some common social, cultural, or historical knowledge or if the game has some deliberate or accidental features that enable their expectations to converge. Here is a very extreme example of multiplicity of Nash equilibria in a coordination game. Two players are asked to write down, simultaneously and independently, the share that each wants of a total prize of \$100. If the amounts that they write down add up to \$100 or less, each player is given what she wrote. If the two add up to more than \$100, neither gets anything. For any x, one player writing x and the other writing $(100 - x)$ is a Nash equilibrium. Thus the game has an (almost) infinite range of Nash equilibria. But in practice, 50 : 50 emerges as a focal point. This social norm of equality or fairness seems so deeply ingrained as to be almost an instinct; players who choose 50 say that it is the obvious answer. To be a true focal point, not only should it be obvious to each, but everyone should know that it is obvious to each, and everyone should know that . . . ; in other words, its obviousness should be common knowledge. That need not always be the case, as we see when we consider a situation in which one player is a woman from an enlightened and egalitarian society who believes that 50 : 50 is obvious and the other is a man from a patriarchal society who believes it is obvious that, in any matter of division, a man should get three times as much as a woman. Then each will do what is obvious to her or him, and they will end up with nothing, because neither's obvious solution is obvious as common knowledge to both.

The existence of focal points is often a matter of coincidence, and creating them where none exist is basically an art that requires a lot of attention to the historical and cultural context of a game and not merely its mathematical description. This bothers many game theorists, who would prefer the outcome to depend only on an abstract specification of a game—players and their strategies should be identified by numbers without any external associations. We disagree. We think that historical and cultural contexts are just as important to a game as its purely mathematical description, and, if such context helps in selecting a unique outcome from multiple Nash equilibria, that is all to the better.

In Chapter 6, we will see that sequential-move games can have multiple Nash equilibria. There, we introduce the requirement of *credibility* that enables us to select a particular equilibrium; it turns out that this one is in fact the rollback equilibrium of Chapter 3. In more complex games with information asymmetries or additional complications, other restrictions called **refinements** have been developed to identify and rule out Nash equilibria that are unreasonable in some way. In Chapter 9, we consider one such refinement process that selects an outcome called a *perfect Bayesian equilibrium.* The motivation for each refinement is often specific to a particular type of game. A refinement stipulates how players update their information when they observe what moves other players made or failed to make. Each such stipulation is often perfectly reasonable in its context, and in many games it is not difficult to eliminate most of the Nash equilibria and therefore to narrow down the ambiguity in prediction.

The opposite of the criticism that some games may have too many Nash equilibria is that some games may have none at all. We saw an example of this in Section 4.8 and said that, by extending the concept of strategy to random mixtures, Nash equilibrium could be restored. In Chapters 7 and 8 we explain and consider Nash equilibria in mixed strategies. In higher reaches of game theory, there are more esoteric examples of games that have no Nash equilibrium in mixed strategies either. However, this added complication is not relevant for the types of analysis and applications that we deal with in this book, so we do not attempt to address it here.

C. Requirements of Rationality for Nash Equilibrium

Remember that Nash equilibrium can be regarded as a system of the strategy choices of each player and the belief that each player holds about the other players' choices. In equilibrium, (1) the choice of each should give her the best payoff given her belief about the others' choices, and (2) the belief of each player should be correct—that is, her actual choices should be the same as what this player believes them to be. These seem to be natural expressions of the requirements of the mutual consistency of individual rationality. If all players have common knowledge that they are all rational, how can any one of them rationally believe something about others' choices that would be inconsistent with a rational response to her own actions?

To begin to address this question, we consider the three-by-three game in Figure 5.5. Cell-by-cell inspection quickly reveals that it has only one Nash equilibrium—namely, (R2, C2), leading to payoffs (3, 3). In this equilibrium, Row plays R2 because she believes that Column is playing C2. Why does she believe this? Because she knows Column to be rational, Row must simultaneously believe that Column believes that Row is choosing R2, because C2 would not be Column's best choice if she believed Row would be playing either R1 or R3. Thus, the claim goes, in any rational process of formation of beliefs and responses, beliefs would have to be correct.

The trouble with this argument is that it stops after one round of thinking about beliefs. If we allow it to go far enough, we can justify other choice combinations. We can, for example, rationally justify Row's choice of R1. To do so, we note that R1 is Row's best choice if she believes Column is choosing C3. Why does she believe this? Because she believes that Column believes that Row is playing R3. Row justifies this belief by thinking that Column believes that Row

		COLUMN		
		C1	C2	C3
ROW	R1	0, 7	2, 5	7, 0
	R2	5, 2	3, 3	5, 2
	R3	7, 0	2, 5	0, 7

FIGURE 5.5 Justifying Choices by Chains of Beliefs and Responses

believes that Column is playing C1, believing that Row is playing R1, believing in turn . . . This is a chain of beliefs, each link of which is perfectly rational.

Thus rationality alone does not justify Nash equilibrium. There are more sophisticated arguments of this kind that do justify a special form of Nash equilibrium in which players can condition their strategies on a publicly observable randomization device. But we leave that to more advanced treatments. In the next section, we develop a simpler concept that captures what is logically implied by the players' common knowledge of their rationality alone.

4 RATIONALIZABILITY

What strategy choices in games can be justified on the basis of rationality alone? In the matrix of Figure 5.5, we can justify any pair of strategies, one for each player, by using the same type of logic as that used in Section 3.C. In other words, we can justify any one of the nine logically conceivable combinations. Thus rationality alone does not give us any power to narrow down or predict outcomes at all. Is this a general feature of all games? No. For example, if a strategy is dominated, rationality alone can rule it out of consideration. And when players recognize that other players, being rational, will not play dominated strategies, iterated elimination of dominated strategies can be performed on the basis of common knowledge of rationality. Is this the best that can be done? No. Some more ruling out of strategies can be done, by using a property slightly stronger than being dominated in pure strategies. This property identifies strategies that are **never a best response.** The set of strategies that survive elimination on this ground are called **rationalizable,** and the concept itself is known as **rationalizability.**

Why introduce this additional concept, and what does it do for us? As for why, it is useful to know how far we can narrow down the possible outcomes of a game based on the players' rationality alone, without invoking correctness of expectations about the other player's actual choice. It is sometimes possible to figure out that the other player *will not* choose some available action or actions, even when it is not possible to pin down the single action that she *will* choose. As for what it achieves, that depends on the context. In some cases rationalizability may not narrow down the outcomes at all. This was so in the three-by-three example of Figure 5.5. In some cases it narrows down the possibilities to some extent, but not all the way down to the Nash equilibrium if the game has a unique one, or to the set of Nash equilibria if there are several. An example of such a situation is the four-by-four enlargement of the previous example, considered in Section 4.A. In some other cases, the narrowing down may go all the way to the Nash equilibrium; in these cases we have a more powerful justification for the Nash equilibrium that relies on rationality alone, without assuming correctness of expectations. The quantity competition example of Section 4.B is an example in which the rationalizability argument takes us all the way to the game's unique Nash equilibrium.

A. Applying the Concept of Rationalizability

Consider the game in Figure 5.6, which is the same as Figure 5.5 but with an additional strategy for each player.[17] We just indicated that nine of the strategy combinations that pick one of the first three strategies for each of the players can be justified by a chain of beliefs about each other's beliefs. That remains true in this enlarged matrix. But can R4 and C4 be justified in this way?

Could Row ever believe that Column would play C4? Such a belief would have to be justified by Column's beliefs about Row's choice. What might Column believe about Row's choice that would make C4 Column's best response? Nothing. If Column believes that Row would play R1, then Column's best choice is C1. If Column believes that Row will play R2, then Column's best choice is C2. If Column believes that Row will play R3, then C3 is Column's best choice. And, if Column believes that Row will play R4, then C1 and C3 are tied for her best choice. Thus C4 is never a best response for Column.[18] This means that Row, knowing Column to be rational, can never attribute to Column any belief about Row's choice that would justify Column's choice of C4. Therefore Row should never believe that Column would choose C4.

Note that, although C4 is never a best response, it is not dominated by any of C1, C2, and C3. For Column, C4 does better than C1 against Row's R3, better than

		COLUMN			
		C1	C2	C3	C4
ROW	R1	0, 7	2, 5	7, 0	0, 1
	R2	5, 2	3, 3	5, 2	0, 1
	R3	7, 0	2, 5	0, 7	0, 1
	R4	0, 0	0, −2	0, 0	10, −1

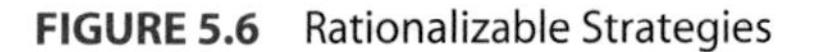
FIGURE 5.6 Rationalizable Strategies

[17]This example is taken from the original article that developed the concept of rationalizability. See Douglas Bernheim, "Rationalizable Strategic Behavior," *Econometrica,* vol. 52, no. 4 (July 1984), pp. 1007–1028. See also Andreu Mas-Colell, Michael Whinston, and Jerry Green, *Microeconomic Theory* (New York: Oxford University Press, 1995), pp. 242–245.

[18]Note that in each case the best choice is strictly better than C4 for Column. Thus C4 is never even tied for a best response. We can distinguish between weak and strong senses of never being a best response just as we distinguished between weak and strong dominance. Here, we have the strong sense.

C2 against Row's R4, and better than C3 against Row's R1. If a strategy *is* dominated, it also can never be a best response. Thus "never a best response" is a more general concept than "dominated." Eliminating strategies that are never a best response may be possible even when eliminating dominated strategies is not. So eliminating strategies that are never a best response can narrow down the set of possible outcomes more than can elimination of dominated strategies.[19]

The elimination of "never best response" strategies can also be carried out iteratively. Because a rational Row can never believe that a rational Column will play C4, a rational Column should foresee this. Because R4 is Row's best response only against C4, Column should never believe that Row will play R4. Thus R4 and C4 can never figure in the set of rationalizable stratgies. The concept of rationalizability does allow us to narrow down the set of possible outcomes of this game to this extent.

If a game has a Nash equilibrium, it is rationalizable and in fact can be sustained by a simple one-round system of beliefs, as we saw in Section 3.C. But more generally, even if a game does not have a Nash equilibrium, it may have rationalizable outcomes. Consider the two-by-two game obtained from Figure 5.5 or Figure 5.6 by retaining just the strategies R1 and R3 for Row and C1 and C3 for Column. It is easy to see that it has no Nash equilibrium in pure strategies. But all four outcomes are rationalizable with the use of exactly the chain of beliefs, constructed earlier, that went around and around these strategies.

Thus the concept of rationalizability provides a possible way of solving games that do not have a Nash equilibrium. And more important, the concept tells us how far we can narrow down the possibilities in a game on the basis of rationality alone.

B. Rationalizability Can Take Us All the Way to Nash Equilibrium

In some games, iterated elimination of never-best-response strategies can narrow things down all the way to Nash equilibrium. Note we said *can,* not *must.* But if it does, that is useful because in these games we can strengthen the case for Nash equilibrium by arguing that it follows purely from the players' rational thinking about each other's thinking. Interestingly, one class of games that can be solved in this way is very important in economics. This class consists of competition between firms that choose the quantities that they produce, knowing that the total quantity that is put on the market will determine the price.

We illustrate a game of this type in the context of a small coastal town. It has two fishing boats that go out every evening and return the following morning

[19]In Chapter 8, we will see that in two-player games, a strategy that is never a best response can be dominated by a *mixture* of the other strategies. Therefore, in two-player games that allow mixed strategies, the two kinds of elimination become equivalent.

to put their night's catch on the market. The game is played out in an era before modern refrigeration, so all the fish has to be sold and eaten the same day. Fish are quite plentiful in the ocean near the town, so the owner of each boat can decide how much to catch each night. But each knows that, if the total that is brought to the market is too large, the glut of fish will mean a low price and low profits.

Specifically, we suppose that, if one boat brings R barrels and the other brings S barrels of fish to the market, the price P (measured in ducats per barrel) will be $P = 60 - (R + S)$. We also suppose that the two boats and their crews are somewhat different in their fishing efficiency. Fishing costs the first boat 30 ducats per barrel and the second boat 36 ducats per barrel.

Now we can write down the profits of the two boat owners, U and V, in terms of their strategies R and S:

$$U = [(60 - R - S) - 30]R = (30 - S)\,R - R^2,$$
$$V = [(60 - R - S) - 36]S = (24 - R)\,S - S^2.$$

With these payoff expressions, we construct best-response curves and find the Nash equilibrium. As in our price competition example from Section 1, each player's payoff is a quadratic function of his own strategy, holding the strategy of the other player constant. Therefore the same mathematical methods we develop there and in the Appendix can be applied.

The first boat's best response R should maximize U for each given value of the other boat's S. With the use of calculus, this means that we should differentiate U with respect to R, holding S fixed, and set the derivative equal to zero, which gives

$$(30 - R) - 2R = 0; \quad \text{so} \quad R = 15 - S/2.$$

The noncalculus approach uses the result that the U-maximizing value of $R = B/(2C)$ where in this case $B = 30 - S$ and $C = 1$. This gives $R = (30 - S)/2$, or $R = 15 - S/2$.

Similarly, the best-response equation of the second boat is found by choosing S to maximize V for each fixed R, yielding

$$S = (24 - R)/2; \quad \text{so} \quad S = 12 - R/2.$$

The Nash equilibrium is found by solving the two best-response equations jointly for R and S, which is easy to do. So we just state the results:[20] quantities are $R = 12$ and $S = 6$; price is $P = 42$; and profits are $U = 144$ and $V = 36$.

[20]Although they are incidental to our purpose, some interesting properties of the solution are worth pointing out. The quantities differ because the costs differ; the more efficient (lower-cost) boat gets to sell more. The cost and quantity differences together imply even bigger differences in the resulting profits. The cost advantage of the first boat over the second is only 20%, but it makes four times as much profit as the second boat.

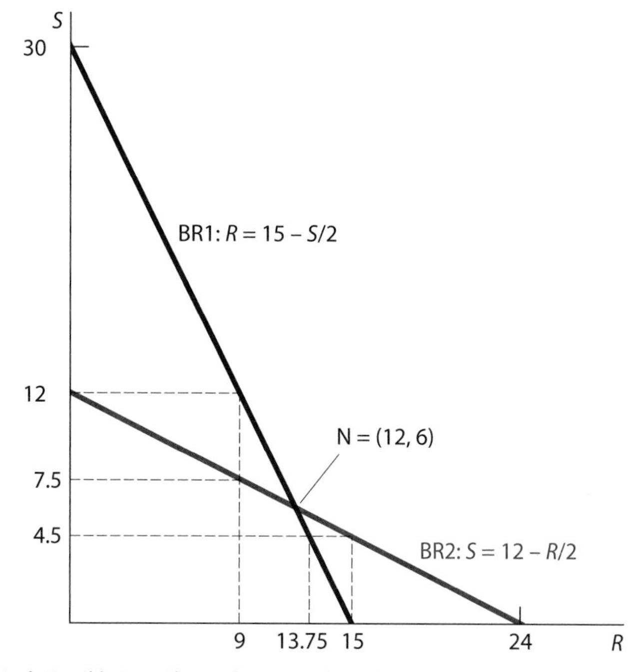

FIGURE 5.7 Nash Equilibrium Through Rationalizability

Figure 5.7 shows the two fishermen's best-response curves (labeled BR1 and BR2 with the equations displayed) and the Nash equilibrium (labeled N with its coordinates displayed) at the intersection of the two curves. Figure 5.7 also shows how the players' beliefs about each other's choices can be narrowed down by iteratively eliminating strategies that are never best responses.

What values of S can the first owner rationally believe the second owner will choose? That depends on what the second owner thinks the first owner will produce. But no matter what this might be, the whole range of the second owner's best responses is between 0 and 12. So the first owner cannot rationally believe that the second owner will choose anything else; all negative choices of S (obviously) and all choices of S greater than 12 (less obviously) are eliminated. Similarly, the second owner cannot rationally think that the first owner will produce anything less than 0 or greater than 15.

Now take this to the second round. When the first owner has restricted the second owner's choices of S to the range between 0 and 12, her own choices of R are restricted to the range of best responses to S's range. The best response to $S = 0$ is $R = 15$, and the best response to $S = 12$ is $R = 15 - 12/2 = 9$. Because BR1 has a negative slope throughout, the whole range of R allowed at this round of thinking is between 9 and 15. Similarly, the second owner's choice of S is restricted to the range of best responses to R between 0 and 15—namely, values between $S = 12$ and $S = 12 - 15/2 = 4.5$. Figure 5.7 shows these restricted ranges on the axes.

The third round of thinking narrows the ranges further. Because R must be at least 9 and BR2 has a negative slope, S can be at most the best response to 9—namely, $S = 12 - 9/2 = 7.5$. In the second round, S was already shown to be at least 4.5. Thus S is now restricted to be between 4.5 and 7.5. Similarly, because S must be at least 4.5, R can be at most $15 - 4.5/2 = 12.75$. In the second round, R was shown to be at least 9, so now it is restricted to the range from 9 to 12.75.

This succession of rounds can be carried on as far as you like, but it is already evident that the successive narrowing of the two ranges is converging on the Nash equilibrium, $R = 12$ and $S = 6$. Thus the Nash equilibrium is the only outcome that survives the iterated elimination of strategies that are never best responses.[21] We know that in general the rationalizability argument need not narrow down the outcomes of a game to its Nash equilibria, so this is a special feature of this example. Actually, the process works for an entire class of games; it will work for any game that has a unique Nash equilibrium at the intersection of downward-sloping best-response curves.[22]

This argument should be carefully distinguished from an older one based on a succession of best responses. The old reasoning proceeded as follows. Start at any strategy for one of the players—say, $R = 18$. Then the best response of the other is $S = 12 - 18/2 = 3$. The best response of R to $S = 3$ is $R = 15 - 3/2 = 13.5$. In turn, the best response of S to $R = 13.5$ is $12 - 13.5/2 = 5.25$. Then, in its turn, the best R against this S is $R = 15 - 5.25/2 = 12.375$. And so on.

The chain of best responses in the old argument also converges to the Nash equilibrium. But the argument is flawed. The game is played once with simultaneous moves. It is not possible for one player to respond to what the other player has chosen, then have the first player respond back again, and so on. If such dynamics of actual play were allowed, would the players not foresee that the other is going to respond and so do something different in the first place?

The rationalizability argument is different. It clearly incorporates the fact that the game is played only once and with simultaneous moves. All the thinking regarding the chain of best responses is done in advance, and all the successive rounds of thinking and responding are purely conceptual. Players are not responding to actual choices but are merely calculating those choices that will never be made. The dynamics are purely in the minds of the players.

[21]This example can also be solved by iteratively eliminating dominated strategies, but proving dominance is harder and needs more calculus, whereas the never-best-response property is obvious from Figure 5.7, so we use the simpler argument.

[22]A similar argument works with upward-sloping best-response curves, such as those in the pricing game of Figure 5.1, for narrowing the range of best responses starting at low prices. Narrowing from the higher end is possible only if there is some obvious starting point. This starting point might be a very high price that can never be exceeded for some externally enforced reason—if, for example, people simply do not have the money to pay prices beyond a certain level.

SUMMARY

When players in a simultaneous-move game have a continuous range of actions to choose, best-response analysis yields mathematical *best-response rules* that can be solved simultaneously to obtain Nash equilibrium strategy choices. The best-response rules can be shown on a diagram in which the intersection of the two curves represents the Nash equilibrium. Firms choosing price or quantity from a large range of possible values and political parties choosing campaign advertising expenditure levels are examples of games with *continuous strategies.*

The results of laboratory tests of the Nash equilibrium concept show that a common cultural background is essential for coordinating in games with multiple equilibria. Repeated play of some games shows that players can learn from experience and begin to choose strategies that approach Nash equilibrium choices. Further, predicted equilibria are accurate only when the experimenters' assumptions match the true preferences of players. Real-world applications of game theory have helped economists and political scientists, in particular, to understand important consumer, firm, voter, legislature, and government behaviors.

Theoretical criticisms of the Nash equilibrium concept have argued that the concept does not adequately account for risk, that it is of limited use because many games have multiple equilibria, and that it cannot be justified on the basis of rationality alone. In many cases, a better description of the game and its payoff structure or a refinement of the Nash equilibrium concept can lead to better predictions or fewer potential equilibria. The concept of *rationalizability* relies on the elimination of strategies that are *never a best response* to obtain a set of *rationalizable* outcomes. When a game has a Nash equilibrium, that outcome will be rationalizable, but rationalizability also allows one to predict equilibrium outcomes in games that have no Nash equilibria.

KEY TERMS

best-response curves (137)
best-response rule (134)
continuous strategy (134)
never a best response (157)
rationalizability (157)
rationalizable (157)
refinement (155)

SOLVED EXERCISES

S1. In the political campaign advertising game in Section 1.B, party L chooses an advertising budget, x (millions of dollars), and party R similarly chooses a budget, y (millions of dollars). We showed there that the best-response rules in that game are $y = 10\sqrt{x} - x$, for party R, and $x = 10\sqrt{y} - x$ for party L.

(a) What is party R's best response if party L spends \$16 million?

(b) Use the specified best-response rules to verify that the Nash equilibrium advertising budgets are $x = y = 25$, or \$25 million.

S2. The bistro game illustrated in Figure 5.1 defines demand functions for meals at Xavier's (Q_x) and Yvonne's (Q_y) as $Q_x = 44 - 2\ P_x + P_y$, and $Q_y = 44 - 2\ P_y + P_x$. Profits for each firm depend in addition on their costs of serving each customer. Suppose here that Yvonne's is able to reduce its costs to a mere \$2 per customer by completely eliminating the wait staff (customers pick up their orders at the counter, and a few remaining employees bus the tables). Xavier's continues to incur a cost of \$8 per customer.

(a) Recalculate the best-response rules and the Nash equilibrium prices for the two firms, given the change in the cost conditions.

(b) Graph the two best-response curves and describe the differences between your graph and Figure 5.1. In particular, which curve has moved and by how much? Explain why these changes occurred in the diagram.

S3. Yuppietown has two food stores, La Boulangerie, which sells bread, and La Fromagerie, which sells cheese. It costs \$1 to make a loaf of bread and \$2 to make a pound of cheese. If La Boulangerie's price is P_1 dollars per loaf of bread and La Fromagerie's price is P_2 dollars per pound of cheese, their respective weekly sales, Q_1 thousand loaves of bread and Q_2 thousand pounds of cheese, are given by the following equations:

$$Q_1 = 14 - P_1 - 0.5P_2, \quad Q_2 = 19 - 0.5P_1 - P_2.$$

(a) For each store, write its profit as a function of P_1 and P_2 (in the exercises that follow, we will call this "the profit function" for brevity). Then find their respective best-response rules. Graph the best-response curves, and find the Nash equilibrium prices in this game.

(b) Suppose that the two stores collude and set prices jointly to maximize the sum of their profits. Find the joint profit-maximizing prices for the stores.

(c) Provide a short intuitive explanation for the differences between the Nash equilibrium prices and those that maximize joint profit. Why is joint profit maximization not a Nash equilibrium?

(d) In this problem, bread and cheese are mutual *complements*. They are often consumed together; that is why a drop in the price of one increases the sales of the other. The products in our bistro example in Section 1.A are *substitutes* for each other. How does this difference explain the differences among your findings for the best-response rules, the Nash equilibrium prices, and the joint profit-maximizing prices in this question, and the corresponding entities in the bistro example in the text?

S4. The game illustrated in Figure 5.3 has a unique Nash equilibrium in pure strategies. However, all nine outcomes in that game are rationalizable. Confirm this assertion, explaining your reasoning for each outcome.

S5. For the game presented in Exercise S5 in Chapter 4, what are the rationalizable strategies for each player? Explain your reasoning.

S6. Section 4.B of this chapter describes a fishing game played in a small coastal town. When the response rules for the two boats have been derived, rationalizability can be used to justify the Nash equilibrium in the game. In the description in the text, we take the process of narrowing down strategies that can never be best responses through three rounds. By the third round, we know that R (the number of barrels of fish brought home by boat 1) must be at least 9, and that S (the number of barrels of fish brought home by boat 2) must be at least 4.5. The narrowing process in that round restricted R to the range between 9 and 12.75 while restricting S to the range between 4.5 and 7.5. Take this process of narrowing through one additional (fourth) round and show the reduced ranges of R and S that are obtained at the end of the round.

S7. Two carts selling coconut milk (from the coconut) are located at 0 and 1, 1 mile apart on the beach in Rio de Janeiro. (They are the only two coconut-milk carts on the beach.) The carts—Cart 0 and Cart 1—charge prices p_0 and p_1, respectively, for each coconut. One thousand beachgoers buy coconut milk, and these customers are uniformly distributed along the beach between carts 0 and 1. Each beachgoer will purchase one coconut milk in the course of her day at the beach and, in addition to the price, each will incur a transport cost of 0.5 times d^2, where d is the distance (in miles) from her beach blanket to the coconut cart. In this system, Cart 0 sells to all of the beachgoers located between 0 and x, and Cart 1 sells to all of the beachgoers located between x and 1, where x is the location of the beachgoer who pays the same total price if she goes to 0 or 1. Location x is then defined by the expression:

$$p_0 + 0.5x^2 = p_1 + 0.5(1 - x)^2.$$

The two carts will set their prices to maximize their bottom-line profit figures, B; profits are determined by revenue (the cart's price times its number of customers) and cost (the carts each incur a cost of \$0.25 per coconut times the number of coconuts sold).

(a) For each cart, determine the expression for the number of customers served as a function of p_0 and p_1. (Recall that Cart 0 gets the customers between 0 and x, or just x, while Cart 1 gets the customers between x and 1, or $1 - x$. That is, cart 0 sells to x customers, where x is measured in thousands, and cart 1 sells to $(1 - x)$ thousand.)

(b) Write the profit functions for the two carts. Find the two best-response rules for each cart as a function of their rival's price.

(c) Graph the best-response rules, and then calculate (and show on your graph) the Nash equilibrium price level for coconut milk on the beach.

S8. Crude oil is transported across the globe in enormous tanker ships called Very Large Crude Carriers (VLCCs). By 2001, more than 92% of all new VLCCs were built in South Korea and Japan. Assume that the price of new VLCCs (in millions of dollars) is determined by the function $P = 180 - Q$, where $Q = q_{Korea} + q_{Japan}$. (That is, assume that only Japan and Korea produce VLCCs, so they are a duopoly.) Assume that the cost of building each ship is \$30 million in both Korea and Japan. That is, $c_{Korea} = c_{Japan} = 30$, where the per-ship cost is measured in millions of dollars.

(a) Write the profit functions for each country in terms of q_{Korea} and q_{Japan} and either c_{Korea} or c_{Japan}. Find each country's best-response function.

(b) Using the best-response functions found in part (a), solve for the Nash equilibrium quantity of VLCCs produced by each country per year. What is the price of a VLCC? How much profit is made in each country?

(c) Labor costs in Korean shipyards are actually much lower than in their Japanese counterparts. Assume now that the cost per ship in Japan is \$40 million and that in Korea it is only \$20 million. Given $c_{Korea} = 20$ and $c_{Japan} = 40$, what is the market share of each country (ie, the percentage of ships that each country sells relative to the total number sold)? What are the profits for each country?

S9. Extending the previous problem, suppose China decides to enter the VLCC construction market. The duopoly now becomes a triopoly, so that although price is still $P = 180 - Q$, quantity is now given by $Q = q_{Korea} + q_{Japan} + q_{China}$. Assume that all three countries have a per-ship cost of \$30 million: $c_{Korea} = c_{Japan} = c_{China} = 30$.

(a) Write the profit functions for each of the three countries in terms of q_{Korea}, q_{Japan}, and q_{China}, and c_{Korea}, c_{Japan}, or c_{China}. Find each country's best-response rule.

(b) Using your answer to part (a), find the quantity produced, the market share captured (see Exercise S8, part (c)), and the profits earned by each country. This will require the solution of three equations in three unknowns.

(c) What happens to the price of a VLCC in the new triopoly relative to the duopoly situation in Exercise S8, part (b)? Why?

S10. Monica and Nancy have formed a business partnership to provide consulting services in the golf industry. They each have to decide how much effort to put into the business. Let m be the amount of effort put into the business by Monica, and n be the amount of effort put in by Nancy.

The joint profits of the partnership are given by $4m + 4n + mn$, in tens of thousands of dollars, and the two partners split these profits equally. However, they must each separately incur the costs of their own effort; the cost to Monica of her effort is m^2, while the cost to Nancy of her effort is n^2 (both measured in tens of thousands of dollars). Each partner must make her effort decision without knowing what effort decision the other player has made.

(a) If Monica and Nancy each put in effort of $m = n = 1$, then what are their payoffs?

(b) If Monica puts in effort of $m = 1$, then what is Nancy's best response?

(c) What is the Nash equilibrium to this game?

S11. Nash equilibrium through rationalizability can be achieved in games with upward-sloping best-response curves if the rounds of eliminating never-best-response strategies begin with the smallest possible values. Consider the pricing game between Xavier's Tapas Bar and Yvonne's Bistro that is illustrated in Figure 5.1. Use Figure 5.1 and the best-response rules from which it is derived to begin rationalizing the Nash equilibrium in that game. Start with the lowest possible prices for the two firms and describe (at least) two rounds of narrowing the set of rationalizable prices toward the Nash equilibrium.

S12. A professor presents the following game to Elsa and her forty-nine classmates. Each of them simultaneously and privately writes down a number between zero and 100 on a piece of paper, and they all hand in their numbers. The professor then computes the mean of these numbers and defines X to be the mean of the students' numbers. The student who submits the number closest to two-thirds of X wins \$50. If multiple students tie, they split the prize equally.

(a) Show that choosing the number 80 is a dominated strategy.

(b) What would the set of best responses be for Elsa if she knew that all of her classmates would submit the number 40? That is, what is the range of numbers for which each number in the range is closer to the winning number than 40?

(c) What would the set of best responses be for Elsa is she knew that all of her classmates would submit the number 10?

(d) Find a symmetric Nash equilibrium to this game. That is, what number is a best response to everyone else submitting that same number?

(e) Which strategies are rationalizable in this game?

UNSOLVED EXERCISES

U1. Diamond Trading Company (DTC), a subsidiary of De Beers, is the dominant supplier of high quality diamonds for the wholesale market. For simplicity, assume that DTC has a monopoly on wholesale diamonds. The quantity that DTC chooses to sell thus has a direct impact on the wholesale price of diamonds. Let the wholesale price of diamonds (in hundreds of dollars) be given by the following inverse demand function: $P = 120 - Q_{DTC}$. Assume that DTC has a cost of 12 (hundred dollars) per high-quality diamond.

(a) Write DTC's profit function in terms of Q_{DTC}, and solve for DTC's profit-maximizing quantity. What will be the wholesale price of diamonds at that quantity? What will DTC's profit be?

Frustrated with DTC's monopoly, several diamond mining interests and large retailers collectively set up a joint venture called Adamantia to act as a competitor to DTC in the wholesale market for diamonds. The wholesale price is now given by $P = 120 - Q_{DTC} - Q_{ADA}$. Assume that Adamantia has a cost of 12 (hundred dollars) per high-quality diamond.

(b) Write the best-response functions for both DTC and Adamantia. What quantity does each wholesaler supply to the market in equilibrium? What wholesale price do these quantities imply? What will the profit of each supplier be in this duopoly situation?

(c) Describe the differences in the market for wholesale diamonds under the duopoly of DTC and Adamantia relative to the monopoly of DTC. What happens to the quantity supplied in the market and the market price when Adamantia enters? What happens to the collective profit of DTC and Adamantia?

U2. There are two movie theaters in the town of Harkinsville: Modern Multiplex, a first-run theater, and Sticky Shoe, which shows movies that have been out for a while at a cheaper price. The demand for Modern Multiplex is given by: $Q_{MM} = 14 - P_{MM} + P_{SS}$, while the demand for Sticky Shoe is: $Q_{SS} = 8 - 2P_{SS} + P_{MM}$, where prices are in dollars and quantities are measured in hundreds of moviegoers. Modern Multiplex has a per-customer cost of \$4, while Sticky Shoe has a per-customer cost of only \$2.

(a) From the demand equations alone, what indicates whether Modern Multiplex and Sticky Shoe offer services that are substitutes or complements?

(b) Write the profit function for each theater in terms of P_{SS} and P_{MM}. Find each theater's best-response rule.

(c) Find the Nash equilibrium price, quantity, and profit for each theater.

(d) What would each theater's price, quantity, and profit be if the two decided to collude to maximize joint profits in this market? Why isn't the collusive outcome a Nash equilibrium?

U3. Fast forward a decade beyond the situation in Exercise **S3**. Yuppietown's demand for bread and cheese has decreased, and the town's two food stores, La Boulangerie and La Fromagerie, have been bought out by a third company: L'Épicerie. It still costs \$1 to make a loaf of bread and \$2 to make a pound of cheese, but the quantities of bread and cheese sold (Q_1 and Q_2 respectively, measured in thousands) are now given by the equations:

$$Q_1 = 8 - P_1 - 0.5P_2, \quad Q_2 = 16 - 0.5P_1 - P_2.$$

Again, P_1 is the price in dollars of a loaf of bread, and P_2 is the price in dollars of a pound of cheese.

(a) Initially, L'Épicerie runs La Boulangerie and La Fromagerie as if they were separate firms, with independent managers who each try to maximize their own profit. What are the Nash equilibrium quantities, prices, and profits for the two divisions of L'Épicerie, given the new quantity equations?

(b) The owners of L'Épicerie think that they can make more total profit by coordinating the pricing strategies of the two Yuppietown divisions of their company. What are the joint-profit-maximizing prices for bread and cheese under collusion? What quantities do La Boulangerie and La Fromagerie sell of each good, and what is the profit that each division earns separately?

(c) In general, why might companies sell some of their goods at prices below cost? That is, explain a rationale of loss leaders, using your answer from part **(b)** as an illustration.

U4. The coconut-milk carts from Exercise **S7** set up again the next day. Nearly everything is exactly the same as Exercise **S7**: the carts are in the same locations, the number and distribution of beachgoers is identical, and the demand of the beachgoers for exactly one coconut milk each is unchanged. The only difference is that it is a particularly hot day, so that now each beachgoer incurs a higher transport cost of $0.6d^2$. Again, Cart 0 sells to all of the beachgoers located between 0 and x, and Cart 1 sells to all of the beachgoers located between x and 1, where x is the location of the beachgoer who pays the same total price if she goes to 0 or 1. However, now location x is defined by the expression:

$$p_0 + 0.6x^2 = p_1 + 0.6(1 - x)^2.$$

Again, each cart has a cost of \$0.25 per coconut sold.

(a) For each cart, determine the expression for the number of customers served as a function of p_0 and p_1. (Recall that Cart 0 gets the customers between 0 and x, or just x, while Cart 1 gets the customers between x

and 1, or $1 - x$. That is, Cart 0 sells to x customers, where x is measured in thousands, and Cart 1 sells to $(1 - x)$ thousand.)

(b) Write out profit functions for the two carts and find the two best-response rules.

(c) Calculate the Nash equilibrium price level for coconuts on the beach. How does this price compare with the price found in Exercise S7? Why?

U5. The game illustrated in Figure 5.4 has a unique Nash equilibrium in pure strategies. Find that Nash equilibrium, and then show that it is also the unique rationalizable outcome in that game.

U6. What are the rationalizable strategies of the game "Evens or Odds" from Exercise S11 in Chapter 4?

U7. In the fishing-boat game of Section 4, we showed how it is possible for there to be a uniquely rationalizable outcome in continuous strategies that is also a Nash equilibrium. However, this is not always the case; there may be many rationalizable strategies, and not all of them will necessarily be part of a Nash equilibrium.

Returning to the political advertising game of Exercise S1, find the set of rationalizable strategies for party L. (Due to their symmetric payoffs, the set of rationalizable strategies will be the same for party R.) Explain your reasoning.

U8. Intel and AMD, the primary producers of computer central processing units (CPUs), compete with one another in the mid-range chip category (among other categories). Assume that global demand for mid-range chips depends on the quantity that the two firms make, so that the price (in dollars) for mid-range chips is given by $P = 210 - Q$, where $Q = q_{Intel} + q_{AMD}$ and where the quantities are measured in millions. Each mid-range chip costs Intel \$60 to produce. AMD's production process is more streamlined; each chip costs them only \$48 to produce.

(a) Write the profit function for each firm in terms of q_{Intel} and q_{AMD}. Find each firm's best-response rule.

(b) Find the Nash equilibrium price, quantity, and profit for each firm.

(c) **(Optional)** Suppose Intel acquires AMD, so that it now has two separate divisions with two different production costs. The merged firm wishes to maximize total profits from the two divisions. How many chips should each division produce? (Hint: You may need to think carefully about this problem, rather than blindly applying mathematical techniques.) What is the market price and the total profit to the firm?

U9. Return to the VLCC triopoly game of Exercise S9. In reality, the three countries do not have identical production costs. China has been gradually entering the VLCC construction market for several years, and its production costs started out rather high due to lack of experience.

(a) Solve for the triopoly quantities, market shares, price, and profits for the case where the per-ship costs are \$20 million for Korea, \$40 million for Japan, and \$60 million for China ($c_{Korea} = 20$, $c_{Japan} = 40$, and $c_{China} = 60$).

After it gains experience and adds production capacity, China's per-ship cost will decrease dramatically. Since labor is even cheaper in China than Korea, eventually the per-ship cost will be even lower in China than it is in Korea.

(b) Repeat part (a) with the adjustment that China's per-ship cost is \$16 million ($c_{Korea} = 20$, $c_{Japan} = 40$, and $c_{China} = 16$).

U10. Return to the story of Monica and Nancy from Exercise **S10**. After some additional professional training, Monica is more productive on the job, so that the joint profits of their company are now given by $5m + 4n + mn$, in tens of thousands of dollars. Again, m is the amount of effort put into the business by Monica, n is the amount of effort put in by Nancy, and the costs are m^2 and n^2 to Monica and Nancy respectively (in tens of thousands of dollars).

The terms of their partnership still require that the joint profits be split equally, despite the fact that Monica is more productive. Assume that their effort decisions are made simultaneously.

(a) What is Monica's best response if she expects Nancy to put in an effort of $n = \frac{4}{3}$?

(b) What is the Nash equilibrium to this game?

(c) Compared to the old Nash equilibrium found in Exercise **S10**, part (c), does Monica now put in more, less, or the same amount of effort? What about Nancy?

(d) What are the final payoffs to Monica and Nancy in the new Nash equilibrium (after splitting the joint profits and accounting for their costs of effort)? How do they compare to the payoffs to each of them under the old Nash equilibrium? In the end, who receives more benefit from Monica's additional training?

U11. A professor presents a new game to Elsa and her forty-nine classmates (similar to the situation in Exercise **S12**). As before, each of the students simultaneously and privately writes down a number between zero and 100 on a piece of paper, and the professor computes the mean of these numbers and calls it X. This time the student who submits the number closest to $(\frac{2}{3}) \times (X + 9)$ wins \$50. Again, if multiple students tie, they split the prize equally.

(a) Find a symmetric Nash equilibrium to this game. That is, what number is a best response to everyone else submitting the same number?

(b) Show that choosing the number 5 is a dominated strategy. (Hint: what would class average X have to be for the target number to be 5?)

(c) Show that choosing the number 90 is a dominated strategy.

(d) What are all of the dominated strategies?

(e) Suppose Elsa believes that none of her classmates will play the dominated strategies found in part (d). Given these beliefs what strategies are never a best response for Elsa?

(f) Which strategies do you think are rationalizable in this game? Explain your reasoning.

U12. **(Optional—requires calculus)** Recall the political campaign advertising example from Section 1.B concerning parties L and R. In that example, when L spends \$$x$ million on advertising and R spends \$$y$ million, L gets a share $x/(x + y)$ of the votes and R gets $y/(x + y)$. We also mentioned that two types of asymmetries can arise between the parties in that model. One party—say, R—may be able to advertise at a lower cost or R's advertising dollars may be more effective in generating votes than L's. To allow for both possibilities, we can write the payoff functions of the two parties as

$$V_L = \frac{x}{x + ky} - x \quad \text{and} \quad V_R = \frac{ky}{x + ky} - cy, \text{ where } k > 0 \text{ and } c > 0.$$

These payoff functions show that R has an advantage in the relative effectiveness of its ads when k is high and that R has an advantage in the cost of its ads when c is low.

(a) Use the payoff functions to derive the best-response functions for R (which chooses y) and L (which chooses x).

(b) Use your calculator or your computer to graph these best-response functions when $k = 1$ and $c = 1$. Compare the graph with the one for the case in which $k = 1$ and $c = 0.8$. What is the effect of having an advantage in the cost of advertising?

(c) Compare the graph from part (b), when $k = 1$ and $c = 1$ with the one for the case in which $k = 2$ and $c = 1$. What is the effect of having an advantage in the effectiveness of advertising dollars?

(d) Solve the best-response functions that you found in part (a), jointly for x and y, to show that the campaign advertising expenditures in Nash equilibrium are:

$$x = \frac{ck}{(c + k)^2} - x \quad \text{and} \quad y = \frac{k}{(c + k)^2}.$$

(e) Let $k = 1$ in the equilibrium spending-level equations and show how the two equilibrium spending levels vary with changes in c (i.e., interpret the signs of dx/dc and dy/dc). Then let $c = 1$ and show how the two equilibrium spending levels vary with changes in k (i.e., interpret the signs of dx/dk and dy/dk). Do your answers support the effects that you observed in parts (b) and (c) of this exercise?

■

Appendix: Finding a Value to Maximize a Function

Here we develop in a simple way the method for choosing a variable X to obtain the maximum value of a variable that is a function of it, say $Y = F(X)$. Our applications will mostly be to cases where the function is quadratic, such as $Y = A + BX - CX^2$. For such functions we derive the formula $X = B/(2C)$ that was stated and used in the chapter. We develop the general idea using calculus, and then offer an alternative approach that does not use calculus but applies only to the quadratic function.*

The calculus method tests a value of X for optimality by seeing what happens to the value of the function for other values on either side of X. If X does indeed maximize $Y = F(X)$, then the effect of increasing or decreasing X should be a drop in the value of Y. Calculus gives us a quick way to perform such a test.

Figure 5A.1 illustrates the basic idea. It shows the graph of a function $Y = F(X)$, where we have used a function of the type that fits our application, even though the idea is perfectly general. Start at any point P with coordinates (X, Y) on the graph. Consider a slightly different value of X, say $(X + h)$. Let k be the resulting change in $Y = F(X)$, so the point Q with coordinates $(X + h, Y + k)$ is also on the graph. The slope of the chord joining P to Q is the ratio k / h. If this ratio is positive, then h and k have the same sign; as X increases, so does Y. If the ratio is negative, then h and k have opposite signs; as X increases, Y decreases.

If we now consider smaller and smaller changes h in X, and the corresponding smaller and smaller changes k in Y, the chord PQ will approach the tangent to the graph at P. The slope of this tangent is the limiting value of the ratio k / h. It is called the derivative of the function $Y = F(X)$ at the point X. Symbolically, it is written as $F'(X)$ or dY/dX. Its sign tells us whether the function is increasing or decreasing at precisely the point X.

For the quadratic function in our application, $Y = A + BX - CX^2$ and

$$Y + k = A + B(X + h) - C(X + h)^2.$$

Therefore, we can find an expression for k as follows:

$$k = [A + B(X + h) - C(X + h)^2] - [A + BX - CX^2]$$

$$= Bh - C[(X + h)^2 - X^2] = Bh - C[X^2 + 2Xh + h^2 - X^2]$$

$$= (B - 2CX)\,h - Ch^2.$$

*Needless to say, we give only the briefest, quickest treatment, leaving out all issues of functions that don't have derivatives, functions that are maximized at an extreme point of the interval over which they are defined, and so on. Some readers will know all we say here; some will know much more. Others who want to find out more should refer to any introductory calculus textbook.

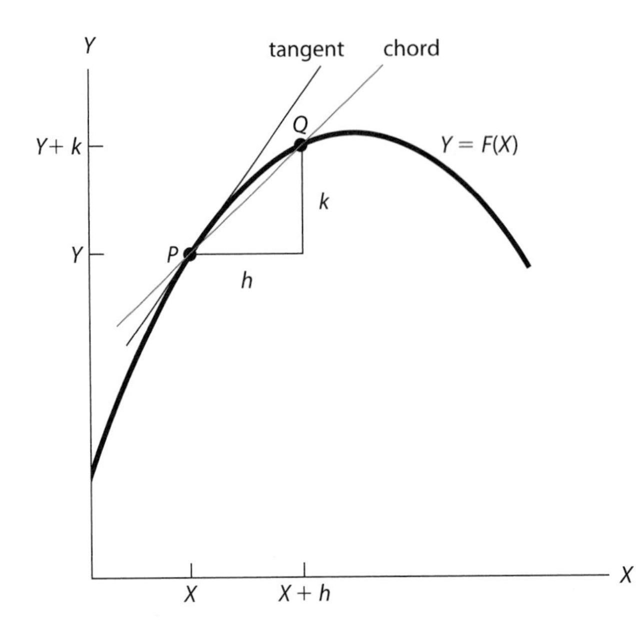

FIGURE 5A.1 Derivative of a Function Illustrated

Then $k/h = (B - 2CX) - Ch$. In the limit as h goes to zero, $k/h = (B - 2CX)$. This last expression is then the derivative of our function.

Now we use the derivative to find a test for optimality. Figure 5A.2 illustrates the idea. The point M yields the highest value of $Y = F(X)$. The function increases as we approach the point M from the left and decreases after we have passed to the right of M. Therefore the derivative $F'(X)$ should be positive for values of X smaller than M and negative for values of X larger than M. By continuity, the derivative precisely at M should be zero. In ordinary language, the graph of the function should be flat where it peaks.

In our quadratic example, the derivative is: $F'(X) = B - 2CX$. Our optimality test implies that the function is optimized when this is zero, or at $X = B/(2C)$. This is exactly the formula given in the chapter.

One additional check needs to be performed. If we turn the whole figure upside down, M is the minimum value of the upside-down function, and at this trough the graph will also be flat. So for a general function $F(X)$, setting $F'(X) = 0$ might yield an X that gives its minimum rather than the maximum. How do we distinguish the two possibilities?

At a maximum, the function will be increasing to its left and decreasing to its right. Therefore the derivative will be positive for values of X smaller than the purported maximum, and negative for larger values. In other words, the derivative, itself regarded as a function of X, will be decreasing at this point. A decreasing function has a negative derivative. Therefore the derivative of the derivative, what is called the second derivative of the original function, written as $F''(X)$ or d^2Y/dX^2,

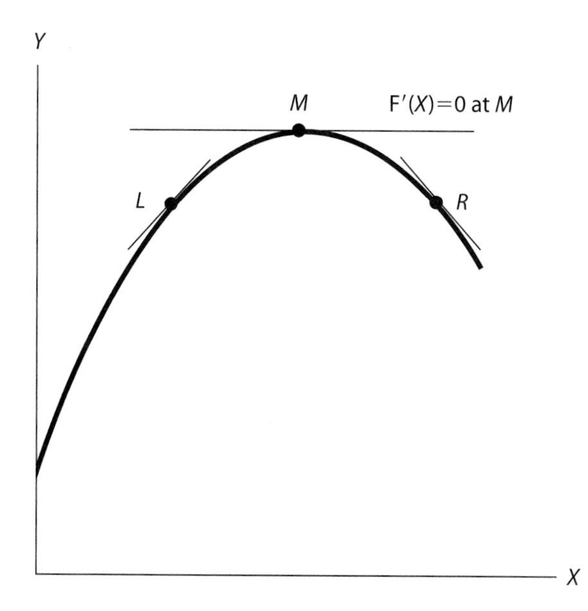

FIGURE 5A.2 Optimum of a Function

should be negative at a maximum. Similar logic shows that the second derivative should be positive at a minimum; that is what distinguishes the two cases.

For the derivative $F'(X) = B - 2CX$ of our quadratic example, applying the same h, k procedure to $F'(X)$ as we did to $F(X)$ shows $F''(X) = -2C$. This is negative so long as C is positive, which we assumed when stating the problem in the chapter. The test $F'(X) = 0$ is called the first-order condition for maximization of $F(X)$, and $F''(X) < 0$ is the second-order condition.

To fix the idea further, let us apply it to the specific example of Xavier's best response that we considered in the chapter. We had the expression

$$\Pi_x = -8(44 + P_y) + (16 + 44 + P_y)P_x - 2(P_x)^2.$$

This is a quadratic function of P_x (holding the other restaurant's price, P_y, fixed). Our method gives its derivative:

$$\frac{d\Pi_x}{dP_x} = (60 + p_y) - 4p_x.$$

The first-order condition for P_x to maximize Π_x is that this derivative should be zero. Setting it equal to zero and solving for P_x gives the same equation as derived in Section 5.1.A. (The second-order condition is $d^2\Pi_x/dP^2_x < 0$, which is satisfied because the second-order derivative is just –4.)

We hope you will regard the calculus method as simple enough and that you will have occasion to use it again in a few places later, for example, in Chapter 12 on collective action. But if you find it too difficult, here is a noncalculus alternative method that works for quadratic functions. Rearrange terms to write the function as

$$Y = A + BX - CX^2$$

$$= A + B^2/(4C) - B^2/(4C) + BX - CX^2$$

$$= A + \frac{B^2}{4C} - C\left(\frac{B^2}{4C^2} - 2\frac{B}{C} + X^2\right)$$

$$= A + \frac{B^2}{4C} - C\left(\frac{B}{2C} - X\right)^2.$$

In the final form of the expression, X appears only in the last term, where a square involving it is being subtracted (remember $C > 0$). The whole expression is maximized when this subtracted term is made as small as possible, which happens when $X = B/(2C)$. Voila!

This method of "completing the square" works for quadratic functions and therefore will suffice for most of our uses. It also avoids calculus. But we must admit it smacks of magic. Calculus is more general and more methodical. It repays a little study many times over.

Combining Sequential and Simultaneous Moves

In Chapter 3 we considered games of purely sequential moves; Chapters 4 and 5 dealt with games of purely simultaneous moves. We developed concepts and techniques of analysis appropriate to the pure game types—trees and rollback equilibrium for sequential moves, payoff tables and Nash equilibrium for simultaneous moves. In reality, however, many strategic situations contain elements of both types of interaction. Also, although we used game trees (extensive forms) as the sole method of illustrating sequential-move games and game tables (strategic forms) as the sole method of illustrating simultaneous-move games, we can use either form for any type of game.

In this chapter, we examine many of these possibilities. We begin by showing how games that combine sequential and simultaneous moves can be solved by combining trees and payoff tables and by combining rollback and Nash equilibrium analysis in appropriate ways. Then we consider the effects of changing the nature of the interaction in a particular game. Specifically, we look at the effects of changing the rules of a game to convert sequential play into simultaneous play and vice versa and of changing the order of moves in sequential play. This topic gives us an opportunity to compare the equilibria found by using the concept of rollback, in a sequential-move game, with those found by using the Nash equilibrium concept, in the simultaneous version of the same game. From this comparison, we extend the concept of Nash equilibria to sequential-play games. It turns out that the rollback equilibrium is a special case, usually called a refinement, of these Nash equilibria.

1 GAMES WITH BOTH SIMULTANEOUS AND SEQUENTIAL MOVES

As mentioned several times thus far, most real games that you will encounter will be made up of numerous smaller components. Each of these components may entail simultaneous play or sequential play, so the full game requires you to be familiar with both. The most obvious examples of strategic interactions containing both sequential and simultaneous parts are those between two (or more) players over an extended period of time. You may play a number of different simultaneous-play games against your roommate, for example, in the course of a week; your play, as well as hers, in previous situations is important in determining how each of you decide to act in the next "round." Also, many sporting events, interactions between competing firms in an industry, and political relationships are sequentially linked series of simultaneous-move games. Such games are analyzed by combining the tools presented in Chapter 3 (trees and rollback) and in Chapters 4 and 5 (payoff tables and Nash equilibria).[1] The only difference is that the actual analysis becomes more complicated as the number of moves and interactions increases.

A. Two-Stage Games and Subgames

Our main illustrative example for such situations includes two would-be telecom giants, CrossTalk and GlobalDialog. Each can choose whether to invest $10 billion in the purchase of a fiber-optic network. They make their investment decisions simultaneously. If neither chooses to make the investment, that is the end of the game. If one invests and the other does not, then the investor has to make a pricing decision for its telecom services. It can choose either a high price, which will attract 60 million customers, from each of whom it will make an operating profit of $400, or a low price, which will attract 80 million customers, from each of whom it will make an operating profit of $200. If both firms acquire fiber-optic networks and enter the market, then their pricing choices become a second simultaneous-move game. Each can choose either the high or the low price. If both choose the high price, they will split the total market equally; so each will get 30 million customers and an operating profit of $400 from each. If both choose the low price, again they will split the total market equally; so each will get 40 million customers and an operating profit of $200 from each. If one chooses the high price and the other the low price, then the low-price

[1]Sometimes the simultaneous part of the game will have equilibria in mixed strategies, when the tools we develop in Chapters 7 and 8 will be required. We mention this possibility in this chapter where relevant and give you an opportunity to use such methods in exercises for the later chapters.

firm will get all the 80 million customers at that price, and the high-price firm will get nothing.

The interaction between CrossTalk and GlobalDialog forms a two-stage game. Of the four combinations of the simultaneous-move choices at the first (investment) stage, one ends the game, two lead to a second-stage (pricing) decision by just one player, and the fourth leads to a simultaneous-move (pricing) game at the second stage. We show this game pictorially in Figure 6.1.

Regarded as a whole, Figure 6.1 illustrates a game tree, but one that is more complex than the trees in Chapter 3. You can think of it as an elaborate "tree house" with multiple levels. The levels are shown in different parts of the same two-dimensional figure, as if you are looking down at the tree from a helicopter positioned directly above it.

The first-stage game is represented by the payoff table in the top-left quadrant of Figure 6.1. You can think of it as the first floor of the tree house. It has four "rooms." The room in the northwest corner corresponds to the "Don't invest" first-stage moves of both firms. If the firms' decisions take the game to this room, there are no further choices to be made, so we can think of it being like a terminal node of a tree in Chapter 3 and show the payoffs in the cell of the table;

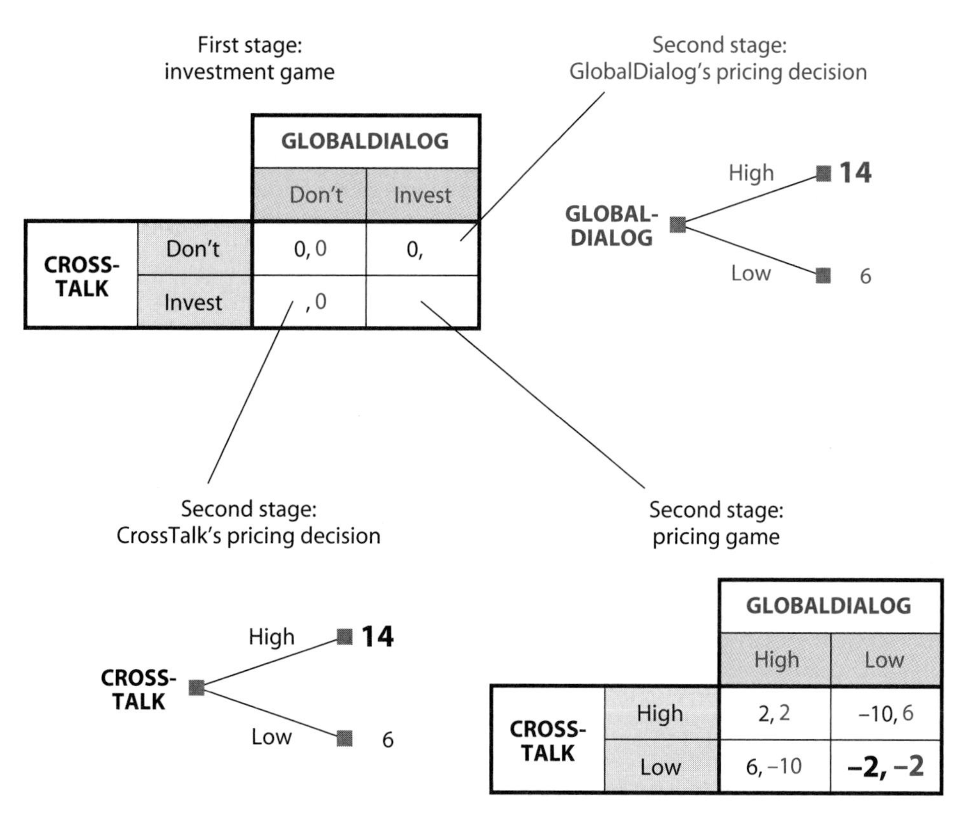

FIGURE 6.1 Two-Stage Game Combining Sequential and Simultaneous Moves

both firms get 0. However, all of the other combinations of actions for the two firms lead to rooms that lead to further choices; so we cannot yet show the payoffs in those cells. Instead, we show branches leading to the second floor. The northeast and southwest rooms show only the payoff to the firm that has not invested; the branches leading from each of these rooms take us to single-firm pricing decisions in the second stage. The southeast room leads to a multiroom second-floor structure within the tree house, which represents the second-stage pricing game that is played if both firms have invested in the first stage. This second-floor structure has four rooms corresponding to the four combinations of the two firms' pricing moves.

All of the second-floor branches and rooms are like terminal nodes of a game tree, so we can show the payoffs in each case. Payoffs here consist of each firm's operating profits minus the previous investment costs; payoff values are written in billions of dollars.

Consider the branch leading to the southwest corner of Figure 6.1. The game arrives in that corner if CrossTalk is the only firm that has invested. Then, if it chooses the high price, its operating profit is $400 $\times$ 60 million = $24 billion; after subtracting the $10 billion investment cost, its payoff is $14 billion, which we write as 14. In the same corner, if CrossTalk chooses the low price, then its operating profit is $200 $\times$ 80 million = $16 billion, yielding the payoff 6 after accounting for its original investment. In this situation, GlobalDialog's payoff is 0, as shown in the southwest room of the first floor of our tree. Similar calculations for the case in which GlobalDialog is the only firm to invest give us the payoffs shown in the northeast corner of Figure 6.1; again, the payoff of 0 for CrossTalk is shown in the northeast room of the first-stage game table.

If both firms invest, both play the second-stage pricing game illustrated in the southeast corner of the figure. When both choose the high price in the second stage, each gets operating profit of $400 $\times$ 30 million (half of the market), or $12 billion; after subtracting the $10 billion investment cost, each is left with a net profit of $2 billion, or a payoff of 2. If both firms choose the low price in the second stage, each gets operating profit of $200 $\times$ 40 million = $8 billion, and, after subtracting the $10 billion investment cost, each is left with a net loss of $2 billion, or a payoff of -2. Finally, if one firm charges the high price and the other firm the low price, then the low-price firm has operating profit of $200 $\times$ 80 million = $16 billion, leading to the payoff 6, while the high-price firm gets no operating profit and simply loses its $10 billion investment, for a payoff of -10.

As with any multistage game in Chapter 3, we must solve this game backward, starting with the second-stage game. In the two single-firm decision problems, we see at once that the high-price policy yields the higher payoff. We highlight this by showing that payoff in a larger-size type.

The second-stage pricing game has to be solved by using methods developed in Chapter 4. It is immediately evident, however, that this game is a prisoners' dilemma. Low is the dominant strategy for each firm; so the outcome is the room in the southeast corner of the second-stage game table; each firm gets payoff -2.[2] Again, we show these payoffs in a larger type size to highlight the fact that they are the payoffs obtained in the second-stage equilibrium.

Rollback now tells us that each first-stage configuration of moves should be evaluated by looking ahead to the equilibrium of the second-stage game (or the optimum second-stage decision) and the resulting payoffs. We can therefore substitute the payoffs that we have just calculated into the previously empty or partly empty rooms on the first floor of our tree house. This substitution gives us a first floor with known payoffs, shown in Figure 6.2.

Now we can use the methods of Chapter 4 to solve this simultaneous-move game. You should immediately recognize the game in Figure 6.2 as a chicken game. It has two Nash equilibria, each of which entails one firm choosing Invest and the other choosing Don't. The firm that invests makes a huge profit; so each firm prefers the equilibrium in which it is the investor while the other firm stays out. In Chapter 4, we briefly discussed the ways in which one of the two equilibria might get selected. We also pointed out the possibility that each firm might try to get its preferred outcome, with the result that both of them invest and both lose money. Indeed, this is what seems to have happened in the real-life play of this game. In Chapter 7, we investigate this type of game further, showing that it has a third Nash equilibrium, in mixed strategies.

Analysis of Figure 6.2 shows that the first-stage game in our example does not have a unique Nash equilibrium. This problem is not too serious, because we can leave the solution ambiguous to the extent that was done in the preceding paragraph. Matters would be worse if the second-stage game did not have a unique equilibrium. Then it would be essential to specify the precise process by which an outcome gets selected so that we could figure out the second-stage payoffs and use them to roll back to the first stage.

The second-stage pricing game shown in the table in the bottom-right quadrant of Figure 6.1 is one part of the complete two-stage game. However, it is also

		GLOBALDIALOG	
		Don't	Invest
CROSSTALK	Don't	0, 0	0, 14
	Invest	14, 0	−2, −2

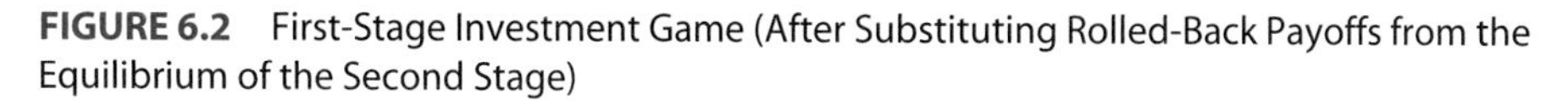

FIGURE 6.2 First-Stage Investment Game (After Substituting Rolled-Back Payoffs from the Equilibrium of the Second Stage)

[2]As is usual in a prisoners' dilemma, if the firms could successfully collude and charge high prices, both could get the higher payoff of 2. But this outcome is not an equilibrium, because each firm is tempted to cheat to try to get the much higher payoff of 6.

a full-fledged game in its own right, with a fully specified structure of players, strategies, and payoffs. To bring out this dual nature more explicitly, it is called a **subgame** of the full game.

More generally, a subgame is the part of a multimove game that begins at a particular node of the original game. The tree for a subgame is then just that part of the tree for the full game that takes this node as its root, or initial, node. A multimove game has as many subgames as it has decision nodes.

B. Configurations of Multistage Games

In the multilevel game illustrated in Figure 6.1, each stage consists of a simultaneous-move game. However, that may not always be the case. Simultaneous and sequential components may be mixed and matched in any way. We give two more examples to clarify this point and to reinforce the ideas introduced in the preceding section.

The first example is a slight variation of the CrossTalk–GlobalDialog game. Suppose one of the firms—say, GlobalDialog—has already made the $10 billion investment in the fiber-optic network. CrossTalk knows of this investment and now has to decide whether to make its own investment. If CrossTalk does not invest, then GlobalDialog will have a simple pricing decision to make. If CrossTalk invests, then the two firms will play the second-stage pricing game already described. The tree for this multistage game has conventional branches at the initial node and has a simultaneous-move subgame starting at one of the nodes to which these initial branches lead. The complete tree is shown in Figure 6.3.

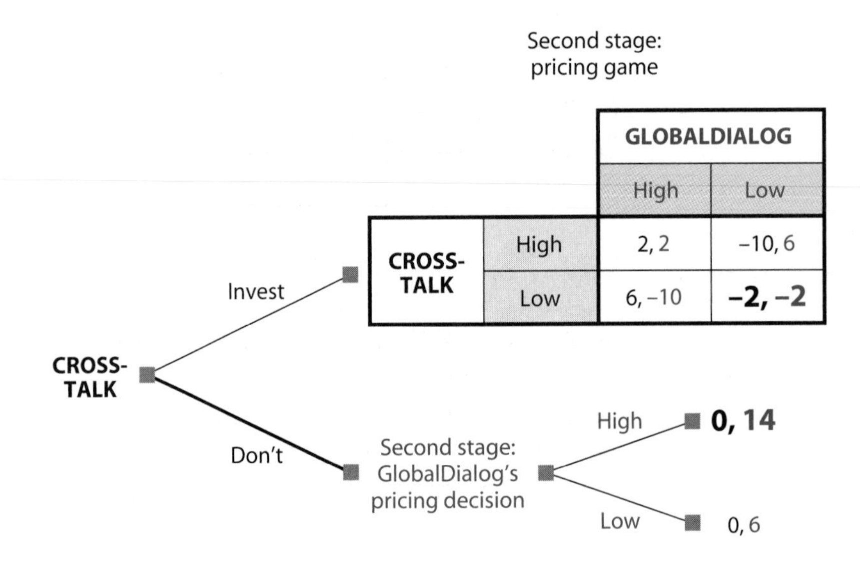

FIGURE 6.3 Two-Stage Game When One Firm Has Already Invested

When the tree has been set up, it is easy to analyze the game. We show the rollback analysis in Figure 6.3 by using large type for the equilibrium payoffs that result from the second-stage game or decision and a thicker branch for CrossTalk's first-stage choice. In words, CrossTalk figures out that, if it invests, the ensuing prisoners' dilemma of pricing will leave it with payoff −2, whereas staying out will get it 0. Thus it prefers the latter. GlobalDialog gets 14 instead of the −2 that it would have gotten if CrossTalk had invested, but CrossTalk's concern is to maximize its own payoff and not to ruin GlobalDialog deliberately.

This analysis does raise the possibility, though, that GlobalDialog may try to get its investment done quickly before CrossTalk makes its decision so as to ensure its most preferred outcome from the full game. And CrossTalk may try to beat GlobalDialog to the punch in the same way. In Chapter 10, we study some methods, called strategic moves, that may enable players to secure such advantages.

Our second example comes from football. Before each play, the coach for the offense chooses the play that his team will run; simultaneously, the coach for the defense sends his team out with instructions on how they should align themselves to counter the offense. Thus these moves are simultaneous. Suppose the offense has just two alternatives, a safe play and a risky play, and the defense may align itself to counter either of them. If the offense has planned to run the risky play and the quarterback sees the defensive alignment that will counter it, he can change the play at the line of scrimmage. And the defense, hearing the change, can respond by changing its own alignment. Thus we have a simultaneous-move game at the first stage, and one of the combination of choices of moves at this stage leads to a sequential-move subgame. Figure 6.4 shows the complete tree.

This is a zero-sum game, and we show only the offense's payoffs, measured in the number of yards that they expect to gain. The safe play gets 2 yards, even if the defense is ready for it; if the defense is not ready for it, the safe play does not do much better, gaining 6 yards. The risky play, if it catches the defense unready to cover it, gains 30 yards. But if the defense is ready for the risky play, the offense loses 10 yards. We show this payoff of −10 for the offense at the terminal node where the offense does not change the play. If the offense changes the play (back to safe), the payoffs are 2 if the defense responds and 6 if it does not; these payoffs are the same as those that arise when the offense plans the safe play from the start.

We show the chosen branches in the sequential subgame as thick lines in Figure 6.4. It is easy to see that, if the offense changes its play, the defense will respond to keep the offense's gain to 2 rather than 6 and that the offense should change the play to get 2 rather than −10. Rolling back, we should put the resulting payoff, 2, in the bottom-right cell of the simultaneous-move game of the first stage. Then we see that this game has no Nash equilibrium in pure strategies.

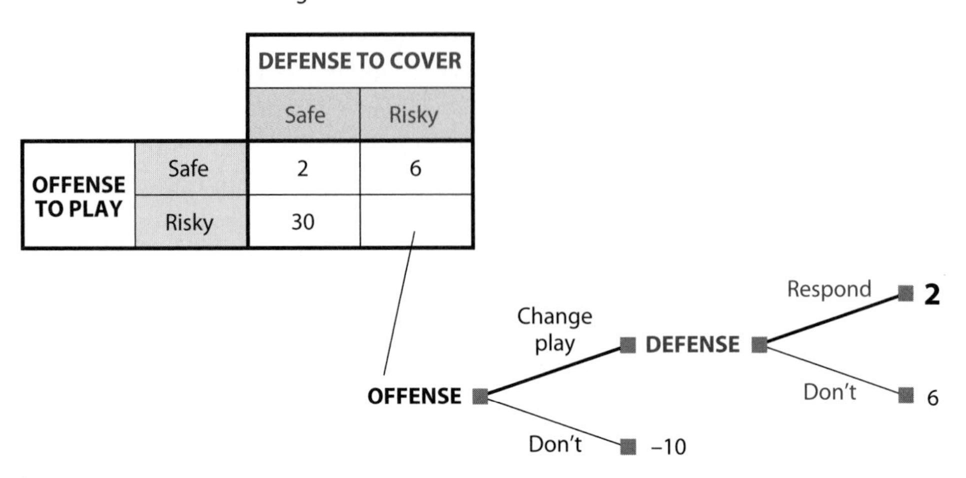

FIGURE 6.4 Simultaneous-Move First Stage Followed by Sequential Moves

The reason is the same as that in the tennis game of Chapter 4, Section 8; one player (defense) wants to match the moves (align to counter the play that the offense is choosing) while the other (offense) wants to unmatch moves (catch the defense in the wrong alignment). In Chapter 7, we show how to calculate the mixed-strategy equilibrium of such a game. It turns out that the offense should choose the risky play with probability 1/8, or 12.5%.

2 CHANGING THE ORDER OF MOVES IN A GAME

The games considered in preceding chapters were presented as either sequential or simultaneous in nature. We used the appropriate tools of analysis to predict equilibria in each type of game. In Section 1 of this chapter, we discussed games with elements of both sequential and simultaneous play. These games required both sets of tools to find solutions. But what about games that could be played either sequentially or simultaneously? How would changing the play of a particular game and thus changing the appropriate tools of analysis alter the expected outcomes?

The task of turning a sequential-play game into a simultaneous one requires changing only the timing or observability with which players make their choices of moves. Sequential-move games become simultaneous if the players cannot observe moves made by their rivals before making their own choices. In that case, we would analyze the game by searching for a Nash equilibrium rather than for a rollback equilibrium. Conversely, a simultaneous-move game could become sequential if one player were able to observe the other's move before choosing her own.

Any changes to the rules of the game can also change its outcomes. Here, we illustrate a variety of possibilities that arise owing to changes in different types of games.

A. Changing Simultaneous-Move Games into Sequential-Move Games

I. NO CHANGE IN OUTCOME Certain games have the same outcomes in the equilibria of both simultaneous and sequential versions and regardless of the order of play in the sequential-play game. This result generally arises only when both or all players have dominant strategies. We show that it holds for the prisoners' dilemma.

Consider the prisoners' dilemma game of Chapter 4, in which a husband and wife are being questioned regarding their roles in a crime. The simultaneous version of that game, reproduced in Figure 6.5a, can be redrawn as either of the sequential-play games shown in Figure 6.5b and c. As in Figure 4.4, the payoff numbers indicate years in jail; so low numbers are better than high ones. In Figure 6.5b, Husband chooses his strategy before Wife does; so she knows what he has chosen before making her own choice; in Figure 6.5c the roles are reversed.

The Nash equilibrium of the prisoners' dilemma game in Figure 6.5a is for each player to confess (or to defect from cooperating with the other). Using rollback to solve the sequential versions of the game, illustrated in Figure 6.5b and c, we see that the second player does best to confess if the first has confessed (10 rather than 25 years in jail) and the second player also does best to confess if the first has denied (1 year rather than 3 years of jail). Given these choices by the second player, the first player does best to confess (10 rather than 25 years in jail). The equilibrium entails 10 years of jail for both players regardless of which player moves first. Thus, the equilibrium is the same in all three versions of this game.

II. FIRST-MOVER ADVANTAGE A first-mover advantage may emerge when the rules of a game are changed from simultaneous to sequential play. At a minimum, if the simultaneous-move version has multiple equilibria, the sequential-move version enables the first mover to choose his preferred outcome. We illustrate such a situation with the use of chicken, the game in which two teenagers drive toward each other in their cars, both determined not to swerve. We reproduce the strategic form of Figure 4.14 in Figure 6.6a and two extensive forms, one for each possible ordering of play, in Figure 6.6b and c.

Under simultaneous play, the two outcomes in which one player swerves (is "chicken") and the others goes straight (is "tough") are both pure-strategy Nash

(a) Simultaneous play

		WIFE	
		Confess (Defect)	Deny (Cooperate)
HUSBAND	Confess (Defect)	10 yr, 10 yr	1 yr, 25 yr
	Deny (Cooperate)	25 yr, 1 yr	3 yr, 3 yr

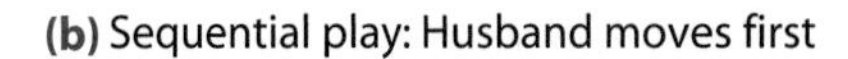

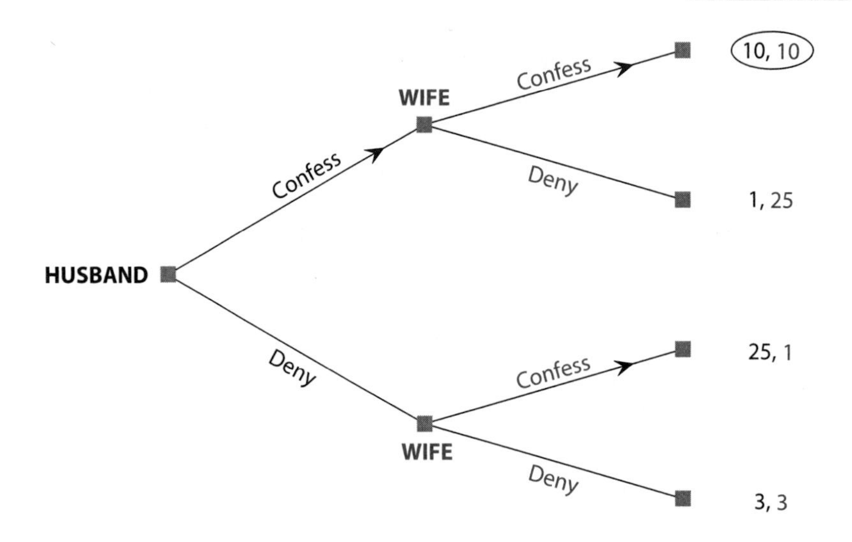

(c) Sequential moves: Wife moves first

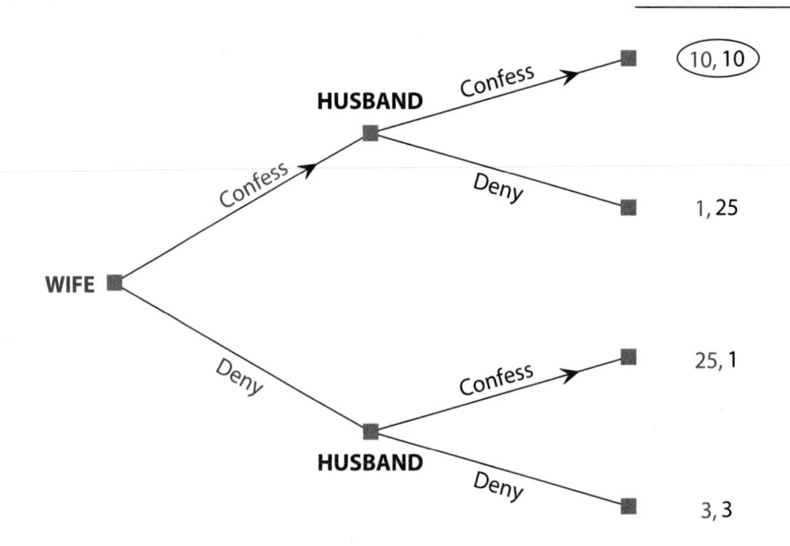

FIGURE 6.5 Three Versions of the Prisoners' Dilemma Game

(a) Simultaneous play

		DEAN	
		Swerve (Chicken)	Straight (Tough)
JAMES	Swerve (Chicken)	0, 0	−1, 1
	Straight (Tough)	1, −1	−2, −2

(b) Sequential play: James moves first

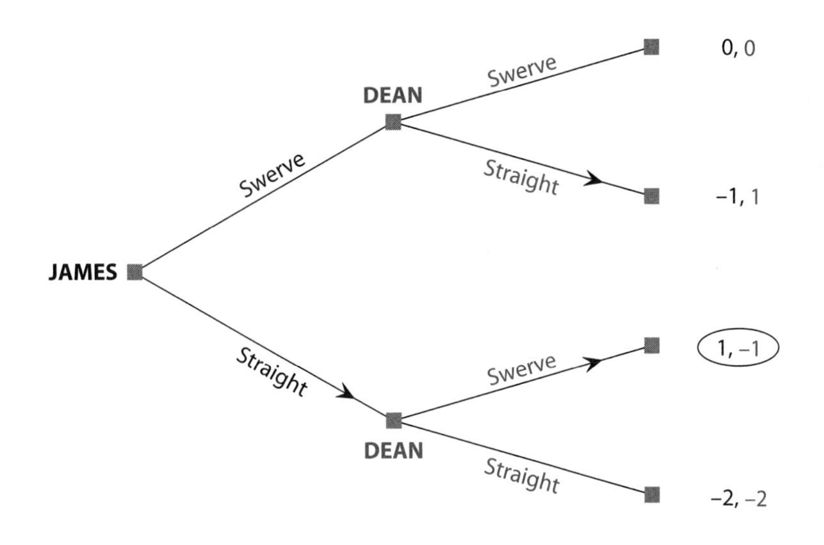

(c) Sequential play: Dean moves first

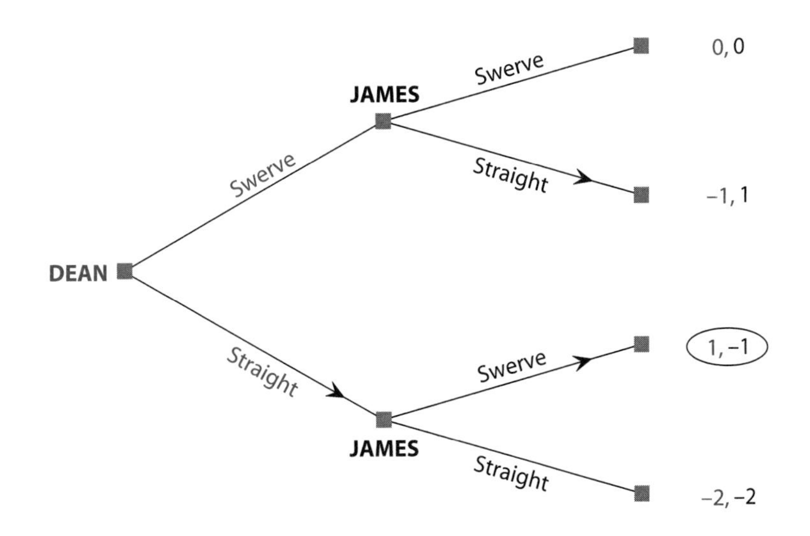

FIGURE 6.6 Chicken in Simultaneous- and Sequential-Play Versions

equilibria. Without specification of some historical, cultural, or other convention, neither has a claim to be a focal point. Our analysis in Chapter 4 suggested that coordinated play could help the players in this game, perhaps through an agreement to alternate between the two equilibria.

When we alter the rules of the game to allow one of the players the opportunity to move first, there are no longer two equilibria. Rather, we see that the second mover's equilibrium strategy is to choose the action opposite that chosen by the first mover. Rollback then shows that the first mover's equilibrium strategy is Straight. We see in Figure 6.6b and c that allowing one person to move first and to be observed making the move results in a single rollback equilibrium in which the first mover gets a payoff of 1, while the second mover gets a payoff of −1. The actual play of the game becomes almost irrelevant under such rules, which may make the sequential version uninteresting to many observers. Although teenagers might not want to play such a game with the rule change, the strategic consequences of the change are significant.

III. SECOND-MOVER ADVANTAGE In other games, a second-mover advantage may emerge when simultaneous play is changed into sequential play. This can be illustrated using the tennis game of Chapter 4. Recall that, in that game, Evert is planning the location of her return while Navratilova considers where to cover. The version considered earlier assumed that both players were skilled at disguising their intended moves until the very last moment so that they moved at essentially the same time. If Evert's movement as she goes to hit the ball belies her shot intentions, however, then Navratilova can react and move second in the game. In the same way, if Navratilova leans toward the side that she intends to cover before Evert actually hits her return, then Evert is the second mover. Figure 6.7 shows all three possibilities. The simultaneous-move version is Figure 4.15 reproduced as Figure 6.7a; the two orderings of the sequential-play game are Figure 6.7b and c.

The simultaneous-play version of this game has no equilibrium in pure strategies. In each ordering of the sequential version, however, there is a unique rollback equilibrium outcome; the equilibrium differs, depending on who moves first. If Evert moves first, then Navratilova chooses to cover whichever direction Evert chooses and Evert opts for a down-the-line shot. Each player is expected to win the point half the time in this equilibrium. If the order is reversed, Evert chooses to send her shot in the opposite direction from that which Navratilova covers; so Navratilova should move to cover crosscourt. In this case, Evert is expected to win the point 80% of the time. The second mover does better by being able to respond optimally to the opponent's move.

We return to the simultaneous version of this game in Chapter 7. There we show that it does have a Nash equilibrium in mixed strategies. In that equilibrium, Evert succeeds on average 62% of the time. Her success rate in the

(a) Simultaneous play

		NAVRATILOVA	
		DL	CC
EVERT	DL	50	80
	CC	90	20

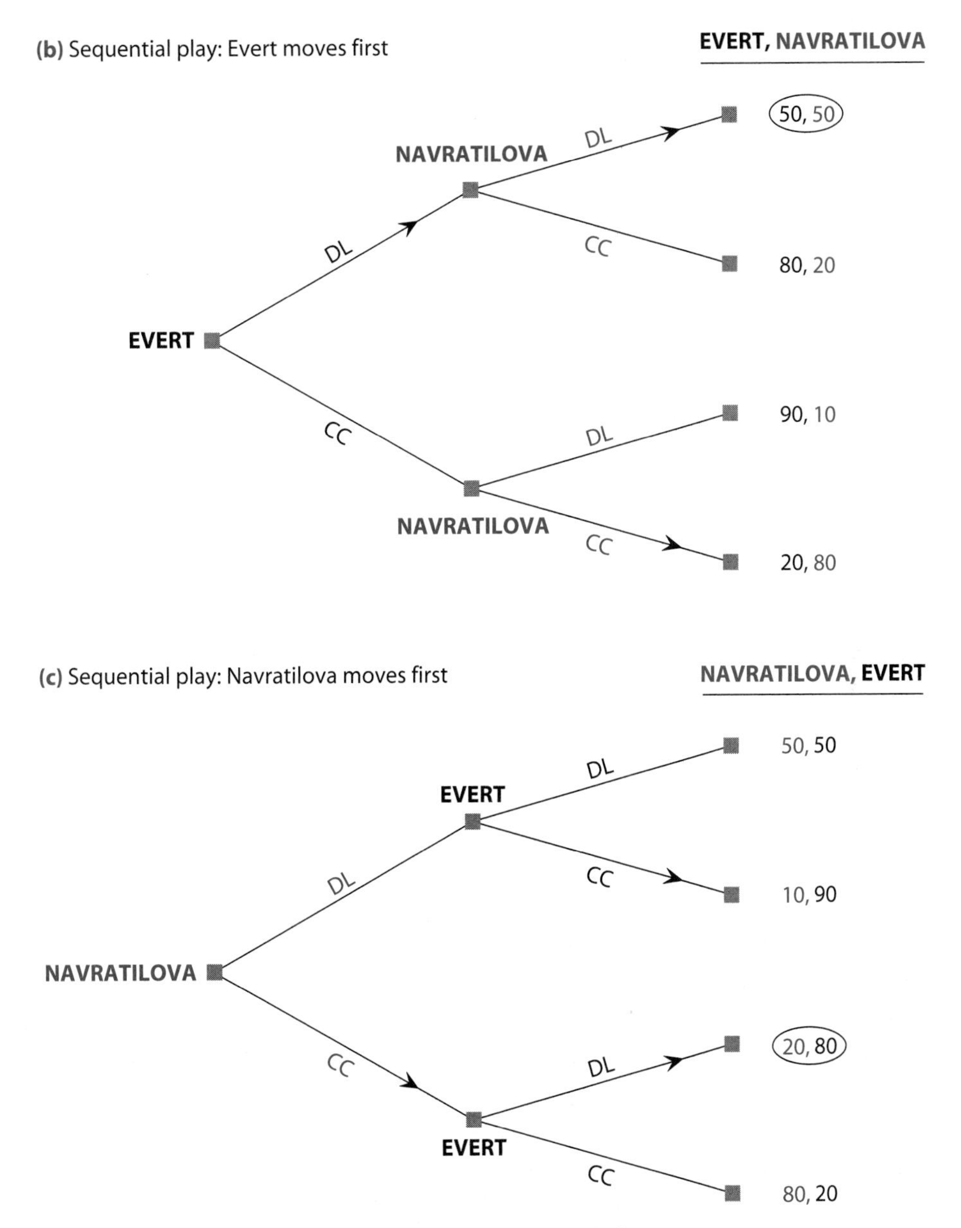

FIGURE 6.7 Tennis Game in Simultaneous- and Sequential-Play Versions

mixed-strategy equilibrium of the simultaneous game is thus better than the 50% that she gets by moving first but is worse than the 80% that she gets by moving second, in the two sequential-move versions.

IV. BOTH PLAYERS MAY DO BETTER That a game may have a first-mover or a second-mover advantage, which is suppressed when moves have to be simultaneous but emerges when an order of moves is imposed, is quite intuitive. Somewhat more surprising is the possibility that both players may do better under one set of rules of play than under another. We illustrate this possibility by using the game of monetary and fiscal policies played by the Federal Reserve and the Congress. In Chapter 4, we studied this game with simultaneous moves; we reproduce the payoff table (Figure 4.5) as Figure 6.8a and show the two sequential-move versions as Figure 6.8b and c. For brevity, we write the strategies as Balance and Deficit instead of Budget Balance and Budget Deficit for the Congress and as High and Low instead of High Interest Rates and Low Interest Rates for the Fed.

In the simultaneous-move version, the Congress has a dominant strategy (Deficit), and the Fed, knowing this, chooses High, yielding payoffs of 2 to both players. Almost the same thing happens in the sequential version where the Fed moves first. The Fed foresees that, for each choice it might make, the Congress will respond with Deficit. Then High is the better choice for Fed, yielding 2 instead of 1.

But the sequential-move version where the Congress moves first is different. Now the Congress foresees that, if it chooses Deficit, the Fed will respond with High, whereas, if it chooses Balance, the Fed will respond with Low. Of these two developments, the Congress prefers the latter, where it gets payoff 3 instead of 2. Therefore the rollback equilibrium with this order of moves is for the Congress to choose a balanced budget and the Fed to respond with low interest rates. The resulting payoffs, 3 for the Congress and 4 for the Fed, are better for both players than those of the other two versions.

The difference between the two outcomes is even more surprising because the better outcome obtained in Figure 6.8c results from the Congress choosing Balance, which is its dominated strategy in Figure 6.8a. To resolve the apparent paradox, one must understand more precisely the meaning of dominance. For Deficit to be a dominant strategy, it must be better than Balance from the Congress's perspective for *each given* choice of the Fed. This type of comparison between Deficit and Balance is relevant in the simultaneous-move game because there the Congress must make a decision without knowing the Fed's choice. The Congress must think through, or formulate a belief about, the Fed's action, and choose its best response to that. In our example, this best response is always Deficit for the Congress. The concept of dominance is also relevant with sequential moves if the Congress moves second, because then it knows what the

(a) Simultaneous moves

		FEDERAL RESERVE	
		Low interest rates	High interest rates
CONGRESS	Budget balance	3, 4	1, 3
	Budget deficit	4, 1	2, 2

(b) Sequential moves: Fed moves first

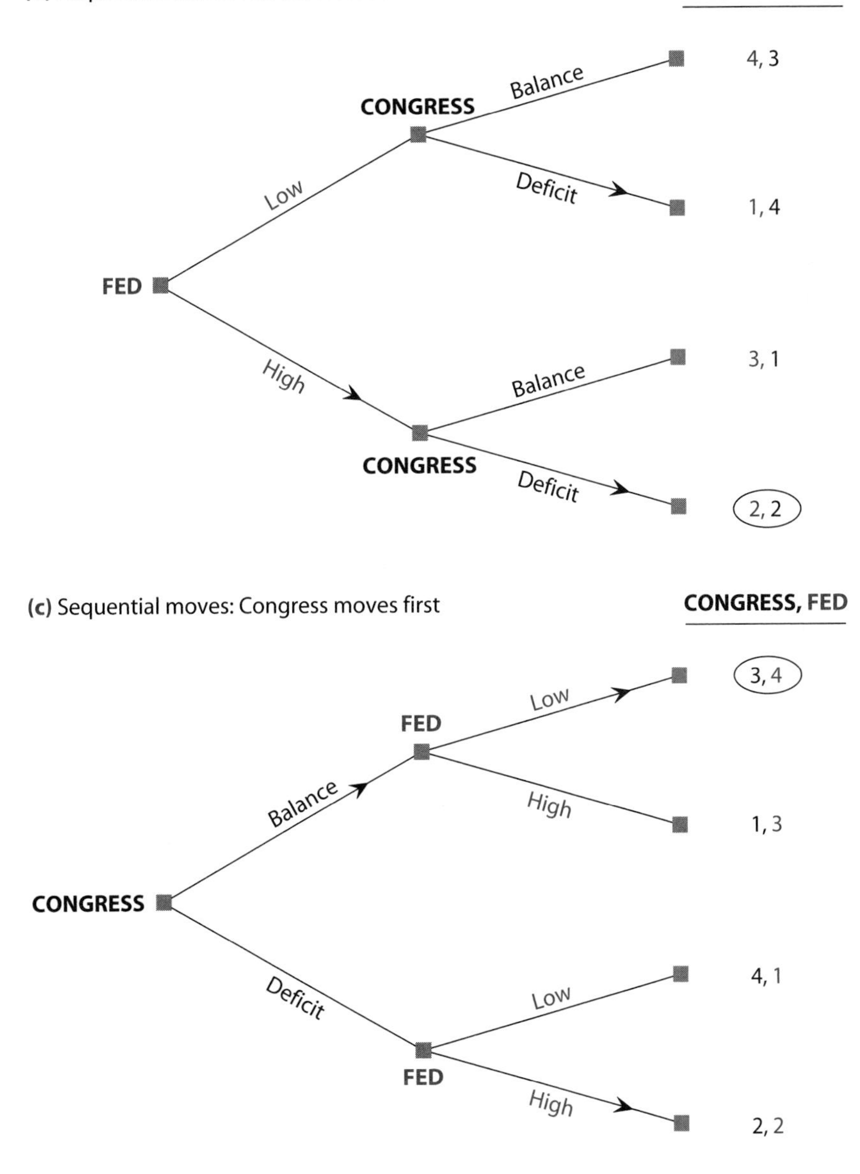

FIGURE 6.8 Three Versions of the Monetary–Fiscal Policy Game

Fed has already done and merely picks its best response, which is always Deficit. However, if the Congress moves first, it cannot take the Fed's choice as *given.* Instead, it must recognize how the Fed's second move will be affected by its own first move. Here it knows that the Fed will respond to Deficit with High and to Balance with Low. The Congress is then left to choose between these two alternatives; its most preferred outcome of Deficit and Low becomes irrelevant because it is precluded by the Fed's response.

The idea that dominance may cease to be a relevant concept for the first mover reemerges in Chapter 10. There we consider the possibility that one player or the other may deliberately change the rules of a game to become the first mover. Players can alter the outcome of the game in their favor in this way.

Suppose that the two players in our current example could choose the order of moves in the game. In this case, they would agree that the Congress should move first. Indeed, when budget deficits and inflation threaten, the chairs of the Federal Reserve in testimony before various congressional committees often offer such deals; they promise to respond to fiscal discipline by lowering interest rates. But it is often not enough to make a verbal deal with the other player. The technical requirements of a first move—namely, that it be observable to the second mover and not reversible thereafter—must be satisfied. In the context of macroeconomic policies, it is fortunate that the legislative process of fiscal policy in the United States is both very visible and very slow, whereas monetary policy can be changed quite quickly in a meeting of the Federal Reserve Board. Therefore the sequential play where the Congress moves first and the Fed moves second is quite realistic.

B. Other Changes in the Order of Moves

The preceding section presented various examples in which the rules of the game were changed from simultaneous play to sequential play. We saw how and why such rule changes can change the outcome of a game. The same examples also serve to show what happens if the rules are changed in the opposite direction, from sequential to simultaneous moves. Thus, if a first- or a second-mover advantage exists with sequential play, it can be lost under simultaneous play. And if a specific order benefits both players, then losing the order can hurt both.

The same examples also show us what happens if the rules are changed to reverse the order of play while keeping the sequential nature of a game unchanged. If there is a first-mover or a second-mover advantage, then the player who shifts from moving first to moving second may benefit or lose accordingly, with the opposite change for the other player. And if one order is in the common interests of both, then an externally imposed change of order can benefit or hurt them both.

3 CHANGE IN THE METHOD OF ANALYSIS

Game trees are the natural way to display sequential-move games, and payoff tables the natural representation of simultaneous-move games. However, each technique can be adapted to the other type of game. Here we show how to translate the information contained in one illustration to an illustration of the other type. In the process, we develop some new ideas that will prove useful in subsequent analysis of games.

A. Illustrating Simultaneous-Move Games by Using Trees

Consider the game of the passing shot in tennis as originally described in Chapter 4, where the action is so quick that moves are truly simultaneous, as shown in Figure 6.7a. But suppose we want to show the game in extensive form—that is, by using a tree. We show how this can be done in Figure 6.9.

To draw the tree in the figure, we must choose one player—say, Evert—to make her choice at the initial node of the tree. The branches for her two choices, DL and CC, then end in two nodes, at each of which Navratilova makes her choices. However, because the moves are actually simultaneous, Navratilova must choose without knowing what Evert has picked. That is, she must choose without knowing whether she is at the node following Evert's DL branch or the one following Evert's CC branch. Our tree diagram must in some way show this lack of information on Navratilova's part.

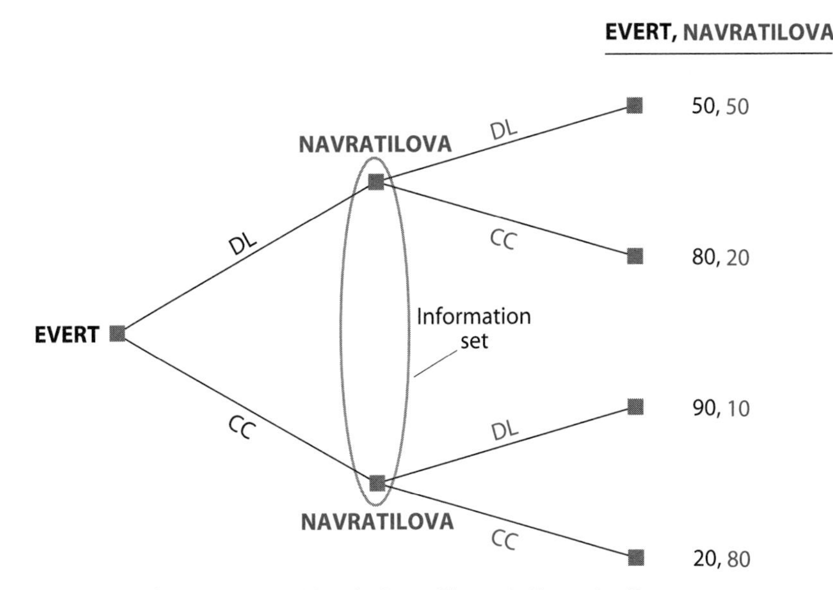

FIGURE 6.9 Simultaneous-Move Tennis Game Shown in Extensive Form

We illustrate Navratilova's strategic uncertainty about the node from which her decision is being made by drawing an oval to surround the two relevant nodes. (An alternative is to connect them by a dotted line; a dotted line is used to distinguish it from the solid lines that represent the branches of the tree.) The nodes within this oval or balloon are called an **information set** for the player who moves there. Such a set indicates the presence of imperfect information for the player; she cannot distinguish between the nodes in the set, given her available information (because she cannot observe the row player's move before making her own). As such, her strategy choice from within a single information set must specify the same move at all the nodes contained in it. That is, Navratilova must choose either DL at both the nodes in this information set or CC at both of them. She cannot choose DL at one and CC at the other, as she could in Figure 6.7b, where the game had sequential moves and she moved second.

Accordingly, we must adapt our definition of strategy. In Chapter 3, we defined a strategy as a complete plan of action, specifying the move that a player would make at each *node* where the rules of the game specified that it was her turn to move. We should now more accurately redefine a strategy as a complete plan of action, specifying the move that a player would make at each *information set* at whose nodes the rules of the game specify that it is her turn to move.

The concept of an information set is also relevant when a player faces external uncertainty about some conditions that affect his decision, rather than about another player's moves. For example, a farmer planting a crop is uncertain about the weather during the growing season, although he knows the probabilities of various alternative possibilities from past experience or meteorological forecasts. We can regard the weather as a random choice of an outside player, Nature, who has no payoffs but merely chooses according to known probabilities.[3] We can then enclose the various nodes corresponding to Nature's moves into an information set for the farmer, constraining the farmer's choice to be the same at all of these nodes. Figure 6.10 illustrates this situation.

Using the concept of an information set, we can formalize the concepts of perfect and imperfect information in a game, which we introduced in Chapter 2 (Section 2.D). A game has perfect information if it has neither strategic nor external uncertainty, which will happen if it has no information sets enclosing two or more nodes. Thus a game has perfect information if all of its information sets consist of singleton nodes.

[3]Some people believe that Nature is actually a malevolent player who plays a zero-sum game with us, so its payoffs are higher when ours are lower. For example, it is more likely to rain if we have forgotten to bring an umbrella. We understand such thinking, but it does not have real statistical support.

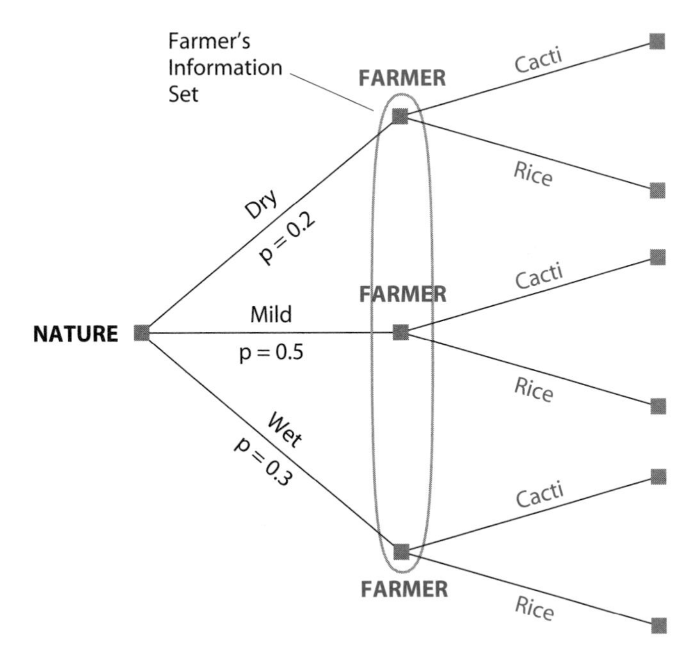

FIGURE 6.10 Nature and Information Sets

Although this representation is conceptually simple, it does not provide any simpler way of solving the game. Therefore we use it only occasionally, where it conveys some point more simply. Some examples of game illustrations using information sets can be found later in Chapters 9 and 15.

B. Showing and Analyzing Sequential-Move Games in Strategic Form

Consider now the sequential-move game of monetary and fiscal policy from Figure 6.8c, in which the Congress has the first move. Suppose we want to show it in normal or strategic form—that is, by using a payoff table. The rows and the columns of the table are the strategies of the two players. We must therefore begin by specifying the strategies.

For the Congress, the first mover, listing its strategies is easy. There are just two moves—Balance and Deficit—and they are also the two strategies. For the second mover, matters are more complex. Remember that a strategy is a complete plan of action, specifying the moves to be made at each node where it is a player's turn to move. Because the Fed gets to move at two nodes (and because we are supposing that this game actually has sequential moves and so the two nodes are not confounded into one information set) and can choose either Low or High at each node, there are four combinations of its choice patterns. These combinations are (1) Low if Balance, High if Deficit (we write this as "L if B, H if

		FED			
		L if B, H if D	H if B, L if D	Low always	High always
CONGRESS	Balance	3, 4	1, 3	3, 4	1, 3
	Deficit	2, 2	4, 1	4, 1	2, 2

FIGURE 6.11 Sequential-Move Game of Monetary and Fiscal Policy in Strategic Form

D" for short); (2) High if Balance, Low if Deficit ("H if B, L if D" for short); (3) Low always; and (4) High always.

We show the resulting two-by-four payoff matrix in Figure 6.11. The last two columns are no different from those for the two-by-two payoff matrix for the game under simultaneous-move rules (Figure 6.8a). This is because, if the Fed is choosing a strategy in which it makes the same move always, it is just as if the Fed were moving without taking into account what the Congress had done; it is as if their moves were simultaneous. But calculation of the payoffs for the first two columns, where the Fed's second move does depend on the Congress's first move, needs some care.

To illustrate, consider the cell in the first row and the second column. Here the Congress is choosing Balance, and the Fed is choosing "H if B, L if D." Given Congress's choice, the Fed's actual choice under this strategy is High. Then the payoffs are those for the Balance and High combination—namely, 1 for Congress and 3 for the Fed.

Cell-by-cell inspection quickly shows that the game has two pure-strategy Nash equilibria, which we show by shading the cells gray. One is in the top-left cell, where the Congress's strategy is Balance and the Fed's is "L if B, H if D," and so the Fed's actual choice is L. This outcome is just the rollback equilibrium of the sequential-move game. But there is another Nash equilibrium in the bottom-right cell, where the Congress chooses Deficit and the Fed chooses "High always." As always in a Nash equilibrium, neither player has a clear reason to deviate from the strategies that lead to this outcome. The Congress would do worse by switching to Balance, and the Fed could do no better by switching to any of its other three strategies, although it could do just as well with "L if B, H if D."

The sequential-move game, when analyzed in its extensive form, produced just one rollback equilibrium. But when analyzed in its normal or strategic form, it has two Nash equilibria. What is going on?

The answer lies in the different nature of the logic of Nash and rollback analyses. Nash equilibrium requires that neither player have a reason to deviate, given the strategy of the other player. However, rollback does not take the strategies of later movers as given. Instead, it asks what would be optimal to do if the opportunity to move actually arises.

In our example, the Fed's strategy of "High always" does not satisfy the criterion of being optimal if the opportunity to move actually arises. If the Congress chose Deficit, then High is indeed the Fed's optimal response. However, if the Congress chose Balance and the Fed had to respond, it would want to choose Low, not High. So "High always" does not describe the Fed's optimal response in all possible configurations of play and cannot be a rollback equilibrium. But the logic of Nash equilibrium does not impose such a test, instead regarding the Fed's "High always" as a strategy that the Congress could legitimately take as given. If it does so, then Deficit is the Congress's best response. And, conversely, "High always" is one best response of the Fed to the Congress's Deficit (although it is tied with "L if B, H if D"). Thus the pair of strategies "Deficit" and "High always" are mutual best responses and constitute a Nash equilibrium, although they do not constitute a rollback equilibrium.

We can therefore think of rollback as a further test, supplementing the requirements of a Nash equilibrium and helping to select from among multiple Nash equilibria of the strategic form. In other words, it is a refinement of the Nash equilibrium concept.

To state this idea somewhat more precisely, recall the concept of a subgame. At any one node of the full game tree, we can think of the part of the game that begins there as a subgame. In fact, as successive players make their choices, the play of the game moves along a succession of nodes, and each move can be thought of as starting a subgame. The equilibrium derived by using rollback corresponds to one particular succession of choices in each subgame and gives rise to one particular path of play. Certainly, other paths of play are consistent with the rules of the game. We call these other paths **off-equilibrium paths,** and we call any subgames that arise along these paths **off-equilibrium subgames,** for short.

With this terminology, we can now say that the equilibrium path of play is itself determined by the players' expectations of what would happen if they chose a different action—if they moved the game to an off-equilibrium path and started an off-equilibrium subgame. Rollback requires that all players make their best choices in *every* subgame of the larger game, whether or not the subgame lies along the path to the ultimate equilibrium outcome.

Strategies are complete plans of action. Thus a player's strategy must specify what she will do in each eventuality, or each and every node of the game, whether on or off the equilibrium path, where it is her turn to act. When one such node arrives, only the plan of action starting there—namely, the part of the full strategy that pertains to the subgame starting at that node—is pertinent. This part is called the **continuation** of the strategy for that subgame. Rollback requires that the equilibrium strategy be such that its continuation in every subgame is optimal for the player whose turn it is to act at that node, whether or not the node and the subgame lie on the equilibrium path of play.

Return to the monetary policy game with the Congress moving first, and consider the second Nash equilibrium that arises in its strategic form. Here the path of play is for the Congress to choose Deficit and the Fed to choose High. On the equilibrium path, High is indeed the Fed's best response to Deficit. The Congress's choice of Balance would be the start of an off-equilibrium path. It leads to a node where a rather trivial subgame starts—namely, a decision by the Fed. The Fed's purported equilibrium strategy "High always" asks it to choose High in this subgame. But that is not optimal; this second equilibrium is specifying a nonoptimal choice for an off-equilibrium subgame.

In contrast, the equilibrium path of play for the Nash equilibrium in the upper-left corner of Figure 6.11 is for the Congress to choose Balance and the Fed to follow with Low. The Fed is responding optimally on the equilibrium path. The off-equilibrium path would have the Congress choosing Deficit, and the Fed, given its strategy of "L if B, H if D," would follow with High. It is optimal for the Fed to respond to Deficit with High, so the strategy remains optimal off the equilibrium path, too.

The requirement that continuation of a strategy remain optimal under all circumstances is important because the equilibrium path itself is the result of players' thinking strategically about what would happen if they did something different. A later player may try to achieve an outcome that she would prefer by threatening the first mover that certain actions would be met with dire responses or by promising that certain other actions would be met with nice responses. But the first mover will be skeptical of the **credibility** of such threats and promises. The only way to remove that doubt is to check if the stated responses would actually be optimal if the need arose. If the responses are not optimal, then the threats or promises are not credible, and the responses would not be observed along the equilibrium path of play.

The equilibrium found by using rollback is called a **subgame-perfect equilibrium (SPE).** It is a set of strategies (complete plans of action), one for each player, such that, at every node of the game tree, whether or not the node lies along the equilibrium path of play, the continuation of the same strategy in the subgame starting at that node is optimal for the player who takes the action there. More simply, an SPE requires players to use strategies that constitute a Nash equilibrium in every subgame of the larger game.

In fact, as a rule, in games with finite trees and perfect information, where players can observe every previous action taken by all other players so that there are no multiple nodes enclosed in one information set, rollback finds the unique (except for trivial and exceptional cases of ties) subgame-perfect equilibrium of the game. Consider: if you look at any subgame that begins at the last decision node for the last player who moves, the best choice for that player is the one that gives her the highest payoff. But that is precisely the action chosen with the use of rollback. As players move backward through the game tree, rollback eliminates all unreasonable strategies, including incredible threats or promises, so that the collection of actions ultimately selected is the SPE. Therefore, for the purposes of this book, subgame perfectness is just a fancy name for rollback. At more advanced levels of game theory, where games include complex information structures and information sets, subgame perfectness becomes a richer notion.

4 THREE-PLAYER GAMES

We have restricted the discussion so far in this chapter to games with two players and two moves each. But the same methods also work for some larger and more general examples. We now illustrate this by using the street-garden game of Chapter 3. Specifically, we (1) change the rules of the game from sequential to simultaneous moves and then (2) keep the moves sequential but show

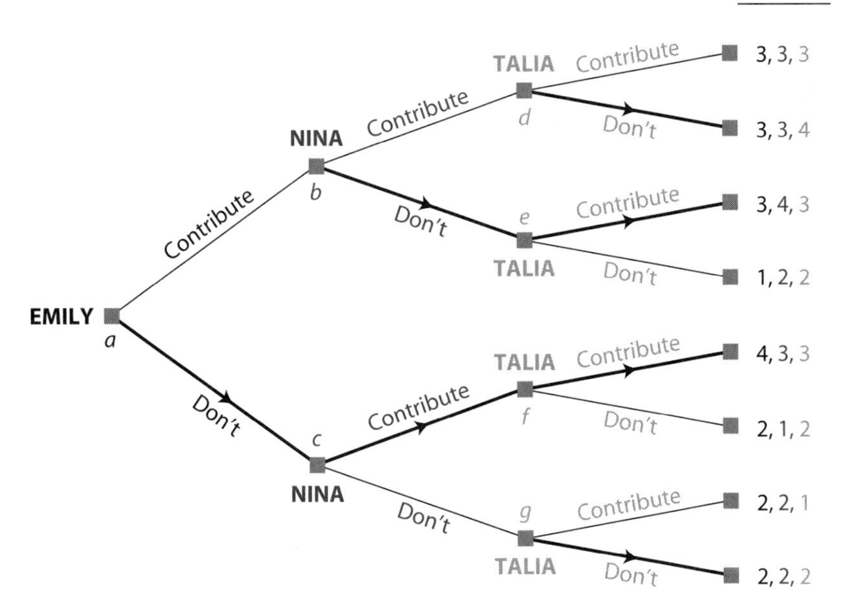

FIGURE 6.12 The Street-Garden Game with Sequential Moves

and analyze the game in its strategic form. First we reproduce the tree of that sequential-move game (Figure 3.6) as Figure 6.12 here and remind you of the rollback equilibrium.

The equilibrium strategy of the first mover (Emily) is simply a move, "Don't contribute." The second mover chooses from among four possible strategies (choice of two responses at each of two nodes) and chooses the strategy "Don't contribute (D) if Emily has chosen her Contribute, and Contribute (C) if Emily has chosen her Don't contribute," or, more simply, "D if C, C if D," or even more simply "DC." Talia has 16 available strategies (choice of two responses at each of four nodes), and her equilibrium strategy is "D following Emily's C and Nina's C, C following their CD, C following their DC, and D following their DD," or "DCCD" for short.

Remember, too, the reason for these choices. The first mover has the opportunity to choose Don't, knowing that the other two will recognize that the nice garden won't be forthcoming unless they contribute and that they like the nice garden sufficiently strongly that they will contribute.

Now we change the rules of the game to make it a simultaneous-move game. (In Chapter 4, we solved a simultaneous-move version with somewhat different payoffs; here we keep the payoffs the same as in Chapter 3.) The payoff matrix is in Figure 6.13. Cell-by-cell inspection shows very easily that there are four Nash equilibria.

In three of the Nash equilibria of the simultaneous-move game, two players contribute, while the third does not. These equilibria are similar to the rollback equilibrium of the sequential-move game. In fact, each one corresponds to the rollback equilibrium of the sequential game with a particular order of play. Further, any given order of play in the sequential-move version of this game leads to the same simultaneous-move payoff table.

But there is also a fourth Nash equilibrium here, where no one contributes. *Given* the specified strategies of the other two—namely, Don't contribute—any one player is powerless to bring about the nice garden and therefore chooses

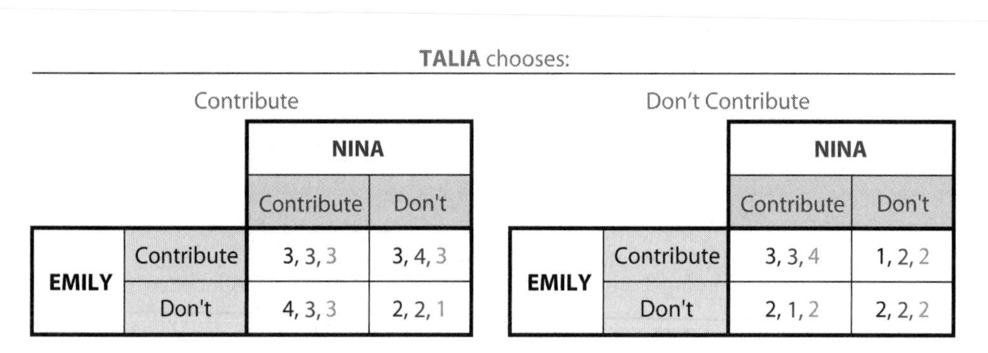

TALIA chooses:

Contribute

		NINA	
		Contribute	Don't
EMILY	Contribute	3, 3, 3	3, 4, 3
	Don't	4, 3, 3	2, 2, 1

Don't Contribute

		NINA	
		Contribute	Don't
EMILY	Contribute	3, 3, 4	1, 2, 2
	Don't	2, 1, 2	2, 2, 2

FIGURE 6.13 The Street-Garden Game with Simultaneous Moves

not to contribute as well. Thus, in the change from sequential to simultaneous moves, the first-mover advantage has been lost. Multiple equilibria arise, only one of which retains the original first mover's high payoff.

Next we return to the sequential-move version—Emily first, Nina second, and Talia third—but show the game in its normal or strategic form. In the sequential-move game, Emily has 2 pure strategies, Nina has 4, and Talia has 16; so this means constructing a payoff table that is 2 by 4 by 16. With the use of the same conventions as we used for three-player tables in Chapter 4, this particular game would require a table with 16 "pages" of two-by-four payoff tables. That would look too messy; so we opt instead for a reshuffling of the players. Let Talia be the row player, Nina be the column player, and Emily be the page

	EMILY							
	Contribute				Don't			
	NINA				**NINA**			
TALIA	CC	CD	DC	DD	CC	CD	DC	DD
CCCC	3, 3, 3	3, 3, 3	3, 4, 3	3, 4, 3	3, 3, 4	1, 2, 2	3, 3, 4	1, 2, 2
CCCD	3, 3, 3	3, 3, 3	3, 4, 3	3, 4, 3	3, 3, 4	2, 2, 2	3, 3, 4	2, 2, 2
CCDC	3, 3, 3	3, 3, 3	3, 4, 3	3, 4, 3	2, 1, 2	1, 2, 2	2, 1, 2	1, 2, 2
CDCC	3, 3, 3	3, 3, 3	2, 2, 1	2, 2, 1	3, 3, 4	1, 2, 2	3, 3, 4	1, 2, 2
DCCC	4, 3, 3	4, 3, 3	3, 4, 3	3, 4, 3	3, 3, 4	1, 2, 2	3, 3, 4	1, 2, 2
CCDD	3, 3, 3	3, 3, 3	3, 4, 3	3, 4, 3	2, 1, 2	2, 2, 2	2, 1, 2	2, 2, 2
CDDC	3, 3, 3	3, 3, 3	2, 2, 1	2, 2, 1	2, 1, 2	1, 2, 2	2, 1, 2	1, 2, 2
DDCC	4, 3, 3	4, 3, 3	2, 2, 1	2, 2, 1	3, 3, 4	1, 2, 2	3, 3, 4	1, 2, 2
CDCD	3, 3, 3	3, 3, 3	2, 2, 1	2, 2, 1	3, 3, 4	2, 2, 2	3, 3, 4	2, 2, 2
DCDC	4, 3, 3	4, 3, 3	3, 4, 3	3, 4, 3	2, 1, 2	1, 2, 2	2, 1, 2	1, 2, 2
DCCD	4, 3, 3	4, 3, 3	3, 4, 3	3, 4, 3	3, 3, 4	2, 2, 2	3, 3, 4	2, 2, 2
CDDD	3, 3, 3	3, 3, 3	2, 2, 1	2, 2, 1	2, 1, 2	2, 2, 2	2, 1, 2	2, 2, 2
DCDD	4, 3, 3	4, 3, 3	3, 4, 3	3, 4, 3	2, 1, 2	2, 2, 2	2, 1, 2	2, 2, 2
DDCD	4, 3, 3	4, 3, 3	2, 2, 1	2, 2, 1	3, 3, 4	2, 2, 2	3, 3, 4	2, 2, 2
DDDC	4, 3, 3	4, 3, 3	2, 2, 1	2, 2, 1	2, 1, 2	1, 2, 2	2, 1, 2	1, 2, 2
DDDD	4, 3, 3	4, 3, 3	2, 2, 1	2, 2, 1	2, 1, 2	2, 2, 2	2, 1, 2	2, 2, 2

FIGURE 6.14 Street-Garden Game in Strategic Form

player. Then "all" that is required to illustrate this game is the 16 by 4 by 2 game table shown in Figure 6.14. The order of payoffs still corresponds to our earlier convention in that they are listed row, column, page player; in our example, that means the payoffs are now listed in the order Talia, Nina, and Emily.

As in the monetary–fiscal policy game between the Fed and the Congress, there are multiple Nash equilibria in the simultaneous street-garden game. (In Exercise S8, we ask you to find them all.) But there is only one subgame-perfect equilibrium, corresponding to the rollback equilibrium found in Figure 6.13. Although cell-by-cell inspection finds all of the Nash equilibria, iterated elimination of dominated strategies can reduce the number of reasonable equilibria for us here. This process works because elimination identifies those strategies that include noncredible components (such as "High always" for the Fed in Section 3.B). As it turns out, such elimination can take us all the way to the unique subgame-perfect equilibrium.

In Figure 6.14, we start with Talia and eliminate all of her (weakly) dominated strategies. This step eliminates all but the strategy listed in the eleventh row of the table, DCCD, which we have already identified as Talia's rollback equilibrium strategy. Elimination can continue with Nina, for whom we must compare outcomes from strategies across both pages of the table. To compare her CC to CD, for example, we look at the payoffs associated with CC in *both pages* of the table and compare these payoffs with the similarly identified payoffs for CD. For Nina, the elimination process leaves only her strategy DC; again, this is the rollback equilibrium strategy found for her above. Finally, Emily has only to compare the two remaining cells associated with her choice of Don't and Contribute; she gets the highest payoff when she chooses Don't and so makes that choice. As before, we have identified her rollback equilibrium strategy.

The unique subgame-perfect outcome in the game table in Figure 6.14 thus corresponds to the cell associated with the rollback equilibrium strategies for each player. Note that the process of iterated elimination that leads us to this subgame-perfect equilibrium is carried out by considering the players in reverse order of the actual play of the game. This order conforms to the order in which player actions are considered in rollback analysis and therefore allows us to eliminate exactly those strategies, for each player, that are not consistent with rollback. In so doing, we eliminate all of the Nash equilibria that are not subgame-perfect.

SUMMARY

Many games include multiple components, some of which entail simultaneous play and others of which entail sequential play. In *two-stage* (and multistage) games, a "tree house" can be used to illustrate the game; this construction allows the identification of the different stages of play and the ways in which those stages are linked together. Full-fledged games that arise in later stages of play are called *subgames* of the full game.

Changing the rules of a game to alter the timing of moves may or may not alter the equilibrium outcome of a game. Simultaneous-move games that are changed to make moves sequential may have the same outcome (if both players have dominant strategies), may have a first-mover or second-mover advantage, or may lead to an outcome in which both players are better off. The sequential version of a simultaneous game will generally have a unique rollback equilibrium even if the simultaneous version has no equilibrium or multiple equilibria. Similarly, a sequential-move game that has a unique rollback equilibrium may have several Nash equilibria when the rules are changed to make the game a simultaneous-move game.

Simultaneous-move games can be illustrated in a game tree by collecting decision nodes in *information sets* when players make decisions without knowing at which specific node they find themselves. Similarly, sequential-move games can be illustrated by using a game table; in this case, each player's full set of strategies must be carefully identified. Solving a sequential-move game from its strategic form may lead to many possible Nash equilibria. The number of potential equilibria can be reduced by using the criteria of *credibility* to eliminate some strategies as possible equilibrium strategies. This process leads to the *subgame-perfect equilibrium (SPE)* of the sequential-move game. These solution processes also work for games with additional players.

KEY TERMS

continuation (197)
credibility (198)
information set (194)
off-equilibrium path (197)
off-equilibrium subgame (197)
subgame (182)
subgame-perfect equilibrium (SPE) (198)

SOLVED EXERCISES

S1. Consider the simultaneous-move game with two players that has no Nash equilibrium in pure strategies, illustrated in Figure 4.15. If the game were transformed into a sequential-move game, would you expect that game to exhibit a first-mover advantage, a second-mover advantage, or neither? Explain your reasoning.

S2. Consider the game represented by the game tree below. The first mover, Player 1, may move either Up or Down, after which Player 2 may move either Left or Right. Payoffs for the possible outcomes appear below. Reexpress this game in strategic (table) form. Find all of the pure-strategy Nash equilibria in the game. If there are multiple equilibria, indicate which one is subgame-perfect. For those equilibria that are not subgame-perfect, identify the reason (the source of the lack of credibility).

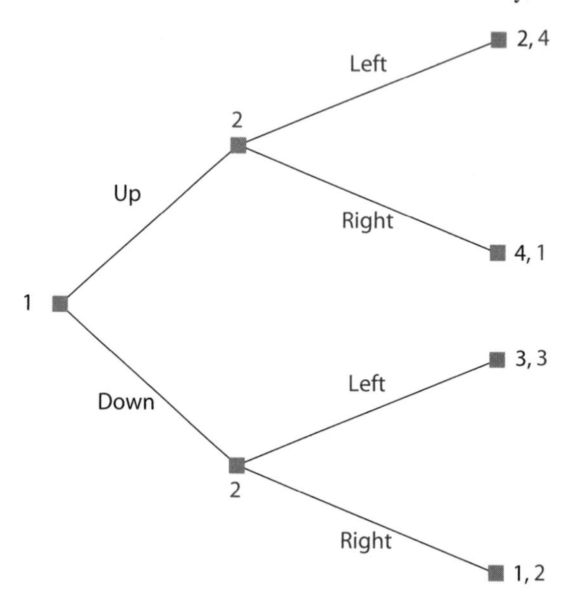

S3. Consider the Airbus–Boeing game in Exercise S4 in Chapter 3. Show that game in strategic form and locate all of the Nash equilibria. Which one of the equilibria is subgame-perfect? For those equilibria that are not subgame-perfect, identify the source of the lack of credibility.

S4. Return to the two-player game tree in part (a) of Exercise S2 in Chapter 3.
(a) Write the game in strategic form, making Scarecrow the row player and Tinman the column player.
(b) Find the Nash equilibrium.

S5. Return to the two-player game tree in part (b) of Exercise S2 in Chapter 3.
(a) Write the game in strategic form. (Hint: Refer to your answer to Exercise S2 of Chapter 3.) Find all of the Nash equilibria. There will be many.
(b) For the equilibria that you found in part (a) that are not subgame-perfect, identify the credibility problems.

S6. Return to the three-player game tree in part (c) of Exercise S2 in Chapter 3.
(a) Draw the game table. Make Scarecrow the row player, Tinman the column player, and Lion the page player. (Hint: Refer to your answer to Exercise S2 of Chapter 3.) Find all of the Nash equilibria. There will be many.

(b) For the equilibria that you found in part (a) that are not subgame-perfect, identify the credibility problems.

S7. Consider a simplified baseball game played between a pitcher and a batter. The pitcher chooses between throwing a fastball or a curve, while the batter chooses which pitch to anticipate. The batter has an advantage if he correctly anticipates the type of pitch. In this constant-sum game, the batter's payoff is the probability that the batter will get a base hit. The pitcher's payoff is the probability that the batter fails to get a base hit, which is simply one minus the payoff of the batter. There are four potential outcomes:

(i) If a pitcher throws a fastball, and the batter guesses fastball, the probability of a hit is 0.300.

(ii) If the pitcher throws a fastball, and the batter guesses curve, the probability of a hit is 0.200.

(iii) If the pitcher throws a curve, and the batter guesses curve, the probability of a hit is 0.350.

(iv) If the pitcher throws a curve, and the batter guesses fastball, the probability of a hit is 0.150.

Suppose that the pitcher is "tipping" his pitches. This means that the pitcher is holding the ball, positioning his body, or doing something else in a way that reveals to the batter which pitch he is going to throw. For our purposes, this means that the pitcher-batter game is a sequential game in which the pitcher announces his pitch choice before the batter has to choose his strategy.

(a) Draw this situation, using a game tree.

(b) Suppose that the pitcher knows he is tipping his pitches but can't stop himself from doing so. Thus, the pitcher and batter are playing the game you just drew. Find the rollback equilibrium of this game.

(c) Now change the timing of the game, so that the batter has to reveal his action (perhaps by altering his batting stance) before the pitcher chooses which pitch to throw. Draw the game tree for this situation, and find the rollback equilibrium.

Now assume that the tips of each player occur so quickly that neither opponent can react to them, so that the game is in fact simultaneous.

(d) Draw a game tree to represent this simultaneous game, indicating information sets where appropriate.

(e) Draw the game table for the simultaneous game. Is there a Nash equilibrium in pure strategies? If so, what is it?

S8. The street-garden game analyzed in Section 4 of this chapter has a 16-by-4-by-2 game table when the sequential-move version of the game is expressed in strategic form, as in Figure 6.14. There are *many* Nash equilibria to be found in this table.

(a) Use best-response analysis to find all of the Nash equilibria in the table in Figure 6.14.

(b) Identify the subgame-perfect equilibrium from among your set of all Nash equilibria. Other equilibrium outcomes look identical to the subgame-perfect one—they entail the same payoffs for each of the three players—but they arise after different combinations of strategies. Explain how this can happen. Describe the credibility problems that arise in the nonsubgame-perfect equilibria.

S9. As it appears in the text, Figure 6.1 represents the two-stage game between CrossTalk and GlobalDialog with a combination of tables and trees. Instead, represent the entire two-stage game in a single, very large game tree. Be careful to label which player makes the decision at each node, and remember to draw information sets between nodes where necessary.

S10. Recall the mall location game in Exercise S8 in Chapter 3. That three-player sequential game has a game tree that is similar to the one for the street-garden game, shown in Figure 6.12.

(a) Draw the tree for the mall location game. How many strategies does each store have?

(b) Illustrate the game in strategic form and find all of the pure-strategy Nash equilibria in the game.

(c) Use iterated dominance to find the subgame-perfect equilibrium. (Hint: Reread the last two paragraphs of Section 4.)

S11. The rules of the mall location game, analyzed in Exercise **S10** above, specify that when all three stores request space in Urban Mall, the two bigger (more prestigious) stores get the available spaces. The original version of the game also specifies that the firms move sequentially in requesting mall space.

(a) Suppose that the three firms make their location requests simultaneously. Draw the payoff table for this version of the game and find all of the Nash equilibria. Which one of these equilibria do you think is most likely to be played in practice? Explain.

Now suppose that when all three stores simultaneously request Urban Mall, the two spaces are allocated by lottery, giving each store an equal chance of getting into Urban Mall. With such a system, each would have a two-thirds probability (or a 66.67% chance) of getting into Urban Mall when all three had requested space there, and a one-third probability (33.33% chance) of being alone in the Rural Mall.

(b) Draw the game table for this new version of the simultaneous-play mall location game. Find all of the Nash equilibria of the game. Which one of these equilibria do you think is most likely to be played in practice? Explain.

(c) Compare and contrast the equilibria found in part (b) with the equilibria found in part (a). Do you get the same Nash equilibria? Why or why not?

S12. Return to the game of Monica and Nancy in Exercise S10 of Chapter 5. Assume that Monica and Nancy choose their effort levels sequentially instead of simultaneously. Monica commits to her choice of effort first, and on observing this decision, Nancy commits to her own effort.

(a) What is the subgame-perfect equilibrium to the game where the joint profits are $4m + 4n + mn$, the effort costs to Monica and Nancy are m^2 and n^2 respectively, and Monica commits to an effort level first?

(b) Compare the payoffs of Monica and Nancy with those found in Exercise S10 of Chapter 5. Does this game have a first-mover or a second-mover advantage? Explain.

S13. Extending Exercise S12, Monica and Nancy need to decide which (if either) of them will commit to an effort level first. To do this, each of them simultaneously writes on a separate slip of paper whether or not she will commit first. If they both write "yes" or they both write "no," they choose effort levels simultaneously, as in Exercise S10 in Chapter 5. If Monica writes "yes" and Nancy writes "no," then Monica commits to her move first, as in Exercise S12. If Monica writes "no" and Nancy writes "yes," then Nancy commits to her move first.

(a) Use the payoffs to Monica and Nancy in Exercise S12 above as well as in Exercise S10 in Chapter 5 to construct the game table for the first-stage paper-slip decision game. (Hint: Note the symmetry of the game.)

(b) Find the pure-strategy Nash equilibria of this first-stage game.

UNSOLVED EXERCISES

U1. Consider a game in which there are two players, A and B. Player A moves first and chooses either Up or Down. If A chooses Up, the game is over, and each player gets a payoff of 2. If A moves Down, then B gets a turn and chooses between Left and Right. If B chooses Left, both players get 0; if B chooses Right, A gets 3 and B gets 1.

(a) Draw the tree for this game, and find the subgame-perfect equilibrium.

(b) Show this sequential-play game in strategic form, and find all of the Nash equilibria. Which is or are subgame-perfect? Which is or are not? If any are not, explain why.

(c) What method of solution could be used to find the subgame-perfect equilibrium from the strategic form of the game? (Hint: Refer to the last two paragraphs of Section 4.)

U2. Return to the two-player game tree in part (a) of Exercise U2 in Chapter 3.
   (a) Write the game in strategic form, making Albus the row player and Minerva the column player. Find all of the Nash equilibria.
   (b) For the equilibria you found in part (a) of this exercise that are not subgame-perfect, identify the credibility problems.

U3. Return to the two-player game tree in part (b) of Exercise U2 in Chapter 3.
   (a) Write the game in strategic form. Find all of the Nash equilibria.
   (b) For the equilibria you found in part (a) that are not subgame-perfect, identify the credibility problems.

U4. Return to the two-player game tree in part (c) of Exercise U2 in Chapter 3.
   (a) Draw the game table. Make Albus the row player, Minerva the column player, and Severus the page player. Find all of the Nash equilibria.
   (b) For the equilibria you found in part (a) that are not subgame-perfect, identify the credibility problems.

U5. Consider the cola industry, in which Coke and Pepsi are the two dominant firms. (To keep the analysis simple, just forget about all the others.) The market size is $8 billion. Each firm can choose whether to advertise. Advertising costs $1 billion for each firm that chooses it. If one firm advertises and the other doesn't, then the former captures the whole market. If both firms advertise, they split the market 50:50 and pay for the advertising. If neither advertises, they split the market 50:50 but without the expense of advertising.
   (a) Write the payoff table for this game, and find the equilibrium when the two firms move simultaneously.
   (b) Write the game tree for this game (assume that it is played sequentially), with Coke moving first and Pepsi following.
   (c) Is either equilibrium in parts (a) and (b) better from the joint perspective of Coke and Pepsi? How could the two firms do better?

U6. Along a stretch of a beach are 500 children in five clusters of 100 each. (Label the clusters A, B, C, D, and E in that order.) Two ice-cream vendors are deciding simultaneously where to locate. They must choose the exact location of one of the clusters.

If there is a vendor in a cluster, all 100 children in that cluster will buy an ice cream. For clusters without a vendor, 50 of the 100 children are willing to walk to a vendor who is one cluster away, only 20 are willing to walk to a vendor two clusters away, and no children are willing to walk the distance of three or more clusters. The ice cream melts quickly, so the walkers cannot buy for the nonwalkers.

If the two vendors choose the same cluster, each will get a 50% share of the total demand for ice cream. If they choose different clusters, then those children (locals or walkers) for whom one vendor is closer than the other will go to the closer one, and those for whom the two are equidistant will split 50% each. Each vendor seeks to maximize her sales.

(a) Construct the five-by-five payoff table for the vendor location game; the entries stated here will give you a start and a check on your calculations:

If both vendors choose to locate at A, each sells 85 units.

If the first vendor chooses B and the second chooses C, the first sells 150 and the second sells 170.

If the first vendor chooses E and the second chooses B, the first sells 150 and the second sells 200.

(b) Eliminate dominated strategies as far as possible.
(c) In the remaining table, locate all pure-strategy Nash equilibria.
(d) If the game is altered to one with sequential moves, where the first vendor chooses her location first and the second vendor follows, what are the locations and the sales that result from the subgame-perfect equilibrium? How does the change in the timing of moves here help players resolve the coordination problem in part (c)?

U7. Return to the game among the three lions in the Roman Colosseum in Exercise S7 in Chapter 3.
(a) Write out this game in strategic form. Make Lion 1 the row player, Lion 2 the column player, and Lion 3 the page player.
(b) Find the Nash equilibria for the game. How many did you find?
(c) You should have found Nash equilibria that are not subgame-perfect. For each of those equilibria, which lion is making a noncredible threat? Explain.

U8. Now assume that the mall location game (from Exercises S8 in Chapter 3 and S10 in this chapter) is played sequentially but with a different order of play: Big Giant, then Titan, then Frieda's.
(a) Draw the new game tree.
(b) What is the subgame-perfect equilibrium of the game? How does it compare to the subgame-perfect equilibrium for Exercise S8 in Chapter 3?
(c) Now write the strategic form for this new version of the game.
(d) Find all of the Nash equilibria of the game. How many are there? How does this compare with the number of equilibria from Exercise S10 in this chapter?

U9. Return to the game of Monica and Nancy in Exercise U10 of Chapter 5. Assume that Monica and Nancy choose their effort levels sequentially instead of simultaneously. Monica commits to her choice of effort first. On observing this decision, Nancy commits to her own effort.

(a) What is the subgame-perfect equilibrium to the game where the joint profits are $5m + 4n + mn$, the effort costs to Monica and Nancy are m^2 and n^2 respectively, and Monica commits to an effort level first?

(b) Compare the payoffs of Monica and Nancy with those found in Exercise U10 of Chapter 5. Does this game have a first-mover or second-mover advantage?

(c) Using the same joint profit function as in part (a), find the subgame-perfect equilibrium for the game where *Nancy* must commit first to an effort level.

U10. In an extension of Exercise **U9**, Monica and Nancy need to decide which (if either) of them will commit to an effort level first. To do this, each of them simultaneously writes on a separate slip of paper whether or not she will commit first. If they both write "yes" or they both write "no," they choose effort levels simultaneously, as in Exercise U10 in Chapter 5. If Monica writes "yes" and Nancy writes "no," they play the game in part (a) of Exercise **U9**, above. If Monica writes "no" and Nancy writes "yes," they play the game in part (c).

(a) Use the payoffs to Monica and Nancy in parts (b) and (c) in Exercise U9 above as well as those in Exercise U10 in Chapter 5 to construct the game table for the first-stage paper-slip decision game.

(b) Find the pure-strategy Nash equilibria of this first-stage game.

U11. In the faraway town of Saint James two firms, Bilge and Chem, compete in the soft-drink market (Coke and Pepsi aren't in this market yet). They sell identical products, and since their good is a liquid, they can easily choose to produce fractions of units. Since they are the only two firms in this market, the price of the good (in dollars), P, is determined by $P = (30 - Q_B - Q_C)$, where Q_B is the quantity produced by Bilge and Q_C is the quantity produced by Chem (each measured in liters). At this time both firms are considering whether to invest in new bottling equipment that will lower their variable costs.

(i) If firm j decides *not* to invest, its cost will be $C_j = Q_j^2 / 2$, where j stands for either B (Bilge) or C (Chem).

(ii) If a firm decides to invest, its cost will be $C_j = 20 + Q_j^2 / 6$, where j stands for either B (Bilge) or C (Chem). This new cost function reflects the fixed cost of the new machines (20) as well as the lower variable costs.

The two firms make their investment choices simultaneously, but the payoffs in this investment game will depend on the subsequent duopoly games that arise. The game is thus really a two-stage game: decide to invest, and then play a duopoly game.

(a) Suppose both firms decide to invest. Write the profit functions in terms of Q_B and Q_C for the two firms. Use these to find the Nash equilibrium of the quantity-setting game. What are the equilibrium quantities and profits for both firms? What is the market price?

(b) Now suppose both firms decide not to invest. What are the equilibrium quantities and profits for both firms? What is the market price?

(c) Now suppose that Bilge decides to invest, and Chem decides not to invest. What are the equilibrium quantities and profits for both firms? What is the market price?

(d) Write out the two-by-two game table of the investment game between the two firms. Each firm has two strategies: Investment and No Investment. The payoffs are simply the profits found in parts (a), (b), and (c). (Hint: Note the symmetry of the game.)

(e) What is the subgame-perfect equilibrium of the overall two-stage game?

U12. Two French aristocrats, Chevalier Chagrin and Marquis de Renard, fight a duel. Each has a pistol loaded with one bullet. They start 10 steps apart and walk toward each other at the same pace, 1 step at a time. After each step, either may fire his gun. When one shoots, the probability of scoring a hit depends on the distance. After k steps it is $k/5$, and so it rises from 0.2 after the first step to 1 (certainty) after 5 steps, at which point they are right up against one another. If one player fires and misses while the other has yet to fire, the walk must continue even though the bulletless one now faces certain death; this rule is dictated by the code of the aristocracy. Each gets a payoff of 21 if he himself is killed and 1 if the other is killed. If neither or both are killed, each gets 0.

This is a game with five sequential steps and simultaneous moves (shoot or not shoot) at each step. Find the rollback (subgame-perfect) equilibrium of this game.

Hint: Begin at step 5, when the duelists are right up against one another. Set up the two-by-two table for the simultaneous-move game at this step, and find its Nash equilibrium. Now move back to step 4, where the probability of scoring a hit is 4/5, or 0.8, for each. Set up the two-by-two table for the simultaneous-move game at this step, correctly specifying in the appropriate cell what happens in the future. For example, if one shoots and misses, but the other does not shoot, then the other will wait until step 5 and score a sure hit. If neither shoots, then the game will go to the next step, for which you have already found the equilibrium. Using all this information, find the payoffs in the two-by-two table of step 4, and find the Nash equilibrium at this step. Work backward in the same way through the rest of the steps to find the Nash equilibrium strategies of the full game.

U13. Describe an example of business competition that is similar in structure to the duel in Exercise U12.

Simultaneous-Move Games with Mixed Strategies I: Two-by-Two Games

IN OUR STUDY of simultaneous-move games in Chapter 4, we came across a class of games that the solution methods described there could not solve; in fact, games in that class have no Nash equilibria in pure strategies. To predict outcomes for such games, we need an extension of our concepts of strategies and equilibria. This is to be found in the randomization of moves, which is the focus of this chapter and the next.

Consider the tennis-point game from the end of Chapter 4. This game is zero sum; the interests of the two tennis players are purely in mutual conflict. Evert wants to hit her passing shot to whichever side—down the line (DL) or crosscourt (CC)—is not covered by Navratilova, whereas Navratilova wants to cover the side to which Evert hits her shot. In Chapter 4, we pointed out that in such a situation, any systematic choice by Evert will be exploited by Navratilova to her own advantage and therefore to Evert's disadvantage. Conversely, Evert can exploit any systematic choice by Navratilova. To avoid being thus exploited, each player wants to keep the other guessing, which can be done by acting unsystematically or randomly.

However, randomness doesn't mean choosing each shot half the time, or alternating between the two. The latter would itself be a systematic action open to exploitation, and a 60-40 or 75-25 random mix may be better than 50-50 depending on the situation. In this chapter we develop methods for calculating the best mix and discuss how well this theory helps us understand actual play in such games.

Our method for calculating the best mix can also be applied to non-zero-sum games. However, in such games the players' interests can partially coincide, so when player B exploits A's systematic choice to her own advantage, it is not necessarily to A's disadvantage. Therefore the logic of keeping the other player guessing is weaker or even absent altogether in non-zero-sum games. We will discuss whether and when mixed-strategy equilibria make sense in such games.

Throughout this chapter, we limit the discussion to two-by-two games and to the most direct method for calculating a mixed-strategy equilibrium in order to present the basic ideas in the simplest possible setting. Many of the concepts and methods we develop here continue to be valid in more general games; most important, our discussion in Sections 6 and 7 of how to mix strategies and whether mixing is observed in reality is perfectly general. However, we leave to Chapter 8 the general analysis of best responses in mixed strategies and of mixed-strategy equilibria for games with more than two pure strategies.

1 WHAT IS A MIXED STRATEGY?

When players choose to act unsystematically, they pick from among their pure strategies in some random way. In the tennis-point game, Navratilova and Evert each choose from two initially given pure strategies, DL and CC. We call a random mixture of these two pure strategies a mixed strategy.

Such mixed strategies cover a whole continuous range. At one extreme, DL could be chosen with probability 1 (for sure), meaning that CC is never chosen (probability 0); this "mixture" is just the pure strategy DL. At the other extreme, DL could be chosen with probability 0 and CC with probability 1; this "mixture" is the same as pure CC. In between is the whole set of possibilities: DL chosen with probability 75% (0.75) and CC with probability 25% (0.25); or both chosen with probabilities 50% (0.5) each; or DL with probability 1/3 (33.33 . . . %) and CC with probability 2/3 (66.66 . . . %); and so on.[1]

The payoffs from a mixed strategy are defined as the corresponding probability-weighted averages of the payoffs from its constituent pure

[1]When a chance event has just two possible outcomes, people often speak of the odds in favor of or against one of the outcomes. If the two possible outcomes are labeled A and B, and the probability of A is p so that the probability of B is $(1 - p)$, then the ratio $p/(1 - p)$ gives the odds in favor of A, and the reverse ratio $(1 - p)/p$ gives the odds against A. Thus, when Evert chooses CC with probability 0.25 (25%), the odds against her choosing CC are 3 to 1, and the odds in favor of it are 1 to 3. This terminology is often used in betting contexts, so those of you who misspent your youth in that way will be more familiar with it. However, this usage does not readily extend to situations in which three or more outcomes are possible, so we avoid its use here.

strategies. For example, in the tennis game of Section 4.8, against Navratilova's DL, Evert's payoff from DL is 50 and from CC is 90. Therefore the payoff of Evert's mixture (0.75 DL, 0.25 CC) against Navratilova's DL is $0.75 \times 50 + 0.25 \times 90 = 37.5 + 22.5 = 60$. This is Evert's **expected payoff** from this particular mixed strategy.[2]

You may have noticed that mixed strategies are very similar to the continuous strategies we studied in Chapter 5. In fact, they are just special kinds of continuously variable strategies. With the possibility of mixing made available, each player can now choose from among all conceivable mixtures of the basic or pure strategies initially specified (which, as we just saw, include pure strategies as extreme special cases).

The notion of Nash equilibrium also extends easily to include mixed strategies. Nash equilibrium is defined as a list of mixed strategies, one for each player, such that the choice of each is her best choice, in the sense of yielding the highest expected payoff for her, given the mixed strategies of the others. Allowing for mixed strategies in a game solves the problem of possible nonexistence of Nash equilibrium, which we encountered for pure strategies, automatically and almost entirely. Nash's celebrated theorem shows that, under very general circumstances (which are broad enough to cover all the games that we meet in this book and many more besides), a Nash equilibrium in mixed strategies exists.

At this broadest level, therefore, incorporating mixed strategies into our analysis does not entail anything different from the general theory of continuous strategies developed in Chapter 5. However, the special case of mixed strategies does bring with it several special conceptual as well as methodological matters, and therefore deserves separate study.

2 UNCERTAIN ACTIONS: MIXING MOVES TO KEEP THE OPPONENT GUESSING

We begin with the tennis example of Section 4.8, which did not have a Nash equilibrium in pure strategies. We show how the extension to mixed strategies remedies this deficiency, and we interpret the resulting equilibrium as one in which each player keeps the other guessing.

[2]Game theory assumes that players will calculate and try to maximize their expected payoffs when probabilistic mixtures of strategies or outcomes are included. We consider this further in the Appendix to this chapter, but for now we proceed to use it, with just one important note. The word *expected* in "expected payoff" is a technical term from probability and statistics. It merely denotes a probability-weighted average. It does not mean this is the payoff that the player should expect in the sense of regarding it as her right or entitlement.

A. The Benefit of Mixing

We reproduce in Figure 7.1 the payoff matrix of Figure 4.15 with both players' payoffs included. Even though this is a constant-sum game, we now show the payoffs for the two players separately, because it is more intuitive to think of each player as trying to get a higher payoff for herself. In this game, if Evert always chooses DL, Navratilova will then cover DL and hold Evert's payoff down to 50. Similarly, if Evert always chooses CC, Navratilova will choose to cover CC and hold Evert down to 20. If Evert can only choose one of her two basic (pure) strategies, and Navratilova can predict that choice, Evert's better (or less bad) pure strategy will be DL, yielding her a payoff of 50. (This argument follows the minimax reasoning of Chapter 4, Section 5.)

But suppose Evert is not restricted to using only pure strategies and can choose a mixed strategy, perhaps one in which the probability of playing DL on any one occasion is 75%, or 0.75; this makes her probability of playing CC 25%, or 0.25. Using the method outlined in Section 1, we can calculate Navratilova's expected payoff against this mixture as

$$0.75 \times 50 + 0.25 \times 10 = 37.5 + 2.5 = 40 \text{ if she covers DL, and}$$
$$0.75 \times 20 + 0.25 \times 80 = 15 + 20 = 35 \text{ if she covers CC.}$$

If Evert chooses this mixture, the expected payoffs show that Navratilova can best exploit it by covering DL.

When Navratilova chooses DL to best exploit Evert's 75-25 mix, her choice works to Evert's disadvantage because this is a zero-sum game. Evert's expected payoffs are

$$0.75 \times 50 + 0.25 \times 90 = 37.5 + 22.5 = 60 \text{ if Navratilova covers DL, and}$$
$$0.75 \times 80 + 0.25 \times 20 = 60 + 5 = 65 \text{ if Navratilova covers CC.}$$

By choosing DL, Navratilova holds Evert down to 60 rather than 65. But notice that Evert's payoff with the mixture is still better than the 50 she would get by playing purely DL, or the 20 she would get by playing purely CC.[3]

The 75-25 mix does leave Evert's strategy open to some exploitation by Navratilova. By choosing to cover DL she can hold Evert down to a lower expected payoff than when she chooses CC. Ideally, Evert would like to find a mix that would be exploitation proof. To find such a mix requires taking a more general approach to describing Evert's mixed strategy. Specifically, we denote the

[3]Not every mixed strategy will perform better than the pure strategies. For example, if Evert mixes 50:50 between DL and CC, Navratilova can hold Evert's expected payoff down to 50, exactly the same as from pure DL. And a mixture that attaches a probability of less than 30% to DL will be worse for Evert than pure DL. We ask you to verify these statements as a useful exercise to acquire the skill of calculating expected payoffs and comparing strategies.

		NAVRATILOVA	
		DL	CC
EVERT	DL	50, 50	80, 20
	CC	90, 10	20, 80

FIGURE 7.1 Payoffs of Both Players in the Tennis Point

probability of Evert choosing DL by the algebraic symbol p (so the probability of choosing CC is $1 - p$) and find the condition for this mixture to be exploitation proof (in terms of p). Solving the condition for p provides the answer we seek.

So suppose Evert chooses DL with probability p and CC with probability $(1 - p)$. We will refer to this mixture as Evert's p-mix for short. Against the p-mix, Navratilova's expected payoffs are

$$50p + 10(1 - p) \text{ if she covers DL, and}$$
$$20p + 80(1 - p) \text{ if she covers CC.}$$

For Evert's strategy to be exploitation proof, these two expected payoffs should be equal. That implies $50p + 10(1 - p) = 20p + 80(1 - p)$; or $30p = 70(1 - p)$; or $100p = 70$; or $p = 0.7$. Thus Evert's exploitation-proof mix uses DL with probability 70% and CC with probability 30%. Evert's expected payoff from this mixed strategy is

$$50 \times 0.7 + 90 \times 0.3 = 35 + 27 = 62 \text{ if Navratilova covers DL, and also}$$
$$80 \times 0.7 + 20 \times 0.3 = 56 + 6 = 62 \text{ if Navratilova covers CC.}$$

This expected payoff is better than the 50 that Evert would get if she used the pure strategy DL, and better than the 60 from the 75-25 mixture. We know this mixture is exploitation proof, but does it ensure Evert the best possible expected payoff?

To answer this question, we first consider how Navratilova will respond to various choices of p by Evert. To see this explicitly, we construct a diagram, Figure 7.2, showing Navratilova's expected payoffs from covering DL and from covering CC for all of Evert's possible p-mixes. Evert's choice of p for her p-mix is shown on the horizontal axis in Figure 7.2; Navratilova's expected payoffs are on the vertical axis. In the figure are two lines, one corresponding to Navratilova's choice of pure DL and the other to her choice of pure CC. The equations of the lines are given by the two expressions above—that is, Navratilova's expected payoff from covering DL $= 50p + 10(1 - p) = 10 + 40p$ and her expected payoff from covering CC $= 20p + 80(1 - p) = 80 - 60p$.

The upward sloping straight line in Figure 7.2 shows Navratilova's expected payoff from her DL against Evert's p-mix. The intuition for this shape is that

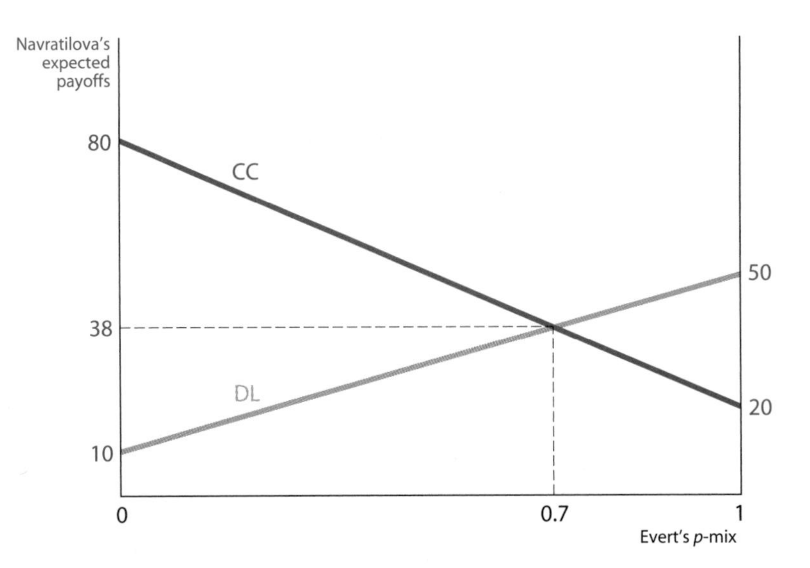

FIGURE 7.2 Navratilova's Expected Payoffs Against Evert's p-mixes

Navratilova does better by covering DL (so her expected payoff is higher) when Evert is more likely to use DL (or when p is higher). The falling straight line in the figure depicts Navratilova's expected payoff from her CC against Evert's p-mix. Navratilova fares worse as the likelihood that Evert uses DL rises.

If Navratilova were to use a mixture of DL and CC instead of a pure strategy, her expected payoff would be a weighted average of the payoffs from the two pure strategies. Thus her expected payoff from choosing a mixed strategy would lie somewhere between the two expected payoff lines in Figure 7.2. (As we did for Evert, we will use an algebraic symbol, this time a q, to represent the probability with which Navratilova covers DL. Then we will refer to the mixture putting probability q on DL and probability $1 - q$ on CC as Navratilova's q-mix, for short.)

It is useful to calculate the value of p at the point of intersection of the two lines in Figure 7.2. That p satisfies $10 + 40p = 80 - 60p$, or $p = 0.7$. The expected payoff for Navratilova at this value of p is the same on both lines: it is $50 \times 0.7 + 10 \times 0.3 = 35 + 3 = 38$ when calculated using the line for DL, and $20 \times 0.7 + 80 \times 0.3 = 14 + 24 = 38$ when calculated using the line for CC.

We can now identify the strategy that is best for Navratilova against each possible value of Evert's p. This is just the strategy that yields her the highest expected payoff in each case. For $p=0$, for example, the diagram shows that Navratilova gets a higher payoff when she plays CC, namely 80, than the payoff she would get by playing DL, namely 10. If Navratilova were to play a 75-25 mixture of DL and CC against Evert's $p = 0$, then Navratilova's expected payoff would be a weighted average of the pure strategy payoffs, or $80 \times 0.75 + 10 \times 0.25 = 62.5$. If her payoff from pure CC is higher than her payoff from pure DL, then her payoff from pure CC is also higher than that from any such average. Therefore her pure CC is not only better for Navratilova than her pure DL, it is also her best choice among all her strategies, pure and mixed.

This continues to be true for all values of p to the left of the intersection point in the diagram. For all such values of p, pure CC is better than DL for Navratilova when played against Evert's p-mix and is also better than any mixed strategy of her own. Pure CC is Navratilova's best response for this range of values of p. Similarly, to the right of the point of intersection, pure DL is her best response.[4]

Against Evert's specific p-mix at the point of intersection, Navratilova does equally well (expected payoff 38) from her CC and DL. Because the expected payoff associated with a mixture of her CC and DL lies "between" the CC and DL lines, it follows that she also does equally well with any q-mix of her own. The weighted average of 38 and 38 is 38, no matter what the weights used for averaging. Thus, Navratilova's best response to Evert's choice of $p = 0.7$ can be a q-mix for any q in the entire range from 0 to 1.

We can summarize Navratilova's best-response rule as follows:

If $p < 0.7$, choose pure CC ($q = 0$).
If $p = 0.7$, all values of q in the range from 0 to 1 are equal best responses.
If $p > 0.7$, choose pure DL ($q = 1$).

To see the intuition behind this rule, note that p is the probability with which Evert hits her passing shot down the line. Navratilova wants to prepare for Evert's shot. Therefore if p is low (Evert is more likely to go crosscourt), Navratilova does better to defend crosscourt. If p is high (Evert is more likely to go down the line), Navratilova does better to defend down the line. For a critical value of p in between, the two choices are equally good for Navratilova. This break-even point is not $p = 0.5$ because the consequences (measured by the payoff numbers of the various combinations) are not symmetric for the two choices.

Recall that our purpose in drawing Figure 7.2 was to help answer the question of whether Evert's exploitation-proof mixture, $p = 0.7$, is her best mixed strategy, in the sense of giving her the highest expected payoff. The answer is yes. We can see this by looking at the expected payoff that Navratilova can get when she makes her best response to each of Evert's p-mixes. In Figure 7.2, this is shown by the higher of the two lines, which corresponds to Navratilova's coverage of DL or CC. The best-response expected payoffs make up a V-shaped curve. The point of intersection of the two lines in the figure is the lowest point of the V. When Evert chooses $p = 0.7$, Navratilova gets her worst possible expected payoff, and therefore Evert gets her best possible expected payoff.

[4]If, in some numerical problem you are trying to solve, the expected payoff lines for the pure strategies do not intersect, that would indicate that one pure strategy was best for all of the opponent's mixtures. Then this player's best response would always be that pure strategy.

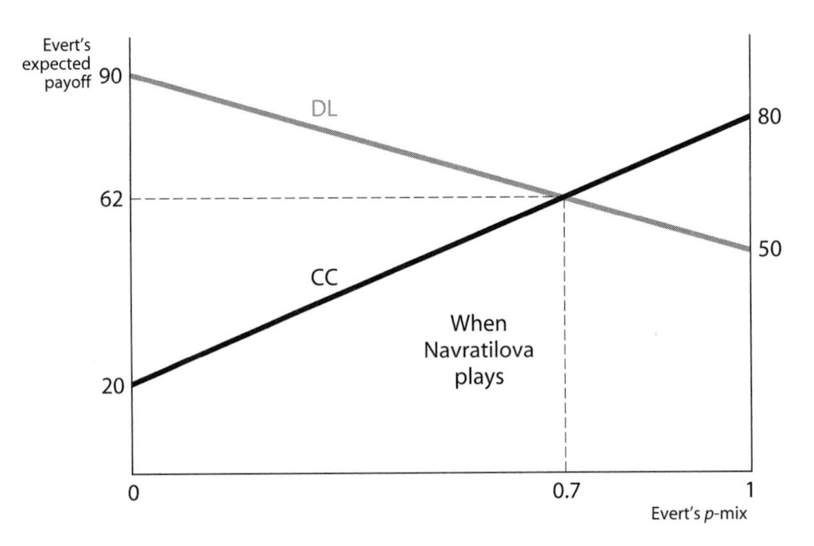

FIGURE 7.3 Navratilova's Best-Response Calculation Using Evert's Payoffs

We can see this even more clearly by graphing Evert's own expected payoffs for each of her p-mixes. We just have to remember that in the constant-sum game, Navratilova does better when Evert gets low payoffs. Figure 7.3 puts Evert's p-mix on the horizontal axis as in Figure 7.2, but Evert's own expected payoffs are on the vertical axis. Since Evert's payoffs are 100 minus Navratilova's payoffs, Figure 7.3 is a vertically inverted version of Figure 7.2. We show two lines again, one corresponding to Navratilova's choice of pure CC and the other for her choice of pure DL. Here, the CC line is rising from left to right. This corresponds to the idea that Evert's expected payoff rises as she uses her DL shot more often (higher p) against Navratilova's covering purely CC. Similarly, the DL line is falling from left to right, as does Evert's expected payoff from using DL more often against an opponent who is always covering for that shot. We see that the point of intersection of the expected payoff lines is again at $p = 0.7$. To the left of this p value, the CC line is lower than the DL line. Therefore, CC is better than DL for Navratilova when p is low; it entails a lower expected payoff for Evert. To the right of $p = 0.7$, the DL line is lower than the CC line, so DL is better than CC for Navratilova. And the expected payoff at the point of intersection is 62 for Evert, which is just 100 minus the 38 that we found for Navratilova in the analysis based on her payoffs. Further, for each of Evert's p-mixes, Navratilova's best response will hold Evert down to the lower of the expected payoffs along the two lines. These best responses make an inverted V in this case, and Evert's highest expected payoff comes at the vertex of the inverted V, that is, at the point of intersection of the two straight lines. Thus, the exploitation-proof mix is indeed Evert's best choice. Our results are the same as we found by using Navratilova's own payoffs; only the basis of the analysis is slightly different.

Let us recapitulate the argument for finding Evert's exploitation-proof mixture. For each p that Evert could choose for her p-mix, we find the lowest payoff to which Navratilova can hold her down by choosing the appropriate response. Then we find the p that gives the highest of these lowest payoffs. This matches exactly the minimax reasoning that we developed for pure strategies in Chapter 4, Section 5, and it has the same justification here. Navratilova is going to choose the strategy that is best for her. The game is zero sum; what is best for Navratilova is worst for Evert. Therefore Evert would be correct to entertain a pessimistic belief about Navratilova's best responses, and that is exactly what minimax reasoning does. We develop this idea in a little more detail in subsection C below.

We can find Navratilova's best mixture by using exactly the same reasoning. First you will need to identify Evert's best choice of p against each of Navratilova's q-mixes. We leave the analysis to you but provide the result. Evert's best-response rule is:

If $q < 0.6$, choose pure DL ($p = 1$).
If $q = 0.6$, all values of p from 0 to 1 are equal best responses.
If $q > 0.6$, choose pure CC ($p = 0$).

To understand the intuition, observe that q is the probability of Navratilova defending down the line. Evert wants to hit her passing shot to the side that Navratilova is not defending. Therefore, when q is low, Evert does better to go down the line and, when q is high, she does better to go crosscourt. Following the same reasoning as we gave above from Evert's perspective, $q = 0.6$ is Navratilova's best mixture, immune to exploitation by Evert.

B. Equilibrium in Mixed Strategies

When each player chooses her best, exploitation-proof, mixture, the outcome will be an equilibrium in mixed strategies. In the next chapter we will verify formally that it is indeed a Nash equilibrium in continuous strategies, regarding the mixture probabilities as the continuous variables in the choices of the two players' strategies. There we will also develop more general theory of mixed strategies and their equilibria when each player has several pure strategies. Here we merely state the procedure you have to follow to calculate such an equilibrium in two-by-two games.

Let the Row player's general mixed strategy be her p-mix, putting probability p on playing her first pure strategy and probability $(1 - p)$ on her second pure strategy. Given this p-mix, calculate the algebraic expressions for the Column player's expected payoffs from each of her pure strategies. As in Section 2.A, these expressions will depend on the parameter p. Then equate the two expressions, and solve the resulting equation for p. This p defines the Row player's optimal, exploitation-proof mixture.

Similarly, for the Column player, let her q-mix put probability q on playing her first pure strategy and probability $(1 - q)$ on her second pure strategy. Find the q that defines the Column player's best q-mix by equating the expressions for the expected payoffs the Row player would get from each of her pure strategies against this q-mix.[5]

We illustrate this method using the tennis point example. We have already seen that against Evert's p-mix, Navratilova's expected payoff is $50p + 10(1 - p)$ if she covers DL, and $20p + 80(1 - p)$ if she covers CC. Equating these two we have

$$50p + 10(1 - p) = 20p + 80(1 - p),$$
$$\text{or}\quad 30p = 70(1 - p), \quad \text{or} \quad 100p = 70, \quad \text{or} \quad p = 0.7.$$

Against Navratilova's q-mix, Evert's expected payoffs are similarly found using Figure 7.1; they are $50q + 80(1 - q)$ if Evert plays DL, and $90q + 20(1 - q)$ if she plays CC. Equating the two, we have

$$50q + 80(1 - q) = 90q + 20(1 - q),$$
$$\text{or}\quad 60(1 - q) = 40q, \quad \text{or} \quad 100q = 60, \quad \text{or} \quad q = 0.6.$$

These values, $p = 0.7$ and $q = 0.6$, describe the equilibrium in mixed strategies in the tennis point game.

Each player's equilibrium (best) mixture probability is found by equating the other player's expected payoffs for her two pure strategies. Thus, the equilibrium mixture of each player satisfies the condition that the other player should be indifferent between her two pure strategies. We call this the **opponent's indifference property** of mixed strategy equilibria.

C. The Minimax Method

In Chapter 4, Section 5, we developed the minimax reasoning for zero-sum games in which players used only pure strategies. That approach can be used for zero-sum games with no equilibrium in pure strategies as well, if we expand the set of strategies to include mixtures for the two players.

In Figure 7.4, we show Evert's payoffs in the tennis point game with pure strategies alone, but augmented to show row minima and column maxima, as we did in Figure 4.8 for the football game. Evert's minimum payoffs (success values) for each of her strategies are shown at the far right of the table: 50 for the DL row and 20 for the CC row. The maximum of these minima, or Evert's maximin, is 50. This is the best that Evert can guarantee for herself by using pure strategies,

[5]If your solution yields values of p or q that are outside the range for probabilities—that is, if one or both of the solutions is either less than 0 or greater than 1—then the game has an equilibrium in pure strategies. Inspect the original payoff matrix to find it.

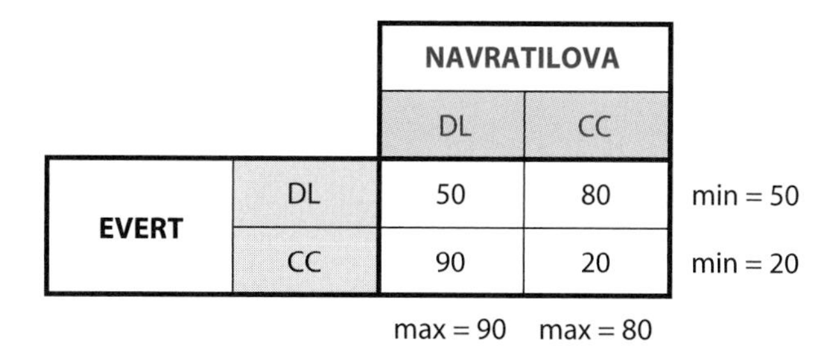

		NAVRATILOVA		
		DL	CC	
EVERT	DL	50	80	min = 50
	CC	90	20	min = 20
		max = 90	max = 80	

FIGURE 7.4 Minimax Analysis of the Tennis Point with Pure Strategies

knowing that Navratilova responds in her own best interests. To achieve this guaranteed payoff, Evert would play DL.

At the foot of each column, we show the maximum of Evert's success percentages attainable in that column. This value represents the worst payoff that Navratilova can get if she plays the pure strategy corresponding to that column. The smallest of these maxima, or the minimax, is 80. This is the best payoff that Navratilova can guarantee for herself by using pure strategies, knowing that Evert responds in her own best interests. To achieve this, Navratilova would play CC.

The maximin and minimax values for the two players are not the same: Evert's maximin success percentage (50) is less than Navratilova's minimax (80). As we explained in Chapter 4, this shows that the game has no equilibrium in the (pure) strategies available to the players in this analysis. In this game, each player can achieve a better outcome—a higher maximin value for Evert and a lower minimax value for Navratilova—by choosing a suitable random mixture of DL and CC.

Let us expand the sets of strategies available to players to include randomization of moves or mixed strategies. We do the analysis here from Evert's perspective; that for Navratilova is similar, and we leave it for you as an exercise. First, in Figure 7.5 we expand the payoff table of Figure 7.4 by adding a row for Evert's p-mix, where she plays DL with probability p and CC with probability $(1 - p)$. We show the expected payoffs of this mix against each of Navratilova's

		NAVRATILOVA		
		DL	CC	
EVERT	DL	50	80	min = 50
	CC	90	20	min = 20
	p-mix	$50p + 90(1 - p)$	$80p + 20(1 - p)$	min = ?

FIGURE 7.5 Minimax Analysis of the Tennis Point with Evert's Mixed Strategies

pure strategies, DL and CC. The row minimum for the p-mix is then just the smaller of these two expressions. But which expression is smaller depends on the value of p, so we leave it as a question mark there.

To find the correct value for p, turn to Figure 7.3, which graphs each of the two payoff expressions from the last row of the table in Figure 7.5 against p. When $p < 0.7$, the minimum payoff occurs along the line corresponding to Navratilova's CC, and when $p > 0.7$, it occurs along the line corresponding to Navratilova's DL. The maximum of these minima for Evert occurs at $p = 0.7$, namely the peak of the inverted V formed by the lower segments of the two lines.[6] Evert's expected payoff at this point is 62. Thus suitable mixing enables Evert to raise her maximin from 50 to 62.

A similar analysis done from Navratilova's perspective shows that suitable mixing (choice of an exploitation-proof q) enables her to lower her minimax. We already calculated this mix; it is $q = 0.6$. The resulting expected payoff to Evert is $50 \times 0.6 + 80 \times 0.4 = 30 + 32 = 62$. Thus Navratilova's best mixing enables her to lower her minimax from 80 to 62.

When the two best mixes are pitted against each other, Evert's maximin equals Navratilova's minimax, and we have a Nash equilibrium in mixed strategies. The equality of maximin and minimax for optimal mixes is a general property of zero-sum games and was proved by John von Neumann and Oskar Morgenstern in 1945.

3 NASH EQUILIBRIUM AS A SYSTEM OF BELIEFS AND RESPONSES

When the moves in a game are simultaneous, neither player can respond to the other's actual choice. Instead, each takes her best action in light of what she thinks the other might be choosing at that instant. In Chapter 4, we called such thinking a player's belief about the other's strategy choice. We then interpreted Nash equilibrium as a configuration where such beliefs are correct, so each chooses her best response to the actual actions of the other. This concept proved useful for understanding the structures and outcomes of many important types of games, most notably zero-sum games and minimax strategies, dominance and the prisoners' dilemma, and focal points and various coordination games, as well as chicken.

[6]We could have replaced the question mark in the table by the whole graph of the inverted V, but that would be complicated to do and difficult to read.

However, in Chapter 4 we considered only pure-strategy Nash equilibria. Therefore a hidden assumption went almost unremarked—namely, that each player was sure or confident in her belief that the other would choose a particular pure strategy. Now that we are considering more general mixed strategies, the concept of belief requires a corresponding reinterpretation.

Players may be unsure about what others might be doing. In the coordination game in Chapter 5, in which Harry wanted to meet Sally, he might be unsure whether she would go to Starbucks or Local Latte, and his belief might be that there was a 50–50 chance that she would go to either one. And in the tennis example, Evert might recognize that Navratilova was trying to keep her (Evert) guessing and would therefore be unsure of which of Navratilova's available actions she would play. In Chapter 2, Section 4, we labeled this as strategic uncertainty, and in Chapter 4 we mentioned that such uncertainty can give rise to mixed-strategy equilibria. Now we develop this idea more fully.

It is important, however, to distinguish between being unsure and having incorrect beliefs. For example, in the tennis example, Navratilova cannot be sure of what Evert is choosing on any one occasion. But she can still have correct beliefs about Evert's mixture—namely, about the probabilities with which Evert chooses between her two pure strategies. Having correct beliefs about mixed actions means knowing or calculating or guessing the correct probabilities with which the other player chooses from among her underlying basic or pure actions. In the equilibrium of our example, it turned out that Evert's equilibrium mixture was 70% DL and 30% CC. If Navratilova believes that Evert will play DL with 70% probability and CC with 30% probability, then her belief, although uncertain, will be correct in equilibrium.

Thus we have an alternative and mathematically equivalent way to define Nash equilibrium in terms of beliefs: each player forms beliefs about the probabilities of the mixture that the other is choosing and chooses her own best response to this. A Nash equilibrium in mixed strategies occurs when the beliefs are correct, in the sense just explained.

In the next section, we consider mixed strategies and their Nash equilibria in non-zero-sum games. In such games, there is no general reason that the other player's pursuit of her own interests should work against your interests. Therefore it is not in general the case that you would want to conceal your intentions from the other player, and there is no general argument in favor of keeping the other player guessing. However, because moves are simultaneous, each player may still be subjectively unsure of what action the other is taking and therefore may have uncertain beliefs that in turn lead her to be unsure about how she should act. This can lead to mixed-strategy equilibria, and their interpretation in terms of subjectively uncertain but correct beliefs proves particularly important.

4 MIXING IN NON-ZERO-SUM GAMES

The same mathematical method used to find mixed-strategy equilibria in zero-sum games—namely, exploitation-proofness or the opponent's indifference property—can be applied to non-zero-sum games as well, and it can reveal mixed strategy equilibria in some of them. However, in such games the players' interests may coincide to some extent. Therefore the fact that the other player will exploit your systematic choice of strategy to her advantage need not work out to your disadvantage, as was the case with zero-sum interactions. In a coordination game of the kind we studied in Chapter 4, for example, the players are better able to coordinate if each can rely on the other's acting systematically; random actions only increase the risk of coordination failure. As a result, mixed-strategy equilibria have a weaker rationale, and sometimes no rationale at all, in non-zero-sum games. Here we examine mixed-strategy equilibria in some prominent non-zero-sum games and discuss their relevance, or lack thereof.

A. Will Harry Meet Sally? Assurance, Pure Coordination, and Battle of the Sexes

We illustrate mixing in non-zero-sum games by using the assurance version of the meeting game. For your convenience, we reproduce its table (Figure 4.12) as Figure 7.6 below. We consider the game from Sally's perspective first. If she is confident that Harry will go to Starbucks, she also should go to Starbucks. If she is confident that Harry will go to Local Latte, so should she. But if she is unsure about Harry's choice, what is her own best choice?

To answer this question, we must give a more precise meaning to the uncertainty in Sally's mind. (The technical term for this uncertainty, in the theory of probability and statistics, is her subjective uncertainty. In the context where the uncertainty is about another player's action in a game, it is also strategic uncertainty; recall the distinctions we discussed in Chapter 2, Section 2.D.) We gain precision by stipulating the probability with which Sally thinks Harry will

		SALLY	
		Starbucks	Local Latte
HARRY	Starbucks	1, 1	0, 0
	Local Latte	0, 0	2, 2

FIGURE 7.6 Assurance Game

choose one café or the other. The probability of his choosing Local Latte can be any real number between 0 and 1 (that is, between 0% and 100%). We can cover all possible cases by using algebra, letting the symbol p denote the probability (in Sally's mind) that Harry chooses Starbucks; the variable p can take on any real value between 0 and 1. Then $(1 - p)$ is the probability (again in Sally's mind) that Harry chooses Local Latte. In other words, we describe Sally's strategic uncertainty as follows: she thinks that Harry is using a mixed strategy, mixing the two pure strategies, Starbucks and Local Latte, in proportions or probabilities p and $(1 - p)$ respectively. We call this mixed strategy Harry's p-mix, even though for the moment it is purely an idea in Sally's mind.

Given her uncertainty, Sally can calculate the expected payoffs from her actions when they are played against her belief about Harry's p-mix. If she chooses Starbucks, it will yield her $1 \times p + 0 \times (1 - p) = p$. If she chooses Local Latte, it will yield her $0 \times p + 2 \times (1 - p) = 2 \times (1 - p)$. When p is high, $p > 2(1 - p)$, so that Sally is fairly sure that Harry is going to Starbucks, then she does better by also going to Starbucks. Similarly, if p is low, $p < 2(1 - p)$ and if Sally is fairly sure that Harry is going to Local Latte, then she does better by going to Local Latte. If $p = 2(1 - p)$, or $3p = 2$, or $p = 2/3$, the two choices give Sally the same expected payoff. Therefore if she believes that $p = 2/3$, she might be unsure about her own choice, so she might dither between the two.

Harry can figure this out, and that makes him unsure about Sally's choice. Thus Harry also faces subjective strategic uncertainty. Suppose in his mind Sally will choose Starbucks with probability q and Local Latte with probability $(1 - q)$. Similar reasoning shows that Harry should choose Starbucks if $q > 2/3$ and Local Latte if $q < 2/3$. If $q = 2/3$, he will be indifferent between the two actions and unsure about his own choice.

Now we have the basis for a mixed-strategy equilibrium with $p = 2/3$ and $q = 2/3$. In such an equilibrium, these p and q values are simultaneously the actual mixture probabilities and the subjective beliefs of each player about the other's mixture probabilities. The correct beliefs sustain each player's own indifference between the two pure strategies and therefore each player's willingness to mix between the two. This matches exactly the concept of a Nash equilibrium as a system of self-fulfilling beliefs and responses described in Section 3.

The key to finding the mixed-strategy equilibrium is that Sally is willing to mix between her two pure strategies only if her subjective uncertainty about Harry's choice is just right—that is, if the value of p in Harry's p-mix is just right. Algebraically, this idea is borne out by solving for the equilibrium value of p by using the equation $p = 2(1 - p)$, which ensures that Sally gets the same expected payoff from her two pure strategies when each is matched against Harry's p-mix. When the equation holds in equilibrium, it is as if Harry's mixture probabilities are doing the job of keeping Sally indifferent. We emphasize the "as if" because in this game, Harry has no reason to keep Sally indifferent; the outcome is merely a property of the equilibrium. Still, the general idea is worth remembering: in a mixed-strategy Nash equilibrium, each person's mixture probabilities keep the other player indifferent between her pure strategies. We called this the opponent's indifference method in the zero-sum discussion above, and now we see that it remains valid even in non-zero-sum games.

However, the mixed-strategy equilibrium has some very undesirable properties in the assurance game. First, it yields both players rather low expected payoffs. The formulas for Sally's expected payoffs from her two actions, p and $2(1 - p)$, both equal 2/3 when $p = 2/3$. Similarly, Harry's expected payoffs against Sally's equilibrium q-mix for $q = 2/3$ are also both 2/3. Thus, each player gets 2/3 in the mixed-strategy equilibrium. In Chapter 4 we found two pure strategy equilibria for this game; even the worse of them (both choosing Starbucks) yields the players 1 each, and the better one (both choosing Local Latte) yields them 2 each.

The reason the two players fare so badly in the mixed-strategy equilibrium is that when they choose their actions independently and randomly, they create a significant probability of going to different places; when that happens, they do not meet, and each gets a payoff of 0. Harry and Sally fail to meet if one goes to Starbucks and the other goes to Local Latte, or vice versa. The probability of this happening when both are using their equilibrium mixtures is $2 \times (2/3) \times (1/3) = 4/9$.[7] Similar problems exist in the mixed-strategy equilibria of most non-zero-sum games.

A second undesirable property of the mixed-strategy equilibrium here is that it is very fragile. If either player departs ever so slightly from the exact values $p = 2/3$ or $q = 2/3$, the best choice of the other tips to one pure strategy. Once one player chooses a pure strategy, then the other also does better by choosing the same pure strategy, and play moves to one of the two pure-strategy equilibria. This instability of mixed-strategy equilibria is common to many non-zero-sum games. However, some important non-zero-sum games do have mixed-strategy equilibria that are not so fragile. One example considered later in this chapter and in Chapter 13 is the mixed-strategy equilibrium in Chicken, which has an interesting evolutionary interpretation.

Given the analysis of the mixed-strategy equilibrium in the assurance version of the meeting game, you can now probably guess the mixed-strategy

[7]The probability that each chooses Starbucks in equilibrium is 2/3. The probability that each chooses Local Latte is 1/3. The probability that one chooses Starbucks while the other chooses Local Latte is $2/3 \times 1/3$. But that can happen two different ways (once when Harry chooses Starbucks and Sally chooses Local Latte, and again when the choices are reversed) so the total probability of not meeting is $2 \times 2/3 \times 1/3$. See the Appendix to this chapter for more details on the algebra of probabilities.

equilibria for the related non-zero-sum meeting games. In the pure-coordination version (see Figure 4.11), the payoffs from meeting in the two cafés are the same; so the mixed-strategy equilibrium will have $p = 1/2$ and $q = 1/2$. In the battle-of-the-sexes variant (see Figure 4.13), Sally prefers to meet at Local Latte because her payoff is 2 rather than the 1 that she gets from meeting at Starbucks. Her decision hinges on whether her subjective probability of Harry's going to Starbucks is greater than or less than 2/3. (Sally's payoffs here are similar to those in the assurance version, so the critical p is the same.) Harry prefers to meet at Starbucks, so his decision hinges on whether his subjective probability of Sally's going to Starbucks is greater than or less than 1/3. Therefore the mixed-strategy Nash equilibrium has $p = 2/3$ and $q = 1/3$.

B. Will James Meet Dean? Chicken

The non-zero-sum game of Chicken also has a mixed-strategy equilibrium that can be found using the same method developed above, although its interpretations are slightly different. Recall that this is a game between James and Dean, who are trying to *avoid* a meeting; the game table, originally introduced in Figure 4.14, is reproduced here as Figure 7.7.

If we introduce mixed strategies, James's p-mix will entail a probability p of swerving and a probability $1 - p$ of going straight. Against that p-mix, Dean gets $0 \times p - 1 \times (1 - p) = p - 1$ if he chooses Swerve and $1 \times p - 2 \times (1 - p) = 3p - 2$ if he chooses Straight. Comparing the two, we see that Dean does better by choosing swerve when $p - 1 > 3p - 2$, or when $2p < 1$, or when $p < 1/2$, that is, when p is low and James is more likely to choose Straight. Conversely, when p is high and James is more likely to choose Swerve, then Dean does better by choosing Straight. If James's p-mix has p exactly equal to 1/2, then Dean is indifferent between his two pure actions; he is therefore equally willing to mix between the two. Similar analysis of the game from James's perspective when considering his options against Dean's q-mix yields the same results. Therefore $p = 1/2$ and $q = 1/2$ is a mixed-strategy equilibrium of this game.

		DEAN	
		Swerve (Chicken)	Straight (Tough)
JAMES	Swerve (Chicken)	0, 0	−1, 1
	Straight (Tough)	1, −1	−2, −2

FIGURE 7.7 Chicken

The properties of this equilibrium have some similarities but also some differences when compared with the mixed-strategy equilibria of the meeting game. Here, each player's expected payoff in the mixed-strategy equilibrium is low ($-1/2$). This is bad, as was the case in the meeting game, but unlike in that game, the mixed-strategy equilibrium payoff is not worse for both players than either of the two pure-strategy equilibria. In fact, because player interests are somewhat opposed here, each player will do strictly better in the mixed-strategy equilibrium than in the pure-strategy equilibrium that entails his choosing Swerve.

This mixed-strategy equilibrium is again unstable, however. If James increases his probability of choosing Straight to just slightly above 1/2, this change tips Dean's choice to pure Swerve. Then (Straight, Swerve) becomes the pure-strategy equilibrium. If James instead lowers his probability of choosing Straight slightly below 1/2, Dean chooses Straight, and the game goes to the other pure-strategy equilibrium.[8]

5 GENERAL DISCUSSION OF MIXED-STRATEGY EQUILIBRIA

Now that we have seen how to find mixed-strategy equilibria in both zero-sum and non-zero-sum games, it is worthwhile to consider some additional features of these equilibria. In particular, we highlight in this section some general properties of mixed-strategy equilibria. We also introduce you to some results that seem counterintuitive at first, until you fully analyze the game in question.

A. Weak Sense of Equilibrium

The opponent's indifference property described in Section 2 implies that in a mixed-strategy equilibrium, each player gets the same expected payoff from each of her two pure strategies, and therefore also gets the same expected payoff from any mixture between them. Thus mixed-strategy equilibria are Nash equilibria only in a weak sense. When one player is choosing her equilibrium mix, the other has no positive reason to deviate from her own equilibrium mix. But she would not do any worse if she chose another mix or even one of her pure strategies. Each player is indifferent between her pure strategies, or indeed between any mixture of them, so long as the other player is playing her correct (equilibrium) mix. This is also a very general property of mixed-strategy Nash equilibria.

[8]In Chapter 13 we consider a different kind of stability, namely evolutionary stability. The question in the evolutionary context is whether a stable mix of Straight and Swerve choosers can arise and persist in a population of Chicken players. The answer is yes, and the proportions of the two types are exactly equal to the probabilities of playing each action in the mixed-strategy equilibrium. Thus, we derive a new and different motivation for that equilibrium in this game.

This property seems to undermine the basis for mixed-strategy Nash equilibria as the solution concept for games. Why should a player choose her appropriate mixture when the other player is choosing her own? Why not just do the simpler thing by choosing one of her pure strategies? After all, the expected payoff is the same. The answer is that to do so would not be a Nash equilibrium; it would not be a stable outcome, because then the other player would not choose to use her mixture. For example, if Harry chooses pure Starbucks in the assurance version of the meeting game, then Sally can get a higher payoff in equilibrium (1 instead of 2/3) by switching from her 50-50 mix to her pure Starbucks as well.

B. Counterintuitive Changes in Mixture Probabilities in Zero-Sum Games

Games with mixed-strategy equilibria may exhibit some features that seem counterintuitive at first glance. The most interesting of them is the change in the equilibrium mixes that follow a change in the structure of a game's payoffs. To illustrate, we return to Evert and Navratilova and their tennis point.

Suppose that Navratilova works on improving her skills covering down the line to the point where Evert's success using her DL strategy against Navratilova's covering DL drops to 30% from 50%. This improvement in Navratilova's skill alters the payoff table, including the mixed strategies for each player, from that illustrated in Figure 7.1. We present the new table in Figure 7.8.

The only change from the table in Figure 7.1 has occurred in the upper-left-hand cell, where our earlier 50 for Evert is now a 30 and the 50 for Navratilova is now a 70. This change in the payoff table does not lead to a game with a pure-strategy equilibrium, because the players still have opposing interests; Navratilova still wants their choices to coincide, and Evert still wants their choices to differ. We still have a game in which mixing will occur.

But how will the equilibrium mixes in this new game differ from those calculated in Section 2? At first glance, many people would argue that Navratilova should cover DL more often now that she has gotten so much better at doing so. Thus, the assumption is that her equilibrium q-mix should be more heavily weighted toward DL and her equilibrium q should be higher than the 0.6 calculated before.

		NAVRATILOVA	
		DL	CC
EVERT	DL	30, 70	80, 20
	CC	90, 10	20, 80

FIGURE 7.8 Changed Payoffs in the Tennis Point

But when we calculate Navratilova's q-mix by using the condition of Evert's indifference between her two pure strategies, we get $30q + 80(1 - q) = 90q + 20(1 - q)$, or $q = 0.5$. The actual equilibrium value for q, 50%, has exactly the opposite relation to the original q of 60% than what many people's intuition predicts.

Although the intuition seems reasonable, it misses an important aspect of the theory of strategy: the interaction between the two players. Evert will also be reassessing her equilibrium mix after the change in payoffs, and Navratilova must take the new payoff structure *and* Evert's behavior into account when determining her new mix. Specifically, because Navratilova is now so much better at covering DL, Evert uses CC more often in her mix. To counter that, Navratilova covers CC more often, too.

We can see this more explicitly by calculating Evert's new mixture. Her equilibrium p must equate Navrtilova's expected payoff from covering DL, $30p + 90(1 - p)$, with her expected payoff from covering CC, $80p + 20(1 - p)$. So we have $30p + 90(1 - p) = 80p + 20(1 - p)$, or $90 - 60p = 20 + 60p$, or $120p = 70$. Thus, Evert's p must be $7/12$, which is 0.583, or 58.3%. Comparing this new equilibrium p with the original 70% calculated in Section 2 shows that Evert has significantly decreased the number of times she sends her shot DL in response to Navratilova's improved skills. Evert has taken into account the fact that she is now facing an opponent with better DL coverage, and so she does better to play DL less frequently in her mixture. By virtue of this behavior, Evert makes it better for Navratilova also to decrease the frequency of her DL play. Evert would now exploit any other choice of mix by Navratilova, in particular a mix heavily favoring DL.

So is Navratilova's skill improvement wasted? No, but we must judge it properly—not by how often one strategy or the other gets used but by the resulting payoffs. When Navratilova uses her new equilibrium mix with $q = 0.5$, Evert's success percentage from either of her pure strategies is $(30 \times 0.5) + (80 \times 0.5) = (90 \times 0.5) + (20 \times 0.5) = 55$. This is less than Evert's success percentage of 62 in the original example. Thus, Navratilova's average payoff also rises from 38 to 45, and she does benefit by improving her DL coverage.

Unlike the counterintuitive result that we saw when we considered Navratilova's strategic response to the change in payoffs, we see here that her response is absolutely intuitive when considered in light of her expected payoff. In fact, players' expected payoff responses to changed payoffs can never be counterintuitive, although strategic responses, as we have seen, can be.[9] The most interesting aspect of this counterintuitive outcome in players' strategic responses is the message that it sends to tennis players and to strategic game players more

[9]For a general theory of the effect that changing the payoff in a particular cell has on the equilibrium mixture and the expected payoffs in equilibrium, see Vincent Crawford and Dennis Smallwood, "Comparative Statics of Mixed-Strategy Equilibria in Noncooperative Games," *Theory and Decision*, vol. 16 (May 1984), pp. 225–232.

generally. The result here is equivalent to saying that Navratilova should improve her down-the-line coverage so that she does not have to use it so often.

There are other similar examples of possibly counterintuitive results, but they require slightly more ease with algebra than many readers may have. Therefore we postpone them to the next, optional, chapter (Section 8.6).

6 HOW TO USE MIXED STRATEGIES IN PRACTICE

There are several important things to remember when finding or using a mixed strategy in a zero-sum game. First, to use a mixed strategy effectively in such a game, a player needs to do more than calculate the equilibrium percentages with which to use each of her actions. Indeed, in our tennis-point game, Evert cannot simply play DL seven-tenths of the time and CC three-tenths of the time by mechanically rotating seven shots down the line and three shots cross court. Why not? Because mixing your strategies is supposed to help you benefit from the element of surprise against your opponent. If you use a recognizable pattern of plays, your opponent is sure to discover it and exploit it to her advantage.

The lack of a pattern means that, after any history of choices, the probability of choosing DL or CC on the next turn is the same as it always was. If a run of several successive DLs happens by chance, there is no sense in which CC is now "due" on the next turn. In practice, many people mistakenly think otherwise, and therefore they alternate their choices too much compared with what a truly random sequence of choices would require; they produce too few runs of identical successive choices. However, detecting a pattern from observed actions is a tricky statistical exercise that the opponents may not be able to perform while playing the game. As we will see in Section 7, analysis of data from grand-slam tennis finals found that servers alternated their serves too much, but receivers were not able to detect this departure from true randomization.

However, to make sure that your opponent cannot exploit your mixed strategy, you need a truly random pattern of actions on each play of the game. For example, you may want to rely on a computer's ability to generate random numbers for you, from which you can determine your appropriate choice of action. If the computer generates numbers between 1 and 100 and you want to mix pure strategies A and B in a 60-40 split, you may decide to play A for any random number between 1 and 60 and to play B for any random number between 61 and 100. Similarly, you could employ a device such as the color-coded spinners provided in many children's games. For the same 60-40 mixture, you would color 60% of the circle on the spinner in blue, for example, and 40% in red; the first 216 degrees of the circle would be blue, the remaining 144 red. Then you would spin the spinner arrow and play A if it landed in blue but B if it landed in red. You can use the second hand on a watch as the same type of device, but it is important that your watch not be so accurate and synchronized that your opponent can use the same watch and figure out what you are going to do.

The importance of avoiding a predictable system of randomization is clearest in ongoing interactions of a zero-sum nature. Because of the diametrically opposed interests of the players in such games, your opponent always benefits from exploiting your choice of action to the greatest degree possible. Thus, if you play the same game against each other on a regular basis, she will always be on the lookout for ways to break the code that you are using to randomize your moves. If she can do so, she has a chance to improve her payoffs in future plays of the game. Mixing is still justified in single-meet (sometimes called one-shot) zero-sum games because the benefit of tactical surprise remains important.

Finally, players must understand and accept the fact that the use of mixed strategies guards you against exploitation, and gives the best possible expected payoff against an opponent who is making her best choices, but that it is only a probabilistic average. On particular occasions, you can get poor outcomes. For example, the long pass on third down with a yard to go, intended to keep the defense honest, may fail on any specific occasion. If you use a mixed strategy in a situation in which you are responsible to a higher authority, therefore, you may need to plan ahead for this possibility. You may need to justify your use of such a strategy ahead of time to your coach or your boss, for example. They need to understand why you have adopted your mixture and why you expect it to yield you the best possible payoff on average, even though it might yield an occasional low payoff as well. Even such advance planning may not work to protect your "reputation," though, and you should prepare yourself for criticism in the face of a bad outcome.

7 EVIDENCE ON MIXING

A. Zero-Sum Games

Early researchers who performed laboratory experiments were generally dismissive of mixed strategies. To quote Douglas Davis and Charles Holt, "Subjects in experiments are rarely (if ever) observed flipping coins, and when told ex post that the equilibrium involves randomization, subjects have expressed surprise and skepticism."[10] When the predicted equilibrium entails mixing two or more pure strategies, experimental results do show some subjects in the group pursuing one of the pure strategies and others pursuing another, but this does not

[10]Douglas D. Davis and Charles A. Holt, *Experimental Economics* (Princeton: Princeton University Press, 1993), p. 99.

constitute true mixing by an individual player. When subjects play zero-sum games repeatedly, individual players often choose different pure strategies over time. But they seem to mistake alternation for randomization, that is, they switch their choices more often than true randomization would require.

Later research has found somewhat better evidence for mixing in zero-sum games. When laboratory subjects are allowed to acquire a lot of experience, they do appear to learn mixing in zero-sum games. However, departures from equilibrium predictions remain significant. To quote Camerer, "The overall picture is that mixed equilibria do not provide bad guesses about how people behave, on average."[11]

An instance of randomization in practice comes from Malaya in the late 1940s.[12] The British army escorted convoys of food trucks to protect the trucks from communist terrorist attacks. The terrorists could either launch a large-scale attack or create a smaller sniping incident intended to frighten the truck drivers and keep them from serving again. The British escort could be either concentrated or dispersed throughout the convoy. For the army, concentration was better to counter a full-scale attack, and dispersal was better against sniping. For the terrorists, a full-scale attack was better if the army escort was dispersed, and sniping was better if the escort was concentrated. This zero-sum game has only a mixed-strategy equilibrium. The escort commander, who had never heard of game theory, made his decision as follows. Each morning, as the convoy was forming, he took a blade of grass and concealed it in one of his hands, holding both hands behind his back. Then he asked one of his troops to guess which hand held the blade, and he chose the form of the convoy according to whether the man guessed correctly. Although the precise payoff numbers are difficult to judge and therefore we cannot say whether 50-50 was the right mixture, the officer had correctly figured out the need for true randomization and the importance of using a fresh randomization procedure every day to avoid falling into a pattern or making too much alternation between the choices

The best evidence in support of mixed strategies in zero-sum games comes from sports, especially from professional sports, in which players accumulate a great deal of experience of such games, and their intrinsic desire to win is buttressed by large financial gains from winning.

Mark Walker and John Wooders examined the serve-and-return play of top-level players at Wimbledon.[13] They model this interaction as a game with two players, the server and the receiver, in which each player has two pure strategies.

[11]For a detailed account and discussion see Chapter 3 of Colin F. Camerer, *Behvioral Game Theory* (Princeton: Princeton University Press, 2003). The quote is from p. 468 of this book.

[12]R. S. Beresford and M. H. Peston, "A Mixed Strategy in Action," *Operations Research*, vol. 6, no. 4 (December 1955), pp. 173–176.

[13]Mark Walker and John Wooders, "Minimax Play at Wimbledon," *American Economic Review*, vol. 91, no. 5 (December 2001), pp. 1521–1538.

The server can serve to the receiver's forehand or backhand, and the receiver can guess to which side the serve will go and move that way. Because serves are so fast at the top levels of men's singles, the receiver cannot react after observing the actual direction of the serve; rather, the receiver must move in anticipation of the serve's direction. Thus, this game has simultaneous moves. Further, because the receiver wants to guess correctly and the server wants to wrong-foot the receiver, this interaction has a mixed-strategy equilibrium.

If the tennis players are using their equilibrium mixtures in the serve-and-return game, the server should win the point with the same probability whether he serves to the receiver's forehand or backhand. An actual tennis match contains a hundred or more points played by the same two players; thus there is enough data to test whether this implication holds. Walker and Wooders tabulated the results of serves in 10 matches. Each match contains four kinds of serve-and-return combinations: A serving to B and vice versa, combined with service from the right or the left side of the court (Deuce or Ad side). Thus they had data on 40 serving situations and found that in 39 of them the server's success rates with forehand and backhand serves were equal to within acceptable limits of statistical error.

The top-level players must have had enough general experience playing the game, as well as particular experience playing against the specific opponents, to have learned the general principle of mixing and the correct proportions to mix against the specific opponents. However, in one respect the servers' choices departed from true mixing. To achieve the necessary unpredictability, there should be no pattern of any kind in a sequence of serves: the choice of side for each serve should be independent of what has gone before. As we said in reference to the practice of mixed strategies, players can alternate too much, not realizing that alternation is a pattern just as much as repeating the same action a few times would be a pattern. And indeed, the data show that the tennis servers alternated too much. But the data also indicate that this departure from true mixing was not enough for the opponents to pick up and exploit.

Penalty kicks in soccer are another good context in which to study mixed strategies. Two such studies find firm support for predictions of the theory.

Kickers usually kick with the inside of the foot. Therefore the natural direction of kicking for a right-footed kicker is to the goalie's right; for a left-footed kicker it is to the goalie's left. For simplicity of writing we will refer to the natural side as "Right." So the choices are Left and Right for each player. When the goalie chooses Right, it means covering the kicker's natural side. Using a large data set from professional soccer leagues in Europe, Ignacio Palacios-Huerta constructed the payoff table of the kicker's average success probabilities shown in Figure 7.9.[14]

[14]See "Professionals Play Minimax," by Ignacio Palacios-Huerta, *Review of Economics Studies*, vol. 70, no. 20 (2003), pp. 395–415.

		GOALIE	
		Left	Right
KICKER	Left	58	95
	Right	93	70

FIGURE 7.9 Soccer Penalty Kick Success Probabilities in European Major Leagues

This game is similar to the tennis-point game and similarly has only a mixed-strategy equilibrium. Using the opponent's indifference property, it is easy to calculate that the kicker *should* choose Left 38.3% of the time and Right 61.7% of the time. This mixture achieves a success rate of 79.6% no matter what the goalie chooses. The goalie *should* choose the probabilities of covering her Left and Right to be 41.7 and 58.3 respectively; this mixture holds the kicker down to a success rate of 79.6%.

What actually happens? Kickers choose Left 40.0% of times, and goalies choose Left 41.3% of the times. These values are startlingly close to the theoretical predictions. The chosen mixtures are almost exploitation proof. The kickers' mix achieves a success rate of 79.0% against the goalie's Left and 80% against the goalie's Right. The goalies' mix holds kickers down to 79.3% if they choose Left and 79.7% if they choose Right.

In an earlier paper, Pierre-Andre Chiappori, Timothy Groseclose, and Steven Levitt used similar data and found similar results.[15] They also analyzed the whole sequence of choices of each kicker and goalie and did not even find too much alternation. Thus these findings suggest that behavior in soccer penalty kicks is even closer to true mixing than behavior in the tennis serve-and-return game.

B. Non-Zero-Sum Games

Laboratory experiments on games with mixed strategies in non-zero-sum games yield even more negative results than experiments involving mixing in zero-sum games. This is not surprising. As we have seen, in such games the property that each player's equilibrium mixture keeps her opponent indifferent among her pure strategies is a logical property of the equilibrium. Unlike in zero-sum games, in general each player in a non-zero-sum game has no positive or

[15]Pierre-André Chiappori, Timothy Groseclose, and Steven Levitt, "Testing Mixed Strategy Equilibria When Players are Heterogeneous: The Case of Penalty Kicks in Soccer," *American Economic Review*, vol. 92, no. 4 (September 2002), pp. 1138–1151.

purposive reason to keep the other players indifferent. Then the reasoning underlying the mixture calculations is more difficult for players to comprehend and learn. This shows up in their behavior.

In a group of experimental subjects playing a non-zero-sum game, we may see some pursuing one pure strategy and others pursuing another. This type of mixing in the population, although it does not fit the theory of mixed-strategy equilibria, does have an interesting evolutionary interpretation, which we examine in Chapter 13.

Other experimental issues concern subjects who play the game many times. When collecting evidence on play in non-zero-sum games, it is important to rotate or randomize each subject's opponents to avoid tacit cooperation in repeated interactions. In such experiments, players change their actions from one play to the next. But each player's mixture probabilities should be generated using the other player's indifference conditon; therefore these probabilities should not change when a player's own payoffs change. (You will find more on this in Chapter 8, Section 6.B.) But in fact they do.[16] Thus the changes of action from one play to the next may not be true mixing, but some other kind of experimentation.

The overall conclusion is that you should interpret and use mixed-strategy equilibria in non-zero-sum games with, at best, considerable caution.

SUMMARY

Zero-sum games in which one player prefers a coincidence of actions and the other prefers the opposite often have no Nash equilibrium in pure strategies. In these games, each player wants to be unpredictable and thus uses a mixed strategy that specifies a probability distribution over her set of pure strategies. Each player's equilibrium mixture probabilities are calculated using the *opponent's indifference property,* namely that the opponent should get equal *expected payoffs* from all her pure strategies when facing the first player's equilibrium mix. In zero-sum games each player wants to keep the other indifferent in this way, since any clear advantage for the opponent would only work to the first player's disadvantage.

Non-zero-sum games can also have mixed-strategy equilibria that can be calculated from the opponent's indifference property. But here the motivation

[16]Jack Ochs, "Games with Unique Mixed-Strategy Equilibria: An Experimental Study," *Games and Economic Behavior,* vol. 10, no. 1 (July 1995), pp. 202–217.

for keeping the opponent indifferent is weaker or missing; therefore such equilibria have less appeal and are often unstable.

When using mixed strategies, players should remember that their system of randomization should not be predictable in any way. Most important, they should avoid excessive alternation of actions.

Mixed strategies are a special case of continuous strategies but have additional matters that deserve separate study. Mixed-strategy equilibria can be interpreted as outcomes in which each player has correct beliefs about the probabilities with which the other player chooses from among her underlying pure actions. And mixed-strategy equilibria may have some counterintuitive properties when payoffs for players change. Laboratory experiments show only weak support for the use of mixed strategies. But mixed-strategy equilibria give good predictions in many zero-sum situations in sports played by experienced professionals.

KEY TERMS

expected payoff (215) **opponent's indifference property** (222)

SOLVED EXERCISES

S1. "When a two-by-two game has a mixed-strategy equilibrium, a player's equilibrium mixture is designed to yield her the same expected payoff when used against each of the other player's pure strategies." True or false? Explain and give an example of a game that illustrates your answer.

S2. Consider the following game:

		COLUMN	
		Safe	Risky
ROW	Safe	4, 4	4, 1
	Risky	1, 4	6, 6

(a) Which game does this most resemble: tennis, assurance, or chicken? Explain.

(b) Find all of this game's Nash equilibria.

S3. The following table illustrates the money payoffs associated with a two-person simultaneous-play game.

		COLUMN	
		Left	Right
ROW	Up	1, 16	4, 6
	Down	2, 20	3, 40

(a) Find the Nash equilibrium in mixed strategies for this game.

(b) What are the players' expected payoffs in this equilibrium?

(c) The two players jointly get the most money when Row plays Down. However, in the equilibrium, Row does not always play Down. Why not? Can you think of ways in which a more cooperative outcome can be sustained?

S4. Recall Exercise S6 from Chapter 4, about an old lady looking for help crossing the street. Only one person is needed to help her; more are okay but no better than one. You and I are the two people in the vicinity who can help; each has to choose simultaneously whether to do so. Each of us will get pleasure worth 3 from her success (no matter who helps her). But each one who goes to help will bear a cost of 1, this being the value of our time taken up in helping. You were asked to set this up as a game and to write the payoff table in Exercise S6 of Chapter 4. If you did that exercise, you also found all of the pure-strategy Nash equilibria of the game. Now find the mixed-strategy equilibrium of this game.

S5. Revisit the tennis game in Section 2.A of this chapter. Recall that the mixed-strategy Nash equilibrium found in that section had Evert playing DL with probability 0.7, while Navratilova played DL with probability 0.6.

Later in the match Evert injures herself, so her DL shots are much slower and easier for Navratilova to defend. The payoffs are now given by the table:

		NAVRATILOVA	
		DL	CC
EVERT	DL	30	60
	CC	90	20

(a) Relative to the game before her injury (see Figure 7.1), Evert's payoffs are reduced when she plays DL and Navratilova plays either DL or CC. Overall, DL seems much less attractive to Evert than before. Would you expect Evert to play DL more or less, or stay the same? Explain.

(b) Find each player's equilibrium mixture for the game above. What is the expected value of the game to Evert?

(c) How do the equilibrium mixtures found in part (b) compare with those of the original game? Explain why each has changed or hasn't changed.

S6. Undeterred by their experiences wih chicken so far (see Section 4.B), James and Dean decide to increase the excitement (and the stakes) by starting their cars farther apart. This way they can keep the crowd in suspense longer, and they'll be able to accelerate to even higher speeds before they may or may not be involved in a much more serious collision. The new game table thus has a higher penalty for collision:

		DEAN	
		Swerve	Straight
JAMES	Swerve	0, 0	−1, 1
	Straight	1, −1	−10, −10

(a) What is the mixed-strategy Nash equilibrium for this more dangerous version of chicken? Do James and Dean play Straight more or less often than the game shown in Figure 7.7?

(b) What is the expected payoff to each player in the mixed-strategy equilibrium found in part (a)?

(c) James and Dean decide to play the chicken game repeatedly (say, in front of different crowds of reckless youths). Moreover, because they don't want to collide, they collude. They alternate between the two pure-strategy equilibria, so that half the time they play (Swerve, Straight) and half the time they play (Straight, Swerve). Assuming they play an even number of games, what is the average payoff to each of them when they alternate between the two pure-strategy equilibria? Is this better or worse than they can expect from playing the mixed-strategy equilibrium? Why?

(d) After several weeks of not playing chicken as in part (c), James and Dean agree to play again. However, it has been so long since their last meeting that each of them has entirely forgotten which pure-strategy Nash equilibrium they played last time. Worse still, they don't realize this until they're revving their engines moments before starting the game, and—sadly—they live decades before the advent of cell phones. Instead of playing the mixed-strategy Nash equilibrium, each of them tosses a separate coin to decide which strategy to play. What is the expected payoff to James and Dean when each mixes 50-50 in this game? How does this compare with their expected payoff when they play their equilibrium mixtures? Explain why these payoffs are the same or different from those found in part (c).

S7. Consider the following game:

		COLUMN	
		Left	Right
ROW	Up	2, 1	−1, 4
	Down	0, 3	3, 2

(a) Show that this game has no pure-strategy Nash equilibrium.
(b) Find the unique mixed-strategy Nash equilibrium to this game.
(c) If the payoffs for the cell (Up, Left) were changed to (4, 1), would Row's equilibrium p-mixture increase, decrease, or remain the same? Explain your answer.
(d) What would happen to the equilibrium if the payoffs for the (Up, Left) cell were instead changed to (−1, 1)?

S8. Exercise S7 in Chapter 6 introduced a simplified version of baseball, and part (e) pointed out that the simultaneous-move game has no Nash equilibrium in pure strategies. This is because pitchers and batters have conflicting goals. Pitchers want to get the ball *past* batters, but batters want to *connect* with pitched balls. The game table is as follows:

		PITCHER	
		Throw fastball	Throw curve
BATTER	Anticipate fastball	0.30	0.20
	Anticipate curve	0.15	0.35

(a) Find the mixed-strategy Nash equilibrium to this simplified baseball game.
(b) What is each player's expected payoff for the game?

S9. Extending Exercise S8, the pitcher wants to improve his expected payoff in the mixed-strategy equilibrium of this game by slowing down his fastball, thereby making it more similar to a curve ball. Assume that slowing down the pitched fastball changes the payoff to the hitter in the "anticipate fastball/throw fastball" cell from 0.30 to 0.25, and the pitcher's payoff adjusts accordingly. Can this modification improve the pitcher's expected payoff as desired? Explain carefully how you determine the answer here and show your work. Also, explain *why* slowing the fastball can or cannot improve the pitcher's expected payoff in the game.

S10. In the last minute of a football game, the home team is down by 5 points. The home team has the ball, and it's third down and goal from the rival team's 20-yard line. The home team thus has two chances (third down and fourth down) to move the ball a total of 20 yards and win the game. If it fails, the rival team will win. For each team, the payoff for winning the football game is 1, and the payoff for losing is 0.

On each down the home team can choose to run either a 10-yard play or a 20-yard play. For its part, the rival team can anticipate (and prepare for) either the 10-yard play or the 20-yard play. The success rate of the home team's play given each team's strategy is as follows:

		RIVAL TEAM	
		Anticipate 10-yard play	Anticipate 20-yard play
HOME TEAM	10-yard play	80%	100%
	20-yard play	100%	50%

The success of a particular play is all or nothing. If it succeeds, it yields exactly the number of yards intended, but if it fails, the home team gains 0 yards.

As in Chapter 6, this game combines simultaneous and sequential moves. There are three possible outcomes on third down: the home team can gain 0 yards if its play fails, it can gain 10 yards if a 10-yard play succeeds, or it can gain 20 yards and win the game if a 20-yard play succeeds. On fourth down (if necessary), the home team will thus either have 10 more yards or 20 more yards to go. To solve the larger two-down game, we use rollback and start with the fourth down.

(a) Suppose the home team's third-down play failed, so that there are still 20 yards to go on fourth down. What strategy must the home team play in this situation? Given that, what's the rival team's best response?

(b) Given your answer in part (a), what is the home team's expected payoff when there are 20 yards to go on fourth down?

(c) Suppose the home team ran a successful 10-yard play on third down, so that there are 10 yards to go on fourth down. Since the end zone is 10 yards deep, note that the home team has both strategies available. What is the mixed-strategy Nash equilibrium on fourth down in this situation?

(d) Given your answer in part (c), what is the home team's expected payoff when there are 10 yards to go on fourth down?

Using the expected payoffs for fourth down with 20 yards to go and fourth down with 10 yards to go as calculated above, we now roll back and look at what the home team might do on third down. We construct a table for the simultaneous-move game at third down and 20.

(e) What are the expected payoffs for each team when the home team runs the 20-yard play on third down while the rival team anticipates the 20-yard play? (Use your answer to part (b) and remember that there is a 50% success rate for the 20-yard play when the rival team anticipates the 20-yard play.)

(f) What are the expected payoffs to each team when the home team runs the 10-yard play on third down and the rival team anticipates the 10-yard play on that down? (Use your answer in part (d) and remember that there is an 80% success rate for the 10-yard play when the rival team anticipates the 10-yard play.)

(g) What are the expected payoffs to each team when the home team runs the 10-yard play on third down while the rival team anticipates the 20-yard play?

(h) Now construct the game table for third down with 20 yards to go. (Use your answers from parts (e), (f), and (g).

(i) What are the equilibrium p-mix and q-mix for each team on third down?

(j) What is the expected payoff to the home team for the overall two-stage game?

S11. The recalcitrant James and Dean are playing their more dangerous variant of chicken again (see Exercise S6). They've noticed that their payoff for being perceived as "tough" varies with the size of the crowd. The larger the crowd on hand, the more glory and praise each receives from driving straight when his opponent swerves. Smaller crowds, of course, have the opposite effect. Let $k > 0$ be the payoff for appearing "tough." The game may now be represented as:

		DEAN	
		Swerve	Straight
JAMES	Swerve	0, 0	−1, k
	Straight	k, −1	−10, −10

(a) Expressed in terms of k, with what probability does each driver play Swerve in the mixed-strategy Nash equilibrium? Do James and Dean play Swerve more or less often as k increases?

(b) In terms of k, what is the expected value of the game to each player when both are playing the mixed-strategy Nash equilibrium found in part (a)?

(c) At what value of k do both James and Dean mix 50-50 in the mixed-strategy equilibrium?

(d) How large must k be for the average payoff to be positive under the alternating scheme discussed in part (c) of Exercise S6?

S12. Consider the following zero-sum game:

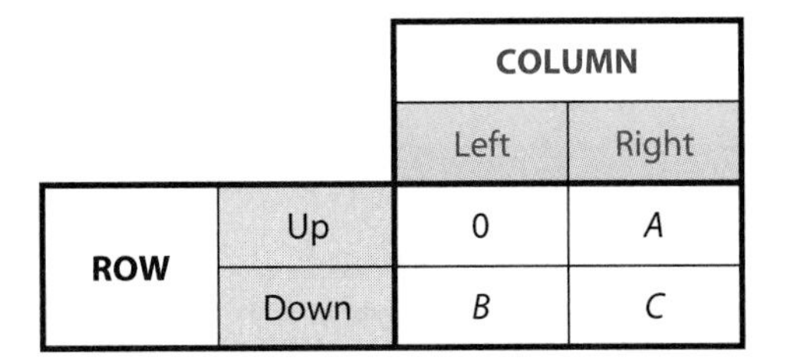

		COLUMN	
		Left	Right
ROW	Up	0	A
	Down	B	C

The entries are the Row player's payoffs, and the numbers A, B, and C are all positive. What other relations among these numbers (for example, $A < B < C$) must be valid for each of the following cases to arise?

(a) At least one of the players has a dominant strategy.

(b) Neither player has a dominant strategy, but there is a Nash equilibrium in pure strategies.

(c) There is no Nash equilibrium in pure strategies, but there is one in mixed strategies.

(d) Given that case (c) holds, write a formula for Row's probability of choosing Up. Call this probability p, and write it as a function of A, B, and C.

S13. **(Optional)** Return to the game between Evert and Navratilova as shown in Figure 7.1. Suppose that Evert is risk-averse, as discussed in Appendix 2 to this chapter, so that she dislikes uncertainty in outcomes. In particular, suppose that Evert has a square-root utility function, so that her utility equals the square root of the payoff listed in the table, and suppose that Navratilova remains risk neutral, so that her utility equals her payoff. Suppose that the players know each other's utility functions, and each player wishes to maximize her expected utility.

(a) Find the mixed-strategy Nash equilibrium of this game.

(b) How did the two players' mixing proportions change, relative to the original case where both players were risk neutral? Explain why this might have happened.

UNSOLVED EXERCISES

U1. Find the Nash equilibria in mixed strategies for the following games.

(a)

		COLUMN	
		Left	Right
ROW	Up	4	−1
	Down	1	2

(b)

		COLUMN	
		Left	Right
ROW	Up	3	2
	Down	1	4

U2. Exercise S11 in Chapter 4 introduced the game Evens or Odds, which has no Nash equilibrium in pure strategies. There is, however, an equilibrium in mixed strategies.

(a) If Anne plays 1 (that is, she puts in one finger) with probability p, what is the expected payoff to Bruce from playing 1, in terms of p? What is his expected payoff from playing 2?

(b) What level of p will make Bruce indifferent between playing 1 and playing 2?

(c) If Bruce plays 1 with probability q, what level of q will make Anne indifferent between playing 1 and playing 2?

(d) Write the mixed-strategy equilibrium of this game. What is the expected payoff of the game to each player?

U3. In football the offense can either run the ball or pass the ball, whereas the defense can either anticipate (and prepare for) a run or anticipate (and prepare for) a pass. The defense wants to guess correctly to reduce the yards gained by the offense, whereas the offense wants its opponents to guess incorrectly so that it can gain more yards. Assume that the expected payoffs (in yards) for the two teams on any given down are as follows:

		DEFENSE	
		Anticipate Run	Anticipate Pass
OFFENSE	Run	1	5
	Pass	9	−3

(a) Show that this game has no pure-strategy Nash equilibrium.
(b) Find the unique mixed-strategy Nash equilibrium to this game.
(c) Explain why the mixture used by the offense is different from the mixture used by the defense.
(d) How many yards is the offense expected to gain per down in equilibrium?

U4. On the eve of a problem-set due date, a professor receives an e-mail from one of her students, who claims to be stuck on one of the problems after working on it for more than an hour. The professor would rather help the student if he has sincerely been working at the problem, but she would rather not render aid if the student is just fishing for hints. Given the timing of the request, she could simply pretend not to have read the e-mail until later. Obviously, the student would rather receive help whether or not he has been working on the problem. But if help isn't coming, he would rather be working instead of slacking, since the problem set *is* due the next day. Assume the payoffs are as follows:

		STUDENT	
		Work and ask for help	Slack and fish for hints
PROFESSOR	Help student	3, 3	−1, 4
	Ignore e-mail	−2, 1	0, 0

(a) What is the mixed-strategy Nash equilibrium to this game?
(b) What is the expected payoff to each of the players?

U5. Return again to the tennis rivals Evert and Navratilova, discussed in Section 2.A. Months later, they meet again in a new tournament. Evert has healed from her injury (see Exercise S5), but during that same time Navratilova has worked very hard on improving her defense against DL serves. The payoffs of the game are now given by the table:

		NAVRATILOVA	
		DL	CC
EVERT	DL	25	80
	CC	90	20

(a) Find each player's equilibrium mixture for the game above.

(b) What happened to Evert's p-mixture compared to the game presented in Section 2.A? Why?

(c) What is the expected value of the game to Evert? Why is it different from the expected value of the original game in Section 2.A?

U6. Section 4.A of this chapter discussed mixing in the battle of the sexes game between Harry and Sally.

(a) What do you expect to happen to the equilibrium values of p and q found in the chapter if Sally decides she really likes Local Latte a lot more than Starbucks, so that the payoffs in the Local Latte, Local Latte cell are now 1, 3? Explain your reasoning.

(b) Now find the new mixed-strategy equilibrium. What are the new equilibrium values of p and q? How do they compare with those of the original game?

(c) What is the expected payoff to each player in the new mixed-strategy equilibrium?

(d) Do you think Harry and Sally might play the mixed-strategy equilibrium in this new version of the game? Explain why or why not.

U7. Consider the following variant of chicken, in which James's payoff from being "tough" when Dean is "chicken" is 2, rather than 1.

		DEAN	
		Swerve	Straight
JAMES	Swerve	0, 0	−1, 1
	Straight	2, −1	−2, −2

(a) Find the mixed-strategy equilibrium in this game, including the expected payoffs for the players.

(b) Compare the results with those of the original game in Section 4.B of this chapter. Is Dean's probability of playing Straight (being tough) higher now than before? What about James's probability of playing Straight?

(c) What has happened to the two players' expected payoffs? Are these differences in the equilibrium outcomes paradoxical in light of the new payoff structure? Explain how your findings can be understood in light of the opponent's-indifference principle.

U8. Lucy offers to play the following game with Charlie: "Let us show pennies to each other, each choosing either heads or tails. If we both show heads, I pay you \$3. If we both show tails, I pay you \$1. If the two don't match, you pay me \$2." Charlie reasons as follows. "The probability of both heads is 1-4, in which case I get \$3. The probability of both tails is 1-4, in which case I get \$1. The probability of no match is 1-2, and in that case I pay \$2. So it is a fair game." Is he right? If not, (a) why not, and (b) what is Lucy's expected profit from the game?

U9. Consider the following game:

		COLUMN	
		Yes	No
ROW	Yes	x, x	0, 1
	No	1, 0	1, 1

(a) For what values of x does this game have a unique Nash equilibrium? What is that equilibrium?

(b) For what values of x does this game have a mixed-strategy Nash equilibrium? With what probability, expressed in terms of x, does each player play Yes in this mixed-strategy equilibrium?

(c) For the values of x found in part (b), is the game an example of an assurance game, a game of chicken, or a game similar to tennis? Explain.

U10. Consider the following game:

		COLUMN	
		L	R
ROW	U	3, 1	2, 2
	D	0, 2	2, 3

(a) Find and describe all pure-strategy Nash equilibria.

(b) If Row plays U with probability p and Column plays L with probability q, demonstrate that $p = 0.75$, $q = 0$ is a mixed-strategy Nash equilibrium for this game.

(c) Is $p = 0.4$, $q = 0$ a mixed-strategy equilibrium for this game? Explain why or why not.

(d) Is $p = 1$, $q = 0.5$ a mixed-strategy equilibrium for this game? Explain why or why not.

(e) How many mixed-strategy equilibria does this game have? Explain.

U11. Consider another simplified version of baseball. The pitcher can throw either a fastball or a curveball; the batter can either swing at the pitch or take (not swing). These choices are simultaneous for each pitch. On the first pitch, if the batter swings at a curveball or takes a fastball, he strikes out and gets a 0. If the batter swings at a fastball, he has a probability of 0.75 of hitting a home run and getting a 1, and a probability of hitting a fly ball and getting a 0. If the batter takes a curveball, there is a second pitch.

On the second pitch, the first three combinations (swing at a curveball, take a fastball, and swing at a fastball) work as before; if the batter takes a curveball on the second pitch, he walks and earns 0.25.

This is a zero-sum game; the batter tries to maximize his expected score (his probability-weighted average payoff), and the pitcher tries to minimize the batter's expected score. Note that the two pitches constitute a sequential-move game, whereas each individual pitch is a simultaneous-move game.

(a) Use the techniques of Chapter 6 to draw a game tree to represent this two-stage game.

(b) Solve this game using rollback; construct a table of payoffs for the second pitch, and use them to determine the table of payoffs for the first pitch. Show that on the first pitch, the batter should take with a probability of 0.8.

(c) What is the pitcher's strategy in the subgame-perfect equilibrium?

(d) What is the batter's expected score in this equilibrium?

(e) Explain intuitively why the batter's probability of swinging is so small.

U12. **(Optional)** Exercises S5 and U5 demonstrate that in zero-sum games such as the Evert-Navratilova tennis rivalry, changes in a player's payoffs can sometimes lead to unexpected or unintuitive changes to her equilibrium mixture. But what happens to the expected value of the game? Consider the following general form of a two-player zero-sum game:

		COLUMN	
		L	R
ROW	U	a	b
	D	c	d

Assume that there is no Nash equilibrium in pure strategies, and assume that a, b, c, and d are all greater than or equal to zero. Can an *increase* in any one of a, b, c, or d lead to a *lower* expected value of the game for Row? If not, prove why not. If so, provide an example.

U13. **(Optional)** Return to the game in Exercise S3. Suppose that both players are risk averse, as discussed in Appendix 2 of this chapter. In particular, suppose that each player has a square-root utility function, and that each player knows the other's utility function. Each player wants to maximize his expected utility.

(a) Find the mixed-strategy Nash equilibrium of this game.
(b) How did the two players' mixing proportions change relative to Exercise S3, where both players were risk neutral? Explain why this might have happened.

■

Appendix: Probability and Expected Utility

To calculate the expected payoffs and mixed-strategy equilibria of games in this chapter, we had to do some simple manipulation of probabilities. Some simple rules govern calculations involving probabilities. Many of you may be familiar with them, but we give a brief statement and explanation of the basics here by way of reminder or remediation, as appropriate. We also state how to calculate expected values of random numerical values.

We also consider the expected utility approach to calculating expected payoffs. When the outcomes of your action in a particular game are not certain, either because your opponent is mixing strategies or because of some uncertainty in nature, you may not want to maximize your expected monetary payoff as we have generally assumed in our analysis to this point; rather, you may want to give some attention to the *riskiness* of the payoffs. As mentioned in Chapter 2, such situations can be handled by using the expected values (which are probability-weighted averages) of an appropriate nonlinear rescaling of the monetary payoffs. We offer here a brief discussion of how this can be done.

You should certainly read this material, but to get real knowledge and mastery of it, the best thing to do is to use it. The chapters to come, especially Chapters 8, 9, 13, and 14, will give you plenty of opportunity for practice.

1 THE BASIC ALGEBRA OF PROBABILITIES

The basic intuition about the probability of an event comes from thinking about the frequency with which this event occurs by chance among a larger set of possibilities. Usually any one element of this larger set is just as likely to occur by chance as any other, so finding the probability of the event in which we are interested is simply a matter of counting the elements corresponding to that event and dividing by the total number of elements in the whole large set.[1]

In any standard deck of 52 playing cards, for instance, there are four suits (clubs, diamonds, hearts, and spades) and 13 cards in each suit (ace through 10 and the face cards—jack, queen, king). We can ask a variety of questions about the likelihood that a card of a particular suit or value—or suit *and* value—might be drawn from this deck of cards: How likely are we to draw a spade? How likely are we to draw a black card? How likely are we to draw a 10? How likely are we to draw the queen of spades? and so on. We would need to know something about the calculation and manipulation of probabilities to answer such questions. If we had two decks of cards, one with blue backs and one with green backs, we could ask even more complex questions ("How likely are we to draw one card from each deck and have them both be the jack of diamonds?"), but we would still use the algebra of probabilities to answer them.

In general, a **probability** measures the likelihood of a particular event or set of events occurring. The likelihood that you draw a spade from a deck of cards is just the probability of the event "drawing a spade." Here the large set has 52 elements—the total number of equally likely possibilities—and the event "drawing a spade" corresponds to a subset of 13 particular elements. Thus you have 13 chances out of the 52 to get a spade, which makes the probability of getting a spade in a single draw equal to $13/52 = 1/4 = 25\%$. To see this another way, consider the fact that there are four suits of 13 cards each, so your chance of drawing a card from any particular suit is one out of four, or 25%. If you made a large number of such draws (each time from a complete deck), then out of 52 times you will not always draw exactly 13 spades; by chance you may draw a few more or a few less. But the chance averages out over different such occasions—over different sets of 52 draws. Then the probability of 25% is the average of the frequencies of spades drawn in a large number of observations.[2]

[1]When we say "by chance," we simply mean that a systematic order cannot be detected in the outcome or that it cannot be determined by using available scientific methods of prediction and calculation. Actually, the motions of coins and dice are fully determined by laws of physics, and highly skilled people can manipulate decks of cards but, for all practical purposes, coin tosses, rolls of dice, or card shuffles are devices of chance that can be used to generate random outcomes. However, randomness can be harder to achieve than you think. For example, a perfect shuffle, where a deck of cards is divided exactly in half and then interleaved by dropping cards one at a time alternately from each, may seem a good way to destroy the initial order of the deck. But Cornell mathematician Persi Diaconis has shown that, after eight of the shuffles, the original order is fully restored. For slightly imperfect shuffles that people carry out in reality, he finds that some order persists through six, but randomness suddenly appears on the seventh! See "How to Win at Poker, and Other Science Lessons," *The Economist,* October 12, 1996. For an interesting discussion of such topics, see Deborah J. Bennett, *Randomness* (Cambridge: Harvard University Press, 1998), chaps. 6–9.

[2]Bennett, *Randomness,* chaps. 4 and 5, offers several examples of such calculations of probabilities.

The algebra of probabilities simply develops such ideas in general terms and obtains formulas that you can then apply mechanically instead of having to do the thinking from scratch every time. We will organize our discussion of these probability formulas around the types of questions that one might ask when drawing cards from a standard deck (or two: blue backed and green backed).[3] This method will allow us to provide both specific and general formulas for you to use later. You can use the card-drawing analogy to help you reason out other questions about probabilities that you encounter in other contexts. One other point to note: In ordinary language, it is customary to write probabilities as percentages, but the algebra requires that they be written as fractions or decimals; thus instead of 25% the mathematics works with 13/52 or 0.25. We will use one or the other, depending on the occasion; be aware that they mean the same thing.

A. The Addition Rule

The first questions that we ask are: If we were to draw one card from the blue deck, how likely are we to draw a spade? And how likely are we to draw a card that is not a spade? We already know that the probability of drawing a spade is 25% because we determined that earlier. But what is the probability of drawing a card that is not a spade? It is the same likelihood of drawing a club or a diamond or a heart instead of a spade. It should be clear that the probability in question should be larger than any of the individual probabilities of which it is formed; in fact, the probability is 13/52 (clubs) + 13/52 (diamonds) + 13/52 (hearts) = 0.75. The *or* in our verbal interpretation of the question is the clue that the probabilities should be added together, because we want to know the chances of drawing a card from any of those three suits.

We could more easily have found our answer to the second question by noting that not getting a spade is what happens the other 75% of the time. Thus the probability of drawing "not a spade" is 75% (100% − 25%) or, more formally, 1 − 0.25 = 0.75. As is often the case with probability calculations, the same result can be obtained here by two different routes, entailing different ways of thinking about the event for which we are trying to find the probability. We will see other examples of this later in this Appendix, where it will become clear that the different methods of calculation can sometimes require vastly different amounts of effort. As you develop experience, you will discover and remember the easy ways or shortcuts. In the meantime, be comforted that each of the different routes, when correctly followed, leads to the same final answer.

[3]If you want a more detailed exposition of the following addition and multiplication rules, as well as more exercises to practice these rules, we recommend David Freeman, Robert Pisani, and Robert Purves, *Statistics*, 3rd ed. (New York: Norton, 1998), chaps. 13 and 14.

To generalize our preceding calculation, we note that, if you divide the set of events, X, in which you are interested into some number of subsets, $Y, Z, \ldots,$ none of which overlap (in mathematical terminology, such subsets are said to be **disjoint**), then the probabilities of each subset occurring must sum to the probability of the full set of events; if that full set of events includes all possible outcomes, then its probability is 1. In other words, if the occurrence of X requires the occurrence of *any one* of several disjoint $Y, Z, \ldots,$ then the probability of X is the sum of the separate probabilities of $Y, Z, \ldots$ Using Prob(X) to denote the probability that X occurs and remembering the caveats on X (that it requires any one of $Y, Z, \ldots$) and on $Y, Z, \ldots$ (that they must be disjoint), we can write the **addition rule** in mathematical notation as $\text{Prob}(X) = \text{Prob}(Y) + \text{Prob}(Z) + \cdots$.

EXERCISE Use the addition rule to find the probability of drawing two cards, one from each deck, such that the two cards have identical faces.

B. The Modified Addition Rule

Our analysis in Section A of this appendix covered only situations in which a set of events could be broken down into disjoint, nonoverlapping subsets. But suppose we ask, What is the likelihood, if we draw one card from the blue deck, that the card is either a spade *or* an ace? The *or* in the question suggests, as before, that we should be adding probabilities, but in this case the two categories "spade" and "ace" are not mutually exclusive, because one card, the ace of spades, is in both subsets. Thus "spade" and "ace" are not disjoint subsets of the full deck. So if we were to sum only the probabilities of drawing a spade (13/52) and of drawing an ace (4/52), we would get 17/52. This would suggest that we had 17 different ways of finding either an ace or a spade when in fact we have only 16—there are 13 spades (including the ace) and three additional aces from the other suits. The incorrect answer, 17/52, comes from counting the ace of spades twice. To get the correct probability in the nondisjoint case, then, we must subtract the probability associated with the overlap of the two subsets. The probability of drawing an ace or a spade is the probability of drawing an ace plus the probability of drawing a spade minus the probability of drawing the overlap, the ace of spades; that is, $13/52 + 4/52 - 1/52 = 16/52 = 0.31$.

To make this more general, if you divide the set of events, X, in which you are interested into some number of subsets $Y, Z, \ldots,$ which may overlap, then the sum of the probabilities of each subset occurring minus the probability of the overlap yields the probability of the full set of events. More formally, the **modified addition rule** states that, if the occurrence of X requires the occurrence of any one of the nondisjoint Y and Z, then the probability of X is the sum of the separate probabilities of Y and Z minus the probability that *both* Y and Z occur: $\text{Prob}(X) = \text{Prob}(Y) + \text{Prob}(Z) - \text{Prob}(Y \text{ and } Z)$.

EXERCISE Use the modified addition rule to find the probability of drawing two cards, one from each deck, and getting at least one face card.

C. The Multiplication Rule

Now we ask, What is the likelihood that when we draw two cards, one from each deck, both of them will be spades? This event occurs if we draw a spade from the blue deck *and* a spade from the green deck. The switch from *or* to *and* in our interpretation of what we are looking for indicates a switch in mathematical operations from addition to multiplication. Thus the probability of two spades, one from each deck, is the product of the probabilities of drawing a spade from each deck, or $(13/52) \times (13/52) = 1/16 = 0.0625$, or 6.25%. Not surprisingly, we are much less likely to get two spades than we were in Section A to get one spade. (Always check to make sure that your calculations accord in this way with your intuition regarding the outcome.)

In much the same way as the addition rule requires events to be disjoint, the multiplication rules requires them to be independent; if we break down a set of events, X, into some number of subsets $Y, Z, \ldots$, those subsets are independent if the occurrence of one does not affect the probability of the other. Our events—a spade from the blue deck and a spade from the green deck—satisfy this condition of independence; that is, drawing a spade from the blue deck does nothing to alter the probability of getting a spade from the green deck. If we were drawing both cards from the same deck, however, then after we had drawn a spade (with a probability of 13/52), the probability of drawing another spade would no longer be 13/52 (in fact, it would be 12/51); drawing one spade and then a second spade from the *same* deck are not **independent events**.

The formal statement of the **multiplication rule** tells us that, if the occurrence of X requires the simultaneous occurrence of *all* the several independent $Y, Z, \ldots$, then the probability of X is the *product* of the separate probabilities of $Y, Z, \ldots$: $\text{Prob}(X) = \text{Prob}(Y) \times \text{Prob}(Z) \times \ldots$.

EXERCISE Use the multiplication rule to find the probability of drawing two cards, one from each deck, and getting a red card from the blue deck and a face card from the green deck.

D. The Modified Multiplication Rule

What if we are asking about the probability of an event that depends on two nonindependent occurrences? For instance, suppose that we ask, What is the likelihood that with one draw we get a card that is both a spade *and* an ace? If we think about this for a moment, we realize that the probability of this event is just the probability of drawing a spade *and* the probability that our card is an ace *given that* it is a spade. The probability of drawing a spade is $13/52 = 1/4$, and the probability of drawing an ace, given that we have a spade, is 1/13. The *and* in our question tells us to take the product of these two probabilities: $(13/52)(1/13) = 1/52$.

We could have gotten the same answer by realizing that our question was the same as asking, What is the likelihood of drawing the ace of spades? The calculation of that probability is straightforward; only 1 of the 52 cards is the ace of spades, so the probability of drawing it must be 1/52. As you see, how you word the question affects how you go about looking for an answer.

In the technical language of probabilities, the probability of a particular event occurring (such as getting an ace), given that another event has already occurred (such as getting a spade) is called the **conditional probability** of drawing an ace, for example, conditioned on having drawn a spade. Then the formal statement of the **modified multiplication rule** is that, if the occurrence of X requires the occurrence of both Y and Z, then the probability of X equals the product of two things: (1) the probability that Y alone occurs, and (2) the probability that Z occurs given that Y has already occurred, or the *conditional probability* of Z, conditioned on Y having already occurred: $\text{Prob}(X) = \text{Prob}(Y \text{ alone}) \times \text{Prob}(Z \text{ given } Y)$.

A third way would be to say that the probability of drawing an ace is 4/52, and the conditional probability of the suit being a spade, given that the card is an ace, is 1/4; so the overall probability of getting an ace of spades is $(4/52) \times 1/4$. More generally, using the terminology just introduced, we have $\text{Prob}(X) = \text{Prob}(Z)\ \text{Prob}(Y \text{ given } Z)$.

> EXERCISE Use the modified multiplication rule to find the probability that, when you draw two cards from a deck, the second card is the jack of hearts.

E. The Combination Rule

We could also ask questions of an even more complex nature than we have tried so far, in which it becomes necessary to use both the addition (or modified addition) and the multiplication (or modified multiplication) rules simultaneously. We could ask, What is the likelihood, if we draw one card from each deck, that we draw *at least one* spade? As usual, we could approach the calculation of the necessary probability from several angles, but suppose that we come at it first by considering all of the different ways in which we could draw at least one spade when drawing one card from each deck. There are three possibilities: either we could get one spade from the blue deck and none from the green deck ("spade *and* none") *or* we could get no spade from the blue deck and a spade from the green deck ("none *and* spade") *or* we could get a spade from each deck ("spade *and* spade"); our event requires that one of these three possibilities occurs, each of which entails the occurrence of both of two independent events.

It should be obvious now, by using the *or*s and *and*s as guides, how to calculate the necessary probability. We find the probability of each of the three possible ways of getting at least one spade (which entails three products of two probabilities each) and sum these probabilities together: $(1/4 \times 3/4) + (3/4 \times 1/4) + (1/4 \times 1/4) = 7/16 = 43.75\%$.

The second approach entails recognizing that "at least one spade" and "not any spades" are disjoint events; together they constitute a sure thing. Therefore the probability of "at least one spade" is just 1 minus the probability of "not any spades." And the event "not any spades" occurs only if the blue card is not a spade 3/4 *and* the green card is not a spade 3/4; so its probability is $3/4 \times 3/4 = 9/16$. The probability of "at least one spade" is then $1 - 9/16 = 7/16$, as we found in the preceding paragraph.

Finally, we can formally state the **combination rule** for probabilities: if the occurrence of X requires the occurrence of *exactly one* of a number of *disjoint* $Y, Z, \ldots$, the occurrence of Y requires that of *all* of a number of *independent* $Y_1, Y_2, \ldots$ the occurrence of Z requires that of *all* of a number of *independent* $Z_1, Z_2, \ldots$, and so on, then the probability of X is the sum of the probabilities of $Y, Z, \ldots$, which are the products of the probabilities $Y_1, Y_2, \ldots, Z_1, Z_2, \ldots$: or

$$\begin{aligned} \text{Prob}(X) &= \text{Prob}(Y) + \text{Prob}(Z) + \cdots \\ &= \text{Prob}(Y_1) \times \text{Prob}(Y_2) \times \cdots + \text{Prob}(Z_1) \times \text{Prob}(Z_2) \times \cdots + \cdots \end{aligned}$$

EXERCISE Suppose we now have a third (orange) deck of cards. Find the probability of drawing at least one spade when you draw one card from each of the three decks.

F. Expected Values

If a numerical magnitude (such as money winnings or rainfall) is subject to chance and can take on any one of n possible values $X_1, X_2, \ldots, X_n$ with respective probabilities $p_1, p_2, \ldots, p_n$, then the **expected value** is defined as the weighted average of all its possible values using the probabilities as weights; that is, as $p_1X_1 + p_2X_2 + \cdots + p_nX_n$. For example, suppose you bet on the toss of two fair coins. You win \$5 if both coins come up heads, \$1 if one shows heads and the other tails, and nothing if both come up tails. Using the rules for manipulating probabilities discussed earlier in this section, you can see that the probabilities of these events are, respectively, 0.25, 0.50, and 0.25. Therefore your expected winnings are $(0.25 \times \$5) + (0.50 \times \$1) + (0.25 \times \$0) = \$1.75$.

In game theory, the numerical magnitudes that we need to average in this way are payoffs, measured in numerical ratings, or money, or, as we will see later in this appendix, utilities. We will refer to the expected values in each context appropriately, for example, as *expected payoffs* or **expected utilities**.

2 ATTITUDES TOWARD RISK AND EXPECTED UTILITY

In Chapter 2, we pointed out a difficulty about using probabilities to calculate the average or expected payoff for players in a game. Consider a game where players gain or lose money, and suppose we measure payoffs simply in money amounts. If a player has a 75% chance of getting nothing and a 25% chance of getting $100, then the expected payoff is calculated as a *probability-weighted average;* the expected payoff is the average of the different payoffs with the probabilities of each as weights. In this case, we have $0 with a probability of 75%, which yields $0.75 \times 0 = 0$ on average, added to $100 with a probability of 25%, which yields $0.25 \times 100 = 25$ on average. That is the same payoff as the player would get from a simple nonrandom outcome that guaranteed him $25 every time he played. People who are indifferent between two alternatives with the same average monetary value but different amounts of risk are said to be **risk-neutral.** In our example, one prospect is riskless ($25 for sure), while the other is risky, yielding either $0 with a probability of 0.75 or $100 with a probability of 0.25, for the same average of $25. In contrast are **risk-averse** people—those who, given a pair of alternatives each with the same average monetary value, would prefer the less risky option. In our example, they would rather get $25 for sure than face the risky $100-or-nothing prospect and, given the choice, would pick the safe prospect. Such risk-averse behavior is quite common; we should therefore have a theory of decision making under uncertainty that takes it into account.

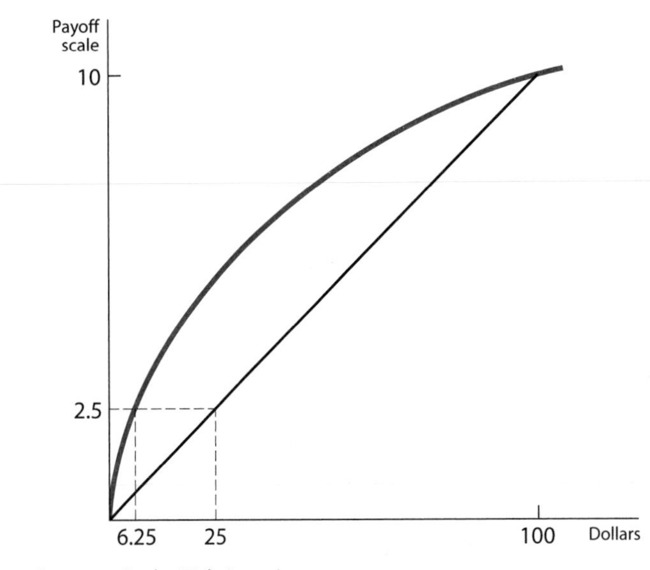

FIGURE 7A.1 Concave Scale: Risk Aversion

We also said in Chapter 2 that a very simple modification of our payoff calculation can get us around this difficulty. We said that we could measure payoffs not in money sums but by using a nonlinear rescaling of the dollar amounts. Here we show explicitly how that rescaling can be done and why it solves our problem for us.

Suppose that, when a person gets D dollars, we define the payoff to be something other than just D, perhaps $\sqrt{D}$. Then the payoff number associated with \$0 is 0, and that for \$100 is 10. This transformation does not change the way in which the person rates the two payoffs of \$0 and \$100; it simply rescales the payoff numbers in a particular way.

Now consider the risky prospect of getting \$100 with probability 0.25 and nothing otherwise. After our rescaling, the expected payoff (which is the average of the two payoffs with the probabilities as weights) is $(0.75 \times 0) + (0.25 \times 10) = 2.5$. This expected payoff is equivalent to the person's getting the dollar amount whose square root is 2.5; because $2.5 = \sqrt{6.25}$, a person getting \$6.25 for sure would also receive a payoff of 2.5. In other words, the person with our square-root payoff scale would be just as happy getting \$6.25 for sure as he would getting a 25% chance at \$100. This indifference between a guaranteed \$6.25 and a 1 in 4 chance of \$100 indicates quite a strong aversion to risk; this person is willing to give up the difference between \$25 and \$6.25 to avoid facing the risk. Figure 7A.1 shows this nonlinear scale (the square root), the expected payoff, and the person's indifference between the sure prospect and the gamble.

What if the nonlinear scale that we use to rescale dollar payoffs is the cube root instead of the square root? Then the payoff from \$100 is 4.64, and the

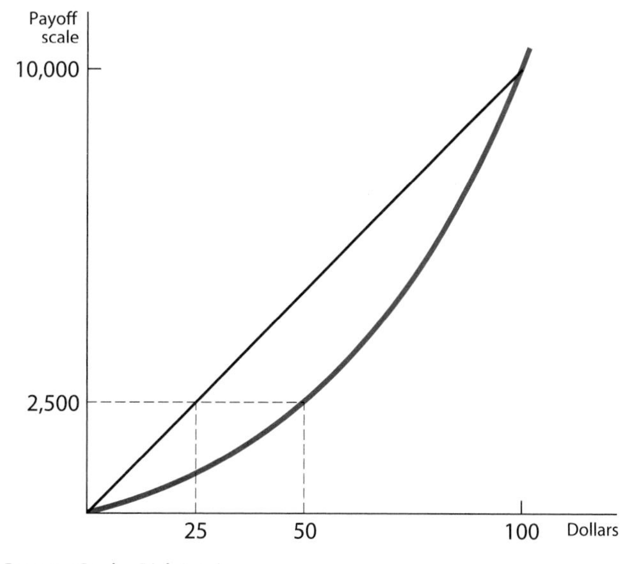

FIGURE 7A.2 Convex Scale: Risk Loving

expected payoff from the gamble is $(0.75 \times 0) + (0.25 \times 4.64) = 1.16$, which is the cube root of 1.56. Therefore a person with this payoff scale would accept only $1.56 for sure instead of a gamble that has a money value of $25 on average; such a person is extremely risk-averse indeed. (Compare a graph of the cube root of *x* with a graph of the square root of *x* to see why this should be so.)

And what if the rescaling of payoffs from *x* dollars is done by using the function x^2? Then the expected payoff from the gamble is $(0.75 \times 0) + (0.25 \times 10{,}000) = 2{,}500$, which is the square of 50. Therefore a person with this payoff scale would be indifferent between getting $50 for sure and the gamble with an expected money value of only $25. This person must be a risk lover because he is not willing to give up any money to get a reduction in risk; on the contrary, he must be given an extra $25 in compensation for the loss of risk. Figure 7A.2 shows the nonlinear scale associated with a function such as x^2.

So by using different nonlinear scales instead of pure money payoffs, we can capture different degrees of risk-averse or risk-loving behavior. A concave scale like that of Figure 7A.1 corresponds to risk aversion, and a convex scale like that of Figure 7A.2 to risk-loving behavior. You can experiment with different simple nonlinear scales—for example, logarithms, exponentials, and other roots and powers—to see what they imply about attitudes toward risk.[4]

This method of evaluating risky prospects has a long tradition in decision theory; it is called the expected utility approach. The nonlinear scale that gives payoffs as functions of money values is called the **utility function;** the square root, cube root, and square functions referred to earlier are simple examples. Then the mathematical expectation, or probability-weighted average, of the utility values of the different money sums in a random prospect is called the expected utility of that prospect. And different random prospects are compared with one another in terms of their expected utilities; prospects with higher expected utility are judged to be better than those with lower expected utility.

Almost all of game theory is based on the expected utility approach, and it is indeed very useful, although is not without flaws. We will adopt it in this book, leaving more detailed discussions to advanced treatises.[5] However, we will indicate the difficulties that it leaves unresolved by means of a simple example in Chapter 8.

[4]Additional information on the use of expected utility and risk attitudes of players can be found in many intermediate microeconomic texts; for example, Hal Varian, *Intermediate Microeconomics,* 7th ed. (New York: Norton, 2006), ch. 12; Walter Nicholson and Christopher Snyder, *Microeconomic Theory,* 10th ed. (New York: Dryden Press, 2008), ch. 7.

[5]See R. Duncan Luce and Howard Raiffa, *Games and Decisions* (New York: Wiley, 1957), chap. 2 and app. 1, for an exposition; and Mark Machina, "Choice Under Uncertainty: Problems Solved and Unsolved," *Journal of Economic Perspectives,* vol. 1, no. 1 (Summer 1987), pp. 121–154, for a critique and alternatives. Although decision theory based on these alternatives has made considerable progress, it has not yet influenced game theory to any significant extent.

SUMMARY

The *probability* of an event is the likelihood of its occurrence by chance from among a larger set of possibilities. Probabilities can be combined by using some rules. The *addition rule* says that the probability of any one of a number of *disjoint* events occurring is the sum of the probabilities of these events; the *modified addition rule* generalizes the addition rule to overlapping events. According to the *multiplication rule,* the probability that all of a number of *independent events* will occur is the product of the probabilities of these events; the *modified multiplication rule* generalizes the multiplication rule to allow for lack of independence, by using *conditional probabilities.*

Judging consequences by taking expected monetary payoffs assumes *risk-neutral* behavior. *Risk aversion* can be allowed, by using the *expected-utility* approach, which requires the use of a *utility function,* which is a concave rescaling of monetary payoffs, and taking its probability-weighted average as the measure of expected payoff.

KEY TERMS

addition rule (254)
combination rule (257)
conditional probability (256)
disjoint (254)
expected utility (257)
expected value (257)
independent events (255)
modified addition rule (254)
modified multiplication rule (256)
multiplication rule (255)
probability (252)
risk-averse (258)
risk-neutral (258)
utility function (260)

Simultaneous-Move Games with Mixed Strategies II: Some General Theory

In Chapter 7 we developed the basic concept of a mixed strategy and a method for finding mixed-strategy equilibria in two-by-two games based on the *opponent's indifference property.* That analysis will suffice for readers who wish only to acquire basic familiarity with the subject of mixing. But more systematic study of games with mixed-strategy equilibria will give a deeper understanding to those who plan to go on to further study of game theory. We provide that deeper analysis in this chapter.

We should say quite frankly that most of this chapter is more difficult than the rest of the book. But we do not apologize for including this material. First, it has an important place in the history of game theory. Von Neumann and Morgenstern's pioneering classic book and Luce and Raiffa's equally classic textbook[1] devote more than a third of their space to the theory of mixed strategies. Second, this topic has its own constituency. Just as many of our other chapters will have special appeal to students whose interests lie in economics, business, political science, or evolutionary biology, this chapter is for students who come from a more mathematical background. Although only a small amount of college mathematics is needed, readers will require a familiarity with algebraic notation

[1]See John von Neumann and Oskar Morgenstern, *Theory of Games and Economic Behavior* (Princeton, NJ: Princeton University Press, 1944; 2nd ed., 1947; 3rd ed., 1953); R. Duncan Luce and Howard Raiffa, *Games and Decisions* (New York: John Wiley and Sons), 1957.

and logic to appreciate fully the material presented here. Those who lack the interest or the preparation for this material can omit this chapter or cover only parts of it without loss of continuity.

In addition to providing a more general approach to solving two-by-two games with mixed strategies, we consider larger games in this chapter as well—those in which players have three or more pure strategies initially available. A substantive new question arises in such games: When will a player's equilibrium mix include all of his pure strategies, and when will it include only a subset? Finally, the basic result regarding general mixed-strategy equilibria, due to von Neumann and Morgenstern, can claim to be the first true theorem of game theory. We describe that general result in Section 7 and introduce the other ideas through numerical examples in the earlier sections.

1 BEST-RESPONSE ANALYSIS

When we developed the concept of a mixed strategy in Section 7.1, we pointed out that a mixture is a special kind of continuous strategy. The probability of using one of the pure strategies is the continuous variable that characterizes the strategy. In Chapter 7, we then developed the opponent's indifference property as the method for calculating equilibrium-mixture probabilities. But when we studied Nash equilibria with continuous strategies in Chapter 5, we developed a different solution method, namely best-response analysis. Now we reconcile the apparent difference in approaches by using best-response analysis to find Nash equilibria in mixed strategies. Although the opponent's indifference property remains a quick method of calculating mixture probabilities, best-response analysis is the more general method that locates all Nash equilibria (if the game has multiple equilibria), pure and mixed.

A. Best-Response Analysis of the Tennis Point

We develop best-response analysis of games with mixed strategies using the same tennis-point example that we used for developing the concept of mixed strategies and their equilibria in Chapter 7. For your convenience we reproduce as Figure 8.1 the payoff table from Figure 7.1, but we now explicitly include a third, mixed, strategy for each player. For Evert, we refer to the mixture as her p-mix, in which we assign the general probability p to choosing her first strategy, DL; we do likewise for Navratilova and her q-mix. The expressions for the expected payoffs when one player's mixture faces a pure strategy of the other player were derived in Section 7.2. The algebraic expression for the bottom right cell where mix meets mix is more complicated, but we won't need it, so we leave that cell blank.

		NAVRATILOVA		
		DL	CC	q-mix
EVERT	DL	50, 50	80, 20	$50q + 80(1-q)$, $50q + 20(1-q)$
	CC	90, 10	20, 80	$90q + 20(1-q)$, $10q + 80(1-q)$
	p-mix	$50p + 90(1-p)$, $50p + 10(1-p)$	$80p + 20(1-p)$, $20p + 80(1-p)$	

FIGURE 8.1 Expected Payoffs for General Mixtures in the Tennis Point

In Section 7.2 we also derived Navratilova's best response to Evert's p-mix. We restate it here:

If $p < 0.7$, choose pure CC ($q = 0$).
If $p = 0.7$, all values of q in the range from 0 to 1 are equal best responses.
If $p > 0.7$, choose pure DL ($q = 1$).

Just as a reminder, if p is low (Evert is quite likely to choose CC), Navratilova does better by covering CC; if p is high (Evert is quite likely to choose DL), Navratilova does better by covering DL; for a critical value of p in between, namely 0.7, Navratilova gets the same expected payoff from either of her pure strategies and therefore also the same from any mixture of the two.

We show this best response graphically in the left-hand panel of Figure 8.2. Navratilova is choosing q for each given value of Evert's p; thus q is a function of p. Therefore p is on the horizontal axis and q is on the vertical axis. We know that, for $p < 0.7$, Navratilova does better by choosing pure CC ($q = 0$); this segment of her best-response curve is the horizontal solid (green) line along the bottom edge of the graph. For p > 0.7, Navratilova does better by choosing pure DL ($q = 1$); this segment of her best-response curve is the horizontal solid line along the top edge of the graph. For $p = 0.7$, Navratilova does equally well with all of her choices, pure and mixed, and so any value of q between 0 and 1 (inclusive) is a best response; the vertical solid line in the graph at $p = 0.7$ shows her best responses for this choice by Evert. Navratilova's best responses are then shown by the three separate line segments joined end to end and shown in green. As we did for general continuous strategies in Chapter 5, we call this construction Navratilova's *best-response curve*. It is conceptually the same as the best-response curves that we drew in Chapter 5; the only difference is that, for mixed strategies, the curve has this special shape. Because it actually consists of three straight-line segments, "curve" is a misnomer, but it is the standard general terminology in this context.

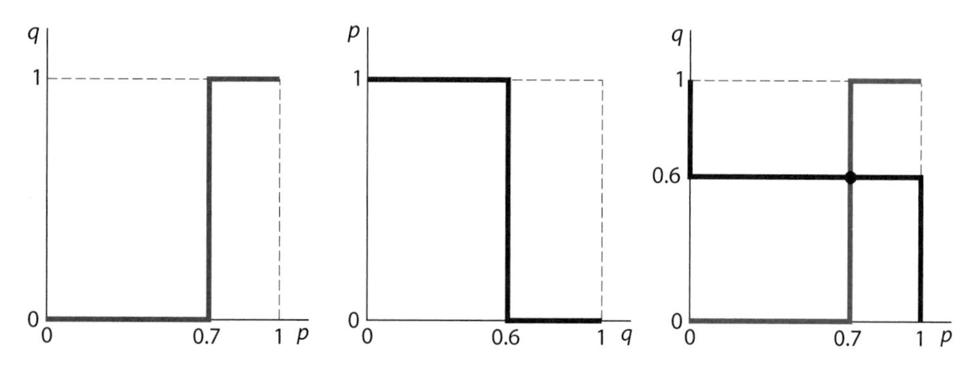

FIGURE 8.2 Best-Response Curves and Nash Equilibrium

Similarly, we recall Evert's best-response rule from Section 7.2:

If $q < 0.6$, choose pure DL ($p = 1$).
If $q = 0.6$, all values of p from 0 to 1 are equal best responses.
If $q > 0.6$, choose pure CC ($p = 0$).

And we depict her best response in the middle panel of Figure 8.2. Here, Evert's best p values are determined in relation to Navratilova's various possible choices for q; so q is on the horizontal axis and p on the vertical axis. For $q < 0.6$, Evert does better playing pure DL ($p = 1$); for $q > 0.6$, she does better with pure CC ($p = 0$); for $q = 0.6$, she does equally well with all choices, pure or mixed. The thick black curve, consisting of three line segments joined end to end, is Evert's best-response curve.

The right-hand panel in Figure 8.2 combines the other two panels by reflecting the middle graph across the diagonal (45° line) so that p is on the horizontal axis and q on the vertical axis and then superimposing this graph on the left-hand graph. Now the green and black curves meet at the point $p = 0.7$ and $q = 0.6$. Here each player's mixture choice is a best response to the other's choice, so we see clearly the derivation of our Nash equilibrium in mixed strategies.

This picture also shows that the best-response curves do not have any other common points. Thus the mixed-strategy equilibrium in this example is the unique Nash equilibrium in the game. What is more, this representation includes pure strategies as special cases corresponding to extreme values of p and q. So we can also see that the best-response curves do not have any points in common at any of the sides of the square where each value of p and q equals either 0 or 1; thus we have another way to verify that the game does not have any pure-strategy equilibria. The mixed-strategy equilibrium in this example is the unique Nash equilibrium in the game.

Observe that each player's best response is a pure strategy for almost all values of her opponent's mixture. Thus Navratilova's best response is pure CC for all of Evert's choices of $p < 0.7$, and it is pure DL for all of Evert's choices of $p > 0.7$. Only for the one crucial value $p = 0.7$ is Navratilova's best response a mixed strategy, as is represented by the vertical part of her three-segment best-response curve in the left panel of Figure 8.2. Similarly, only for the one crucial value $q = 0.6$ of Navratilova's mixture is Evert's best response a mixed strategy—namely, the horizontal segment of her best-response curve in the middle panel of Figure 8.2. But these seemingly exceptional or rare strategies are just the ones that emerge in the equilibrium.

These special values of p and q have an important feature in common. Evert's equilibrium p is where Navratilova's best-response curve has its vertical segment; so Navratilova is indifferent among all her strategies, pure and mixed. Navratilova's equilibrium q is where Evert's best-response curve has its horizontal segment; so Evert is indifferent among all her strategies, pure and mixed. Each player's equilibrium mixture is such that the other player is indifferent among all her mixes. Thus the opponent's indifference property is reconfirmed by our best-response analysis.

The best-response-curve method thus provides a very complete analysis of the game. Like the cell-by-cell inspection method, which examines all the cells of a pure-strategy game, the best-response-curve method is the one to use when we want to locate all of the equilibria, whether in pure or mixed strategies, that a game might have. The best-response-curve diagram can show both types of equilibria in the same place. It could also be used to show equilibria in which one player uses a mixed strategy and the other player uses a pure strategy. Such hybrids occur only in exceptional cases; we give some examples in Section 2.B.

B. Best-Response Analysis in Non-Zero-Sum Games

Best-response analysis can also be used for non-zero-sum games and yields all equilibria, in mixed as well as pure strategies. We illustrate this case using the assurance game, leaving the other two classic games we discussed in Chapter 7 (battle of the sexes and chicken) for you as exercises.

We reproduce the payoff matrix of the assurance version of the meeting game (Figure 4.12 or 7.4) as Figure 8.3, but as in Figure 8.1 for the tennis point, we show an added row and an added column corresponding to Harry's p-mix and Sally's q-mix.[2]

We depict Sally's payoffs from her two pure strategies in relation to Harry's p-mix, and her best-response rule, in the two panels of Figure 8.4. When $p < 2/3$, Sally's best response is pure Local Latte ($q = 0$). The intuition is simple; if Harry is not very likely to go to Starbucks, neither should

[2]Following standard practice, p and q represent the probabilities of choosing the first strategy (Starbucks).

		SALLY		
		Starbucks	Local Latte	q-mix
HARRY	Starbucks	1, 1	0, 0	q, q
	Local Latte	0, 0	2, 2	$2(1-q)$, $2(1-q)$
	p-mix	p, p	$2(1-p)$, $2(1-p)$	

FIGURE 8.3 Expected Payoffs for General Mixtures in the Assurance Game

Sally. Similarly, when $p > 2/3$, Sally's best response is pure Starbucks ($q = 1$). When $p = 2/3$, all values of q are equally good for Sally, so her best response is any combination of Local Latte and Starbucks. This portion of her best-response curve is shown by the vertical straight line from $q = 0$ to $q = 1$ at $p = 2/3$ in the right panel of Figure 8.4.

We find the mixed-strategy equilibrium by superimposing the two best-response curves, as we did for the tennis-point game in Section 1.A above. Because the payoffs of the two players are symmetric, Harry's best-response curve will look just like Sally's with the two axes interchanged. Figure 8.5 shows both of the best-response curves at the same time. Sally's is the thick green curve; Harry's is the thick black curve.

The two best-response curves meet at three points. One is at the top right where $p = 1$ and $q = 1$. This point corresponds to each player choosing Starbucks

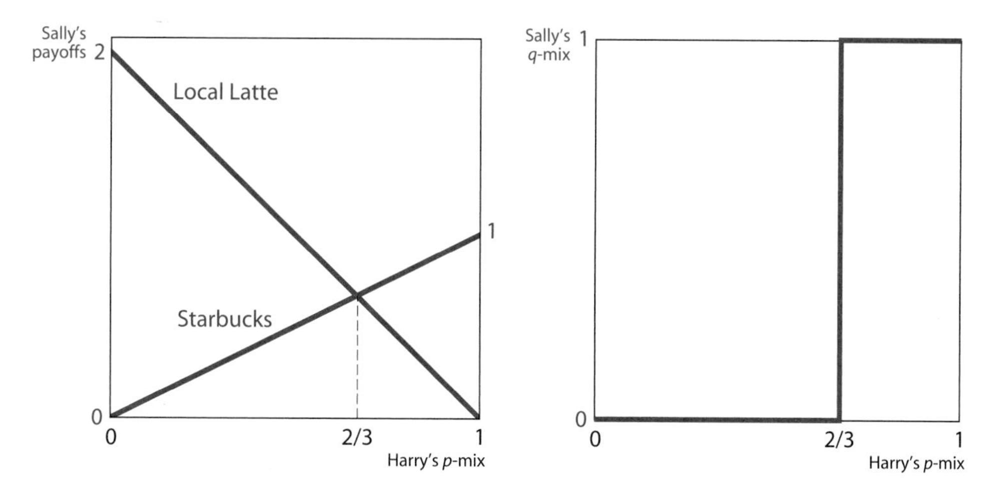

FIGURE 8.4 Sally's Best Responses

for sure, in the correct subjective belief that the other is doing likewise. This is a self-sustaining state of affairs. The second meeting point is at the bottom left, where $p = 0$ and $q = 0$. Here each is not putting any probability on going to Starbucks—that is, is going to Local Latte for sure—in the correct subjective belief that the other is doing likewise. This also is a self-sustaining situation. These are just the two pure-strategy Nash equilibria for this game that we found in Chapter 4.

But there is a third meeting point in Figure 8.5, where $p = 2/3$ and $q = 2/3$. This intersection is just the mixed-strategy equilibrium that we calculated using the opponent's indifference property in Section 7.4.A. As promised, best-response analysis gives us all three of the Nash equilibria for this game, whether in pure or mixed strategies, in one go.

2 MIXING WHEN ONE PLAYER HAS THREE OR MORE PURE STRATEGIES

The discussion of mixed strategies to this point has been confined to games in which each player has only two pure strategies, as well as mixes between them. In many strategic situations, each player has available a larger number of pure strategies, and we should be ready to calculate equilibrium mixes for those cases as well. However, these calculations get complicated quite quickly. For truly complex games, we would turn to a computer to find the mixed-strategy equilibrium. But for some small games, it is possible to calculate equilibria by hand quite easily. The calculation process gives us a better understanding of how the equilibrium works than can be obtained just from looking at a computer-generated solution. Therefore in this section and the next one we solve some larger games.

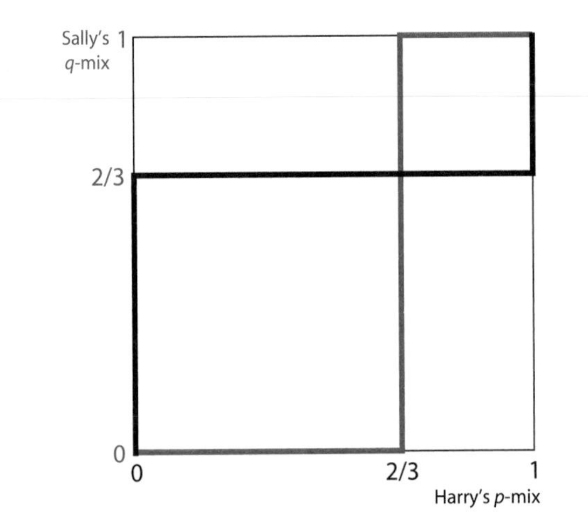

FIGURE 8.5 Mixed-Strategy Equilibrium in the Assurance Game

Here we consider zero-sum games in which one of the players has only two pure strategies, whereas the other has more. In such games, we find that the player who has three (or more) pure strategies typically uses only two of them in equilibrium. The others do not figure in his mix; they get zero probabilities. We must determine which ones are used and which ones are not.[3]

Our example is that of the tennis-point game augmented by giving Evert a third type of return. In addition to going down the line or crosscourt, she now can consider using a lob (a slower but higher and longer return). The equilibrium depends on the payoffs of the lob against each of Navratilova's two defensive stances. We begin with the case that is most likely to arise and then consider a coincidental or exceptional case.

A. A General Case

Evert now has three pure strategies in her repertoire: DL, CC, and Lob. We leave Navratilova with just two pure strategies, Cover DL or Cover CC. The payoff table for this new game can be obtained by adding a Lob row to the table in Figure 8.1. The result is shown in Figure 8.6. Now that you are more familiar with mixing in constant-sum games, we show only Evert's payoffs, and ask you to remember that Navratilova chooses her strategies so as to achieve smaller expected payoffs for Evert.

The payoffs in the first three rows of the table are straightforward. When Evert uses her pure strategies DL and CC, her payoffs against Navratilova's pure strategies or the q-mix are exactly as in Figure 8.1. The third row also is analogous. When Evert uses Lob, we assume that her success percentages against Navratilova's DL and CC are, respectively, 70% and 60%. When Navratilova uses

		NAVRATILOVA		
		DL	CC	q-mix
EVERT	DL	50	80	$50q + 80(1 - q)$
	CC	90	20	$90q + 20(1 - q)$
	Lob	70	60	$70q + 60(1 - q)$
	p-mix	$50p_1 + 90p_2 + 70(1 - p_1 - p_2)$	$80p_1 + 20p_2 + 60(1 - p_1 - p_2)$	

FIGURE 8.6 Payoff Table for Tennis Point with Lob

[3]Even when a player has only two pure strategies, he may not use one of them in equilibrium. The other player then generally finds one of his strategies to be better against the one that the first player does use. In other words, the equilibrium "mixtures" collapse to the special case of pure strategies. But when one or both players have three or more strategies, we can have a genuinely mixed-strategy equilibrium where some of the pure strategies go unused.

her q-mix, using DL a fraction q of the time and CC a fraction $(1 - q)$ of the time, Evert's expected payoff from Lob is $70q + 60(1 - q)$; therefore that is the entry in the cell where Evert's Lob meets Navratilova's q-mix.

The really new feature here is the last row of the table. Evert now has three pure strategies, so she must now consider three different actions in her mix. The mix cannot be described by just one number p. Rather, we suppose that Evert plays DL with probability p_1 and CC with probability p_2, leaving Lob to get the remaining probability, $1 - p_1 - p_2$. Thus we need two numbers, p_1 and p_2, to define Evert's p-mix. Each of them, being a probability, must be between 0 and 1. Moreover, the two must add to something no more than 1; that is, they must satisfy the condition $p_1 + p_2 \leq 1$, because the probability $(1 - p_1 - p_2)$ of using Lob must be nonnegative.

Using this characterization of Evert's p-mix, then, we see that her expected payoff, when Navratilova plays her pure strategy DL, is given by $50p_1 + 90p_2 + 70(1 - p_1 - p_2)$. This is the entry in the first cell of the last row of the table in Figure 8.6. Evert's expected payoff from using her p-mix against Navratilova's CC is similarly $80p_1 + 20p_2 + 60(1 - p_1 - p_2)$. We do not show the expression for the payoff of mix against mix, because it is too long and we do not need it for our calculations.

Technically, before we begin looking for a mixed-strategy equilibrium, we should verify that there is no pure-strategy equilibrium. This is easy to do, however, so we leave it to you and turn to mixed strategies.

The easiest way to solve for a mixed-strategy equilibrium in a constant-sum game where one player has just two pure strategies, and the other has any number, is to use the minimax method from the perspective of the player who has just two pure strategies; here that player is Navratilova.[4] This approach works because Navratilova's mixture can be specified by using just one variable—namely, the single probability (q) used to define her mixed strategy. That probability, of choosing DL in this case, fully specifies her mixed strategy; after q is known, the probability of choosing CC is simply $(1 - q)$.

Figure 8.7 shows Evert's expected payoffs (success percentages) from playing each of her pure strategies DL, CC, and Lob as the q in Navratilova's q-mix varies over its full range from 0 to 1. These graphs are just those of the expressions in the right-hand column of Figure 8.6. Given the usual worst-case assumption that is appropriate in zero-sum games, Navratilova's calculation of her minimax strategy is as follows. For each q, if Navratilova were to choose that q-mix in equilibrium, Evert's best response would be to choose the strategy that gives her (Evert) the highest payoff. Evert's best response, which is also the worst-case

[4]For many amusing uses of this method, see John D. Williams, *The Compleat Strategyst* (New York: McGraw-Hill, 1954; reprint, New York: Dover, 1986).

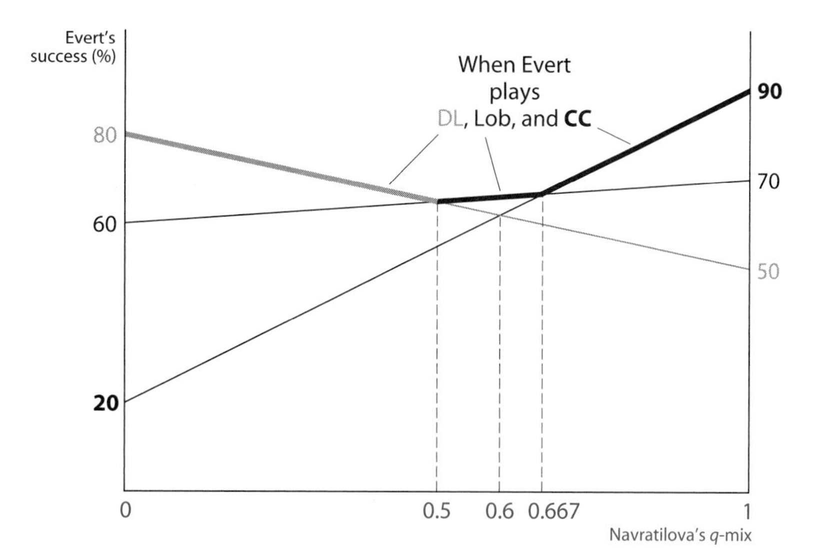

FIGURE 8.7 Diagrammatic Solution for Navratilova's q-Mix

scenario for Navratilova, is then shown on the highest of the three lines at that q. We show this set of worst-case outcomes with the thicker lines in Figure 8.7; these outcomes are formed from the *upper envelope* of the three payoff lines. Navratilova's optimal choice of q will make Evert's payoff as low as possible (thereby making her own as large as possible) from this set of worst-case outcomes.

To be more precise about Navratilova's optimal choice of q, we must calculate the coordinates of the kink points in the line showing her worst-case outcomes. The value of q at the left-most kink in this line makes Evert indifferent between DL and Lob. That q must equate the two payoffs from DL and Lob when used against the q-mix. Setting those two expressions equal gives us $50q + 80(1 - q) = 70q + 60(1 - q)$ and $q = 20/40 = 1/2$. This first kink is thus at $q = 0.5$, or 50%. Evert's expected payoff at this point is $50 \times 0.5 + 80 \times 0.5 = 70 \times 0.5 + 60 \times 0.5 = 65$. At the second (rightmost) kink, Evert is indifferent between CC and Lob. Thus the q value at this kink is the one that equates the CC and Lob payoff expressions. Setting $90q + 20(1 - q) = 70q + 60(1 - q)$, we find $q = 40/60$; the rightmost kink is at $q = 0.667$, or 66.7%. Here, Evert's expected payoff is $90 \times 0.667 + 20 \times 0.333 = 70 \times 0.667 + 60 \times 0.333 = 66.67$.

Now we can explicitly describe Evert's best responses to Navratilova's different choices of q. Evert's best response is DL when $q < 0.5$, CC when $q > 0.667$, and Lob when $0.5 < q < 0.667$. As usual, Evert's best response is pure for most values of q. When $q = 0.5$, Evert is indifferent between DL and Lob and therefore equally indifferent between those two pure strategies and any mixture of them.

When $q = 0.667$, she is indifferent between CC and Lob and therefore equally indifferent between those two pure strategies and any mixture of them.

Figure 8.7 also shows that, of all the worst-case scenarios for Navratilova, the best (or least bad) occurs at the left kink, where $q = 0.5$ and Evert's expected payoff is 65. The thick line shows all the maxima (for each q), and this point represents the minimum among them; this is Navratilova's minimax. At this point, Evert achieves the smallest of the payoffs associated with choosing her best response to each q that Navratilova might pick. Therefore, at $q = 0.5$, Navratilova achieves the largest of her worst-case payoffs, and she should choose this q in equilibrium.

When Navratilova chooses $q = 0.5$, Evert is indifferent between DL and Lob, and either of these choices gives her a better payoff than does CC. Therefore Evert will not use CC at all in equilibrium. CC will be an unused strategy in her equilibrium mix.

Now we can proceed with the equilibrium analysis as if this were a game with just two pure strategies for each player: DL and CC for Navratilova, and DL and Lob for Evert. We are back in familiar territory. Therefore we leave the calculation to you and just tell you the result. Evert's optimal mixture in this game entails her using DL with probability 0.25 and Lob with probability 0.75. Evert's expected payoff from this mixture, taken against Navratilova's DL and CC, respectively, is

$$50 \times 0.25 + 70 \times 0.75 = 80 \times 0.25 + 60 \times 0.75 = 65.$$

This payoff is Evert's maximin value, and it equals Navratilova's minimax, in conformity with the general result on mixed-strategy Nash equilibrium in zero-sum games. Thus, in equilibrium Evert mixes her DL and Lob with probabilities 0.25 and 0.75. Navratilova mixes her DL and CC with probabilities 0.5 each. The maximin (minimax) payoff to each is 65.

We could not have started our analysis with this two-by-two game, because we did not know in advance which of her three strategies Evert would not use. But we can be confident that in the general case there will be one such strategy. When the three expected payoff lines take the most general positions, they intersect pair by pair rather than all crossing at a single point. Then the upper envelope has the shape that we see in Figure 8.7. Its lowest point is defined by the intersection of the payoff lines associated with two of the three strategies. The payoff from the third strategy lies below the intersection at this point, so the player choosing among the three strategies does not use that third one.

B. Exceptional Cases

The positions and intersections of the three lines of Figure 8.7 depend on the payoffs specified for the pure strategies. We chose the payoffs for that particular game to show a general configuration of the lines. But if the payoffs stand in very specific relationships to each other, we can get some exceptional configurations with different results. We describe the possibilities here but leave it to you to redraw the diagrams for these cases.

First, if Evert's payoffs from Lob against Navratilova's DL and CC are equal, then the line for Lob is horizontal, and a whole range of q-values achieve Navratilova's minimax. For example, if the two payoffs in the Lob row of the table in Figure 8.6 are 70 each, then it is easy to calculate that the left kink in a revised Figure 8.7 would be at $q = 1/3$ and the right kink at $q = 5/7$. For any q in the range from 1/3 to 5/7, Evert's best response is Lob, and we get an unusual equilibrium in which Evert plays a pure strategy and Navratilova mixes. Further, Navratilova's equilibrium mixture probabilities are indeterminate within the range from $q = 1/3$ to $q = 5/7$.

Second, if Evert's payoffs from Lob against Navaratilova's DL and CC are lower than those of Figure 8.7 by just the right amounts (or those of the other two strategies are higher by just the right amounts), all three lines can meet in one point. For example, if the payoffs of Evert's Lob are 66 and 56 against Navratilova's DL and CC, respectively, instead of 70 and 60, then for $q = 0.6$ Evert's expected payoff from the Lob becomes $66 \times 0.6 + 56 \times 0.4 = 39.6 + 22.6 = 62$, the same as that from DL and CC when $q = 0.6$. Then Evert is indifferent among all three of her strategies when $q = 0.6$ and is equally willing to mix among them.

In this special case, Evert's equilibrium mixture probabilities are not fully determinate. Rather, a whole range of mixtures, including some where all three strategies are used, can do the job of keeping Navratilova indifferent between her DL and CC and therefore willing to mix. However, to achieve her minimax, Navratilova must use the mixture with $q = 0.6$. If she does not, Evert's best response will be to switch to one of her pure strategies, and this will work to Navratilova's detriment. We do not dwell on the determination of the precise range over which Evert's equilibrium mixtures can vary, because this case can only arise for exceptional combinations of the payoff numbers and is therefore relatively unimportant.

C. Case of Domination by a Mixed Strategy

What if Evert's payoffs from using her Lob against Navratilova's DL and CC are even lower than the values that make all three lines intersect in one point? Figure 8.8 illustrates such a payoff matrix.

When we graph the expected payoff lines from Evert's three pure strategies against Navratilova's choice of q, we now find that Lob has shifted down from its position in Figure 8.7. Figure 8.9 shows the new configuration of lines and the determination of Navratilova's minimax. The calculations that give us the positions of the three lines labeled DL, CC, and Lob, and their intersections follow

		NAVRATILOVA	
		DL	CC
EVERT	DL	50	80
	CC	90	20
	Lob	75	30

FIGURE 8.8 Tennis Point When Lob Is Never a Best Response

the same procedures as before; so we omit the details. The line labeled Mix is explained soon.

It is clear that, with these numbers, Lob is not a very good strategy for Evert. In fact, the line showing the payoffs from Lob lies everywhere below either DL or CC—and so below the upper envelope of those lines—as well as below the point of intersection of the DL and CC lines. Thus, for each q in Navratilova's mix, at least one of the pure strategies DL or CC gives Evert a higher payoff than does Lob. Figure 8.9 shows that if $q < 0.667$, DL is a better response for Evert than Lob; if $q > 0.4$, CC is better than Lob; and when $0.4 < q < 0.667$, both DL and CC are better than Lob. In other words, Lob is *never the best response* for Evert.

However, it is also true that Lob is not dominated by either DL or CC. If $q < 0.4$, Lob does better than CC, whereas if $q > 0.667$, Lob does better than

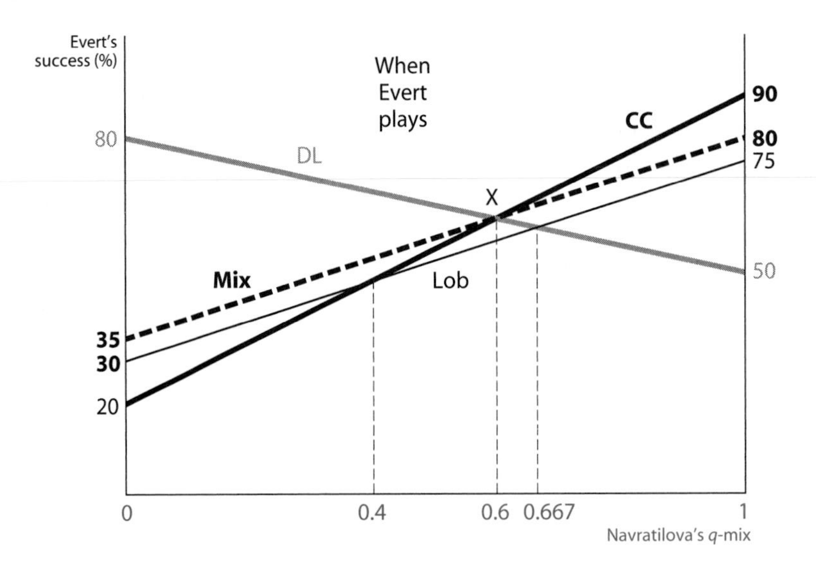

FIGURE 8.9 Domination by a Mixed Strategy

DL. Thus it appears that Lob cannot be eliminated from Evert's consideration by using dominance reasoning. Instead, we use reasoning below that is related to the concept of rationalizability from Chapter 5.

But now that we are allowing mixed strategies, we can consider mixtures of DL and CC as well. Can one such mixture dominate Lob? Yes. Consider a mixture between DL and CC with respective probabilities p and $(1 - p)$. Its payoff against Navratilova's pure DL is $50p + 90(1 - p)$, and its payoff against Navratilova's pure CC is $80p + 20(1 - p)$. Therefore its payoff against Navratilova's q-mix is $[50p + 90(1 - p)]\, q + [80p + 20(1 - p)](1 - q) = [50q + 80(1 - q)]p + [90q + 20(1 - q)](1 - p)$. This expression is an average of the heights of the DL and CC lines, with proportions p and $(1 - p)$.

In other words, Evert's expected payoff from her p-mix between DL and CC against Navratilova's q-mix is a straight line that averages the two separate DL and CC lines. Naturally, the average line must pass through the point of intersection of the two separate DL and CC lines; if, for some q, the two pure strategies DL and CC give equal payoffs, so must any mixture of them.

If $p = 1$ in Evert's mixture, she plays pure DL and the line for her mixture is just the DL line; if $p = 0$, the mixture line is the CC line. As p ranges from 0 to 1, the line for the p-mix rotates from the pure CC to the pure DL line, always passing through the point of intersection at $q = 0.6$. Among all such lines corresponding to different values of p, one will be parallel to the line for Lob. This line is shown in Figure 8.9 as the fourth (dashed) line labeled Mix. Because the line for Lob lies below the point of intersection of DL and CC, it also lies entirely below this parallel line representing a particular mixture of DL and CC. Then, no matter what q may be, Lob yields a lower expected payoff than this particular mixed strategy. In other words, Lob is dominated by this mixed strategy.

It remains to find the probabilities that define the mixture that dominates Lob for Evert. By construction, the Mix line is parallel to the Lob line and passes through the point at which $q = 0.6$ and the expected payoff is 62. Then it is easy to calculate the vertical coordinates of the Mix line. We use the fact that the slope of Lob $= (75 - 30)/(1 - 0) = 45$. Then the vertical coordinate of the Mix line when $q = 0$ is given by v in the equation $45 = (62 - v)/(0.6 - 0)$, or $v = 35$; similarly, when $q = 1$, the vertical coordinate solves $45 = (v - 62)/(1 - 0.6)$, which gives us $v = 80$. For any value of q between 0 and 1, the vertical coordinate of the Mix line is then $80 \times q + 35 \times (1 - q)$. Now compare this result with the expression for the general p-mix line just derived. For the two to be consistent, the part of that p-mix expression that multiplied q—$50p + 90(1 - p)$—must equal 80, and $80p + 20(1 - p)$ must equal 35. Both of them imply that $p = 0.25$.

Thus Evert's mixture of DL with probability 0.25 and CC with probability 0.75 yields the Mix line drawn in Figure 8.9. This mixed strategy gives her a better expected payoff than does her Lob, for each and every value of Navratilova's q. In other words, this particular mixture dominates Lob from Evert's perspective.[5]

Now we have shown that if a strategy is never a best response, then we can find a mixed strategy that dominates it.[6] In the process, we have expanded the scope of the concept of dominance to include domination by mixed strategies. The converse also is true; if a strategy is dominated by another strategy, albeit a mixed one, it can never be a best response to any of the other player's strategies. We can then use all of the other concepts associated with dominance that were developed in Chapter 4, but now allowing for mixed strategies also. We can do successive or iterated elimination of strategies, pure or mixed, that are dominated by other strategies, pure or mixed. If the process leaves just one strategy for each player, the game is dominance solvable and we have found a Nash equilibrium. More typically, the process only narrows down the range of strategies. In Chapter 5, we defined as *rationalizable* the set of strategies that remain after iterated elimination of pure strategies that are never best responses. Now we see that in two-player games we can think of rationalizable strategies as the set that survives after doing all possible iterated elimination of strategies that are dominated by other pure or mixed strategies.[7]

3 MIXING WHEN BOTH PLAYERS HAVE THREE STRATEGIES

As we saw in our two-by-three strategy example in the preceding section, a player mixing among three pure strategies can choose the probabilities of any two of them independently (as long as they are nonnegative and add up to no more than 1); then the probability of the third must equal 1 minus the sum of the probabil-

[5]We constructed the parallel line to guarantee dominance. Other lines through the point of intersection of the DL and CC lines also can dominate Lob as long as their slopes are not too different from that of the Lob line.

[6]In the example of Section 5.4 (Figure 5.6), we saw that Column's strategy C4 is never a best response but it is not dominated by any of the pure strategies C1, C2, or C3. Now we know that we can look for domination by a mixed strategy. In that game, it is easy to see that C4 is (strictly) dominated by an equal-probability mixture of C1 and C3.

[7]This equivalence between "never a best response" and "dominated by a mixed strategy" works fine in two-player games, but an added complication arises in many-player games. Consider a game with three players, A, B, and C. One of A's strategies—say, A1—may never be a best response to any pure strategies or independently mixed strategies of B and C, but A1 may fail to be dominated by any other pure or mixed strategy for A. However, if A1 is never a best response to any pure strategies or arbitrarily correlated mixed strategies of B and C, then it must be dominated by another of A's pure or mixed strategies. A complete treatment of this requires more advanced game theory, so we merely mention it. See Andreu Mas-Colell, Michael Whinston, and Jerry R. Green, *Microeconomic Theory* (New York: Oxford University Press, 1995), pp. 242–245 and 262–263.

ities of the other two. Thus we need two variables to specify a mix.[8] When both players have three strategies, we cannot find a mixed-strategy equilibrium without doing two-variable algebra. In many cases, such algebra is still manageable.

A. Full Mixture of All Strategies

Consider a simplified representation of a penalty kick in soccer. Suppose the kicker has just three pure strategies: kick to the left, right, or center. (As in Chapter 7, left and right refer to the goalie's left or right, for a right-footed kicker.) Then he can mix among these strategies, with probabilities denoted by p_L, p_R, and p_C, respectively. Any two of them can be taken to be the independent variables and the third expressed in terms of them. If p_L and p_R are made the two independent choice variables, then $p_C = 1 - p_L - p_R$. The goalie also has three pure strategies—namely, move to the kicker's left (the goalie's own right), move to the kicker's right, or continue to stand in the center—and can mix among them with probabilities q_L, q_R, and q_C, two of which can be chosen independently.

Best-response analysis also works in this context. The goalie would choose his two independent variables, say (q_L, q_R), as his best response to the kicker's two, (p_L, p_R), and vice versa. The Nash equilibrium occurs when each is choosing his best response to the other's mix. However, because each is choosing two magnitudes in response to the other's two, we can't use best-response graphs, because they would have to be in four dimensions.

Instead, we use the principle of the opponent's indifference, which enables us to focus on one player's mixture probabilities. They should be such that the other is indifferent among all the pure strategies that constitute his mixture. This gives us a set of equations that can be solved for the mixture probabilities. In the soccer example, the kicker's (p_L, p_R) would satisfy two equations expressing the requirement that the goalie's expected payoff from using his left should equal that from using his right and that the goalie's expected payoff from using his right should equal that from using his center. (Then the equality of expected payoffs from left and center follows automatically and is not a separate equation.) With more pure strategies, the number of the probabilities to be solved for and the number of equations that they must satisfy also increase.

Figure 8.10 shows the payoff matrix with the kicker as the Row player and the goalie as the Column player. (Unlike in the example in Chapter 7, these are not real data, but similar rounded numbers to simplify calculations.) Because the kicker wants to maximize the percentage probabilities that he successfully scores a goal and the goalie wants to minimize the probability that he lets the

[8]More generally, if a player has N pure strategies, then her mix has $(N - 1)$ independent variables, or "degrees of freedom of choice."

		GOALIE		
		Left	Center	Right
KICKER	Left	45	90	90
	Center	85	0	85
	Right	95	95	60

FIGURE 8.10 Mixing in Soccer Penalty Kick When All Pure Strategies Are Used

goal through, this is a zero-sum game. We show the payoffs from the kicker's (the Row player's) perspective. For example, if the kicker kicks to his left while the goalie moves to the kicker's left (the top-left-corner cell), we suppose that the kicker still succeeds 45% of the time. But if the kicker kicks to his right and the goalie goes to the kicker's left, then the kicker has a 90% chance of scoring; we suppose a 10% probability that he might kick wide or too high. If the kicker kicks to his right (bottom row of the matrix), his probabilities of success are 95% if the goalie guesses wrong and 60% if the goalie guesses correctly. You can experiment with different payoff numbers that you think might be more appropriate.

It is easy to verify that the game has no equilibrium in pure strategies. So suppose the kicker is mixing with probabilities p_L, p_R, and $p_C = 1 - p_L - p_R$. Against each of the goalie's pure strategies, this mixture yields the following payoffs:

$$\begin{aligned}
&\text{Left:} && 45p_L + 85p_C + 95p_R = 45p_L + 85(1 - p_L - p_R) + 95p_R \\
&\text{Center:} && 90p_L + 0p_C + 95p_R = 90p_L + 95p_R \\
&\text{Right:} && 90p_L + 85p_C + 60p_R = 90p_L + 85(1 - p_L - p_R) + 60p_R
\end{aligned}$$

The goalie wants these numbers to be as small as possible. But in a mixed-strategy equilibrium, the kicker's mixture must be such that the goalie is indifferent among his pure strategies. Therefore all three of these expressions must be equal in equilibrium.

Equating the Left and Right expressions and simplifying, we have $45p_L = 35p_R$, or $p_R = (9/7)p_L$. Next, equate the Center and Right expressions and simplify, by using the link between p_L and p_R just obtained. This gives

$$\begin{aligned}
&90p_L + 95(9p_L/7) = 90p_L + 85[1 - p_L - (9p_L/7)] + 60(9p_L/7), \\
&\text{or} \quad [85 + 120(9/7)]\, p_L = 85, \\
&\text{which yields } p_L = 0.355.
\end{aligned}$$

Then we get $p_R = 0.355(9/7) = 0.457$, and finally $p_C = 1 - 0.355 - 0.457 = 0.188$. The kicker's payoff from this mixture against any of the goalie's pure strategies and therefore against any mixture of them can then be calculated by using any of the preceding three payoff lines; the result is 75.4.

The goalie's mixture probabilities can be found by writing down and solving the equations for the kicker's indifference among his three pure strategies against the goalie's mixture. We will do this in detail for a slight variant of the same game in Section 3.B, so we omit the details here and just give you the answer: $q_L = 0.325$, $q_R = 0.561$, and $q_C = 0.113$. The kicker's payoff from any of his pure strategies when played against the goalie's equilibrium mixture is 75.4. Note that this payoff is the same as the number found when calculating the kicker's mix; this is just the maximin = minimax property of zero-sum games.

Now we can interpret the findings. The kicker does better with his pure Right than his pure Left, both when the goalie guesses correctly ($60 > 45$) and when he guesses incorrectly ($95 > 90$). (Presumably the kicker is left-footed and can kick harder to his right.) Therefore the kicker chooses his Right with greater probability and, to counter that, the goalie chooses Right with the highest probability, too. However, the kicker should not and does not choose his pure-strategy Right; if he did so, the goalie would then choose his own pure-strategy Right, too, and the kicker's payoff would be only 60, less than the 75.4 that he gets in equilibrium.

B. Equilibrium Mixtures with Some Strategies Unused

In the preceding equilibrium, the probabilities of using Center in the mix are quite low for each player. The (Center, Center) combination would result in a sure save and the kicker would get a really low payoff—namely, 0. Therefore the kicker puts a low probability on this choice. But then the goalie also should put a low probability on it, concentrating on countering the kicker's more likely choices. But if the kicker gets a sufficiently high payoff from choosing Center when the goalie chooses Left or Right, then the kicker will choose Center with some positive probability. If the kicker's payoffs in the Center row were lower, he might then choose Center with zero probability; if so, the goalie would similarly put zero probability on Center. The game would reduce to one with just two basic pure strategies, Left and Right, for each player.

We show such a variant of the soccer game in Figure 8.11. The only difference in payoffs between this variant and the original game of Figure 8.10 is that the kicker's payoffs from (Center, Left) and (Center, Right) have been lowered even further, from 85 to 70. This might be because this kicker has the habit of kicking too high and therefore missing the goal when aiming for the center. Let us try to calculate the equilibrium here by using the same methods as in Section 3.A. This time we do it from the goalie's perspective; we try to find his mixture probabilities q_L, q_R, and q_C by using the condition that the kicker should be indifferent among all three of his pure strategies when played against this mixture.

		GOALIE		
		Left	Center	Right
KICKER	Left	45	90	90
	Center	70	0	70
	Right	95	95	60

FIGURE 8.11 Mixing in Soccer Penalty Kick When Not All Pure Strategies Are Used

The kicker's payoffs from his pure strategies are

Left: $45q_L + 90q_C + 90q_R = 45q_L + 90(1 - q_L - q_R) + 90q_R = 45q_L + 90(1 - q_L)$
Center: $70q_L + 0q_C + 70q_R = 70q_L + 70q_R$
Right: $95q_L + 95q_C + 60q_R = 95q_L + 95(1 - q_L - q_R) + 60\,q_R = 95(1 - q_R) + 60q_R$

Equating the Left and Right expressions and simplifying, we have $90 - 45q_L = 95 - 35q_R$, or $35q_R = 5 + 45q_L$. Next, equate the Left and Center expressions and simplify to get $90 - 45q_L = 70q_L + 70q_R$, or $115q_L + 70q_R = 90$. Substituting for q_R from the first of these equations (after multiplying through by 2 to get $70q_R = 10 + 90q_L$) into the second yields $205q_L = 80$, or $q_L = 0.390$. Then using this value for q_L in either of the equations gives $q_R = 0.644$. Finally, we use both of these values to obtain $q_C = 1 - 0.390 - 0.644 = -0.034$. Because probabilities cannot be negative, something has obviously gone wrong.

To understand what happens in this example, start by noting that Center is now a poorer strategy for the kicker than it was in the original version of the game, where his probability of choosing it was already quite low. But the concept of the opponent's indifference, expressed in the equations that led to the solution, means that the kicker has to be kept willing to use this poor strategy. That can happen only if the goalie is using his best counter to the kicker's Center—namely, the goalie's own Center—sufficiently infrequently. And in this example that logic has to be carried so far that the goalie's probability of Center has to become negative.

As pure algebra, the solution that we derived may be fine, but it violates the requirement of probability theory and real-life randomization that probabilities be nonnegative. The best that can be done in reality is to push the goalie's probability of choosing Center as low as possible—namely, to zero. But that leaves the kicker unwilling to use his own Center. In other words, we get a situation in which each player is not using one of his pure strategies in his mixture, that is, each is using it with zero probability.

Can there then be an equilibrium in which each player is mixing between his two remaining strategies—namely, Left and Right? If we regard this reduced two-by-two game in its own right, we can easily find its mixed-strategy equilibrium. With all the practice that you have had so far, it is safe to leave the details to you and to state the result:

Kicker's mixture probabilities: $p_L = 0.4375$, $p_R = 0.5625$;
Goalie's mixture probabilities: $q_L = 0.3750$, $q_R = 0.6250$;
Kicker's expected payoff (success percentage): 73.13.

We found this result by simply removing the two players' Center strategies from consideration on intuitive grounds. But we must check that it is a genuine equilibrium of the full three-by-three game. That is, we must check that neither player finds it desirable to bring in his third strategy, given the mixture of two strategies chosen by the other player.

When the goalie is choosing this particular mixture, the kicker's payoff from pure Center is $0.375 \times 70 + 0.625 \times 70 = 70$. This payoff is less than the 73.13 that he gets from either of his pure Left or pure Right or any mixture between the two, so the kicker does not want to bring his Center strategy into play. When the kicker is choosing the two-strategy mixture with the preceding probabilities, his payoff against the goalie's pure Center is $0.4375 \times 90 + 0.5625 \times 95 =$ 92.8. This number is higher than the 73.13 that the kicker would get against the goalie's pure Left or pure Right or any mixture of the two and is therefore worse for the goalie. Thus the goalie does not want to bring his Center strategy into play either. The equilibrium that we found for the two-by-two game is indeed an equilibrium of the three-by-three game.

To allow for the possibility that some strategies may go unused in an equilibrium mixture, we must modify or extend the "opponent's indifference" principle. Each player's equilibrium mix should be such that the other player is indifferent among all the strategies *that are actually used in his equilibrium mix.* The other player is not indifferent between these and his unused strategies; he prefers the ones used to the ones unused. In other words, against the opponent's equilibrium mix, all of the strategies used in your own equilibrium mix should give you the same expected payoff, which in turn should be higher than what you would get from any of your unused strategies. This is called the principle of *complementary slackness;* we consider it in greater generality in the next section, where the reason for this strange-sounding name will become clearer.

4 MORE COUNTERINTUITIVE PROPERTIES OF MIXED STRATEGIES

In Chapter 7, Section 5, we pointed out, discussed, and explained some features of mixed-strategy equilibria that appear counterintuitive at first sight. We postponed the discussion of some other counterintuitive outcomes

		OPPONENT EXPECTS	
		P	R
YOU PLAY	P	*c*	*b*
	R	*a*	*d*

FIGURE 8.12 Table of Success Probabilities of Risky and Percentage Plays

because they required somewhat more mathematics. We take these additional examples up now.

A. Risky and Safe Choices in Zero-Sum Games

In sports, there are always some strategies that are relatively safe; they do not fail disastrously even if anticipated by the opponent but do not do very much better even if unanticipated. Other strategies are risky; they do brilliantly if the other side is not ready for them but fail miserably if the other side is ready. Thus in football, on third down with a yard to go, a run up the middle is safe and a long pass is risky. The following example incorporates this idea of safe–versus–risky strategies. In addition, although most of our examples use illustrative numbers for payoffs, here we change that practice to emphasize the generality of the problem. Therefore, we let the payoffs be general algebraic symbols, subject only to some conditions concerning the basic strategy being considered.

Consider any zero-sum game in which you have two pure strategies. Let us call the relatively safe strategy (the percentage play) P and the riskier strategy R. The opponent has two pure strategies that we also call P and R; his P is his best response to your P, as is his R to your R. Figure 8.12 shows the table of probabilities that your play succeeds; these are not your payoffs. The sense of "safe" and "risky" is captured by requiring $a > b > c > d$. The risky play does really well if the opponent is not prepared for it (your success probability is a) but really badly if he is (your success probability is d); the percentage play does moderately well in either case (you succeed with probability b or c) but a little worse if the opponent expects it ($c < b$).

Let your payoff or utility be W if your play succeeds and L if it fails. A "really big occasion" is when W is much bigger than L. Note that W and L are not necessarily money amounts, so they can be utilities that capture any aversion to risk, as explained in the Appendix to Chapter 7. Now we can write down the table of expected payoffs from the various strategy combinations as in Figure 8.13. Note how this table is constructed. For example, if you play P and your

		OPPONENT EXPECTS	
		P	R
YOU PLAY	P	$cW + (1 - c)L$	$bW + (1 - b)L$
	R	$aW + (1 - a)L$	$dW + (1 - d)L$

FIGURE 8.13 Payoff Table with Risky and Percentage Plays

opponent expects R, then you get utility W with probability b and utility L with probability $(1 - b)$; your expected payoff is $bW + (1 - b)L$. This game is zero-sum; so in each cell your opponent's payoffs are just the negative of yours.

In the mixed-strategy equilibrium, your probability p of choosing P is defined by the opponent's indifference property; therefore

$$p[cW + (1 - c)L] + (1 - p)[aW + (1 - a)L] =$$
$$p[bW + (1 - b)L] + (1 - p)[dW + (1 - d)L]$$

This equation simplifies to $p = (a - d)/[(a - d) + (b - c)]$. Because $(b - c)$ is small in relation to $(a - d)$, we see that p is close to 1. That is exactly why the strategy P is called the percentage play; it is the normal play in these situations, and the risky strategy R is played only occasionally to keep the opponent guessing or, in football commentators' terminology, "to keep the defense honest."

The interesting part of this result is that the expression for p is completely independent of W and L. That is, the theory says that you should mix the percentage play and the risky play in exactly the same proportions on a big occasion as you would on a minor occasion. This runs against the intuition of many people. They think that the risky play should be engaged in less often when the occasion is more important. Throwing a long pass on third down with a yard to go may be fine on an ordinary Sunday afternoon in October, but doing so in the Super Bowl is too risky.

So which is right: theory or intuition? We suspect that readers will be divided on this issue. Some will think that the sports commentators are wrong and will be glad to have found a theoretical argument to refute their claims. Others will side with the commentators and argue that bigger occasions call for safer play. Still others may think that bigger risks should be taken when the prizes are bigger, but even they will find no support in the theory, which says that the size of the prize or the loss should make no difference to the mixture probabilities.

On many previous occasions when discrepancies between theory and intuition arose, we argued that the discrepancies were only apparent, that they were the result of failing to make the theory sufficiently general or rich enough to capture all the features of the situation that created the intuition, and that improving the theory removed the discrepancy. This one is different; the problem is fundamental to the calculation of payoffs from mixed strategies as probability-weighted averages or expected payoffs. And almost all of existing game theory has this starting point.[9]

B. Counterintuitive Changes in Mixture Probabilities for Non-Zero-Sum Games

In Chapter 7, Section 5.C, we described a counterintuitive property of mixed strategies in zero-sum games. If a player improves the payoffs from one of his pure strategies, the probability of using that strategy in the equilibrium mixture can go down. Here we demonstrate an even more general and more surprising result in general non-zero-sum games. One player's equilibrium mixture probabilities depend only on the other player's payoffs, not on his own. Consider a general two-by-two non-zero-sum game with the payoff table shown in Figure 8.14. In actual games, the payoffs would be actual numbers and the strategies would have particular names. In this example, we again use general algebraic symbols for payoffs so that we can examine how the probabilities of the equilibrium mixtures depend on them. Similarly, we use arbitrary generic labels for the strategies.

Suppose the game has a mixed-strategy equilibrium in which Row plays Up with probability p and Down with probability $(1 - p)$. To guarantee that Column also mixes in equilibrium, Row's p-mix must keep Column indifferent between

		COLUMN	
		Left	Right
ROW	Up	a, A	b, B
	Down	c, C	d, D

FIGURE 8.14 General Algebraic Payoff Matrix for Two-by-Two Non-Zero-Sum Game

[9]Vincent P. Crawford, "Equilibrium Without Independence," *Journal of Economic Theory,* vol. 50, no. 1 (February 1990), pp. 127–154; and James Dow and Sergio Werlang, "Nash Equilibrium Under Knightian Uncertainty," *Journal of Economic Theory,* vol. 64, no. 2 (December 1994), pp. 305–324, are among the few research papers that suggest alternative foundations for game theory. And our exposition of this problem in the first edition of this book inspired an article that uses such new methods on it: Simon Grant, Atsushi Kaji, and Ben Polak, "Third Down and a Yard to Go: Recursive Expected Utility and the Dixit-Skeath Conundrum," *Economic Letters,* vol. 73, no. 3 (December 2001), pp. 275–286. Unfortunately, it uses more advanced concepts than those available at the introductory level of this book.

his two pure strategies, Left and Right. Equating Column's expected payoffs from these two strategies when played against Row's mixture, we have[10]

$$pA + (1 - p)C = pB + (1 - p)D, \quad \text{or}$$
$$p = (D - C)/[(A - B) + (D - C)]$$

The surprising thing about the expression for p is not what it contains, but what it does *not* contain. None of Row's own payoffs, a, b, c, or d, appear on the right-hand side. Row's mixture probabilities are totally independent of his own payoffs!

Similarly, the equilibrium probability q of Column playing Left is given by

$$q = (d - b)/[(a - c) + (d - b)].$$

Column's equilibrium mixture also is determined independently of his own payoffs.

The surprising or counterintuitive aspect of these results is resolved if you remember the general principle of the opponent's indifference. Because each player's mixture probabilities are solved by requiring the opponent to be indifferent between his pure strategies, it is natural that these probabilities should depend on the opponent's payoffs, not on one's own. But remember also that it is only in zero-sum games that a player has a genuine reason to keep the opponent indifferent. There, any clear preference of the opponent for one of his pure strategies would work to one's own disadvantage. In non-zero-sum games, the opponent's indifference does not have any such purposive explanation; it is merely a logical property of equilibrium in mixed strategies.

5 MIXING AMONG ANY NUMBER OF STRATEGIES: GENERAL THEORY

We conclude this chapter with some general theory of mixed-strategy equilibria, to unify all of the ideas introduced in the various examples in Chapters 7 and 8 so far. Such general theory unavoidably requires some algebra and some abstract thinking. Readers unprepared for such mathematics or averse to it can omit this section without loss of continuity.

Suppose the Row player has available the pure strategies $R_1, R_2, \ldots, R_m$, and the Column player has strategies $C_1, C_2, \ldots C_n$. Write the Row player's payoff from the strategy combination (i, j) as A_{ij}, and Column's as B_{ij}, where the index i ranges from 1 to m, and the index j ranges from 1 to n. We allow each player

[10]For there to be a mixed-strategy equilibrium, the probability p must be between 0 and 1. This requires that $(A - B)$ and $(D - C)$ have the same sign; if A is bigger than B, then D must be bigger than C, and, if A is smaller than B, then D must be smaller than C. (Otherwise, one of the pure strategies, Left or Right, would dominate the other.)

to mix the available pure strategies. Suppose the Row player's p-mix has probabilities P_i and the Column player's q-mix has Q_j. All these probabilities must be nonnegative, and each set must add to 1, so

$$P_1 + P_2 + \cdots + P_m = 1 = Q_1 + Q_2 + \cdots + Q_n.$$

We write V_i for Row's expected payoff from using his pure strategy i against Column's q-mix. Using the reasoning that we have already seen in several examples, we have

$$V_i = A_{i1}Q_1 + A_{i2}Q_2 + \cdots + A_{in}Q_n = \sum_{j=1}^{n} A_{ij}\, Q_j,$$

where the last expression on the right uses the mathematical notation for summation of a collection of terms. When Row plays his p-mix and it is matched against Column's q-mix, Row's expected payoff is

$$P_1V_1 + \cdots + P_mV_m = \sum_{i=1}^{m} P_iV_i = \sum_{i=1}^{m}\sum_{j=1}^{n} P_iA_{ij}Q_j.$$

The Row player chooses his p-mix to maximize this expression.

Similarly, writing W_j for Column's expected payoff when his pure strategy j is pitted against Row's p-mix, we have

$$W_j = P_1B_{1j} + P_2B_{2j} + \cdots + P_mB_{mj} = \sum_{i=1}^{m} P_iB_{ij}.$$

Pitting mix against mix, we have Column's expected payoff:

$$Q_1W_1 + \cdots + Q_nW_n = \sum_{j=1}^{n} Q_jW_j = \sum_{i=1}^{m}\sum_{j=1}^{n} P_iB_{ij}Q_j,$$

and he chooses his q-mix to maximize this expression.

We have a Nash equilibrium when each player simultaneously chooses his best mix, given that of the other. That is, Row's equilibrium p-mix should be his best response to Column's equilibrium q-mix, and vice versa. Let us begin by finding Row's best-response rule. That is, let us temporarily fix Column's q-mix and consider Row's choice of his p-mix.

Suppose that, against Column's given q-mix, Row has $V_1 > V_2$. Then Row can increase his expected payoff by shifting some probability from strategy R_2 to R_1; that is, Row reduces his probability P_2 of playing R_2 and increases the probability P_1 of playing R_1 by the same amount. Because the expressions for V_1 and V_2 do not include any of the probabilities P_i at all, this is true no matter what the original values of P_1 and P_2 were. Therefore Row should reduce the probability P_2 of playing R_2 as much as possible—that is, all the way to zero.

The idea generalizes immediately. Row should rank the V_i in descending order. At the top there may be just one strategy, in which case it should be the only one used; that is, Row should then use a pure strategy. Or there may be a tie among two or more strategies at the top, in which case Row should mix solely among these strategies and not use any of the others.[11] When there is such a tie, all mixtures between these strategies give Row the same expected payoff. Therefore this consideration alone does not fix Row's equilibrium p-mix. We show later how, in a way that may seem somewhat strange at first sight, Column's indifference condition does that job.

The same argument applies to Column. He should use only that pure strategy which gives him the highest W_j or should mix only among those of his pure strategies C_j whose W_j are tied at the top. If there is such a tie, then all mixtures are equally good from Column's perspective; the probabilities of the mix are not fixed by this consideration alone.

In general, for most values of $(Q_1, Q_2, \ldots, Q_n)$ that we hold fixed in Column's q-mix, Row's $V_1, V_2, \ldots, V_m$ will not have any ties at the top, and therefore Row's best response will be a pure strategy. Conversely, Column's best response will be one of his pure strategies for most values of $(P_1, P_2, \ldots, P_m)$ that we hold fixed in Row's p-mix. We saw this several times in the examples of Chapters 7 and 8; for example, in Figure 8.2, for most values of p in Row's p-mix, the best q for Column was either 0 or 1, and vice versa. For only one critical value of Row's p was it optimal for Column to mix (choose any q between 0 and 1), and vice versa.

All of these conditions—ties at the top, and worse outcomes from the other strategies—constitute the complicated set of equations and inequalities that, when simultaneously satisfied, defines the mixed-strategy Nash equilibrium of the game. To understand it better, suppose for the moment that we have done all the work and found which strategies are used in the equilibrium mix. We can always relabel the strategies so that Row uses, say, the first g pure strategies, $R_1, R_2, \ldots, R_g$, and does not use the remaining $(m - g)$ pure strategies, $R_{g+1}, R_{g+2}, \ldots, R_m$, while Column uses his first h pure strategies, $C_1, C_2, \ldots, C_h$, and does not use the remaining $(n - h)$ pure strategies, $C_{h+1}, C_{h+2}, \ldots, C_n$. Write V for the tied value of Row's top expected payoffs V_i and, similarly, W for the tied value of Column's top expected payoffs W_i. Then the equations and inequalities can be written as follows. First, for each player, we set the probabilities of the unused strategies equal to zero and require those of the used strategies to sum to 1:

$$P_1 + P_2 + \cdots + P_g = 1, \quad P_{g+1} = P_{g+2} = \cdots P_m = 0 \tag{8.1}$$

and

$$Q_1 + Q_2 + \cdots + Q_h = 1, \quad Q_{h+1} = Q_{h+2} = \cdots Q_n = 0 \tag{8.2}$$

[11]In technical mathematical terms, the expression $\Sigma_i P_i V_i$ is *linear* in the P_i; therefore its maximum must be at an extreme point of the set of permissible P_i.

Next we set Row's expected payoffs for the pure strategies that he uses equal to the top tied value:

$$V = A_{i1}\,Q_1 + A_{i2}\,Q_2 + \cdots + A_{ih}Q_h \quad \text{for } i = 1, 2, \ldots, g, \tag{8.3}$$

and note that his expected payoffs from his unused strategies must be smaller (that is why they are unused):

$$V > A_{i1}\,Q_1 + A_{i2}\,Q_2 + \cdots + A_{ih}Q_h \quad \text{for } i = g+1, g+2, \ldots, n. \tag{8.4}$$

Next, we do the same for Column, writing W for his top tied payoff value:

$$W = P_1\,B_{1j} + P_2B_{2j} + \cdots + P_gB_{gj} \quad \text{for } j = 1, 2, \ldots, h, \tag{8.5}$$

and

$$W > P_1B_{1j} + P_2B_{2j} + \cdots + P_gB_{gj} \quad \text{for } j = h+1, h+2, \ldots, n. \tag{8.6}$$

To find the equilibrium, we must take this whole system; regard the choice of g and h as well as the probabilities $P_1, P_2, \ldots, P_g$, and $Q_1, Q_2, \ldots, Q_h$ as unknowns; and attempt to solve for them.

There is always the exhaustive search method. Try a particular selection of g and h; that is, choose a particular set of pure strategies as candidates for use in equilibrium. Then take Eqs. (8.1) and (8.5) as a set of $(h + 1)$ simultaneous linear equations regarding $P_1, P_2, \ldots, P_g$ and W as $(g + 1)$ unknowns; solve for them; and check if the solution satisfies all the inequalities in Eq. (8.6). Similarly, take Eqs. (8.2) and (8.3) as a set of $(g + 1)$ simultaneous linear equations in the $(h + 1)$ unknowns $Q_1, Q_2, \ldots, Q_h$ and V; solve for them; and check if the solution satisfies all the inequalities in Eq. (8.4). If all these things check out, we have found an equilibrium. If not, take another selection of pure strategies as candidates and try again. There is only a finite number of selections. Row can use $(2^m - 1)$ possible pure strategies in his mix and Column can use $(2^n - 1)$ possible pure strategies in his mix. Therefore the process must end successfully after a finite number of attempts.

When m and n are reasonably small, exhaustive search is manageable. Even then, shortcuts suggest themselves in the course of the calculation for each specific problem. Thus, in the second variant of our soccer penalty kick, the way in which the attempted solution with all strategies used failed told us which strategy to discard in the next attempt.

Even for moderately large problems, however, solutions based on exhaustive search or ad hoc methods become too complex. That is when one must resort to more systematic computer searches or algorithms. What these computer algorithms do is to search simultaneously for solutions to two linear maximization (or linear programming, in the terminology of decision theory) problems: given a q-mix, and therefore all the V_i values, choose a p-mix to maximize

Row's expected payoff $\Sigma_i P_i V_i$; and given a p-mix and therefore all the W_i values, choose a q-mix to maximize Column's expected payoff $\Sigma_j Q_j W_j$. However, for a typical q-mix, all the V_i values will be unequal. If Column were to play this q-mix in an actual game, Row would not mix but would instead play just the one pure strategy that gave him the highest V_i. But in our numerical solution method we should not adjust Row's strategy in this drastic fashion. If we did, then at the next step of our algorithm, Column's best q-mix also would change drastically, and Row's chosen pure strategy would no longer look so good. Instead, the algorithm should take a more gradual step, adjusting the p-mix a little bit to improve Row's expected payoff. Then, with the use of this new p-mix for Row, the algorithm should adjust Column's q-mix a little bit to improve his expected payoff. Then back again to another adjustment in the p-mix. The method proceeds in this way until no improvements can be found; that is the equilibrium.

We do not need the details of such procedures, but the general ideas that we have developed above already tell us a lot about equilibrium. Here are some important lessons of this kind.

1. We solve for Row's equilibrium mix probabilities $P_1, P_2, \ldots, P_g$ from Eqs. (8.1) and (8.5). The former is merely the adding-up requirement for probabilities. The more substantive equation is (8.5), which gives the conditions under which Column gets the same payoff from all the pure strategies that he uses against the p-mix. It might seem puzzling that Row adjusts his mix so as to keep Column indifferent when Row is concerned about his own payoffs, not Column's. Actually the puzzle is only apparent. We derived those conditions [Eq. (8.5)] by thinking about Column's choice of his q-mix, motivated by concerns about his own payoffs. We argued that Column would use only those strategies that gave him the best (tied) payoffs against Row's p-mix. This is the requirement embodied in Eq. (8.5). Even though it appears *as if* Row is deliberately choosing his p-mix so as to keep Column indifferent, the actual force that produces this outcome is Column's own purposive strategic choice.

In Chapter 7, we gave the name "the opponent's indifference property" to the idea that each player's indifference conditions constitute the equations that can be solved for the other player's equilibrium mix. We now have a proof of this principle for general games, zero-sum and non-zero-sum.

2. However, in the zero-sum case, the idea that each player chooses his mixture to keep the other indifferent is not just an *as if* matter; there is some genuine reason that a player should behave in this way. When the game is zero-sum, we have a natural link between the two players' payoffs: $B_{ij} = -A_{ij}$ for all i and j, and then similar relations hold among all the combinations and expected payoffs, too. Therefore we can multiply Eqs. (8.5) and (8.6) by -1 to write them in

terms of Row's payoffs rather than Column's. We write these "zero-sum versions" of the conditions as Eqs. (8.5z) and (8.6z):

$$V = P_1 A_{1j} + P_2 A_{2j} + \cdots + P_g A_{gj} \quad \text{for } j = 1, 2, \ldots, h \qquad (8.5z)$$

and

$$V < P_1 A_{1j} + P_2 A_{2j} + \cdots + P_g A_{gj} \quad \text{for } j = h + 1, h + 2, \ldots, n. \qquad (8.6z)$$

(Note that multiplying by -1 to go from Eq. (8.6) to Eq. (8.6z) reverses the direction of the inequality.)

Of these, Eqs. (8.5) and (8.5z) tell us that as long as Row is using his equilibrium mix, Column (and therefore Row, too, in this zero-sum game) gets the same payoff from any of the pure strategies that he actually uses in equilibrium. Column cannot do any better for himself—and therefore in this zero-sum game cannot cause any harm to Row—by choosing one of those strategies rather than another. What is more, Eq. (8.6z) tells us that were Column to use any of the other strategies, Row would do even better. In other words, these conditions tell us that Column cannot exploit Row's equilibrium mix. Thus we see in a more general setting the purposive role of mixing in zero-sum games that we saw in the examples of Chapter 7; we also see more explicitly why it works only for zero-sum games.

3. Now we return to the general, non-zero-sum, case. Note that the system comprising Eqs. (8.1) and (8.5) has $(h + 1)$ linear equations and $(g + 1)$ unknowns. In general, such a system has no solution if $h > g$, has exactly one solution if $h = g$, and has many solutions if $h < g$. Conversely, the system comprising Eqs. (8.2) and (8.3) has $(g + 1)$ linear equations and $(h + 1)$ unknowns. In general, such a system has no solution if $g > h$, has exactly one solution if $g = h$, and has multiple solutions if $g < h$. Because in equilibrium we want both systems to be satisfied, in general we need $g = h$. Thus in a mixed-strategy equilibrium, the two players use equal numbers of pure strategies.

We keep on saying "in general" because exceptions can arise for fortuitous combinations of coefficients and right-hand sides of equations. In particular, it is possible for too many equations in too few unknowns to have solutions. This is just what happens in the "exceptional cases" mentioned in Section 2.B of this chapter.

4. We observe a very particular relation between the use of strategies and their payoffs. Row uses strategies P_1 to P_g with positive probabilities, and Eq. (8.3) shows that he gets exactly the payoff V when any one of these pure strategies is played against Column's equilibrium mix. For the remaining pure strategies in Row's armory, Eq. (8.4) shows that they yield a lower payoff than V when played against Column's equilibrium mix, and then they are not used; that is, P_{h+1} to P_m are all zero. In other words, for any i, it is *impossible* to have

$$\text{both } V > A_{i1} Q_1 + A_{i2} Q_2 + \cdots + A_{ih} Q_h \quad \text{and} \quad P_i > 0.$$

At least one of these inequalities must collapse into equality. This is known as the principle of **complementary slackness,** and it is of great importance in the general theory of games and equilibria, as well as in mathematical optimization (programming).

5. Back to the zero-sum case. When both players choose their equilibrium mix, Row's expected payoff is

$$V = \sum_{i=1}^{m} \sum_{j=1}^{n} P_i A_{ij} Q_j,$$

and Column's is just the negative of this payoff. Moreover, the equilibrium comes about when Row, for the given q-mix, chooses his p-mix to maximize this expression, and simultaneously Column, for the given p-mix, chooses his q-mix to maximize the negative of the same expression, or to minimize the same expression. If we regard the expression as a function of all the P_i and the Q_j, therefore, and graph it in a sufficiently high-dimensional space, it will look like a saddle. The front-to-back cross section of a saddle looks like a valley or a U, with its minimum at the middle, while the side-to-side cross section looks like a peak or an inverted U, with its maximum at the middle. If each player has just two pure strategies, the p-mix and q-mix can each be described by a single number—say, the probability of choosing the first strategy. (For each player, the probability of choosing his second pure strategy is then just one minus that of choosing his first pure strategy.) We can then draw a graph in three dimensions, where the x and y axes are in a horizontal plane and the z axis points vertically upward. The p-mix is shown along the x-axis, the q-mix along the y-axis, and the value V along the z-axis. The cross section of this

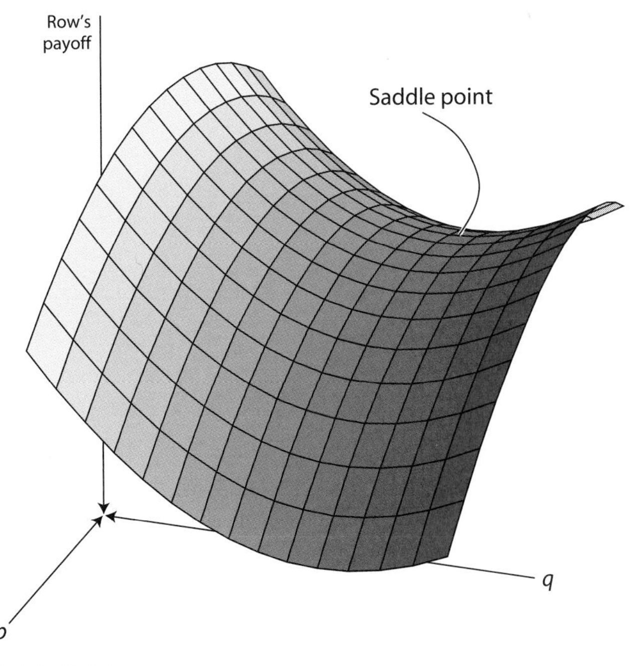

FIGURE 8.15 Saddle Point

saddle-shaped surface along the x direction will show the maximization of V with respect to p, therefore a peak. And the cross section along the y direction will show the minimization of V with respect to q, therefore a valley. Thus the graph will look like a saddle, as illustrated in Figure 8.15. Such an equilibrium is called a **saddle point.**

The value of V in equilibrium—that is, the simultaneous maximum with respect to the P_i and the minimum with respect to the Q_j—is called the *minimax* value of the zero-sum game. The idea of such an equilibrium, as well as the formulation of the conditions such as Eqs. (8.5z) and (8.6z) that define it, was the first important achievement of game theory and appeared in the work of von Neumann and Morgenstern in the 1940s. It is called their minimax theorem.

SUMMARY

Best-response analysis can be used to solve for mixed-strategy equilibria. The best-response-curve diagram can be used to show all mixed-strategy as well as pure-strategy equilibria of a game.

If one player has three strategies and the other has only two, the player with three available pure strategies will generally use only two in her equilibrium mix. In some exceptional cases, equilibrium mixtures may also be indeterminate.

When one or both players have three (or more) strategies, equilibrium mixed strategies may put positive probability on all pure strategies or may include only a subset of the pure strategies. All strategies that are actively used in the mixture yield equal expected payoff against the opponent's equilibrium mix; all the ones unused yield lower expected payoff. This is the principle of complementary slackness.

KEY TERMS

complementary slackness (290) **saddle point** (292)

SOLVED EXERCISES

S1. "When a zero-sum game has a mixed-strategy equilibrium, a player's equilibrium mixture is designed to yield her the same expected payoff when used against each of the other player's pure strategies." True or false? Explain.

S2. Sections 1.A and 1.B illustrate how to use a best-response graph to find all of the Nash equilibria of the tennis-point game and the assurance game respectively. For the battle of the sexes game from Figure 4.13, graph the best responses of Harry and Sally on a p-q coordinate plane. Label all of the Nash equilibria.

S3. Revisit Exercise U9 from Chapter 7. For parts (a) and (b) below, graph the best-response curves of Row and Column on a p-q coordinate plane. On the graphs, identify and label the pure-strategy Nash equilibria and any mixed-strategy Nash equilibria.

(a) Let $x = 3$.

(b) Let $x = 1$.

S4. Recall the exceptional cases of the three-by-two tennis-point game described in Section 2.B.

(a) Draw the diagram of Evert's success rate relative to Navratilova's q-mix (similar to Figure 8.7) for the case where Evert's payoff from playing Lob is 70 when Navratilova plays either DL or CC. Use this diagram to explain why it is an equilibrium for Evert to play pure Lob and for Navratilova to mix 50-50.

(b) Describe all possible Nash equilibria for the game given in part (a).

(c) Draw the diagram of Evert's success rate as a function of Navratilova's q-mix (similar to Figure 8.7) for the case where the the payoffs of Evert's Lob are 66 and 56 against Navratilova's DL and CC, respectively. Use this diagram to illustrate that Navratilova's equilibrium q must be 0.6.

(d) As the text notes, there are a number of different Nash equilibria for the game in part (c) involving different mixtures for Evert. Write down the indifference equation for Navratilova in terms of p_1 and p_2. (Hint: Look at Figure 8.6.) Note that this equation can be satisfied by many values of p_1 and p_2. Find the largest and smallest possible values for p_1 and p_2.

S5. Consider the following game:

		OLLISTAN	
		Laurel	Hardy
KARL	Groucho	9	2
	Harpo	7	5
	Chico	5	6
	Zeppo	4	9
	Gummo	1	8

(a) On a single graph, plot the expected payoffs from each of Karl's strategies as a function of Ollistan's q-mix.

(b) Which strategies appear on the upper envelope of the graph in part (a)?
(c) Which strategies are never a best response for Karl? Why?
(d) What is the mixed-strategy Nash equilibrium of this game?

S6. Consider the following game:

		PROFESSOR PLUM		
		Revolver	Knife	Wrench
MRS. PEACOCK	Conservatory	1, 3	2, −2	0, 6
	Ballroom	3, 1	1, 4	5, 0

(a) Graph the expected payoffs from each of Professor Plum's strategies as a function of Mrs. Peacock's p-mix.
(b) Over what range of p does Revolver yield a higher expected payoff for Professor Plum than Knife?
(c) Over what range of p does Revolver yield a higher expected payoff than Wrench?
(d) Which pure strategies will Professor Plum use in his equilibrium mixture? Why?
(e) What is the mixed-strategy Nash equilibrium of this game?

S7. Find all Nash equilibria of the following game. (Hint: Look for dominated strategies.)

		GREEK			
		α	β	γ	δ
ROMAN	A	6	−1	5	4
	B	5	−2	2	1
	C	3	7	4	8

S8. Find all Nash equilibria of the Roman-Greek game in Exercise S7 when the payoff to (A, γ) changes from 5 to 3.

S9. Many of you will be familiar with the children's game rock–paper–scissors. In rock–paper–scissors, two people simultaneously choose either "rock," "paper," or "scissors," usually by putting their hands into the shape of one of the three choices. The game is scored as follows. A person choosing Scissors beats a person choosing Paper (because scissors cut paper). A person choosing Paper beats a person choosing Rock (because paper covers rock). A person choosing Rock beats a person choosing Scissors (because rock breaks scissors). If two players choose the same object, they tie. Suppose that each individual play of the game is worth 10 points. The following matrix shows the possible outcomes in the game:

		PLAYER 2		
		Rock	Scissors	Paper
PLAYER 1	Rock	0	10	−10
	Scissors	−10	0	10
	Paper	10	−10	0

(a) Find the mixed-strategy equilibrium of this rock–paper–scissors game.

(b) Suppose that Player 2 announced that she would use a mixture in which her probability of choosing Rock would be 40%, her probability of choosing Scissors would be 30%, and her probability of Paper, 30%. What is Player 1's best response to this strategy choice by Player 2? Explain why your answer makes sense, given your knowledge of mixed strategies.

S10. Harry and Sally have planned to go out to eat tonight. Harry strongly prefers pasta, which Sally finds disagreeable. Sally really likes a particular sandwich place, but Harry's not a fan. Another option is a buffet, which would allow each of them to get a lower-quality version of the food they'd most like to eat. Harry and Sally enjoy one another's company and would rather eat together than apart, but they are abysmally poor communicators. Once again they have neglected to decide beforehand where to go, and they have both absentmindedly allowed their cell-phone batteries to die. Their payoffs are as follows:

		SALLY		
		Pasta	Sandwich	Buffet
HARRY	Pasta	5, 1	0, 0	0, 0
	Sandwich	0, 0	1, 5	0, 0
	Buffet	0, 0	0, 0	2, 2

The pure-strategy Nash equilibria are easy to see, but this game also has multiple mixed-strategy Nash equilibria.

(a) Show that it is a mixed-strategy equilibrium when Harry plays Pasta 5/6 of the time, Sandwich 1/6 of the time, and Buffet never, whereas Sally plays Pasta 1/6 of the time, Sandwich 5/6 of the time, and Buffet never. What is the expected payoff to each player in this equilibrium?

(b) What mixed-strategy equilibrium results when both Harry and Sally mix only over Pasta and Buffet? What is the expected payoff to each player in this equilibrium? Explain why Sally would never play Sandwich if she knew that Harry was playing his mixed strategy for this equilibrium.

(c) What mixed-strategy equilibrium results when both Harry and Sally mix only over Sandwich and Buffet? What is the expected payoff to each player in this equilibrium? Explain why Harry would never play Pasta if he knew that Sally was playing her mixed strategy for this equilibrium.

(d) There is also a mixed-strategy equilibrium where both Harry and Sally play all three of their strategies with positive probability. What is it? What is the expected payoff to each player?

(e) Which of the seven equilibria of this game is focal? Explain your reasoning.

S11. Recall the game between ice-cream vendors on a beach from Exercise U6 in Chapter 6. In that game, we found two asymmetric pure-strategy equilibria. There is also a symmetric mixed-strategy equilibrium to the game.

(a) Write down the five-by-five table for the game.

(b) Eliminate dominated strategies, and explain why they should not be used in the equilibrium.

(c) Use your answer to part (b) to help you find the mixed-strategy equilibrium to the game.

S12. Suppose that the soccer penalty-kick game of Section 5.A in this chapter is expanded to include a total of six distinct strategies for the kicker: to shoot high and to the left (HL), low and to the left (LL), high and in the center (HC), low and in the center (LC), high right (HR), and low right (LR). The goalkeeper continues to have three strategies: to move to the kicker's left (L) or right (R) or to stay in the center (C). The kicker's success percentages are shown in the following table.

		GOALIE		
		L	C	R
KICKER	HL	0.50	0.85	0.85
	LL	0.40	0.95	0.95
	HC	0.85	0	0.85
	LC	0.70	0	0.70
	HR	0.85	0.85	0.50
	LR	0.95	0.95	0.40

These payoffs incorporate the following information. Shooting high runs some risk of missing the goal even if the goalie goes the wrong way (hence $0.85 < 0.95$). If the goalie guesses correctly, she has a better chance of collecting or deflecting a low shot than a high one (hence $0.40 < 0.50$). And if the shot is to the center while the goalie goes to one side, she has a better chance of using her feet to deflect a low shot than a high one (hence $0.70 < 0.85$).

In this problem, you will verify the following mixed-strategy equilibrium of this game. The goalie uses L and R each 42.2% of the time, and C 15.6% of the time. The kicker uses LL and LR each 37.8% of the time, and HC 24.4% of the time.

(a) Given the goalie's proposed mixed strategy, compute the expected payoff to the kicker for each her six pure strategies. (Use only three significant digits, in order to keep things simple.)

(b) Use your answer to part (a) to explain why the kicker's proposed mixed strategy is a best response to the goalie's proposed mixed strategy.

(c) Given the kicker's proposed mixed strategy, compute the expected payoff to the goalie for each her three pure strategies. (Again, use only three significant digits, in order to keep things simple.)

(d) Use your answer to part (a) to explain why the goalie's proposed mixed strategy is a best response to the kicker's proposed mixed strategy.

(e) Using your previous answers, explain why the proposed strategies are indeed a Nash equilibrium.

(f) Compute the equilibrium payoff to the kicker.

S13. **(Optional)** Recall the three-player game among Marta's sons in Exercise S9 of Chapter 4. In that game, we found three asymmetric Nash equilibria in pure strategies. In this exercise, you will find a symmetric equilibrium in mixed strategies. Note that with three players, we need three different variables (p, q, and r) to stand for the mixing probabilities. We next need to understand how to compute expected payoffs for one player when both of the other players are mixing.

(a) Suppose that Bernardo plays Yes with probability q and No with probability $1 - q$. Further suppose that Carlos independently plays Yes with probability r and No with probability $1 - r$. Then from Arturo's perspective, what is the probability that Bernardo plays Yes *and* Carlos plays No? (Hint: See Section 1.C of the Appendix to Chapter 7.)

(b) What is Arturo's expected payoff from playing Yes, in terms of Bernardo's q and Carlos's r?

(c) Write down an indifference equation for Arturo in terms of q and r.

(d) Write down the indifference equations for Bernardo and Carlos.

(e) Solve the system of three (nonlinear) equations in three unknowns to find the mixed-strategy equilibrium.

S14. **(Optional)** Recall Exercise S12 of Chapter 4, which was based on the bar scene from the film *A Beautiful Mind.* Here we consider the mixed-strategy equilibria of that game when played by $n > 2$ young men.

(a) Begin by considering the symmetric case in which each of the n young men independently goes after the solitary blonde with some probability P. This probability is determined by the condition that each young man should be indifferent between the pure strategies Blonde and Brunette, given that everyone else is mixing. What is the condition that guarantees the indifference of each player? What is the equilibrium value of P in this game?

(b) There are also some asymmetric mixed-strategy equilibria in this game. In these equilibria, $m < n$ young men each go for the blonde with probability Q, and the remaining $n - m$ young men go after the brunettes. What is the condition that guarantees that each of the m young men is indifferent, given what everyone else is doing? What condition must hold so that the remaining $n - m$ players don't want to switch from the pure strategy of choosing a brunette? What is the equilibrium value of Q in the asymmetric equilibrium?

UNSOLVED EXERCISES

U1. For the chicken game from Figure 4.14, graph the best responses of James and Dean on a p-q coordinate plane. Label all of the Nash equilibria.

U2. Revisit Exercise U10 from Chapter 7.

(a) Graph the best-response curves of Row and Column on a p-q coordinate plane.

(b) Identify and label the pure-strategy Nash equilibria.

(c) Identify and label the set of mixed-strategy Nash equilibria.

U3. (a) Find all pure-strategy Nash equilibria of the following non-zero-sum game.

		COLUMN			
		A	B	C	D
ROW	1	1, 1	2, 2	3, 4	9, 3
	2	2, 5	3, 3	1, 2	7, 1

(b) Now find a mixed-strategy equilibrium of the game. What are the players' expected payoffs in the equilibrium?

U4. Consider the following game:

		PROFESSOR PLUM		
		Revolver	Knife	Wrench
MRS. PEACOCK	Conservatory	1, 3	2, −2	0, 6
	Ballroom	3, 2	1, 4	5, 0

(a) Graph the expected payoffs from each of Professor Plum's strategies as a function of Mrs. Peacock's p-mix.

(b) Which strategies will Professor Plum use in his equilibrium mixture? Why?

(c) What is the mixed-strategy Nash equilibrium of this game?

(d) Note that this game is only slightly different from the game in Exercise S6. How are the two games different? Explain why you intuitively think the equilibrium outcome has changed from Exercise S6.

U5. Find all Nash equilibria of the Roman-Greek game in Exercise S7 when the payoff to (A, γ) changes from 5 to 4.

U6. Find all Nash equilibria of the Roman-Greek game in Exercise S7 when the payoff to (A, γ) changes from 5 to 2.

U7. Consider the following game:

		MAXWELL		
		Air	Sea	Land
JAMES	Air	0, 3	2, 0	1, 7
	Sea	2, 4	0, 6	2, 0
	Land	1, 3	2, 4	0, 3

(a) Does this game have a pure-strategy Nash equilibrium? If so, what is it?

(b) Find a mixed-strategy equilibrium to this game.

(c) Actually, this game has two mixed-strategy equilibria. Find the one you didn't find in part (b). (Hint: In one of these equilibria, one of the players plays a mixed strategy, whereas the other plays a pure strategy.)

U8. Consider a slightly different version of rock–paper–scissors in which Player 1 has an advantage. If Player 1 picks Rock and Player 2 picks Scissors, Player 1 wins 20 points from Player 2 (rather than 10). The new payoff matrix is:

		PLAYER 2		
		Rock	Scissors	Paper
PLAYER 1	Rock	0	20	−10
	Scissors	−10	0	10
	Paper	10	−10	0

(a) What is the mixed-strategy equilibrium in this version of the game?

(b) Compare your answer here with your answer for the mixed-strategy equilibrium in Exercise S9. How can you explain the differences in the equilibrium strategy choices?

U9. Section 2.C of Chapter 1 mentioned the story of two chemistry students at Duke who had opted to party hard instead of studying for their final. In the hopes of obtaining a makeup final at a later date, they lied about getting a flat tire on their return trip. Their professor agreed to the makeup, but the students were unpleasantly surprised by the second (and last) question, worth 90 points: "Which tire?" The students hadn't previously decided this part of their story, and they can't communicate during the exam.

(a) Write the game table for the tire-guessing game. (Note that each student has four pure strategies.) Giving the same answer yields a payoff of 90 for each student, while all other outcomes are worth a payoff of 0 to each.

(b) How many pure-strategy Nash equilibria are there? What are they?

As in the restaurant-choice game between Harry and Sally in Exercise S10, the number of zeroes in the table makes it relatively straightforward to find all of the game's mixed-strategy equilibria. There are eleven of them.

(c) There is one mixed-strategy Nash equilibrium where the two students play each of their four strategies with positive probability. What is this equilibrium? What is the expected value to each student from playing this equilibrium?

(d) How many mixed-strategy equilibria are there where each student plays three of the four strategies with positive probability (and never plays the fourth one)? What are they? What is the expected value to each student from playing each of these equilibria?

(e) How many mixed-strategy equilibria are there where each student plays two of the four strategies with positive probability (and never plays the other two)? What are they? What is the expected value to each student from playing each of these equilibria?

(f) If the students can't coordinate on one of the pure-strategy equilibria, can they at least improve their expected payoffs by mixing over a coordinated subset of their strategies? Explain.

U10. Barry and Neill sit down to play a relatively simple card game. They each hold one card of each suit. Each privately selects one of the suits and pushes the card of that suit toward the middle of the table. If the cards are both diamonds, then Barry wins. If the cards match suit but are not diamonds, then Neill wins. If the cards do not match suit and neither card is a diamond, then Barry wins. If the cards do not match suit and one is a diamond, then Neill wins. That is, the payoff table is as follows:

		NEILL			
		Club	Heart	Spade	Diamond
BARRY	Club	0, 1	1, 0	1, 0	0, 1
	Heart	1, 0	0, 1	1, 0	0, 1
	Spade	1, 0	1, 0	0, 1	0, 1
	Diamond	0, 1	0, 1	0, 1	1, 0

(a) What is the mixed-strategy Nash equilibrium of this game?

(b) What is the expected value of this game to each player?

U11. Recall the duel game between Renard and Chagrin in Exercise U12 in Chapter 6. Remember that the duelists start 10 steps apart and walk toward one another at the same pace, 1 step at a time, and either may fire his gun after each step. When one duelist shoots, the probability of scoring a hit depends on the distance; after k steps, it is $k/5$. Each gets a payoff of –1 if he himself is killed and 1 if the other duelist is killed. If neither or both are killed, each gets zero. Now, however, suppose that the duelists have guns with silencers. If one duelist fires and misses, the other does not know that this has happened and cannot follow the strategy of then holding his fire until the final step to get a sure shot. Each must formulate a strategy at the outset that is not conditional on the other's intermediate actions. Thus we have a simultaneous-move game, with strategies of the form "Shoot after n steps if still alive." Each player has five such strategies corresponding to the five steps that can be taken toward his adversary in the duel.

(a) The five-by-five payoff table for this game is shown below. Demonstrate how to calculate the payoffs for Row 2 of the table.

		CHAGRIN				
		1	2	3	4	5
RENARD	1	0	−0.12	−0.28	−0.44	−0.6
	2	0.12	0	0.04	−0.08	−0.2
	3	0.28	−0.04	0	0.28	0.2
	4	0.44	0.08	−0.28	0	0.6
	5	0.6	0.2	−0.2	−0.6	0

The mixed-strategy Nash equilibrium involves each player playing strategies 2, 3, and 5, with proportions 5/11, 5/11, and 1/11, respectively. Strategies 1 and 4 go unused. You will now verify that this is a Nash equilibrium.

(b) Compute the expected payoff to each of the five strategies for Renard, given Chagrin's proposed equilibrium mixture.

(c) Explain how your answer to part (b) demonstrates that the proposed mixed-strategy equilibrium really is a Nash equilibrium.

(d) What is the expected payoff for each player in equilibrium?

U12. **(Optional)** Recall the game from Exercise S10 in Chapter 4, where Larry, Moe, and Curly can choose to buy tickets toward a prize worth \$30. We found six pure-strategy Nash equilibria in that game. In this problem you will find a symmetric equilibrium in mixed strategies.

(a) Eliminate the weakly dominated strategy for each player. Explain why a player would never use this weakly dominated strategy in his equilibrium mixture.

(b) Find the equilibrium in mixed strategies.

U13. **(Optional)** Revisit the three-player version of evens or odds played by Roxanne, Sara, and Ted in Exercise U11 of Chapter 4. In addition to the two pure-strategy Nash equilibria found in that problem, the game also has a mixed-strategy equilibrium.

(a) Find the mixed-strategy Nash equilibrium of the game.

(b) Once again, is this a fair game? Explain.

U14. **(Optional)** Find all Nash equilibria of the Roman-Greek game in Exercise S7 when the payoff to (A, γ) changes from 5 to 1.

U15. **(Optional)** Revist the soccer-shootout problem from Exercise S12 to see what happens to the equilibrium mixtures when the payoff in a particular cell is changed slightly from the original game table. You may wish to use software such as Gambit (freely available at http://gambit.sourceforge.net/) to compute the answers to these questions. You can also try to compute the answers by hand, but this is harder because of the number of possibilities to consider.

What is the new mixed-strategy equilibrium, given each of the following changes to the original payoff matrix? What is the expected payoff to the kicker? Compare your answers in each case with the answers for the original game.

(a) Change the payoff for (HC, C) to 0.10, making it slightly more likely for the shooter to score on a high-center shot.

(b) Change the payoff for (HL, L) to 0.70, making it more likely for the shooter to score on a high-left shot.

(c) Change the payoff for (LC, C) to 0.50, making it more likely for the shooter to score on a low-center shot.

PART THREE

■

Some Broad Classes of Games and Strategies

Uncertainty and Information

IN CHAPTER 2 we mentioned different ways in which uncertainty can arise in a game—external and strategic—and ways in which players can have limited information about aspects of the game—imperfect and incomplete, symmetric and asymmetric. We have already encountered and analyzed some of these. Most notably, in simultaneous-move games, each player does not know the actions the other is taking; this is strategic uncertainty. In Chapter 6 we saw that strategic uncertainty gives rise to asymmetric and imperfect information, because the different actions taken by one player must be lumped into one information set for the other player. In Chapters 4 and 7 we saw how such strategic uncertainty is handled by having each player formulate beliefs about the other's action (including beliefs about the probabilities with which different actions may be taken when mixed strategies are played) and by applying the concept of Nash equilibrium, in which such beliefs are confirmed. In this chapter we focus on some further ways in which uncertainty and informational limitations arise in games.

We begin by examining various strategies that individuals and societies can use for coping with the imperfect information generated by external uncertainty or risk. Recall that external uncertainty is about matters outside any player's control but affecting the payoffs of the game; weather is a simple example. We discussed attitudes toward risk and the methodology for evaluating risky situations in the Appendix to Chapter 7. Here we show the basic ideas behind diversification, or spreading, of risk by an individual player and pooling of risk by multiple players. These strategies can benefit everyone, although the division of total gains among the participants can be unequal; therefore these situations contain a mixture of common interest and conflict.

We then consider the informational limitations that often exist in situations with strategic interdependence. Information in a game is *complete* only if all of the rules of the game—the strategies available to all players and the payoffs of each player as functions of the strategies of all players—are fully known by all players, and moreover, are common knowledge among them. By this exacting standard, most games in reality have *incomplete information.* Moreover, the incompleteness is usually *asymmetric*: each player knows his own capabilities and payoffs much better than he knows those of other players. As we pointed out in Chapter 2, manipulation of the information becomes an important dimension of strategy in such games. In this chapter, we will discuss when information can or cannot be communicated verbally in a credible manner. We will also examine other strategies designed to convey or conceal one's own information and to elicit another player's information. We spoke briefly of some such strategies—namely, screening and signaling—in Chapters 1 and 2; here, we study those in more detail.

Of course, players in many games would also like to manipulate the actions of others. Managers would like their workers to work hard and well; insurance companies would like their policyholders to exert care to reduce the risk that is being insured. If information were perfect, the actions would be observable. Workers' pay could be made contingent on the quality and quantity of their effort; payouts to insurance policyholders could be made contingent on the care they exercised. But in reality these actions are difficult to observe; that creates a situation of imperfect asymmetric information, commonly called **moral hazard**. Then the counterparties in these games have to devise various indirect methods to give incentives to influence others' actions in the right direction.

The study of the topic of information and its manipulation in games has been very active and important in recent decades. It has shed new light on many previously puzzling matters in economics, such as the nature of incentive contracts, the organization of companies, markets for labor and for durable goods, government regulation of business, and myriad others.[1] More recently, political scientists have used the same concepts to explain phenomena such as the relation of tax- and expenditures-policy changes to elections, as well as the delegation of legislation to committees. These ideas have also spread to biology, where evolutionary game theory explains features such as the peacock's large and ornate tail as a signal. Perhaps even more important, you will recognize the

[1]The pioneers of the theory of asymmetric information in economics—George Akerlof, Michael Spence, and Joseph Stiglitz—received the 2001 Nobel Prize in economics for these contributions.

important role that signaling and screening play in your daily interaction with family, friends, teachers, coworkers, and so on, and you will be able to improve your strategies in these games.

Although the study of information clearly goes well beyond consideration of external uncertainty and the basic concepts of signaling and screening, we focus only on those few topics in this chapter. We will return to the analysis of information and its manipulation in Chapter 14, however. There we will use the methods developed here to study the design of mechanisms to provide incentives to and elicit information from other players who have some private information.

1 IMPERFECT INFORMATION: DEALING WITH RISK

Imagine that you are a farmer subject to the vagaries of weather. If the weather is good for your crops, you will have an income of \$160,000. If it is bad, your income will be only \$40,000. The two possibilities are equally likely (probability 1/2, or 0.5, or 50% each). Therefore your average or expected income is \$100,000 ($= 1/2 \times 160{,}000 + 1/2 \times 40{,}000$), but there is considerable risk around this average value.

What can you do to reduce the risk that you face? You might try a crop that is less subject to the vagaries of weather, but suppose you have already done all such things that are under your individual control. Then you might be able to reduce your income risk further by getting someone else to accept some of the risk. Of course you must give the other person something else in exchange. This quid pro quo usually takes one of two forms: cash payment, or a mutual exchange or sharing of risks.

A. Sharing of Risk

We begin with an analysis of the possibility of risk sharing for mutual benefit. Suppose you have a neighbor who faces a similar risk but gets good weather exactly when you get bad weather and vice versa. (Suppose you live on opposite sides of an island, and rain clouds visit one side or the other but not both.) In technical jargon, *correlation* is a measure of alignment between any two uncertain quantities—in this discussion, between one person's risk and another's. Thus we would say that your neighbor's risk is totally **negatively correlated** with yours. Then your combined income is \$200,000, no matter what the weather: it is totally risk free. You can enter into a contract that gets each of you \$100,000 for sure: you promise to give him \$60,000 in years when you are lucky, and he promises to give you \$60,000 in years when he is lucky. You have eliminated your risks by combining them.

Currency swaps provide a good example of negative correlation of risk in real life. A U.S. firm exporting to Europe gets its revenues in euros, but it is interested in its dollar profits, which depend on the fluctuating euro-dollar exchange rate. Conversely, a European firm exporting to the United States faces similar uncertainty about its profits in euros. When the euro falls relative to the dollar, the U.S. firm's euro revenues convert into fewer dollars, and the European firm's dollar revenues convert into more euros. The opposite happens when the euro rises relative to the dollar. Thus fluctuations in the exchange rate generate negatively correlated risks for the two firms. Both can reduce these risks by contracting for an appropriate swap of their revenues.

Even without such perfect negative correlation, risk sharing has some benefit. Return to your role as an island farmer and suppose you and your neighbor face risks that are independent from each other, as if the rain clouds could toss a separate coin to decide which one of you to visit. Then there are four possible outcomes, each with a probability of 1/4. The incomes you and your neighbor earn in these four cases are illustrated in panel a of Figure 9.1. However, suppose the two of your were to make a contract to share and share alike; then your incomes would be those shown in panel b of Figure 9.1. Although your average (expected) income in each table is $100,000, without the sharing contract, you each would have $160,000, or $40,000 with probabilities of 1/2 each. With the contract, you each would have $160,000 with probability 1/4, $100,000 with probability 1/2, and $40,000 with probability 1/4. Thus, for each of you, the contract has reduced the probabilities of the two extreme outcomes from 1/2 to 1/4 and increased the probability of the middle outcome from 0 to 1/2. In other words, the contract has reduced the risk for each of you.

In fact, as long as your incomes are not totally **positively correlated**—that is, as long as your luck does not move in perfect tandem—you can both reduce your risks by sharing them. And if there are more than two of you with some degree of independence in your risks, then the law of large numbers makes

		NEIGHBOR	
		Lucky	Not
YOU	Lucky	160,000, 160,000	160,000, 40,000
	Not	40,000, 160,000	40,000, 40,000

(a) Without sharing

		NEIGHBOR	
		Lucky	Not
YOU	Lucky	160,000, 160,000	100,000, 100,000
	Not	100,000, 100,000	40,000, 40,000

(b) With sharing

FIGURE 9.1 Sharing Income Risk

possible even greater reduction in the risk of each. That is exactly what insurance companies do: by combining the similar but independent risks of many people, an insurance company is able to compensate any one of them when he suffers a large loss. It is also the basis of portfolio diversification: by dividing your wealth among many different assets with different kinds and degrees of risk, you can reduce your total exposure to risk.

However, such arrangements for risk sharing depend on public observability of outcomes and enforcement of contracts. Otherwise each farmer has the temptation to pretend to have suffered bad luck, or simply to renege on the deal and refuse to share when he has good luck. An insurance company may similarly falsely deny claims, but its desire to maintain is reputation in ongoing business may check such reneging.

Here we consider another issue. In the discussion above, we simply assumed that sharing meant equal shares. That seems natural, because you and your farmer-neighbor are in identical situations. But you may have different strategic skills and opportunities, and one may be able to do better than the other in bargaining or contracting.

To understand this, we must recognize the basic reason that farmers want to make such sharing arrangements, namely, that they are averse to risk. As we saw in the Appendix to Chapter 7, attitudes toward risk can be captured by using nonlinear scales to convert money incomes into "utility" numbers. In that Appendix, we used the square root as a simple example of such a scale that reflects risk aversion; we continue to do so here.

When you bear the full risk of getting $160,000 or $40,000 with probabilities 1/2 each, your expected (probability-weighted average) utility is

$$1/2 \times \sqrt{160{,}000} + 1/2 \times \sqrt{40{,}000} = 1/2 \times 400 + 1/2 \times 200 = 300.$$

The riskless income that will give you the same utility is the number whose square root is 300, that is, $90,000. This is less than the average money income you have, namely $100,000. The difference, $10,000, is the maximum money sum you would be willing to pay as a price for eliminating the risk in your income entirely. Your neighbor faces a risk of equal magnitude, so if he has the same utility scale, he is also willing to pay the same maximum amount to eliminate all of his risk.

Consider the situation where your risks are perfectly negatively correlated, so that the sum of your two incomes is $200,000 no matter what. You make your neighbor the following offer: I will pay you $90,001 − $40,000 = $50,001 when your luck is bad, if you pay me $160,000 − $90,001 = $69,999 when your luck is good. That leaves your neighbor with $90,001 whether his luck is good or bad ($160,000 − $69,999 in the former situation and $40,000 + $50,001 in the latter situation). He prefers this situation to facing the risk. When his luck is good, yours is bad; you have $40,000 of your own but receive $69,999 from him for a total of $100,999. When his luck is bad, yours is good; you have $160,000 of your own but pay him $50,001, leaving you with $100,999. You have also eliminated your own risk. Both of you are made better off by this deal, but you have collared almost all the gain.

Of course your neighbor could have made you the opposite offer. And a whole range of intermediate offers, involving more equitable sharing of the gains from risk sharing, is also conceivable. Which of these will prevail? That depends on the parties' bargaining power, as we will see in more detail in Chapter 18; the full range of mutually beneficial risk-sharing outcomes will correspond to the efficient frontier of negotiation in the bargaining game between the players.

B. Paying to Reduce Risk

Now we consider the possibility of trading of risks for cash. Suppose you are the farmer facing the same risk as before. But now your neighbor has a sure income of $100,000. You face a lot of risk, and he faces none. He may be willing to take a little of your risk for a price that is agreeable to both of you. We just saw that $10,000 is the maximum "insurance premium" you would be willing to pay to get rid of your risk completely. Would your neighbor accept this as payment for eliminating your risk? In effect, he is taking over control of his riskless income plus your risky income, that is, $100,000 + $160,000 = $260,000 if your luck is good and $100,000 + $40,000 = $140,000 if your luck is bad. He gives you $90,000 in either eventuality, thus leaving him with $170,000 or $50,000 with equal probabilities. His expected utility is then

$$1/2 \times \sqrt{170{,}000} + 1/2 \times \sqrt{50{,}000} = 1/2 \times 412.31 + 1/2 \times 223.61 = 317.96.$$

His utility if he did not trade with you would be $\sqrt{100{,}000} = 316.23$, so the trade makes him just slightly better off. The range of mutually beneficial deals in this case is very narrow, so the outcome is almost determinate, but there is not much scope for mutual benefit if you aim to trade all of your risk away.

What about a partial trade? Suppose you pay him x if your luck is good, and he pays you y if your luck is bad. For this to raise expected utilities for both of you, we need both of the following inequalities to hold:

$$1/2 \times \sqrt{160{,}000 - x} + 1/2 \sqrt{40{,}000 + y} > 300,$$

$$1/2 \times \sqrt{100{,}000 + x} + 1/2 \times \sqrt{100{,}000 - y} > \sqrt{100{,}000}.$$

As an example, suppose $y = 10{,}000$. Then the second inequality yields $x > 10{,}526.67$, and the first yields $x < 18{,}328.16$. The first value for x is the minimum payment he requires from you to be willing to make the trade, and the second value for x is the maximum you are willing to pay to him to have him assume your risk. Thus there is a substantial range for mutually beneficial trade and bargaining.

What if your neighbor is risk neutral, that is, concerned solely with expected monetary magnitudes? Then the deal must satisfy

$$1/2 \times (100{,}000 + x) + 1/2 \times (100{,}000 - y) > 100{,}000,$$

or simply $x > y$, to be acceptable to him. Almost-full insurance, where you pay him \$60,001 if your luck is good and he pays you \$59,999 if your luck is bad, is possible. This is the situation where you reap all the gain from the trade in risks.

If your "neighbor" is actually an insurance company, the company can be close to risk neutral because it is combining numerous such risks and is owned by well-diversified investors for each of whom this business is only a small part of their total risk. Then the fiction of a friendly, risk-neutral, good neighbor can become a reality. And if insurance companies compete for your business, the insurance market can offer you almost complete insurance at a price that leaves almost all of the gain with you.

Common to all such arrangements is the idea that mutually beneficial deals can be struck whereby, for a suitable price, someone facing less risk takes some of the risk off the shoulders of someone else who faces more. In fact, the idea that a price and a market for risk exist is the basis for almost all of the financial arrangements in a modern economy. Stocks and bonds, as well as all of the complex financial instruments, such as derivatives, are just ways of spreading risk to those who are willing to bear it for the lowest asking price. Many people think these markets are purely forms of gambling. In a sense, they are. But those who start out with the least risk take the gambles, perhaps because they have already diversified in the way that we saw earlier. And the risk is sold or shed by those who are initially most exposed to it. This enables the latter to be more adventurous in their enterprises than they would be if they had to bear all of the risk themselves. Thus financial markets promote entrepreneurship by facilitating risk trading.

Here we have only considered sharing of a given total risk. In practice, people may be able to take actions to reduce that total risk: a farmer can guard crops against frosts, and a car owner can drive more carefully to reduce the risk of an accident. If such actions are not publicly observable, the game will be one of imperfect information, raising the problem of moral hazard that we mentioned in the introduction: people who are well insured will lack the incentive to reduce the risk they face. We will look at such problems, and the design of mechanisms to cope with them, in Chapter 14.

C. Manipulating Risk in Contests

The farmers above faced risk due to the weather rather than from any actions of their own or of other farmers. If the players in a game can affect the risk they or others face, then they can use such manipulation of risk strategically. A prime example is contests such as research and development races between companies to develop and market new information technology or biotech products; many sports contests have similar features.

The outcome of sports and related contests is determined by a mixture of skill and chance. You win if

$$\text{Your skill} + \text{your luck} > \text{rival's skill} + \text{rival's luck}$$

or

$$\text{Your luck} - \text{rival's luck} > \text{rival's skill} - \text{your skill}.$$

Denote the left-hand side by the symbol *L;* it measures your "luck surplus." *L* is an uncertain magnitude; suppose its probability distribution is a normal, or bell, curve, as illustrated by the black curve in Figure 9.2. At any point on the horizontal axis, the height of the curve represents the probability that *L* takes on that value. Thus the area under this curve between any two points on the horizontal axis equals the probability that *L* lies between those points. Suppose your rival has more skill, so you are an underdog. Your "skill deficit," which equals the difference between your rival's skill and your skill, is therefore positive, as shown by the point *S*. You win if your luck surplus, *L*, exceeds your skill deficit, *S*. Therefore the area under the curve to the right of the point *S*, which is shaded in gray in Figure 9.2, represents your probability of winning. If you make the situation chancier, the bell curve will be flatter, like the green curve in Figure 9.2, because the probability of relatively high and low values of *L* increases while the probability of moderate values decreases. Then the area under the curve to the right of *S* also increases. In Figure 9.2, the area under the original bell curve is shown

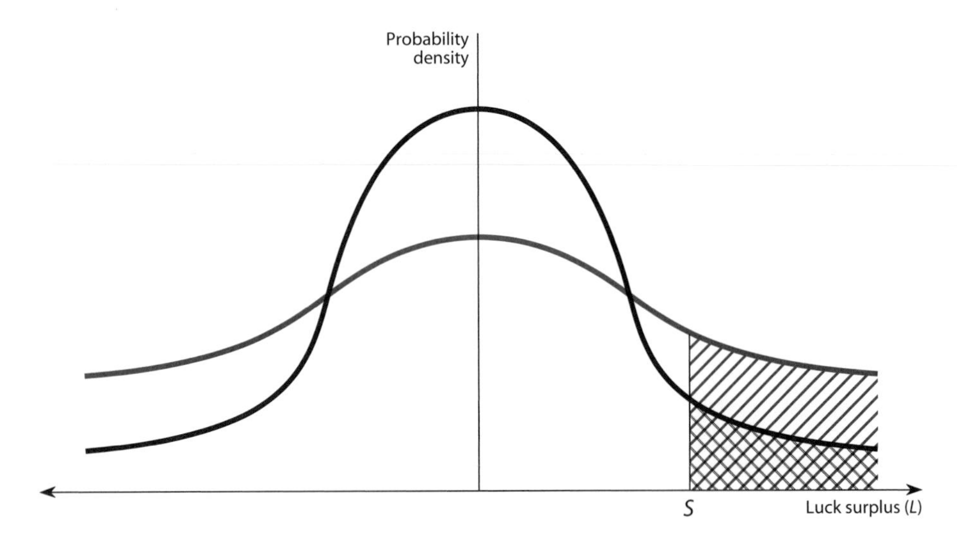

FIGURE 9.2 The Effect of Greater Risk on the Chances of Winning

by gray shading, and the larger area under the flatter bell curve by the green hatching. As the underdog, you should therefore adopt a strategy that flattens the curve. Conversely, if you are the favorite, you should try to reduce the element of chance in the contest.

Thus we should see underdogs or those who have fallen behind in a long race try unusual or risky strategies: it is their only chance to get level or ahead. On the other hand, favorites or those who have stolen a lead will play it safe. A practical piece of advice based on this principle: if you want to challenge someone who is a better player than you to a game of tennis, choose a windy day.

You may stand to benefit by manipulating not just the amount of risk in your strategy, but also the correlation between the risks. The player who is ahead will try to choose a correlation as high and as positive as possible: then, whether his own luck is good or bad, the luck of his opponent will be the same and his lead protected. Conversely, the player who is behind will try to find a risk as uncorrelated with that of his opponent as possible. It is well known that in a two-sailboat race, the boat that is behind should try to steer differently from the boat ahead, and the boat ahead should try to imitate all the tack of the one behind.[2]

2 ASYMMETRIC INFORMATION: BASIC IDEAS

In many games, one or some of the players may have an advantage of knowing with greater certainty what has happened or what will happen. Such advantages, or asymmetries of information, are common in actual strategic situations. At the most basic level, each player may know his own preferences or payoffs—for example, risk tolerance in a game of brinkmanship, patience in bargaining, or peaceful or warlike intentions in international relations—quite well but those of the other players much more vaguely. The same is true for a player's knowledge of his own innate characteristics (such as the skill of an employee or the riskiness of an applicant for auto or health insurance). And sometimes the actions available to one player—for example, the weaponry and readiness of a country—are not fully known to other players. Finally, some actual outcomes (such as the actual dollar value of loss to an insured homeowner in a flood or an earthquake) may be observed by one player but not by others.

By manipulating what the other players know about your abilities and preferences, you can affect the equilibrium outcome of a game. Therefore such

[2]Avinash Dixit and Barry Nalebuff, *Thinking Strategically* (New York: Norton, 1991), give a famous example of the use of this strategy in sailboat racing. For a more general theoretical discussion, see Luis Cabral, "R&D Competition When the Firms Choose Variance," *Journal of Economics and Management Strategy*, vol. 12, no. 1 (Spring 2003), pp. 139–150.

manipulation of asymmetric information itself becomes a game of strategy. You may think that each player will always want to conceal his own information and elicit information from the others, but that is not so. Here is a list of various possibilities, with examples. The better-informed player may want to do one of the following:

1. *Conceal information* or *reveal misleading information.* When mixing moves in a zero-sum game, you don't want the other player to see what you have done; you bluff in poker to mislead others about your cards.
2. *Reveal selected information truthfully.* When you make a strategic move, you want others to see what you have done so that they will respond in the way you desire. For example, if you are in a tense situation but your intentions are not hostile, you want others to know this credibly, so that there will be no unnecessary fight.

Similarly, the less-informed player may want to do one of the following:

1. *Elicit information* or *filter truth from falsehood.* An employer wants to find out the skill of a prospective employee and the effort of a current employee. An insurance company wants to know an applicant's risk class, the amount of a claimant's loss, and any contributory negligence by the claimant that would reduce its liability.
2. *Remain ignorant.* Being unable to know your opponent's strategic move can immunize you against his commitments and threats. Top-level politicians or managers often benefit from having such "credible deniability."

In most cases, we will find that words alone do not suffice to convey credible information; rather, actions speak louder than words. Even actions may not convey information credibly if they are too easily performed by any random player. In general, however, the less-informed players should pay attention to what a better-informed player does, not to what he says. And knowing that the others will interpret actions in this way, the better-informed player should in turn try to manipulate his actions for their information content.

When you are playing a strategic game, you may find that you have information that other players do not. You may have information that is "good" (for yourself) in the sense that, if the other players knew this information, they would alter their actions in a way that would increase your payoff. You know that you are a nonsmoker, for example, and should qualify for lower life-insurance premiums. Or you may have "bad" information whose disclosure would cause others to act in a way that would hurt you. You cheated your way through college, for example, and don't deserve to be admitted to a prestigious law school. You know that others will infer your information from your actions. Therefore you try to think of, and take, actions that will induce them to believe your information is good. Such actions are called **signals,** and the strategy of using them is called **signaling.** Conversely, if others are likely to conclude that your information is bad, you may be able to stop them from making this inference by confusing them. This strategy, called **signal jamming,** is typically a mixed strategy, because the randomness of mixed strategies makes inferences imprecise.

If other players know more than you do or take actions that you cannot directly observe, you can use strategies that reduce your informational disadvantage. The strategy of making another player act so as to reveal his information is called **screening,** and specific methods used for this purpose are called **screening devices.**[3]

Because a player's private information often consists of knowledge of his own abilities or preferences, it is useful to think of players who come to a game possessing different private information as different **types.** When credible signaling works, in the equilibrium of the game the less-informed players will be able to infer the information of the more-informed ones correctly from the actions; the law school, for example, will admit only the truly qualified applicants. Another way to describe the outcome is to say that in equilibrium the different types are correctly revealed or separated. Therefore we call this a **separating equilibrium.** In some cases, however, one or more types may successfully mimic the actions of other types, so that the uninformed players cannot infer types from actions and cannot identify the different types; insurance companies, for example, may offer only one kind of life insurance policy. Then, in equilibrium we say the types are pooled together, and we call this a **pooling equilibrium.** When studying games of incomplete information, we will see that identifying the kind of equilibrium that occurs is of primary importance.

3 DIRECT COMMUNICATION, OR "CHEAP TALK"

The simplest way to convey information to others would seem to be to tell them; likewise, the simplest way to elicit information would seem to be to ask. But in a game of strategy, players should be aware that others may not tell the truth and, likewise, that their own assertions may not be believed by others. That is, the *credibility* of mere words may be questionable. It is a common saying that talk is cheap; indeed, direct communication has zero or negligible *direct* cost.

[3]A word of warning: Don't confuse screening with signal jamming. In ordinary language, the word *screening* can have different meanings. The one used in game theory is that of testing or scrutinizing. Thus a less-informed player uses screening to find out what a better-informed player knows. For the alternative sense of screening—that is, concealing—the game-theoretic term is signal jamming. Thus a better-informed player uses a signal-jamming action to prevent the less-informed player from correctly inferring the truth from the action (that is, from screening the better-informed player).

However, it can *indirectly* affect the outcome and payoffs of a game by changing one players's beliefs about another player's actions, or by influencing the selection of one equilibrium out of multiple equilibria. Direct communication that has no direct cost has come to be called *cheap talk* by game theorists, and the equilibrium achieved by using direct communication is termed a **cheap talk equilibrium.**

A. Perfectly Aligned Interests

Direct communication of information works well if the players' interests are well aligned. The assurance game first introduced in Chapter 4 provides the most extreme example of this. We reproduce its payoff table (Figure 4.12) as Figure 9.3.

The interests of Harry and Sally are perfectly aligned in this game; they both want to meet, and both prefer meeting in Local Latte. The problem is that the game is played noncooperatively; they are making their choices independently, without knowledge of what the other is choosing. But suppose that Harry is given an opportunity to send a message to Sally (or Sally is given an opportunity to ask a question and Harry replies) before their choices are made. If Harry's message (or reply; we will not keep repeating this) is, "I am going to Local Latte," Sally has no reason to think he is lying.[4] If she believes him, she should choose Local Latte, and if he believes she will believe him, it is equally optimal for him to choose Local Latte, making his message truthful. Thus direct communication very easily achieves the mutually preferable outcome. This is indeed the reason that, when we considered this game in Chapter 4, we had to construct an elaborate scenario in which such communication was infeasible; recall that the two were in separate classes until the last minute before their meeting and did not have their cell phones.

		SALLY	
		Starbucks	Local Latte
HARRY	Starbucks	1, 1	0, 0
	Local Latte	0, 0	2, 2

FIGURE 9.3 Assurance

[4]This reasoning assumes that Harry's payoffs are as stated, and that this fact is common knowledge between the two. If Sally suspects that Harry wants her to go to Local Latte so he can go to Starbucks to meet another girlfriend, her strategy will be different! Analysis of games of asymmetric information thus depends on how many different possible "types" of players are actually conceivable.

Let us examine the outcome of allowing direct communication in the assurance game more precisely in game-theoretic terms. We have created a two-stage game. In the first stage, only Harry acts, and his action is his message to Sally. In the second stage, the original simultaneous-move game is played. In the full two-stage game, we have a rollback equilibrium where the strategies (complete plans of action) are as follows. The second-stage action plans for both players are: "If Harry's first-stage message was 'I am going to Starbucks,' then choose Starbucks; if Harry's first-stage message was 'I am going to Local Latte,' then choose Local Latte." (Remember that players in sequential games must specify *complete* plans of action.) The first-stage action for Harry is to send the message "I am going to Local Latte." Verification that this is indeed a rollback equilibrium of the two-stage game is easy, and we leave it to you.

However, this equilibrium where cheap talk "works" is not the only rollback equilibrium of this game. Consider the following strategies: The second-stage action plan for each player is to go to Starbucks regardless of Harry's first-stage message; and Harry's first-stage message can be anything. We can verify that this also is indeed a rollback equilibrium. Regardless of Harry's first-stage message, if one player is going to Starbucks, then it is optimal for the other player to go there also. Thus, in each of the second-stage subgames that could arise—one after each of the two messages that Harry could send—both choosing Starbucks is a Nash equilibrium of the subgame. Then, in the first stage, Harry, knowing his message is going to be disregarded, is indifferent about which message he sends.

The cheap talk equilibrium—where Harry's message is not disregarded—yields higher payoffs, and we might normally think that it would be the one selected as a focal point. However, there may be reasons of history or culture that favor the other equilibrium. For example, for some reasons quite extraneous to this particular game, Harry may have a reputation for being totally unreliable. He might be a compulsive practical joker or just absent minded. Then people might generally disregard his statements and, knowing this to be the usual state of affairs, Sally might not believe this particular one.

Such problems exist in all communication games. They always have alternative equilibria where the communication is disregarded and therefore irrelevant. Game theorists call these **babbling equilibria.** Having noted that they exist, however, we will focus on the cheap talk equilibria, where communication does have some effect.

B. Totally Conflicting Interests

The credibility of direct communication depends on the degree of alignment of players' interests. As a dramatic contrast with the assurance game example, consider a game where the players' interests are totally in conflict—namely, a

		NAVRATILOVA	
		DL	CC
EVERT	DL	50	80
	CC	90	20

FIGURE 9.4 Tennis Point

zero-sum game. A good example is the tennis point of Figure 4.15; we reproduce its payoff matrix as Figure 9.4. Remember that the payoffs are Evert's success percentages. Remember also that this game has only a mixed-strategy Nash equilibrium (derived in Chapter 7); Evert's expected payoff in this equilibrium is 62.

Now suppose that we construct a two-stage game. In the first stage, Evert is given an opportunity to send a message to Navratilova. In the second stage, the simultaneous-move game of Figure 9.4 is played. What will be the rollback equilibrium?

It should be clear that Navratilova will not believe any message she receives from Evert. For example, if Evert's message is, "I am going to play DL," and Navratilova believes her, then Navratilova should choose to cover DL. But if Evert thinks that Navratilova will cover DL, then Evert's best choice is CC. At the next level of thinking, Navratilova should see through this and not believe the assertion of DL.

But there is more. Navratilova should not believe that Evert would do exactly the opposite of what she says either. Suppose Evert's message is, "I am going to play DL," and Navratilova thinks, "She is just trying to trick me, and so I will take it that she will play CC." This will lead Navratilova to choose to cover CC. But if Evert thinks that Navratilova will disbelieve her in this simple way, then Evert should choose DL after all. And Navratilova should see through this, too.

Thus Navratilova's disbelief should mean that she should just totally disregard Evert's message. Then the full two-stage game has only the babbling equilibrium. The two players' actions in the second stage will be simply those of the original equilibrium, and Evert's first-stage message can be anything. This is true of all zero-sum games.

C. Partially Aligned Interests

But what about more general games in which there is a mixture of conflict and common interest? Whether direct communication is credible in such games depends on how the two aspects of conflict and cooperation mix when players' interests are only partially aligned. Thus, we should expect to see both cheap talk and babbling equilibria in games of this type.

Consider games with multiple equilibria where one player prefers one equilibrium and the other prefers the other equilibrium, but both prefer either of the

		SALLY	
		Starbucks	Local Latte
HARRY	Starbucks	2, 1	0, 0
	Local Latte	0, 0	1, 2

FIGURE 9.5 Battle of the Sexes

equilibria to some other outcome. One example is the battle of the sexes; we reproduce its payoff table (Figure 4.13) as Figure 9.5.

Suppose Harry is given an opportunity to send a message. Then the two-stage game has a rollback equilibrium where he sends the message "I am going to Starbucks," and the second-stage action plan for both players is to choose the location identified in Harry's message. Here, there is a cheap talk equilibrium, and the opportunity to send a message can enable a player to select his preferred outcome.[5]

Another example of a game with partially aligned payoffs comes from a situation that you may have already experienced, or, if not, soon will when you start to earn and invest. When your stockbroker recommends that you should buy a particular stock, he may be doing so as part of developing a long-run relationship with you for the steady commissions that your business will bring him, or he may be touting a loser that his firm wants to get rid of for a quick profit. You have to guess the relative importance of these two possibilities in his payoffs.

Suppose there are just two possibilities: the stock may have good prospects or bad ones. If the former, you should buy it; if the latter, sell it, or *sell short.* Figure 9.6 shows your payoffs in each of the eventualities given the two possibilities, Good and Bad, and your two actions, Buy and Sell. Note that this is not a payoff matrix of a game; the columns are not the choices of a strategic player.

		THE STOCK IS	
		Good	Bad
YOU	Buy	1	−1
	Sell	−1	1

FIGURE 9.6 Your Payoffs from Your Investment Decisions

[5]What about the possibility that, in the second stage, the action plans are for both players to choose exactly the opposite of the location in Harry's message? That, too, is a Nash equilibrium of the second-stage subgame, so there would seem to be a "perverse" cheap talk equilibrium. In this situation, Harry's optimal first-stage action will be to say, "I am going to Local Latte," so he can still get his preferred outcome.

		THE STOCK IS	
		Good	Bad
YOU	Buy	1	$-1+X$
	Sell	-1	1

FIGURE 9.7 Your Broker's Payoffs from Your Investment Decisions

The broker knows whether the stock is actually Good or Bad; you don't. He can give you a recommendation of Buy, or Sell. Should you follow his advice? That should depend on your broker's payoffs in these same situations. Suppose the broker's payoffs are a mixture of two considerations. One is the long-term relationship with you; that part of his payoffs is just a replica of yours. But he also gets an extra kickback X from his firm if he can persuade you to buy a bad stock that the firm happens to own and is eager to unload on you. Then his payoffs are as shown in Figure 9.7.

Can there be a cheap talk equilibrium with truthful communication? In this example, the broker sends you a message in the first stage of the game. That message will be Buy or Sell, depending on his observation of whether the stock is Good or Bad. At the second stage, you make your choice of Buy or Sell, depending on his message. So we are looking for an equilibrium where his strategy is honesty (say Buy if Good, say Sell if Bad), and your strategy is to follow his recommendation. We have to test whether the strategy of each player is optimal given that of the other.

Given that the broker is sending honest messages, obviously it is best for you to follow the advice. Given that you are following the advice, what about the broker's strategy? Suppose he knows the stock is Good. If he sends the message Buy, you will buy the stock and his payoff will be – 1; if he says Sell, you will sell and his payoff will be – 1. So the "say Buy if Good" part of his strategy is indeed optimal for him. Now suppose he knows the stock to be Bad. If he says Sell, you will sell and he will get 1. If he says Buy, you will buy and he will get $-1 + X$. So honesty is optimal for him in this situation if $1 > -1 + X$, or if $X < 2$. Direct communication from your broker is credible, and there is a cheap talk equilibrium in this game as long as his extra payoff from selling you a loser is not "too large."

However, if $X > 2$, then the broker's best response to your strategy of following his advice is to say Buy regardless of the truth. But if he is doing that, then following his advice is no longer your optimal strategy. You have to disregard his message and fall back on your own prior estimate of whether the stock is Good or Bad. In this case, only the babbling equilibrium is possible.

In these examples, the available messages were simple binary ones—Starbucks or Local Latte and Buy or Sell. What happens when richer messages are possible? For example, suppose that the broker could send you a number g, representing his estimate of the rate of growth of the stock price, and this number could range over a whole continuum. Now, as long as the broker gets some extra benefit if you buy a bad stock that he recommends, he has some incentive to exaggerate g. Therefore fully accurate truthful communication is no longer possible. But partial revelation of the truth may be possible. That is, the continuous range of growth rates may split into intervals—say, from 0% to 1%, from 1% to 2%, and so on—such that the broker finds it optimal to tell you truthfully into which of these intervals the actual growth rate falls and you find it optimal to accept this advice and take your optimal action on its basis. However, we must leave further explanation of this idea to more advanced treatments.[6]

4 ADVERSE SELECTION, SIGNALING, AND SCREENING

A. Adverse Selection and Market Failure

In many games, one of the players knows something pertinent to the outcomes that the other players don't know. An employer knows much less about the skills of a potential employee than does the employee himself; vaguer but important matters such as work attitude and collegiality are even harder to observe. An insurance company knows much less about the health or driving skills of someone applying for medical or auto insurance than does the applicant. The seller of a used car knows a lot about the car from long experience; a potential buyer can at best get a little information by inspection.

In such situations, direct communication will not credibly signal information. Unskilled workers will claim to have skills to get higher-paid jobs; people who are bad risks will claim good health or driving habits to get lower insurance premiums; owners of bad cars will assert that their cars run fine and have given them no trouble in all the years they have owned them. The other parties to the transactions will be aware of the incentives to lie and will not trust the information conveyed by the words.

[6] The seminal paper that developed this theory of partial communication is by Vincent Crawford and Joel Sobel, "Strategic Information Transmission," *Econometrica*, vol. 50, no. 6 (November 1982), pp. 1431–1452. An elementary exposition and survey of further work is in Joseph Farrell and Matthew Rabin, "Cheap Talk," *Journal of Economic Perspectives*, vol. 10, no. 3 (Summer 1996), pp. 103–118.

What if the less-informed parties in these transactions have no way of obtaining the pertinent information at all? In other words, to use the terminology introduced in Section 2 above, suppose that no credible screening devices nor signals are available. If an insurance company offers a policy that costs 5 cents for each dollar of coverage, then the policy will be especially attractive to people who know that their own risk (of illness or a car crash) exceeds 5%. Of course, some people who know their risk to be lower than 5% will still buy the insurance because they are risk averse. But the pool of applicants for this insurance policy will have a larger proportion of the poorer risks than the proportion of these risks in the population as a whole. The insurance company will selectively attract an unfavorable, or adverse, group of customers. This phenomenon is very common in transactions involving asymmetric information and is known as **adverse selection.** (This term in fact originated within the insurance industry.)

Potential consequences of adverse selection for market transactions were dramatically illustrated by George Akerlof in a paper that became the starting point of economic analysis of asymmetric information situations and won him a Nobel Prize in 2001.[7] We use his example to introduce you to the effects that adverse selection may have.

Think of the market in 2009 for a specific kind of used car, say a 2006 Citrus. Suppose that in use these cars have proved to be either largely trouble free and reliable or have had many things go wrong. The usual slang name for the latter type is "lemon," so for contrast let us call the former type "orange."

Suppose that each owner of an orange Citrus values it at $12,500; he is willing to part with it for a price higher than this but not for a lower price. Similarly, each owner of a lemon Citrus values it at $3,000. Suppose that potential buyers are willing to pay more than these values for each type. If a buyer could be confident that the car he was buying was an orange, he would be willing to pay $16,000 for it; if the car was a known lemon, he would be willing to pay $6,000. Since the buyers value each type of car more than do the original owners, it benefits everyone if all the cars are traded. The price for an orange can be anywhere between $12,500 and $16,000; that for a lemon anywhere between $3,000 and $6,000. For definiteness, we will suppose that there is a limited stock of such cars and a larger number of potential buyers. Then the buyers, competing with each other, will drive the price up to their full willingness to pay. The prices will be $16,000 for an orange and $6,000 for a lemon—if each type could be identified with certainty.

But information about the quality of any specific car is not symmetric between the two parties to the transaction. The owner of a Citrus knows perfectly

[7]George Akerlof, "The Market for Lemons: Qualitative Uncertainty and the Market Mechanism," *Quarterly Journal of Economics*, vol. 84, no. 3 (August 1970), pp. 488–500.

well whether it is an orange or a lemon. Potential buyers don't, and the owner of a lemon has no incentive to disclose the truth. For now, we confine our analysis to the private used-car market in which laws requiring truthful disclosure are either nonexistent or hard to enforce. We also assume away any possibility that the potential buyer can observe something that tells him whether the car is an orange or a lemon; similarly, the car owner has no way to indicate the type of car he owns. Thus, for this example, we consider the effects of the information asymmetry alone without allowing either side of the transaction to signal or screen.

When buyers cannot distinguish between oranges and lemons, there cannot be distinct prices for the two types in the market. There can be just one price, p, for a Citrus; the two types—oranges and lemons—must be pooled. Whether efficient trade is possible under such circumstances will depend on the proportions of oranges and lemons in the population. We suppose that oranges are a fraction f of used Citruses, and lemons the remaining fraction $(1 - f)$.

Even though buyers cannot verify the quality of an individual car, they can know the proportion of good cars in the population as a whole, for example, from newspaper reports, and we assume this to be the case. If all cars are being traded, a potential buyer will expect to get a random selection, with probabilities f and $(1 - f)$ of getting an orange and a lemon, respectively. The expected value of the car purchased is $16{,}000 \times f + 6{,}000 \times (1 - f) = 6{,}000 + 10{,}000 \times f$. He will buy such a car if its expected value exceeds the price he is asked to pay, that is, if $6{,}000 + 10{,}000 f > p$.

Now consider the point of view of the seller. The owners know whether their cars are oranges or lemons. The owner of a lemon is willing to sell it as long as the price exceeds its value to him, that is, if $p > 3{,}000$. But the owner of an orange requires $p > 12{,}500$. If this condition for an orange owner to sell is satisfied, so is the sell condition for a lemon owner.

To meet the requirements for all buyers and sellers to want to make the trade, therefore, we need $6{,}000 + 10{,}000 \times f > p > 12{,}500$. If the fraction of oranges in the population satisfies $6{,}000 + 10{,}000 \times f > 12{,}500$, or $f > 0.65$, a price can be found that does the job; otherwise there cannot be efficient trade. If $6{,}000 + 10{,}000 f < 12{,}500$ (leaving out the exceptional and unlikely case where the two are just equal), owners of oranges are unwilling to sell at the maximum price the potential buyers are willing to pay. We then have adverse selection in the set of used cars put up for sale; no oranges will appear in the market at all. The potential buyers will recognize this, will expect to get a lemon for sure, and will pay at most $6,000. The owners of lemons will be happy with this outcome, so lemons will trade. But the market for oranges will collapse completely due to the asymmetric information. The outcome will be a kind of Gresham's law, where bad cars drive out the good.

Because the lack of information makes it impossible to get a reasonable price for an orange, the owners of oranges will want a way to convince the buyers that their cars are the good type. They will want to signal their type. The trouble is that the owners of lemons would also like to pretend that their cars are oranges, and to this end can imitate most of the signals that owners of oranges might attempt to use. Michael Spence, who developed the concept of signaling and shared the 2001 Nobel Prize for information economics with Akerlof and Stiglitz, summarizes the problems facing our orange owners in his pathbreaking book on signaling: "Verbal declarations are costless and therefore useless. Anyone can lie about why he is selling the car. One can offer to let the buyer have the car checked. The lemon owner can make the same offer. It's a bluff. If called, nothing is lost. Besides, such checks are costly. Reliability reports from the owner's mechanic are untrustworthy. The clever nonlemon owner might pay for the checkup but let the purchaser choose the inspector. The problem for the owner, then, is to keep the inspection cost down. Guarantees do not work. The seller may move to Cleveland, leaving no forwarding address."[8]

In reality, the situation is not so hopeless as Spence implies. People and firms that regularly sell used cars as a business can establish a reputation for honesty and profit from this reputation by charging a markup. (Of course, some used car dealers are unscrupulous.) Some buyers are knowledgeable about cars; some buy from personal acquaintances and can therefore verify the history of the car they are buying. And in other markets it is harder for bad types to mimic the actions of good types, so credible signaling will be viable. For a specific example of such a situation, consider the possibility that education can signal skill. Then it may be hard for the unskilled to acquire enough education to be mistaken for highly skilled people. The key requirement for education to separate the types is that education should be sufficiently more costly for the truly unskilled to acquire than for the truly skilled. To show how and when signaling can successfully separate types, therefore, we turn to the labor market.

B. Signaling in the Labor Market

Many of you expect that when you graduate you will work for an elite firm in finance or computing. These firms have two kinds of jobs. One kind requires high quantitative and analytical skills and capacity for hard work and offers high pay in return. The other kind of jobs are semiclerical, lower-skill, lower-pay jobs. Of course, you want the job with higher pay. You know your own qualities and skills

[8] A. Michael Spence, *Market Signaling: Information Transfer in Hiring and Related Screening Processes* (Cambridge, Mass.: Harvard University Press, 1974), pp. 93–94. The present authors apologize on behalf of Spence to any residents of Cleveland who may be offended by any unwarranted suggestion that that's where shady sellers of used cars go!

far better than your prospective employer does. If you are highly skilled, you want your employer to know this about you and he also wants to know. He can test and interview you, but what he can find out by these methods is limited by the available time and resources. You can tell him how skilled you are but mere assertions about your qualifications are not credible. More objective evidence is needed, both for you to offer and for your employer to seek out.

What items of evidence can the employer seek, and what can you offer? Recall from Section 2 of this chapter that your prospective employer will use *screening devices* to identify your qualities and skills. You will use *signals* to convey your information about those same qualities and skills. Sometimes similar or even identical devices can be used for either signaling or screening.

In this instance, if you have selected (and passed) particularly tough and quantitative courses in college, your course choices can be credible evidence of your capacity for hard work in general and of your quantitative and logical skills in particular. Let us consider the role of course choice as a screening device.

To keep things simple, suppose college students are of just two types when it comes to the qualities most desired by employers: A (able) and C (challenged). Potential employers in finance or computing are willing to pay $160,000 a year to a type A, and $60,000 to a type C. Other employment opportunities yield the A types a salary of $125,000 and the C types a salary of $30,000. These are just the numbers in the used-car example in Section 4.A above, but multiplied by a factor of 10 better to suit the reality of the job-market example. And just as in the used-car example where we supposed there was fixed supply and numerous potential buyers, we suppose here that there are many potential employers who have to compete with each other for a limited number of job candidates, so they have to pay the maximum amount that they are willing to pay. Because employers cannot directly observe any particular job applicant's type, they have to look for other credible means to distinguish among them.[9]

Suppose the types differ in their tolerance for taking a tough course rather than an easy one in college. Each type must sacrifice some party time or other activities to take a tougher course, but this sacrifice is smaller or easier to bear for the A types than it is for the C types. Suppose the A types regard the cost of each such course as equivalent to $3,000 a year of salary, while the C types regard it as $15,000 a year of salary. Can an employer use this differential to screen his applicants and tell the A types from the C types?

Consider the following hiring policy: anyone who has taken a certain number, n, or more of the tough courses will be regarded as an A and paid $160,000,

[9]You may wonder whether the fact that the two types have different outside opportunities can be used to distinguish between them. For example, an employer may say, "Show me an offer of a job at $125,000, and I will accept you as type A and pay you $160,000." However, such a competing offer can be forged or obtained in cahoots with someone else, so it is not reliable.

and anyone who has taken fewer than n will be regarded as a C and paid $60,000. The aim of this policy is to create natural incentives whereby only the A types will take the tough courses, and the C types will not. Neither wants to take more of the tough courses than he has to, so the choice is between taking n to qualify as an A or giving up and settling for being regarded as a C, in which case he may as well not take any of the tough courses and just coast through college.

To succeed, such a policy must satisfy two kinds of conditions. The first set of conditions requires that the policy gives each type of job applicant the incentive to make the choice that the firm wants him to make. In other words, the policy should be compatible with the incentives of the workers; therefore the relevant conditions are called **incentive-compatibility conditions.** The second kind of conditions ensure that, with such an incentive-compatible choice, the workers get a better (at least, no worse) payoff from these jobs than they would get in their alternative opportunities. In other words, the workers should be willing to participate in this firm's offer; therefore the relevant conditions are called the **participation conditions.** We will develop these conditions in the labor market context now. Similar conditions will appear in other examples later in this chapter and again in Chapter 14, where we develop the general theory of mechanism design.

[1] INCENTIVE COMPATIBILITY The criterion that employers devise to distinguish an A from a C—namely, the number of tough courses taken—should be sufficiently strict that the C types do not bother to meet it but not so strict as to discourage even the A types from attempting it. The correct value of n must be such that the true C types prefer to settle for being revealed as such and getting $60,000, rather than incurring the extra cost of imitating the A type's behavior. That is, we need the policy to be incentive compatible for the C types, so [10]

$$60{,}000 \geq 160{,}000 - 15{,}000\,n, \quad \text{or} \quad 15\,n \geq 100, \text{ or } n \geq 6.67.$$

Similarly, the condition that the true A types prefer to prove their type by taking n tough courses is

$$160{,}000 - 3{,}000\,n \geq 60{,}000, \text{ or } 3n \leq 100, \text{ or } n \leq 33.33.$$

These incentive-compatibility conditions or, equivalently, **incentive-compatibility constraints,** align the job applicant's incentives with the employer's desires, or make it optimal for the applicant to reveal the truth about his skill

[10]We require merely that the payoff from choosing the option intended for one's type be at least as high as that from choosing a different option, not that it be strictly greater. However, it is possible to approach the outcome of this analysis as closely as one wants while maintaining a strict inequality, so nothing substantial hinges on this assumption.

through his action. The n satisfying both constraints, because it is required to be an integer, must be at least 7 and at most 33.[11] The latter is not realistically relevant in this example, as an entire college program is typically thirty-two courses, but in other examples it might matter.

What makes it possible to meet both conditions is the *difference* in the costs of taking tough courses between the two types: the cost is sufficiently lower for the "good" type that the employers wish to identify. When the constraints are met, the employer can use a policy to which the two types will respond differently, thereby revealing their types. This is called **separation of types** based on **self-selection.**

We did not assume here that the tough courses actually imparted any additional skills or work habits that might convert C types into A types. In our scenario, the tough courses serve only the purpose of identifying the persons who already possess these attributes. In other words, they have a pure screening function.

In reality, education does increase productivity. But it also has the additional screening or signaling function of the kind described here. In our example, we found that education might be undertaken solely for the latter function; in reality, the corresponding outcome is that education is carried further than is justified by the extra productivity alone. This extra education carries an extra cost—the cost of the information asymmetry.

[2] PARTICIPATION When the incentive-compatibility conditions for the two types of jobs in this firm are satisfied, the A types take n tough courses and get a payoff of $160{,}000 - 3{,}000\,n$, and the C types take no tough courses and get a payoff of 60,000. For the types to be willing to make these choices instead of taking their alternative opportunities, the participation conditions must be satisfied as well. So we need

$$160{,}000 - 3{,}000\,n \geq 125{,}000, \text{ and } 60{,}000 \geq 30{,}000.$$

The C types' participation condition is trivially satisfied in this example (although that may not be the case in other examples); the A types' participation condition requires $n \leq 11.67$, or, since n must be an integer, $n \leq 11$. Here, any n that satisfies the A types' participation constraint of $n \leq 11$ also satifies their incentive compatibility constraint of $n \leq 33$, so the latter becomes logically redundant, regardless of its realistic irrelevance.

[11] If in some other context the corresponding choice variable is not required to be an integer—for example, if it is a sum of money or an amount of time—then a whole continuous range will satisfy both incentive-compatibility constraints.

The full set of conditions that are required to achieve separation of types in this labor market is then $7 \leq n \leq 11$. This restriction on possible values of n combines the incentive-compatibility condition for the C types and the participation condition for the A types. The participation condition for the C types and the incentive-compatibility condition for the A types in this example are automatically satisfied when the other conditions hold.

When the requirement of taking enough tough courses is used for screening, the A types bear the cost. Assuming that only the minimum needed to achieve separation is used—namely, $n = 7$—the cost to each A type has the monetary equivalent of $7 \times \$3{,}000 = \$21{,}000$. This is the cost, in this context, of the information asymmetry. It would not exist if a person's type could be directly and objectively identified. Nor would it exist if the population consisted solely of A types. The A types have to bear this cost because there are some C types in the population from whom they (or their prospective employers) seek to distinguish themselves.[12]

Rather than having the A types bear this cost, might it be better not to bother with the separation of types at all? With the separation, A types get a salary of $160,000 but suffer a cost, the monetary equivalent of $21,000, in taking the tough courses; thus their net money-equivalent payoff is $139,000. And C types get the salary of $60,000. What happens to the two types if they are not separated?

If employers do not use screening devices, they have to treat every applicant as a random draw from the population and pay all the same salary. This is called **pooling of types,** or simply **pooling** when the sense is clear.[13] In a competitive job market, the common salary under pooling will be the population average of what the types are worth to an employer, and this average will depend on the proportions of the types in the population. For example, if 60% of the population is type A and 40% is type C, then the common salary with pooling will be

$$0.6 \times \$160{,}000 + 0.4 \times \$60{,}000 = \$120{,}000.$$

The A types will then prefer the situation with separation because it yields $139,000 instead of the $120,000 with pooling. But if the proportions are 80% A and 20% C, then the common salary with pooling will be $140,000, and the A types will be worse off under separation than they would be under pooling. The C types are always better off under pooling. The existence of the A types in the population means that the common salary with pooling will always exceed the C types' separation salary of $60,000.

[12]In the terminology of economics, the C types in this example inflict a *negative external effect* on the A types. We will develop this concept in Chapter 12, Section 5.

[13]It is the opposite of *separation of types,* described above where players differing in their characteristics get different outcomes, so the outcome reveals the type perfectly.

However, even if both types prefer the pooling outcome, it cannot be an equilibrium when many employers or workers compete with each other in the screening or signaling process. Suppose the population proportions are 80-20 and there is an initial situation with pooling where both types are paid $140,000. An employer can announce that he will pay $144,000 for someone who takes just one tough course. Relative to the initial situation, the A types will find it worthwhile because their cost of taking the course is only $3,000 and it raises their salary by $4,000, whereas C types will not find it worthwhile because their cost, $15,000, exceeds the benefit, $4,000. Because this particular employer selectively attracts the A types, each of whom is worth $160,000 to him but is paid only $144,000, he makes a profit by deviating from the pooling salary package.

But his deviation starts a process of adjustment by competing employers, and that causes the old pooling situation to collapse. As A types flock to work for him, the pool available to the other employers is of lower average quality, and eventually they cannot afford to pay $140,000 anymore. As the salary in the pool is lowered, the differential between that salary and the $144,000 offered by the deviating employer widens to the point where the C types also find it desirable to take that one tough course. But then the deviating employer must raise his requirement to two courses and must increase the salary differential to the point where two courses become too much of a burden for the C types but the A types find it acceptable. Other employers who would like to hire some A types must use similar policies to attract them. This process continues until the job market reaches the separating equilibrium described earlier.

Even if the employers did not take the initiative to attract As rather than Cs, a type A earning $140,000 in a pooling situation might take a tough course, take his transcript to a prospective employer, and say, "I have a tough course on my transcript, and I am asking for a salary of $144,000. This should be convincing evidence that I am type A; no type C would make you such a proposition." Given the facts of the situation, the argument is valid, and the employer should find it very profitable to agree: the employee, being type A, will generate $160,000 for the employer but get only $144,000 in salary. Other A types can do the same. This starts the same kind of cascade that leads to the separating equilibrium. The only difference is in who takes the initiative. Now the type A workers choose to get the extra education as credible proof of their type; it becomes a case of signaling rather than screening.

The general point is that, even though the pooling outcome may be better for all, they are not choosing the one or the other in a cooperative, binding process. They are pursuing their own individual interests, which lead them to the separating equilibrium. This is like a prisoners' dilemma game with many players, and therefore there is something unavoidable about the cost of the information asymmetry.

We have considered an example with only two types, but the analysis generalizes immediately. Suppose there are several types: A, B, C, . . . , ranked in an order that is at the same time decreasing in their worth to the employer and increasing in the costs of extra education. Then it is possible to set up a sequence of requirements of successively higher and higher levels of education, such that the very worst type needs none, the next-worst type needs the lowest level, the type third from the bottom needs the next higher level, and so on, and the types will self-select the level that identifies them.

To finish this discussion, we provide one further point, or perhaps a word of warning, regarding signaling. You are the informed party and have available an action that would credibly signal good information (information whose credible transmission would work to your advantage). If you fail to send that signal, you will be assumed to have bad information. In this respect, signaling is like playing chicken: if you refuse to play, you have already played and lost.

You should keep this in mind when you have the choice between taking a course for a letter grade or on a pass/fail basis. The whole population in the course spans the whole spectrum of grades; suppose the average is B. A student is likely to have a good idea of his own abilities. Those reasonably confident of getting an A+ have a strong incentive to take the course for a letter grade. When they have done so, the average of the rest is less than B, say, B−, because the top end has been removed from the distribution. Now, among the rest, those expecting an A have a strong incentive to choose the letter-grade option. That in turn lowers the average of the rest. And so on. Finally, the pass/fail option is chosen by only those anticipating Cs and Ds. A strategically smart reader of a transcript (a prospective employer or the admissions officer for a professional graduate school) will be aware that the pass/fail option will be selected mainly by students in the lower portion of the grade distribution; such a reader will therefore interpret a Pass as a C or a D, not as the class-wide average B.

5 EQUILIBRIA IN SIGNALING GAMES

Our analysis so far in this chapter has covered the general concept of incomplete information as well as the specific strategies of screening and signaling; we have also seen the possible outcomes of separation and pooling that can arise when these strategies are being used. We saw how adverse selection could arise in a market where many car owners and buyers came together and how signals and screening devices would operate in an environment where many employers and employees meet each other. However, we have not specified and solved a game in which just two players with differential information confront one another. Here we develop an example to show how that can be done. We will see that either separating or pooling can be an equilibrium and that a new type of **partially revealing** or **semiseparating equilibrium** can emerge.

In this section, we analyze a game of market entry with asymmetric information; the players are two auto manufacturers, Tudor and Fordor. Tudor Auto Corporation currently enjoys a monopoly in the market for a particular kind of automobile, say a nonpolluting, fuel-efficient compact car. An innovator, Fordor, has a competing concept and is deciding whether to enter the market. But Fordor does not know how tough a competitor Tudor will prove to be. Specifically, Tudor's production cost, unknown to Fordor, may be high or low. If it is high, Fordor can enter and compete profitably; if it is low, Fordor's entry and development costs cannot be recouped by subsequent operating profits, and it will make a net loss if it enters.

The two firms interact in a sequential game. In the first stage of the game (period 1), Tudor sets a price (high or low, for simplicity) knowing that it is the only manufacturer in the market. In the next stage, Fordor makes its entry decision. Payoffs, or profits, are determined based on the market price of the automobile relative to each firm's production costs and, for Fordor, entry and development costs as well.

Tudor would of course prefer that Fordor not enter the market. It might therefore try to use its price in the first stage of the game as a signal of its cost. A low-cost firm would charge a lower price than would a high-cost firm. Tudor might therefore hope that if it keeps its period-1 price low, Fordor will interpret this as evidence that Tudor's cost is low and will stay out. (Once Fordor has given up and is out of the picture, in later periods Tudor can jack its price back up.) Just as a poker player might bet on a poor hand, hoping that the bluff will succeed and the opponent will fold, Tudor might try to bluff Fordor into staying out. Of course Fordor is a strategic player and is aware of this possibility. The question is whether Tudor can bluff successfully in an equilibrium of their game. The answer depends on the probability that Tudor is genuinely low cost and on Tudor's cost of bluffing. We consider different cases below and show the resulting different equilibria.

In all the cases, the per-unit costs and prices are expressed in thousands of dollars, and the numbers of cars sold are expressed in hundreds of thousands, so the profits are measured in hundreds of millions. This will help us write the payoffs and tables in a relatively compact form that is easy to read. We calculate those payoffs using the same type of analysis that we used for the restaurant pricing game of Chapter 5, assuming that the underlying relationship between the price charged (P) and the quantity demanded (Q) is given by $P = 25 - Q$.[14] To enter the market, Fordor must incur an up-front cost of 40 (this payment is in

[14]We do not supply the full calculations necessary to generate the profit-maximizing prices and the resulting firm profits in each case. You may do so on your own for extra practice, using the methods learned in Chapter 5.

the same units as profits, or hundreds of millions, so the actual figure is $4 billion) to build its plant, launch an ad campaign, and so on. If it enters the market, its cost for producing and delivering each of its cars to the market will always be 10 (thousand dollars).

A. Separating Equilibrium

Tudor could be either a lumbering, old firm with a high unit production cost of 15 (thousand dollars), or a nimble, efficient producer with a lower unit cost. To start, we suppose that the lower cost is 5; this cost is less than what Fordor can achieve. Suppose further that Tudor can achieve the lower unit cost with probability 0.4, or 40% of the time; therefore it has high unit cost with probability 0.6, or 60% of the time.[15]

Fordor's choices in the entry game will depend on how much information it has about Tudor's costs. We assume that Fordor knows the two possible levels of cost, and therefore can calculate the profits associated with each case (as we do below). In addition, Fordor will form some belief about the probability that Tudor is the low-cost type. We are assuming that the structure of the game is common knowledge to both players. Therefore although Fordor does not know the type of the specific Tudor it is facing, Fordor's prior belief exactly matches the probability with which Tudor has the lower unit cost; that is, Fordor's belief is that the probability of facing a low-cost Tudor is 40%.

If Tudor's cost is high, 15 (thousand), then under conditions of unthreatened monopoly it will maximize its profit by pricing its car at 20 (thousand). At that price it will sell 5 (hundred thousand) units and make a profit of 25 ($= 5 \times (20 - 15)$ hundred million, or 2.5 billion). If Fordor enters and the two compete, then the Nash equilibrium of their duopoly game will yield operating profits of 3 to Tudor and 45 to Fordor. The operating profit exceeds Fordor's up-front cost of entry (40), so Fordor would choose to enter and earn a net profit of 5 if it knew Tudor to be high cost.

If Tudor's cost is low, 5, then in unthreatened monopoly it will price its car at 15, selling 10 and making a profit of 100. In the equilibrium following the entry of Fordor, the operating profits will be 69 for Tudor and 11 for Fordor. The 11 is less than Fordor's cost of entry. Therefore it would not enter and avoid incurring a loss of 29 if it knew Tudor to be low cost.

If Tudor is actually high cost, but wants Fordor to think that it is low cost, Tudor must mimic the action of the low-cost type; that is, it has to price at 15. But that price equals its cost in this case; it will make zero profit. Will this sacri-

[15]Tudor's probability of having low unit cost could be denoted with an algebraic parameter, z. The equilibrium will be the same regardless of the value of z, as you will be asked to show in Exercise S5 at the end of this chapter.

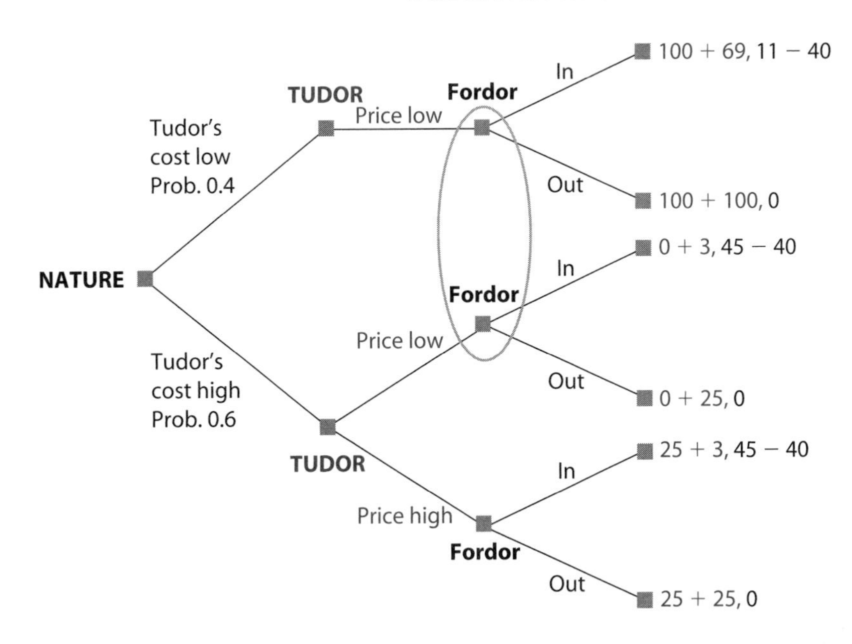

FIGURE 9.8 Extensive Form of Entry Game: Tudor's Low Cost Is 5

fice of initial profit give Tudor the benefit of scaring Fordor off and enjoying the benefits of being a monopoly in subsequent periods?

We show the full game in extensive form in Figure 9.8. Note that we use the fictitious player called Nature, introduced in Chapter 3, to choose Tudor's cost type at the start of the game. Then Tudor makes its pricing decision. We assume that if Tudor has low cost, it will not choose a high price.[16] But if Tudor has high cost, it may choose either the high price or the low price if it wants to bluff. Fordor cannot tell apart the two situations in which Tudor prices low; therefore its entry choices at these two nodes are enclosed in one information set, as explained in Section 3 of Chapter 6. Fordor must choose either In at both or Out at both.

At each terminal node, the first payoff entry (in green) is Tudor's profit, and the second entry (in black) is Fordor's profit. Tudor's profit is added over two periods, the first period when it is the sole producer, and the second period when

[16]This seems obvious: why choose a price different from the profit-maximizing price? Charging the high price when you have low cost not only sacrifices some profit in period 1 (if the low-cost Tudor charges 20, its sales will drop by so much that it will make a profit of only 75 instead of the 100 it gets by charging 15), but also increases the risk of entry and so lowers period-2 profits as well (competing with Fordor, the low-cost Tudor would have a profit of only 69 instead of the 100 it gets under monopoly). However, game theorists have found strange equilibria where a high period-1 price for Tudor is perversely interpreted as evidence of low cost, and they have applied great ingenuity in ruling out these equilibria. We leave out these complications, but refer interested readers to In-Koo Cho and David Kreps, "Signaling Games and Stable Equilibria," *Quarterly Journal of Economics*, vol. 102, no. 2 (May 1987), pp. 179–222.

		FORDOR	
		Regardless (II)	Conditional (OI)
Tudor	Bluff (LL)	$169 \times 0.4 + 3 \times 0.6 = 69.4$, $-29 \times 0.4 + 5 \times 0.6 = 14.6$	$200 \times 0.4 + 25 \times 0.6 = 95$, 0
	Honest (LH)	$169 \times 0.4 + 28 \times 0.6 = 84.4$, $-29 \times 0.4 + 5 \times 0.6 = -8.6$	$200 \times 0.4 + 28 \times 0.6 = 96.8$, $5 \times 0.6 = 3$

FIGURE 9.9 Strategic Form of Entry Game: Tudor's Low Cost Is 5

it may be a monopolist or a duopolist, depending on whether Fordor enters. Fordor's profit covers only the second period and is non-zero only when it has chosen to enter.

Using one step of rollback analysis, we see that Fordor will choose In at the bottom node where Tudor has chosen the high price, because $45 - 40 = 5 > 0$. Therefore we can prune the Out branch at that node. Then each player has just two strategies (complete plans of action). For Tudor the strategies are Bluff: choose the low price in period 1 regardless of cost (LL in the shorthand notation of Chapter 3), and Honest: choose the low price in period 1 if cost is low and the high price if cost is high (LH). For Fordor, the two strategies are Regardless: enter irrespective of Tudor's period-1 price (II, for In-In) and Conditional: enter only if Tudor's period-1 price is high (OI).

We can now show the game in strategic (normal) form. Figure 9.9 shows each player with two possible strategies; payoffs in each cell are the expected profits to each firm, given the probability (40%) that Tudor's cost is low.

This is a simple dominance-solvable game. For Tudor, Honest dominates Bluff. And Fordor's best response to Tudor's dominant strategy of Honest is Conditional. Thus (Honest, Conditional) is the only (subgame-perfect) Nash equilibrium of the game.

The equilibrium found in Figure 9.9 is separating. The two cost types of Tudor charge different prices in period 1. This action reveals Tudor's type to Fordor, which then makes its entry decision appropriately.

The key to understanding why Honest is the dominant strategy for Tudor can be found in the comparison of its payoffs against Fordor's Conditional strategy. These are the outcomes when Tudor's bluff "works": Fordor enters if Tudor charges the high price in period 1 and stays out if Tudor charges the low price in period 1. If Tudor is truly low cost, then its payoffs against Fordor playing Conditional are the same whether it chooses Bluff or Honest. But when Tudor is actually high cost, the results are different.

If Fordor's strategy is Conditional and Tudor is high cost, Tudor can use Bluff successfully. However, even the successful bluff will be too costly. If Tudor charged its best monopoly (Honest) price in period 1, it would make a profit of 25; the bluffing low price reduces this period-1 profit drastically, in this instance all the way to zero. The higher monopoly price in period 1 would encourage Fordor's entry and reduce period-2 profit for Tudor, from the monopoly level of 25 to the duopoly level of 3. But Tudor's period-2 benefit from charging the low (Bluff) price and keeping Fordor out (25 − 3 = 22) is less than the period-1 cost imposed by bluffing and giving up its monopoly profits (25 − 0 = 25). As long as there is some positive probability that Tudor is high cost, then, the benefits from choosing Honest will outweigh those from choosing Bluff, even when Fordor's choice is Conditional.

If the low price were not so low, then a truly high-cost Tudor would sacrifice less by mimicking the low-cost type. In such a case, Bluff might be a more profitable strategy for a high-cost Tudor. We consider exactly this possibility in the analysis below.

B. Pooling Equilibrium

Let us now suppose that the lower of the production costs for Tudor is 10 per car instead of 5. With this cost change, the high-cost Tudor still makes profit of 25 under monopoly if it charges its profit-maximizing price of 20. But the low-cost Tudor now charges 17.5 as a monopolist (instead of 15) and makes a profit of 56. If the high-cost type mimics the low-cost type and also charges 17.5, its profit is now 19 (rather than the zero it earned in this case before); the loss of profit from bluffing is now much smaller: 25 − 19 = 6, rather than 25. If Fordor enters, then the two firms' profits in their duopoly game are 3 for Tudor and 45 for Fordor if Tudor has high costs (as in the previous section). Duopoly profits are now 25 for each firm if Tudor has low costs; in this situation, Fordor and the low-cost Tudor have identical unit costs of 10.

Suppose again that the probability of Tudor being the low-cost type is 40% (0.4) and Fordor's belief about the low-cost probability is correct. The new game tree is shown in Figure 9.10. Because Fordor will still choose In when Tudor prices High, the game again collapses to one in which each player has exactly two complete strategies; those strategies are the same ones we described in Section 5.A above. The payoff table for the normal form of this game is then the one illustrated in Figure 9.11.

This is another dominance-solvable game. Here it is Fordor with a dominant strategy, however; it will always choose Conditional. And given the dominance of Conditional, Tudor will choose Bluff. Thus (Bluff, Conditional) is the unique (subgame-perfect) Nash equilibrium of this game. In all other cells of the table, one firm gains by deviating to its other action. We leave it to you to think about the intuitive explanations of why each of these deviations is profitable.

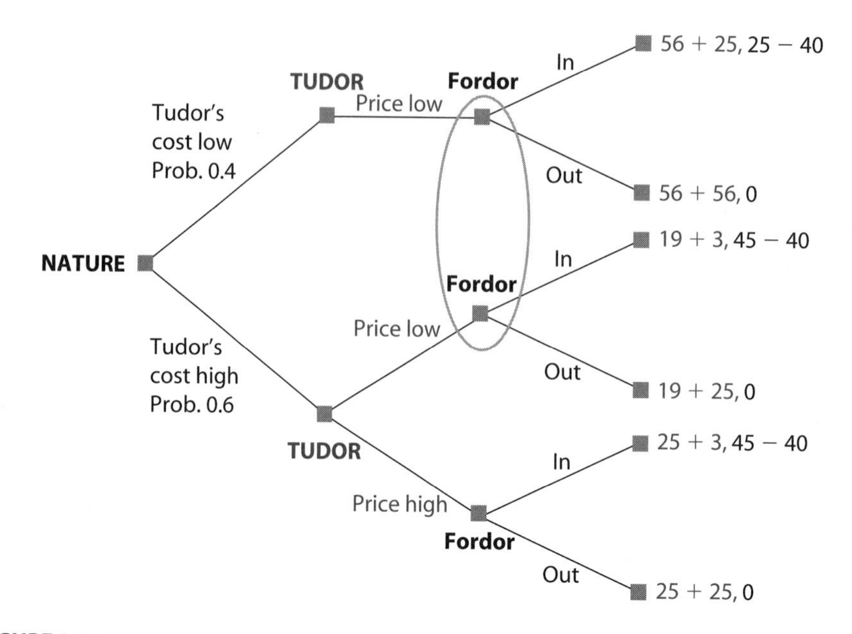

FIGURE 9.10 Extensive Form of Entry Game: Tudor's Low Cost Is 10

The equilibrium found using Figure 9.11 involves pooling. Both cost types of Tudor charge the same (low) price and, seeing this, Fordor stays out. When both types of Tudor charge the same price, observation of that price does not convey any information to Fordor. Its estimate of the probability of Tudor's cost being low stays at 0.4, and it calculates its expected profit from entry to be $-3 < 0$, so it does not enter. Even though Fordor knows full well that Tudor is bluffing in equilibrium, the risk of calling the bluff is too great because the probability of Tudor's cost actually being low is sufficiently great.

What if this probability were smaller—say, 0.1—and Fordor was aware of this fact? If all the other numbers remain unchanged, then Fordor's expected profit from its Regardless strategy is $-15 \times 0.1 + 5 \times 0.9 = 4.5 - 1.5 = 3 > 0$. Then Fordor will enter no matter what price Tudor charges, and Tudor's bluff will not work. Such a situation results in a new kind of equilibrium; we consider its features below.

		FORDOR	
		Regardless (II)	Conditional (OI)
Tudor	Bluff (LL)	$81 \times 0.4 + 22 \times 0.6 = 45.6$, $-15 \times 0.4 + 5 \times 0.6 = -3$	$112 \times 0.4 + 44 \times 0.6 = 71.2$, 0
	Honest (LH)	$81 \times 0.4 + 28 \times 0.6 = 49.2$, $-15 \times 0.4 + 5 \times 0.6 = -3$	$112 \times 0.4 + 28 \times 0.6 = 61.6$, $5 \times 0.6 = 3$

FIGURE 9.11 Strategic Form of Entry Game: Tudor's Low Cost Is 10

C. Semiseparating Equilibrium

Here we consider the outcomes in the entry game when Tudor's probability of achieving the low production cost of 10 is small, only 10% (0.1). All of the cost and profit numbers are the same as in the previous section; only the probabilities have changed. Therefore we do not show the game tree (Figure 9.10) again. We show only the payoff table as Figure 9.12.

In this new situation, the game illustrated in Figure 9.12 has no equilibrium in pure strategies. From (Bluff, Regardless), Tudor gains by deviating to Honest; from (Honest, Regardless), Fordor gains by deviating to Conditional; from (Honest, Conditional), Tudor gains by deviating to Bluff; and from (Bluff, Conditional), Fordor gains by deviating to Regardless. Once again we leave it to you to think about the intuitive explanations of why each of these deviations is profitable.

So now we need to look for an equilibrium in mixed strategies. We suppose Tudor mixes Bluff and Honest with probabilities p and $(1 - p)$, respectively. Similarly, Fordor mixes Regardless and Conditional with probabilities q and $(1 - q)$, respectively. Tudor's p-mix must keep Fordor indifferent between its two pure strategies of Regardless and Conditional; therefore we need

$$3\,p + 3\,(1 - p) = 0\,p + 4.5\,(1 - p), \text{ or } 4.5\,(1 - p) = 3, \text{ or } 1 - p = 2/3, \text{ or } p = 1/3.$$

And Fordor's q-mix must keep Tudor indifferent between its two pure strategies of Bluff and Honest; therefore we need

$$27.9\,q + 50.8\,(1 - q) = 33.3\,q + 36.4\,(1 - q), \text{ or } 5.4\,q = 14\,(1 - q), \text{ or}$$

$$q = 14.4/19.8 = 16/22 = 0.727.$$

The mixed-strategy equilibrium of the game then entails Tudor playing Bluff one-third of the time and Honest two-thirds of the time, while Fordor plays Regardless sixteen twenty-seconds of the time and Conditional six twenty-seconds of the time.

In this equilibrium, the Tudor types are only partially separated. The low-cost-type Tudor always prices Low in period 1, but the high-cost-type mixes and

		FORDOR	
		Regardless (II)	Conditional (OI)
Tudor	Bluff (LL)	$81 \times 0.1 + 22 \times 0.9 = 27.9$, $-15 \times 0.1 + 5 \times 0.9 = 3$	$112 \times 0.1 + 44 \times 0.9 = 50.8$, 0
	Honest (LH)	$81 \times 0.1 + 28 \times 0.9 = 33.3$, $-15 \times 0.1 + 5 \times 0.9 = 3$	$112 \times 0.1 + 28 \times 0.9 = 36.4$, $5 \times 0.9 = 4.5$

FIGURE 9.12 Strategic Form of Entry Game: Tudor's Low Cost Is 10 with Probability 0.1

will also charge the low price one-third of the time. If Fordor observes a high price in period 1, it can be sure that Tudor has high cost; in that case, it will always enter. But if Fordor observes a low price, it does not know whether it faces a truly low-cost Tudor or a bluffing, high-cost Tudor. Then Fordor also plays a mixed strategy, entering 72.7% of the time. Thus a high price conveys full information, but a low price conveys only partial information about Tudor's type. Therefore this kind of equilibrium is labeled *semiseparating.*

To understand better the mixed strategies of each firm and the semiseparating equilibrium, consider how Fordor can use the partial information conveyed by Tudor's low price. If Fordor sees the low price in period 1, it will use this observation to update its belief about the probability that Tudor is low cost; it does this updating using Bayes' theorem.[17] The table of calculations is shown as Figure 9.13; this table is similar to Figure 9.A.1 in the Appendix.

The table shows the possible types of Tudor in the rows and the prices Fordor observes in the columns. The values in the cells represent the overall probability that a Tudor of the type shown in the corresponding row chooses the price shown in the corresponding column (incorporating Tudor's equilibrium mixture probability); the final row and column show the total probabilities of each type and of observing each price, respectively.

Using Bayes' rule, when Fordor observes Tudor charging a low period-1 price, it will revise its belief about the probability of Tudor being low cost by taking the probability that a low-cost Tudor is charging the low price (the 0.1 in the top left cell) and dividing that by the total probability of the two types of Tudor choosing the low price (0.4, the column sum in the left column). This calculation yields Fordor's updated belief about the probability that Tudor has low costs to be $0.1 / 0.4 = 0.25$. Then Fordor also updates its expected profit from entry to be $-15 \times 0.25 + 5 \times 0.75 = 0$. Thus Tudor's equilibrium mixture is exactly right for making Fordor indifferent between entering and not entering when it sees the

		TUDOR'S PRICE		Sum of row
		Low	High	
TUDOR'S COST	Low	0.1	0	0.1
	High	$0.9 \times 1/3 = 0.3$	$0.9 \times 2/3 = 0.6$	0.9
Sum of column		0.4	0.6	

FIGURE 9.13 Applying Bayes' Theorem to the Entry Game

[17]We provide a thorough explanation of Bayes' theorem in the Appendix to this chapter. Here, we simply apply the analysis found there to our entry game.

low period-1 price. This outcome is exactly what is needed to keep Tudor willing to mix in the equilibrium.

The original probability 0.1 of Tudor being low cost was too low to deter Fordor from entering. Fordor's revised probability of 0.25, after observing the low price in period 1, is higher. Why? Precisely because the high-cost type Tudor is not always bluffing. If it were, then the low price would convey no information at all. Fordor's revised probability would equal 0.1 in that case, whereupon it would enter. But when the high-cost type Tudor bluffs only sometimes, a low price is more likely to be indicative of low cost.

We developed the equilibria in this entry game in an intuitive way, but we now look back and think systematically about the nature of those equilibria. In each case, we first ensured that each player's (and each type's) strategy was optimal, given the strategies of everyone else; we applied the Nash concept of equilibrium. Second, we ensured that players drew the correct inference from their observations; this required a probability calculation using Bayes' theorem, most explicitly in the semiseparating equilibrium. The combination of concepts necessary to identify equilibria in such asymmetric information games justifies giving them the label **Bayesian-Nash equilibria.** Finally, although this was a rather trivial part of this example, we did a little bit of rollback, or subgame perfectness, reasoning. The use of rollback justifies calling the equilibruim **perfect Bayesian** as well. Our example was a simple instance of all of these equilibrium concepts; you will meet some of them again in slightly more sophisticated forms in later chapters, and in much fuller contexts in further studies of game theory.

6 EVIDENCE ABOUT SIGNALING AND SCREENING

As we saw in Section 5, the characterization and solution of Bayesian-Nash equilibria for games of signaling and screening entail some quite subtle concepts and computations. Should we expect players to perform such calculations correctly? How realistic are these equilibria as descriptions of the outcomes of these games?

There is ample evidence that people are very bad at performing calculations that include probabilities and are especially bad at conditioning probabilities on new information.[18] These calculations are exactly the ones that one must

[18]Deborah J. Bennett, *Randomness* (Cambridge, Mass.: Harvard University Press, 1998), pp. 2–3 and Chapter 10. See also Paul Hoffman, *The Man Who Loved Only Numbers* (New York: Hyperion, 1998), pp. 233–240, for an entertaining account of how several probability theorists, as well as the brilliant and prolific mathematician Paul Erdös, got a very simple probability problem wrong and even failed to understand their error when it was explained to them.

perform to update one's information on the basis of observed actions. Therefore we should be justifiably suspicious of equilibria that depend on the players' doing so. Relative to this expectation, the findings of economists who have conducted laboratory experiments of signaling games are encouraging. Some surprisingly subtle refinements of Bayesian-Nash and perfect Bayesian equilibria are successfully observed, even though these refinements require not only updating of information by observing actions along the equilibrium path but also deciding how one would infer information from off-equilibrium actions that should never have been taken in the first place. However, the verdict of the experiments is not unanimous; much seems to depend on the precise details of the laboratory design of the experiment.[19]

Although the equilibria of signaling and screening games can be quite subtle and complex, the basic idea of the role of signaling or screening to elicit information is very simple: players of different "types" (that is, possessing different information about their own characteristics or about the game and its payoffs more generally) should find it optimal to take different actions so that their actions truthfully reveal their types. One can point to numerous practical applications of this idea. Here are some examples; once you start thinking along these lines you should be able to come up with dozens more.

1. Insurance companies usually offer a spectrum or menu of policies, with different provisions for deductible amounts and coinsurance, for which they charge different premiums. Those customers who know themselves to be particularly susceptible to risk prefer the policies with low deductibles and coinsurance, whereas those who know themselves to be less risky are more willing to take the policies stipulating that they have to bear more of the losses. Thus the different risk classes have different optimal actions, and the insurance companies use their clients' self-selection to screen and elicit the risk class of any particular client.
2. Venture capitalists are looking for inventors with good ideas, but how are they to judge the quality of any particular idea? They look for the inventor's willingness to "put his money where his mouth is"; they look for inventors who will take up as large an equity participation in the venture as their financial situation permits. Thus the willingness to bear a part of any loss from the project becomes a credible indicator of the inventor's belief that his project will turn a profit.
3. The quality of durable goods such as cars and computers is hard to evaluate for consumers, but each manufacturer has a much better idea of the quality of his own product. And if a product is of higher quality, it is less costly for the maker to offer a longer or better warranty on it. Therefore

[19]Douglas D. Davis and Charles A. Holt, *Experimental Economics* (Princeton: Princeton University Press, 1995), review and discuss these experiments in their Chapter 7.

warranties can serve as signals of quality, and consumers are intuitively quite aware of this when they make their purchase decisions.

4. Finally, here is an example from biology.[20] In many species of birds, the males have very elaborate and heavy plumage that females find attractive. One should expect the females to seek genetically superior males so that their offspring will be better equipped to survive and attract mates in their turn. But why does elaborate plumage indicate such desirable genetic qualities? One would think it would be a handicap, making the bird more visible to predators (including human hunters), and less mobile and therefore less able to evade these predators. Why do females choose these handicapped males? The answer comes from the conditions for credible signaling. Although heavy plumage is a handicap, it is less of a handicap to a male who is genetically sufficiently superior in qualities such as strength and speed. The weaker the male, the harder it is for him to produce and maintain plumage of a given quality. Thus it is precisely the heaviness of the plumage that makes it a credible signal of the male's quality.

As you can see from these examples and others, such as the dating story in Chapter 1, information asymmetry is everywhere, and strategies to deal with it are an important part not only of the science of game theory, but also of the art of strategy in everyday life. We hope that you will be intrigued by these ideas and will want to learn more than we can offer you at the elementary level of this book. For this purpose, we suggest the following additional readings:

From sociology: Erving Goffman, *The Presentation of Self in Everyday Life,* revised edition (New York: Anchor Books, 1959).

From biology: Matt Ridley, *The Red Queen: Sex and the Evolution of Human Behavior* (New York: Penguin, 1995).

From economics: A. Michael Spence, *Market Signaling* (Cambridge, Mass.: Harvard University Press, 1974).

And, finally, also from economics, two graduate-level textbooks, strictly for the very ambitious who want to know the higher reaches of game theory: Drew Fudenberg and Jean Tirole, *Game Theory* (Cambridge, Mass.: MIT Press, 1991), Chapters 6–8; and David Kreps, *A Course in Microeconomic Theory* (Princeton: Princeton University Press, 1990), Chapters 13, 16, and 17.

[20]Matt Ridley, *The Red Queen: Sex and the Evolution of Human Behavior* (New York: Penguin, 1995), p. 148.

SUMMARY

When facing imperfect or incomplete information, game players with different attitudes toward risk or different amounts of information can engage in strategic behavior to control and manipulate the risk and information in a game. Players can reduce their risk with payment schemes or by sharing the risk with others, although the latter is complicated by *moral hazard* and *adverse selection*. Risk can sometimes be manipulated to a player's benefit, depending on the circumstances within the game.

Players with private information may want to conceal or reveal that information, while those without the information try to elicit it or avoid it. Actions speak louder than words in the presence of asymmetric information. To reveal information, a credible *signal* is required. *Signaling* works only if the signal action entails different costs to players with different information. To obtain information, when questioning is not sufficient to elicit truthful information, a *screening* scheme that looks for a specific action may be required. Screening works only if the *screening device* induces others to reveal their *types* truthfully; there must be *incentive compatibility* to get *separation*. At times, credible signaling or screening may not be possible; then the equilibrium can entail *pooling* or there can be a complete collapse of the market or transaction for one of the types.

In the equilibrium of a game with asymmetric information, players must not only use their best actions given their information, but must also draw correct inferences (update their information) by observing the actions of others. This type of equilibrium is known as a *Bayesian-Nash equilibrium*. When the further requirement of optimality at all nodes (as in rollback analysis) must be imposed, the equilibrium becomes a *perfect Bayesian equilibrium*. The outcome of such a game may entail pooling, separation, or *partial separation*, depending on the specifics of the payoff structure and the specified updating processes used by players. In some parameter ranges, such games may have multiples types of perfect Bayesian equilibria.

The evidence on players' abilities to achieve perfect Bayesian equilibria seems to suggest that, despite the difficult probability calculations necessary, such equilibria are often observed. Different experimental results appear to depend largely on the design of the experiment. Many examples of signaling and screening games can be found in ordinary situations such as the labor market or in the provision of insurance.

KEY TERMS

adverse selection (324)
babbling equilibrium (319)
Bayesian-Nash equilibrium (341)
cheap talk equilibrium (318)
incentive-compatability conditions (constraints) (328)
moral hazard (308)
negatively correlated (309)
partially revealing equilibrium (332)
participation condition (constraint) (328)
perfect Bayesian equilibrium (341)
pooling (of types) (330)
pooling equilibrium (317)
positively correlated (310)
screening (317)
screening device (317)
self-selection (329)
semiseparating equilibrium (332)
separation (of types) (329)
separating equilibrium (317)
signal (316)
signaling (317)
signal jamming (317)
type (317)

SOLVED EXERCISES

S1. In the risk-trading example in Section 1, you had a risky income that was \$160,000 with good luck (probability 0.5) and \$40,000 with bad luck (probability 0.5). When your neighbor had a sure income of \$100,000, we derived a scheme in which you could eliminate all of your risk while raising his expected utility slightly. Assume that the utility of each of you is still the square root of the respective income. Now, however, let the probability of good luck be 0.6. Consider a contract that leaves you with exactly \$100,000 when you have bad luck. Let x be the payment that you make to your neighbor when you have good luck.

(a) What is the minimum value of x (to the nearest penny) such that your neighbor slightly prefers to enter into this kind of contract rather than no contract at all?

(b) What is the maximum value of x (to the nearest penny) for which this kind of contract gives you a slightly higher expected utility than no contract at all?

S2. A local charity has been given a grant to serve free meals to the homeless in its community, but it is worried that its program might be exploited by nearby college students, who are always on the lookout for a free meal. Both a homeless person and a college student receive a payoff of 10 for a free meal. The cost of standing in line for the meal is t^2 / 320 for a homeless person and t^2 / 160 for a college student, where t is the amount of time in line measured in minutes. Assume that the staff of the charity cannot observe the true type of those coming for free meals.

(a) What is the minimum wait time t that will achieve separation of types?

(b) After a while, the charity finds that it can successfully identify and turn away college students half of the time. College students who are turned away receive no free meal and, further, incur a cost of 5 for their time and embarrassment. Will the partial identification of college students reduce or increase the answer in part (a)? Explain.

S3. Consider the used-car market for the 2006 Citrus described in Section 4.A. There is now a surge in demand for used Citruses; buyers would now be willing to pay up to $18,000 for an orange and $8,000 for a lemon. All else remains identical to the example in Section 4.A.

(a) What price would buyers be willing to pay for a 2006 Citrus of unknown type if the fraction of oranges in the population, f, were 0.6?

(b) Will there be a market for oranges if $f = 0.6$? Explain.

(c) What price would buyers be willing to pay if f were 0.2?

(d) Will there be a market for oranges if $f = 0.2$? Explain.

(e) What is the minimum value of f such that the market for oranges does not collapse?

(f) Explain why the increase in the buyers' willingness to pay changes the threshold value of f, where the market for oranges collapses.

S4. Suppose electricians come in two types: competent and incompetent. Both types of electricians can get certified, but for the incompetent types certification takes extra time and effort. Competent ones have to spend C months preparing for the certification exam; incompetent ones take twice as long. Certified electricians can earn 100 (thousand dollars) each year working on building sites for licensed contractors. Uncertified electricians can earn only 25 (thousand dollars) each year in freelance work; licensed contractors won't hire them. Each type of electrician gets a payoff equal to $\sqrt{S} - M$, where S is the salary measured in thousands of dollars and M is the number of months spent getting certified. What is the range of values of C for which a competent electrician will choose to signal with this device but an incompetent one will not?

S5. Return to the Tudor-Fordor example in Section 5.A, when Tudor's low per-unit cost is 5. Let z be the probability that Tudor actually has a low per-unit cost.

(a) Rewrite the table in Figure 9.9 in terms of z.

(b) How many pure-strategy equilibria are there when $z = 0$? Explain.

(c) How many pure-strategy equilibria are there when $z = 1$? Explain.

(d) Show that the Nash equilibrium of this game is always a separating equilibrium for any value of z between 0 and 1 (inclusive).

S6. Looking at Tudor and Fordor again, assume that the old, established company Tudor is risk averse, whereas the would-be entrant Fordor (which is planning to finance its project through venture capital) is risk neutral. That is, Tudor's utility is always the square root of its total profit over both periods. Fordor's utility is simply the amount of its profit—if any—during the second period. Assume that Tudor's low per-unit cost is 5, as in Section 5.A.

(a) Redraw the extensive-form game shown in Figure 9.8, giving the proper payoffs for a risk-averse Tudor.

(b) Let the probability that Tudor is low cost, z, be 0.4. Will the equilibrium be separating, pooling, or semiseparating? (Hint: Use a table equivalent to Figure 9.9.)

(c) Repeat part (b) with $z = 0.1$.

S7. Return to a risk-neutral Tudor, but with a low per-unit cost of 6 (instead of 5 or 10 as in Section 5). If Tudor's cost is low, 6, then it will earn 90 in a profit-maximizing monopoly. If Fordor enters, Tudor will earn 59 in the resulting duopoly while Fordor earns 13. If Tudor is actually high cost (i.e., its per-unit cost is 15) and prices as if it were low cost (i.e., with a per-unit cost of 6), then it earns 5 in a monopoly situation.

(a) Draw a game tree for this game equivalent to Figure 9.8 or 9.10 in the text, changing the appropriate payoffs.

(b) Write the normal form of this game, assuming that the probability that Tudor is low price is 0.4.

(c) What is the equilibrium of the game? Is it separating, pooling, or semiseparating? Explain why.

S8. Felix and Oscar are playing a simplified version of poker. Each makes an initial bet of 8 dollars. Then each separately draws a card, which may be High or Low with equal probabilities. Each sees his own card but not that of the other.

Then Felix decides whether to Pass or to Raise (bet an additional 4 dollars). If he chooses to pass, the two cards are revealed and compared. If the outcomes are different, the one who has the High card collects the whole pot. The pot has 16 dollars, of which the winner himself contributed 8, so his winnings are 8 dollars. The loser's payoff is -8 dollars. If the outcomes are the same, the pot is split equally and each gets his 8 dollars back (payoff 0).

If Felix chooses Raise, then Oscar has to decide whether to Fold (concede) or See (match with his own additional 4 dollars). If Oscar chooses Fold, then Felix collects the pot irrespective of the cards. If Oscar chooses

See, then the cards are revealed and compared. The procedure is the same as that in the preceding paragraph, but the pot is now bigger.

(a) Show the game in extensive form. (Be careful about information sets.)

If the game is instead written in the normal form, Felix has four strategies: (1) Pass always (PP for short), (2) Raise always (RR), (3) Raise if his own card is High and Pass if it is Low (RP), and (4) the other way round (PR). Similarly, Oscar has four strategies: (1) Fold always (FF), (2) See always (SS), (3) See if his own card is High and Fold if it is Low (SF), and (4) the other way round (FS).

(b) Show that the table of payoffs to Felix is as follows:

		OSCAR			
		FF	SS	SF	FS
FELIX	PP	0	0	0	0
	RR	8	0	1	7
	RP	2	1	0	3
	PR	6	−1	1	4

(In each case you will have to take an expected value by averaging over the consequences for each of the four possible combinations of the card draws.)

(c) Eliminate dominated strategies as far as possible. Find the mixed-strategy equilibrium in the remaining table and the expected payoff to Felix in the equilibrium.

(d) Use your knowledge of the theory of signaling and screening to explain intuitively why the equilibrium has mixed strategies.

S9. Felix and Oscar are playing another simplified version of poker called Stripped-Down Poker. Both make an initial bet of one dollar. Felix (and only Felix) draws one card, which is either a King or a Queen with equal probability (there are four Kings and four Queens). Felix then chooses whether to Fold or to Bet. If Felix chooses to Fold, the game ends, and Oscar receives Felix's dollar in addition to his own. If Felix chooses to Bet, he puts in an additional dollar, and Oscar chooses whether to Fold or to Call.

If Oscar Folds, Felix wins the pot (consisting of Oscar's initial bet of one dollar and two dollars from Felix). If Oscar Calls, he puts in another dollar to match Felix's bet, and Felix's card is revealed. If the card is a King, Felix wins the pot (two dollars from each of the roommates). If it is a Queen, Oscar wins the pot.

(a) Show the game in extensive form. (Be careful about information sets.)

(b) How many strategies does each player have?

(c) Show the game in strategic form, where the payoffs in each cell reflect the expected payoffs given each player's respective strategy.

(d) Eliminate dominated strategies, if any. Find the equilibrium in mixed strategies. What is the expected payoff to Felix in equilibrium?

S10. Wanda works as a waitress and consequently has the opportunity to earn cash tips that are not reported by her employer to the Internal Revenue Service. Her tip income is rather variable. In a good year (G), she earns a high income, so her tax liability to the IRS is $5,000. In a bad year (B), she earns a low income, and her tax liability to the IRS is $0. The IRS knows that the probability of her having a good year is 0.6, and the probability of her having a bad year is 0.4, but it doesn't know for sure which outcome has resulted for her this tax year.

In this game, first Wanda decides how much income to report to the IRS. If she reports high income (H), she pays the IRS $5,000. If she reports low income (L), she pays the IRS $0. Then the IRS has to decide whether to audit Wanda. If she reports high income, they do not audit, because they automatically know they're already receiving the tax payment Wanda owes. If she reports low income, then the IRS can either audit (A) or not audit (N). When the IRS audits, it costs the IRS $1,000 in administrative costs, and also costs Wanda $1,000 in the opportunity cost of the time spent gathering bank records and meeting with the auditor. If the IRS audits Wanda in a bad year (B), then she owes nothing to the IRS, although she and the IRS have each incurred the $1,000 auditing cost. If the IRS audits Wanda in a good year (G), then she has to pay the $5,000 she owes to the IRS, in addition to her and the IRS each incurring the cost of auditing.

(a) Suppose that Wanda has a good year (G), but she reports low income (L). Suppose the IRS then audits her (A). What is the total payoff to Wanda, and what is the total payoff to the IRS?

(b) Which of the two players has an incentive to bluff (that is, to give a false signal) in this game? What would bluffing consist of?

(c) Show this game in extensive form. (Be careful about information sets.)

(d) How many pure strategies does each player have in this game? Explain your reasoning.

(e) Write down the strategic-form game matrix for this game. Find all of the Nash equilibria to this game. Identify whether the equilibria you find are separating, pooling, or semiseparating.

(f) Let x equal the probability that Wanda has a good year. In the original version of this problem, we had $x = 0.6$. Find a value of x such that Wanda always reports low income in equilibrium.

(g) What is the full range of values of x for which Wanda always reports low income in equilibrium?

S11. The design of a health-care system concerns matters of information and strategy at several points. The users—potential and actual patients—have better information about their own state of health, lifestyle, and so forth—than the insurance companies can find out. The providers—doctors, hospitals, and so forth—know more about what the patients need than do either the patients themselves or the insurance companies. Doctors also know more about their own skills and efforts, and hospitals about their own facilities. Insurance companies may have some statistical information about outcomes of treatments or surgical procedures from their past records. But outcomes are affected by many unobservable and random factors, so the underlying skills, efforts, or facilities cannot be inferred perfectly from observation of the outcomes. The pharmaceutical companies know more about the efficacy of drugs than do the others. As usual, the parties do not have natural incentives to share their information fully or accurately with others. The design of the overall scheme must try to face these matters and find the best feasible solutions.

Consider the relative merits of various payment schemes—fee for service versus capitation fees to doctors, comprehensive premiums per year versus payment for each visit for patients, and so forth—from this strategic perspective. Which are likely to be most beneficial to those seeking health care? To those providing health care? Think also about the relative merits of private insurance and coverage of costs from general tax revenues.

S12. In a television commercial for a well-known brand of instant cappuccino, a gentleman is entertaining a lady friend at his apartment. He wants to impress her and offers her cappuccino with dessert. When she accepts, he goes into the kitchen to make the instant cappuccino—simultaneously tossing take-out boxes into the trash and faking the noises made by a high-class (and expensive) espresso machine. As he is doing so, a voice comes from the other room: "I want to see the machine . . . ."

Use your knowledge of games of asymmetric information to comment on the actions of these two people. Pay attention to their attempts to use signaling and screening, and point out specific instances of each strategy. Offer an opinion about which player is the better strategist.

S13. **(Optional, requires Appendix)** In the genetic test example, suppose the test comes out negative (*Y* is observed). What is the probability that the person does not have the defect (*B* exists)? Calculate this probability by applying Bayes' rule, and then check your answer by doing an enumeration of the 10,000 members of the population.

S14. **(Optional, requires Appendix)** Return to the example of the 2006 Citrus in Section 4.A. The two types of Citrus—the reliable orange and the hapless lemon—are outwardly indistinguishable to a buyer. In the example, if the fraction f of oranges in the Citrus population is less than 0.65, the seller of an orange will not be willing to part with the car for the maximum price buyers are willing to pay, so the market for oranges collapses.

But what if a seller has a costly way to signal her car's type? Although oranges and lemons are in nearly every respect identical, the defining difference between the two is that lemons break down much more frequently. Knowing this, owners of oranges make the following proposal. On a buyer's request, the seller will in one day take a 500-mile round-trip drive in the car. (Assume this trip will be verifiable via odometer readings and a time-stamped receipt from a gas station 250 miles away.) For the sellers of both types of Citrus, the cost of the trip in fuel and time is \$0.50 per mile (that is, \$250 for the 500-mile trip). However, with probability q a lemon attempting the journey will break down. If a car breaks down, the cost is \$2 per mile of the total length of the attempted road trip (that is, \$1,000). Additionally, breaking down will be a sure sign that the car is a lemon, so a Citrus that does so will sell for only \$6,000.

Assume that the fraction of oranges in the Citrus population, f, is 0.6. Also, assume that the probability of a lemon breaking down, q, is 0.5 and that owners of lemons are risk neutral.

(a) Use Bayes' rule to determine $f_{updated}$, the fraction of Citruses that have successfully completed a 500-mile road trip that are oranges. Assume that all Citrus owners attempt the trip. Is $f_{updated}$ greater than or less than f? Explain why.

(b) Use $f_{updated}$ to determine the price, $p_{updated}$, that buyers are willing to pay for a Citrus that has successfully completed the 500-mile road trip.

(c) Will an owner of an orange be willing to make the road trip and sell her car for $p_{updated}$? Why or why not?

(d) What is the expected payoff of attempting the road trip to the seller of a lemon?

(e) Would you describe the outcome of this market as pooling, separating, or semiseparating? Explain.

UNSOLVED EXERCISES

U1. Jack is a talented investor, but his earnings vary considerably from year to year. In the coming year he expects to earn either \$250,000 with good luck or \$90,000 with bad luck. Somewhat oddly, given his chosen profession, Jack is risk averse, so that his utility is equal to the square root of his income. The probability of Jack's having good luck is 0.5.

(a) What is Jack's expected utility for the coming year?

(b) What amount of certain income would yield the same level of utility for Jack as the expected utility in part (a)?

Jack meets Janet, whose situation is identical in every respect. She's an investor who will earn $250,000 in the next year with good luck and $90,000 with bad, she's risk averse with square-root utility, and her probability of having good luck is 0.5. Crucially, it turns out that Jack and Janet invest in such a way that their luck is completely independent. They agree to the following deal. Regardless of their respective luck, they will always pool their earnings and then split them equally.

(c) What are the four possible luck-outcome pairs, and what is the probability of reaching each one?

(d) What is the expected utility for Jack or Janet under this arrangement?

(e) What amount of certain income would yield the same level of utility for Jack and Janet as in part (d)?

Incredibly, Jack and Janet then meet Chrissy, who is also identical to Jack and Janet with respect to her earnings, utility, and luck. Chrissy's probability of good luck is independent from either Jack's or Janet's. After some discussion, they decide that Chrissy should join the agreement of Jack and Janet. All three of them will pool their earnings and then split them equally three ways.

(f) What are the eight possible luck-outcome triplets, and what is the probability of reaching each of them?

(g) What is the expected utility for each of the investors under this expanded arrangement?

(h) What amount of certain income would yield the same level of utility as in part (g) for these risk-averse investors?

U2. Consider again the case of the 2006 Citrus. Almost all cars depreciate over time, and so it is with the Citrus. Every month that passes, all sellers of Citruses—regardless of type—are willing to accept $100 less than they were the month before. Also, with every passing month buyers are maximally willing to pay to $400 less for an orange than they were the previous month and $200 less for a lemon. Assume that the example in the text takes place in month 0. Eighty percent of the Citruses are oranges, and this proportion never changes.

(a) Fill out three versions of the following table for month 1, month 2, and month 3:

	Willingness to accept of sellers	Willingness to pay of buyers
Orange		
Lemon		

(b) Graph the willingness to accept of the sellers of oranges over the next twelve months. On the same figure, graph the price that buyers are willing to pay for a Citrus of unknown type (given that the proportion of oranges is 0.8). (Hint: Make the vertical axis range from 10,000 to 14,000.)

(c) Is there a market for oranges in month 3? Why or why not?

(d) In what month does the market for oranges collapse?

(e) If owners of lemons experienced no depreciation (that is, they were never willing to accept anything less than $3,000), would this affect the timing of the collapse of the market for oranges? Why or why not? In what month does the market for oranges collapse in this case?

(f) If buyers experienced no depreciation for a lemon (that is, they were always willing to pay up to $6,000 for a lemon), would this affect the timing of the collapse of the market for oranges? Why or why not? In what month does the market for oranges collapse in this case?

U3. An economy has two types of jobs, Good and Bad, and two types of workers, Qualified and Unqualified. The population consists of 60% Qualified and 40% Unqualified. In a Bad job, either type of worker produces 10 units of output. In a Good job, a Qualified worker produces 100 units, and an Unqualified worker produces 0. There is enough demand for workers that for each type of job, companies must pay what they expect the appointee to produce.

Companies must hire each worker without observing his type and pay him before knowing his actual output. But Qualified workers can signal their qualification by getting educated. For a Qualified worker, the cost of getting educated to level n is $n^2/2$, whereas for an Unqualified worker, it is n^2. These costs are measured in the same units as output, and n must be an integer.

(a) What is the minimum level of n that will achieve separation?

(b) Now suppose the signal is made unavailable. Which kind of jobs will be filled by which kinds of workers and at what wages? Who will gain and who will lose from this change?

U4. You are the Dean of the Faculty at St. Anford University. You hire Assistant Professors for a probationary period of seven years, after which they come up for tenure and are either promoted and gain a job for life, or turned down, in which case they must find another job elsewhere.

Your Assistant Professors come in two types, Good and Brilliant. Any types worse than Good have already been weeded out in the hiring process, but you cannot directly distinguish between Good and Brilliant types. Each individual Assistant Professor knows whether he or she is Brilliant or merely Good. You would like to tenure only the Brilliant types.

The payoff from a tenured career at St. Anford is $2 million; think of this as the expected discounted present value of salaries, consulting fees, and book royalties, plus the monetary equivalent of the pride and joy that the faculty member and his or her family would get from being tenured at St. Anford. Anyone denied tenure at St. Anford will get a faculty position at Boondocks College, and the present value of that career is $0.5 million.

Your faculty can do research and publish the findings. But each publication requires effort and time and causes strain on the family; all these are costly to the faculty member. The monetary equivalent of this cost is $30,000 per publication for a Brilliant Assistant Professor, and $60,000 per publication for a Good one. You can set a minimum number, N, of publications that an Assistant Professor must produce in order to achieve tenure.

(a) Without doing any math, describe, as completely as you can, what would happen in a separating equilibrium to this game.

(b) There are two potential types of pooling outcomes to this game. Without doing any math, describe what they would look like, as completely as you can.

(c) Now please go ahead and do some math. What is the set of possible N that will accomplish your goal of screening the Brilliant professors out from the merely Good ones?

U5. Return to the Tudor-Fordor problem from Section 5.B, when Tudor's low per-unit cost is 10. Let z be the probability that Tudor actually has a low per-unit cost.

(a) Rewrite the table in Figure 9.11 in terms of z.

(b) How many pure-strategy equilibria are there when $z = 0$? What type of equilibrium (separating, pooling, or semiseparating) occurs when $z = 0$? Explain.

(c) How many pure-strategy equilibria are there when $z = 1$? What type of equilibrium (separating, pooling, or semiseparating) occurs when $z = 1$? Explain.

(d) What is the lowest value of z such that there is a pooling equilibrium?

(e) Explain intuitively why the pooling equilibrium cannot occur when the value of z is too low.

U6. Assume that Tudor is risk averse, with square-root utility over its total profit (see Exercise S6), and that Fordor is risk neutral. Also, assume that Tudor's low per-unit cost is 10, as in Section 5.B.

(a) Redraw the extensive-form game shown in Figure 9.10, giving the proper payoffs for a risk-averse Tudor.

(b) Let the probability that Tudor is low cost, z, be 0.4. Will the equilibrium be separating, pooling, or semiseparating? (Hint: Use a table equivalent to Figure 9.11.)

(c) Repeat part (b) with $z = 0.1$.

(d) **(Optional)** Will Tudor's risk aversion change the answer to part (d) of Exercise U5? Explain why or why not.

U7. Return to the situation in Exercise S7, where Tudor's low per-unit cost is 6.

(a) Write the normal form of this game in terms of z, the probability that Tudor is low price.

(b) What is the equilibrium when $z = 0.1$? Is it separating, pooling, or semiseparating?

(c) Repeat part (b) for $z = 0.2$.

(d) Repeat part (b) for $z = 0.3$.

(e) Compare your answers in parts (b), (c), and (d) of this problem with part (d) of Exercise U5. When Tudor's low cost is 6 instead of 10, can pooling equilibria be achieved at lower values of z? Or are higher values of z required for pooling equilibria to occur? Explain intuitively why this is the case.

U8. Corporate lawsuits may sometimes be signaling games. Here is one example. In 2003, AT&T filed suit against eBay, alleging that its Billpoint and PayPal electronic-payment systems infringed on AT&T's 1994 patent on "mediation of transactions by a communications system."

Let's consider this situation from the point in time when the suit was filed. In response to this suit, as in most patent-infringement suits, eBay can offer to settle with AT&T without going to court. If AT&T accepts eBay's settlement offer, there will be no trial. If AT&T rejects eBay's settlement offer, the outcome will be determined by the court.

The amount of damages claimed by AT&T is not publicly available. Let's assume that AT&T is suing for $300 million. In addition, let's assume that if the case goes to trial, the two parties will incur court costs (paying lawyers and consultants) of $10 million each.

Because eBay is actually in the business of processing electronic payments, we might think that eBay knows more than AT&T does about its probability of winning the trial. For simplicity, let's assume that eBay knows for sure whether it will be found innocent (i) or guilty (g) of patent infringement. From AT&T's point of view, there is a 25% chance that eBay is guilty (g) and a 75% chance that eBay is innocent (i).

Let's also suppose that eBay has two possible actions: a generous settlement offer (G) of $200 million or a stingy settlement offer (S) of $20 million. If eBay offers a generous settlement, assume that AT&T will accept, thus avoiding a costly trial. If eBay offers a stingy settlement, then AT&T must decide whether to accept (A) and avoid a trial, or reject and take the case to court (C). In the trial, if eBay is guilty, it must pay AT&T $300 million in addition to paying all the court costs. If eBay is found innocent, it will pay AT&T nothing, and AT&T will pay all the court costs.

(a) Show the game in extensive form. (Be careful to label information sets correctly.)

(b) Which of the two players has an incentive to bluff (that is, to give a false signal) in this game? What would bluffing consist of? Explain your reasoning.

(c) Write the strategic-form game matrix for this game. Find all of the Nash equilibria to this game. What are the expected payoffs to each player in equilibrium?

U9. For the Stripped-Down Poker game that Felix and Oscar play in Exercise S9, what does the mix of Kings and Queens have to be for the game to be fair? That is, what fraction of Kings will make the expected payoff of the game zero for both players?

U10. Bored with Stripped-Down Poker, Felix and Oscar now make the game more interesting by adding a third card type: Jack. Four Jacks are added to the deck of four Kings and four Queens. All rules remain the same as before, except for what happens when Felix Bets and Oscar Calls. When Felix Bets and Oscar Calls, Felix wins the pot if he has a King, they "tie" and each gets his money back if Felix is holding a Queen, and Oscar wins the pot if the card is a Jack.

(a) Show the game in extensive form. (Be careful to label information sets correctly.)

(b) How many pure strategies does Felix have in this game? Explain your reasoning.

(c) How many pure strategies does Oscar have in this game? Explain your reasoning.

(d) Represent this game in strategic form. This should be a matrix of *expected* payoffs for each player, given a pair of strategies.

(e) Find the unique pure-strategy Nash equilibrium of this game.

(f) Would you call this a pooling equilibrium, a separating equilibrium, or a semiseparating equilibrium?

(g) In equilibrium, what is the expected payoff to Felix of playing this game? Is it a fair game?

U11. Consider Spence's job-market signaling model with the following specifications. There are two types of workers, 1 and 2. The productivities of the two types, as functions of the level of education E, are

$$W_1(E) = E,\ W_2(E) = 1.5\,E.$$

The costs of education for the two types, as functions of the level of education, are

$$C_1(E) = E^2 / 2 \text{ and } C_2(E) = E^2 / 3.$$

Each worker's utility equals his or her income minus the cost of education. Companies that seek to hire these workers are perfectly competitive in the labor market.

(a) If types are public information (observable and verifiable), find expressions for the levels of education, incomes, and utilities of the two types of workers.

Now suppose each worker's type is his or her private information.

(b) Verify that if the contracts of part (a) are attempted in this situation of information asymmetry, then type 2 does not want to take up the contract intended for type 1, but type 1 does want to take up the contract intended for type 2, so "natural" separation cannot prevail.

(c) If we leave the contract for type 1 as in part (a), what is the range of contracts (education-wage pairs) for type 2 that can achieve separation?

(d) Of the possible separating contracts, which one do you expect to prevail? Give a verbal but not a formal explanation for your answer.

(e) Who gains or loses from the information asymmetry? How much?

U12. "Mr. Robinson pretty much concludes that business schools are a sifting device—M.B.A. degrees are union cards for yuppies. But perhaps the most important fact about the Stanford business school is that all meaningful sifting occurs before the first class begins. No messy weeding is done within the walls. 'They don't want you to flunk. They want you to become a rich alum who'll give a lot of money to the school.' But one wonders: If corporations are abdicating to the Stanford admissions office the responsibility for selecting young managers, why don't they simply replace their personnel departments with Stanford admissions officers, and eliminate the spurious education? Does the very act of throwing away a lot of money and two years of one's life demonstrate a commitment to business that employers find appealing?" (From the review by Michael Lewis of Peter Robinson's *Snapshots from Hell: The Making of an MBA*, in the *New York Times,* May 8, 1994, Book Review section.) What answer to Lewis's question can you give, based on our analysis of strategies in situations of asymmetric information?

U13. **(Optional, requires Appendix)** An auditor for the IRS is reviewing Wanda's latest tax return (see Exercise **S10**), on which she reports having had a bad year. Assume that Wanda is playing according to her equilibrium strategy and that the auditor knows this.

(a) Using Bayes' rule, find the probability that Wanda had a good year given that she reports having had a bad year.

(b) Explain why the answer in part (a) is more or less than the baseline probability of having a good year, 0.6.

U14. **(Optional, requires Appendix)** Return to Exercise S14. Assume, reasonably, that the probability of a lemon's breaking down increases over the length of the road trip. Specifically, let $q = m / (m + 500)$, where m is the number of miles in the round trip.

(a) Find the minimum integer number of miles, m, necessary to avoid the collapse of the market for oranges. That is, what is the smallest m such that the seller of an orange is willing to sell her car at the market price for a Citrus that has successfully completed the road trip? (Hint: Remember to calculate f_{updated} and p_{updated}.)

(b) What is the minimum integer number of miles, m, necessary to achieve complete separation between functioning markets for oranges and lemons? That is, what is the smallest m such that the owner of a lemon will never decide to attempt the road trip?

■

Appendix: Inferring Probabilities from Observing Consequences

When players have different amounts of information in a game, they will try to use some device to ascertain their opponents' private information. As we saw in Section 3 of this chapter, it is sometimes possible for direct communication to yield a cheap talk equilibrium. But more often, players will need to determine one another's information by observing one another's actions. They then must estimate the probabilities of the underlying information by using those actions or their observed consequences. This estimation requires some relatively sophisticated manipulation of the rules of probability, and we examine this process in detail here.

The rules given in Section 1 of the Appendix to Chapter 7 for manipulating and calculating the probability of events, particularly the combination rule, prove useful in our calculations of payoffs when individual players are differently informed. In games of asymmetric information, players try to find out the other's information by observing their actions. Then they must draw inferences about the likelihood of—estimate the probabilities of—the underlying information by exploiting the actions or consequences that are observed.

The best way to understand this is by example. Suppose 1% of the population has a genetic defect that can cause a disease. A test that can identify this genetic defect has a 99% accuracy: when the defect is present, the test will fail to detect it 1% of the time, and the test will also falsely find a defect when none is present 1% of the time. We are interested in determining the probability that a person with a positive test result really has the defect. That is, we cannot directly observe the person's genetic defect (underlying condition), but we can observe the results of the test for that defect (consequences)—except that the test is not a perfect indicator of the defect. How certain can we be, given our observations, that the underlying condition does in fact exist?

We can do a simple numerical calculation to answer the question for our particular example. Consider a population of 10,000 persons in which 100 (1%) have the defect and 9,900 do not. Suppose they all take the test. Of the 100 persons with the defect, the test will be (correctly) positive for 99. Of the 9,900 without the defect, it will be (wrongly) positive for 99. That is 198 positive test results of which one-half are right and one-half are wrong. If a random person receives a positive test result, it is just as likely to be because the test is indeed right as because the test is wrong, so the risk that the defect is truly present for a person with a positive result is only 50%. (That is why tests for rare conditions must be designed to have especially low error rates of generating "false positives.")

For general questions of this type, we use an algebraic formula called **Bayes' theorem** to help us set up the problem and do the calculations. To do so, we generalize our example, allowing for two alternative underlying conditions, A and B (genetic defect or not, for example), and two observable consequences, X and Y (positive or negative test result, for example). Suppose that, in the absence of any information (over the whole population), the probability that A exists is p, so the probability that B exists is $(1 - p)$. When A exists, the chance of observing X is a, so the chance of observing Y is $(1 - a)$. (To use the language that we developed in the Appendix to Chapter 7, a is the probability of X conditional on A, and $(1 - a)$ is the probability of Y conditional on A.) Similarly, when B exists, the chance of observing X is b, so the chance of observing Y is $(1 - b)$.

This description shows us that four alternative combinations of events could arise: (1) A exists and X is observed, (2) A exists and Y is observed, (3) B exists and X is observed, and (4) B exists and Y is observed. Using the modified multiplication rule, we find the probabilities of the four combinations to be, respectively, pa, $p(1 - a)$, $(1 - p)b$, and $(1 - p)(1 - b)$.

Now suppose that X is observed: a person has the test for the genetic defect and gets a positive result. Then we restrict our attention to a subset of the four preceding possibilities—namely, the first and third, both of which include the observation of X. These two possibilities have a total probability of $pa + (1 - p)b$; this is the probability that X is observed. Within this subset of outcomes in which X is observed, the probability that A *also* exists is just pa, as we have already seen. So we know how likely we are to observe X alone and how likely it is that both X and A exist.

But we are more interested in determining how likely it is that A exists, given that we have observed X—that is, the probability that a person has the genetic defect, given that the test is positive. This calculation is the trickiest one. Using the modified multiplication rule, we know that the probability of both A and X happening equals the product of the probability that X alone happens times the probability of A conditional on X; it is this last probability that we are after. Using the formulas for "A and X" and for "X alone," which we just calculated, we get:

$$\text{Prob}(A \text{ and } X) = \text{Prob}(X \text{ alone}) \times \text{Prob}(A \text{ conditional on } X)$$

$$pa = [\,pa + (1-p)b\,] \times \text{Prob}(A \text{ conditional on } X)$$

$$\text{Prob}(A \text{ conditional on } X) = \frac{pa}{pa + (1-p)b}.$$

This formula gives us an assessment of the probability that A has occurred, given that we have observed X (and have therefore conditioned everything on this fact). The outcome is known as *Bayes' theorem* (or rule or formula).

In our example of testing for the genetic defect, we had $\text{Prob}(A) = p = 0.01$, $\text{Prob}(X \text{ conditional on } A) = a = 0.99$, and $\text{Prob}(X \text{ conditional on } B) = b = 0.01$. We can substitute these values into Bayes' formula to get

$$\text{Probability defect exists given that test is positive} = \text{Prob}(A \text{ conditional on } X)$$

$$\text{Probability defect exists given that test is positive} = \text{Prob}(A \text{ conditional on } X)$$

$$= \frac{(0.01)(0.99)}{(0.01)(0.99) + (1-0.01)(0.01)}$$

$$= \frac{0.0099}{0.0099 + 0.0099}$$

$$= 0.5$$

The probability algebra employing Bayes' rule confirms the arithmetical calculation that we used earlier, which was based on an enumeration of all of the possible cases. The advantage of the formula is that, once we have it, we can apply it mechanically; this saves us the lengthy and error-susceptible task of enumerating every possibility and determining each of the necessary probabilities.

We show Bayes' rule in Figure 9A.1 in tabular form, which may be easier to remember and to use than the preceding formula. The rows of the table show the alternative true conditions that might exist, for example, "genetic defect" and "no genetic defect." Here, we have just two, A and B, but the method generalizes immediately to any number of possibilities. The columns show the observed events—for example, "test positive" and "test negative."

Each cell in the table shows the overall probability of that combination of the true condition and the observation; these are just the probabilities for the four alternative combinations listed above. The last column on the right shows

		OBSERVATION		Sum of row
		X	Y	
TRUE CONDITION	A	pa	$p(1-a)$	p
	B	$(1-p)b$	$(1-p)(1-b)$	$1-p$
Sum of column		$pa+(1-p)b$	$p(1-a)+(1-p)(1-b)$	

FIGURE 9A.1 Bayes' Rule

the sum across the first two columns for each of the top two rows. This sum is the total probability of each true condition (so, for instance, A's probability is p, as we have seen). The last row shows the sum of the first two rows in each column. This sum gives the probability that each observation occurs. For example, the entry in the last row of the X column is the total probability that X is observed, either when A is the true condition (a true positive in our genetic test example) or when B is the true condition (a false positive).

To find the probability of a particular condition, given a particular observation, then, Bayes' rule says that we should take the entry in the cell corresponding to the combination of that condition and that observation and divide it by the column sum in the last row for that observation. As an example, Prob (B given X) = $(1-p)b/[pa+(1-p)b]$.

SUMMARY

If players have asymmetric information in a game, they may try to infer probabilities of hidden underlying conditions from observing actions or the consequences of those actions. *Bayes' theorem* provides a formula for inferring such probabilities.

KEY TERMS

Bayes' theorem (359)

Strategic Moves

A GAME IS specified by the choices or moves available to the players, the order, if any, in which they make those moves, and the payoffs that result from all logically possible combinations of all the players' choices. In Chapter 6, we saw how changing the order of moves from sequential to simultaneous or vice versa can alter the game's outcomes. Adding or removing moves available to a player or changing the payoffs at some terminal nodes or in some cells of the game table also can change outcomes. Unless the rules of a game are fixed by an outside authority, each player has the incentive to manipulate them to produce an outcome that is more to his own advantage. Devices to manipulate a game in this way are called **strategic moves,** which are the subject of this chapter.

A strategic move changes the rules of the original game to create a new two-stage game. In this sense, strategic moves are similar to the direct communications of information that we examined in Chapter 9. With strategic moves, though, the second stage is the original game, often with some alteration of the order of moves and the payoffs; there was no such alteration in our games with direct communication. The first stage in a game with strategic moves specifies how you will act in the second stage. Different first-stage actions correspond to different strategic moves, and we classify them into three types: *commitments, threats,* and *promises.* The aim of all three is to alter the outcome of the second-stage game to your own advantage. Which, if any, suits your purpose depends on the context. But most important, any of the three works only if the other player believes that at the second stage you will indeed do what you declared at the first stage. In other words, the *credibility* of the strategic move is open to question. Only a credible strategic move will have the desired effect and, as was often the case in Chapter 9, mere declarations are not enough. At the first stage, you must take some ancillary actions that lend credibility to your declared second-stage actions. We will study both the kinds of second-stage actions that work to your benefit and the first-stage ancillary moves that make them credible.

You are probably more familiar with the use and credibility of strategic moves than you might think. Parents, for instance, constantly attempt to influence the behavior of their children by using threats ("no dessert unless you finish your vegetables") and promises ("you will get the new racing bike at the end of the term if you maintain at least a B average in school"). And children know very well that many of these threats and promises are not credible; much bad behavior can escape the threatened punishment if the child sweetly promises not to do that again, even though the promise itself may not be credible. Furthermore, when the children get older and become concerned with their own appearance, they find themselves making commitments to themselves to exercise and diet; many of these commitments also turn out to lack credibility. All of these devices—commitments, threats, and promises—are examples of strategic moves. Their purpose is to alter the actions of another player, perhaps even your own future self, at a later stage in a game. But they will not achieve this purpose unless they are credible. In this chapter, we will use game theory to study systematically how to use such strategies and how to make them credible.

Be warned, however, that credibility is a difficult and subtle matter. We can offer you some general principles and an overall understanding of how strategic moves can work—a science of strategy. But actually making them work depends on your specific understanding of the context, and your opponent may get the better of you by having a better understanding of the concepts or the context or both. Therefore the use of strategic moves in practice retains a substantial component of art. It also entails risk, particularly when using the strategy of **brinkmanship,** which can sometimes lead to disasters. You can have success as well as fun trying to put these ideas into practice, but note our disclaimer and warning: Use such strategies at your own risk.

1 A CLASSIFICATION OF STRATEGIC MOVES

Because the use of strategic moves depends so critically on the order of moves, to study them we need to know what it means to "move first." Thus far we have taken this concept to be self-evident, but now we need to make it more precise. It has two components. First, your action must be **observable** to the other player; second, it must be **irreversible.**

Consider a strategic interaction between two players, A and B, in which A's move is made first. If A's choice is not observable to B, then B cannot respond to it, and the mere chronology of action is irrelevant. For example, suppose A and B are two companies bidding in an auction. A's committee meets in secret on Monday to determine its bid; B's committee meets on Tuesday; the bids are mailed separately to the auctioneer and opened on Friday. When B makes its decision, it does not know what A has done; therefore the moves are strategically the same as if they were simultaneous.

If A's move is not irreversible, then A might pretend to do one thing, lure B into responding, and then change its own action to its own advantage. B should anticipate this ruse and not be lured; then it will not be responding to A's choice. Once again, in the true strategic sense A does not have the first move.

Considerations of observability and irreversibility affect the nature and types of strategic moves as well as their credibility. We begin with a taxonomy of strategic moves available to players.

A. Unconditional Strategic Moves

Let us suppose that player A is the one making a strategic observable and irreversible move in the first stage of the game. He can declare: "In the game to follow, I will make a particular move, X." This declaration says that A's future move is unconditional; A will do X irrespective of what B does. Such a statement, if credible, is tantamount to changing the order of the game at stage 2 so that A moves first and B second, and A's first move is X. This strategic move is called a **commitment.**

If the previous rules of the game at the second stage already have A moving first, then such a declaration would be irrelevant. But if the game at the second stage has simultaneous moves or if A is to move second there, then such a declaration, if credible, can change the outcome because it changes B's beliefs about the consequences of his actions. Thus a commitment is a simple seizing of the first-mover advantage when it exists.

In the street-garden game of Chapter 3, three women play a sequential-move game in which each must decide whether to contribute toward the creation of a public flower garden on their street; two or more contributors are necessary for the creation of a pleasant garden. The rollback equilibrium entails the first player (Emily) choosing not to contribute while the other players (Nina and Talia) do contribute. By making a credible commitment not to contribute, however, Talia (or Nina) could alter the outcome of the game. Even though she does not get her turn to announce her decision until after Emily and Nina have made theirs public, Talia could let it be known that she has sunk all of her savings (and energy) into a large house-renovation project, and so she will have absolutely nothing left to contribute to the street garden. Then Talia essentially commits herself not to contribute regardless of Emily's and Nina's decisions, before

Emily and Nina make those decisions. In other words, Talia changes the game to one in which she is in effect the first mover. You can easily check that the new rollback equilibrium entails Emily and Nina both contributing to the garden and the equilibrium payoffs are 3 to each of them but 4 to Talia—the equilibrium outcome associated with the game when Talia moves first. Several more detailed examples of commitments are given in the following sections.

B. Conditional Strategic Moves

Another possibility for A is to declare at the first stage: "In the game to follow, I will respond to your choices in the following way. If you choose Y_1, I will do Z_1; if you do Y_2, I will do Z_2, . . . " In other words, A can use a move that is conditional on B's behavior; we call this type of move a **response rule** or *reaction function.* A's statement means that, in the game to be played at the second stage, A will move second, but how he will respond to B's choices at that point is already predetermined by A's declaration at stage 1. For such declarations to be meaningful, A must be physically able to wait to make his move at the second stage until after he has observed what B has irreversibly done. In other words, at the second stage, B should have the true first move in the double sense just explained.

Conditional strategic moves take different forms, depending on what they are trying to achieve and how they set about achieving it. When A wants to stop B from doing something, we say that A is trying to deter B, or to achieve **deterrence;** when A wants to induce B to do something, we say that A is trying to compel B, or to achieve **compellence.** We return to this distinction later. Of more immediate interest is the method used in pursuit of either of these aims. If A declares, "*Unless* your action (or inaction, as the case may be) conforms to my stated wish, I will respond in a way that will *hurt* you," that is, a **threat.** If A declares, "If your action (or inaction, as the case may be) conforms to my stated wish, I will respond in a way that will *reward* you," that is, a **promise.** "Hurt" and "reward" are measured in terms of the payoffs in the game itself. When A hurts B, A does something that lowers B's payoff; when A rewards B, A does something that leads to a higher payoff for B. Threats and promises are the two conditional strategic moves on which we focus our analysis.

To understand the nature of these strategies, consider the dinner game mentioned earlier. In the natural chronological order of moves, first the child decides whether to eat his vegetables, and then the parent decides whether to give the child dessert. Rollback analysis tells us the outcome: the child refuses to eat the vegetables, knowing that the parent, unwilling to see the child hungry and unhappy, will give him the dessert. The parent can foresee this outcome, however, and can try to alter it by making an initial move—namely, by stating a conditional response rule of the form "no dessert unless you finish your vegetables." This declaration constitutes a threat. It is a first move in a pregame, which fixes how you will make your second move in the actual game to follow. If the child believes the threat, that alters the child's rollback calculation. The child "prunes" that branch of the game tree in which the parent serves dessert even if the child has not finished his vegetables. This may alter the child's behavior; the parent hopes that it will make the child act as the parent wants him to. Similarly, in the "study game," the promise of the bike may induce a child to study harder.

2 CREDIBILITY OF STRATEGIC MOVES

We have already seen that payoffs to the other player can be altered by one player's strategic move, but what about the payoffs for the player making that move? Player A gets a higher payoff when B acts in conformity with A's wishes. But A's payoff also may be affected by his own response. In regard to a threat, A's threatened response if B does not act as A would wish may have consequences for A's own payoffs: the parent may be made unhappy by the sight of the unhappy child who has been denied dessert. Similarly, in regard to a promise, rewarding B if he does act as A would wish can affect A's own payoff: the parent who rewards the child for studying hard has to incur the monetary cost of the gift but is happy to see the child's happiness on receiving the gift and even happier about the academic performance of the child.

This effect on A's payoffs has an important implication for the efficacy of A's strategic moves. Consider the threat. If A's payoff is actually increased by carrying out the threatened action, then B reasons that A will carry out this action even if B fulfills A's demands. Therefore B has no incentive to comply with A's wishes, and the threat is ineffective. For example, if the parent is a sadist who enjoys seeing the child go without dessert, then the child thinks, "I am not going to get dessert anyway, so why eat the vegetables?"

Therefore an essential aspect of a threat is that it should be *costly* for the threatener to carry out the threatened action. In the dinner game, the parent must prefer to give the child dessert. Threats in the true strategic sense have the innate property of imposing some cost on the threatener, too; they are threats of *mutual harm.*

In technical terms, a threat fixes your strategy (response rule) in the subsequent game. A strategy must specify what you will do in each eventuality along the game tree. Thus, "no dessert if you don't finish your vegetables" is an incomplete specification of the strategy; it should be supplemented by "and dessert if you do." Threats generally don't specify this latter part. Why not? Because the second part of the strategy is automatically understood; it is implicit. And for the threat to work, this second part of the strategy—the *implied promise* in this case—has to be automatically credible, too.

Thus the threat "no dessert if you don't finish your vegetables" carries with it an implicit promise of "dessert if you do finish your vegetables." This promise also should be credible if the threat is to have the desired effect. In our example, the credibility of the implicit promise is automatic when the parent prefers to see the child get and enjoy his dessert. In other words, the implicit promise is automatically credible precisely when the threatened action is costly for the parent to carry out.

To put it yet another way, a threat carries with it the stipulation that you will do something if your wishes are not met that, if those circumstances actually arise, you will regret having to do. Then why make this stipulation at the first stage? Why tie your own hands in this way when it might seem that leaving one's options open would always be preferable? Because in the realm of game theory, having more options is not always preferable. In regard to a threat, your lack of freedom in the second stage of the game has strategic value. It changes other players' expectations about your future responses, and you can use this change in expectations to your advantage.

A similar effect arises with a promise. If the child knows that the parent enjoys giving him gifts, he may expect to get the racing bike anyway on some occasion in the near future—for example, an upcoming birthday. Then the promise of the bike has little effect on the child's incentive to study hard. To have the intended strategic effect, the promised reward must be so costly to provide that the other player would not expect you to hand over that reward anyway. (This is a useful lesson in strategy that you can point out to your parents: the rewards that they promise must be larger and more costly than what they would give you just for the pleasure of seeing you happy.)

The same is true of unconditional strategic moves (commitments, too). In bargaining, for example, others know that, when you have the freedom to act, you also have the freedom to capitulate; so a "no concessions" commitment can secure you a better deal. If you hold out for 60% of the pie and the other party offers you 55%, you may be tempted to take it. But if you can credibly assert in advance that you will not take less than 60%, then this temptation does not arise and you can do better than you otherwise would.

Thus it is in the very nature of strategic moves that after the fact—that is, when the stage 2 game actually requires it—you do not want to carry out the action that you had stipulated you would take. This is true for all types of strategic moves and it is what makes credibility so problematic. You have to do something at the first stage to create credibility—something that convincingly tells the other player that you will not give in to the temptation to deviate from the stipulated action when the time comes—in order for your strategic move to work. That is why giving up your own freedom to act can be strategically beneficial. Alternatively, credibility can be achieved by changing your own payoffs in the second-stage game in such a way that it becomes truly optimal for you to act as you declare.

Thus there are two general ways of making your strategic moves credible: (1) remove from your own set of future choices the other moves that may tempt you or (2) reduce your own payoffs from those temptation moves so that the stipulated move becomes the actual best one. In the sections that follow, we first elucidate the mechanics of strategic moves, assuming them to be credible. We make some comments about credibility as we go along but postpone our general analysis of credibility until the last section of the chapter.

3 COMMITMENTS

We studied the game of chicken in Chapter 4 and found two pure-strategy Nash equilibria. Each player prefers the equilibrium in which he goes straight and the other person swerves.[1] We saw in Chapter 6 that, if the game were to have sequential rather than simultaneous moves, the first mover would choose Straight, leaving the second to make the best of the situation by settling for Swerve rather than causing a crash. Now we can consider the same matter from another perspective. Even if the game itself has simultaneous moves, if one player can make a strategic move—create a first stage in which he makes a credible declaration about his action in the chicken game itself, which is to be played at the second stage—then he can get the same advantage afforded a first mover by making a commitment to act tough (choose Straight).

Although the point is simple, we outline the formal analysis to develop your understanding and skill, which will be useful for later, more complex ex-

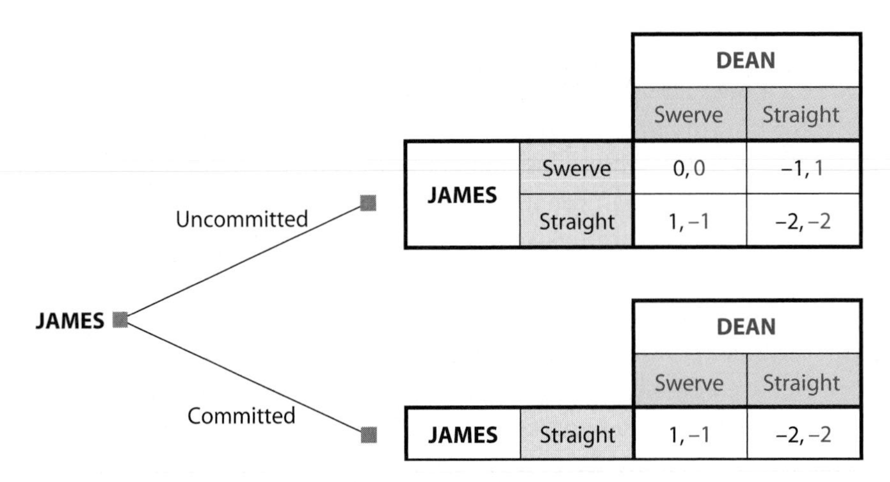

FIGURE 10.1 Chicken: Commitment by Restricting Freedom to Act

[1]We saw in Chapter 7 and will see again in Chapter 13 that the game has a third equilibrium, in mixed strategies, in which both players do quite poorly.

amples. Remember our two players, James and Dean. Suppose James is the one who has the opportunity to make a strategic move. Figure 10.1 shows the tree for the two-stage game. At the first stage, James has to decide whether to make a commitment. Along the upper branch emerging from the first node, he does not make the commitment. Then at the second stage the simultaneous-move game is played, and its payoff table is the familiar one shown in Figures 4.14 and 6.6. This second-stage game has multiple equilibria, and James gets his best payoff in only one of them. Along the lower branch, James makes the commitment. Here, we interpret this commitment to mean giving up his freedom to act in such a way that Straight is the only action available to James at this stage. Therefore the second-stage game table has only one row for James, corresponding to his declared choice of Straight. In this table, Dean's best action is Swerve; so the equilibrium outcome gives James his best payoff. Therefore, at the first stage, James finds it optimal to make the commitment; this strategic move ensures his best payoff, while not committing leaves the matter uncertain.

How can James make this commitment credibly? Like any first move, the commitment move must be (1) irreversible and (2) visible to the other player. People have suggested some extreme and amusing ideas. James can disconnect the steering wheel of the car and throw it out of the window so that Dean can see that James can no longer Swerve. (James could just tie the wheel so that it could no longer be turned, but it would be more difficult to demonstrate to Dean that the wheel was truly tied and that the knot was not a trick one that could be undone quickly.) These devices simply remove the Swerve option from the set of choices available to James in the stage 2 game, leaving Straight as the only thing he can do.

More plausibly, if such games are played every weekend, James can acquire a general reputation for toughness that acts as a guarantee of his action on any one day. In other words, James can alter his own payoff from swerving by subtracting an amount that represents the loss of reputation. If this amount is large enough—say, 3—then the second-stage game when James has made the commitment has a different payoff table. The complete tree for this version of the game is shown in Figure 10.2.

Now, in the second stage with commitment, Straight has become truly optimal for James; in fact, it is his dominant strategy in that stage. Dean's optimal strategy is then Swerve. Looking ahead to this outcome at stage 1, James sees that he gets 1 by making the commitment (changing his own stage 2 payoffs), while without the commitment he cannot be sure of 1 and may do much worse. Thus a rollback analysis shows that James should make the commitment.

Both (or all) can play the game of commitment, so success may depend both on the speed with which you can seize the first move and on the credibility with which you can make that move. If there are lags in observation, the two may even make incompatible simultaneous commitments: each disconnects his steering wheel and tosses it out of the window just as he sees the other's wheel come flying out, and then the crash is unavoidable.

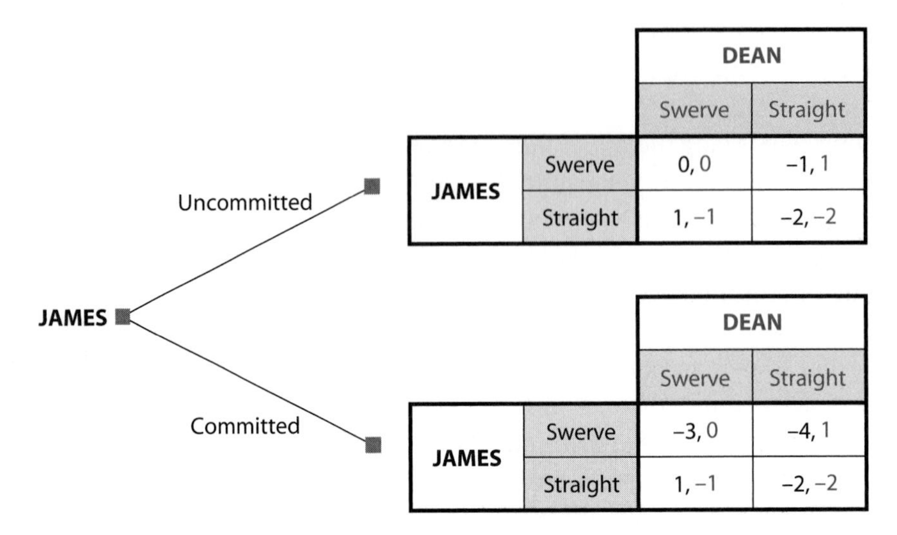

FIGURE 10.2 Chicken: Commitment by Changing Payoffs

Even if one of the players has the advantage in making a commitment, the other player can defeat the first player's attempt to do so. The second player could demonstrably remove his ability to "see" the other's commitment, for example, by cutting off communication.

Games of chicken may be a 1950s anachronism, but our second example is perennial and familiar. In a class, the teacher's deadline enforcement policy can be Weak or Tough, and the students' work can be Punctual or Late. Figure 10.3 shows this game in the strategic form. The teacher does not like being tough; for him the best outcome (a payoff of 4) is when students are punctual even when he is weak; the worst (1) is when he is tough but students are still late. Of the two intermediate strategies, he recognizes the importance of punctuality and rates (Tough, Punctual) better than (Weak, Late). The students most prefer the outcome (Weak, Late), where they can party all weekend without suffering any penalty for the late assignment. (Tough, Late) is the worst for them, just as it is for the teacher. Between the intermediate ones, they prefer (Weak, Punctual) to (Tough, Punctual) because they have higher self-esteem if they can think that

		STUDENT	
		Punctual	Late
TEACHER	Weak	4, 3	2, 4
	Tough	3, 2	1, 1

FIGURE 10.3 Payoff Table for Class Deadline Game

they acted punctually of their own volition rather than because of the threat of a penalty.[2]

If this game is played as a simultaneous-move game or if the teacher moves second, Weak is dominant for the teacher, and then the student chooses Late. The equilibrium outcome is (Weak, Late), and the payoffs are (2, 4). But the teacher can achieve a better outcome by committing at the outset to the policy of Tough. We do not draw a tree as we did in Figures 10.1 and 10.2. The tree would be very similar to that for the preceding chicken case, and so we leave it for you to draw. Without the commitment, the second-stage game is as before, and the teacher gets a 2. When the teacher is committed to Tough, the students find it better to respond with Punctual at the second stage, and the teacher gets a 3.

The teacher commits to a move different from what he would do in simultaneous play or indeed, his best second move if the students moved first. This is where strategic thinking enters. The teacher has nothing to gain by declaring that he will have a Weak enforcement regime; the students expect that anyway in the absence of any declaration. To gain advantage by making a strategic move, he must commit not to follow what would be his equilibrium strategy of the simultaneous-move game. This strategic move changes the students' expectations and therefore their action. Once they believe the teacher is really committed to tough discipline, they will choose to turn in their assignments punctually. If they tested this out by being late, the teacher would like to forgive them, maybe with an excuse to himself, such as "just this once." The existence of this temptation to shift away from your commitment is what makes its credibility problematic.

Even more dramatic, in this instance the teacher benefits by making a strategic move that commits him to a dominated strategy. He commits to choosing Tough, which is dominated by Weak. The choice of Tough gets the teacher a 3 if the student chooses Punctual and a 1 if the student chooses Late, whereas if the teacher had chosen Weak, his corresponding payoffs would have been 4 and 2. If you think it paradoxical that one can gain by choosing a dominated strategy, you are extending the concept of dominance beyond the proper scope of its validity. Dominance entails either of two calculations: (1) After the other player does something, how do I respond, and is some choice best (or worst), given all possibilities? (2) If the other player is simultaneously doing action X, what is best (or worst) for me, and is this the same for all the X actions that the other could be choosing? Neither is relevant when you are moving first. Instead, you must look

[2]You may not regard these specific rankings of outcomes as applicable either to you or to your own teachers. We ask you to accept them for this example, whose main purpose is to convey some *general ideas* about commitment in a simple way. The same disclaimer applies to all the examples that follow.

ahead to how the other will respond. Therefore the teacher does not compare his payoffs in vertically adjacent cells of the table (taking the possible actions of the students one at a time). Instead, he calculates how the students will react to each of his moves. If he is committed to Tough, they will be Punctual, but if he is committed to Weak (or uncommitted), they will be Late, so the only pertinent comparison is that of the top-right cell with the bottom left, of which the teacher prefers the latter.

To be credible, the teacher's commitment must be everything a first move has to be. First, it must be made before the other side makes its move. The teacher must establish the ground rules of deadline enforcement before the assignment is due. Next, it must be observable—the students must know the rules by which they must abide. Finally, and perhaps most important, it must be irreversible—the students must know that the teacher cannot, or at any rate will not, change his mind and forgive them. A teacher who leaves loopholes and provisions for incompletely specified emergencies is merely inviting imaginative excuses accompanied by fulsome apologies and assertions that "it won't happen again."

The teacher might achieve credibility by hiding behind general university regulations; this simply removes the Weak option from his set of available choices at stage 2. Or, as is true in the chicken game, he might establish a reputation for toughness, changing his own payoffs from Weak by creating a sufficiently high cost of loss of reputation.

4 THREATS AND PROMISES

We emphasize that threats and promises are *response rules*: your actual future action is conditioned on what the other players do in the meantime, but your freedom of future action is constrained to following the stated rule. Once again, the aim is to alter the other players' expectations and therefore their actions in a way favorable to you. Tying yourself to a rule that you would not want to follow if you were completely free to act at the later time is an essential part of this process. Thus the initial declaration of intention must be credible. Once again, we will elucidate some principles for achieving credibility of these moves, but we remind you that their actual implementation remains largely an art.

Remember the taxonomy given in Section 1. A *threat* is a response rule that leads to a bad outcome for the other players if they act contrary to your interests. A *promise* is a response rule by which you offer to create a good outcome for the other players if they act in a way that promotes your own interests. Each of these responses may aim either to stop the other players from doing something that they would otherwise do (*deterrence*) or to induce them to do something that they would otherwise not do (*compellence*). We consider these features in turn.

A. Example of a Threat: U.S.–Japan Trade Relations

Our example comes from a hardy perennial of U.S. international economic policy—namely, trade friction with Japan. Each country has the choice of keeping its own markets open or closed to the other's goods. They have somewhat different preferences regarding the outcomes.

Figure 10.4 shows the payoff table for the trade game. For the United States, the best outcome (a payoff of 4) comes when both markets are open; this is partly because of its overall commitment to the market system and free trade and partly because of the benefit of trade with Japan itself—U.S. consumers get high-quality cars and consumer electronics products, and U.S. producers can export their agricultural and high-tech products. Similarly, its worst outcome (payoff 1) occurs when both markets are closed. Of the two outcomes when only one market is open, the United States would prefer its own market to be open, because the Japanese market is smaller, and loss of access to it is less important than the loss of access to Hondas and video games.

As for Japan, for the purpose of this example we accept the protectionist, producer-oriented picture of Japan, Inc. Its best outcome is when the U.S. market is open and its own is closed; its worst is when matters are the other way around. Of the other two outcomes, it prefers that both markets be open, because its producers then have access to the much larger U.S. market.[3]

Both sides have dominant strategies. No matter how the game is played—simultaneously or sequentially with either move order—the equilibrium outcome is (Open, Closed), and the payoffs are (3, 4). This outcome also fits well the common American impression of how the actual trade policies of the two countries work.

		JAPAN	
		Open	Closed
UNITED STATES	Open	4, 3	3, 4
	Closed	2, 1	1, 2

FIGURE 10.4 Payoff Table for U.S.–Japan Trade Game

[3]Again, we ask you to accept this payoff structure as a vehicle for conveying the ideas. You can experiment with the payoff tables to see what difference that would make to the role and effectiveness of the strategic moves.

Japan is already getting its best payoff in this equilibrium and so has no need to try any strategic moves. The United States, however, can try to get a 4 instead of a 3. But in this case an ordinary unconditional commitment will not work. Japan's best response, no matter what commitment the United States makes, is to keep its market closed. Then the United States does better for itself by committing to keep its own market open, which is the equilibrium without any strategic moves anyway.

But suppose the United States can choose the following conditional response rule: "We will close our market if you close yours." The situation then becomes the two-stage game shown in Figure 10.5. If the United States does not use the threat, the second stage is as before and leads to the equilibrium in which the U.S. market is open and it gets a 3, whereas the Japanese market is closed and it gets a 4. If the United States does use the threat, then at the second stage only Japan has freedom of choice; given what Japan does, the United States then merely does what its response rule dictates. Therefore, along this branch of the tree, we show only Japan as an active player and write down the payoffs to the two parties: If Japan keeps its market closed, the United States closes its own, and the United States gets a 1 and Japan gets a 2. If Japan keeps its market open, then the United States threat has worked, it is happy to keep its own market open, and it gets a 4, while Japan gets a 3. Of these two possibilities, the second is better for Japan.

Now we can use the familiar rollback reasoning. Knowing how the second stage will work in all eventualities, it is better for the United States to deploy its threat at the first stage. This threat will result in an open market in Japan, and the United States will get its best outcome.

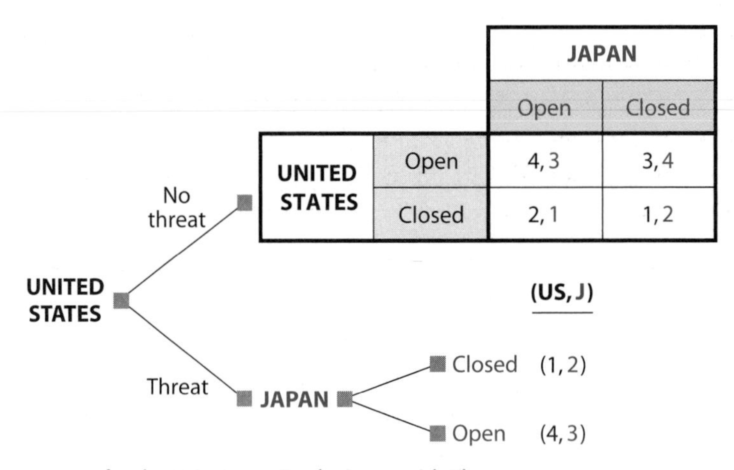

FIGURE 10.5 Tree for the U.S.–Japan Trade Game with Threat

Having described the mechanics of the threat, we now point out some of its important features.

1. When the United States deploys its threat credibly, Japan doesn't follow its dominant strategy Closed. Again, the idea of dominance is relevant only in the context of simultaneous moves or when Japan moves second. Here, Japan knows that the United States will take actions that depart from *its* dominant strategy. In the payoff table, Japan is looking at a choice between just two cells, the top left and the bottom right, and, of those two, it prefers the latter.

2. Credibility of the threat is problematic because, if Japan puts it to the test by keeping its market closed, the United States faces the temptation to refrain from carrying out the threat. In fact, if the threatened action were the best U.S. response after the fact, then there would be no need to make the threat in advance (but the United States might issue a *warning* just to make sure that the Japanese understand the situation). The strategic move has a special role exactly because it locks a player into doing something other than what it would have wanted to do after the fact. As explained earlier, a threat in the true strategic sense is necessarily costly for the threatener to carry out; the threatened action would inflict *mutual* harm.

3. The conditional rule "We will close our market if you close yours" does not completely specify the U.S. *strategy.* To be complete, it needs an additional clause indicating what the United States will do in response to an open Japanese market: "and we will keep our market open if you keep yours open." This additional clause, the implicit promise, is really part of the threat, but it does not need to be stated explicitly, because it is automatically credible. Given the payoffs of the second-stage game, it is in the best interests of the United States to keep its market open if Japan keeps its market open. If that were not the case, if the United States would respond by keeping its market closed even when Japan kept its own market open, then the implicit promise would have to be made explicit and somehow made credible. Otherwise, the U.S. threat would become tantamount to the unconditional commitment "We will keep our market closed," and that would not elicit the desired response from Japan.

4. The threat, when credibly deployed, results in a change in Japan's action. We can regard this as deterrence or compellence, depending on the status quo. If the Japanese market is initially open, and the Japanese are considering a switch to protectionism, then the threat deters them from that action. But if the Japanese market is initially closed, then the threat compels them to open it. Thus whether a strategic move is deterrent or compellent depends on the status quo. The distinction may seem to be a matter of semantics, but in practice the credibility of a move and the way that it works are importantly affected by this distinction. We return to this matter later in the chapter.

5. Here are a few ways in which the United States can make its threat credible. First, it can enact a law that mandates the threatened action under the right circumstances. This removes the temptation action from the set of available choices at stage 2. Some reciprocity provisions in the World Trade Organization agreements have this effect, but the procedures are very slow and uncertain. Second, it can delegate fulfillment to an agency such as the U.S. Commerce Department that is captured by U.S. producers who would like to keep our markets closed and so reduce the competitive pressure on themselves. This changes the U.S. payoffs at stage 2—replacing the true U.S. payoffs by those of the Commerce Department—with the result that the threatened action becomes truly optimal. (The danger is that the Commerce Department will then retain a protectionist stance even if Japan opens its market; gaining credibility for the threat may lose credibility for the implied promise.)

6. If a threat works, it doesn't have to be carried out. So its cost to you is immaterial. In practice, the danger that you may have miscalculated or the risk that the threatened action will take place by error even if the other player complies is a strong reason to refrain from using threats more severe than necessary. To make the point starkly, the United States could threaten to pull out of defensive alliances with Japan if it didn't buy our rice and semiconductors, but that threat is "too big" and too risky for the United States ever to carry out; therefore it is not credible.

But sometimes a range of threats is not available from which a player can choose one that is, on the one hand, sufficiently big that the other player fears it and alters his action in the way that the first player desires and, on the other hand, not so big as to be too risky for the first player ever to carry out and therefore lacking credibility. If the only available threat is too big, then a player can reduce its size by making its fulfillment a matter of chance. Instead of saying, "If you don't open your markets, we will refuse to defend you in the future," the United States can say to Japan, "If you don't open your markets, the relations between our countries will deteriorate to the point where Congress may refuse to allow us to come to your assistance if you are ever attacked, even though we do have an alliance." In fact, the United States can deliberately foster sentiments that raise the probability that Congress will do just that, so the Japanese will feel the danger more vividly. A threat of this kind, which creates a risk but not a certainty of the bad outcome, is called brinkmanship. It is an extremely delicate and even dangerous variant of the strategic move. We will study brinkmanship in greater detail in Chapter 15.

7. Japan gets a worse outcome when the United States deploys its threat than it would without this threat, so it would like to take strategic actions that defeat or disable U.S. attempts to use the threat. For example, suppose its market is currently closed, and the United States is attempting compellence. The Japanese can accede in principle but stall in practice, pleading unavoidable delays for assembling the necessary political consensus to legislate the market

opening, then delays for writing the necessary administrative regulations to implement the legislation, and so on. Because the United States does not want to go ahead with its threatened action, at each point it has the temptation to accept the delay. Or Japan can claim that its domestic politics makes it difficult to open all markets fully; will the United States accept the outcome if Japan keeps just a few of its industries protected? It gradually expands this list, and at any point the extra small step is not enough cause for the United States to unleash a trade war. This device of defeating a compellent threat by small steps, or "slice by slice," is called **salami tactics.**

B. Example of a Promise: The Restaurant Pricing Game

We now illustrate a promise by using the restaurant pricing game of Chapter 5. We saw in Chapter 5 that the game is a prisoners' dilemma, and we simplify it here by supposing that only two choices of price are available: the jointly best price of $26 or the Nash equilibrium price of $20. The profits for each restaurant in this version of the game can be calculated by using the functions in Section 1 of Chapter 5; the results are shown in Figure 10.6. Without any strategic moves, the game has the usual equilibrium in dominant strategies in which both stores charge the low price of 20, and both get lower profits than they would if they both charged the high price of 26.

If either side can make the credible promise "I will charge a high price if you do," the cooperative outcome is achieved. For example, if Xavier's makes the promise, then Yvonne's knows that its choice of 26 will be reciprocated, leading to the payoff shown in the lower-right cell of the table, and that its choice of 20 will bring forth Xavier's usual action—namely, 20—leading to the upper-left cell. Between the two, Yvonne's prefers the first and therefore chooses the high price.

The analysis can be done more properly by drawing a tree for the two-stage game in which Xavier's has the choice of making or not making the promise at the first stage. We omit the tree, partly so that you can improve your understanding of the process by constructing it yourself and partly to show how such detailed analysis becomes unnecessary as one becomes familiar with the ideas.

		YVONNE'S BISTRO	
		20 (low)	26 (high)
XAVIER'S TAPAS	20 (low)	288, 288	360, 216
	26 (high)	216, 360	324, 324

FIGURE 10.6 Payoff Table for Restaurant Prisoners' Dilemma ($100s per month)

The credibility of Xavier's promise is open to doubt. To respond to what Yvonne's does, Xavier's must arrange to move second in the second stage of the game; correspondingly, Yvonne's must move first in stage 2. Remember that a first move is an irreversible and observable action. Therefore, if Yvonne's moves first and prices high, it leaves itself vulnerable to Xavier's cheating, and Xavier's is very tempted to renege on its promise to price high when it sees Yvonne's in this vulnerable position. Xavier's must somehow convince Yvonne's that it will not give in to the temptation to charge a low price when Yvonne's charges a high price.

How can it do so? Perhaps Xavier's owner can leave the pricing decision in the hands of a local manager, with clear written instructions to reciprocate with the high price if Yvonne's charges the high price. Xavier's owner can invite Yvonne's to inspect these instructions, after which he leaves on a solo round-the-world sailing trip so that he cannot rescind them. (Even then, Yvonne's management may be doubtful—Xavier might secretly carry a telephone or a laptop computer on board.) This scenario is tantamount to removing the cheating action from the choices available to Xavier's at stage 2.

Or Xavier's restaurant can develop a reputation for keeping its promises, in business and in the community more generally. In a repeated relationship, the promise may work because reneging on the promise once may cause future cooperation to collapse. In essence, an ongoing relationship means splitting the game into smaller segments, in each of which the benefit from reneging is too small to justify the costs. In each such game, then, the payoff from cheating is altered by the cost of collapse of future cooperation.[4]

We saw earlier that every threat has an implicit attached promise. Similarly, every promise has an implicit attached threat. In this case, the threat is "I will charge the low price if you do." It does not have to be stated explicitly, because it is automatically credible—it describes Xavier's best response to Yvonne's low price.

There is also an important difference between a threat and a promise. If a threat is successful, it doesn't have to be carried out and is then costless to the threatener. Therefore a threat can be bigger than what is needed to make it effective (although making it too big may be too risky, even to the point of losing its credibility as suggested earlier). If a promise is successful in altering the other's action in the desired direction, then the promisor has to deliver what he had promised, and so it is costly. In the preceding example, the cost is simply giving up the opportunity to cheat and get the highest payoff; in other instances where the promisor offers an actual gift or an inducement to the other, the cost may

[4]In Chapter 11, we will investigate in great detail the importance of repeated or ongoing relationships in attempts to reach the cooperative outcome in a prisoners' dilemma.

be more tangible. In either case, the player making the promise has a natural incentive to keep its size small—just big enough to be effective.

C. Example Combining Threat and Promise: Joint U.S.–China Political Action

When we considered threats and promises one at a time, the explicit statement of a threat included an implicit clause of a promise that was automatically credible, and vice versa. There can, however, be situations in which the credibility of both aspects is open to question; then the strategic move has to make both aspects explicit and make them both credible.

Our example of an explicit-threat-and-promise combination comes from a context in which multiple nations must work together toward some common goal in dealing with a dangerous situation in a neighboring country. Specifically, we consider an example of the United States and China contemplating whether to take action to compel North Korea to give up its nuclear weapons programs. We show in Figure 10.7 the payoff table for the United States and China when each must choose between action and inaction.

Each country would like the other to take on the whole burden of taking action against the North Koreans; so the top-right cell has the best payoff for China (4), and the bottom-left cell is best for the United States. The worst situation for the United States is where no action is taken, because it finds the increased threat of nuclear war in that case to be unacceptable. For China, however, the worst outcome arises when it takes on the whole burden of action, because the costs of action are so high. Both regard a joint involvement as the second best (a payoff of 3). The United States assigns a payoff of 2 to the situation in which it is the only one to act. And for China, a payoff of 2 is assigned to the case in which no action is taken.

Without any strategic moves, the intervention game is dominance solvable. Inaction is the dominant strategy for China, and then Action is the best choice for the United States. The equilibrium outcome is the top-right cell, with payoffs of 2 for the United States and 4 for China. Because China gets its best outcome, it has no reason to try any strategic moves. But the United States can try to do better than a 2.

		CHINA	
		Action	Inaction
UNITED STATES	Action	3, 3	2, 4
	Inaction	4, 1	1, 2

FIGURE 10.7 Payoff Table for U.S.–China Political Action Game

What strategic move will work to improve the equilibrium payoff for the United States? An unconditional move (commitment) will not work, because China will respond with "Inaction" to either first move by the United States. A threat alone ("We won't take action unless you do") does not work, because the implied promise ("We will if you do") is not credible—if China does act, the United States would prefer to back off and leave everything to China, getting a payoff of 4 instead of the 3 that would come from fulfilling the promise. A promise alone won't work: because China knows that the United States will intervene if China does not, an American promise of "We will intervene if you do" becomes tantamount to a simple commitment to intervene; then China can stay out and get its best payoff of 4.

In this game, an explicit promise from the United States must carry the implied threat "We won't take action if you don't," but that threat is not automatically credible. Similarly, America's explicit threat must carry the implied promise "We will act if you do," but that is not automatically credible, either. Therefore the United States has to make both the threat and the promise explicit. It must issue the combined threat-cum-promise "We will act if, and only if, you do." It needs to make both clauses credible. Usually such credibility has to be achieved by means of a treaty that covers the whole relationship, not just with agreements negotiated separately when each incident arises.

5 SOME ADDITIONAL TOPICS

A. When Do Strategic Moves Help?

We have seen several examples in which a strategic move brings a better outcome to one player or another, compared with the original game without such moves. What can be said in general about the desirability of such moves?

An unconditional move—a commitment—need not always be advantageous to the player making it. In fact, if the original game gives the advantage to the second mover, then it is a mistake to commit oneself to move in advance, thereby effectively becoming the first mover.

The availability of a conditional move—threat or promise—can never be an actual disadvantage. At the very worst, one can commit to a response rule that would have been optimal after the fact. However, if such moves bring one an actual gain, it must be because one is choosing a response rule that in some eventualities specifies an action different from what one would find optimal at that later time. Thus whenever threats and promises bring a positive gain, they do so precisely when (one might say precisely because) their credibility is inherently questionable and must be achieved by some specific credibility "device."

We have mentioned some such devices in connection with each earlier example and will later discuss the topic of achieving credibility in greater generality.

What about the desirability of being on the receiving end of a strategic move? It is never desirable to let the other player threaten you. If a threat seems likely, you can gain by looking for a different kind of advance action—one that makes the threat less effective or less credible. We will consider some such actions shortly. However, it is often desirable to let the other player make promises to you. In fact, both players may benefit when one can make a credible promise, as in the prisoners' dilemma example of restaurant pricing earlier in this chapter, in which a promise achieved the cooperative outcome. Thus it may be in the players' mutual interest to facilitate the making of promises by one or both of them.

B. Deterrence Versus Compellence

In principle, either a threat or a promise can achieve either deterrence or compellence. For example, a parent who wants a child to study hard (compellence) can promise a reward (a new racing bike) for good performance in school or can threaten a punishment (a strict curfew the following term) if the performance is not sufficiently good. Similarly, a parent who wants the child to keep away from bad company (deterrence) can try either a reward (promise) or a punishment (threat). In practice, the two types of strategic moves work somewhat differently, and that will affect the ultimate decision regarding which to use. Generally, deterrence is better achieved by a threat and compellence by a promise. The reason is an underlying difference of timing and initiative.

A deterrent threat can be passive—you don't need to do anything so long as the other player doesn't do what you are trying to deter. And it can be static—you don't have to impose any time limit. Thus you can set a trip wire and then leave things up to the other player. So the parent who wants the child to keep away from bad company can say, "If I ever catch you with X again, I will impose a 7 P.M. curfew on you for a whole year." Then the parent can sit back to wait and watch; only if the child acts contrary to the parent's wishes does the parent have to act on her threat. Trying to achieve the same deterrence by a promise would require more complex monitoring and continual action: "At the end of each month in which I know that you did not associate with X, I will give you $25."

Compellence must have a deadline or it is pointless—the other side can defeat your purpose by procrastinating or by eroding your threat in small steps (salami tactics). This makes a compellent threat harder to implement than a compellent promise. The parent who wants the child to study hard can simply say, "Each term that you get an average of B or better, I will give you CDs or games worth $500." The child will then take the initiative in showing the parent each time he has fulfilled the conditions. Trying to achieve the same thing by a threat—"Each term that your average falls below B, I will take away one of your computer games"—will require the parent to be much more vigilant and active. The child will postpone bringing the grade report or will try to hide the games.

The concepts of reward and punishment are relative to those of some status quo. If the child has a perpetual right to the games, then taking one away is a punishment; if the games are temporarily assigned to the child on a term-by-term basis, then renewing the assignment for another term is a reward. Therefore you can change a threat into a promise or vice versa by changing the status quo. You can use this change to your own advantage when making a strategic move. If you want to achieve compellence, try to choose a status quo such that what you do when the other player acts to comply with your demand becomes a reward, and so you are using a compellent promise. To give a rather dramatic example, a mugger can convert the threat "If you don't give me your wallet, I will take out my knife and cut your throat" into the promise "Here is a knife at your throat; as soon as you give me your wallet I will take it away." But if you want to achieve deterrence, try to choose a status quo such that, if the other player acts contrary to your wishes, what you do is a punishment, and so you are using a deterrent threat.

6 ACQUIRING CREDIBILITY

We have emphasized the importance of credibility of strategic moves throughout, and we accompanied each example with some brief remarks about how credibility could be achieved in that particular context. Devices for achieving credibility are indeed often context specific, and there is a lot of art to discovering or developing such devices. Some general principles can help you organize your search.

We pointed out two broad approaches to credibility: (1) reducing your own future freedom of action in such a way that you have no choice but to carry out the action stipulated by your strategic move and (2) changing your own future payoffs in such a way that it becomes optimal for you to do what you stipulate in your strategic move. We now elaborate some practical methods for implementing each of these approaches.

A. Reducing Your Freedom of Action

I. AUTOMATIC FULFILLMENT Suppose at stage 1 you relinquish your choice at stage 2 and hand it over to a mechanical device or similar procedure or mechanism that is programmed to carry out your committed, threatened, or promised action under the appropriate circumstances. You demonstrate to the other player that you have done so. Then he will be convinced that you have no freedom to change your mind, and your strategic move will be credible. The **doomsday device,** a nuclear explosive device that would detonate and contaminate the whole world's atmosphere if the enemy launched a nuclear attack, is the best-known example, popularized by the early 1960s movies *Fail Safe* and *Dr. Strangelove*. Luckily, it remained in the realm of fiction. But automatic procedures that retaliate with import tariffs if another country tries to subsidize its exports to your country (countervailing duties) are quite common in the arena of trade policy.

II. DELEGATION A fulfillment device does not even have to be mechanical. You could delegate the power to act to another person or to an organization that is required to follow certain preset rules or procedures. In fact, that is how the countervailing duties work. They are set by two agencies of the U.S. government—the Commerce Department and the International Trade Commission—whose operating procedures are laid down in the general trade laws of the country.

An agent should not have his own objectives that defeat the purpose of his strategic move. For example, if one player delegates to an agent the task of inflicting threatened punishment and the agent is a sadist who enjoys inflicting punishment, then he may act even when there is no reason to act—that is, even when the second player has complied. If the second player suspects this, then the threat loses its effectiveness, because the punishment becomes a case of "damned if you do and damned if you don't."

Delegation devices are not complete guarantees of credibility. Even the doomsday device may fail to be credible if the other side suspects that you control an override button to prevent the risk of a catastrophe. And delegation and mandates can always be altered; in fact, the U.S. government has often set aside the stipulated countervailing duties and reached other forms of agreements with other countries so as to prevent costly trade wars.

III. BURNING BRIDGES Many invaders, from Xenophon in ancient Greece to William the Conqueror in England to Cortés in Mexico, are supposed to have deliberately cut off their own army's avenue of retreat to ensure that it will fight hard. Some of them literally burned bridges behind them, while others burned ships, but the device has become a cliche. Its most recent users in military contexts may have been the Japanese *kamikaze* pilots in World War II, who took only enough fuel to reach the U.S. naval ships into which they were to ram their airplanes. The principle even appears in the earliest known treatise on war, in a commentary attributed to Prince Fu Ch'ai: "Wild beasts, when they are at bay, fight desperately. How much more is this true of men! If they know there is no alternative they will fight to the death."[5]

Related devices are used in other high-stakes games. Although the European Monetary Union could have retained separate currencies and merely fixed

[5]Sun Tzu, *The Art of War*, trans. Samuel B. Griffith (Oxford: Oxford University Press, 1963), p. 110.

the exchange rates among them, a common currency was adopted precisely to make the process irreversible and thereby give the member countries a much greater incentive to make the union a success. (In fact, it is the extent of the necessary commitment that has kept some nations, Great Britain in particular, from agreeing to be part of the European Monetary Union.) It is not totally impossible to abandon a common currency and go back to separate national ones; it is just inordinately costly. If things get really bad inside the Union, one or more countries may yet choose to get out. As with automatic devices, the credibility of burning bridges is not an all-or-nothing matter, but one of degree.

IV. CUTTING OFF COMMUNICATION If you send the other player a message demonstrating your commitment and at the same time cut off any means for him to communicate with you, then he cannot argue or bargain with you to reverse your action. The danger in cutting off communication is that, if both players do so simultaneously, then they may make mutually incompatible commitments that can cause great mutual harm. Additionally, cutting off communication is harder to do with a threat, because you have to remain open to the one message that tells you whether the other player has complied and therefore whether you need to carry out your threat. In this age, it is also quite difficult for a person to cut himself off from all contact.

But players who are large teams or organizations can try variants of this device. Consider a labor union that makes its decisions at mass meetings of members. To convene such a meeting takes a lot of planning—reserving a hall, communicating with members, and so forth—and several weeks of time. A meeting is convened to decide on a wage demand. If management does not meet the demand in full, the union leadership is authorized to call a strike and then it must call a new mass meeting to consider any counteroffer. This process puts management under a lot of time pressure in the bargaining; it knows that the union will not be open to communication for several weeks at a time. Here, we see that cutting off communication for extended periods can establish some degree of credibility, but not absolute credibility. The union's device does not make communication totally impossible; it only creates several weeks of delay.

B. Changing Your Payoffs

I. REPUTATION You can acquire a **reputation** for carrying out threats and delivering on promises. Such a reputation is most useful in a repeated game against the same player. It is also useful when playing different games against different players, if each of them can observe your actions in the games that you play with others. The circumstances favorable to the emergence of such a reputation are the same as those for achieving cooperation in the prisoners' dilemma, and for the same reasons. The greater the likelihood that the interaction will continue and the greater the concern for the future relative to the present, the more likely the players will be to sacrifice current temptations for the sake of future gains. The players will therefore be more willing to acquire and maintain reputations.

In technical terms, this device links different games, and the payoffs of actions in one game are altered by the prospects of repercussions in other games. If you fail to carry out your threat or promise in one game, your reputation suffers and you get a lower payoff in other games. Therefore when you consider any one of these games, you should adjust your payoffs in it to take into consideration such repercussions on your payoffs in the linked games.

The benefit of reputation in ongoing relationships explains why your regular car mechanic is less likely to cheat you by doing an unnecessary or excessively costly or shoddy repair than is a random garage that you go to in an emergency. But what does your regular mechanic actually stand to gain from acquiring this reputation if competition forces him to charge a price so low that he makes no profit on any deal? His integrity in repairing your car must come at a price—you have to be willing to let him charge you a little bit more than the rates that the cheapest garage in the area might advertise.

The same reasoning also explains why, when you are away from home, you might settle for the known quality of a restaurant chain instead of taking the risk of going to an unknown local restaurant. And a department store that expands into a new line of merchandise can use the reputation that it has acquired in its existing lines to promise its customers the same high quality in the new line.

In games where credible promises by one or both parties can bring mutual benefit, the players can agree and even cooperate in fostering the development of reputation mechanisms. But if the interaction ends at a known finite time, there is always the problem of the endgame.

In the Middle East peace process that started in 1993 with the Oslo Accord, the early steps, in which Israel transferred some control over Gaza and small isolated areas of the West Bank to the Palestinian Authority and in which the latter accepted the existence of Israel and reduced its anti-Israel rhetoric violence, continued well for a while. But as the final stages of the process approached, mutual credibility of the next steps became problematic, and by 1998 the process stalled. Sufficiently attractive rewards could have come from the outside; for example, the United States or Europe could have given both parties contingent offers of economic aid or prospects of expanded commerce to keep the process going. The United States offered Egypt and Israel large amounts of aid in this way to achieve the Camp David Accords in 1978. But such rewards were not offered in the more recent situation and, at the date of this writing, prospects for progress do not look bright.

II. DIVIDING THE GAME INTO SMALL STEPS Sometimes a single game can be divided into a sequence of smaller games, thereby allowing the reputation mechanism to come into effect. In home-construction projects, it is customary to pay by installments as the work progresses. In the Middle East peace process, Israel would never have agreed to a complete transfer of the West Bank to the Palestinian Authority in one fell swoop in return for a single promise to recognize Israel and cease the terrorism. Proceeding in steps has enabled the process to go at least part of the way. But this again illustrates the difficulty of sustaining momentum as the endgame approaches.

III. TEAMWORK Teamwork is yet another way to embed one game into a larger game to enhance the credibility of strategic moves. It requires a group of players to monitor each another. If one fails to carry out a threat or a promise, others are required to inflict punishment on him; failure to do so makes them in turn vulnerable to similar punishment by others, and so on. Thus a player's payoffs in the larger game are altered in a way that makes adhering to the team's creed credible.

Many universities have academic honor codes that act as credibility devices for students. Examinations are not proctored by the faculty; instead, students are required to report to a student committee if they see any cheating. Then the committee holds a hearing and hands out punishment, as severe as suspension for a year or outright expulsion, if it finds the accused student guilty of cheating. Students are very reluctant to place their fellow students in such jeopardy. To stiffen their resolve, such codes include the added twist that failure to report an observed infraction is itself an offense against the code. Even then, the general belief is that the system works only imperfectly. A poll conducted at Princeton University last year found that only a third of students said that they would report an observed infraction, especially if they knew the guilty person.

IV. IRRATIONALITY Your threat may lack credibility because the other player knows that you are rational and that it is too costly for you to follow through with your threatened action. Therefore others believe you will not carry out the threatened action if you are put to the test. You can counter this problem by claiming to be irrational so that others will believe that your payoffs are different from what they originally perceived. Apparent irrationality can then turn into strategic rationality when the credibility of a threat is in question. Similarly, apparently irrational motives such as honor or saving face may make it credible that you will deliver on a promise even when tempted to renege.

The other player may see through such **rational irrationality.** Therefore if you attempt to make your threat credible by claiming irrationality, he will not readily believe you. You will have to acquire a reputation for irrationality, for example, by acting irrationally in some related game. You could also use one of the strategies discussed in Chapter 9 and do something that is a credible signal of irrationality to achieve an equilibrium in which you can separate from the falsely irrational.

V. CONTRACTS You can make it costly to yourself to fail to carry out a threat or to deliver on a promise by signing a **contract** under which you have to pay a sufficiently large sum in that eventuality. If such a contract is written with sufficient clarity that it can be enforced by a court or some outside authority, the change in payoffs makes it optimal to carry out the stipulated action, and the threat or the promise becomes credible.

In regard to a promise, the other player can be the other party to the contract. It is in his interest that you deliver on the promise, so he will hold you to the contract if you fail to fulfill the promise. A contract to enforce a threat is more problematic. The other player does not want you to carry out the threatened action and will not enforce the contract unless he gets some longer-term benefit in associated games from being subject to a credible threat in this one. Therefore in regard to a threat, the contract has to be with a third party. But when you bring in a third party and a contract merely to ensure that you will carry out your threat if put to the test, the third party does not actually benefit from your failure to act as stipulated. The contract thus becomes vulnerable to any renegotiation that would provide the third-party enforcer with some positive benefits. If the other player puts you to the test, you can say to the third party, "Look, I don't want to carry out the threat. But I am being forced to do so by the prospect of the penalty in the contract, and you are not getting anything out of all this. Here is a real dollar in exchange for releasing me from the contract." Thus the contract itself is not credible; therefore neither is the threat. The third party must have its own longer-term reasons for holding you to the contract, such as wanting to maintain its reputation, if the contract is to be renegotiation proof and therefore credible.

Written contracts are usually more binding than verbal ones, but even verbal ones may constitute commitments. When George H. W. Bush said, "Read my lips; no new taxes," in the presidential campaign of 1988, the American public took this promise to be a binding contract; when Bush reneged on it in 1990, the public held that against him in the election of 1992.

VI. BRINKMANSHIP In the U.S.–Japan trade-policy game, we found that a threat might be too "large" to be credible. If a smaller but effective threat cannot be found in a natural way, the size of the large threat can be reduced to a credible level by making its fulfillment a matter of chance. The United States cannot credibly say to Japan, "If you don't keep your markets open to U.S. goods, we will not defend you if the Russians or the Chinese attack you." But it can credibly say, "If you don't keep your markets open to U.S. goods, the relations between our countries will deteriorate, which will create the risk that, if you are faced with an invasion, Congress at that time will not sanction U.S. military involvement in your aid." As mentioned earlier, such deliberate creation of risk is called brinkmanship. This is a subtle idea, difficult to put into practice. Brinkmanship is best understood by seeing it in operation, and the detailed case study of the Cuban missile crisis in Chapter 15 serves just that purpose.

We have described several devices for making one's strategic moves credible and examined how well they work. In conclusion, we want to emphasize a feature common to the entire discussion. Credibility in practice is not an all-or-nothing matter but one of degree. Even though the theory is stark—rollback analysis shows either that a threat works or that it does not—practical application must recognize that between these polar extremes lies a whole spectrum of possibility and probability.

7 COUNTERING YOUR OPPONENT'S STRATEGIC MOVES

If your opponent can make a commitment or a threat that works to your disadvantage, then, before he actually does so, you may be able to make a strategic countermove of your own. You can do so by making his future strategic move less effective, for example, by removing its irreversibility or undermining its credibility. In this section, we examine some devices that can help achieve this purpose. Some are similar to devices that the other side can use for its own needs.

A. Irrationality

Irrationality can work for the would-be receiver of a commitment or a threat just as well as it does for the other player. If you are known to be so irrational that you will not give in to any threat and will suffer the damage that befalls you when your opponent carries out that threat, then he may as well not make the threat in the first place, because having to carry it out will only end up hurting him, too. Everything that we said earlier about the difficulties of credibly convincing the other side of your irrationality holds true here as well.

B. Cutting Off Communication

If you make it impossible for the other side to convey to you the message that it has made a certain commitment or a threat, then your opponent will see no point in doing so. Thomas Schelling illustrates this possibility with the story of a child who is crying too loudly to hear his parent's threats.[6] Thus it is pointless

[6]Thomas C. Schelling, *The Strategy of Conflict* (Oxford: Oxford University Press, 1960), p. 146.

for the parent to make any strategic moves; communication has effectively been cut off.

C. Leaving Escape Routes Open

If the other side can benefit by burning bridges to prevent its retreat, you can benefit by dousing those fires or perhaps even by constructing new bridges or roads by which your opponent can retreat. This device was also known to the ancients. Sun Tzu said, "To a surrounded enemy, you must leave a way of escape." The intent is not actually to allow the enemy to escape. Rather, "show him there is a road to safety, and so create in his mind the idea that there is an alternative to death. Then strike."[7]

D. Undermining Your Opponent's Motive to Uphold His Reputation

If the person threatening you says, "Look, I don't want to carry out this threat, but I must because I want to maintain my reputation with others," you can respond, "It is not in my interest to publicize the fact that you did not punish me. I am only interested in doing well in this game. I will keep quiet; both of us will avoid the mutually damaging outcome; and your reputation with others will stay intact." Similarly, if you are a buyer bargaining with a seller and he refuses to lower his price on the grounds that "if I do this for you, I would have to do it for everyone else," you can point out that you are not going to tell anyone else. This may not work; the other player may suspect that you would tell a few friends who would tell a few others, and so on.

E. Salami Tactics

Salami tactics are devices used to whittle down the other player's threat in the way that a salami is cut—one slice at a time. You fail to comply with the other's wishes (whether for deterrence or compellence) to a very small degree so that it is not worth the other's while to carry out the comparatively more drastic and mutually harmful threatened action just to counter that small transgression. If that works, you transgress a little more, and a little more again, and so on.

You know this perfectly well from your own childhood. Schelling[8] gives a wonderful description of the process:

> Salami tactics, we can be sure, were invented by a child. . . . Tell a child not to go in the water and he'll sit on the bank and submerge his bare feet; he is not yet "in" the water. Acquiesce, and he'll stand up; no more of him is in the

[7]Sun Tzu, *The Art of War,* pp. 109–110.

[8]Thomas C. Schelling, *Arms and Influence* (New Haven: Yale University Press, 1966), pp. 66–67.

> water than before. Think it over, and he'll start wading, not going any deeper. Take a moment to decide whether this is different and he'll go a little deeper, arguing that since he goes back and forth it all averages out. Pretty soon we are calling to him not to swim out of sight, wondering whatever happened to all our discipline.

Salami tactics work particularly well against compellence, because they can take advantage of the *time* dimension. When your mother tells you to clean up your room "or else," you can put off the task for an extra hour by claiming that you have to finish your homework, then for a half day because you have to go to football practice, then for an evening because you can't possibly miss *The Simpsons* on TV, and so on.

To counter the countermove of salami tactics you must make a correspondingly graduated threat. There should be a scale of punishments that fits the scale of noncompliance or procrastination. This can also be achieved by gradually raising the risk of disaster, another application of brinkmanship.

SUMMARY

Actions taken by players to fix the rules of later play are known as *strategic moves*. These first moves must be *observable* and *irreversible* to be true first moves, and they must be credible if they are to have their desired effect of altering the equilibrium outcome of the game. *Commitment* is an unconditional first move used to seize a first-mover advantage when one exists. Such a move usually entails committing to a strategy that would not have been one's equilibrium strategy in the original version of the game.

Conditional first moves such as *threats* and *promises* are *response rules* designed either to *deter* rivals' actions and preserve the status quo or to *compel* rivals' actions and alter the status quo. Threats carry the possibility of mutual harm but cost nothing if they work; threats that create only the risk of a bad outcome fall under the classification of *brinkmanship*. Promises are costly only to the maker and only if they are successful. Threats can be arbitrarily large, although excessive size compromises credibility, but promises are usually kept just large enough to be effective. If the implicit promise (or threat) that accompanies a threat (or promise) is not credible, players must make a move that combines both a promise and a threat and see to it that both components are credible.

Credibility must be established for any strategic move. There are a number of general principles to consider in making moves credible and a number of specific devices that can be used to acquire credibility. They generally work either by reducing your own future freedom to choose or by altering your own payoffs from future actions. Specific devices of this kind include establishing a *reputation,* using teamwork, demonstrating apparent irrationality, burning bridges, and making *contracts,* although the acquisition of credibility is often context specific. Similar devices exist for countering strategic moves made by rival players.

KEY TERMS

brinkmanship (363)
commitment (364)
compellence (365)
contract (387)
deterrence (365)
doomsday device (382)
irreversible (363)
observable (363)
promise (365)
rational irrationality (386)
reputation (384)
response rule (365)
salami tactics (377)
strategic moves (389)
threat (365)

SOLVED EXERCISES

S1. "One could argue that the size of a promise is naturally bounded, while in principle a threat can be arbitrarily severe so long as it is credible (and error free)." First, briefly explain why the statement is true. Despite the truth of the statement, players might find that an arbitrarily severe threat might not be to their advantage. Explain why the latter statement is also true.

S2. For each of the following three games, answer these questions:

(a) What is the equilibrium if neither player can use any strategic moves?

(b) Can one player improve his payoff by using a strategic move (commitment, threat, or promise) or a combination of such moves? If so, which player makes what strategic move(s)?

(i)

		COLUMN	
		Left	Right
ROW	Up	0, 0	2, 1
	Down	1, 2	0, 0

(ii)

		COLUMN	
		Left	Right
ROW	Up	4, 3	3, 4
	Down	2, 1	1, 2

(iii)

		COLUMN	
		Left	Right
ROW	Up	4, 1	2, 2
	Down	3, 3	1, 4

S3. In the classic film *Mary Poppins,* the Banks children are players in a strategic game with a number of different nannies. In their view of the world, nannies are inherently harsh, and playing tricks on nannies is great fun. That is, they view themselves as playing a game in which the nanny moves first, showing herself to be either Harsh or Nice, and the children move second, choosing to be either Good or Mischievous. The nanny prefers to have Good children to take care of but is also inherently harsh, and so she gets her highest payoff of 4 from (Harsh, Good) and her lowest payoff of 1 from (Nice, Mischievous), with (Nice, Good) yielding 3 and (Harsh, Mischievous) yielding 2. The children similarly most prefer to have a Nice nanny and then to be Mischievous; they get their highest two payoffs when the nanny is Nice (4 if Mischievous, 3 if Good) and their lowest two payoffs when the nanny is Harsh (2 if Mischievous, 1 if Good).

(a) Draw the game tree for this game and find the subgame-perfect equilibrium in the absence of any strategic moves.

(b) In the film, before the arrival of Mary Poppins, the children write their own ad for a new nanny in which they state: "If you won't scold and dominate us, we will never give you cause to hate us; we won't hide your spectacles so you can't see, put toads in your bed, or pepper in your tea." Use the tree from part **(a)** to argue that this statement constitutes a promise. What would the outcome of the game be if the children keep their promise?

(c) What is the implied threat that goes with the promise in part **(b)**? Is that implied threat automatically credible? Explain your answer.

(d) How could the children make the promise in part **(b)** credible?

(e) Is the promise in part (b) compellent or deterrent? Explain your answer by referring to the status quo in the game—namely, what would happen in the absence of the strategic move.

S4. The following is an interpretation of the rivalry between the United States and the Soviet Union for geopolitical influence during the 1970s and 1980s.[9] Each side has the choice of two strategies: Aggressive and Restrained. The Soviet Union wants to achieve world domination, so being Aggressive is its dominant strategy. The United States wants to prevent the Soviet Union from achieving world domination; it will match Soviet aggressiveness with aggressiveness, and restraint with restraint. Specifically, the payoff table is:

		SOVIET UNION	
		Restrained	Aggressive
UNITED STATES	Restrained	4, 3	1, 4
	Aggressive	3, 1	2, 2

For each player, 4 is best and 1 is worst.

(a) Consider this game when the two countries move simultaneously. Find the Nash equilibrium.

(b) Next consider three different and alternative ways in which the game could be played with sequential moves: (i) The United States moves first, and the Soviet Union moves second. (ii) The Soviet Union moves first, and the United States moves second. (iii) The Soviet Union moves first, and the United States moves second, but the Soviet Union has a further move in which it can change its first move. For each case, draw the game tree and find the subgame-perfect equilibrium.

(c) What are the key strategic matters (commitment, credibility, and so on) for the two countries?

S5. Consider the following games. In each case, (i) identify which player can benefit from making a strategic move, (ii) identify the nature of the strategic move appropriate for this purpose, (iii) discuss the conceptual and practical difficulties that will arise in the process of making this move credible, and (iv) discuss whether and how the difficulties can be overcome.

[9]We thank political science professor Thomas Schwartz at UCLA for the idea for this exercise.

(a) The other countries of the European Monetary Union (France, Germany, and so on) would like Britain to join the common currency and the common central bank.

(b) The United States would like North Korea to stop exporting missiles and missile technology to countries such as Iran and would like China to join the United States in working toward this aim.

(c) The United Auto Workers would like U.S. auto manufacturers not to build plants in Mexico and would like the U.S. government to restrict imports of autos made abroad.

UNSOLVED EXERCISES

U1. In a scene from the movie *Manhattan Murder Mystery,* Woody Allen and Diane Keaton are at a hockey game in Madison Square Garden. She is obviously not enjoying herself, but he tells her: "Remember our deal. You stay here with me for the entire hockey game, and next week I will come to the opera with you and stay until the end." Later, we see them coming out of the Met into the deserted Lincoln Center Plaza while inside the music is still playing. Keaton is visibly upset: "What about our deal? I stayed to the end of the hockey game, and so you were supposed to stay till the end of the opera." Allen answers: "You know I can't listen to too much Wagner. At the end of the first act, I already felt the urge to invade Poland." Comment on the strategic choices made here by using your knowledge of the theory of strategic moves and credibility.

U2. Consider a game between a parent and a child. The child can choose to be good (G) or bad (B); the parent can punish the child (P) or not (N). The child gets enjoyment worth a 1 from bad behavior, but hurt worth -2 from punishment. Thus a child who behaves well and is not punished gets a 0; one who behaves badly and is punished gets $1 - 2 = -1$; and so on. The parent gets -2 from the child's bad behavior and -1 from inflicting punishment.

(a) Set up this game as a simultaneous-move game, and find the equilibrium.

(b) Next, suppose that the child chooses G or B first and that the parent chooses its P or N after having observed the child's action. Draw the game tree and find the subgame-perfect equilibrium.

(c) Now suppose that before the child acts, the parent can commit to a strategy. For example, the threat "P if B" ("If you behave badly, I will punish you"). How many such strategies does the parent have? Write the table for this game. Find all pure-strategy Nash equilibria.

(d) How do your answers to parts (b) and (c) differ? Explain the reason for the difference.

U3. The general strategic game in Thucydides' history of the Peloponnesian War has been expressed in game-theoretic terms by Professor William Charron of St. Louis University.[10] Athens had acquired a large empire of coastal cities around the Aegean as part of its leadership role in defending the Greek world from Persian invasions. Sparta, fearing Athenian power, was contemplating war against Athens. If Sparta decided against war, Athens would have to decide whether to retain or relinquish its empire. But Athens in turn feared that if it gave independence to the cities, they could choose to join Sparta in a greatly strengthened alliance against Athens and receive very favorable terms from Sparta for doing so. Thus there are three players, Sparta, Athens, and Small cities, who move in this order. There are four outcomes, and the payoffs are as follows (4 being best):

Outcome	Sparta	Athens	Small cities
War	2	2	2
Athens retains empire	1	4	1
Small cities join Sparta	4	1	4
Small cities stay independent	3	3	3

(a) Draw the game tree and find the rollback equilibrium. Is there another outcome that is better for all players?

(b) What strategic move or moves could attain the better outcome? Discuss the credibility of such moves.

U4. It is possible to reconfigure the payoffs in the game in Exercise S3 so that the children's statement in their ad is a threat, rather than a promise.

(a) Redraw the tree from part (a) of Exercise S3 and fill in payoffs for both players so that the children's statement becomes a *threat* in the full technical sense.

(b) Define the status quo in your game, and determine whether the threat is deterrent or compellent.

(c) Explain why the threatened action is not automatically credible, given your payoff structure.

(d) Explain why the implied promise *is* automatically credible.

[10] William C. Charron, "Greeks and Games: Forerunners of Modern Game Theory," *Forum for Social Economics*, vol. 29, no. 2 (Spring 2000), pp. 1–32.

(e) Explain why the children would want to make a threat in the first place, and suggest a way in which they might make their threatened action credible.

U5. Answer the questions in Exercise S5 for the following situations:

(a) The students at your university or college want to prevent the administration from raising tuition.

(b) Most participants, as well as outsiders, want to achieve a durable peace in Afghanistan, Iraq, Israel, and Palestine.

(c) Nearly all nations of the world want Iran to shut down its nuclear program.

U6. Write a brief description of a game in which you have participated, entailing strategic moves such as a commitment, threat, or promise and paying special attention to the essential aspect of credibility. Provide an illustration of the game if possible, and explain why the game that you describe ended as it did. Did the players use sound strategic thinking in making their choices?

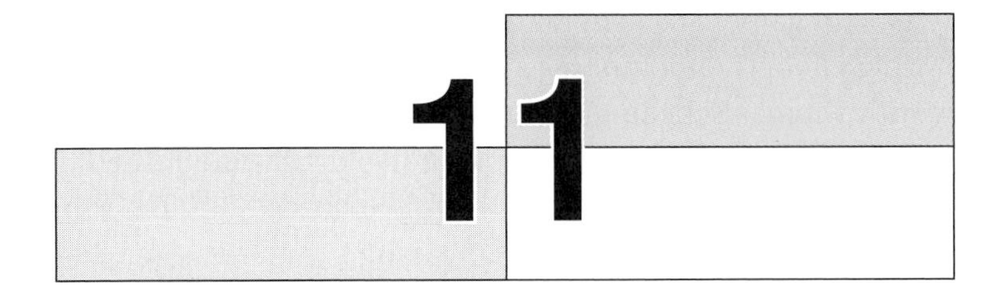

The Prisoners' Dilemma and Repeated Games

IN THIS CHAPTER, we continue our study of broad classes of games with an analysis of the prisoners' dilemma game. It is probably *the* classic example of the theory of strategy and its implications for predicting the behavior of game players, and most people who learn only a little bit of game theory learn about it. Even people who know *no* game theory may know the basic story behind this game or they may have at least heard that it exists. The prisoners' dilemma is a game in which each player has a dominant strategy, but the equilibrium that arises when all players use their dominant strategies provides a worse outcome for every player than would arise if they all used their dominated strategies instead. The paradoxical nature of this equilibrium outcome leads to several more complex questions about the nature of the interactions that only a more thorough analysis can hope to answer. The purpose of this chapter is to provide that additional thoroughness.

We already considered the prisoners' dilemma in Section 3 of Chapter 4. There we took note of the curious nature of the equilibrium that is actually a "bad" outcome for the players. The "prisoners" can find another outcome that both prefer to the equilibrium outcome, but they find it difficult to bring about. The focus of this chapter is the potential for achieving that better outcome. That is, we consider whether and how the players in a prisoners' dilemma can attain and sustain their mutually beneficial cooperative outcome, overcoming their separate incentives to defect for individual gain. We first review the standard prisoners' dilemma game and then develop four categories of solutions. The first and most important method of solution consists of repetition of the standard one-shot game. The general theory of repeated games was the contribution for which Robert Aumann was awarded the 2005 Nobel Prize in Economics (jointly with Thomas Schelling). As usual at this introductory level, we look at a few simple examples of this general theory. Two other potential solutions rely on penalty (or reward) schemes and on the role of leadership. The fourth incorporates asymmetric information into a finitely repeated dilemma game. As we consider each potential solution, the importance of the costs of defecting and the benefits of cooperation will become clear.

This chapter concludes with a discussion of some of the experimental evidence regarding the prisoners' dilemma as well as several examples of actual dilemmas in action. Experiments generally put live players in a variety of prisoners' dilemma–type games and show some perplexing as well as some more predictable behavior; experiments conducted with the use of computer simulations yield additional interesting outcomes. Our examples of real-world dilemmas that end the chapter are provided to give a sense of the diversity of situations in which prisoners' dilemmas arise and to show how, in at least one case, players may be able to create their own solution to the dilemma.

1 THE BASIC GAME (REVIEW)

Before we consider methods for avoiding the "bad" outcome in the prisoners' dilemma, we briefly review the basics of the game. Recall our example from Chapter 4 of the husband and wife suspected of murder. Each is interrogated separately and can choose to confess to the crime or to deny any involvement. The payoff matrix that they face was originally presented as Figure 4.4 and is reproduced here as Figure 11.1. The numbers shown indicate years in jail; therefore low numbers are better for both players.

Both players here have a dominant strategy. Each does better to confess, regardless of what the other player does. The equilibrium outcome entails both players deciding to confess and each getting 10 years in jail. If they both had

		WIFE	
		Confess (Defect)	Deny (Cooperate)
HUSBAND	Confess (Defect)	10 yr, 10 yr	1 yr, 25 yr
	Deny (Cooperate)	25 yr, 1 yr	3 yr, 3 yr

FIGURE 11.1 Payoffs for the Standard Prisoners' Dilemma

chosen to deny any involvement, however, they would have been better off, with only 3 years of jail time to serve.

In any prisoners' dilemma game, there is always a *cooperative strategy* and a *cheating* or *defecting strategy*. In Figure 11.1, Deny is the cooperative strategy; both players using that strategy yields the best outcome for the players. Confess is the cheating or defecting strategy; when the players do not cooperate with one another, they choose to Confess in the hope of attaining individual gain at the rival's expense. Thus, players in a prisoners' dilemma can always be labeled, according to their choice of strategy, as either *defectors* or *cooperators*. We will use this labeling system throughout the discussion of potential solutions to the dilemma.

We want to emphasize that, although we speak of a cooperative strategy, the prisoners' dilemma game is noncooperative in the sense explained in Chapter 2—namely, the players make their decisions and implement their choices individually. If the two players could discuss, choose, and play their strategies jointly—as if, for example, the prisoners were in the same room and could give a joint answer to the question of whether they were both going to confess—there would be no difficulty about their achieving the outcome that both prefer. The essence of the questions of whether, when, and how a prisoners' dilemma can be resolved is the difficulty of achieving a cooperative (jointly preferred) outcome through noncooperative (individual) actions.

2 SOLUTIONS I: REPETITION

Of all the mechanisms that can sustain cooperation in the prisoners' dilemma, the best known and the most natural is **repeated play** of the game. Repeated or ongoing relationships between players imply special characteristics for the games that they play against one another. In the prisoners' dilemma, this result plays out in the fact that each player fears that one instance of defecting will lead to a collapse of cooperation in the future. If the value of future cooperation is large and exceeds what can be gained in the short term by defecting, then the long-term individual interests of the players can automatically and tacitly keep them from defecting, without the need for any additional punishments or enforcement by third parties.

We consider the meal-pricing dilemma faced by the two restaurants, Xavier's Tapas and Yvonne's Bistro, introduced in Chapter 5. For our purposes here, we have chosen to simplify that game by supposing that only two choices of price are available: the jointly best (collusive) price of $26 or the Nash equilibrium price of $20. The payoffs (profits measured in hundreds of dollars per month) for each restaurant can be calculated by using the quantity (demand) functions in Section 1.A of Chapter 5; these payoffs are shown in Figure 11.2. As in any

		YVONNE'S BISTRO	
		20 (Defect)	26 (Cooperate)
XAVIER'S TAPAS	20 (Defect)	288, 288	360, 216
	26 (Cooperate)	216, 360	324, 324

FIGURE 11.2 Prisoners' Dilemma of Pricing ($100s per month)

prisoners' dilemma, each store has a dominant strategy to defect and price its meals at $20, although both stores would prefer the outcome in which each cooperates and charges the higher price of $26 per meal.

Let us start our analysis by supposing that the two restaurants are initially in the cooperative mode, each charging the higher price of $26. If one restaurant—say, Xavier's—deviates from this pricing strategy, it can increase its profit from 324 to 360 (from $32,400 to $36,000) for one month. But then cooperation has dissolved and Xavier's rival, Yvonne's, will see no reason to cooperate from then on. Once cooperation has broken down, presumably permanently, the profit for Xavier's is 288 each month instead of the 324 it would have been if Xavier's had never defected in the first place. By gaining 36 ($3,600) in one month of defecting, Xavier's gives up 36 ($3,600) each month thereafter by destroying cooperation. Even if the relationship lasts as little as three months, it seems that defecting is not in Xavier's best interest. A similar argument can be made for Yvonne's. Thus, if the two restaurants competed on a regular basis for at least three months, it seems that we might see cooperative behavior and high prices rather than the defecting behavior and low prices predicted by theory for the one-shot game.

A. Finite Repetition

But the solution of the dilemma is not actually that simple. What if the relationship did last exactly three months? Then strategic restaurants would want to analyze the full three-month game and choose their optimal pricing strategies. Each would use rollback to determine what price to charge each month. Starting their analyses with the third month, they would realize that, at that point, there was no future relationship to consider. Each restaurant would find that it had a dominant strategy to defect. Given that, there is effectively no future to consider in the second month either. Each player knows that there will be mutual defecting in the third month, and therefore both will defect in the second month; defecting is the dominant strategy in month 2 also. Then the same argument applies to the first month as well. Knowing that both will defect in months 2 and 3 anyway, there is no future value of cooperation in the first month. Both players defect right from the start, and the dilemma is alive and well.

This result is very general. As long as the relationship between the two players in a prisoners' dilemma game lasts a fixed and known length of time, the dominant-strategy equilibrium with defecting should prevail in the last period of play. When the players arrive at the end of the game, there is never any value to continued cooperation, and so they defect. Then rollback predicts mutual defecting all the way back to the very first play. However, in practice, players in finitely repeated prisoners' dilemma games show a lot of cooperation; more on this to come.

B. Infinite Repetition

Analysis of the finitely repeated prisoners' dilemma shows that even repetition of the game cannot guarantee the players a solution to their dilemma. But what would happen if the relationship did not have a predetermined length? What if the two restaurants expected to continue competing with one another indefinitely? Then our analysis must change to incorporate this new aspect of their interaction, and we will see that the incentives of the players change also.

In repeated games of any kind, the sequential nature of the relationship means that players can adopt strategies that depend on behavior in preceding plays of the games. Such strategies are known as **contingent strategies,** and several specific examples are used frequently in the theory of repeated games. Most contingent strategies are **trigger strategies.** A player using a trigger strategy plays cooperatively as long as her rival(s) do so, but any defection on their part "triggers" a period of **punishment,** of specified length, in which she plays noncooperatively in response. Two of the best-known trigger strategies are the grim strategy and tit-for-tat. The **grim strategy** entails cooperating with your rival until such time as she defects from cooperation; once a defection has occurred, you punish your rival (by choosing the Defect strategy) on every play for the rest of the game.[1] **Tit-for-tat (TFT)** is not so harshly unforgiving as the grim strategy and is famous (or infamous) for its ability to solve the prisoners' dilemma without requiring permanent punishment. Playing TFT means choosing, in any specified period of play, the action chosen by your rival in the preceding period of play. Thus, when playing TFT, you cooperate with your rival if she cooperated during the most recent play of the game and defect (as punishment) if your rival defected. The punishment phase lasts only as long as your rival continues to defect; you will return to cooperation one period after she chooses to do so.

[1]Defecting as retaliation under the requirements of a trigger strategy is often termed *punishing* to distinguish it from the original decision to deviate from cooperation.

Let us consider how play might proceed in the repeated restaurant pricing game if one of the players uses the contingent strategy tit-for-tat. We have already seen that if Xavier's Tapas defects one month, it could add 36 to its profits (360 instead of 324). But if Xavier's rival is playing TFT, then such defecting would induce Yvonne's Bistro to punish Xavier's the next month in retaliation. At that point, Xavier's has two choices. One option is to continue to defect by pricing at $20, and to endure Yvonne's continued punishment according to TFT; in this case, Xavier's loses 36 (288 rather than 324) for every month thereafter in the foreseeable future. This option appears quite costly. But Xavier's *could* get back to cooperation, too, if it so desired. By reverting to the cooperative price of $26 after one month's defection, Xavier's would incur only one month's punishment from Yvonne's. During that month, Xavier's would suffer a loss in profit of 108 (216 rather than the 324 that would have been earned without any defection). In the second month after Xavier's defection, both restaurants could be back at the cooperative price earning 324 each month. This one-time defection yields an extra 36 in profit but costs an additional 108 during the punishment, also apparently quite costly to Xavier's.

It is important to realize here, however, that Xavier's extra $36 from defecting is gained in the first month. Its losses are ceded in the future. Therefore the relative importance of the two depends on the relative importance of the present versus the future. Here, because payoffs are calculated in dollar terms, an objective comparison can be made. Generally, money (or profit) that is earned today is better than money that is earned later because, even if you do not need (or want) the money until later, you can invest it now and earn a return on it until you need it. So Xavier's should be able to calculate whether it is worthwhile to defect, on the basis of the total rate of return on its investment (including capital gains and/or dividends and/or interest, depending on the type of investment). We use the symbol r to denote this rate of return. Thus one dollar invested generates r dollars of interest and/or dividends and/or capital gains, or 100 dollars generate $100r$, therefore the rate of return is sometimes also said to be $100r\%$.

Note that we can calculate whether it is in Xavier's interest to defect because the firms' payoffs are given in dollar terms, rather than as simple ratings of outcomes, as in some of the games in earlier chapters (the street-garden game in Chapters 3 and 6, for example). This means that payoff values in different cells are directly comparable; a payoff of 4 (dollars) is twice as good as a payoff of 2 (dollars) here, whereas a payoff of 4 is not necessarily exactly twice as good as a payoff of 2 in any two-by-two game in which the four possible outcomes are ranked from 1 (worst) to 4 (best). As long as the payoffs to the players are given in measurable units, we can calculate whether defecting in a prisoners' dilemma game is worthwhile.

I. IS IT WORTHWHILE TO DEFECT ONLY ONCE AGAINST A RIVAL PLAYING TFT? One of Xavier's options when playing repeatedly against a rival using TFT is to defect just once from a cooperative outcome and then to return to cooperating. This particular strategy gains the restaurant 36 in the first month (the month during which it defects) but loses it 108 in the second month. By the third month, cooperation is restored. Is defecting for only one month worth it?

We cannot directly compare the 36 gained in the first month with the 108 lost in the second month, because the additional money value of time must be incorporated into the calculation. That is, we need a way to determine how much the 108 lost in the second month is worth during the first month. Then we can compare that number with 36 to see whether defecting once is worthwhile. What we are looking for is the **present value (PV)** of 108, or how much in profit earned this month (in the present) is equivalent to (has the same value as) the 108 earned next month. We need to determine the number of dollars earned this month that, with interest, would give us 108 next month; we call that number PV, the present value of 108.

Given that the (monthly) total rate of return is r, getting PV this month and investing it until next month yields a total next month of PV + rPV, where the first term is the principal being paid back and the second term is the return (interest or dividend or capital gain). When the total is exactly 108, then PV equals the present value of 108. Setting PV + rPV = 108 yields a solution for PV:

$$\text{PV} = \frac{108}{1 + r}.$$

For any value of r, we can now determine the exact number of dollars that, earned this month, would be worth 108 next month.

From the perspective of Xavier's Tapas, the question remains whether the gain of 36 this month is offset by the loss of 108 next month. The answer depends on the value of PV. Xavier's must compare the gain of 36 with the PV of the loss of 108. To defect once (and then return to cooperation) is worthwhile only if $36 > 108/(1 + r)$. This is the same as saying that defecting once is beneficial only if $36(1 + r) > 108$, which reduces to $r > 2$. Thus Xavier's should choose to defect once against a rival playing TFT only if the monthly total rate of return exceeds 200%. This outcome is very unlikely; for example, prime lending rates rarely exceed 12% per year. This translates into a monthly interest rate of no more than 1% (compounded annually, not monthly), well below the 200% just calculated. Here, it is better for Xavier's to continue cooperating than to try a single instance of defecting when Yvonne's is playing TFT.

II. IS IT WORTHWHILE TO DEFECT FOREVER AGAINST A RIVAL PLAYING TFT? What about the possibility of defecting once and then continuing to defect forever? This second option of

Xavier's gains the restaurant 36 in the first month but loses it 36 in every month thereafter into the future if the rival restaurant plays TFT. To determine whether such a strategy is in Xavier's best interest again depends on the present value of the losses incurred. But this time the losses are incurred over an **infinite horizon** of future months of competition.

We need to figure out the present value of all of the 36s that are lost in future months, add them all up, and compare them with the 36 gained during the month of defecting. The PV of the 36 lost during the first month of punishment and continued defecting on Xavier's part is just $36/(1 + r)$; the calculation is identical with that used in Section 2.B.I to find that the PV of 108 was $108/(1 + r)$. For the next month, the PV must be the dollar amount needed this month that, with two months of **compound interest,** would yield 36 in two months. If the PV is invested now, then in one month the investor would have that principal amount plus a return of rPV, for a total of PV + rPV, as before; leaving this total amount invested for the second month means that at the end of two months, the investor has the amount invested at the beginning of the second month (PV + rPV) plus the return on that amount, which would be r(PV + rPV). The PV of the 36 lost two months from now must then solve the equation: $PV + rPV + r(PV + rPV) = 36$. Working out the value of PV here yields $PV(1 + r)^2 = 36$, or $PV = 36/(1 + r)^2$. You should see a pattern developing. The PV of the 36 lost in the third month of continued defecting is $36/(1 + r)^3$, and the PV of the 36 lost in the fourth month is $36/(1 + r)^4$. In fact, the PV of the 36 lost in the nth month of continued defecting is just $36/(1 + r)^n$. Xavier's loses an infinite sum of 36s, and the PV of each of them gets smaller each month.

More precisely, Xavier's loses the sum, from $n = 1$ to $n = \infty$ (where n labels the months of continued defecting after the initial month), of $36/(1 + r)^n$. Mathematically, it is written as the sum of an infinite number of terms:[2]

$$36/(1 + r) + 36/(1 + r)^2 + 36/(1 + r)^3 + 36/(1 + r)^4 + \cdots.$$

Because r is a rate of return and presumably a positive number, the ratio of $1/(1 + r)$ will be less than 1; this ratio is generally called the **discount factor** and is referred to by the Greek letter δ. With $\delta = 1/(1 + r) < 1$, the mathematical rule for infinite sums tells us that this sum converges to a specific value, in this case $36/r$.

It is now possible to determine whether Xavier's Tapas will choose to defect forever. The restaurant compares its gain of 36 with the PV of all the lost 36s, or $36/r$. Then it defects forever only if $36 > 36/r$, or $r > 1$; defecting forever is beneficial in this particular game only if the monthly rate of return exceeds 100%, an unlikely event. Thus we would not expect Xavier's to defect against a cooperative rival when both are playing tit-for-tat. When both Yvonne's Bistro and Xavier's

[2]The Appendix to this chapter contains a detailed discussion of the solution of infinite sums.

Tapas play TFT, the cooperative outcome in which both price high is a Nash equilibrium of the game. Both playing TFT is a Nash equilibrium, and use of this contingent strategy solves the prisoners' dilemma for the two restaurants.

Remember that tit-for-tat is only one of many trigger strategies that players could use in repeated prisoners' dilemmas. And it is one of the "nicer" ones. Thus if TFT can be used to solve the dilemma for the two restaurants, other, harsher trigger strategies should be able to do the same. The grim strategy, for instance, also can be used to sustain cooperation in this infinitely repeated game and in others.

C. Games of Unknown Length

In addition to considering games of finite or infinite length, we can incorporate a more sophisticated tool to deal with games of unknown length. It is possible that, in some repeated games, players might not know for certain exactly how long their interaction will continue. They may, however, have some idea of the *probability* that the game will continue for another period. For example, our restaurants might believe that their repeated competition will continue only as long as their customers find *prix fixe* menus to be the dining-out experience of choice; if there were some probability each month that *à la carte* dinners would take over that role, then the nature of the game is altered.

Recall that the present value of a loss next month is already worth only $\delta = 1/(1 + r)$ times the amount earned. If in addition there is only a probability p (less than 1) that the relationship will actually continue to the next month, then next month's loss is worth only p times δ times the amount lost. For Xavier's Tapas, this means that the PV of the 36 lost with continued defecting is worth $36 \times \delta$ [the same as $36/(1 + r)$] when the game is assumed to be continuing with certainty but is worth only $36 \times p \times \delta$ when the game is assumed to be continuing with probability p. Incorporating the probability that the game may end next period means that the present value of the lost 36 is smaller, because $p < 1$, than it is when the game is definitely expected to continue (when p is *assumed* to equal 1).

The effect of incorporating p is that we now effectively discount future payoffs by the factor $p \times \delta$ instead of simply by δ. We call this **effective rate of return** R, where $1/(1 + R) = p \times \delta$, and R depends on p and δ as shown:[3]

$$1/(1 + R) = p\delta$$

$$1 = p\delta(1 + R)$$

$$R = \frac{1 - p\delta}{p\delta}.$$

[3]We could also express R in terms of r and p, in which case $R = (1 + r)/p - 1$.

With a 5% actual rate of return on investments ($r = 0.05$, and so $\delta = 1/1.05 = 0.95$) and a 50% chance that the game continues for an additional month ($p = 0.5$), then $R = [1 - (0.5)(0.95)]/(0.5)(0.95) = 1.1$, or 110%.

Now the high rates of return required to destroy cooperation (encourage defection) in these examples seem more realistic if we interpret them as effective rather than actual rates of return. It becomes conceivable that defecting forever, or even once, might actually be to one's benefit if there is a large enough probability that the game will end in the near future. Consider Xavier's decision whether to defect forever against a TFT-playing rival. Our earlier calculations showed that permanent defecting is beneficial only when r exceeds 1, or 100%. If Xavier's faces the 5% actual rate of return and the 50% chance that the game will continue for an additional month, as we assumed in the preceding paragraph, then the effective rate of return of 110% will exceed the critical value needed for it to continue defecting. Thus the cooperative behavior sustained by the TFT strategy can break down if there is a sufficiently large chance that the repeated game might be over by the end of the next period of play—that is, by a sufficiently small value of p.

D. General Theory

We can easily generalize the ideas about when it is worthwhile to defect against TFT-playing rivals so that you can apply them to any prisoners' dilemma game that you encounter. To do so, we use a table with general payoffs (delineated in appropriately measurable units) that satisfy the standard structure of payoffs in the dilemma as in Figure 11.3. The payoffs in the table must satisfy the relation H > C > D > L for the game to be a prisoners' dilemma, where C is the *cooperative* outcome, D is the payoff when both players *defect* from cooperation, H is the *high* payoff that goes to the defector when one player defects while the other cooperates, and L is the *low* payoff that goes to the loser (the cooperator) in the same situation.

In this general version of the prisoners' dilemma, a player's one-time gain from defecting is $(H - C)$. The single-period loss for being punished while you return to cooperation is $(C - L)$, and the per-period loss for perpetual defect-

		COLUMN	
		Defect	Cooperate
ROW	Defect	*D*, *D*	*H*, *L*
	Cooperate	*L*, *H*	*C*, *C*

FIGURE 11.3 General Version of the Prisoners' Dilemma

ing is $(C - D)$. To be as general as possible, we will allow for situations in which there is a probability $p < 1$ that the game continues beyond the next period and so we will discount payoffs using an effective rate of return of R per period. If $p = 1$, as would be the case when the game is guaranteed to continue, then $R = r$, the simple interest rate used in our preceding calculations. Replacing r with R, we find that the results attained earlier generalize almost immediately.

We found earlier that a player defects exactly once against a rival playing TFT if the one-time gain from defecting $(H - C)$ exceeds the present value of the single-period loss from being punished (the PV of $C - L$). In this general game, that means that a player defects once against a TFT-playing opponent only if $(H - C) > (C - L)/(1 + R)$, or $(1 + R)(H - C) > C - L$, or

$$R > \frac{C - L}{H - C} - 1.$$

Similarly, we found that a player defects forever against a rival playing TFT only if the one-time gain from defecting exceeds the present value of the infinite sum of the per-period losses from perpetual defecting (where the per-period loss is $C - D$). For the general game, then, a player defects forever against a TFT-playing opponent only if $(H - C) > (C - D)/R$, or

$$R > \frac{C - D}{H - C}.$$

The three critical elements in a player's decision to defect, as seen in these two expressions, are the immediate gain from defection $(H - C)$, the future losses from punishment ($C - L$ or $C - D$ per period of punishment), and the value of the effective rate of return (R, which measures the importance of the present relative to the future). Under what conditions on these various values do players find it attractive to defect from cooperation?

First, assume that the values of the gains and losses from defecting are fixed. Then changes in R determine whether a player defects, and defection is more likely when R is large. Large values of R are associated with small values of p and small values of δ (and large values of r), so defection is more likely when the probability of continuation is low or the discount factor is low (or the interest rate is high). Another way to think about it is that defection is more likely when the future is less important than the present or when there is little future to consider; that is, defection is more likely when players are impatient or when they expect the game to end quickly.

Second, consider the case in which the effective rate of return is fixed, as is the one-period gain from defecting. Then changes in the per-period losses associated with punishment determine whether defecting is worthwhile. Here it is smaller values of $C - L$ or $C - D$ that encourage defection. In this case, defection is more likely when punishment is not very severe.[4]

Finally, assume that the effective rate of return and the per-period losses associated with punishment are held constant. Now players are more likely to defect when the gains, $H - C$, are high. This situation is more likely when defecting garners a player large and immediate benefits.

This discussion also highlights the importance of the detection of defecting. Decisions about whether to continue along a cooperative path depend on how long defecting might be able to go on before it is detected, on how accurately it is detected, and on how long any punishment can be made to last before an attempt is made to revert back to cooperation. Although our model does not incorporate these considerations explicitly, if defecting can be detected accurately and quickly, its benefit will not last long, and the subsequent cost will have to be paid more surely. Therefore the success of any trigger strategy in resolving a repeated prisoners' dilemma depends on how well (both in speed and accuracy) players can detect defecting. This is one reason that the TFT strategy is often considered dangerous; slight errors in the execution of actions or in the perception of those actions can send players into continuous rounds of punishment from which they may not be able to escape for a long time, until a slight error of the opposite kind occurs.

You can use all of these ideas to guide you in when to expect more cooperative behavior between rivals and when to expect more defecting and cutthroat actions. If times are bad and an entire industry is on the verge of collapse, for example, so that businesses feel that there is no future, competition may become fiercer (less cooperative behavior may be observed) than in normal times. Even if times are temporarily good but are not expected to last, firms may want to make a quick profit while they can, so cooperative behavior might again break down. Similarly, in an industry that emerges temporarily because of a quirk of fashion and is expected to collapse when fashion changes, we should expect less cooperation. Thus a particular beach resort might become the place to go, but all the hotels there will know that such a situation cannot last, and so they cannot afford to collude on pricing. If, on the other hand, the shifts in fashion are among products made by an unchanging group of companies in long-term relationships with each other, cooperation might persist. For example, even if all the children want cuddly bears one year and Power Ranger

[4]The costs associated with defection may also be smaller if information transmission is not perfect, as might be the case if there are many players, and so difficulties might arise in identifying the defector and in coordinating a punishment scheme. Similarly, gains from defection may be larger if rivals cannot identify a defection immediately.

action figures the next, collusion in pricing may occur if the same small group of manufacturers makes both items.

In Chapter 12, we will look in more detail at prisoners' dilemmas that arise in games with many players. We examine when and how players can overcome such dilemmas and achieve outcomes better for them all.

3 SOLUTIONS II: PENALTIES AND REWARDS

Although repetition is the major vehicle for the solution of the prisoners' dilemma, there are also several others that can be used to achieve this purpose. One of the simplest ways to avert the prisoners' dilemma in the one-shot version of the game is to inflict some direct **penalty** on the players when they defect. When the payoffs have been altered to incorporate the cost of the penalty, players may find that the dilemma has been resolved.[5]

Consider the husband–wife dilemma from Section 1. If only one player defects, the game's outcome entails one year in jail for the defector and 25 years for the cooperator. The defector, though, getting out of jail early, might find the cooperator's friends waiting outside the jail. The physical harm caused by those friends might be equivalent to an additional 20 years in jail. If so, and if the players account for the possibility of this harm, then the payoff structure of the original game has changed.

The "new" game, with the physical penalty included in the payoffs, is illustrated in Figure 11.4. With the additional 20 years in jail added to each player's sentence when one player confesses while the other denies, the game is completely different.

A search for dominant strategies in Figure 11.4 shows that there are none. A cell-by-cell check then shows that there are now two pure-strategy Nash equilibria. One of them is the (Confess, Confess) outcome; the other is the (Deny,

		WIFE	
		Confess	Deny
HUSBAND	Confess	10 yr, 10 yr	21 yr, 25 yr
	Deny	25 yr, 21 yr	3 yr, 3 yr

FIGURE 11.4 Prisoners' Dilemma with Penalty for the Lone Defector

[5]Note that we get the same type of outcome in the repeated-game case considered in Section 2.

Deny) outcome. Now each player finds that it is in his or her best interest to cooperate if the other is going to do so. The game has changed from being a prisoners' dilemma to an assurance game, which we studied in Chapter 4. Solving the new game requires selecting an equilibrium from the two that exist. One of them—the cooperative outcome—is clearly better than the other from the perspective of both players. Therefore it may be easy to sustain it as a focal point if some convergence of expectations can be achieved.

Notice that the penalty in this scenario is inflicted on a defector only when his or her rival does *not* defect. However, stricter penalties can be incorporated into the prisoners' dilemma, such as penalties for *any* confession. Such discipline typically must be imposed by a third party with some power over the two players, rather than by the other player's friends, because the friends would have little authority to penalize the first player when their associate also defects. If both prisoners are members of a special organization (such as a gang or a crime mafia) and the organization has a standing rule of never confessing to the police under penalty of extreme physical harm, the game changes again to the one illustrated in Figure 11.5.

Now the equivalent of an additional 20 years in jail is added to *all* payoffs associated with the Confess strategy. (Compare Figures 11.5 and 11.1.) In the new game, each player has a dominant strategy, as in the original game. The difference is that the change in the payoffs makes Deny the dominant strategy for each player. And (Deny, Deny) becomes the unique pure-strategy Nash equilibrium. The stricter penalty scheme achieved with third-party enforcement makes defecting so unattractive to players that the cooperative outcome becomes the new equilibrium of the game.

In larger prisoners' dilemma games, difficulties arise with the use of penalties. In particular, if there are many players and some uncertainty exists, penalty schemes may be more difficult to maintain. It becomes harder to decide whether actual defecting is taking place or it's just bad luck or a mistaken move. In addition, if there really is defecting, it is often difficult to determine the identity of the defector from among the larger group. And if the game is one shot, there is no opportunity in the future to correct a penalty that is too severe or to inflict a penalty once a defector has been identified. Thus penalties may be less

		WIFE	
		Confess	Deny
HUSBAND	Confess	30 yr, 30 yr	21 yr, 25 yr
	Deny	25 yr, 21 yr	3 yr, 3 yr

FIGURE 11.5 Prisoners' Dilemma with Penalty for Any Defecting

successful in large one-shot games than in the two-person game we consider here. We study prisoners' dilemmas with a large number of players in greater detail in Chapter 12.

A further interesting possibility arises when a prisoners' dilemma that has been solved with a penalty scheme is considered in the context of the larger society in which the game is played. It might be the case that, although the dilemma equilibrium outcome is bad for the players, it is actually good for the rest of society or for some subset of persons within the rest of society. If so, social or political pressures might arise to try to minimize the ability of players to break out of the dilemma. When third-party penalties are the solution to a prisoners' dilemma, as is the case with crime mafias that enforce a no-confession rule, for instance, society can come up with its own strategy to reduce the effectiveness of the penalty mechanism. The Federal Witness Protection Program is an example of a system that has been set up for just this purpose. The U.S. government removes the threat of penalty in return for confessions and testimony in court.

Similar situations can be seen in other prisoners' dilemmas, such as the pricing game between our two restaurants. The equilibrium there entailed both firms charging the low price of $20 even though they enjoy higher profits when charging the higher price of $26. Although the restaurants want to break out of this "bad" equilibrium—and we have already seen how the use of trigger strategies can help them do so—their customers are happier with the low price offered in the Nash equilibrium of the one-shot game. The customers then have an incentive to try to destroy the efficacy of any enforcement mechanism or solution process the restaurants might use. For example, because some firms facing prisoners' dilemma pricing games attempt to solve the dilemma through the use of a "meet the competition" or "price matching" campaign, customers might want to press for legislation banning such policies. We analyze the effects of such price-matching strategies in Section 7.B.

Just as a prisoners' dilemma can be resolved by penalizing defectors, it can also be resolved by rewarding cooperators. Because this solution is more difficult to implement in practice, we mention it only briefly.

The most important question is who is to pay the rewards. If it is a third party, that person or group must have sufficient interest of its own in the cooperation achieved by the prisoners to make it worth its while to pay out the rewards. A rare example of this occurred when the United States brokered the Camp David accords between Israel and Egypt by offering large promises of aid to both.

If the rewards are to be paid by the players themselves to each other, the trick is to make the rewards contingent (paid out only if the other player cooperates) and credible (guaranteed to be paid if the other player cooperates). Meeting these criteria requires an unusual arrangement; for example, the player making the promise should deposit the sum in advance in an escrow account held by an honorable and neutral third party, who will hand the sum over to the other player if she cooperates or return it to the promisor if the other defects. An end-of-chapter exercise shows how this type of arrangement can work.

4 SOLUTIONS III: LEADERSHIP

The third method of solution for the prisoners' dilemma pertains to situations in which one player takes on the role of leader in the interaction. In most examples of the prisoners' dilemma, the game is assumed to be symmetric. That is, all the players stand to lose (and gain) the same amount from defecting (and cooperation). However, in actual strategic situations, one player may be relatively "large" (a leader) and the other "small." If the size of the payoffs is unequal enough, so much of the harm from defecting may fall on the larger player that she acts cooperatively, even while knowing that the other will defect. Saudi Arabia, for example, played such a role as the "swing producer" in OPEC (Organization of Petroleum Exporting Countries) for many years; to keep oil prices high, it cut back on its output when one of the smaller producers, such as Libya, expanded.

As with the OPEC example, **leadership** tends to be observed more often in games between nations than in games between firms or individual persons. Thus our example for a game in which leadership may be used to solve the prisoners' dilemma is one played between countries. Imagine that the populations of two countries, Dorminica and Soporia, are threatened by a disease, Sudden Acute Narcoleptic Episodes (SANE). This disease strikes 1 person in every 2,000, or 0.05% of the population, and causes the victim to fall into a deep sleep state for a year.[6] There are no aftereffects of the disease, but the cost of a worker being removed from the economy for a year is $32,000. Each country has a population of 100 million workers, so the expected number of cases in each is 50,000 ($0.0005 \times 100{,}000{,}000$), and the expected cost of the disease is $1.6 billion to each ($50{,}000 \times 32{,}000$). The total expected cost of the disease worldwide—that is, in both Dorminica and Soporia—is then $3.2 billion.

Scientists are confident that a crash research program costing $2 billion will lead to a vaccine that is 100% effective. Comparing the cost of the research program with the worldwide cost of the disease shows that, from the perspective of the entire population, the research program is clearly worth pursuing. However, the government in each country must consider whether to fund the full research program on its own. They make this decision separately, but their

[6]Think of Rip Van Winkle or of Woody Allen in the movie *Sleeper*, but the duration is much shorter.

		SOPORIA	
		Research	No Research
DORMINICA	Research	−2, −2	−2, 0
	No Research	0, −2	−1.6, −1.6

FIGURE 11.6 Payoffs for Equal-Population SANE Research Game ($billions)

decisions affect the outcomes for both countries. Specifically, if only one government chooses to fund the research, the population of the other country can access the information and use the vaccine without cost. But each government's payoff depends only on the costs incurred by its own population.

The payoff matrix for the noncooperative game between Dorminica and Soporia is shown in Figure 11.6. Each country chooses from two strategies, Research and No Research; payoffs show the costs to the countries, in billions of dollars, of the various strategy combinations. It is straightforward to verify that this game is a prisoners' dilemma and that each country has a dominant strategy to do no research.

But now suppose that the populations of the two countries are unequal, with 150 million in Dorminica and 50 million in Soporia. Then, if no research is funded by either government, the cost to Dorminica of SANE will be $2.4 billion ($0.0005 \times 150{,}000{,}000 \times 32{,}000$) and the cost to Soporia will be $0.8 billion ($0.0005 \times 50{,}000{,}000 \times 32{,}000$). The payoff matrix changes to the one illustrated in Figure 11.7.

In this version of the game, No Research is still the dominant strategy for Soporia. But Dorminica's best response is now Research. What has happened to change Dorminica's choice of strategy? Clearly, the answer lies in the unequal distribution of the population in this revised version of the game. Dorminica now stands to suffer such a large portion of the total cost of the disease that it finds it worthwhile to do the research on its own. This is true even though Dorminica knows full well that Soporia is going to be a free rider and get a share of the full benefit of the research.

		SOPORIA	
		Research	No Research
DORMINICA	Research	−2, −2	−2, 0
	No Research	0, −2	−2.4, −0.8

FIGURE 11.7 Payoffs for Unequal-Population SANE Research Game ($billions)

The research game in Figure 11.7 is no longer a prisoners' dilemma. Here we see that the dilemma has, in a sense, been "solved" by the size asymmetry. The larger country chooses to take on a leadership role and provide the benefit for the whole world.

Situations of leadership in what would otherwise be prisoners' dilemma games are common in international diplomacy. The role of leader often falls naturally to the biggest or most well established of the players, a phenomenon labeled "the exploitation of the great by the small."[7] For many decades after World War II, for instance, the United States carried a disproportionate share of the expenditures of our defense alliances such as NATO and maintained a policy of relatively free international trade even when our partners, such as Japan and Europe, were much more protectionist. In such situations, it might be reasonable to suggest further that a large or well-established player may accept the role of leader because its own interests are closely tied to those of the players as a whole; if the large player makes up a substantial fraction of the whole group, such a convergence of interests would seem unmistakable. The large player would then be expected to act more cooperatively than might otherwise be the case.

5 SOLUTIONS IV: ASYMMETRIC INFORMATION

The final solution method we consider is one in which asymmetric information is introduced into a finitely repeated prisoners' dilemma. We saw in Section 2.A how an attempt to resolve the dilemma by repeated play would unravel by rollback reasoning if there were a fixed, finite number of plays. In actual play, however, even when players know exactly how long their interaction will last, they are able to sustain cooperation for quite a while; it unravels near the end when only a few rounds are left. When asked about their reasoning for cooperating in the early rounds, the players will usually say something such as, "I was willing to try and see if the other player was nice, and when this proved to be the case, I continued to cooperate until the time came to take advantage of the other's niceness." Of course the other player may not have been genuinely nice, but thinking along similar lines. As long as there is some chance that players in the dilemma are nice rather than selfish, it may pay even a selfish player to pretend to be nice. She can reap the higher payoffs from cooperation for a while and then also hope to exploit the gains from double crossing near the end of the sequence of plays. In this section, we will show how to explain such behavior more rigorously. If the above intuition suffices to satisfy your

[7]Mancur Olson, *The Logic of Collective Action* (Cambridge: Harvard University Press, 1965), p. 29.

curiosity about this solution, you can skip the rest of this section without loss of continuity.

A. General Expropriation Game

Note that this will be a game of asymmetric information. Players are of two types, selfish and nice. Each player knows his own type but not the type of the other player. Each is trying to infer the other's type from his actions. We solved such a game in Chapter 9, Section 5, where Fordor tried to infer Tudor's cost type from its choice of price. The same methods of analysis will work here, although the situation we have described above involves both players simultaneously trying to infer the other's type. Because the analysis of such a situation would get quite complicated, we will explain the ideas in a somewhat simpler example, in which only one player has the choice between being selfish and being nice. This type of game is sometimes called a one-person dilemma,[8] and is sometimes called a game of holdup or opportunism.[9]

Let us consider a specific situation in which a firm is deciding whether to invest in an emerging economy. The investment will entail an up-front cost of $1 billion, and will then yield an operating profit of $2 billion. It will also create spillover benefits to the country where the investment is located (the "host" country) of $500 million. We will show all monetary amounts in billions, so these payoff numbers will be –1, 2, and 0.5, respectively.

After the investment is made, the host country's government will be tempted to change the rules so that it can collect the whole profit of 2 (billion) in addition to the spillover benefit of 0.5 (billion). That is, it can leave things as they are, accepting its payoff of 0.5, or it can expropriate the full profits from the firm's investment, thereby gaining itself a payoff of 2.5. The game tree in Figure 11.8 shows the host country's choices as *E* (for expropriate) and *NE* (for not expropriate); it has the opportunity to make this choice only after the firm has chosen to invest (*I*) rather than not to invest (*NI*).

A host country could achieve the expropriation outcome by nationalizing the local operation without compensating the foreign investor. Such expropriation of foreign investment has occurred quite often in history but is relatively rare these days. More common are indirect and partial expropriations that use changes in tax rules, limits on repatriation of profits, and so on. To keep matters simple we assume here that the expropriation of profit from the investing firm

[8] The most notable use of this terminology occurs in the works of Avner Greif; see his book *Institutions and the Path to the Modern Economy: Lessons from Medieval Trade* (New York: Cambridge University Press, 2006).

[9] These concepts were developed and used by Oliver Williamson; see his book *The Economic Institutions of Capitalism* (New York: Free Press, 1987).

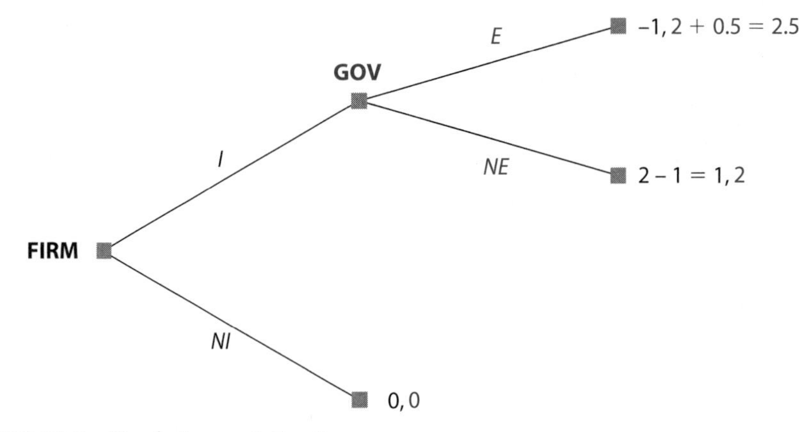

FIGURE 11.8 Simple Expropriation Game

by the host country is total. When applying the theory in other contexts, you will have to change the details to fit the specific situation.

In a single play of the game, then, rollback analysis of the game tree in Figure 11.8 shows that the host government will expropriate any available profits if the firm chooses to invest. Anticipating this choice, the firm will therefore choose not to invest. Similarly, when both players have full information about the other's possible and actual choices, the rollback equilibrium of the finitely repeated version of this game will entail no investment in any period; firms will not invest because they expect all profits to be expropriated. Just as in the finitely repeated prisoners' dilemma of Section 2.A, there will be no cooperation in equilibrium.

But what if we introduce an information asymmetry into this game? Specifically, assume that (host) governments come in two types, Opportunistic and Honorable, or O type and H type for short. Unable to distinguish the government's type, the firm must make its decision about whether to invest without knowing if the government is an O type or an H type. The former type of government will expropriate whenever that choice yields it a higher expected payoff than not expropriating; the latter type will never expropriate. Letting p denote the probability of the government being Honorable, we show the tree for the asymmetric information version of the expropriation game in Figure 11.9. There, in a single play of the game, the O type government will expropriate and the H type will not; therefore the firm's expected payoff from investing will be $2 \times p + 0 \times (1 - p) - 1 = 2p - 1$. The firm will invest if this expected payoff exceeds the expected payoff from not investing, 0; the firm will invest if $p > 1/2$.

Next, suppose the game is played repeatedly, but a fixed finite number of times and with no discounting across periods. The same host government will play in all periods, and its type will not change from one period to the next. Each period, a *new* firm gets the opportunity to make an investment. It observes whether firms invested in previous periods, and if so, whether the government

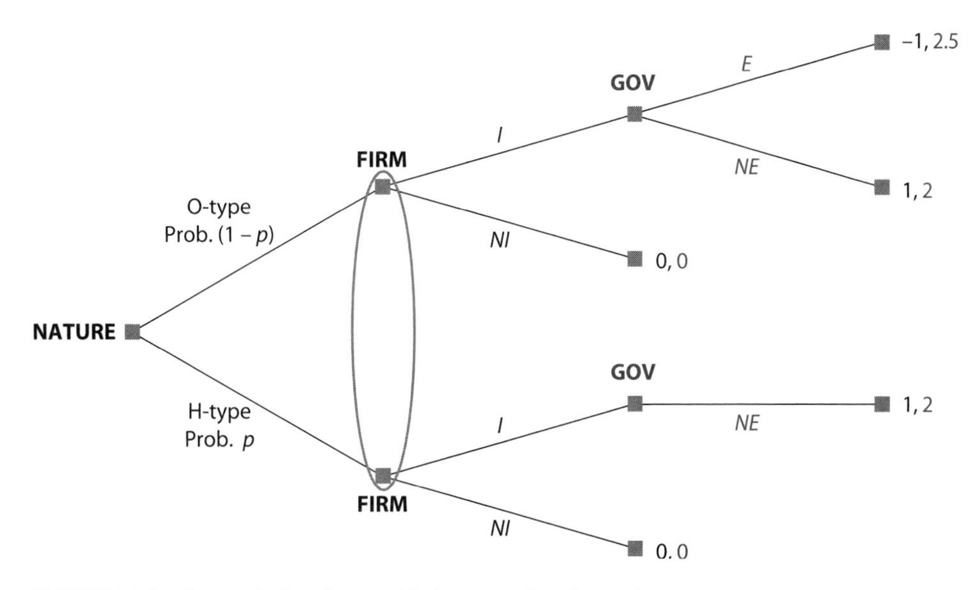

FIGURE 11.9 Expropriation Game with Asymmetric Information

expropriated. The prior probability held by the firm in the very first play of the game is that the government is H type with probability p.

In the repeated asymmetric information game, we will look for the following properties in the equilibrium:

1. Each new firm calculates an updated probability of the government being H type, using the previous period firm's prior belief along with the observed actions in that period and applying Bayes' rule. Its choice of whether to invest or not (*I* or *NI* for short) is optimal, given this updated probability.

2. The O-type government's decision whether to expropriate or not (*E* or *NE* for short) is optimal at all nodes in all periods, with the government recognizing the effect this choice will have on the probability calculations and actions of firms in future periods. An equilibrium that satisfies this properties will be a perfect Bayesian equilibrium (PBE) as defined in Chapter 9, Section 5.

B. Twice-Repeated Game with Asymmetric Information

Begin by considering the asymmetric information expropriation game when it is repeated for just two periods in total. To avoid confusion with our analysis of repeated games in Section 2.B, we use alphabetic, rather than numeric, labels here. The last period in actual time is labeled period Z; period Y is the one before that, and so on.

The firm's prior probability of the government being H type when entering the first period of play (period Y) is $p_Y = p$. Write p_Z for the prior probability of facing an H-type government when going into period Z (the second and final period of play). At this point, p_Z is just our notation for this probability; we will have to solve for its actual value as part of finding the equilibrium for the game.

We know already that in the period-Z game, the equilibrium actions for the government are to play *E* if it is O type and to play *NE* if it is H type. The firm plays *I* if $p_Z > 1/2$, or *NI* if $p_Z < 1/2$. It is indifferent between the two actions, and therefore willing to randomize between them if $p = 1/2$.

Now consider the period-Y game, which is the one played first in actual time. The possibilities for equilibrium play in that period will depend on the underlying value, p. We therefore distinguish three cases (the second of which will further subdivide) and consider each separately.

I. CASE I: $p > 1/2$ As in all situations, the H-type government will play *NE*. Given $p > 1/2$, the period-Y firm (firm Y) would play *I* even if the O-type government was playing *E*. Thus, it has a dominant strategy to play *I*. (It does best to play *I* against the H-type government and against the O type, regardless of the choice made by that government.) The O-type government has two possible strategies however. One, in which the O type plays *E*, would lead to separation, whereas the other, in which it plays *NE*, would result in pooling. We consider each possibility individually.

[1] Separation: We know that an H-type government plays *NE*, and firm Y plays *I*. Suppose an O-type government plays *E* in period *Y*; then the government's action would reveal its type. Can this set of strategies generate a separating equilibrium?

Given the strategies described, the firm investing in the second period (firm Z) will see that firm Y had been expropriated. Firm Z would then conclude that the government was O type for sure and would update the probability of it being H type to $p_Z = 0$. Therefore firm Z would not invest, so the O-type government would get 2.5 in period *Y* and 0 in period *Z*.

But what if the O-type government were to deviate and play *NE* in period *Y* instead? Observing *NE* in period *Y*, firm Z would update the probability of the government being H type to $p_Z = 1$. (Remember that in Nash equilibrium the firm will take the governments' equilibrium strategies as given, so it will believe that a government playing *NE* must be type H.) Therefore firm Z would play *I*, at which point the O-type government could play *E*. The government would then get 0.5 in period *Y* and 2.5 in period *Z*. This total payoff of 3 is better then the total of 2.5 the government gets by using its specified strategy, *E*, so that original strategy cannot be optimal. (Remember that we are not discounting across periods.) So we cannot get a separating equilibrium in the case of $p > 1/2$.

[2] Pooling: The second possibility is that the H-type government plays *NE*, firm Y plays *I*, and the O-type government also plays *NE*. Then the O type's action in

period *Y* is the same as the H type's. Can these strategies constitute a pooling equilibrium?

Because both types of governments take the same action in period Y, no new information regarding type emerges for use by firm Z. Its Bayesian updating will lead to $p_Z = p_Y = p$. So firm Z will play *I* (because $p > 1/2$), and the O-type government will play *E*. Thus the O type gets 0.5 in period *Y* and 2.5 in period *Z* for a total of 3 when it follows the stipulated strategy.

If the O type were to deviate from its stated strategy and play *E* in period *Y*, this raises a question of how firm Z would update. The most natural assumption is that a choice of *E* would be interpreted as a sure indicator of O type, since *E* is not even a strategy available to the H type. Then firm Z, observing *E* in period *Y*, would play *NI*. The O-type government would get 2.5 in period *Y* and 0 in period *Z*. This total payoff is worse than what the O type gets from playing *NE* as stipulated, so the deviation is unprofitable. Pooling in the period-*Y* game is a perfect Bayesian equilibrium.

In this two-play case, the first play (period *Y*) has a different outcome from the single-play version of the game. Investment takes place, and profits are *not* expropriated by either type of government. (With just one play and with $p > 1/2$, investment would take place, but the type-O government would expropriate it.) Many observers would regard this outcome of the twice-played game as better than the single-play game, because actions are honorable, even though the sum of the players' payoffs is the same in both versions.

II. CASE II: $p < 1/2$ Again, the H-type government always plays *NE*. But with $p < 1/2$, it is no longer the case that firm Y has a dominant strategy to play *I*. Nor does it have a dominant strategy to play *NI*; if the O-type government pools and plays *NE*, firm Y's strategy *NI* could be part of a Nash equilibrium in period *Y*. Thus, we will have to consider all four possible combinations of pure strategies for firm Y (playing either *I* or *NI*) and the O-type government (playing either *E* or *NE*) to see which set or sets can be equilibria.

[1] Separation with investment: Consider first the set of strategies in which the H-type government plays *NE*, the O type plays *E*, and firm Y plays *I*. Although these strategies would result in a separation of types for the governments, they cannot be an equilibrium. Firm Y would have negative expected profit (because $p < 1/2$) and it would deviate to *NI*.

[2] Separation without investment: Now suppose that the H-type government plays *NE*, the O-type government plays *E*, and firm Y plays *NI*. Again, this would lead to separation if the set of strategies constitutes an equilibrium.

In this situation, firm Z gains no information about government type from the actions in period *Y* because no investment occurs. Thus, firm Z will also play

NI. Then the O type government's choices are irrelevant in both periods *Y* and *Z*, and it is indifferent between *E* and *NE*.

This makes the specified strategies a Nash equilibrium in period *Y*, but we do not have a perfect Bayesian equilibrium of the two-period game. Consider the off-equilibrium node where firm Y has played *I*. If the O-type government plays its stipulated equilibrium action *E*, that choice will reveal its type to firm Z, which will then play *NI*. So the O type's payoff would be 2.5 in period *Y*, and 0 in period *Z*. If instead the O type deviates to play *NE*, firm Z, which takes equilibrium strategies as given, will believe that the government is H type; that is, it will update to $p_Z = 1$. Therefore firm Z will invest. The O-type government can then expropriate and get a payoff of 2.5 in period *Z* to add to its payoff of 0.5 in period *Y*. The total payoff of 3 exceeds the payoff of 2.5 from playing *E*. This deviation is profitable to the O type, and even though this is true only when firm Y also deviates, it means that we cannot have a *perfect* Bayesian equilibrium of this type.

[3] Pooling with investment: Here we consider the possibility that both types of governments play *NE* while firm Y plays *I*. If these strategies are an equilibrium, we would have pooling of the two types of governments.

Given the stipulated strategies,the O-type government gets 0.5 in period *Y*. Because both governments play *NE*, there is no new information revealed for firm Z. Its updating leaves it with $p_Z = p_Y = p < 1/2$. Then firm Z plays *NI*, and the government gets 0 in period *Z*. The total payoff to the O-type government is 0.5 over the two periods.

If the O-type government were to deviate to *E* in period *Y*, it would get a payoff of 2.5 in that period. Its type would be revealed to firm Z, however, which would update to $p_Z = 0$ and therefore play *NI*. The government would get 0 in period *Z* and a total of 2.5 over the two periods. The O-type government's deviation is then profitable, and the originally stated strategies cannot be a Nash equilibrium.

[4] Pooling without investment: Our last possible set of pure strategies entails both types of governments playing *NE* and firm Y playing *NI*. These strategies cannot be an equilibrium, however, because firm Y would benefit by switching to *I*.

This analysis of the case of $p < 1/2$ shows that none of the four combinations of pure strategies for firm Y and the O-type government generate an equilibrium. With all of these pure-strategy combinations ruled out, we have to consider an equilibrium that entails mixing. With mixed strategies, we may be able to generate a *semiseparating equilibrium*.

[5] Semiseparation (with investment): Here we consider a possible equilibrium in which firm Y plays *I* while the O-type government mixes in period *Y*, playing *NE* with probability q_Y and *E* with probability $(1 - q_Y)$. In period *Z*, firm Z mixes, playing *I* with probability r_Z and *NI* with probability $(1 - r_Z)$, while the government

		GOVERNMENT ACTION		Sum of row
		NE	E	
GOV. TYPE	H	p	0	$p_Y = p$
	O	$(1 - p)q_Y$	$(1 - p)(1 - q_Y)$	$1 - p$
Sum of column		$p + (1 - p)q_Y$	$(1 - p)(1 - q_Y)$	

FIGURE 11.10 Applying Bayes' Rule to the Expropriation Game

reverts to its true type; the O-type government plays E if the firm has invested. The values for q_Y and r_Z will be determined as part of our analysis of the equilibrium conditions. These conditions are the standard "opponent's-indifference" conditions; each player's mixture must keep the other indifferent between its pure actions.

In order for there to be an equilibrium with mixing by firm Z, the O-type government's period-*Y* mixture must keep firm Z indifferent between *I* and *NI*. For that, firm Z's Bayesian updating must yield $p_Z = 1/2$. What does this mean for the O type's choice of q_Y? To answer this question, we need to consider the probability table of types and actions illustrated in Figure 11.10. This table is similar to the ones we created in Chapter 9 when we explained Bayes' theorem in the Appendix and in the bluffing game of Section 5. Note that in the table the probability of observing an O-type government playing *NE* just equals the probability that the government is O type $(1 - p)$ times the probability that the O type chooses *NE* in its mixture (q_Y). The probability of observing an O type playing *E* is calculated similarly.

We can now use the table to determine how firm Z will update its probability that the government is H type. If firm Y's investment meets the government response *NE*, then Bayes' theorem states that the posterior probability of the government being H type (that is, firm Z's updated prior) will equal the probability of observing an H type playing *NE* divided by the sum of the probabilities associated with observing *NE*. The posterior probability is then $p/[p + (1 - p)q_Y]$.

Recall that to ensure mixing by firm Z, we need its updated probability that the government is H type to equal 1/2. Therefore we need

$$\frac{p}{p + (1 - p)q_Y} = \frac{1}{2} \quad \text{or} \quad 2p = p + (1 - p)q_Y \quad \text{or} \quad \frac{p}{1 - p} = q_Y.$$

(Note that $p < 1/2$ ensures that $q_Y < 1$ will hold.) This condition specifies the appropriate level of q_Y for the O-type government's period-*Y* mixing.

Now we need to determine the correct mixture for firm Z. Its mix must keep the O-type government indifferent between *E* and *NE* in period *Y* (and therefore willing to mix in period *Y*). If the O-type government chooses *E* in period *Y*, it earns of payoff of 2.5 in that period but reveals its type. That revelation leads firm Z to play *NI* and the government gets a payoff of 0 in period *Z*, for a total payoff of 2.5. If the O type plays *NE* in period *Y*, it gets 0.5 in that period, and then firm Z will mix in period *Z*. With firm Z's mixing, the O-type government will get a payoff of 2.5 in period *Z* with probability r_Z (the probability that firm Z plays *I*) for a total payoff across the two periods of $0.5 + 2.5\ r_Z$. To keep the O-type government indifferent between *E* and *NE* in period *Y*, firm Z will want to choose the r_Z that equates these two payoffs. So firm Z needs $2.5 = 0.5 + 2.5\ r_Z$, or $r_Z = 0.8$.

We now have calculated equilibrium values for both q_Y and r_Z, but all of our analysis assumed that firm Y would choose *I*. If firm Y chose *NI*, there would be no action for the period-*Y* government to take and nothing to reveal its type even probabilistically. But we do need to verify that this assumption is valid.

To do so, we must consider firm Y's expected profit from investing in period *Y*. We know that firm Y's investment will not be expropriated if it meets an H-type government (probability *p*) or an O-type government choosing *NE* (probability $(1 - p)q_Y = (1 - p) \times p/(1 - p) = p$, using the solution above for q_Y). The total probability of meeting a government that will play *NE* is then $2p$. So firm Y gets expected profits of $(2p \times 2) - 1 = 4p - 1$. This expected profit is positive when $4p - 1 > 0$ or when $p > 1/4$. Therefore firm Y will invest if $p > 1/4$, and there will be a semiseparating equilibrium with the mixture probabilities calculated above. Note that the condition $p > 1/4$ is weaker than the $p > 1/2$ that was required to induce investment in the single-play version of this game. Thus repetition, even just two periods, increases the possibility of the good or cooperative outcome.

III. CASE III: $p = 1/2$ In this final case, firm Y will be indifferent between investing and not investing. This case is exceptional, being just on the borderline between the case of $p > 1/2$ (where we found a pooling equilibrium in which firm Y invests and neither type of government expropriates) and the case $1/2 > p > 1/4$ (where we found a semiseparating equilibrium in which firm Y invests and the O-type government randomizes between *E* and *NE*). As *p* rises to 1/2 in the range of the semiseparating equilibria, the probability of the O-type government choosing *NE*, $q_Y = p/(1 - p)$, rises to 1. So the two cases on either side of $p = 1/2$ converge to the same outcome. Therefore we will regard the case $p = 1/2$ as a limiting case of the first two and we will not go into its details separately.

C. Thrice-Repeated Game

Our analysis in Section 5.B showed that going to a twice-repeated version of the expropriation game increased the likelihood that the cooperative outcome would be observed in equilibrium. We now consider additional repetitions, starting specifically with the case of a three-period game. Counting backward again, the first period of play will be labeled period X, with periods Y and Z being the second and third (or final) periods, as before. Here we can show that the equilibrium has the following features:

Case a: If $p > 1/4$, there will be a pooling equilibrium in period X where firm X invests and even an opportunistic government does not expropriate.

Case b: If $1/4 > p > 1/8$, there will be a semiseparating equilibrium where firm X (the first to play) invests, the O-type government randomizes in its response, and firm Y also randomizes.

Note that the range of values of p where investment takes place and is not expropriated in the first period of play has expanded geometrically (in powers of 1/2) with the increase in the number of repetitions. This pattern would continue if we were to add more repetitions of the game.

The details of the analysis verifying the equilibrium strategies are similar to those of the twice-repeated case, so we omit most of them. But we want to emphasize and check two key issues in the thrice-repeated case.

First, we need to verify the optimality of nonexpropriation in Case a; it must be optimal for the O-type government to play *NE* in period X (the first period of play) when the initial probability is $p_X = p > 1/4$. If the O-type government does play *NE*, it will get 0.5 in period X (remember, firm X plays I). This action pools it with the H-type government, so it reveals no new information about type to firm Y. The game in period Y therefore has the same $p_Y = p > 1/4$. Our analysis in Section 5.B above showed that the O-type government's total payoff over periods Y and Z is 2.5 when $1/2 > p_Y > 1/4$ and 3 when $p_Y > 1/2$. Therefore, over the three periods the O-type government gets 3 when $1/2 > p_Y > 1/4$ and 3.5 when $p_Y > 1/2$. If it deviated and chose E in the very first play (period X), it would get 2.5 in that period, but it would reveal its type to firms Y and Z and so get a payoff of 0 thereafter. The deviation from *NE* in period X is therefore not profitable, and pooling in period X is an equilibrium in this case.

Second, we must check the condition from Case b that guarantees that randomization is sustained when $1/4 > p$. The O-type government's period-X (first play) randomization should keep firm Y indifferent about investing. By the analysis for the twice-repeated case, this indifference will be ensured when firm Y's Bayesian updating yields $p_Y = 1/4$. Therefore, as above, we need $p/[p+(1-p)q_X] = 1/4$. (The equilibrium entry probability, r_Y, in firm Y's mixture is similarly calculated to be 0.8.)

Finally, firm X will indeed invest if its expected profit is positive. Firm X's profits are not expropriated with probability p (that it meets an H-type government) plus $(1-p)q_X$ (that it meets an O-type government playing *NE*). Then firm X's expected profit is $[p + (1-p)q_X] \times 2 - 1 = 8p - 1$, where we have made use of the condition defining q_X that was derived in the preceding paragraph. This expected profit is positive, and firm X does invest, if $p > 1/8$. This condition on investment in the first period of play is even weaker than that found in the two-stage game.

Further repetitions will follow the same pattern. If the game is played N times where N is large, there will be a pooling equilibrium with investment and no expropriation in the initial $(N - n)$ periods, where n is defined as the smallest integer that makes $p < (1/2)^n$ true. In the following $(n - 1)$ periods there will be semiseparating equilibria. The O-type government's randomization in one of these following periods may yield expropriation, in which case later firms will not invest. Otherwise, in the last period (period Z) the firm will be indifferent between investing and not investing because its updated p_Z will exactly equal 1/2. (This result follows from the observation of *NE* in the period-Y semiseparating equilibrium.) But in period Z, an O-type government will play *NE* for sure.

In an exercise at the end of this chapter, we will guide you through a more general formulation of this game, with the payoffs and probabilities denoted by algebraic symbols instead of specific numbers, to show that the idea underlying this solution is perfectly general. The corresponding two-sided dilemma game is harder to solve, and we merely refer ambitious readers to the original article.[10]

6 EXPERIMENTAL EVIDENCE

Numerous people have conducted experiments in which subjects compete in prisoners' dilemma games against each other.[11] Such experiments show that cooperation can and does occur in such games, even in repeated versions of known and finite length. Many players start off by cooperating and continue to cooperate for quite a while, as long as the rival player reciprocates. Only in the last few plays of a finite game does defecting seem to creep in. Although this be-

[10]David Kreps, Paul Milgrom, John Roberts, and Robert Wilson, "Rational Cooperation in a Finitely Repeated Prisoner's Dilemma," *Journal of Economic Theory,* vol. 27 (1982), pp. 245–252.

[11]The literature on experiments involving the prisoners' dilemma game is vast. A brief overview is given by Alvin Roth in *The Handbook of Experimental Economics* (Princeton: Princeton University Press, 1995), pp. 26–28. Journals in both psychology and economics can be consulted for additional references. For some examples of the outcomes that we describe, see Kenneth Terhune, "Motives, Situation, and Interpersonal Conflict Within Prisoners' Dilemmas," *Journal of Personality and Social Psychology Monograph Supplement,* vol. 8, no. 30 (1968), pp. 1–24; and R. Selten and R. Stoecker, "End Behavior in Sequences of Finite Prisoners' Dilemma Supergames," *Journal of Economic Behavior and Organization,* vol. 7 (1986), pp. 47–70. Robert Axelrod's *Evolution of Cooperation* (New York: Basic Books, 1984) presents the results of his computer-simulation tournament for the best strategy in an infinitely repeated dilemma.

havior goes against the reasoning of rollback, it can be "profitable" if sustained for a reasonable length of time. The pairs get higher payoffs than would rational, calculating strategists who defect from the very beginning.

Such observed behavior can be rationalized in different ways. Perhaps the players are not sure that the relationship will actually end at the stated time. Perhaps they believe that their reputations for cooperation will carry over to other similar games against the same opponent or other opponents. Perhaps they think it possible that their opponents are naive cooperators, and they are willing to risk a little loss in testing this hypothesis for a couple of plays. If successful, the experiment will lead to higher payoffs for a sufficiently long time.

In some laboratory experiments, players engage in multiple-round games, each round consisting of a given finite number of repetitions. All of the repetitions in any one round are played against the same rival, but each new round is played against a new opponent. Thus there is an opportunity to develop cooperation with an opponent in each round and to "learn" from preceding rounds when devising one's strategy against new opponents as the rounds continue. These situations have shown that cooperation lasts longer in early rounds than in later rounds. This result suggests that the theoretical argument on the unraveling of cooperation, based on the use of rollback, is being learned from experience of the play itself over time as players begin to understand the benefits and costs of their actions more fully. Another possibility is that players learn simply that they want to be the first to defect, and so the timing of the initial defection occurs earlier as the number of rounds played increases.

Suppose you were playing a game with a prisoners' dilemma structure and found yourself in a cooperative mode with the known end of the relationship approaching. When should you decide to defect? You do not want to do so too early, while a lot of potential future gains remain. But you also do not want to leave it until too late in the game, because then your opponent might preempt you and leave you with a low payoff for the period in which she defects. In fact, your decision about when to defect cannot be deterministic. If it were, your opponent would figure it out and defect in the period before you planned to do so. If no deterministic choice is feasible, then the unraveling of cooperation must include some uncertainty, such as mixed strategies, for both players. Many thrillers whose plots hinge on tenuous cooperation among criminals or between informants and police acquire their suspense precisely because of this uncertainty.

Examples of the collapse of cooperation as players near the end of a repeated game are observed in numerous situations in the real world, as well as in the laboratory. The story of a long-distance bicycle (or foot) race is one such example. There may be a lot of cooperation for most of the race, as players take turns leading and letting others ride in their slipstreams; nevertheless, as the finish line looms, each participant will want to make a dash for the tape. Similarly, signs saying "no checks accepted" often appear in stores in college towns each spring near the end of the semester.

Computer-simulation experiments have matched a range of very simple to very complex contingent strategies against each other in two-player prisoners' dilemmas. The most famous of them were conducted by Robert Axelrod at the University of Michigan. He invited people to submit computer programs that specified a strategy for playing a prisoners' dilemma repeated a finite but large number (200) of times. There were 14 entrants. Axelrod held a "league tournament" that pitted pairs of these programs against one another, in each case for a run of the 200 repetitions. The point scores for each pairing and its 200 repetitions were kept, and each program's scores over all its runs against different opponents were added up to see which program did best in the aggregate against all other programs. Axelrod was initially surprised when "nice" programs did well; none of the top eight programs were ever the first to defect. The winning strategy turned out to be the simplest program: Tit-for-tat, submitted by the Canadian game theorist Anatole Rapoport. Programs that were eager to defect in any particular run got the defecting payoff early but then suffered repetitions of mutual defections and poor payoffs. On the other hand, programs that were always nice and cooperative were badly exploited by their opponents. Axelrod explains the success of Tit-for-tat in terms of four properties: it is at once forgiving, nice, provocable, and clear.

In Axelrod's words, one does well in a repeated prisoners' dilemma to abide by these four simple rules: "Don't be envious. Don't be the first to defect. Reciprocate both cooperation and defection. Don't be too clever."[12] Tit-for-tat embodies each of the four ideals for a good, repeated prisoners' dilemma strategy. It is not envious; it does not continually strive to do better than the opponent, only to do well for itself. In addition, Tit-for-tat clearly fulfills the admonitions not to be the first to defect and to reciprocate, defecting only in retaliation to the opponent's preceding defection and always reciprocating in kind. Finally, Tit-for-tat does not suffer from being overly clever; it is simple and understandable to the opponent. In fact, it won the tournament not because it helped players achieve high payoffs in any individual game—the contest was not about "winner takes all"—but because it was always close; it simultaneously encourages cooperation and avoids exploitation, whereas other strategies cannot.

Axelrod then announced the results of his tournament and invited submissions for a second round. Here, people had a clear opportunity to design programs that would beat Tit-for-tat. The result: Tit-for-tat won again! The programs that were cleverly designed to beat it could not beat it by very much, and they did poorly against one another. Axelrod also arranged a tournament

[12]Axelrod, *Evolution of Cooperation,* p. 110.

of a different kind. Instead of a league where each program met each other program once, he ran a game with a whole population of programs, with a number of copies of each program. Each type of program met an opponent randomly chosen from the population. Those programs that did well were given a larger proportion of the population; those that did poorly had their proportion in the population reduced. This was a game of evolution and natural selection, which we will study in greater detail in Chapter 13. But the idea is simple in this context, and the results are fascinating. At first, nasty programs did well at the expense of nice ones. But as the population became nastier and nastier, each nasty program met other nasty programs more and more often, and they began to do poorly and fall in numbers. Then Tit-for-tat started to do well and eventually triumphed.

However, Tit-for-tat has some flaws. Most importantly, it assumes no errors in execution of the strategy. If there is some risk that the player intends to play the cooperative action but plays the defecting action in error, then this action can initiate a sequence of retaliatory defecting actions that locks two Tit-for-tat programs playing one another into a bad outcome; another error is required to rescue them from this sequence. When Axelrod ran a third variant of his tournament, which provided for such random mistakes, Tit-for-tat could be beaten by even "nicer" programs that tolerated an occasional episode of defecting to see if it was a mistake or a consistent attempt to exploit them and retaliated only when convinced that it was not a mistake.[13]

Interestingly, a twentieth-anniversary competition modeled after Axelrod's original contest and run in 2004 and 2005 generated a new winning strategy.[14] Actually, the winner was a set of strategies designed to recognize one another during play so that one would become docile in the face of the other's continued defections. (The authors likened their approach to a situation in which prisoners manage to communicate with each other by tapping on their cell walls.) This collusion meant that some of the strategies submitted by the winning team did very poorly, whereas others did spectacularly well, a testament to the value of working together. Of course Axelrod's contest did not permit multiple submissions, so such strategy sets were ineligible, but the winners of the recent competition argue that with no way to preclude coordination, strategies such as those they submitted should have been able to win the original competition as well.

[13]For a description and analysis of Axelrod's computer simulations from the biological perspective, see Matt Ridley, *The Origins of Virtue* (New York: Penguin Books, 1997), pp. 61, 75. For a discussion of the difference between computer simulations and experiments using human players, see John K. Kagel and Alvin E. Roth, *Handbook of Experimental Economics* (Princeton: Princeton University Press, 1995), p. 29.

[14]See Wendy M. Grossman, "New Tack Wins Prisoner's Dilemma," *Wired,* October 13, 2004. Available at http://www.wired.com/culture/lifestyle/news/2004/10/65317 (accessed 6/14/08).

7 REAL-WORLD DILEMMAS

Games with the prisoners' dilemma structure arise in a surprisingly varied number of contexts in the world. Although we would be foolish to try to show you every possible instance in which the dilemma can arise, we take the opportunity in this section to consider in detail three specific examples from a variety of fields of study. One example comes from evolutionary biology, a field that we will study in greater detail in Chapter 13. A second example describes the policy of "price matching" as a solution to a prisoners' dilemma pricing game. And a final example concerns international environmental policy and the potential for repeated interactions to mitigate the prisoners' dilemma in this situation.

A. Evolutionary Biology

In our first example, we consider a game known as the bowerbirds' dilemma, from the field of evolutionary biology.[15] Male bowerbirds attract females by building intricate nesting spots called bowers, and female bowerbirds are known to be particularly choosy about the bowers built by their prospective mates. For this reason, male bowerbirds often go out on search-and-destroy missions aimed at ruining other males' bowers. While they are out, however, they run the risk of losing their own bower to the beak of another male. The ensuing competition between male bowerbirds and their ultimate choice regarding whether to maraud or guard has the structure of a prisoners' dilemma game.

Ornithologists have constructed a table that shows the payoffs in a two-bird game with two possible strategies, Maraud and Guard. That payoff table is shown in Figure 11.11. GG represents the benefits associated with Guarding when the rival bird also Guards; GM represents the payoff from Guarding when the rival bird is a Marauder. Similarly, MM represents the benefits associated with Marauding when the rival bird also is a Marauder; MG represents the payoff

		BIRD 2	
		Maraud	Guard
BIRD 1	Maraud	MM, MM	MG, GM
	Guard	GM, MG	GG, GG

FIGURE 11.11 Bowerbirds' Dilemma

[15] Larry Conik, "Science Classics: The Bowerbird's Dilemma," *Discover*, October 1994.

from Marauding when the rival bird Guards. Careful scientific study of bowerbird matings led to the discovery that MG > GG > MM > GM. In other words, the payoffs in the bowerbird game have exactly the same structure as the prisoners' dilemma. The birds' dominant strategy is to maraud, but when both choose that strategy, they end up in equilibrium each worse off than if they had both chosen to guard.

In reality, the strategy used by any particular bowerbird is not actually the result of a process of rational choice on the part of the bird. Rather, in evolutionary games, strategies are assumed to be genetically "hardwired" into individual organisms, and payoffs represent reproductive success for the different types. Then equilibria in such games define the type of population that naturalists can expect to observe—all marauders, for instance, if Maraud is a dominant strategy as in Figure 11.11. This equilibrium outcome is not the best one, however, given the existence of the dilemma. In constructing a solution to the bowerbirds' dilemma, we can appeal to the repetitive nature of the interaction in the game. In the case of the bowerbirds, repeated play against the same or different opponents in the course of several breeding seasons can allow you, the bird, to choose a flexible strategy based on your opponent's last move. Contingent strategies such as tit-for-tat can be, and often are, adopted in evolutionary games to solve exactly this type of dilemma. We will return to the idea of evolutionary games and provide detailed discussions of their structure and equilibrium outcomes in Chapter 13.

B. Price Matching

Now we return to a pricing game, in which we consider two specific stores engaged in price competition with each other, using identical price-matching policies. The stores in question, Toys "R" Us and Kmart, are both national chains that regularly advertise prices for name-brand toys (and other items). In addition, each store maintains a published policy that guarantees customers that it will match the advertised price of any competitor on a specific item (model and item numbers must be identical) as long as the customer provides the competitor's printed advertisement.[16]

For the purposes of this example, we assume that the firms have only two possible prices that they can charge for a particular toy (Low or High). In addition, we use hypothetical profit numbers and further simplify the analysis by

[16]The price-matching policy at Toys "R" Us is printed and posted prominently in all stores. A simple phone call confirmed that Kmart has an identical policy. Similar policies are appearing in many industries, including that for credit cards where "interest rate matching" has been observed. See Aaron S. Edlin, "Do Guaranteed-Low-Price Policies Guarantee High Prices, and Can Antitrust Rise to the Challenge?" *Harvard Law Review,* vol. 111, no. 2 (December 1997), pp. 529–575.

		KMART	
		Low	High
TOYS "R" US	Low	2,000, 2,000	4,000, 0
	High	0, 4,000	3,000, 3,000

FIGURE 11.12 Toys "R" Us and Kmart Toy Pricing

assuming that Toys "R" Us and Kmart are the only two competitors in the toy market in a particular city—Billings, Montana, for example.

Suppose, then, that the basic structure of the game between the two firms can be illustrated as in Figure 11.12. If both firms advertise low prices, they split the available customer demand and each earns $2,000. If both advertise high prices, they split a market with lower sales, but their markups end up being large enough to let them each earn $3,000. Finally, if they advertise different prices, then the one advertising a high price gets no customers and earns nothing, whereas the one advertising a low price earns $4,000.

The game illustrated in Figure 11.12 is clearly a prisoners' dilemma. Advertising and selling at a low price is the dominant strategy for each firm, although both would be better off if each advertised and sold at the high price. But as mentioned earlier, each firm actually makes use of a third pricing strategy: a price-matching guarantee to its customers. How does the inclusion of such a policy alter the prisoners' dilemma that would otherwise exist between these two firms?

Consider the effects of allowing firms to choose among pricing low, pricing high, and price matching. The Match strategy entails advertising a high price but promising to match any lower advertised price by a competitor; a firm using Match then benefits from advertising high if the rival firm does so also, but it does not suffer any harm from advertising a high price if the rival advertises a low price. We can see this in the payoff structure for the new game, shown in Figure 11.13. In that table, we see that a combination of one firm playing Low while the other plays Match is equivalent to both playing Low, while a combination of one firm playing High while the other plays Match (or both playing Match) is equivalent to both playing High.

Using our standard tools for analyzing simultaneous-play games shows that High is weakly dominated by Match for both players and that once High is eliminated, Low is weakly dominated by Match also. The resulting Nash equilibrium entails both firms using the Match strategy. In equilibrium, both firms earn $3,000—the profit level associated with both firms pricing high in the original game. The addition of the Match strategy has allowed the firms to emerge from the prisoners' dilemma that they faced when they had only the choice between two simple pricing strategies, Low or High.

		KMART		
		Low	High	Match
TOYS "R" US	Low	2,000, 2,000	4,000, 0	2,000, 2,000
	High	0, 4,000	3,000, 3,000	3,000, 3,000
	Match	2,000, 2,000	3,000, 3,000	3,000, 3,000

FIGURE 11.13 Toys "R" Us and Kmart Toy Pricing

How did this happen? The Match strategy acts as a penalty mechanism. By guaranteeing to match Kmart's low price, Toys "R" Us substantially reduces the benefit that Kmart achieves by advertising a low price while Toys "R" Us is advertising a high price. In addition, promising to meet Kmart's low price hurts Toys "R" Us, too, because the latter has to accept the lower profit associated with the low price. Thus the price-matching guarantee is a method of penalizing both players whenever either one defects. This is just like the crime mafia example discussed in Section 3, except that this penalty scheme—and the higher equilibrium prices that it supports—is observed in markets in virtually every city in the country.

Actual empirical evidence of the detrimental effects of these policies is available but limited, and some research has found evidence of lower prices in markets with such policies.[17] However, more recent experimental evidence does support the collusive effect of price-matching policies. This result should put all customers on alert.[18] Even though stores that match prices promote their policies in the name of competition, the ultimate outcome when all firms use such policies can be better for the firms than if there were no price matching at all, and so customers can be the ones who are hurt.

C. International Environmental Policy: The Kyoto Protocol

Our final example pertains to the international climate control agreement known as the Kyoto Protocol. Negotiated by the United Nations Framework Convention

[17]J. D. Hess and Eitan Gerstner present evidence of increased prices as a result of price-matching policies in "Price-Matching Policies: An Empirical Case," *Managerial and Decision Economics*, vol. 12 (1991), pp. 305–315. Contrary evidence is provided by Arbatskaya, Hviid, and Shaffer, who find that the effect of matching policies is to lower prices; see Maria Arbatskaya, Morten Hviid, and Greg Shaffer, "Promises to Match or Beat the Competition: Evidence from retail Tire Prices," *Advances in Applied Microeconomics*, vol. 8: Oligopoly (New York: JAI Press, 1999), pp. 123–138.

[18]See Subhasish Dugar, "Price-Matching Guarantees and Equilibrium Selection in a Homogeneous Product Market: An Experimental Study," *Review of Industrial Organization*, vol. 30 (2007), pp. 107–119.

		THEM	
		Cut Emissions	Don't Cut
US	Cut Emissions	−1, −1	−20, 0
	Don't Cut	0, −20	−12, −12

FIGURE 11.14 Greenhouse Gas Emissions Game

on Climate Change in 1997 as a tool for reducing greenhouse gas emissions, it went in to effect in 2005 and is due to expire in 2012. Over 170 countries have signed on to the treaty, although the United States is noticeably absent from the list. Ongoing meetings continue to work on a plan for extending the protocol beyond its current end date.

The difficulty in achieving global reduction in greenhouse gas emissions comes in part from the prisoners' dilemma nature of the interaction. Any individual country will have no incentive to reduce its own emissions, knowing that if it does so alone it bears significant costs with little benefit to overall climate change. If others do reduce their emissions the first country cannot be stopped from enjoyiong the benefits of the others' actions.

Consider the emissions reduction problem as a game played between two countries, Us and Them. Estimates generated by the British government's Office on Climate Change suggest that coordinated action may come at a cost of about 1% of GDP per nation, whereas coordinated inaction could cost each nation between 5 and 20% of GDP, perhaps 12% on average.[19] By extension, the cost to cutting emissions on your own may be at the high end of the inaction estimate (20%), but holding back and letting the other country cut emissions could entail virtually no cost to you at all. We can then summarize the situation between Us and Them using the game table in Figure 11.14, where payoffs represent changes in GDP for each country.

The game in Figure 11.14 is indeed a prisoners' dilemma. Both countries have a dominant strategy to refuse to cut their emissions. The single Nash equilibrium occurs when neither country cuts emissions, but they suffer as a group as a result of the ensuing climate change. From this analysis we should expect little or no progress in greenhouse gas emissions reduction.

This interpretation of the problem inherent in the Kyoto Protocol has been challenged by recent research from Michael Liebriech, who argues that the game

[19]See Nicholas Stern, *The Economics of Climate Change: The Stern Review* (Cambridge: Cambridge University Press, 2007).

is not a one-off interaction and that countries repeatedly interact and negotiate additional amendments to the existing agreement.[20] He argues that the iterated nature of this game makes it amenable to solution by way of contingent strategies and that countries should use strategies that embody the four critical properties of TFT as outlined by Axelrod and described in Section 6 above. Specifically, countries are encouraged to employ strategies that are "nice" (signing on to the protocol and beginning emissions reductions), "retaliatory" (employing mechanisms to punish those that do not do their part), "forgiving" (welcoming to those newly accepting the protocol), and "clear" (specifying actions and reactions).

Liebriech assesses the actions of current players, including the European Union, the United States, and developing countries (as a group), and provides some suggestions for improvements. He explains that the European Union does well with nice, forgiving, and clear but not with retaliation, so other countries will do best to defect when interacting with the European Union. One solution would be for the European Union to institute carbon-related import taxes or another retaliatory-type policy for dealing with recalcitrant trade partners. The United States, on the other hand, ranks high on retaliatory and forgiving, given its history of such behavior following the end of the cold war. But it has not been nice or clear, at least on the national level (individual states may behave differently), giving other countries an incentive to retaliate against it quickly and painfully, if possible. The solution is for the United States to make a meaningful commitment to carbon-emission reduction, a standard conclusion in most policy circles. Developing countries are described as not nice (negotiating no carbon limits for themselves), retaliatory, unclear, and quite unforgiving. A more beneficial strategy, argues Liebriech, would be for these countries—particularly China, India, and Brazil—to make clear their commitment to sharing in international efforts to affect climate change; this approach would leave them less subject to retaliation and more likely to benefit from a global improvement in climatic outlook.

The general conclusion is that the process of international carbon emissions reduction does fit the profile of a prisoners' dilemma game. But the future of global greenhouse gas emissions should not be considered a lost cause simply because of the prisoners' dilemma aspects of the one-time interaction. Repeated play among the nations involved in the Kyoto Protocol negotiations make the game amenable to solutions by way of contingent (nice, clear, and forgiving, but also retaliatory) strategies.

[20]Michael Liebriech presents his analysis of the Kyoto Protocol as an iterated prisoners' dilemma in his paper "How to Save the Planet: Be Nice, Retaliatory, Forgiving and Clear," New Energy Finance White Paper, September 11, 2007. Available at www.newenergyfinance.com/docs/Press/NEF-WP_Carbon-Game-Theory_05.pdf (accessed 9/11/08).

SUMMARY

The prisoners' dilemma is probably the most famous game of strategy. Each player has a dominant strategy (to Defect), but the equilibrium outcome is worse for all players than when each uses her dominated strategy (to Cooperate). The best-known solution to the dilemma is *repetition of play.* In a finitely played game, the *present value* of future cooperation is eventually zero, and rollback yields an equilibrium with no cooperative behavior. With infinite play (or an uncertain end date), cooperation can be achieved with the use of an appropriate contingent strategy such as *tit-for-tat (TFT)* or the *grim strategy;* in either case, cooperation is possible only if the present value of cooperation exceeds the present value of defecting. More generally, the prospects of "no tomorrow" or of short-term relationships lead to decreased cooperation among players.

The dilemma can also be "solved" with *penalty* schemes that alter the payoffs for players who defect from cooperation when their rivals are cooperating or when others also are defecting. A third solution method arises if a large or strong player's loss from defecting is greater than the available gain from cooperative behavior on that player's part. Allowing for asymmetric information in the dilemma can lead to some cooperation, even in finitely repeated games.

Experimental evidence suggests that players often cooperate longer than theory might predict. Such behavior can be explained by incomplete knowledge of the game on the part of the players or by their views regarding the benefits of cooperation. Tit-for-tat has been observed to be a simple, nice, provocable, and forgiving strategy that performs very well on the average in repeated prisoners' dilemmas.

Prisoners' dilemmas arise in a variety of contexts. Specific examples from international environmental policy, evolutionary biology, and product pricing show how to explain and predict actual behavior by using the framework of the prisoners' dilemma.

KEY TERMS

compound interest (404)
contingent strategy (401)
discount factor (404)
effective rate of return (405)
grim strategy (401)
infinite horizon (404)
leadership (412)
penalty (409)
present value (PV) (403)
punishment (401)
repeated play (399)
tit-for-tat (TFT) (401)
trigger strategy (401)

SOLVED EXERCISES

S1. "If a prisoners' dilemma is repeated 100 times, and both players know how many repetitions to expect, they are sure to achieve their cooperative outcome." True or false? Explain and give an example of a game that illustrates your answer.

S2. Consider a two-player game between Child's Play and Kid's Korner, each of which produces and sells wooden swing sets for children. Each player can set either a high or a low price for a standard two-swing, one-slide set. If they both set a high price, each receives profits of $64,000 per year. If one sets a low price and the other sets a high price, the low-price firm earns profits of $72,000 per year, while the high-price firm earns $20,000. If they both set a low price, each receives profits of $57,000.

(a) Verify that this game has a prisoners' dilemma structure by looking at the ranking of payoffs associated with the different strategy combinations (both cooperate, both defect, one defects, and so on). What are the Nash-equilibrium strategies and payoffs in the simultaneous-play game if the players meet and make price decisions only once?

(b) If the two firms decide to play this game for a fixed number of periods—say, for 4 years—what would each firm's total profits be at the end of the game? (Don't discount.) Explain how you arrived at your answer.

(c) Suppose that the two firms play this game repeatedly forever. Let each of them use a grim strategy in which they both price high unless one of them "defects," in which case they price low for the rest of the game. What is the one-time gain from defecting against an opponent playing such a strategy? How much does each firm lose, in each future period, after it defects once? If $r = 0.25$ ($\delta = 0.8$), will it be worthwhile for them to cooperate? Find the range of values of r (or δ) for which this strategy is able to sustain cooperation between the two firms.

(d) Suppose the firms play this game repeatedly year after year, neither expecting any change in their interaction. If the world were to end after 4 years, without either firm having anticipated this event, what would each firm's total profits (not discounted) be at the end of the game? Compare your answer here with the answer in part (b). Explain why the two answers are different, if they are different, or why they are the same, if they are the same.

(e) Suppose now that the firms know that there is a 10% probability that one of them may go bankrupt in any given year. If bankruptcy occurs, the repeated game between the two firms ends. Will this knowledge change the firms' actions when $r = 0.25$? What if the probability of a bankruptcy increases to 35% in any year?

S3. A firm has two divisions, each of which has its own manager. Managers of these divisions are paid according to their effort in promoting productivity in their divisions. The payment scheme is based on a comparison of the two outcomes. If both managers have expended "high effort," each earns \$150,000 a year. If both have expended "low effort," each earns "only" \$100,000 a year. But if one of the two managers shows "high effort" whereas the other shows "low effort," the "high effort" manager is paid \$150,000 plus a \$50,000 bonus, but the second ("low effort") manager gets a reduced salary (for subpar performance in comparison with her competition) of \$80,000. Managers make their effort decisions independently and without knowledge of the other manager's choice.

(a) Assume that expending effort is costless to the managers and draw the payoff table for this game. Find the Nash equilibrium of the game and explain whether the game is a prisoners' dilemma.

(b) Now suppose that expending high effort is costly to the managers (such as a costly signal of quality). In particular, suppose that "high effort" costs an equivalent of \$60,000 a year to a manager who chooses this effort level. Draw the game table for this new version of the game and find the Nash equilibrium. Explain whether the game is a prisoners' dilemma and how it has changed from the game in part **(a)**.

(c) If the cost of high effort is equivalent to \$80,000/year, how does the game change from that described in part **(b)**? What is the new equilibrium? Explain whether the game is a prisoners' dilemma and how it has changed from the games in parts **(a)** and **(b)**.

S4. You have to decide whether to invest \$100 in a friend's enterprise, where in a year's time the money will increase to \$130. You have agreed that your friend will then repay you \$120, keeping \$10 for himself. But instead he may choose to run away with the whole \$130. Any of your money that you don't invest in your friend's venture you can invest elsewhere safely at the prevailing rate of interest r, and get $\$100(1 + r)$ next year.

(a) Draw the game tree for this situation and show the rollback equilibrium.

Next, suppose this game is played repeatedly infinitely often. That is, each year you have the opportunity to invest another \$100 in your friend's enterprise, and the agreement is to split the resulting \$130 in the manner already described. From the second year onward, you get to make your decision of whether to invest with your friend in the light of whether he made the agreed repayment the preceding year. The rate of interest between any two successive periods is r, the same as the outside rate of interest and the same for you and your friend.

(b) For what values of r can there be an equilibrium outcome of the repeated game, in which each period you invest with your friend and he repays as agreed?

(c) If the rate of interest is 10% per year, can there be an alternative profit-splitting agreement that is an equilibrium outcome of the infinitely repeated game, where each period you invest with your friend and he repays as agreed?

S5. Recall the example from Exercise S3 in which two division managers' choices of High or Low effort levels determine their salary payments. In part (b) of that exercise, the cost of exerting High effort is assumed to be $60,000 a year. Suppose now that the two managers play the game in part (b) of Exercise S3 repeatedly for many years. Such repetition allows scope for an unusual type of cooperation in which one is designated to choose High effort while the other chooses Low. This cooperative agreement requires that the High-effort manager make a side payment to the Low-effort manager so that their payoffs are identical.

(a) What size side payment guarantees that the final payoffs of the two managers are identical? How much does each manager earn in a year in which the cooperative agreement is in place?

(b) Cooperation in this repeated game entails each manager's choosing her assigned effort level and the High-effort manager making the designated side payment. Defection entails refusing to make the side payment. Under what values of the rate of return can this agreement sustain cooperation in the managers' repeated game?

S6. Consider the game of chicken in Chapter 4, with slightly more general payoffs (Figure 4.14 had $k = 1$):

		DEAN	
		Swerve	Straight
JAMES	Swerve	0, 0	−1, k
	Straight	k, −1	−2, −2

Suppose this game is played repeatedly, every Saturday evening. If $k < 1$, the two players stand to benefit by cooperating to play (Swerve, Swerve) all the time, whereas if $k > 1$, they stand to benefit by cooperating so that one plays Swerve and the other plays Straight, taking turns to go Straight in alternate weeks. Can either type of cooperation be sustained?

S7. Recall the example from Exercise S8 of Chapter 5, where South Korea and Japan compete in the market for production of VLCCs. As in parts (a) and (b) of that exercise, the cost of building ships is $30 (million) in each country, and the demand for ships is $P = 180 - Q$, where $Q = q_{Korea} + q_{Japan}$.

(a) Previously, we found the Nash equilibrium for the game. Now find the collusive outcome. What total quantity should be set by the two countries in order to maximize their joint profit?

(b) Suppose the two countries produce equal quantities of VLCCs, so that they earn equal shares of this collusive profit. How much profit would each country earn? Compare this profit with the amount they would earn in the Nash equilibrium.

(c) Now suppose the two countries are in a repeated relationship. Once per year, they choose production quantities, and each can observe the amount its rival produced in the previous year. They wish to cooperate to sustain the collusive profit levels found in part (b). In any one year, one of them can defect from the agreement. If one of them holds the quantity at the agreed level, what is the best defecting quantity for the other? What are the resulting profits?

(d) Write down a matrix that represents this game as a prisoners' dilemma.

(e) For what interest rates will collusion be sustainable when the two countries use grim (defect forever) strategies?

UNSOLVED EXERCISES

U1. Two people, Baker and Cutler, play a game in which they choose and divide a prize. Baker decides how large the total prize should be; she can choose either $10 or $100. Cutler chooses how to divide the prize chosen by Baker; Cutler can choose either an equal division or a split where she gets 90% and Baker gets 10%. Write down the payoff table of the game and find its equilibria for each of the following situations:

(a) When the moves are simultaneous.

(b) When Baker moves first.

(c) When Cutler moves first.

(d) Is this game a prisoners' dilemma? Why or why not?

U2. Consider a small town that has a population of dedicated pizza eaters but is able to accommodate only two pizza shops, Donna's Deep Dish and Pierce's Pizza Pies. Each seller has to choose a price for its pizza, but for simplicity, assume that only two prices are available: high and low. If a high price is set, the sellers can achieve a profit margin of $12 per pie; the low price yields a profit margin of $10 per pie. Each store has a loyal captive customer base that will buy 3,000 pies per week, no matter what price is charged by either store. There is also a floating demand of 4,000 pies per week. The people who buy these pies are price conscious and will go to the store with the lower price; if both stores charge the same price, this demand will be split equally between them.

(a) Draw the game table for the pizza-pricing game, using each store's profits per week (in thousands of dollars) as payoffs. Find the Nash equilibrium of this game and explain why it is a prisoners' dilemma.

(b) Now suppose that Donna's Deep Dish has a much larger loyal clientele that guarantees it the sale of 11,000 (rather than 3,000) pies a week. Profit margins and the size of the floating demand remain the same. Draw the payoff table for this new version of the game and find the Nash equilibrium.

(c) How does the existence of the larger loyal clientele for Donna's Deep Dish help "solve" the pizza stores' dilemma?

U3. A town council consists of three members who vote every year on their own salary increases. Two Yes votes are needed to pass the increase. Each member would like a higher salary but would like to vote against it herself because that looks good to the voters. Specifically, the payoffs of each are as follows:

Raise passes, own vote is No: 10
Raise fails, own vote is No: 5
Raise passes, own vote is Yes: 4
Raise fails, own vote is Yes: 0

Voting is simultaneous. Write down the (three-dimensional) payoff table, and show that in the Nash equilibrium the raise fails unanimously. Examine how a repeated relationship among the members can secure them salary increases every year if (i) every member serves a 3-year term, (ii) every year in rotation one of them is up for reelection, and (iii) the townspeople have short memories, remembering only the votes on the salary-increase motion of the current year and not those of past years.

U4. Consider the following game, which comes from James Andreoni and Hal Varian at the University of Michigan.[21] A neutral referee runs the game. There are two players, Row and Column. The referee gives two cards to each: 2 and 7 to Row and 4 and 8 to Column. This is common knowledge. Then, playing simultaneously and independently, each player is asked to hand over to the referee either his high card or his low card. The referee hands out payoffs—which come from a central kitty, not from the players' pockets—

[21] James Andreoni and Hal Varian, "Preplay Contacting in the Prisoners' Dilemma," *Proceedings of the National Academy of Sciences*, vol. 96, no. 19 (September 14, 1999), pp. 10933–10938.

that are measured in dollars and depend on the cards that he collects. If Row chooses his Low card, 2, then Row gets $2; if he chooses his High card, 7, then Column gets $7. If Column chooses his Low card, 4, then Column gets $4; if he chooses his High card, 8, then Row gets $8.

(a) Show that the complete payoff table is as follows:

		COLUMN	
		Low	High
ROW	Low	2, 4	10, 0
	High	0, 11	8, 7

(b) What is the Nash equilibrium? Verify that this game is a prisoners' dilemma.

Now suppose the game has the following stages. The referee hands out cards as before; who gets what cards is common knowledge. At stage I, each player, out of his own pocket, can hand over a sum of money, which the referee is to hold in an escrow account. This amount can be zero but cannot be negative. When both have made their Stage I choices, these are publicly disclosed. Then at stage II, the two make their choices of cards, again simultaneously and independently. The referee hands out payoffs from the central kitty in the same way as in the single-stage game before. In addition, he disposes of the escrow account as follows. If Column chooses his high card, the referee hands over to Column the sum that Row put into the account; if Column chooses his low card, Row's sum reverts back to him. The disposition of the sum that Column deposited depends similarly on Row's card choice. All these rules are common knowledge.

(c) Find the rollback (subgame-perfect) equilibrium of this two-stage game. Does it resolve the prisoners' dilemma? What is the role of the escrow account?

U5. Glassworks and Clearsmooth compete in the local market for windshield repairs. The market size (total available profits) is $10 million per year. Each firm can choose whether to advertise on local television. If a firm chooses to advertise in a given year, it costs that firm $3 million. If one firm advertises and the other doesn't, then the former captures the whole market. If both firms advertise, they split the market 50:50. If both firms choose not to advertise, they also split the market 50:50.

(a) Suppose the two windshield-repair firms know they will compete for just one year. Write down the payoff matrix for this game. Find the Nash equilibrium strategies.

(b) Suppose the firms play this game for five years in a row, and they know that at the end of five years, both firms plan to go out of business. What is the subgame-perfect equilibrium for this five-period game? Explain.

(c) What would be a "tit-for-tat" strategy in the game described in part **(b)**?

(d) Suppose the firms play this game repeatedly forever, and suppose that future profits are discounted with an interest rate of 20% per year. Can you find a subgame-perfect equilibrium that involves higher annual payoffs than the equilbrium in part **(b)**? If so, explain what strategies are involved. If not, explain why not.

U6. Consider the pizza stores introduced in Exercise **U2**, Donna's Deep Dish and Pierce's Pizza Pies. Suppose that they are not constrained to choose from only two possible prices, but that they can choose a specific value for price to maximize profits. Suppose further that it costs $3 to make each pizza (for each store) and that experience or market surveys have shown that the relation between sales (Q) and price (P) for each firm is as follows:

$$Q_{\text{Pierce}} = 12 - P_{\text{Pierce}} + 0.5P_{\text{Donna}}.$$

Then profits per week (Y, in thousands of dollars) for each firm are:

$$Y_{\text{Pierce}} = (P_{\text{Pierce}} - 3)\ Q_{\text{Pierce}} = (P_{\text{Pierce}} - 3)\ (12 - P_{\text{Pierce}} + 0.5P_{\text{Donna}}),$$
$$Y_{\text{Donna}} = (P_{\text{Donna}} - 3)\ Q_{\text{Donna}} = (P_{\text{Donna}} - 3)\ (12 - P_{\text{Donna}} + 0.5\ P_{\text{Pierce}}).$$

(a) Use these profit functions to determine each firm's best-response rule, as in Chapter 5, and use the best-response rules to find the Nash equilibrium of this pricing game. What prices do the firms choose in equilibrium? How much profit per week does each firm earn?

(b) If the firms work together and choose a joint best price, P, then the profit of each will be:

$$Y_{\text{Donna}} = Y_{\text{Pierce}} = (P - 3)\ (12 - P + 0.5\ P) = (P - 3)\ (12 - 0.5\ P).$$

What price do they choose to maximize joint profits?

(c) Suppose the two stores are in a repeated relationship, trying to sustain the joint profit-maximizing prices calculated in part **(b)**. They print new menus each month and thereby commit themselves to prices for the whole month. In any one month, one of them can defect from the agreement. If one of them holds the price at the agreed level, what is the best defecting price for the other? What are its resulting profits? For what interest rates will their collusion be sustainable by using grim-trigger strategies?

U7. Now we extend the analysis of Exercise **S7** to allow for defecting in a collusive triopoly. Exercise S9 of Chapter 5 finds the Nash outcome of a VLCC triopoly of Korea, Japan, and China.

(a) Now find the collusive outcome of the triopoly. That is, what total quantity should be set by the three countries collectively in order to maximize their joint profit?

(b) Assume that under the collusive outcome found in part (a), the three countries produce equal quantities of VLCCs, so that each earns an equal share of the collusive profit. How much profit would each country earn? Compare this profit with the amount each earns in the Nash outcome.

(c) Now suppose the three countries are in a repeated relationship. Once per year, they choose production quantities, and each can observe the amount its rivals produced in the previous year. They wish to cooperate to sustain the collusive profit levels found in part (b). In any one year, one of them can defect from the agreement. If the other two countries are expected to produce their share of the collusive outcome found in parts (a) and (b), what is the best defecting quantity for the third to produce? What is the resulting profit for a defecting country when it produces the optimal defecting quantity while the other two produce their collusive quantities?

(d) Of course, the year after one country defects, both of its rivals will also defect. They will all find themselves back at the Nash outcome (permanently, if they use grim-trigger strategies). How much does the defecting country stand to gain in one year of defecting from the collusive outcome? How much will the defecting country then lose in every subsequent year from earning the Nash profit instead of the collusive profit?

(e) For what interest rates will collusion be sustainable if the three countries are using grim-trigger strategies? Is this set of interest rates larger or smaller than that found in the duopoly case discussed in Exercise S7, part (e)? Why?

■

Appendix: Infinite Sums

The computation of present values requires us to determine the current value of a sum of money that is paid to us in the future. As we saw in Section 2 of Chapter 11, the present value of a sum of money—say, x—that is paid to us n months from now is just $x/(1 + r)^n$, where r is the appropriate monthly rate of return. But the present value of a sum of money that is paid to us next month and every following month in the foreseeable future is more complicated to determine. In that case, the payments continue infinitely, and so there is no defined end to the sum of present values that we need to compute. To compute the present value of this flow of payments requires some knowledge of the mathematics of the summation of infinite series.

Consider a player who stands to gain \$36 this month from defecting in a prisoners' dilemma but who will then lose \$36 every month in the future as a result of her choice to continue defecting while her opponent punishes her (using the tit-for-tat, or TFT, strategy). In the first of the future months—the first for which there is a loss and the first for which values need to be discounted—the present value of her loss is $36/(1 + r)$; in the second future month, the present value of the loss is $36/(1 + r)^2$; in the third future month, the present value of the loss is $36/(1 + r)^3$. That is, in each of the n future months that she incurs a loss from defecting, that loss equals $36/(1 + r)^n$.

We could write out the total present value of all of her future losses as a large sum with an infinite number of components,

$$PV = \frac{36}{1+r} + \frac{36}{(1+r)^2} + \frac{36}{(1+r)^3} + \frac{36}{(1+r)^4} + \frac{36}{(1+r)^5} + \frac{36}{(1+r)^6} + \cdots,$$

or we could use summation notation as a shorthand device and instead write

$$PV = \sum_{n=1}^{\infty} \frac{36}{(1+r)^n}.$$

This expression, which is equivalent to the preceding one, is read as "the sum, from n equals 1 to n equals infinity, of 36 over $(1 + r)$ to the nth power." Because 36 is a common factor—it appears in each term of the sum—it can be pulled out to the front of the expression. Thus we can write the same present value as

$$PV = 36 \times \sum_{n=1}^{\infty} \frac{36}{(1+r)^n}.$$

We now need to determine the value of the sum within the present-value expression to calculate the actual present value. To do so, we will simplify our notation by switching to the *discount factor* δ in place of $1/(1 + r)$. Then the sum that we are interested in evaluating is

$$\sum_{n=1}^{\infty} \delta^n.$$

It is important to note here that $\delta = 1/(1 + r) < 1$ because r is strictly positive.

An expert on infinite sums would tell you, after inspecting this last sum, that it converges to the finite value $\delta/(1 - \delta)$.[1] Convergence is guaranteed because increasingly large powers of a number less than 1, δ in this case, become smaller

[1]An infinite series *con*verges if the sum of the values in the series approaches a specific value, getting closer and closer to that value as additional components of the series are included in the sum. The series *di*verges if the sum of the values in the series gets increasingly larger (more negative) with each addition to the sum. Convergence requires that the components of the series get progressively smaller.

and smaller, approaching zero as n approaches infinity. The later terms in our present value, then, decrease in size until they get sufficiently small that the series approaches (but technically never exactly reaches) the particular value of the sum. Although a good deal of more sophisticated mathematics is required to deduce that the convergent value of the sum is $\delta/(1 - \delta)$, proving that this is the correct answer is relatively straightforward.

We use a simple trick to prove our claim. Consider the sum of the first m terms of the series, and denote it by S_m. Thus

$$S_m = \sum_{n=1}^{\infty} \delta^n = \delta + \delta^2 + \delta^3 + \cdots + \delta^{m-1} + \delta^m.$$

Now we multiply this sum by $(1 - \delta)$ to get

$$\begin{aligned} (1 - \delta) S_m &= \delta + \delta^2 + \delta^3 + \cdots + \delta^{m-1} + \delta^m \\ &\quad - \delta^2 - \delta^3 - \delta^4 - \cdots - \delta^m - \delta^{m-1} \\ &= \delta - \delta^{m-1}. \end{aligned}$$

Dividing both sides by $(1 - \delta)$, we have

$$S_m = \frac{\delta - \delta^{m+1}}{1 - \delta}.$$

Finally we take the limit of this sum as m approaches infinity to evaluate our original infinite sum. As m goes to infinity, the value of δ^{m+1} goes to zero because very large and increasing powers of a number less than 1 get increasingly small but stay nonnegative. Thus as m goes to infinity, the right-hand side of the preceding equation goes to $\delta/(1 - \delta)$, which is therefore the limit of S^m as m approaches infinity. This completes the proof.

We need only convert back into r to be able to use our answer in the calculation of present values in our prisoners' dilemma games. Because $\delta = 1/(1 + r)$, it follows that

$$\frac{\delta}{1 - \delta} = \frac{1/(1 + r)}{r/(1 + r)} = \frac{1}{r}.$$

The present value of an infinite stream of \$36s earned each month, starting next month, is then

$$36 \times \sum_{n=1}^{\infty} \frac{1}{(1 + r)^n} = \frac{36}{r}.$$

This is the value that we use to determine whether a player should defect forever in Section 2 of Chapter 11. Notice that incorporating a probability of continuation, $p \leq 1$, into the discounting calculations changes nothing in the summation procedure used here. We could easily substitute R for r in the preceding calculations, and $p\delta$ for the discount factor, δ.

Remember that you need to find present values only for losses (or gains) incurred (or accrued) *in the future*. The present value of \$36 lost today is just \$36. So if you wanted the present value of a stream of losses, all of them \$36, that begins *today*, you would take the \$36 lost today and add it to the present value of the stream of losses in the future. We have just calculated that present value as $36/r$. Thus the present value of the stream of lost \$36s, including the \$36 lost today, would be $36 + 36/r$, or $36[(r + 1)/r]$, which equals $36/(1 - \delta)$. Similarly, if you wanted to look at a player's stream of profits under a particular contingent strategy in a prisoners' dilemma, you would not discount the profit amount earned in the very first period; you would only discount those profit figures that represent money earned in future periods.

Collective-Action Games

THE GAMES AND STRATEGIC SITUATIONS considered in the preceding chapters have usually included only two or three players interacting with each other. Such games are common in our own academic, business, political, and personal lives and so are important to understand and analyze. But many social, economic, and political interactions are strategic situations in which numerous players participate at the same time. Strategies for career paths, investment plans, rush-hour commuting routes, and even studying have associated benefits and costs that depend on the actions of many other people. If you have been in any of these situations, you likely thought something was wrong—too many students, investors, and commuters crowding just where you wanted to be, for example. If you have tried to organize fellow students or your community in some worthy cause, you probably faced frustration of the opposite kind—too few willing volunteers. In other words, multiple-person games in society often seem to produce outcomes that are not deemed satisfactory by many or even all of the people in that society. In this chapter, we will examine such games from the perspective of the theory that we have already developed. We present an understanding of what goes wrong in such situations and what can be done about it.

In the most general form, such many-player games concern problems of **collective action.** The aims of the whole society or collective are best served if its members take some particular action or actions, but these actions are not in the best private interests of those individual members. In other words, the socially optimal outcome is not automatically achievable as the Nash equilibrium of the game. Therefore we must examine how the game can be modified to lead to the optimal outcome or at least to improve on an unsatisfactory Nash equilibrium. To do so, we must first understand the nature of such games. We find that they come in three forms, all of them familiar to you by now: the prisoners' dilemma, chicken, and assurance games. Although our main focus in this chapter is on situations where numerous individuals play such games at the same time, we build on familiar ground by beginning with games between just two players.

1 COLLECTIVE-ACTION GAMES WITH TWO PLAYERS

Imagine that you are a farmer. A neighboring farmer and you can both benefit by constructing an irrigation and flood-control project. The two of you can join together to undertake this project, or one of you might do so on your own. However, after the project has been constructed, the other automatically benefits from it. Therefore each is tempted to leave the work to the other. That is the essence of your strategic interaction, and the difficulty of securing collective action.

In Chapter 4, we encountered a game of this kind: three neighbors were each deciding whether to contribute to a street garden that all of them would enjoy. That game became a prisoners' dilemma in which all three shirked; our analysis here will include an examination of a more general range of possible payoff structures. Also, in the street-garden game, we rated the outcomes on a scale of 1 to 6; when we describe more general games, we will have to consider more general forms of benefits and costs for each player.

Our irrigation project has two important characteristics. First, its benefits are **nonexcludable:** a person who has not contributed to paying for it cannot be prevented from enjoying the benefits. Second, its benefits are **nonrival:** any one person's benefits are not diminished by the mere fact that someone else is also getting the benefit. Economists call such a project a **pure public good;** national defense is often given as an example. In contrast, a pure *private* good is fully excludable and rival: nonpayers can be excluded from its benefits, and if one person gets the benefit, no one else does. A loaf of bread is a good example of a pure private good. Most goods fall somewhere on the two-dimensional spectrum of varying degrees of excludability and rivalness. We will not go any deeper into this taxonomy, but we mention it to help you relate our discussion to what you may encounter in other courses and books.[1]

[1]Public goods are studied in more detail in textbooks on *public economics* such as those by Harvey Rosen and Ted Gayer, *Public Finance,* 8th ed. (Chicago: Irwin/McGraw-Hill, 2007), and Joseph Stiglitz, *Economics of the Public Sector,* 3rd ed. (New York: Norton, 2000).

A. Collective Action as a Prisoners' Dilemma

The costs and the benefits associated with building the irrigation project depend, as do those associated with all collective actions, on which players participate. In turn, the relative size of the costs and benefits determine the structure of the game that is played. Suppose each of you acting alone could complete the project in 7 weeks, whereas if the two of you acted together, it would take only 4 weeks of time from each. The two-person project is also of better quality; each farmer gets benefits worth 6 weeks of work from a one-person project (whether constructed by you or by your neighbor) and 8 weeks' worth of benefit from a two-person project.

More generally, we can write benefits and costs as functions of the number of players participating. So the cost to you of choosing to build the project depends on whether you build it alone or with help; costs can be written as $C(n)$ where cost, C, depends on the number, n, of players participating in the project. Then $C(1)$ would be the cost to you of building the project alone. $C(2)$ would be the cost *to you* of building the project with your neighbor, here $C(1) = 7$ and $C(2) = 4$. Similarly, benefits (B) from the completed project may vary depending on how many (n) participate in its completion. In our example, $B(1) = 6$ and $B(2) = 8$. Note that these benefits are the same for each farmer regardless of participation due to the public-good nature of this particular project.

In this game, each farmer has to decide whether to work toward the construction of the project or not—that is, to shirk. (Presumably, there is a short window of time in which the work must be done, and you could pretend to be called away on some very important family matter at the last minute, as could your neighbor.) Figure 12.1 shows the payoff table of the game, where the numbers measure the values in weeks of work. Payoffs are determined on the basis of the difference between the cost and the benefit associated with each action. So the payoff for choosing Build will be $B(n) - C(n)$ with $n = 1$ if you build alone and wih $n = 2$ if your neighbor also chooses Build. The payoff for choosing Not is just $B(1)$ if your neighbor chooses Build, because you incur no cost if you do not participate in the project.

		NEIGHBOR	
		Build	Not
YOU	Build	4, 4	−1, 6
	Not	6, −1	0, 0

FIGURE 12.1 Collective Action As a Prisoners' Dilemma: Version I

Given the payoff structure in Figure 12.1, your best response if your neighbor does not participate is not to participate either: your benefit from completing the project by yourself (6) is less than your cost (7), for a net payoff of −1, whereas you can get 0 by not participating. Similarly, if your neighbor does participate, then you can reap the benefit (6) from his work at no cost to yourself; this is better for you than working yourself to get the larger benefit of the two-person project (8) while incurring the cost of the work (4), for a net payoff of 4. The general feature of the game is that it is better for you not to participate no matter what your neighbor does; the same logic holds for him. (In this case, each farmer is said to be a **free rider** on his neighbor's effort if he lets the other do all the work and then reaps the benefits all the same.) Thus not building is the dominant strategy for each. But both would be better off if the two were to work together to build (payoff 4) than if neither builds (payoff 0). Therefore the game is a prisoners' dilemma.

We see in this prisoners' dilemma one of the main difficulties that arises in games of collective action. Individually optimal choices—in this case, not to build regardless of what the other farmer chooses—may not be optimal from the perspective of society as a whole, even if the society is made up of just two farmers. The **social optimum** in a collective-action game is achieved when the sum total of the players' payoffs is maximized; in this prisoners' dilemma, the social optimum is the (Build, Build) outcome. Nash-equilibrium behavior of the players does not consistently bring about the socially optimal outcome, however. Hence, the study of collective-action games has focused on methods to improve on observed (generally Nash) equilibrium behavior to move outcomes toward the socially best ones. As we will see, the divergence between Nash equilibrium and socially optimum outcomes appears in every version of collective-action games.

Now consider what the game would look like if the numbers were to change slightly. Suppose the two-person project yields benefits that are not much better than those in the one-person project: 6.3 weeks' worth of work to each farmer. Then each of you gets 6.3 − 4 = 2.3 when both of you build. The resulting payoff table is shown in Figure 12.2. The game is still a prisoners' dilemma and leads to

		NEIGHBOR	
		Build	Not
YOU	Build	2.3, 2.3	−1, 6
	Not	6, −1	0, 0

FIGURE 12.2 Collective Action As a Prisoners' Dilemma: Version II

the equilibrium (Not, Not). However, when both farmers build, the total payoff for both of you is only 4.6. The social optimum occurs when one of you builds and the other does not, in which case together you get payoff $6 + (-1) = 5$. There are two possible ways to get this outcome. Achieving the social optimum in this case then poses a new problem: Who should build and suffer the payoff of -1 while the other is allowed to be a free rider and enjoy the payoff of 6?

B. Collective Action as Chicken

Yet another variation in the numbers of the original prisoners' dilemma game of Figure 12.1 changes the nature of the game. Suppose the cost of the work is reduced so that it becomes better for you to build your own project if your neighbor does not. Specifically, suppose the one-person project requires 4 weeks of work, so $C(1) = 4$, and the two-person project takes 3 weeks from each, so $C(2) = 3$ (to each); the benefits are the same as before. Figure 12.3 shows the payoff matrix resulting from these changes. Now your best response is to shirk when your neighbor works and to work when he shirks. In form, this game is just like a game of chicken, where shirking is the Straight strategy (tough or uncooperative), and working is the Swerve strategy (conciliatory or cooperative).

If this game results in one of its pure-strategy equilibria, the two payoffs sum to 8; this total is less than the total outcome that both players could get if both of them build. That is, neither of the Nash equilibria provides so much benefit to society as a whole as that of the coordinated outcome, which entails both farmers' choosing to build. The social optimum yields a total payoff of 10. If the outcome of the chicken game is its mixed-strategy equilibrium, the two farmers will fare even worse than in either of the pure-strategy equilibria: their expected payoffs will add up to something less than 8 (4, to be precise).

The collective-action chicken game has another possible structure if we make some additional changes to the benefits associated with the project. As with version II of the prisoners' dilemma, suppose the two-person project is not much better than the one-person project. Then each farmer's benefit from the two-person project, $B(2)$, is only 6.3, whereas each still gets a benefit of $B(1) = 6$

		NEIGHBOR	
		Build	Not
YOU	Build	5, 5	2, 6
	Not	6, 2	0, 0

FIGURE 12.3 Collective Action As Chicken: Version I

from the one-person project. We ask you to practice your skill by constructing the payoff table for this game. You will find that it is still a game of chicken—call it chicken II. It still has two pure-strategy Nash equilibria in each of which only one farmer builds, but the sum of the payoffs when both build is only 6.6, whereas the sum when only one farmer builds is 8. The social optimum is for only one farmer to build. Each farmer prefers the equilibrium in which the other builds. This may lead to a new dynamic game in which each waits for the other to build. Or the original game might yield its mixed-strategy equilibrium with its low expected payoffs.

C. Collective Action as Assurance

Finally, let us change the payoffs of the original prisoners' dilemma case in a different way altogether, leaving the benefits of the two-person project and the costs of building as originally set out and reducing the benefit of a one-person project to $B(1) = 3$. This change reduces your benefit as a free rider so much that now if your neighbor chooses Build, your best response also is Build. Figure 12.4 shows the payoff table for this version of the game. This is now an assurance game with two pure-strategy equilibria: one where both of you participate and the other where neither of you does.

As in the chicken II version of the game, the socially optimal outcome here is one of the two Nash equilibria. But there is a difference. In chicken II, the two players differ in their preferences between the two equilibria, either of which achieves the social optimum. In the assurance game, both of them prefer the same equilibrium, and that is the sole socially optimal outcome. Therefore achieving the social optimum should be easier in the assurance game than in chicken.

D. Collective Inaction

Many games of collective action have payoff structures that differ somewhat from those in our irrigation project example. Our farmers find themselves in a situation in which the social optimum generally entails that at least one, if not

		NEIGHBOR	
		Build	Not
YOU	Build	4, 4	−4, 3
	Not	3, −4	0, 0

FIGURE 12.4 Collective Action as an Assurance Game

both, of them participates in the project. Thus the game is one of collective *action*. Other multiplayer games might better be called games of collective *inaction*. In such games, society as a whole prefers that some or all of the individual players do *not* participate or do *not* act. Examples of this type of interaction include choices between rush-hour commuting routes, investment plans, or fishing grounds.

All of these games have the attribute that players must decide whether to take advantage of some common resource, be it a freeway, a high-yielding stock fund, or an abundantly stocked pond. These collective "inaction" games are better known as common-resource games; the total payoff to all players reaches its maximum when players refrain from overusing the common resource. The difficulty associated with not being able to reach the social optimum in such games is known as the "tragedy of the commons," a phrase coined by Garrett Hardin in his paper of the same name.[2]

We supposed above that the irrigation project yielded equal benefits to both you and your farmer-neighbor. But what if the outcome of both farmers' building was that the project used so much water that the farms had too little water for their livestock? Then each player's payoff could be negative when both choose Build, lower than when both choose Not. This would be yet another variant of the prisoners' dilemma we encountered in Section 1.A, in which the socially optimal outcome entails neither farmer's building even though each one still has an individual incentive to do so. Or suppose that one farmer's activity causes harm to the other, as would happen if the only way to prevent one farm from being flooded is to divert the water to the other. Then each player's payoffs could be negative if his neighbor chose Build. Thus another variant of chicken could also arise. In this variant, each of you wants to build when the other does not, whereas it would be collectively better if neither of you did.

Just as the problems pointed out in these examples of both collective action and collective inaction are familiar, the various alternative ways of tackling the problems also follow the general principles discussed in earlier chapters. Before turning to solutions, let us see how the problems manifest themselves in the more realistic setting where several players interact simultaneously in such games.

2 COLLECTIVE-ACTION PROBLEMS IN LARGE GROUPS

In this section, we extend our irrigation-project example to a situation in which a population of N farmers must each decide whether to participate. Here we make use of the notation we introduced above, with $C(n)$ representing the cost

[2] Garrett Hardin, "The Tragedy of the Commons," *Science*, vol. 162 (1968), pp. 1243–1248.

each participant incurs when n of the N total farmers have chosen to participate. Similarly, the benefit to each, regardless of participation, is $B(n)$. Each participant then gets the payoff $P(n) = B(n) - C(n)$, whereas each nonparticipant, or shirker, gets the payoff $S(n) = B(n)$.

Suppose you are contemplating whether to participate or to shirk. Your decision will depend on what the other $(N - 1)$ farmers in the population are doing. In general, you will have to make your decision when the other $(N - 1)$ players consist of n participants and $(N - 1 - n)$ shirkers. If you decide to shirk, the number of participants in the project is still n, so you get a payoff of $S(n)$. If you decide to participate, the number of participants becomes $n + 1$, so you get $P(n + 1)$. Therefore your final decision depends on the comparison of these two payoffs; you will participate if $P(n + 1) > S(n)$, and you will shirk if $P(n + 1) < S(n)$. This comparison holds true for every version of the collective-action game analyzed in Section 1; differences in behavior in the different versions arise because the changes in the payoff structure alter the values of $P(n + 1)$ and $S(n)$.

We can relate the two-person examples of Section 1 to this more general framework. If there are just two people, then $P(2)$ is the payoff to one from building when the other also builds, $S(1)$ is the payoff to one from shirking when the other builds, and so on. Therefore we can generalize the payoff tables of Figures 12.1 through 12.4 into an algebraic form. This general payoff structure is shown in Figure 12.5.

The game illustrated in Figure 12.5 is a prisoners' dilemma if the inequalities

$$P(2) < S(1),\ P(1) < S(0),\ P(2) > S(0)$$

all hold at the same time. The first says that the best response to Build is Not, the second says that the best response to Not also is Not, and the third says that (Build, Build) is jointly preferred to (Not, Not). The dilemma is of Type I if $2P(2) > P(1) + S(1)$, so the total payoff is higher when both build than when only one builds. You can establish similar inequalities concerning these payoffs that yield the other types of games in Section 1.

Return now to the multiplayer version of the game with a general n. Given the payoff functions for the two actions, $P(n + 1)$ and $S(n)$, we can use graphs to

		NEIGHBOR	
		Build	Not
YOU	Build	$P(2), P(2)$	$P(1), S(1)$
	Not	$S(1), P(1)$	$S(0), S(0)$

FIGURE 12.5 General Form of a Two-Person Collective-Action Game

help us determine which type of game we have encountered and its Nash equilibrium. We can also then compare the Nash equilibrium to the game's socially optimal outcome.

Take a specific version of our irrigation project example in which an entire village of 100 farmers must decide which action to take. Suppose that an irrigation project raises the productivity of each farmer's land in proportion to the size of the project; specifically, suppose the benefit to each farmer when n people work on the project is $P(n) = 2\ n$. Suppose also that if you are not working on the project, you can enjoy this benefit and use your time to earn an extra 4 in some other occupation, so $S(n) = 2n + 4$. Remember that your decision about whether to participate in the project depends on the relative magnitudes of $P(n + 1) = 2(n + 1)$ and $S(n) = 2n + 4$. We draw the two separate graphs of these functions for an individual farmer in Figure 12.6, showing n over its full range from 0 to $(N - 1)$ along the horizontal axis and the payoff to the farmer along the vertical axis. If there are currently very few participants (thus mostly shirkers), your choice will depend on the relative locations of $P(n + 1)$ and $S(n)$ on the left end of the graph. Similarly, if there are already many participants, your choice will depend on the relative locations of $P(n + 1)$ and $S(n)$ on the *right* end of the graph.

Because n actually takes on only integer values, each function $P(n + 1)$ and $S(n)$ technically consists only of a discrete set of points rather than a continuous set as implied by our smooth lines. But when N is large, the discrete points are sufficiently close together that we can connect the successive points and show each payoff function as a continuous curve. We also use linear $P(n + 1)$ and $S(n)$

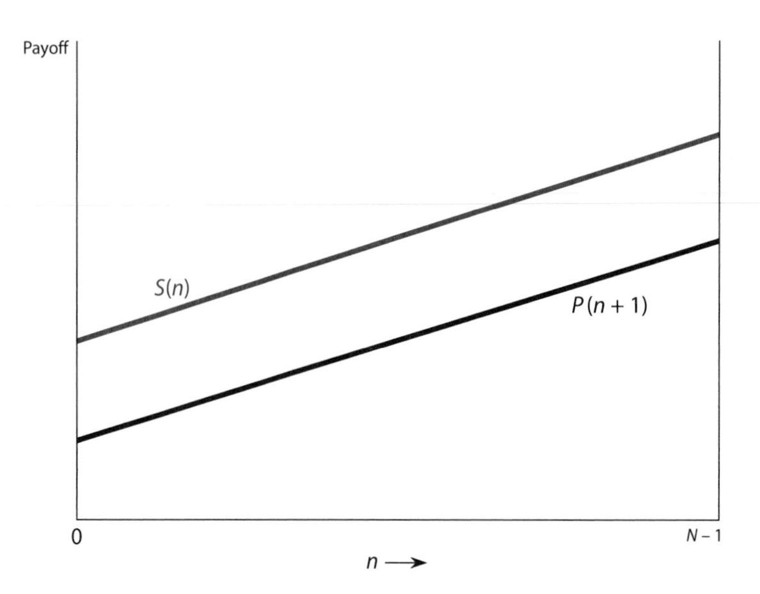

FIGURE 12.6 Multiplayer Prisoners' Dilemma

functions in this section to bring out the basic considerations and will discuss more complicated possibilities later.

Recall that you determine your choice of action by considering the number of current participants in the project, n, and the payoffs associated with each action at that n. Figure 12.6 illustrates a case in which the curve $S(n)$ lies entirely above the curve $P(n + 1)$. Therefore, no matter how many others participate (that is, no matter how large n gets), your payoff is higher if you shirk than if you participate; shirking is your dominant strategy. These payoffs are identical for all players, so everyone has a dominant strategy to shirk. Therefore the Nash equilibrium of the game entails everyone shirking, and the project is not built.

Note that both curves are rising as n increases. For each action you can take, you are better off if more of the others participate. And the left intercept of the $S(n)$ curve is below the right intercept of the $P(n + 1)$ curve, or $S(0) = 4 < P(N) = 102$. This says that if everyone including you shirks, your payoff is less than if everyone including you participates. Everyone would be better off than they are in the Nash equilibrium of the game if the outcome in which everyone participates could be sustained. This makes the game a prisoners' dilemma.

How does the Nash equilibrium found using the curves in Figure 12.6 compare with the social optimum of this game? To answer this question we need a way to describe the total social payoff at each value of n; we do that by using the payoff functions $P(n)$ and $S(n)$ to construct a third function $T(n)$, showing the total payoff to society as a function of n. The total payoff to society when there are n participants consists of the value $P(n)$ for each of the n participants and the value $S(n)$ for each of the $(N - n)$ shirkers:

$$T(n) = nP(n) + (N - n)\,S(n)$$

The social optimum occurs when the allocation of people between participants and shirkers maximizes the total payoff $T(n)$, or at the number of participants—that is, the value of n—that maximizes $T(n)$. To get a better understanding of where this might be, it is convenient to write $T(n)$ differently, rearranging the expression above to get

$$T(n) = NS(n) - n\,[S(n) - P(n)].$$

This version of the total social payoff function shows that we can calculate it as if we gave every one of the N people the shirker's payoff but then removed the shirker's extra benefit $[S(n) - P(n)]$ from each of the n participants.

In collective-action games, as opposed to common-resource games, we normally expect $S(n)$ to increase as n increases. Therefore the first term in this expression, $NS(n)$, also increases as n increases. If the second term does not increase too fast as n increases—as would be the case if the shirker's extra benefit, $[S(n) - P(n)]$, is small and constant—then the effect of the first term dominates in determining the value of $T(n)$.

This is exactly what happens with the total social payoff function for our 100-farmer example. Here $T(n) = n\,P(n) + (N - n)\,S(n)$ becomes $T(n) = n(2n) + (100 - n)\,(2n + 4) = 2n^2 + 200n - 2n^2 + 400 - 4n = 400 + 196n$. In this case, $T(n)$ increases steadily with n and is maximized at $n = N$ when no one shirks.

The large-group version of our two-person example holds the same lesson as above. Society as a whole would be better off if all of the farmers participated in building the irrigation project. But payoffs are such that each farmer has an individual incentive to shirk. The Nash equilibrium of the game is not socially optimal. Figuring out how to achieve the social optimum is one of the most important topics in the study of collective action and one to which we return later in this chapter.

In other situations, $T(n)$ can be maximized for a different value of n, not just at $n = N$. That is, society's aggregate payoff could be maximized by allowing some shirking. Even in the prisoners' dilemma case, it is not automatic that the total payoff function is maximized when n is as large as possible. If the gap between $S(n)$ and $P(n)$ widens sufficiently fast as n increases, then the negative effect of the second term in the expression for $T(n)$ outweighs the positive effect of the first term as n approaches N; then it may be best to let some people shirk—that is, the socially optimal value for n may be less than N.

This type of outcome would arise in our village if $S(n)$ were $4n + 4$, rather than $2n + 4$. Then $T(n) = -2n^2 + 396n + 400$, which is no longer linear in n. In fact, a graphing calculator or some basic calculus shows that this $T(n)$ is maximized at $n = 99$ rather than at $n = 100$ as was true before. The change to the payoff structure has created an inequality in the payoffs—the shirkers fare better than the participants—which adds another dimension of difficulty to society's attempts to resolve the dilemma. How, for example, would the village designate exactly one farmer to be the shirker?

Now we consider some of the other configurations that can arise in the payoffs. For example, when $P(n) = 4n + 36$, so $P(n + 1) = 4n + 40$, and $S(n) = 5n$, the two payoff curves will cross in the figure. This case is illustrated in Figure 12.7. Here, for small values of n, $P(n + 1) > S(n)$, so if few others are participating, your choice is to participate. For large values of n, $P(n + 1) < S(n)$, so if many others are participating, your choice is to shirk. Note the equivalence of these two statements to the idea in the two-person chicken game that "you shirk if your neighbor works and you work if he shirks." This case is indeed that of chicken. More generally, the chicken case occurs when given a choice between two actions, you prefer to do the one that most others are *not* doing.

We can also use Figure 12.7 to determine the location of the Nash equilibrium of this version of the game. Because you choose to participate when n is small and to shirk when n is large, the equilibrium must be some intermediate value of n. Only at that n where the two curves intersect are you indifferent between your two choices. This location represents the equilibrium value of n. In our graph, $P(n + 1) = S(n)$ when $4n + 40 = 5n$ or when $n = 40$; that is the

Nash equilibrium number of farmers from the village who will participate in the irrigation project.

If the two curves intersect at a point corresponding to an integer value of n, then that is the Nash equilibrium number of participants. If that is not the case, then strictly speaking the game has no Nash equilibrium. But in practice, if the current value of n in the population is the integer just to the left of the point of intersection, then one more person will just want to participate, whereas if the current value of n is the integer just to the right of the point of intersection, one person will want to switch to shirking. Therefore the number of participants will stay in a small neighborhood of the point of intersection, and we can justifiably speak of the intersection as the equilibrium in some approximate sense.

The payoff structure illustrated in Figure 12.7 shows both lines positively sloped, although they don't have to be. It is conceivable that the benefit for each person is smaller when more people participate, so the lines could be negatively sloped instead. The important feature of the chicken collective-action game is that when few people are taking one action, it is better for any one person to take that action; when many people are taking one action, it is better for any one person to take the other action.

What is the socially optimal outcome in the chicken form of collective action? If each participant's payoff $P(n)$ increases as the number of participants increases, and if each shirker's payoff $S(n)$ does not become too much greater than the $P(n)$ of each participant, then the total social payoff is maximized when everyone participates. This is the outcome in our example where $T(n) = 536n - n^2$; total social payoff increases in n beyond the value of N (100 here), so $n = N$ is the social optimum.

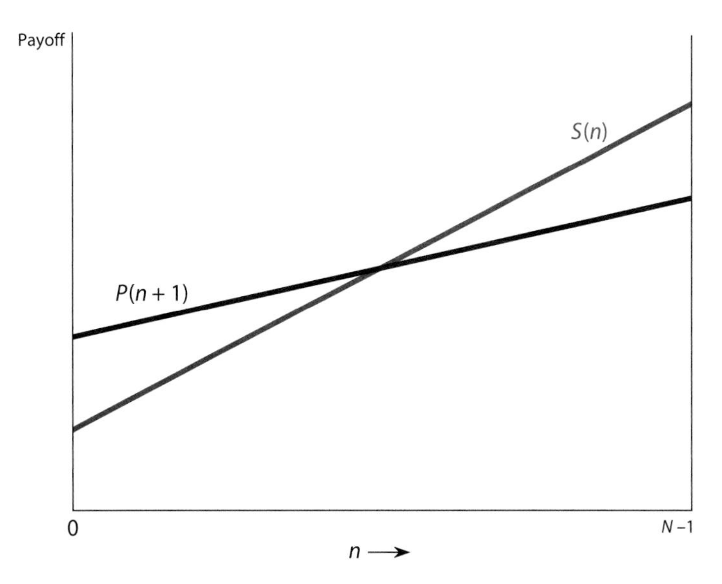

FIGURE 12.7 Multiplayer Chicken

But more generally, some cases of chicken will entail social optima in which it is better to let some shirk. This is exactly the difference between versions I and II of chicken in our example in Section 1. For an exercise, you may try generating a payoff structure that leads to such an outcome for our village farmers. In these more general chicken games, the optimal number of participants could even be smaller than that in the Nash equilibrium. We return to examine the question of the social optimum of all of these versions of the game in greater detail in Section 3.

Finally we consider the third possible type of collective-action game, assurance. Figure 12.8 shows the payoff lines for the assurance case, where we suppose that the village farmers get $P(n + 1) = 4n + 4$ and $S(n) = 2n + 100$. Here $S(n) > P(n + 1)$ for small values of n, so if few others are participating, then you want to shirk, too. But $P(n + 1) > S(n)$ for large values of n, so if many others are participating, then you want to participate too. In other words, unlike chicken, assurance is a collective-action game in which you want to make the choice that the others are making.

Except for the labels, the graph in Figure 12.8 looks nearly identical to that in Figure 12.7. The location of the Nash equilibrium depends critically on the labels associated with the two lines, however. In Figure 12.8, for any initial value of n to the left of the intersection, each farmer will want to shirk, and there will be a Nash equilibrium at $n = 0$ where everyone shirks. But the opposite is true to the right of the intersection. In that portion of the graph, each farmer will want to participate, and there will be a second Nash equilibrium at $n = N$.

Technically, there is also a third Nash equilibrium of this game if the value of n at the intersection is an integer value as it is in our example. There we find that $P(n + 1) = 4n + 4 = 2n + 100 = S(n)$ when $n = 48$. Then if n were exactly 48, we

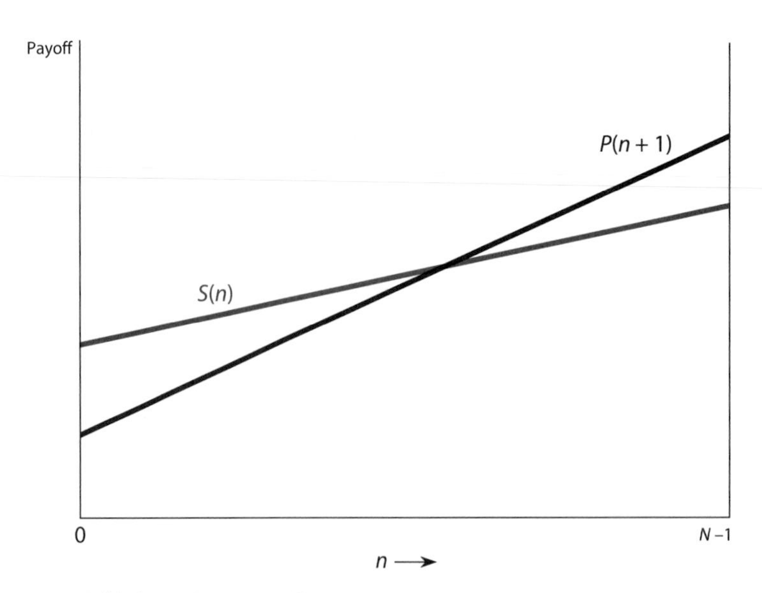

FIGURE 12.8 Multiplayer Assurance Game

would see an outcome in which there were some participants and some shirkers. This situation could be an equilibrium only if the value of n is exactly right. Even then, it would be a highly unstable situation. If any one farmer accidentally joined the wrong group, his choice would alter the incentives for everyone else, driving the game to one of the endpoint equilibria. Those are the two stable Nash equilibria of the game.

The social optimum in this game is fairly easy to see on the graph in Figure 12.8. Because both curves are rising—so each person is better off if more people participate—then clearly the right-hand extreme equilibrium is the better one for society. This is confirmed in our example by noting that $T(n) = 2n^2 + 100n + 10{,}000$, which is increasing in n for all positive values of n; thus the socially optimal value of n is the largest one possible, or $n = N$. In the assurance case, then, the socially optimal outcome is actually one of the stable Nash equilibria of the game. As such, it may be easier to achieve than in some of the other cases. The critical question regarding the social optimum, regardless of whether it represents a Nash equilibrium of the underlying game, is how to bring it about.

So far our examples have focused on relatively small groups of 2 or 100 persons. When the total number of people in the group, N, is very large, however, and any one person makes only a very small difference, then $P(n + 1)$ is almost the same as $P(n)$. Thus, the condition under which any one person chooses to shirk is $P(n) < S(n)$. Expressing this inequality in terms of the benefits and costs of the common project in our example—namely, $P(n) = B(n) - C(n)$ and $S(n) = B(n)$—we see that $P(n)$ (unlike $P(n + 1)$ in our preceding calculations) is *always* less than $S(n)$; individual persons will *always* want to shirk when N is very large. That is why problems of collective provision of public projects in a large group almost always manifest themselves as prisoners' dilemmas. But as we have seen, this result is not necessarily true for smaller groups. Neither is it true for large groups in other contexts such as congestion; we will discuss this case later in this chapter.

In general, we must allow for a broader interpretation of the payoffs $P(n)$ and $S(n)$ than we did in the specific case involving the benefits and the costs of a project. We cannot assume, for example, that the payoff functions will be linear. In fact, in the most general case, $P(n)$ and $S(n)$ can be any functions of n and can intersect many times. Then there can be multiple equilibria, although each can be thought of as representing one of the types described so far.[3] And some games

[3] Several exercises at the end of this chapter present some examples of simple situations with nonlinear payoff curves and multiple equilibria. For a more general analysis and classification of such diagrams, see Thomas Schelling, *Micromotives and Macrobehavior* (New York: Norton, 1978), chap. 7. The theory can be taken further by allowing each player a continuous choice (for example, the number of hours of participation) instead of just a binary choice of whether to participate. Many such situations are discussed in more specialized books on collective action, for example, Todd Sandler, *Collective Action: Theory and Applications* (Ann Arbor: University of Michigan Press, 1993), and Richard Cornes and Todd Sandler, *The Theory of Externalities, Public Goods, and Club Goods*, 2nd ed. (New York: Cambridge University Press, 1996).

will be of the common-resource type as well, so when we allow for completely general games, we will speak of two actions labeled P and S, which have no necessary connotation of "participation" and "shirking" but allow us to continue with the same symbols for the payoffs. Thus, when n players are taking the action P, $P(n)$ becomes the payoff of each player taking the action P, and $S(n)$ becomes that of each player taking the action S.

3 SPILLOVERS, OR EXTERNALITIES

So far we have seen that collective-action games occur in prisoners' dilemma, chicken, and assurance forms. We have also seen that the Nash equilibria in such games rarely yield the socially optimal level of participation (or restraint). And even when the social optimum is a Nash equilibrium, it is usually only one of several equilibria that may arise. Now we delve further into the differences between the individual (or private) incentives in such games and the group (or social) incentives. We also describe more carefully the effects of each individual's decision on other individuals as well as on the collective. This analysis makes explicit why differences in incentives exist, how they are manifested, and how one might go about achieving socially better outcomes than those that arise in Nash equilibrium.

A. Commuting and Spillovers

We start by thinking about a large group of 8,000 commuters who drive every day from a suburb to the city and back. As one of these commuters, you may take either the expressway (action P) or a network of local roads (action S). The local-roads route takes a constant 45 minutes, no matter how many cars are going that way. The expressway takes only 15 minutes when uncongested. But every driver who chooses the expressway increases the time for every other driver on the expressway by 0.005 minutes (about one-quarter of a second).

Measure the payoffs in minutes of time saved—by how much the commute time is less than 1 hour, for instance. Then the payoff to drivers on the local roads, $S(n)$, is a constant $60 - 45 = 15$, regardless of the value of n. But the payoff to drivers on the expressway, $P(n)$, depends on n; in particular, $P(n) = 60 - 15 = 45$ for $n = 0$, but $P(n)$ decreases by 5/1,000 (or 1/200) for every commuter on the expressway. Thus, $P(n) = 45 - 0.005n$.

Suppose that initially 4,000 cars are on the expressway; $n = 4000$. With so many cars on that road, it takes each of them $15 + 4{,}000 \times 0.005 = 15 + 20 = 35$ minutes to commute to work; each gets a payoff of $P(n) = 25$ [which is $60 - 35$, or $P(4{,}000)$]. Now consider the possibility that you, a local-road driver, might decide to switch from driving the local roads to driving on the expressway. Your switch would increase by 1 the value of n and would thereby affect the payoffs of all the other commuters. There would now be 4,001 drivers (including you) on the expressway, and the commute time for each would be 35 and 1/200, or 35.005, minutes; each would now get a payoff of $P(n + 1) = P(4{,}001) = 24.995$. This payoff is higher than the 15 from driving on the local roads. Thus you have a private incentive to make the switch, because for you, $P(n + 1) > S(n)$ $(24.995 > 15)$.

Your switch yields you a *private* gain—because it is privately held by you—equal to the difference between your payoffs before and after the switch; this private gain is $P(n + 1) - S(n) = 9.995$ minutes. Because you are only one person and therefore a small part of the whole group, the gain in payoff that you receive in relation to the total group payoff is small, or *marginal*. Thus we call your gain the **marginal private gain** associated with your switch.

But now the 4,000 other drivers on the expressway each take 0.005 of a minute more as a result of your decision to switch; the payoff to each changes by $P(4{,}001) - P(4{,}000) = -0.005$. Similarly, the drivers on the local roads face a payoff change of $S(4{,}001) - S(4{,}000)$, but this is zero in our example. The cumulative effect on all of these other drivers is $4{,}000 \times -0.005 = -20$ (minutes). Your action, switching from local roads to expressway, has caused this effect on the others' payoffs. Whenever one person's action affects others like this, it is called a **spillover effect, external effect,** or **externality.** Again, because you are but a very small part of the whole group, we should actually call your effect on others the *marginal spillover effect.*

Taken together, the marginal private gain and the marginal spillover effect are the full effect of your switch on the group of commuters, or the overall marginal change in the whole group's or the whole society's payoff. We call this the **marginal social gain** associated with your switch. This "gain" may actually be positive or negative, so the use of the word *gain* is not meant to imply that all switches will benefit the group as a whole. In fact, in our commuting example, the overall marginal social gain is $9.995 - 20 = -10.005$ (minutes). Thus, the overall social effect of your switch is bad; the social payoff is reduced by a total of just over 10 minutes.

B. Spillovers: The General Case

We can describe the effects we observe in the commuting example more generally by returning to our total social payoff function, $T(n)$, where n represents the number of people choosing P, so $N - n$ is the number of people choosing S. Suppose that initially n people have chosen P and that one person switches from S to P. Then the number choosing P increases by 1 to $(n + 1)$, and the number choosing S decreases by 1 to $(N - n - 1)$, so the total social payoff becomes

$$T(n+1) = (n+1)\,P(n+1) + [N-(n+1)]\,S(n+1).$$

The increase in the total social payoff is the difference between $T(n)$ and $T(n+1)$:

$$\begin{aligned} T(n+1) - T(n) &= (n+1)\,P(n+1) + [N-(n+1)]\,S(n+1) - n\,P(n) + (N-n)\,S(n) \\ &= [P(n+1) - S(n)] + n\,[P(n+1) - P(n)] \\ &\quad + [N-(n+1)]\,[S(n+1) - S(n)] \end{aligned} \tag{12.1}$$

after collecting and rearranging terms.

Equation (12.1) describes mathematically the various different effects of one person's switch from S to P that we saw earlier in the commuting example. The equation shows how the marginal social gain is divided into the marginal change in payoffs for the subgroups of the population.

The first of the three terms in Eq. (12.1)—namely, $[P(n+1) - S(n)]$—is the marginal private gain enjoyed by the person who switches. As we saw above, this term is what drives a person's choice, and all such individual choices then determine the Nash equilibrium.

The second and third terms in Eq. (12.1) are just the quantifications of the *spillover effects* of one person's switch on the others in the group. For the n other people choosing P, each sees his payoff change by the amount $[P(n+1) - P(n)]$ when one more person switches to P; this spillover effect is seen in the second group of terms in Eq. (12.1). There are also $N - (n+1)$ (or $N - n - 1$) others still choosing S after the one person switches, and each of these players sees his payoff change by $[S(n+1) - S(n)]$; this spillover effect is shown in the third group of terms in the equation. Of course, the effect that one driver's switch has on the time for any one driver on either route is very small, but, when there are numerous other drivers (that is, when N is large), the full spillover effect can be substantial.

Thus we can rewrite Eq. (12.1) for a general switch of one person from either S to P or P to S as:

Marginal social gain = marginal private gain + marginal spillover effect.

For an example in which one person switches from S to P, we have

$$\begin{aligned} \text{Marginal social gain} &= T(n+1) - T(n), \\ \text{Marginal private gain} &= P(n+1) - S(n), \text{ and} \\ \text{Marginal spillover effect} &= n[P(n+1) - P(n)] + [N-(n+1)]\,[S(n+1) - S(n)]. \end{aligned}$$

USING CALCULUS FOR THE GENERAL CASE Before examining some spillover situations in more detail to see what can be done to achieve socially better outcomes, we restate the general concepts of the analysis in the language of calculus. If you do not know this language, you can omit the remainder of this section without loss of continuity; if you do know it, you will find the alternative statement much simpler to grasp and to use than the algebra employed earlier.

If the total number N of people in the group is very large—say, in the hundreds or thousands—then one person can be regarded as a very small, or infinitesimal, part of this whole. This allows us to treat the number n as a continuous variable. If $T(n)$ is the total social payoff, we calculate the effect of changing n by considering an increase of an infinitesimal marginal quantity dn, instead of a full unit increase from n to $(n + 1)$. To the first order, the change in payoff is $T'(n)dn$, where $T'(n)$ is the derivative of $T(n)$ with respect to n. Using the expression for the total social payoff,

$$T(n) = nP(n) + (N - n)\,S(n),$$

and differentiating, we have

$$\begin{aligned} T'(n) &= P(n) + nP'(n) - S'(n) + (N - n)S'(n) \\ &= [P(n) - S(n)] + nP'(n) + (N - n)S'(n). \end{aligned} \tag{12.2}$$

This is the calculus equivalent of Eq. (12.1). $T'(n)$ represents the marginal social gain. The marginal private gain is $P(n) - S(n)$, which is just the change in the payoff of the person making the switch from S to P. In Eq. (12.1), we had $P(n + 1) - S(n)$ for this change in payoff; now we have $P(n) - S(n)$. This is because the infinitesimal addition of dn to the group of the n people choosing P does not change the payoff to any one of them by a significant amount. However, the total change in their payoff, $nP'(n)$, is sizable and is recognized in the calculation of the spillover effect—it is the second term in Eq. (12.2)—as is the change in the payoff of the $(N - n)$ people choosing S—namely, $(N - n)\,S'(n)$—the third term in Eq. (12.2). These last two terms constitute the marginal-spillover-effect part of Eq. (12.2).

In the commuting problem, we have $P(n) = 45 - 0.005n$, and $S(n) = 15$. Then with the use of calculus, we see that the private marginal gain for each driver who switches to the expressway when n drivers are already using it is $P(n) - S(n) = 30 - 0.005n$. Because $P'(n) = -0.005$ and $S'(n) = 0$, the spillover effect is $n \times (-0.005) + (N - n) \times 0 = -0.005n$, which equals -20 when $n = 4{,}000$. The answer is the same as before, but calculus simplifies the derivation and helps us find the optimum directly.

C. Commuting Revisited: Negative Externalities

A negative externality exists when the action of one person *lowers* others' payoffs; it imposes some extra costs on the rest of society. We saw this in our commuting example, where the marginal spillover effect of one person's switch to the expressway was negative, entailing an extra 20 minutes of drive time for other commuters. But the individual who changes his route to work does not take the spillover—the externality—into account when making his choice. He is motivated only by his own payoffs. (Remember that any guilt that he may suffer from harming others should already be reflected in his payoffs.) He will change his action from S to P as long as this change has a positive marginal *private* gain. He is then made better off by the change.

But society would be better off if the commuter's decision were governed by the marginal *social* gain. In our example, the marginal social gain is negative (-10.005), but the marginal private gain is positive (9.995), so the individual driver makes the switch even though society as a whole would be better off if he did not do so. More generally, in situations with negative externalities, the marginal social gain will be smaller than the marginal private gain due to the existence of the negative spillover effect. Individuals will make decisions based on a cost-benefit calculation that is the wrong one from society's perspective. As a result, individual persons will choose actions with negative spillover effects more often than society would like them to do.

We can use Eq. (12.1) to calculate the precise conditions under which a switch will be beneficial for a particular person versus for society as a whole. Recall that if n people are already using the expressway and another driver is contemplating switching from the local roads to the expressway, he stands to gain from this switch if $P(n + 1) > S(n)$, whereas the total social payoff increases if $T(n + 1) - T(n) > 0$. The private gain is positive if

$$45 - (n + 1) \times 0.005 > 15$$
$$44.995 - 0.005n > 15$$
$$n < 200\,(44.995 - 15) = 5{,}999,$$

whereas the condition for the social gain to be positive is

$$45 - (n + 1) \times 0.005 - 15 - 0.005n > 0$$
$$29.995 - 0.01n > 0$$
$$n < 2{,}999.5.$$

Thus, if given the free choice, commuters will crowd onto the expressway until there are almost 6,000 of them, but all crowding beyond 3,000 reduces the total social payoff. Society as a whole would be best off if the number of commuters on the expressway were kept down to 3,000.

We show this result graphically in Figure 12.9. On the horizontal axis from left to right, we measure the number of commuters using the expressway. On the vertical axis, we measure the payoffs to each commuter when n others are using the expressway. The payoff to each driver on the local road is constant and equal to 15 for all n; this is shown by the horizontal line $S(n)$. The payoff to each driver who switches to the expressway is shown by the line $P(n + 1)$; it falls by 0.005 for each unit increase in n. The two lines meet at $n = 5{,}999$; that is, at the value of n for which $P(n + 1) = S(n)$ or for which the marginal private gain is just zero. Everywhere to the left of this value of n, any one driver on the local roads calculates that he gets a positive gain by switching to the expressway. As some drivers make this switch, the numbers on the expressway increase—the value of n in society rises. Conversely, to the right of the intersection point (that is, for $n > 5{,}999$), $S(n) > P(n + 1)$; so each of the $(n + 1)$ drivers on the expressway stands to gain by switching to the local road. As some do so, the numbers on the expressway decrease and n falls. From the left of the intersection, this process converges to $n = 5{,}999$ and, from the right, it converges to 6,000.

If we had used the calculus approach, we would have regarded 1 as a very small increment in relation to n and graphed $P(n)$ instead of $P(n + 1)$. Then the intersection point would have been at $n = 6{,}000$ instead of at 5,999. As you can see, it makes very little difference in practice. What this means is that we can call $n = 6{,}000$ the Nash equilibrium of the route-choice game when choices are governed by purely individual considerations. Given a free choice, 6,000 of the 8,000 total commuters will choose the expressway, and only 2,000 will drive on the local roads.

But we can also interpret the outcome in this game from the perspective of the whole society of commuters. Society benefits from an increase in the number of commuters, n, on the expressway when $T(n + 1) - T(n) > 0$ and loses from an increase in n when $T(n + 1) - T(n) < 0$. To figure out how to show this on the graph, we express the idea somewhat differently; we rearrange Eq. (12.1) into two pieces, one depending only on P and the other depending only on S:

$$\begin{aligned} T(n+1) - T(n) &= (n+1)\,P(n+1) + [N-(n+1)]\,S(n+1) - nP(n) - [N-n]\,S(n) \\ &= \{P(n+1) + n[P(n+1) - P(n)]\} \\ &\quad - \{S(n) + [N-(n+1)][S(n+1) - S(n)]\}. \end{aligned}$$

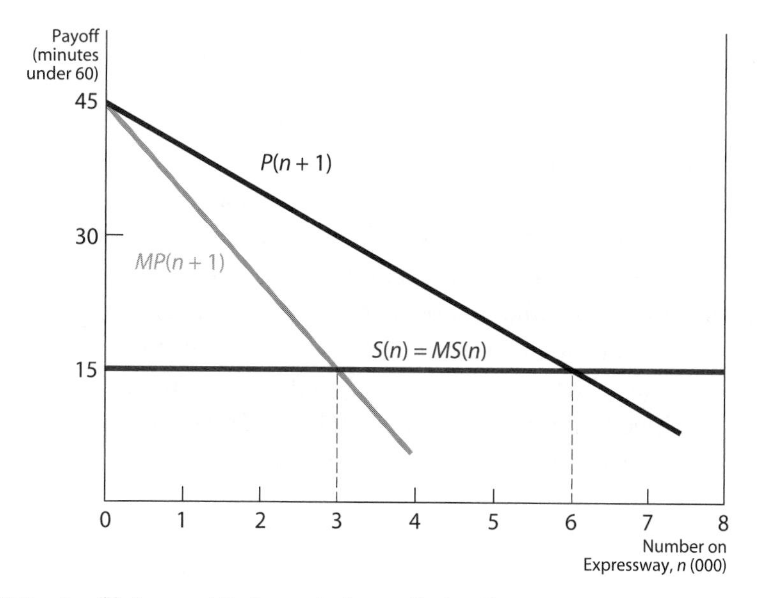

FIGURE 12.9 Equilibrium and Optimum in Route-Choice Game

The expression in the first set of braces is the effect on the payoffs of the set of commuters who choose P; this expression includes the $P(n+1)$ of the switcher and the spillover effect, $n[P(n+1) - P(n)]$, on all the other n commuters who choose P. We call this the marginal social payoff for the P-choosing subgroup, when their number increases from n to $n+1$, or $MP(n+1)$ for short. Similarly, the expression in the second set of braces is the marginal social payoff for the S-choosing subgroup, or $MS(n)$ for short. Then, the full expression for $T(n+1) - T(n)$ tells us that the total social payoff increases when one person switches from S to P (or decreases if the switch is from P to S) if $MP(n+1) > MS(n)$. The total social payoff decreases when one person switches from S to P (or increases when the switch is from P to S) if $MP(n+1) < MS(n)$.

Using our expressions for $P(n+1)$ and $S(n)$ in the commuting example, we have

$$MP(n+1) = 45 - (n+1) \times 0.005 + n \times (-0.005) = 44.995 - 0.01n$$

while $MS(n) = 15$ for all values of n. Figure 12.9 includes graphs of the relations $MP(n+1)$ and $MS(n)$. Note that the $MS(n)$ coincides with $S(n)$ everywhere because the local roads are never congested. But the $MP(n+1)$ curve lies below the $P(n+1)$ curve. Because of the negative spillover, the social gain from one person's switching to the expressway is less than the private gain to the switcher.

The $MP(n+1)$ and $MS(n)$ curves meet at $n = 2{,}999$, or approximately 3,000. To the left of this intersection, $MP(n+1) > MS(n)$, and society stands to gain by allowing one more person on the expressway. To the right, the opposite is true, and society stands to gain by shifting one person from the expressway to the local roads. Thus the socially optimal allocation of drivers is 3,000 on the expressway and 3,000 on the local roads.

If you wish to use calculus, you can write the total payoff for the expressway drivers as $nP(n) = n(45 - 0.005n) = 45n - 0.005n^2$. Then $MP(n+1)$ is the derivative of this with respect to n—namely, $45 - 0.005 \times 2n = 45 - 0.01n$. The rest of the analysis can proceed as before.

How might this society achieve the optimum allocation of its drivers? Different cultures and political groups use different systems, each with its own merits and drawbacks. The society could simply restrict access to the expressway to 3,000 drivers. But how would it choose those 3,000? It could adopt a first-come, first-served rule, but then drivers would race each other to get there early and waste a lot of time. A bureaucratic society could set up criteria based on complex calculations of needs and merits as defined by civil servants; then everyone will undertake some costly activities to meet these criteria. In a politicized society, the important "swing voters" or organized pressure groups or contributors may be favored. In a corrupt society, those who bribe the officials or the politicians may get the preference. A more egalitarian society could allocate the rights to drive on the expressway by lottery or could rotate them from one month to the next. A scheme that lets you drive only on certain days, depending on the last digit of your car's license plate, is an example. But such a scheme is not so egalitarian as it seems, because the rich can have two cars and choose license-plate numbers that will allow them to drive every day.

Many economists prefer a more open system of charges. Suppose each driver on the expressway is made to pay a tax t, measured in units of time. Then the private benefit from using the expressway becomes $P(n) - t$, and the number n in the Nash equilibrium will be determined by $P(n) - t = S(n)$. (Here, we are ignoring the tiny difference between $P(n)$ and $P(n + 1)$, which is possible when N is very large.) We know that the socially optimal value of n is 3,000. Using the expressions $P(n) = 45 - 0.005n$ and $S(n) = 15$, and plugging in 3,000 for n, we find that $P(n) - t = S(n)$—that is, drivers are indifferent between the expressway and the local roads—when $45 - 15 - t = 15$, or $t = 15$. If we value time at the minimum wage of about \$5 an hour, 15 minutes comes to \$1.25. This is the tax or toll that, when charged, will keep the numbers on the expressway down to what is socially optimal.

Note that when 3,000 drivers are on the expressway, the addition of one more increases the time spent by each of them by 0.005 minute, for a total of 15 minutes. This is exactly the tax that each driver is being asked to pay. In other words, each driver is made to pay the cost of the negative spillover that he imposes on the rest of society. This "brings home" to each driver the extra cost of his action and therefore induces him to take the socially optimal action; economists say the individual person is being made to **internalize the externality.** This idea, that people whose actions hurt others are made to pay for the harm that they cause, adds to the appeal of this approach. But the proceeds from the tax are not used to compensate the others directly. If they were, then each expressway user would count on receiving from others just what he pays, and the whole purpose would be defeated. Instead, the proceeds of the tax go into general government revenues, where they may or may not be used in a socially beneficial manner.

Those economists who prefer to rely on markets argue that if the expressway has a private owner, his profit motive will induce him to charge just enough for its use to reduce the number of users to the socially optimal level. An owner knows that if he charges a tax t for each user, the number of users n will be determined by $P(n) - t = S(n)$. His revenue will be $tn = n[P(n) - S(n)]$, and he will act in such a way as to maximize this revenue. In our example, the revenue is $n[45 - 0.005n - 15] = n[30 - 0.005n] = 30n - 0.005n^2$. It is easy to see this revenue is maximized when $n = 3{,}000$. But in this case, the revenue goes into the owner's pocket; most people regard that as a bad solution.

D. Positive Spillovers

Many matters pertaining to positive spillovers or positive externalities can be understood simply as mirror images of those for negative spillovers. A person's private benefits from undertaking activities with positive spillovers are less than society's marginal benefits from such activities. Therefore such actions will be underutilized and their benefits underprovided in the Nash equilibrium. A better outcome can be achieved by augmenting people's incentives; providing those persons whose actions create positive spillovers with a reward just equal to the spillover benefit will achieve the social optimum.

Indeed, the distinction between positive and negative spillovers is to some extent a matter of semantics. Whether a spillover is positive or negative depends on which choice you call P and which you call S. In the commuting example, suppose we called the local roads P and the expressway S. Then one commuter's switch from S to P will reduce the time taken by all the others who choose S, so this action will convey a positive spillover to them. In another example, consider vaccination against some infectious disease. Each person getting vaccinated reduces his own risk of catching the disease (marginal private gain) and reduces the risk of others' getting the disease through him (spillover). If being unvaccinated is called the *S* action, then getting vaccinated has a positive spillover effect. If remaining unvaccinated is called the P action, then the act of remaining unvaccinated has a negative spillover effect. This has implications for the design of policy to bring individual action into conformity with the social optimum. Society can either reward those who get vaccinated or penalize those who fail to do so.

But actions with positive spillovers can have one very important new feature that distinguishes them from actions with negative spillovers—namely, **positive feedback.** Suppose the spillover effect of your choosing P is to increase the payoff to the others who are also choosing P. Then your choice increases the attraction of that action (P) and may induce some others to take it also, setting in train a process that culminates in everyone's taking that action. Conversely, if very few people are choosing P, then it may be so unattractive that they, too, give it up, leading to a situation in which everyone chooses S. In other words, positive feedback can give rise to multiple Nash equilibria, which we now illustrate by using a very real example.

When you buy a computer, you have to choose between one with a Windows operating system and one with an operating system based on Unix, such as Linux. As the number of Unix users rises, the better it will be to purchase such a computer. The system will have fewer bugs because more users will have detected those that exist, more application software will be available, and more experts will be available to help with any problems that arise. Similarly, a Windows-based computer will be more attractive the more Windows users there are. In addition, many computing aficionados would argue that the Unix system is superior. Without necessarily taking a position on that matter, we show what will happen if that is the case. Will individual choice lead to the socially best outcome?

A diagram similar to Figures 12.6 through 12.8 can be used to show the payoffs to an individual computer purchaser of the two strategies, Unix and Windows. As shown in Figure 12.10, the Unix payoff rises as the number of Unix users rises, and the Windows payoff rises as the number of Unix owners falls (the number of Windows users rises). As already explained, the diagram is drawn assuming that the payoff to Unix users when everyone in the population is a Unix user (at the point labeled U) is higher than the payoff to Windows users when everyone in the population is a Windows user (at W).

If the current population has only a small number of Unix users, then the situation is represented by a point to the left of the intersection of the two payoff lines at I, and each individual user finds it better to choose Windows. When there is a larger number of Unix users in the population, placing the society to the right of I, it is better for each person to choose Unix. Thus a mixed population of Unix and Windows users is sustainable as an equilibrium only when the current population has exactly I Unix users; only then will no member of the population have any incentive to switch platforms. And even that situation is unstable. Suppose just one person accidentally makes a different decision. If he switches to Windows, his choice will push the population to the left of I, in which case others will have an incentive to switch to Windows, too. If he switches to Unix, the population point moves to the right of I, creating an incentive for more people to switch to Unix. The cumulative effect of these switches will eventually push the society to an all-Unix or an all-Windows outcome; these are the two stable equilibria of the game.[4]

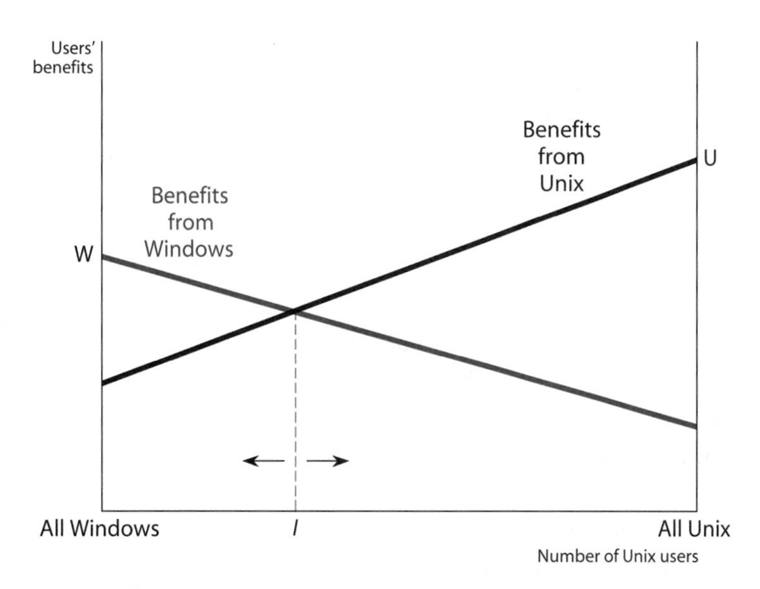

FIGURE 12.10 Payoffs in Operating-System-Choice Game

[4] The term *positive feedback* may create the impression that it is a good thing, but in technical language the term merely characterizes the process and includes no general value judgment about the outcome. In this example, the same positive feedback mechanism could lead to either an all-Unix outcome or an all-Windows outcome; one outcome could be worse than the other.

But which of the two stable equilibria will be achieved in this game? The answer depends on where the game starts. If you look at the configuration of today's computer users, you will see a heavily Windows-oriented population. Thus it seems that because there are so few Unix users (or so many PC users), the world is moving toward the all-Windows equilibrium. Schools, businesses, and private users have become **locked in** to this particular equilibrium as a result of an accident of history. If it is indeed true that Unix provides more benefits to society when used by everyone, then the all-Unix equilibrium should be preferred over the all-Windows one that we are approaching. Unfortunately, although society as a whole might be better off with the change, no individual computer user has an incentive to make a change from the current situation. Only coordinated action can swing the pendulum toward Unix. A critical mass of individual users, more than I in Figure 12.10, must use Unix before it becomes individually rational for others to choose the same operating system.

There are many examples of similar choices of convention being made by different groups of people. The most famous cases are those in which it has been argued, in retrospect, that a wrong choice was made. Advocates claim that steam power could have been developed for greater efficiency than gasoline; it certainly would have been cleaner. Proponents of the Dvorak typewriter/computer keyboard configuration claim that it would be better than the QWERTY keyboard if used everywhere. Many engineers agree that Betamax had more going for it than VHS in the video recorder market. In such cases, the whims of the public or the genius of advertisers help determine the ultimate equilibrium and may lead to a "bad" or "wrong" outcome from society's perspective. Other situations do not suffer from such difficulties. Few people concern themselves with fighting for a reconfiguration of traffic-light colors, for example.[5]

The ideas of positive feedback and lock-in find an important application in macroeconomics. Production is more profitable the higher the level of demand in the economy, which happens when national income is higher. In turn, income is higher when firms are producing more and are therefore hiring more workers. This positive feedback creates the possibility of multiple equilibria, of which the high-production, high-income one is better for society, but individual decisions may lock the economy into the low-production, low-income equilibrium. The better equilibrium could be turned into a focal point by public declaration—"the only thing we have to fear is fear itself"—but the government can also inject demand into the economy to the extent necessary to move it to the better equilibrium. In other words, the possibility of unemployment due

[5] Not everyone agrees that the Dvorak keyboard and the Betamax video recorder were clearly superior alternatives. See two articles by S. J. Liebowitz and Stephen E. Margolis, "Network Externality: An Uncommon Tragedy," *Journal of Economic Perspectives*, vol. 8 (Spring 1994), pp. 146–149, and "The Fable of the Keys," *Journal of Law and Economics*, vol. 33 (April 1990), pp. 1–25.

to a deficiency of aggregate demand—as discussed in the supply-and-demand language of economic theory by the British economist John Maynard Keynes in his well-known 1936 book titled *Employment, Interest, and Money*—can be seen from a game-theoretic perspective as the result of a failure to solve a collective-action problem.[6]

4 A BRIEF HISTORY OF IDEAS

A. The Classics

The problem of collective action has been recognized by social philosophers and economists for a very long time. The seventeenth-century British philosopher Thomas Hobbes argued that society would break down in a "war of all against all" unless it was ruled by a dictatorial monarch, or *Leviathan* (the title of his book). One hundred years later, the French philosopher Jean-Jacques Rousseau described the problem of a prisoners' dilemma in his *Discourse on Inequality.* A stag hunt needs the cooperation of the whole group of hunters to encircle and kill the stag, but any individual hunter who sees a hare may find it better for himself to leave the circle to chase the hare. But Rousseau thought that such problems were the product of civilization and that people in the natural state lived harmoniously as "noble savages." At about the same time, two Scots pointed out some dramatic solutions to such problems. David Hume in his *Treatise on Human Nature* argued that the expectations of future returns of favors can sustain cooperation. Adam Smith's *Wealth of Nations* developed a grand vision of an economy in which the production of goods and services motivated purely by private profit could result in an outcome that was best for society as a whole.[7]

[6] John Maynard Keynes, *Employment, Interest, and Money* (London: Macmillan, 1936). See also John Bryant, "A Simple Rational-Expectations Keynes-type Model," *Quarterly Journal of Economics,* vol. 98 (1983), pp. 525–528, and Russell Cooper and Andrew John, "Coordination Failures in a Keynesian Model," *Quarterly Journal of Economics,* vol. 103 (1988), pp. 441–463, for formal game-theoretic models of unemployment equilibria.

[7] The great old books cited in this paragraph have been reprinted many times in many different versions. For each, we list the year of original publication and the details of one relatively easily accessible reprint. In each case, the editor of the reprinted version provides an introduction that conveniently summarizes the main ideas. Thomas Hobbes, *Leviathan; or the Matter, Form, and Power of Commonwealth Ecclesiastical and Civil,* 1651 (Everyman Edition, London: J. M. Dent, 1973); David Hume, *A Treatise of Human Nature,* 1739 (Oxford: Clarendon Press, 1976); Jean-Jacques Rousseau, *A Discourse on Inequality,* 1755 (New York: Penguin Books, 1984); Adam Smith, *An Inquiry into the Nature and Causes of the Wealth of Nations,* 1776 (Oxford: Clarendon Press, 1976).

The optimistic interpretation persisted, especially among many economists and even several political scientists, to the point where it was automatically assumed that if an outcome was beneficial to a group as a whole, the actions of its members would bring the outcome about. This belief received a necessary rude shock in the mid-1960s when Mancur Olson published *The Logic of Collective Action.* He pointed out that the best collective outcome would not prevail unless it was in each individual person's private interest to perform his assigned action—that is, unless it was a Nash equilibrium. However, Olson did not specify the collective-action game very precisely. Although it looked like a prisoners' dilemma, Olson insisted that it was not necessarily so, and we have already seen that the problem can also take the form of a chicken game or an assurance game.[8]

Another major class of collective-action problems—namely, those concerning the depletion of common-access resources—received attention at about the same time. If a resource such as a fishery or a meadow is open to all, each user will exploit it as much as he can, because any self-restraint on his part will merely make more available for the others to exploit. As we mentioned above, Garrett Hardin wrote a well-known article on this subject titled "The Tragedy of the Commons." Common-resource problems are unlike our irrigation-project game, in which each person has a strong private incentive to free-ride off the efforts of others. In regard to a common resource, each person has a strong private incentive to exploit it to the full, making everyone else pay the social cost that results from the degradation of the resource.

B. Modern Approaches and Solutions

Until recently, many social scientists and most physical scientists took a Hobbesian line on the common-resource problem, arguing that it can be solved only by a government that forces everyone to behave cooperatively. Others, especially economists, retained their Smithian optimism. They argued that placing the resource in proper private ownership, where its benefits can be captured in the form of profit by the owner, will induce the owner to restrain its use in a socially optimal manner. He will realize that the value of the resource (fish or grass, for example) may be higher in the future because less will be available, and therefore he can make more profit by saving some of it for that future.

Nowadays, thinkers from all sides have begun to recognize that collective-action problems come in diverse forms and that there is no uniquely best solution to all of them. They also understand that groups or societies do not stand helpless in the face of such problems, and they devise various ways to

[8] Mancur Olson, *The Logic of Collective Action* (Cambridge: Harvard University Press, 1965).

cope with them. Much of this work has been informed by game-theoretic analysis of repeated prisoners' dilemmas and similar games.[9]

Solutions to collective-action problems of all types must induce individual persons to act cooperatively or in a manner that would be best for the group, even though the person's interests may best be served by doing something else—in particular, taking advantage of the others' cooperative behavior.[10] Humans exhibit much in the way of cooperative behavior. The act of reciprocating gifts and skills at detecting cheating are so common in all societies and throughout history, for example, that there is reason to argue that they may be instincts.[11] But human societies generally rely heavily on purposive social and cultural customs, **norms,** and **sanctions** in inducing cooperative behavior from their individual members. These methods are conscious, deliberate attempts to design the game in order to solve the collective-action problem.[12] We approach the matter of solution methods from the perspective of the type of game being played.

A solution is easiest if the collective-action problem takes the form of an assurance game. Then it is in every person's private interest to take the socially best action if he expects all other persons to do likewise. In other words, the socially optimal outcome is a Nash equilibrium. The only problem is that the same game has other, socially worse, Nash equilibria. Then all that is needed to achieve the best Nash equilibrium and thereby the social optimum is to make it

[9] Prominent in this literature are Michael Taylor, *The Possibility of Cooperation* (New York: Cambridge University Press, 1987); Elinor Ostrom, *Governing the Commons* (New York: Cambridge University Press, 1990); and Matt Ridley, *The Origins of Virtue* (New York: Viking Penguin, 1996).

[10] The problem of the need to attain cooperation and its solutions are not unique to human societies. Examples of cooperative behavior in the animal kingdom have been explained by biologists in terms of the advantage of the gene and of the evolution of instincts. For more, see Chapter 13 and Ridley, *Origins of Virtue.*

[11] See Ridley, *Origins of Virtue,* chaps. 6 and 7.

[12] The social sciences do not have precise and widely accepted definitions of terms such as *custom* and *norm*; nor are the distinctions among such terms always clear and unambiguous. We set out some definitions in this section, but be aware that you may find different usage in other books. Our approach is similar to those found in Richard Posner and Eric Rasmusen, "Creating and Enforcing Norms, with Special Reference to Sanctions," *International Review of Law and Economics,* vol. 19, no. 3 (September 1999), pp. 369–382, and in David Kreps, "Intrinsic Motivation and Extrinsic Incentives," *American Economic Review,* Papers and Proceedings, vol. 87, no. 2 (May 1997), pp. 359–364; Kreps uses the term *norm* for all the concepts that we classify under different names.

Sociologists have a different taxonomy of norms from that of economists; it is based on the importance of the matter (trivial matters such as table manners are called *folkways,* and weightier matters are called *mores*), and on whether the norms are formally codified as *laws.* They also maintain a distinction between *values* and norms, recognizing that some norms may run counter to persons' values and therefore require sanctions to enforce them. This distinction corresponds to ours between customs, internalized norms, and enforced norms. The conflict between individual values and social goals arises for enforced norms but not for customs or *conventions,* as we label them, or for internalized norms. See Donald Light and Suzanne Keller, *Sociology,* 4th ed. (New York: Knopf, 1987), pp. 57–60.

a focal point—that is, to ensure the convergence of the players' expectations on it. Such a convergence can result from a social **custom,** or **convention**—namely, a mode of behavior that finds automatic acceptance because it is in everyone's interest to follow it so long as others are expected to do likewise. For example, if all the farmers, herders, weavers, and other producers in an area want to get together to trade their wares, all they need is the assurance of finding others with whom to trade. Then the custom that the market is held in village X on day Y of every week makes it optimal for everyone to be there on that day.[13]

One complication remains. For the desired outcome to be a focal point, each person must have confidence that all others understand it, which in turn requires that they have confidence that all others understand. . . . In other words, the point must be common knowledge. Usually some prior social action is necessary to ensure that this is true. Publication in a medium that is known by everyone to be sufficiently widely read, and discussion in an inward-facing circle so everyone knows that everyone else was present and paying attention, are some methods used for this purpose.[14]

Our analysis in Section 2 suggested that individual payoffs are often configured in such a way that collective-action problems, particularly of large groups, take the form of a prisoners' dilemma. Not surprisingly, the methods for coping with such problems have received the most attention.

The simplest method attempts to change people's preferences so that the game is no longer a prisoners' dilemma. If individuals get sufficient pleasure from cooperating, or suffer enough guilt or shame when they cheat, they will cooperate to maximize their own payoffs. If the extra payoff from cooperation is conditional—one gets pleasure from cooperating or guilt or shame from cheating if, but only if, many others are cooperating—then the game can turn into an assurance game. In one of its equilibria, everyone cooperates because everyone else does, and in the other, no one cooperates because no one else does. Then the collective action problem is the simpler one of making the better equilibrium the focal point. If the extra payoff from cooperation is unconditional—one gets pleasure from cooperating or guilt or shame from cheating regardless of what the others do—then the game can have a unique equilibrium where

[13]In his study of the emergence of cooperation, *Cheating Monkeys and Citizen Bees* (New York: Free Press, 1999), the evolutionary biologist Lee Dugatkin labels this case "selfish teamwork." He argues that such behavior is likelier to arise in times of crisis, because each person is pivotal at those times. In a crisis, the outcome of the group interaction is likely to be disastrous for everyone if even one person fails to contribute to the group's effort to get out of the dire situation. Thus each person is willing to contribute so long as the others do. We will mention Dugatkin's full classification of alternative approaches to cooperation in Chapter 13 on evolutionary games.

[14]See Michael Chwe, *Rational Ritual: Culture, Coordination, and Common Knowledge* (Princeton, NJ: Princeton University Press, 2001), for a discussion of this issue and numerous examples and applications of it.

everyone cooperates. In many situations it is not even necessary for everyone to have such payoffs. If a substantial proportion of the population does, that may suffice for the desired collective outcome.

Some such prosocial preferences may be innate, hard wired in a biological evolutionary process. But they are more likely to be social or cultural products. Most societies make deliberate efforts to instill prosocial thinking in children during the process of socialization in families and schools. Growth of such preferences is seen in experiments on ultimatum and dictator games of the kind we discussed in Chapter 3. When these experiments are conducted on children of different ages, very young children behave selfishly. By age eight, however, they develop a significant sense of equality. True prosocial preferences develop gradually thereafter, with some relapses, finally to an adult fair-mindedness. Thus a long process of education and experience instills *internalized norms* into people's preferences.[15]

However, people do differ in the extent to which they internalize prosocial preferences, and the process may not go far enough to solve many collective action problems. Most people have sufficiently broad understanding of what the socially cooperative action is in most situations, but individuals retain the personal temptation to cheat. Therefore a system of external sanctions or punishments is needed to sustain the cooperative actions. We call these widely understood but not automatically followed rules of behavior *enforced norms.*

In Chapter 11, we described in detail several methods for achieving a cooperative outcome in prisoners' dilemma games, including repetition, penalties (or rewards), and leadership. In that discussion, we were mainly concerned with two-person dilemmas. The same methods apply to enforcement of norms in collective-action problems in large groups, with some important modifications or innovations.

We saw in Chapter 11 that repetition was the most prominent of these methods; so we focus the most attention on it. Repetition can achieve cooperative outcomes as equilibria of individual actions in a repeated two-person prisoners' dilemma by holding up the prospect that cheating will lead to a breakdown of cooperation. More generally, what is needed to maintain cooperation is the expectation in the mind of each player that his personal benefits from cheating are transitory and that they will quickly be replaced by a payoff lower than that associated with cooperative behavior. For players to believe that cheating is not beneficial from a long-term perspective, cheating should be detected quickly, and the punishment that follows (reduction in future payoffs) should be sufficiently swift, sure, and painful.

[15]Colin Camerer, *Behavioral Game Theory* (Princeton, NJ: Princeton University Press, 2003), pp. 65–67. See also pp. 63–75 for an account of differences in prosocial behavior along different dimensions of demographic characteristics and across different cultures.

A group has one advantage in this respect over a pair of individual persons. The same pair may not have occasion to interact all that frequently, but each of them is likely to interact with *someone* in the group all the time. Therefore B's temptation to cheat A can be countered by his fear that others, such as C, D, and so on, whom he meets in the future will punish him for this action. An extreme case where bilateral interactions are not repeated and punishment must be inflicted on one's behalf by a third party is, in Yogi Berra's well-known saying, "Always go to other people's funerals. Otherwise they won't go to yours."

But a group has some offsetting disadvantages over direct bilateral interaction when it comes to sustaining good behavior in repeated interactions. The required speed and certainty of detection and punishment suffer as the numbers in the group increase. One sees many instances of successful cooperation in small village communities that would be unimaginable in a large city or state.

Start with the detection of cheating, which is never easy. In most real situations, payoffs are not completely determined by the players' actions but are subject to some random fluctuations. Even with two players, if one gets a low payoff, he cannot be sure that the other cheated; it may have been just a bad draw of the random shock. With more people, an additional question enters the picture: If someone cheated, who was it? Punishing someone without being sure of his guilt beyond a reasonable doubt is not only morally repulsive but also counterproductive. The incentive to cooperate gets blunted if even cooperative actions are susceptible to punishment by mistake.

Next, with many players, even when cheating is detected and the cheater identified, this information has to be conveyed sufficiently quickly and accurately to others. For this, the group must be small or else must have a good communication or gossip network. Also, members should not have much reason to accuse others falsely.

Finally, even after cheating is detected and the information spread to the whole group, the cheater's punishment—enforcement of the social norm—has to be arranged. A third person often has to incur some personal cost to inflict such punishment. For example, if C is called on to punish B, who had previously cheated A, C may have to forgo some profitable business that he could have transacted with B. Then the inflicting of punishment is itself a collective-action game and suffers from the same temptation to "shirk," that is, not to participate in the punishment. A society could construct a second-round system of punishments for shirking, but that in turn may be yet another collective-action problem! However, humans seem to have evolved an instinct whereby people get some personal pleasure from punishing cheaters even when they have not themselves been the victims of this particular act of cheating.[16] Interestingly, the

[16]For evidence of such altruistic punishment instinct, see Ernst Fehr and Simon Gächter, "Altruistic Punishment in Humans," *Nature*, vol. 415 (January 10, 2002), pp. 137–140.

notion that "one should impose sanctions, even at personal cost, on violators of enforced social norms" seems itself to have become an internalized norm.[17]

Norms are reinforced by observation of society's general adherence to them, and they lose their force if they are frequently seen to be violated. Before the advent of the welfare state, when those who fell on hard economic times had to rely on help from family or friends or their immediate small social group, the work ethic constituted a norm that held in check the temptation to slacken one's own efforts and become a free rider on the support of others. As government took over the supporting role and unemployment compensation or welfare became an entitlement, this norm of the work ethic weakened. After the sharp increases in unemployment in Europe in the late 1980s and early 1990s, a significant fraction of the population became users of the official support system, and the norm weakened even further.[18]

Different societies or cultural groups may develop different conventions and norms to achieve the same purpose. At the trivial level, each culture has its own set of good manners—ways of greeting strangers, indicating approval of food, and so on. When two people from different cultures meet, misunderstandings can arise. More important, each company or office has its own ways of getting things done. The differences between these customs and norms are subtle and difficult to pin down, but many mergers fail because of a clash of these "corporate cultures."

Next, consider the chicken form of collective-action games. Here, the nature of the remedy depends on whether the largest total social payoff is attained when everyone participates (what we called "chicken version I" in Section 1.B) or when some cooperate and others are allowed to shirk (chicken II). For chicken I, where everyone has the individual temptation to shirk, the problem is much like that of sustaining cooperation in the prisoners' dilemma, and all the earlier remarks for that game apply here, too. Chicken II is different—easier in one respect and harder in another. Once an assignment of roles between participants and shirkers is made, no one has the private incentive to switch: if the other driver is assigned the role of going straight, then you are better off swerving, and the other way around. Therefore, if a custom creates the expectation of

[17]Our distinction between internalized norms and enforced norms is similar to Kreps's distinction between functions (iii) and (iv) of norms (Kreps, "Intrinsic Motivation and Extrinsic Incentives," p. 359). Society can also reward desirable actions just as it can punish undesirable ones. Again, the rewards, financial or otherwise, can be given externally, or players' payoffs can be changed so that they take pleasure in doing the right thing. The two types of rewards can interact; for example, the peerages and knighthoods given to British philanthropists and others who do good deeds for British society are external rewards, but individual persons value them only because respect for knights and peers is a British social norm.

[18]Assar Lindbeck, "Incentives and Social Norms in Household Behavior," *American Economic Review*, Papers and Proceedings, vol. 87, no. 2 (May 1997), pp. 370–377.

an equilibrium, it can be maintained without further social intervention such as sanctions. However, in this equilibrium, the shirkers get higher payoffs than the participants do, and this inequality can create its own problems for the game; the conflicts and tensions, if they are major, can threaten the whole fabric of the society. Often the problem can be solved by repetition. The roles of participants and shirkers can be rotated to equalize payoffs over time.

Sometimes the problem of differential payoffs in version II of the prisoners' dilemma or chicken is "solved," not by restoring equality but by **oppression** or **coercion,** which forces a dominated subset of society to accept the lower payoff and allows the dominant subgroup to enjoy the higher payoff. In many societies throughout history, the work of handling animal carcasses was forced on particular groups or castes in this way. The history of the maltreatment of racial and ethnic minorities and of women provides vivid examples of such practices. Once such a system becomes established, no one member of the oppressed group can do anything to change the situation. The oppressed must get together as a group and act to change the whole system, itself another problem of collective action.

Finally, consider the role of leadership in solving collective-action problems. In Chapter 11, we pointed out that, if the players are of very unequal "size," the prisoners' dilemma may disappear because it may be in the private interests of the larger player to continue cooperation and to accept the cheating of the smaller player. Here we recognize the possibility of a different kind of bigness—namely, having a "big heart." People in most groups differ in their preferences, and many groups have one or a few who take genuine pleasure in expending personal effort to benefit the whole. If there are enough such people for the task at hand, then the collective-action problem disappears. Most schools, churches, local hospitals, and other worthy causes rely on the work of such willing volunteers. This solution, like others before it, is more likely to work in small groups, where the fruits of their actions are more closely and immediately visible to the benefactors, who are therefore encouraged to continue.

C. Applications

In her book *Governing the Commons,* Elinor Ostrom describes several examples of resolution of common-resource problems at local levels. Most of them require taking advantage of features specific to the context in order to set up systems of detection and punishment. A fishing community on the Turkish coast, for example, assigns and rotates locations to its members; the person who is assigned a good location on any given day will naturally observe and report any intruder who tries to usurp his place. Many other users of common resources, including the grazing commons in medieval England, actually restricted access and controlled overexploitation by allocating complex, tacit, but well-understood rights to individual persons. In one sense, this solution bypasses the common-resource problem by dividing up the resource into a number of privately owned subunits.

The most striking feature of Ostrom's range of cases is their immense variety. Some of the prisoners' dilemmas of the exploitation of common-property resources that she examined were solved by private initiative by the group of people actually in the dilemma; others were solved by external public or governmental intervention. In some instances, the dilemma was not resolved at all, and the group remained trapped in the all-shirk outcome. Despite this variety, Ostrom identifies several common features that make it easier to solve prisoners' dilemmas of collective action: (1) it is essential to have an identifiable and stable group of potential participants; (2) the benefits of cooperation have to be large enough to make it worth paying all the costs of monitoring and enforcing the rules of cooperation; and (3) it is very important that the members of the group can communicate with each other. This last feature accomplishes several things. First, it makes the norms clear—everyone knows what behavior is expected, what kind of cheating will not be tolerated, and what sanctions will be imposed on cheaters. Next, it spreads information about the efficacy of the detection of the cheating mechanism, thereby building trust and removing the suspicion that each participant might hold that he is abiding by the rules while others are getting away with breaking them. Finally, it enables the group to monitor the effectiveness of the existing arrangements and to improve on them as necessary. All these requirements look remarkably like those identified in Chapter 11 from our theoretical analysis of the prisoners' dilemma (Sections 2 and 3) and from the observations of Axelrod's tournaments (Section 6).

Ostrom's study of the fishing village also illustrates what can be done if the collective optimum requires different persons to do different things, in which case some get higher payoffs than others. In a repeated relationship, the advantageous position can rotate among the participants, thereby maintaining some sense of equality over time.

Ostrom finds that an external enforcer of cooperation may not be able to detect cheating or impose punishment with sufficient clarity and swiftness. Thus the frequent reaction that centralized or government policy is needed to solve collective-action problems is often proved wrong. Another example comes from village communities or "communes" in late-nineteenth-century Russia. These communities solved many collective-action problems of irrigation, crop rotation, management of woods and pastures, and road and bridge construction and repair in just this way. "The village . . . was not the haven of communal harmony. . . . It was simply that the individual interests of the peasants were often best served by collective activity." Reformers of early twentieth-century czarist governments and Soviet revolutionaries of the 1920s alike failed, partly because the old system had such a hold on the peasants' minds that they resisted anything new, but also because the reformers failed to understand the role that some of the prevailing practices played in solving collective-action problems and thus failed to replace them with equally effective alternatives.[19]

The difference between small and large groups is well illustrated by Avner Greif's comparison of two groups of traders in countries around the Mediterranean Sea in medieval times. The Maghribis were Jewish traders who relied on extended family and social ties. If one member of this group cheated another, the victim informed all the others by writing letters. When guilt was convincingly proved, no one in the group would deal with the cheater. This system worked well on a small scale of trade. But as trade expanded around the Mediterranean, the group could not find sufficiently close or reliable insiders to go to the countries with the new trading opportunities.

In contrast, the Genoese traders established a more official legal system. A contract had to be registered with the central authorities in Genoa. The victim of any cheating or violation of the contract had to take a complaint to the authorities, who carried out the investigation and imposed the appropriate fines on the cheater. This system, with all its difficulties of detection, could be more easily expanded with the expansion of trade.[20] As economies grow and world trade expands, we see a similar shift from tightly linked groups to more arm's-length trading relationships, and from enforcement based on repeated interactions to that of the official law.

The idea that small groups are more successful at solving collective-action problems forms the major theme of Olson's *Logic of Collective Action* (see footnote 8) and has led to an insight important in political science. In a democracy, all voters have equal political rights, and the majority's preference should prevail. But we see many instances in which this does not happen. The effects of policies are generally good for some groups and bad for others. To get its preferred policy adopted, a group has to take political action—lobbying, publicity, campaign contributions, and so on. To do these things, the group must solve a collective-action problem, because each member of the group may hope to shirk and enjoy the benefits that the others' efforts have secured. If small groups are better able to solve this problem, then the policies resulting from the political process will reflect *their* preferences, even if other groups who fail to organize are more numerous and suffer greater losses than the successful groups' gains.

[19]Orlando Figes, *A People's Tragedy: The Russian Revolution 1891–1924* (New York: Viking Penguin, 1997), pp. 89–90, 240–241, 729–730. See also Ostrom, *Governing the Commons*, p. 23, for other instances where external, government-enforced attempts to solve common-resource problems actually made them worse.

[20]Avner Greif, "Cultural Beliefs and the Organization of Society: A Historical and Theoretical Reflection on Collectivist and Individualist Societies," *Journal of Political Economy*, vol. 102, no. 5 (October 1994), pp. 912–950.

The most dramatic example of policies reflecting the preferences of the organized group comes from the arena of trade policy. A country's import restrictions help domestic producers whose goods compete with these imports, but they hurt the consumers of the imported goods and the domestic competing goods alike, because prices for these goods are higher than they would be otherwise. The domestic producers are few in number, and the consumers are almost the whole population; the total dollar amount of the consumers' losses is typically far bigger than the total dollar amount of the producers' gains. Political considerations based on constituency membership numbers and economic considerations of dollar gains and losses alike would lead us to expect a consumer victory in this policy arena; we would expect to see at least a push for the idea that import restrictions should be abolished, but we don't. The smaller and more tightly knit associations of producers are better able to organize for political action than the numerous, dispersed consumers.

More than 70 years ago, the American political scientist E. E. Schattschneider provided the first extensive documentation and discussion of how pressure politics drives trade policy. He recognized that "the capacity of a group for organization has a great influence on its activity," but he did not develop any systematic theory of what determines this capacity.[21] The analysis of Olson and others has improved our understanding of the issue, but the triumph of pressure politics over economics persists in trade policy to this day. For example, in the late 1980s, the U.S. sugar policy cost each of the 240 million people in the United States about $11.50 per year for a total of about $2.75 billion, while it increased the incomes of about 10,000 sugar-beet farmers by about $50,000 each, and the incomes of 1,000 sugarcane farms by as much as $500,000 each, for a total of about $1 billion. The net loss to the U.S. economy was $1.75 billion.[22] Each of the unorganized consumers continues to bear his small share of the costs in silence; many of them are not even aware that each is paying $11.50 a year too much for his sweet tooth.

If this overview of the theory and practice of solving collective-action problems seems diverse and lacking a neat summary statement, that is because the problems are equally diverse, and the solutions depend on the specifics of each problem. The one general lesson that we can provide is the importance of letting the participants themselves devise solutions by using their local knowledge of the situation, their advantage of proximity in monitoring the cooperative or shirking actions of others in the community, and their ability to impose sanctions on shirkers by exploiting various ongoing relationships within the social group.

[21] E. E. Schattschneider, *Politics, Pressures, and the Tariff* (New York: Prentice-Hall, 1935); see especially pp. 285–286.

[22] Stephen V. Marks, "A Reassessment of the Empirical Evidence on the U.S. Sugar Program," in *The Economics and Politics of World Sugar Policies*, ed. Stephen V. Marks and Keith E. Maskus (Ann Arbor: University of Michigan Press, 1993), pp. 79–108.

Finally, a word of caution. You might be tempted to come away from this discussion of collective-action problems with the impression that individual freedom always leads to harmful outcomes that can and must be improved by social norms and sanctions. Remember, however, that societies face problems other than those of collective action; some of them are better solved by individual initiative than by joint efforts. Societies can often get hidebound and autocratic, becoming trapped in their norms and customs and stifling the innovation that is so often the key to economic growth. Collective action can become collective inaction.[23]

5 "HELP!": A GAME OF CHICKEN WITH MIXED STRATEGIES

In the chicken variant of collective-action problems discussed in earlier sections, we looked only at the pure-strategy equilibria. But we know from Chapter 7 that such games have mixed-strategy equilibria, too. In collective-action problems, where each participant is thinking, "It is better if I wait for enough others to participate so that I can shirk; but then again, maybe they won't, in which case I should participate," mixed strategies nicely capture the spirit of such vacillation. Our last story is a dramatic, even chilling application of such a mixed-strategy equilibrium.

In 1964 in New York City (in Kew Gardens, Queens), a woman named Kitty Genovese was killed in a brutal attack that lasted more than half an hour. She screamed through it all and, although her screams were heard by many people and more than 30 actually watched the attack taking place, not one went to help her or even called the police.

The story created a sensation and found several ready theories to explain it. The press and most of the public saw this episode as a confirmation of their belief that New Yorkers—or big-city dwellers or Americans or people more generally—were just apathetic or didn't care about their fellow human beings.

However, even a little introspection or observation will convince you that people do care about the well-being of other humans, even strangers. Social scientists offered a different explanation for what happened, which they labeled **pluralistic ignorance.** The idea behind this explanation is that no one can be sure about what is happening or whether help is really needed and how much. People look to each other for clues or guidance about these matters and try to interpret other people's behavior in this light. If they see that no one else is doing

[23] David Landes, *The Wealth and Poverty of Nations* (New York: Norton, 1998), chaps. 3 and 4, makes a spirited case for this effect.

anything to help, they interpret it as meaning that help is probably not needed, and so they don't do anything either. This explanation has some intuitive appeal but is unsatisfactory in the Kitty Genovese context. There is a very strong presumption that a screaming woman needs help. What did the onlookers think—that a movie was being shot in their obscure neighborhood? If so, where were the lights, the cameras, the director, other crew?

A better explanation would recognize that although each onlooker may experience strong personal loss from Kitty's suffering and get genuine personal pleasure if she were saved, each must balance that against the cost of getting involved. You may have to identify yourself if you call the police; you may then have to appear as a witness, and so on. Thus, we see that each person may prefer to wait for someone else to call and hope to get for himself the free rider's benefit of the pleasure of a successful rescue.

Social psychologists have a slightly different version of this idea of free riding, which they label **diffusion of responsibility.** In this version, the idea is that everyone might agree that help is needed, but they are not in direct communication with each other and so cannot coordinate on who should help. Each person may believe that help is someone else's responsibility. And the larger the group, the more likely it is that each person will think that someone else would probably help, and therefore he can save himself the trouble and the cost of getting involved.

Social psychologists conducted some experiments to test this hypothesis. They staged situations in which someone needed help of different kinds in different places and with different-sized crowds. Among other things, they found that the larger the size of the crowd, the less likely was help to come forth.

The concept of diffusion of responsibility seems to explain this finding, but not quite completely. It claims that the larger the crowd, the less likely is any one person to help. But there are more people, and only one person is needed to act and call the police to secure help. To make it less likely that even one person helps, the chance of any one person helping has to decrease sufficiently fast to offset the increase in the total number of potential helpers. To find out whether it does so requires game-theoretic analysis, which we now supply.[24]

We consider only the aspect of diffusion of responsibility in which action is not consciously coordinated, and we leave aside all other complications of information and inference. Thus we assume that everyone believes the action is needed and is worth the cost.

[24] For a fuller account of the Kitty Genovese story and for the analysis of such situations from the perspective of social psychology, see John Sabini, *Social Psychology*, 2nd ed. (New York: Norton, 1995), pp. 39–44. Our game-theoretic model is based on Thomas Palfrey and Howard Rosenthal, "Participation and the Provision of Discrete Public Goods," *Journal of Public Economics*, vol. 24 (1984), pp. 171–193.

Suppose N people are in the group. The action brings each of them a benefit B. Only one person is needed to take the action; more are redundant. Anyone who acts bears the cost C. We assume that $B > C$; so it is worth any one person's while to act even if no one else is acting. Thus the action is justified in a very strong sense.

The problem is that anyone who takes the action gets the value B and pays the cost C for a net payoff of $(B - C)$, whereas he would get the higher payoff B if someone else took the action. Thus each person has the temptation to let someone else go ahead and to become a free rider on another's effort. When all N people are thinking thus, what will be the equilibrium or outcome?

If $N = 1$, the single person has a simple decision problem rather than a game. He gets $B - C > 0$ if he takes the action and 0 if he does not. Therefore he goes ahead and helps.

If $N > 1$, we have a game of strategic interaction with several equilibria. Let us begin by ruling out some possibilities. With $N > 1$, there cannot be a pure-strategy Nash equilibrium in which all people act, because then any one of them would do better by switching to free ride. Likewise, there cannot be a pure-strategy Nash equilibrium in which no one acts, because *given that no one else is acting* (remember that under the Nash assumption each player takes the others' strategies as given), it pays any one person to act.

There *are* Nash equilibria where exactly one person acts; in fact, there are N such equilibria, one corresponding to each member. But when everyone is making the decision individually in isolation, there is no way to coordinate and designate who is to act. Even if members of the group were to attempt such coordination, they might try to negotiate over the responsibility and not reach a conclusion, at least not in time to be of help. Therefore it is of interest to examine symmetric equilibria in which all members have identical strategies.

We already saw that there cannot be an equilibrium in which all N people follow the same pure strategy. Therefore we should see whether there can be an equilibrium in which they all follow the same mixed strategy. Actually, mixed strategies are quite appealing in this context. The people are isolated, and each is trying to guess what the others will do. Each is thinking, Perhaps I should call the police . . . but maybe someone else will . . . but what if they don't . . . ? Each breaks off this process at some point and does the last thing that he thought of in this chain, but we have no good way of predicting what that last thing is. A mixed strategy carries the flavor of this idea of a chain of guesswork being broken at a random point.

So suppose P is the probability that any one person will not act. If one particular person is willing to mix strategies, he must be indifferent between the two pure strategies of acting and not acting. Acting gets him $(B - C)$ for sure. Not acting will get him 0 if none of the other $(N - 1)$ people act and B if at least one of them does act. Because the probability that any one person fails to act is P and because they are deciding independently, the probability that none of the $(N - 1)$ others acts is P^{N-1}, and the probability that at least one does act is $(1 - P^{N-1})$. Therefore the expected payoff to the one person when he does not act is

$$0 \times P^{N-1} + B(1 - P^{N-1}) = B(1 - P^{N-1}).$$

And that one person is indifferent between acting and not acting when

$$B - C = B(1 - P^{N-1}) \text{ or when } P^{N-1} = C/B \text{ or } P = (C/B)^{1/(N-1)}.$$

Note how this indifference condition of *one* selected player determines the probability with which the *other* players mix their strategies.

Having obtained the equilibrium mixture probability, we can now see how it changes as the group size N changes. Remember that $C/B < 1$. As N increases from 2 to infinity, the power $1/(N - 1)$ decreases from 1 to 0. Then C/B raised to this power—namely, P—increases from C/B to 1. Remember that P is the probability that any one person does not take the action. Therefore the probability of action by any one person—namely, $(1 - P)$—falls from $1 - C/B = (B - C)/B$ to 0.[25]

In other words, the more people there are, the less likely is any one of them to act. This is intuitively true, and in good conformity with the idea of diffusion of responsibility. But it does not yet give us the conclusion that help is less likely to be forthcoming in a larger group. As we said before, help requires action by only one person. Because there are more and more people, each of whom is less and less likely to act, we cannot conclude immediately that the probability of *at least one* of them acting gets smaller. More calculation is needed to see whether this is the case.

Because the N persons are randomizing independently in the Nash equilibrium, the probability Q that *not even one* of them helps is

$$Q = P^N = (C/B)^{N/(N-1)}.$$

As N increases from 2 to infinity, $N/(N - 1)$ decreases from 2 to 1, and then Q increases from $(C/B)^2$ to C/B. Correspondingly, the probability that *at least one* person helps—namely $(1 - Q)$—decreases from $1 - (C/B)^2$ to $1 - C/B$.[26]

So our exact calculation does bear out the hypothesis: the larger the group, the *less* likely is help to be given at all. The probability of provision does not,

[25]Consider the case in which $B = 10$ and $C = 8$. Then P equals 0.8 when $N = 2$, rises to 0.998 when $N = 100$, and approaches 1 as N continues to rise. The probability of action by any one person is $1 - P$, which falls from 0.2 to 0 as N rises from 2 toward infinity.

[26]With the same sample values for B (10) and C (8), this result implies that increasing N from 2 to infinity increases the probability that not even one person helps from 0.64 to 0.8. And the probability that at least one person helps falls from 0.36 to 0.2.

however, reduce to zero even in very large groups; instead it levels off at a positive value—namely, $(B - C)/B$—which depends on the benefit and cost of action to each individual.

We see how game-theoretic analysis sharpens the ideas from social psychology with which we started. The diffusion of responsibility theory takes us part of the way—namely, to the conclusion that any one person is less likely to act when he is part of a larger group. But the desired conclusion—that larger groups are less likely to provide help at all—needs further and more precise probability calculation based on the analysis of individual mixing and the resulting interactive (game) equilibrium.

And now we ask, did Kitty Genovese die in vain? Do the theories of pluralistic ignorance, diffusion of responsibility, and free-riding games still play out in the decreased likelihood of individual action within increasingly large cities? Perhaps not. John Tierney of the *New York Times* has publicly extolled the virtues of "urban cranks."[27] They are people who encourage the civility of the group through prompt punishment of those who exhibit unacceptable behavior—including litterers, noise polluters, and the generally obnoxious boors of society. Such "cranks" are essentially enforcers of a cooperative norm for society. And as Tierney surveys the actions of known "cranks," he reminds the rest of us that "[n]ew cranks must be mobilized! At this very instant, people are wasting time reading while norms are being flouted out on the street. . . . You don't live alone in this world! Have you enforced a norm today?" In other words, we need social norms and some people who have internalized the norm of enforcing norms.

SUMMARY

Multiplayer games generally concern problems of *collective action.* The general structure of collective-action games may be manifested as a prisoners' dilemma, chicken, or an assurance game. The critical difficulty with such games in any form is that the Nash equilibrium arising from individually rational choices may not be the *socially optimal* outcome—the outcome that maximizes the sum of the payoffs of all the players.

In collective-action games, when a person's action has some effect on the payoffs of all the other players, we say that there are *spillovers,* or *externalities.* They can be positive or negative and lead to individually driven outcomes

[27]John Tierney, "The Boor War: Urban Cranks, Unite—Against All Uncivil Behavior. Eggs Are a Last Resort," *New York Times Magazine*, January 5, 1997.

that are not socially optimal. When actions create negative spillovers, they are overused from the perspective of society; when actions create positive spillovers, they are underused. The additional possibility of *positive feedback* exists when there are positive spillovers; in such a case, the game may have multiple Nash equilibria.

Problems of collective action have been recognized for many centuries and discussed by scholars from diverse fields. Several early works professed no hope for the situation, but others offered up dramatic solutions. The most recent treatments of the subject acknowledge that collective-action problems arise in diverse areas and that there is no single optimal solution. Social scientific analysis suggests that social *custom*, or *convention*, can lead to cooperative behavior. Other possibilities for solutions come from the creation of *norms* of acceptable behavior. Some of these norms are *internalized* in individuals' payoffs; others must be *enforced* by the use of *sanctions* in response to the uncooperative behavior. Much of the literature agrees that small groups are more successful at solving collective-action problems than large ones.

In large-group games, *diffusion of responsibility* can lead to behavior in which individual persons wait for others to take action and *free ride* off the benefits of that action. If help is needed, it is less likely to be given at all as the size of the group available to provide it grows.

KEY TERMS

coercion (478)
collective action (446)
convention (474)
custom (474)
diffusion of responsibility (483)
external effect (461)
externality (461)
free rider (449)
internalize the externality (467)
locked in (470)
marginal private gain (461)
marginal social gain (461)
nonexcludable benefits (447)
nonrival benefits (447)
norm (473)
oppression (478)
pluralistic ignorance (482)
positive feedback (468)
pure public good (447)
sanction (473)
social optimum (449)
spillover effect (461)

SOLVED EXERCISES

S1. Suppose that 400 people are choosing between Action X and Action Y. The relative payoffs of the two actions depend on how many of the 400 people choose Action X and how many choose Action Y. The payoffs are as shown in the following diagram, but the vertical axis is not labeled, so you do not know whether the lines show the benefits or the costs of the two actions.

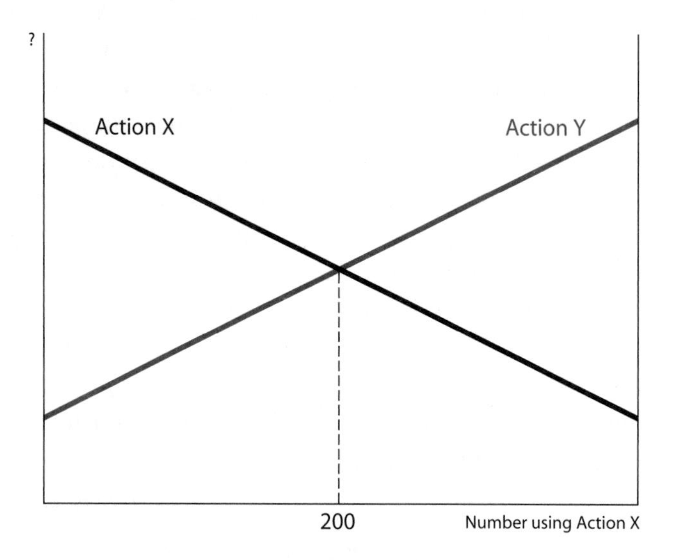

(a) You are told that the outcome in which 200 people choose Action X is an *un*stable equilibrium. If 100 people are currently choosing Action X, would you expect the number of people choosing X to increase or decrease over time? Why?

(b) For the graph to be consistent with the behavior that you described in part (a), should the lines be labeled as indicating the *costs* or *benefits* of Action X and Action Y? Explain your answer.

S2. A group has 100 members. Each person can choose to participate or not participate in a common project. If n of them participate in the project, then each participant derives the benefit $p(n) = n$, and each of the $(100 - n)$ shirkers derives the benefit $s(n) = 4 + 3n$.

(a) Is this an example of a prisoners' dilemma, a game of chicken, or an assurance game?

(b) Write the expression for the total benefit of the group.

(c) Show, either graphically or mathematically, that the maximum total benefit for the group occurs when $n = 74$.

(d) What difficulties will arise in trying to get exactly 74 participants and allowing the remaining 26 to shirk?

(e) How might the group try to overcome these difficulties?

S3. Consider a small geographic region with a total population of 1 million people. There are two towns, Alphaville and Betaville, in which each person can choose to live. For each person, the benefit from living in a town increases for a while with the size of the town (because larger towns have more amenities and so on), but after a point it decreases (because of congestion and so on). If x is the fraction of the population that lives in the same town as you do, your payoff is given by

$$x \text{ if } 0 \leq x \leq 0.4$$
$$0.6 - 0.5x \text{ if } 0.4 < x \leq 1.$$

(a) Draw a graph like Figure 12.10, showing the benefits of living in the two towns, as the fraction living in one versus the other varies continuously from 0 to 1.

(b) Equilibrium is reached either when both towns are populated and their residents have equal payoffs, or when one town—say Betaville—is totally depopulated, and the residents of the other town (Alphaville) get a higher payoff than would the very first person who seeks to populate Betaville. Use your graph to find all such equilibria.

(c) Now consider a dynamic process of adjustment whereby people gradually move toward the town whose residents currently enjoy a larger payoff than do the residents of the other town. Which of the equilibria identified in part (b) will be stable with these dynamics? Which ones will be unstable?

S4. Suppose an amusement park is being built in a city with a population of 100. Voluntary contributions are being solicited to cover the cost. Each citizen is being asked to give \$100. The more people contribute, the larger the park will be and the greater the benefit to each citizen. But it is not possible to keep out the noncontributors; they get their share of this benefit anyway. Suppose that when there are n contributors in the population, where n can be any whole number between 0 and 100, the benefit to each citizen in monetary unit equivalents is n^2 dollars.

(a) Suppose that initially no one is contributing. You are the mayor of the city. You would like everyone to contribute and can use persuasion on some people. What is the minimum number whom you need to persuade before everyone else will join in voluntarily?

(b) Find the Nash equilibria of the game where each citizen is deciding whether to contribute.

S5. Put the idea of Keynesian unemployment described at the end of Section 3.D into a properly specified game, and show the multiple equilibria in a diagram. Show the level of production (national product) on the vertical axis as a function of a measure of the level of demand (national income) on the horizontal axis. Equilibrium is reached when national product equals national income—that is, when the function relating the two cuts the 45° line. For what shapes of the function can there be multiple equilibria? Why might you expect such shapes in reality? Suppose that income increases when current production exceeds current income, and that income decreases when current production is less than current income. In this dynamic process, which equilibria are stable and which ones unstable?

S6. Write a brief description of a strategic game that you have witnessed or participated in that includes a large number of players and in which individual players' payoffs depend on the number of other players and their actions. Try to illustrate your game with a graph if possible. Discuss the outcome of the actual game in light of the fact that many such games have inefficient outcomes. Do you see evidence of such an outcome in your game?

UNSOLVED EXERCISES

U1. Figure 12.5 illustrates the payoffs in a general, two-person, collective-action game. There we showed various inequalities on the algebraic payoffs ($p(1)$, etc.) that made the game a prisoners' dilemma. Now you are asked to find similar inequalities corresponding to other kinds of games:

(a) Under what condition(s) on the payoffs is the two-person game a chicken game? What further condition(s) make the game version I of chicken (as in Figure 12.3)?

(b) Under what condition(s) on the payoffs is the two-person game an assurance game?

U2. A class with 30 students enrolled is given a homework assignment with five questions. The first four are the usual kinds of problems, totalling to 90 points. But the fifth is an interactive game for the class. The question reads: "You can choose whether to answer this question. If you choose to do so, you merely write 'I hereby answer Question 5.' If you choose not to answer Question 5, your score for the assignment will be based on your performance on the first four problems. If you choose to answer Question 5, then your scoring will be as follows: If fewer than half of the students in the class answer Question 5, you get 10 points for Question 5; 10 points will be added to your score on the other four questions to get your total score for the assignment. If half or more than half of the students in the class answer Question 5, you get -10 points; that is, 10 points will be subtracted from your score on the other questions."

(a) Draw a diagram illustrating the payoffs from the two possible strategies, "Answer Question 5" and "Don't Answer Question 5," in relation to the number of other students who answer it. Find the Nash equilibrium of the game.

(b) What would you expect to see happen in this game if it were actually played in a college classroom? Why? Consider two cases: (i) the students make their choices individually with no communication; and (ii) the students make their choices individually but can discuss these choices ahead of time in a discussion forum available on the class Web site.

U3. There are two routes for driving from A to B. One is a freeway, and the other consists of local roads. The benefit of using the freeway is constant and equal to 1.8, irrespective of the number of people using it. Local roads get congested when too many people use this alternative, but if not enough people use it, the few isolated drivers run the risk of becoming victims of crimes. Suppose that when a fraction x of the population is using the local roads, the benefit of this mode to each driver is given by

$$1 + 9x - 10x^2.$$

(a) Draw a graph showing the benefits of the two driving routes as functions of x, regarding x as a continuous variable that can range from 0 to 1.

(b) Identify all possible equilibrium traffic patterns from your graph in part (a). Which equilibria are stable? Which ones are unstable? Why?

(c) What value of x maximizes the total benefit to the whole population?

U4. Suppose a class of 100 students is comparing two careers—lawyer or engineer. An engineer gets take-home pay of $100,000 per year, irrespective of the numbers who choose this career. Lawyers make work for each other, so as the total number of lawyers increases, the income of each lawyer increases—up to a point. Ultimately, the competition between them drives down the income of each. Specifically, if there are N lawyers, each will get $100N - N^2$ thousand dollars a year. The annual cost of running a legal practice (office space, secretary, paralegals, access to online reference services, and so forth) is $800,000. Therefore, each lawyer takes home $100N - N^2 - 800$ thousand dollars a year when there are N of them.

(a) Draw a graph showing the take-home income of each lawyer on the vertical axis and the number of lawyers on the horizontal axis. (Plot a few points—say, for 0, 10, 20, . . . , 90, 100 lawyers. Fit a curve to the points, or use a computer graphics program if you have access to one.)

(b) When career choices are made in an uncoordinated way, what are the possible equilibrium outcomes?

(c) Now suppose the whole class decides how many should become lawyers, aiming to maximize the total take-home income of the whole class. What will be the number of lawyers? (If you can, use calculus, regarding N as a continuous variable. Otherwise you can use graphical methods or a spreadsheet.)

U5. A group of 12 countries is considering whether to form a monetary union. They differ in their assessments of the costs and benefits of this move, but each stands to gain more from joining, and lose more from staying out, when more of the other countries choose to join. The countries are ranked in order of their liking for joining, 1 having the highest preference for joining and 12 the least. Each country has two actions, IN and OUT. Let

$$B(i,n) = 2.2 + n - i$$

be the payoff to country with ranking i when it chooses IN and n others have chosen IN. Let

$$S(i,n) = i - n$$

be the payoff to country with ranking i when it chooses OUT and n others have chosen IN.

(a) Show that for country 1, IN is the dominant strategy.

(b) Having eliminated OUT for country 1, show that IN becomes the dominant strategy for country 2.

(c) Continuing in this way, show that all countries will choose IN.

(d) Contrast the payoffs in this outcome with those where all choose OUT. How many countries are made worse off by the formation of the union?

U6. **(Optional, computer or graphing calculator required.)** Exercise **U2** asks for an asymmetric pure-strategy Nash equilibrium to the game where 30 students decide whether or not to answer the fifth question on their homework assignment. But a symmetric equilibrium—where all players employ the same strategy—also exists in mixed strategies. Suppose each player chooses to answer Question 5 with probability P, and not to answer it with probability $1 - P$. This kind of equilibrium can be found following a procedure similar to that discussed in Section 6, that is, by showing that an individual student is indifferent between answering the question and not answering the question. However, this specific problem is more complicated, because the expected payoff to answering the question depends on the probability that 0 out of 29 of the other students decide to answer the question, the probability that 1 out of 29 of the other students decide to answer the question, the probability that 2 out of 29 of the other students decide to answer the question, and so on. These probabilities are given by the binomial distribution, which you can look up either in a probability textbook or online. Remember that each of the other 29 students will *individually* answer Question 5 with probability P.

(a) Using the binomial distribution, construct an indifference equation that can be solved for the value of P.

(b) Because the indifference equation found in part (a) includes a very high-order polynomial, it may be impossible to solve the equation analytically. Instead, find the approximate solution numerically, using software such as Microsoft Excel, MATLAB, or Mathematica. For example, you can do this in Excel by guessing a value of P in one cell, then setting up a formula in another cell that gives the expected value of answering the question. Change the value in the first cell to zero in on the correct value of P. What is the equilibrium value of P?

Evolutionary Games

WE HAVE SO FAR STUDIED GAMES with many different features—simultaneous and sequential moves, zero-sum and non-zero-sum payoffs, strategic moves to manipulate rules of games to come, one-shot and repeated play, and even games of collective action in which a large number of people play simultaneously. However, one ground rule has remained unchanged in all of the discussions—namely, that all the players in all these games are rational: each player has an internally consistent value system, can calculate the consequences of her strategic choices, and makes the choice that best favors her interests.

In applying this rule, we merely follow the route taken by most of game theory, which was developed mainly by economists. Economics was founded on the dual assumptions of rational behavior and equilibrium. Indeed, these assumptions have proved useful in game theory. We have obtained quite a good understanding of games in which the players participate sufficiently regularly to have learned what their best choices are by experience. The assumptions ensure that a player does not attribute any false naiveté to her rivals and thus does not get exploited by these rivals. The theory also gives some prescriptive guidance to players as to how they *should* play.

However, other social scientists are much more skeptical of the rationality assumption and therefore of a theory built on such a foundation. Economists, too, should not take rationality for granted, as pointed out in Chapter 5. The trouble is finding a feasible alternative. Although we may not wish to impose conscious and perfectly calculating rationality on players, we do not want to abandon the idea that some strategies are better than others. We want good strategies to be rewarded with higher payoffs; we want players to observe or imitate success and to experiment with new strategies; we want good strategies to be used more often and bad strategies less often, as players gain experience playing the game. We find one possible alternative to rationality in the biological theory of evolution and evolutionary dynamics and will study its lessons in this chapter.

1 THE FRAMEWORK

The process of evolution in biology offers a particularly attractive parallel to the theory of games used by social scientists. This theory rests on three fundamentals: heterogeneity, fitness, and selection. The starting point is that a significant part of animal behavior is genetically determined; a complex of one or more genes (**genotype**) governs a particular pattern of behavior, called a behavioral **phenotype.** Natural diversity of the gene pool ensures a heterogeneity of phenotypes in the population. Some behaviors are better suited than others to the prevailing conditions, and the success of a phenotype is given a quantitative measure called its **fitness.** People are used to thinking of this success as meaning the common but misleading phrase "survival of the fittest"; however, the ultimate test of biological fitness is not mere survival, but reproductive success. That is what enables an animal to pass on its genes to the next generation and perpetuate its phenotype. The fitter phenotypes then become relatively more numerous in the next generation than the less fit phenotypes. This process of **selection** is the dynamic that changes the mix of genotypes and phenotypes and perhaps leads eventually to a stable state.

From time to time, chance produces new genetic **mutations.** Many of these mutations produce behaviors (that is, phenotypes) that are ill suited to the environment, and they die out. But occasionally a mutation leads to a new phenotype that is fitter. Then such a mutant gene can successfully **invade** a population—that is, spread to become a significant proportion of the population.

At any time, a population may contain some or all of its biologically conceivable phenotypes. Those that are fitter than others will increase in proportion, some unfit phenotypes may die out, and other phenotypes not currently in the population may try to invade it. Biologists call a configuration of a population and its current phenotypes **evolutionary stable** if the population cannot be invaded successfully by any mutant. This is a static test, but often a more dynamic criterion is applied: a configuration is evolutionary stable if it is the limiting outcome of the dynamics of selection, starting from any arbitrary mixture of phenotypes in the population.[1]

The fitness of a phenotype depends on the relationship of the individual organism to its environment; for example, the fitness of a particular bird depends on the aerodynamic characteristics of its wings. It also depends on the whole complex of the proportions of different phenotypes that exist in the environment—how aerodynamic its wings are relative to those of the rest of its species. Thus the fitness of a particular animal—with its behavioral traits, such as aggression and sociability—depends on whether other members of its species are predominantly aggressive or passive, crowded or dispersed, and so on. For our purpose, this **interaction** between phenotypes within a species is the most interesting aspect of the story. Sometimes an individual member of a species interacts with members of another species; then the fitness of a particular type of sheep, for example, may depend on the traits that prevail in the local population of wolves. We consider this type of interaction as well, but only after we have covered the within-species case.

The biological process of evolution finds a ready parallel in game theory. The behavior of a phenotype can be thought of as a *strategy* of the animal in its interaction with others—for example, whether to fight or to retreat. The difference is that the choice of strategy is not a purposive calculation as it would be in standard game theory; rather, it is a genetically predetermined fixture of the phenotype. The interactions lead to *payoffs* to the phenotypes. In biology, the payoffs measure the evolutionary or reproductive fitness; when we apply these ideas outside of biology, they can have other connotations of success in the social, political, or economic games in question.

The payoffs or fitness numbers can be shown in a payoff table just like that for a standard game, with all conceivable phenotypes of one animal arrayed along the rows of the matrix and those of the other along the columns of the matrix. If more animals interact simultaneously—which is called **playing the field** in biology—the payoffs can be shown by functions like those for collective-action games in Chapter 12. We will consider pair-by-pair matches for most of this chapter and will look at the other case briefly in Section 9.

Because the population is a mix of phenotypes, different pairs selected from it will bring to their interactions different combinations of strategies. The actual quantitative measure of the fitness of a phenotype is the average payoff that it gets in all its interactions with others in the population. Those animals with higher fitness will have greater evolutionary success. The eventual outcome of

[1]The dynamics of phenotypes is driven by an underlying dynamics of genotypes but, at least at the elementary level, evolutionary biology focuses its analysis at the phenotype level and conceals the genetic aspects of evolution. We will do likewise in our exposition of evolutionary games. Some theories at the genotypes level can be found in the materials cited in footnote 2.

the population dynamics will be an evolutionary stable configuration of the population.

Biologists have used this approach very successfully. Combinations of aggressive and cooperative behavior, locations of nesting sites, and many more phenomena that elude more conventional explanations can be understood as the stable outcomes of an evolutionary process of selection of fitter strategies. Interestingly, biologists developed the idea of evolutionary games by using the preexisting body of game theory, drawing from its language but modifying the assumption of conscious maximizing to suit their needs. Now game theorists are in turn using insights from the research on biological evolutionary games to enrich their own subject.[2]

Indeed, the theory of evolutionary games seems a ready-made framework for a new approach to game theory, relaxing the assumption of rational behavior.[3] According to this view of games, individual players have no freedom to choose their strategy at all. Some are "born" to play one strategy, others another. The idea of inheritance of strategies can be interpreted more broadly in applications of the theory other than in biology. In human interactions, a strategy may be embedded in a player's mind for a variety of reasons—not only genetics but also (and probably more important) socialization, cultural background, education, or a rule of thumb based on past experience. The population can consist of a mixture of different people with different backgrounds or experiences that embed different strategies into them. Thus some politicians may be motivated to adhere to certain moral or ethical codes even at the cost of electoral success, whereas others are mainly concerned with their own reelection; similarly, some firms may pursue profit alone, whereas others are motivated by social or ecological objectives. We can call each logically conceivable strategy that can be embedded in this way a phenotype for the population of players in the context being studied.

[2]Robert Pool, "Putting Game Theory to the Test," *Science,* vol. 267 (March 17, 1995), pp. 1591–1593, is a good article for general readers and has many examples from biology. John Maynard Smith deals with such games in biology in his *Evolutionary Genetics* (Oxford: Oxford University Press, 1989), chap. 7, and *Evolution and the Theory of Games* (Cambridge: Cambridge University Press, 1982); the former also gives much background on evolution. Recommended for advanced readers are Peter Hammerstein and Reinhard Selten, "Game Theory and Evolutionary Biology," in *Handbook of Game Theory,* vol. 2, ed. R. J. Aumann and S. Hart (Amsterdam: North Holland, 1994), pp. 929–993; and Jorgen Weibull, *Evolutionary Game Theory* (Cambridge: MIT Press, 1995).

[3]Indeed, applications of the evolutionary perspective need not stop with game theory. The following joke offers an "evolutionary theory of gravitation" as an alternative to Newton's or Einstein's physical theories:

Question: Why does an apple fall from the tree to earth?

Answer: Originally, apples that came loose from trees went in all directions. But only those that were genetically predisposed to fall to the earth could reproduce.

From a population with its heterogeneity of embedded strategies, pairs of phenotypes are repeatedly randomly selected to interact (play the game) with others of the same or different "species." In each interaction, the payoff of each player depends on the strategies of both; this dependence is governed by the usual "rules of the game" and illustrated in the game table or tree. The *fitness* of a particular strategy is defined as its aggregate or average payoff in its pairings with all the strategies in the population. Some strategies have a higher level of fitness than others; in the next generation—that is, the next round of play—these higher-fitness strategies will be used by more players and will proliferate. Strategies with lower fitness will be used by fewer players and will decay or die out. Occasionally, someone may experiment or adopt a previously unused strategy from the collection of those that are logically conceivable. This corresponds to the emergence of a mutant. If the new strategy is fitter than the ones currently being used, it will start to be used by larger proportions of the population. The central question is whether this process of selective proliferation, decay, and mutation of certain strategies in the population will have an evolutionary stable outcome and, if so, what it will be. In regard to the examples just cited, will society end up with a situation in which all politicians are concerned with reelection and all firms with profit? In this chapter, we develop the framework and methods for answering such questions.

Although we use the biological analogy, the reason that the fitter strategies proliferate and the less fit ones die out in socioeconomic games differs from the strict genetic mechanism of biology: players who fared well in the last round will transmit the information to their friends and colleagues playing the next round, those who fared poorly in the last round will observe which strategies succeeded better and will try to imitate them, and some purposive thinking and revision of previous rules of thumb will take place between successive rounds. Such "social" and "educational" mechanisms of transmission are far more important in most strategic games than any biological genetics; indeed, this is how the reelection orientation of legislators and the profit-maximization motive of firms are reinforced. Finally, conscious experimentation with new strategies substitutes for the accidental mutation in biological games.

Evolutionary stable configurations of biological games can be of two kinds. First, a single phenotype may prove fitter than any others, and the population may come to consist of it alone. Such an evolutionary stable outcome is called **monomorphism**—that is, a single (mono) form (morph). In that case, the unique prevailing strategy is called an **evolutionary stable strategy (ESS).** The other possibility is that two or more phenotypes may be equally fit (and fitter than some others not played), so they may be able to coexist in certain proportions. Then the population is said to exhibit **polymorphism**—that is, a multiplicity (poly) of forms (morph). Such a state will be stable if no new phenotype or feasible mutant can achieve a higher fitness against such a population than the fitness of the types that are already present in the polymorphic population.

Polymorphism comes close to the game-theoretic notion of a mixed strategy. However, there is an important difference. To get polymorphism, no individual player need follow a mixed strategy. Each can follow a pure strategy, but the population exhibits a mixture because different individual players pursue different pure strategies.

The whole setup—the population, its conceivable collection of phenotypes, the payoff matrix in the interactions of the phenotypes, and the rule for the evolution of population proportions of the phenotypes in relation to their fitness—constitutes an evolutionary game. An evolutionary stable configuration of the population can be called an *equilibrium* of the evolutionary game.

In this chapter, we develop some of these ideas, as usual through a series of illustrative examples. We begin with symmetric games, in which the two players are on similar footing—for example, two members of the same species competing with each other for food or mates; in a social science interpretation, they could be two elected officials competing for the right to continue in public office. In the payoff table for the game, each can be designated as the row player or the column player with no difference in outcome.

2 PRISONERS' DILEMMA

Suppose a population is made up of two phenotypes. One type consists of players who are natural-born cooperators; they always work toward the outcome that is jointly best for all players. The other type consists of the defectors; they work only for themselves. As an example, we use the restaurant pricing game described in Chapter 5 and presented in a simplified version in Chapter 11. Here, we use the simpler version in which only two pricing choices are available, the jointly best price of $26 or the Nash equilibrium price of $20. A cooperator restaurateur would always choose $26, whereas a defector would always choose $20. The payoffs (profits) of each type in a single play of this discrete dilemma are shown in Figure 13.1, reproduced from Figure 11.2. Here we call the players simply Row and Column because each can be any individual

		COLUMN	
		20 (Defect)	26 (Cooperate)
ROW	20 (Defect)	288, 288	360, 216
	26 (Cooperate)	216, 360	324, 324

FIGURE 13.1 Prisoners' Dilemma of Pricing ($100s per Month)

restaurateur in the population who is chosen at random to compete against another random rival.

Remember that under the evolutionary scenario, no one has the choice between defecting and cooperating; each is "born" with one trait or the other predetermined. Which is the more successful (fitter) trait in the population?

A defecting-type restaurateur gets a payoff of 288 ($28,800 a month) if matched against another defecting type and a payoff of 360 ($36,000 a month) if matched against a cooperating type. A cooperating type gets 216 ($21,600 a month) if matched against a defecting type and 324 ($32,400 a month) if matched against another cooperating type. No matter what the type of the matched rival, the defecting type does better than the cooperating type.[4] Therefore the defecting type has a better expected payoff (and is thus fitter) than does the cooperating type, irrespective of the proportions of the two types in the population.

A little more formally, let x be the proportion of cooperators in the population. Consider any one particular cooperator. In a random draw, the probability that she will meet another cooperator (and get 324) is x and that she will meet a defector (and get 216) is $(1 - x)$. Therefore a typical cooperator's expected payoff is $324x + 216(1 - x)$. For a defector, the probability of meeting a cooperator (and getting 360) is x and that of meeting another defector (and getting 288) is $(1 - x)$. Therefore a typical defector's expected profit is $360x + 288(1 - x)$. Now it is immediately apparent that

$$360x + 288(1 - x) > 324x + 216(1 - x) \text{ for all } x \text{ between 0 and 1.}$$

Therefore a defector has a higher expected payoff and is fitter than a cooperator. This will lead to an increase in the proportion of defectors (a decrease in x) from one "generation" of players to the next, until the whole population consists of defectors.

What if the population initially consists of all defectors? Then in this case no mutant (experimental) cooperator will survive and multiply to take over the population; in other words, the defector population cannot be invaded successfully by mutant cooperators. Even for a very small value of x—that is, when the proportion of cooperators in the population is very small—the cooperators remain less fit than the prevailing defectors, and their population proportion will not increase but will be driven to zero; the mutant strain will die out.

Our analysis shows both that defectors have higher fitness than cooperators and that an all-defector population cannot be invaded by mutant cooperators. Thus the evolutionary stable configuration of the population is monomorphic, consisting of the single strategy or phenotype Defect. We therefore call Defect the evolutionary stable strategy for this population engaged in this dilemma game. Note that Defect is a strictly dominant strategy in the rational behavior

[4]In the rational behavior context of the preceding chapters, we would say that Defect is the strictly dominant strategy.

analysis of this same game. This result is very general: if a game has a strictly dominant strategy, that strategy will also be the ESS.

A. The Repeated Prisoners' Dilemma

We saw in Chapter 11 how a repetition of the prisoners' dilemma permitted consciously rational players to sustain cooperation for their mutual benefit. Let us see if a similar possibility exists in the evolutionary story. Suppose each chosen pair of players plays the dilemma three times in succession. The overall payoff to a player from such an interaction is the sum of what she gets in the three rounds.

Each individual player is still programmed to play just one strategy, but that strategy has to be a complete plan of action. In a game with three moves, a strategy can stipulate an action in the second or third play that depends on what happened in the first or second play. For example, "I will always cooperate no matter what" and "I will always defect no matter what" are valid strategies. But "I will begin by cooperating and continue to cooperate as long as you cooperated on the preceding play; and I will defect in all later plays if you defect in an early play" is also a valid strategy; in fact, this last strategy is just tit-for-tat (TFT).

To keep the initial analysis simple, we suppose in this section that there are just two types of strategies that can possibly exist in the population: always defect (A) and tit-for-tat (T). Pairs are randomly selected from the population, and each selected pair plays the game a specified number of times. The fitness of each player is simply the sum of her payoffs from all the repetitions played against her specific opponent. We examine what happens with two, three, and more generally *n* such repetitions in each pair.

I. TWICE-REPEATED PLAY Figure 13.2 shows the payoff table for the game in which two members of the restaurateur population meet and play against one another exactly twice. If both players are A types, both defect both times, and Figure 13.1 shows that then each gets 288 each time, for a total of 576. If both are T types, defection never starts, and each gets 324 each time, for a total of 648. If one is an A type and the other a T type, then on the first play the A type defects and the T type cooperates, so the former gets 360 and the latter 216. On the second play both defect and get 288. So the A type's total payoff is 360 + 288 = 648, and the T type's total is 216 + 288 = 504.

In the twice-repeated dilemma, we see that A is only weakly dominant. It is easy to see that if the population is all A, then T-type mutants cannot invade, and A is an ESS. But if the population is all T, then A-type mutants cannot do any better than the T types. Does this mean that T must be another ESS, just as it would be a Nash equilibrium in the rational-game-theoretic analysis of this game? The answer is no. If the population is initially all T and a few A mutants enter, then the mutants would meet the predominant T types most of the time and would do as well as T does against another T. But occasionally an A mutant would meet

		COLUMN	
		A	T
ROW	A	576, 576	648, 504
	T	504, 648	648, 648

FIGURE 13.2 Outcomes in the Twice-Repeated Prisoners' Dilemma ($100s)

another A mutant, and in this match she does better than would a T against an A. Thus the mutants have just *slightly* higher fitness than that of a member of the predominant phenotype. This advantage leads to an increase, albeit a slow one, in the proportion of mutants in the population. Therefore an all-T population *can* be invaded successfully by A mutants; T is not an ESS.

Our reasoning relies on two tests for an ESS. First we see if the mutant does better or worse than the predominant phenotype when each is matched against the predominant type. If this primary criterion gives a clear answer, that settles the matter. But if the primary criterion gives a tie, then we use a tie-breaking, or secondary, criterion: does the mutant fare better or worse than a predominant phenotype when each is matched against a mutant? Ties are exceptional and most of the time we do not need the secondary criterion, but it is there in reserve for situations such as the one illustrated in Figure 13.2.[5]

II. THREEFOLD REPETITION Now suppose each matched pair from the (A,T) population plays the game three times. Figure 13.3 shows the fitness outcomes, summed over the three meetings, for each type of player when matched against rivals of each type.

To see how these fitness numbers arise, consider a couple of examples. When two T players meet each other, both cooperate the first time, and therefore both cooperate the second time and the third time as well; both get 324 each time, for a total of 972 each over 3 months. When a T player meets an A player, the latter does well the first time (360 for the A type versus 216 for the T player), but then the T player also defects the second and third times, and each gets 288 in both of those plays (for totals of 936 for A and 792 for T).

The relative fitnesses of the two types depend on the composition of the population. If the population is almost wholly A type, then A is fitter than T (because

[5]This game is just one example of a twice-repeated dilemma. With other payoffs in the basic game, twofold repetition may not have ties. That is so in the husband–wife jail story of Chapter 4. If both the primary and secondary criteria yield ties, neither phenotype satisfies our definition of ESS, and we need to broaden our understanding of what constitutes an equilibrium in the evolutionary game. We consider such a possibility in Section 7 and provide the general theory for dealing with such an outcome in Section 8.

		COLUMN	
		A	T
ROW	A	864, 864	936, 792
	T	792, 936	972, 972

FIGURE 13.3 Outcomes in the Thrice-Repeated Prisoners' Dilemma ($100s)

A types meeting mostly other A types earn 864 most of the time, but T types most often get 792). But if the population is almost wholly T type, then T is fitter than A (because T types earn 972 when they meet mostly other Ts, but A types earn 936 in such a situation). Each type is fitter when it already predominates in the population. Therefore T cannot invade successfully when the population is all A, and vice versa. Now there are two possible evolutionary stable configurations of the population; in one configuration, A is the ESS and, in the other, T is the ESS.

Next consider the evolutionary dynamics when the initial population is made up of a mixture of the two types. How will the composition of the population evolve over time? Suppose a fraction x of the population is T type and the rest, $(1 - x)$, is A type.[6] An individual A player, pitted against various opponents chosen from such a population, gets 936 when confronting a T player, which happens a fraction x of the times, and 864 against another A player, which happens a fraction $(1 - x)$ of the times. This gives an average expected payoff of

$$936x + 864(1 - x) = 864 + 72x$$

for each A player. Similarly, an individual T player gets an average expected payoff of

$$972x + 792(1 - x) = 792 + 180x.$$

Then a T player is fitter than an A player if the former earns more on average; that is, if

$$\begin{aligned} 792 + 180x &> 864 + 72x \\ 108x &> 72 \\ x &> 2/3. \end{aligned}$$

[6]Literally, the fraction of any particular type in the population is finite and can only take on values such as 1/1,000,000, 2/1,000,000, and so on. But, if the population is sufficiently large and we show all such values as points on a straight line, as in Figure 13.4, then these points are very tightly packed together, and we can regard them as forming a continuous line. This amounts to letting the fractions take on any real value between 0 and 1. We can then talk of the population *proportion* of a certain behavioral type. By the same reasoning, if one individual member goes to jail and is removed from the population, her removal does not change the population's proportions of the various phenotypes.

In other words, if more than two-thirds (67%) of the population is already T type, then T players are fitter and their proportion will grow until it reaches 100%. If the population starts with less than 67% T, then A players will be fitter, and the proportion of T players will go on declining until there are 0% of them, or 100% of the A players. The evolutionary dynamics move the population toward one of the two extremes, each of which is a possible ESS. The dynamics leads to the same conclusion as the static test of mutants' invasion. This is a common, although not universal, feature of evolutionary games.

Thus we have identified two evolutionary stable configurations of the population. In each one the population is all of one type (monomorphic). For example, if the population is initially 100% T, then even after a small number of mutant A types arise, the population mix will still be more than 66.66 . . . % T; T will remain the fitter type, and the mutant A strain will die out. Similarly, if the population is initially 100% A, then a small number of T-type mutants will leave the population mix with less than 66.66 . . . % T, so the A types will be fitter and the mutant T strain will die out. And as we saw earlier, experimenting mutants of type N can never succeed in a population of A and T types that is either largely T or largely A.

What if the initial population has exactly 66.66 . . . % T players (and 33.33 . . . % A players)? Then the two types are equally fit. We could call this *poly*morphism. But it is not really a suitable candidate for an evolutionary stable configuration. The population can sustain this delicately balanced outcome only until a mutant of either type surfaces. By chance, such a mutant must arise sooner or later. The mutant's arrival will tip the fitness calculation in favor of the mutant type, and the advantage will accumulate until the ESS with 100% of that type is reached. This is just an application of the secondary criterion for evolutionary stability. We will sometimes loosely speak of such a configuration as an unstable equilibrium, so as to maintain the parallel with ordinary game theory where mutations are not a consideration and a delicately balanced equilibrium can persist. But in the strict logic of the biological process, it is not an equilibrium at all.

This reasoning can be shown in a simple graph that closely resembles the graphs that we drew when calculating the equilibrium proportions in a mixed-strategy equilibrium with consciously rational players. The only difference is that in the evolutionary context, the proportion in which the separate strategies are played is not a matter of choice by any individual player but a property of the whole population, as shown in Figure 13.4. Along the horizontal axis, we measure the proportion x of T players in the population from 0 to 1. We measure fitness along the vertical axis. Each line shows the fitness of one type. The line for the T type starts lower (at 792 compared with 864 for the A-type line) and ends higher (972 against 936). The two lines cross when $x = 0.66. . . .$ To the right of this point, the T type is fitter, so its population proportion

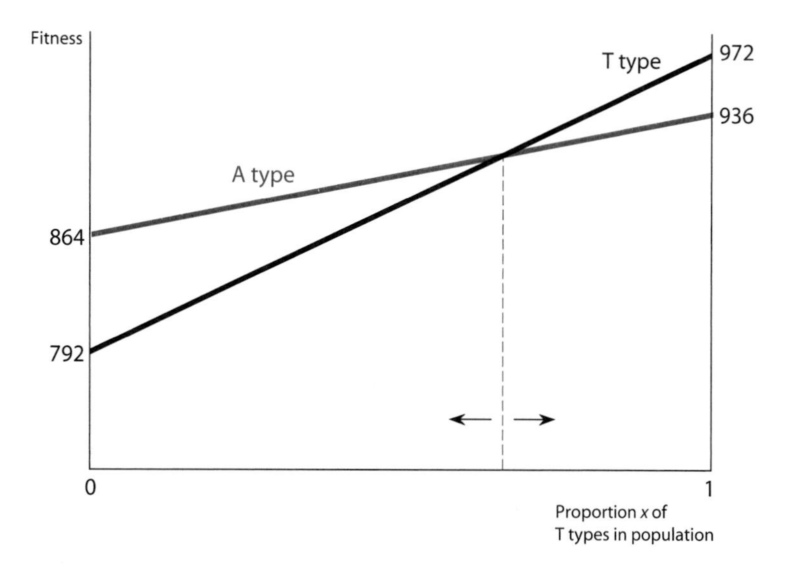

FIGURE 13.4 Fitness Graphs and Equilibria for the Thrice-Repeated Prisoners' Dilemma

increases over time and x *increases* toward 1. Similarly, to the left of this point, the A type is fitter, so its population proportion increases over time and x *decreases* toward 0. Such diagrams often prove useful as visual aids, and we will use them extensively.[7]

B. Multiple Repetitions

What if each pair plays some unspecified number of repetitions of the game? Let us focus on a population consisting of only A and T types in which interactions between random pairs occur n times (where $n > 2$). The table of the total outcomes from playing n repetitions is shown in Figure 13.5. When two A types meet, they always defect and earn 288 every time, so each gets $288n$ in n plays. When two T types meet, they begin by cooperating, and no one is the first to defect, so they earn 324 every time, for a total of $324n$. When an A type meets a T type, on the first play the T type cooperates and the A type defects, and so the A type gets 360 and the T type gets 216; thereafter the T type retaliates against the preceding defection of the A type for all remaining plays, and each gets 288 in all of the remaining $(n - 1)$ plays. Thus the A type earns a total of $360 + 288(n - 1) = 288n + 72$ in n plays against a T type, whereas the T type gets $216 + 288(n - 1) = 288n - 72$ in n plays against an A type.

[7]You should now draw a similar graph for the twice-repeated case. You will see that the A line is above the T line for all values of x less than 1, but the two meet on the right-hand edge of the figure where $x = 1$.

		COLUMN	
		A	T
ROW	A	$288n, 288n$	$288n + 72, 288n - 72$
	T	$288n - 72, 288n + 72$	$324n, 324n$

FIGURE 13.5 Outcomes in the *n*-fold-Repeated Dilemma

If the proportion of T types in the population is x, then a typical A type gets $x(288n + 72) + (1 - x)288n$ on average, and a typical T type gets $x(324n) + (1 - x)(288n - 72)$ on average. Therefore the T type is fitter if

$$x(324n) + (1 - x)(288n - 72) > x(288n + 72) + (1 - x)288n$$
$$36xn > 72$$
$$x > \frac{72}{36n} = \frac{2}{n}.$$

Once again we have two monomorphic ESSs, one all T (or $x = 1$, to which the process converges starting from any $x > 2/n$) and the other all A (or $x = 0$, to which the process converges starting from any $x < 2/n$). As in Figure 13.4, there is also an unstable polymorphic equilibrium at the balancing point $x = 2/n$.

Notice that the proportion of T at the balancing point depends on n; it is smaller when n is larger. When $n = 10$, it is 2/10, or 0.2. So if the population initially is 20% T players, in a situation where each pair plays 10 repetitions, the proportion of T types will grow until they reach 100%. Recall that when pairs played three repetitions ($n = 3$), the T players needed an initial strength of 67% or more to achieve this outcome, and only two repetitions meant that T types needed to be 100% of the population to survive. (We see the reason for this outcome in our expression for the critical value for x, which shows that when $n = 2$, x must be above 1 before the T types are fitter.) Remember, too, that a population consisting of all T players achieves cooperation. Thus cooperation emerges from a larger range of the initial conditions when the game is repeated more times. In this sense, with more repetition, cooperation becomes more likely. What we are seeing is the result of the fact that the value of establishing cooperation increases as the length of the interaction increases.

C. Comparing the Evolutionary and Rational-Player Models

Finally, let us return to the thrice-repeated game illustrated in Figure 13.3 and, instead of using the evolutionary model, consider it played by two consciously rational players. What are the Nash equilibria? There are two in pure strategies, one in which both play A and the other in which both play T. There is also an equilibrium in mixed strategies, in which T is played 67% of the time and A 33% of the time. The first two are just the monomorphic ESSs that we found, and the third is the unstable polymorphic evolutionary equilibrium. In other words, there is a close relation between evolutionary and consciously rational perspectives on games.

That is not a coincidence. An ESS must be a Nash equilibrium of the game played by consciously rational players with the same payoff structure. To see this, suppose the contrary for the moment. If all players using some strategy—call it S—is not a Nash equilibrium, then some other strategy—call it R—must yield a higher payoff for one player when played against S. A mutant playing R will achieve greater fitness in a population playing S and so will invade successfully. Thus S cannot be an ESS. In other words, if all players using S is not a Nash equilibrium, then S cannot be an ESS. This is the same as saying that, if S is an ESS, it must be a Nash equilibrium for all players to use S.

Thus the evolutionary approach provides a backdoor justification for the rational approach. Even when players are not consciously maximizing, if the more successful strategies get played more often and the less successful ones die out and if the process converges eventually to a stable strategy, then the outcome must be the same as that resulting from consciously rational play.

Although an ESS must be a Nash equilibrium of the corresponding rational-play game, the converse is not true. We have seen two examples of this. In the twice-repeated dilemma game of Figure 13.2 played rationally, T would be a Nash equilibrium in the weak sense that if both players choose T, neither has any positive gain from switching to A. But in the evolutionary approach A can arise as a mutation and can successfully invade the T population. And in the thrice-repeated dilemma game of Figures 13.3 and 13.4, rational play would produce a mixed-strategy equilibrium. But the biological counterpart to this mixed-strategy equilibrium, the polymorphic state, can be successfully invaded by mutants and is therefore not a true evolutionary stable equilibrium. Thus the biological concept of stability can help us select from a multiplicity of Nash equilibria of a rationally played game.

There is one limitation of our analysis of the repeated game. At the outset, we allowed just two strategies: A and T. Nothing else was supposed to exist or arise as a mutation. In biology, the kinds of mutations that arise are determined by genetic considerations. In social or economic or political games, the genesis of new strategies is presumably governed by history, culture, and the experience of the players; the ability of people to assimilate and process information and to experiment with different strategies must also play a role. However, the restrictions that we place on the set of strategies that can possibly exist in a particular game have important implications for which of these strategies (if any) can be evolutionary stable. In the thrice-repeated prisoners' dilemma example, if we had allowed for a strategy S that cooperated on the first play and defected on the second and third, then S-type mutants could have successfully invaded an all-T population, so T would not have been an ESS. We develop this possibility further in the exercises at the end of this chapter.

3 CHICKEN

Remember our 1950s youths racing their cars toward one another and seeing who will be the first to swerve to avoid a collision? Now we suppose the players have no choice in the matter: each is genetically hardwired to be either a Wimp (always swerve) or a Macho (always go straight). The population consists of a mixture of the two types. Pairs are picked at random every week to play the game. Figure 13.6 shows the payoff table for any two such players—say, A and B. (The numbers replicate those we used before in Figure 4.14.)

How will the two types fare? The answer depends on the initial population proportions. If the population is almost all Wimps, then a Macho mutant will win and score 1 lots of times, whereas all the Wimps meeting their own types will get mostly zeroes. But if the population is mostly Macho, then a Wimp mutant scores −1, which may look bad but is better than the −2 that all the Machos get. You can think of this appropriately in terms of the biological context and the sexism of the 1950s: in a population of Wimps, a Macho newcomer will show all the rest to be chickens and so will impress all the girls. But if the population consists mostly of Machos, they will be in the hospital most of the time and the girls will have to go for the few Wimps who are healthy.

In other words, each type is fitter when it is relatively rare in the population. Therefore each can successfully invade a population consisting of the other type. We should expect to see both types in the population in equilibrium; that is, we should expect an ESS with a mixture, or polymorphism.

To find the proportions of Wimps and Machos in such an ESS, let us calculate the fitness of each type in a general mixed population. Write x for the fraction of Machos and $(1 - x)$ for the proportion of Wimps. A Wimp meets another Wimp and gets 0 for a fraction $(1 - x)$ of the time and meets a Macho and gets

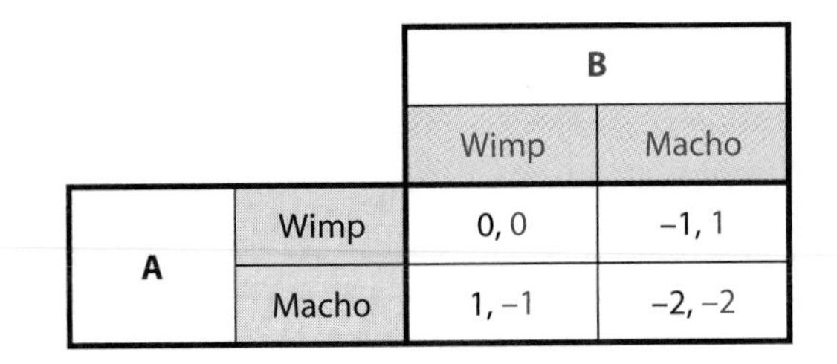

		B	
		Wimp	Macho
A	Wimp	0, 0	−1, 1
	Macho	1, −1	−2, −2

FIGURE 13.6 Payoff Table for Chicken

-1 for a fraction x of the time. Therefore the fitness of a Wimp is $0 \times (1 - x) - 1 \times x = -x$. Similarly, the fitness of a Macho is $1 \times (1 - x) - 2x = 1 - 3x$. The Macho type is fitter if

$$\begin{aligned} 1 - 3x &> -x \\ 2x &< 1 \\ x &< 1/2. \end{aligned}$$

If the population is less than half Macho, then the Machos will be fitter and their proportion will increase. On the other hand, if the population is more than half Macho, then the Wimps will be fitter and the Macho proportion will fall. Either way, the population proportion of Machos will tend toward $1/2$, and this 50–50 mix will be the stable polymorphic ESS.

Figure 13.7 shows this outcome graphically. Each straight line shows the fitness (the expected payoff in a match against a random member of the population) for one type, in relation to the proportion x of Machos. For the Wimp type, this functional relation showing their fitness as a function of the proportion of the Machos is $-x$, as we saw two paragraphs ago. This is the gently falling line that starts at the height 0 when $x = 0$ and goes to -1 when $x = 1$. The corresponding function for the Macho type is $1 - 3x$. This is the rapidly falling line that starts at height 1 when $x = 0$ and falls to -2 when $x = 1$. The Macho line lies above the Wimp line for $x < 1/2$ and below it for $x > 1/2$, showing that the Macho types are fitter when the value of x is small and the Wimps are fitter when x is large.

Now we can compare and contrast the evolutionary theory of this game with our earlier theory of Chapters 4 and 7, which was based on the assumption that the players were conscious rational calculators of strategies. There we found three Nash equilibria: two in pure strategies, where one player goes straight and

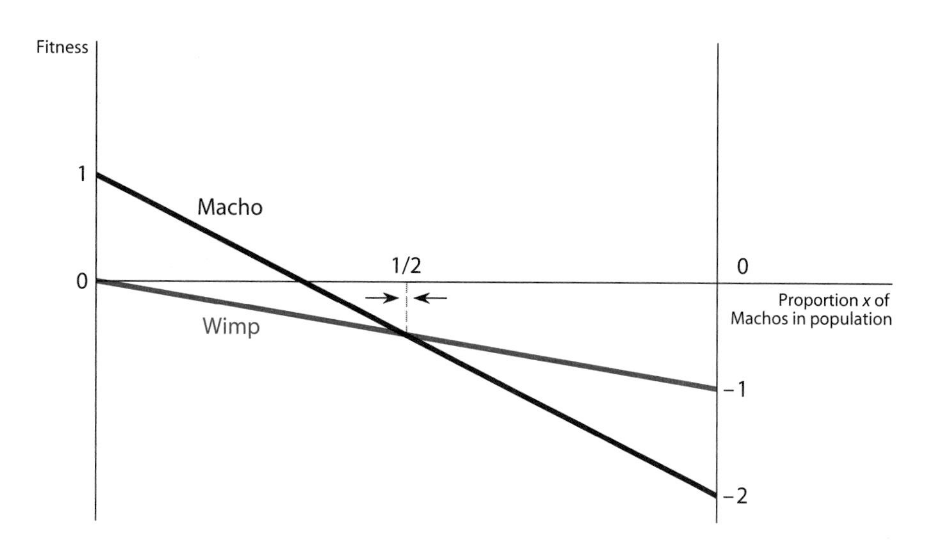

FIGURE 13.7 Fitness Graphs and Polymorphic Equilibrium for Chicken

the other swerves, and one in mixed strategies, where each player goes straight with a probability of 1/2 and swerves with a probability of 1/2.

If the population is truly 100% Macho, then all players are equally fit (or equally unfit). Similarly, in a population of nothing but Wimps, all are equally fit. But these monomorphic configurations are unstable. In an all-Macho population, a Wimp mutant will outscore them and invade successfully.[8] Once some Wimps get established, no matter how few, our analysis shows that their proportion will rise inexorably toward 1/2. Similarly, an all-Wimp population is vulnerable to a successful invasion of mutant Machos, and the process again goes to the same polymorphism. Thus the polymorphic configuration is the only true evolutionary stable outcome.

Most interesting is the connection between the mixed-strategy equilibrium of the rationally played game and the polymorphic equilibrium of the evolutionary game. The mixture proportions in the equilibrium strategy of the former are *exactly the same* as the population proportions in the latter: a 50–50 mixture of Wimp and Macho. But the interpretations differ: in the rational framework, each player mixes his own strategies; in the evolutionary framework, every member of the population uses a pure strategy, but different members use different strategies, and so we see a mixture in the population.[9]

This correspondence between Nash equilibria of a rationally played game and stable outcomes of a game with the same payoff structure when played according to the evolutionary rules is a very general proposition, and we see it in its generality later, in Section 6. Indeed, evolutionary stability provides an additional rationale for choosing one of the many Nash equilibria in such rationally played games.

When we looked at chicken from the rational perspective, the mixed-strategy equilibrium seemed puzzling. It left open the possibility of costly mistakes. Each player went straight one time in two, so one time in four they collided. The pure-strategy equilibria avoided the collisions. At that time, this may have led you to think that there was something undesirable about the mixed-strategy equilibrium, and you may have wondered why we were spending time on it. Now you see the reason. The seemingly strange equilibrium emerges as the stable outcome of a natural dynamic process in which each player tries to improve his payoff against the population that he confronts.

4 THE ASSURANCE GAME

Among the important classes of strategic games introduced in Chapter 4, we have studied prisoners' dilemma and chicken from the evolutionary perspective.

[8] *The Invasion of the Mutant Wimps* could be an interesting science-fiction comedy movie.

[9] There can also be evolutionary stable mixed strategies in which each member of the population adopts a mixed strategy. We investigate this idea further in Section 6.E.

That leaves the assurance game. We illustrated this type of game in Chapter 4 with the story of two undergraduates, Harry and Sally, deciding where to meet for coffee. In the evolutionary context, each player is born liking either Starbucks or Local Latte and the population includes some of each type. Here we assume that pairs of the two types, which we classify generically as men and women, are chosen at random each day to play the game. We denote the strategies now by S (for Starbucks) and L (for Local Latte). Figure 13.8 shows the payoff table for a random pairing in this game; the payoffs are the same as those illustrated earlier in Figure 4.12.

If this were a game played by rational strategy-choosing players, there would be two equilibria in pure strategies: (S, S) and (L, L). The latter is better for both players. If they communicate and coordinate explicitly, they can settle on it quite easily. But if they are making the choices independently, they need to coordinate through a convergence of expectations—that is, by finding a focal point.

The rationally played game has a third equilibrium, in mixed strategies, that we found in Chapter 7. In that equilibrium, each player chooses Starbucks with a probability of 2/3 and Local Latte with a probability of 1/3; the expected payoff for each player is 2/3. As we showed in Chapter 7, this payoff is worse than the one associated with the less attractive of the two pure-strategy equilibria, (S, S), because independent mixing leads the players to make clashing or bad choices quite a lot of the time. Here, the bad outcome (a payoff of 0) has a probability of 4/9: the two players go to different meeting places almost half the time.

What happens when this is an evolutionary game? In the large population, each member is hardwired, either to choose S or to choose L. Randomly chosen pairs of such people are assigned to attempt a meeting. Suppose x is the proportion of S types in the population and $(1 - x)$ is that of L types. Then the fitness of a particular S type—her expected payoff in a random encounter of this kind—is $x \times 1 + (1 - x) \times 0 = x$. Similarly, the fitness of each L type is $x \times 0 + (1 - x) \times 2 = 2(1 - x)$. Therefore the S type is fitter when $x > 2(1 - x)$, or for $x > 2/3$. The L type is fitter when $x < 2/3$. At the balancing point $x = 2/3$, the two types are equally fit.

		WOMEN	
		S	L
MEN	S	1, 1	0, 0
	L	0, 0	2, 2

FIGURE 13.8 Payoff Matrix for the Assurance Game

As in chicken, once again the probabilities associated with the mixed-strategy equilibrium that would obtain under rational choice seem to reappear under evolutionary rules as the population proportions in a polymorphic equilibrium. But now this mixed equilibrium is not stable. The slightest chance departure of the proportion x from the balancing point 2/3 will set in motion a cumulative process that takes the population mix farther away from the balancing point. If x increases from 2/3, the S type becomes fitter and propagates faster, increasing x even more. If x falls from 2/3, the L type becomes fitter and propagates faster, lowering x even more. Eventually x will either rise all the way to 1 or fall all the way to 0, depending on which disturbance occurs. The difference is that in chicken each type was fitter when it was rarer, so the population proportions tended to move away from the extremes and toward a midrange balancing point. In contrast, in the assurance game each type is fitter when it is more numerous; the risk of failing to meet falls when more of the rest of the population is the same type as you—so population proportions tend to move toward the extremes.

Figure 13.9 illustrates the fitness graphs and equilibria for the assurance game; this diagram is very similar to Figure 13.7. The two lines show the fitness of the two types in relation to the population proportion. The intersection of the lines gives the balancing point. The only difference is that, away from the balancing point, the more numerous type is the fitter, whereas in Figure 13.7 it was the less numerous type.

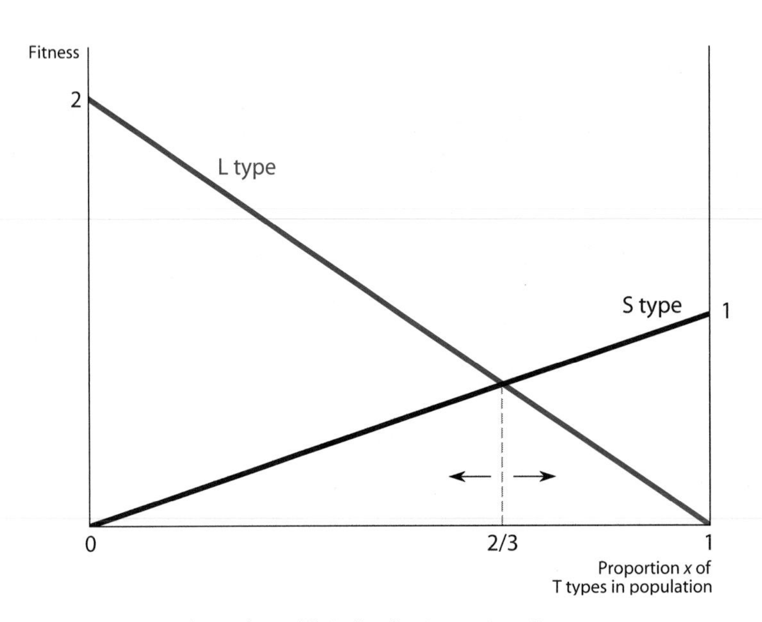

FIGURE 13.9 Fitness Graphs and Equilibria for the Assurance Game

Because each type is less fit when it is rare, only the two extreme monomorphic configurations of the population are possible evolutionary stable states. It is easy to check that both outcomes are ESS according to the static test: an invasion by a small mutant population of the other type will die out because the mutants, being rare, will be less fit. Thus in assurance or coordination games, unlike in chicken, the evolutionary process does not preserve the bad equilibrium, where there is a positive probability that the players choose clashing strategies. However, the dynamics do not guarantee convergence to the better of the two equilibria when starting from an arbitrary initial mixture of phenotypes—where the population ends up depends on where it starts.

5 INTERACTIONS ACROSS SPECIES

A final class of strategic games to consider is that of the battle-of-the-sexes game. In Chapter 4 (Figure 4.13), we saw that the battle of the sexes game looks similar to the assurance game in some respects. We differentiate between the two by assuming here that "men" and "women" are still interested in meeting at either Starbucks or Local Latte—no meeting yields each a payoff of 0—but now each type prefers a different café. Thus a premium remains on taking mutually consistent actions, just as in the assurance game. But the consequences of the two possible mutually consistent actions differ. The types in the assurance game do not differ in their preferences; both prefer (L, L) to (S, S). The players in the battle game differ in theirs: Local Latte gives a payoff of 2 to women and 1 to men, and Starbucks the other way around. These preferences distinguish the two types. In the language of biology, they can no longer be considered random draws from a homogeneous population of animals.[10] Effectively, they belong to different species (as indeed men and women often believe of each other).

To study such games from an evolutionary perspective, we must extend our methodology to the case in which the matches are between randomly drawn members of different species or populations. We develop the battle-of-the-sexes example to illustrate how this is done.

Suppose there is a large population of men and a large population of women. One of each "species" is picked, and the two are asked to attempt a meeting. All men agree among themselves about the valuation (payoffs) of Starbucks, Local Latte, and no meeting. Likewise, all women agree among themselves. But within each population, some members are hard-liners and others are compromisers.

[10]In evolutionary biology, games of this type are labeled "asymmetric" games. Symmetric games are those in which a player cannot distinguish the type of another player simply from observing that player's outward characteristics; in asymmetric games, players can tell each other apart.

A hard-liner will always go to his or her species' preferred café. A compromiser recognizes that the other species wants the opposite and goes to that location, to get along.

If the random draws happen to have picked a hard-liner of one species and a compromiser of the other, the outcome is that preferred by the hard-liner's species. We get no meeting if two hard-liners are paired and, strangely, also if two compromisers are chosen, because they go to each other's preferred café. (Remember, they have to choose independently and cannot negotiate. Perhaps even if they did get together in advance, they would reach an impasse of "No, I insist on giving way to your preference.")

We alter the payoff table in Figure 4.13 as shown in Figure 13.10; what were choices are now interpreted as actions predetermined by type (hard-liner or compromiser).

In comparison with all the evolutionary games studied so far, the new feature here is that the row player and the column player come from different species. Although each species is a heterogeneous mixture of hard-liners and compromisers, there is no reason why the proportions of the types should be the same in both species. Therefore we must introduce two variables to represent the two mixtures and study the dynamics of both.

We let x be the proportion of hard-liners among the men and y that among the women. Consider a particular hard-liner man. He meets a hard-liner woman a proportion y of the time and gets a 0, and he meets a compromising woman the rest of the time and gets a 2. Therefore his expected payoff (fitness) is $y \times 0 + (1 - y) \times 2 = 2(1 - y)$. Similarly, a compromising man's fitness is $y \times 1 + (1 - y) \times 0 = y$. Among men, therefore, the hard-liner type is fitter when $2(1 - y) > y$, or $y < 2/3$. The hard-liner men will reproduce faster when they are fitter; that is, x increases when $y < 2/3$. Note the new, and at first sight surprising, feature of the outcome: the fitness of each type within a given species depends on the proportion of types found in other species. This is not surprising; remember that the games that each species plays are now all against the members of the other species.[11]

		WOMEN	
		Hard-liner	Compromiser
MEN	Hard-liner	0, 0	2, 1
	Compromiser	1, 2	0, 0

FIGURE 13.10 Payoffs in the Battle-of-the-Sexes Game

[11]And this finding supports and casts a different light on the property of mixed-strategy equilibria, that each player's mixture keeps the *other* player indifferent among her pure strategies. Now we can think of it as saying that in a polymorphic evolutionary equilibrium of a two-species game, the proportion of each species' type keeps all the surviving types of the others species equally fit.

Similarly, considering the other species, we have the result that the hard-liner women are fitter; so y increases when $x < 2/3$. To understand the result intuitively, note that it says that the hard-liners of each species do better when the other species does not have too many hard-liners of its own, because then they meet compromisers of the other species quite frequently.

Figure 13.11 shows the dynamics of the configurations of the two species. Each of x and y can range from 0 to 1, so we have a graph with a unit square and x and y on their usual axes. Within that, the vertical line *AB* shows all points where $x = 2/3$, the balancing point at which y neither increases nor decreases. If the current population proportions lie to the left of this line (that is, $x < 2/3$), y is increasing (moving the population proportion of hard-liner women in the vertically upward direction). If the current proportions lie to the right of *AB* ($x > 2/3$), then y is decreasing (motion vertically downward). Similarly, the horizontal line *CD* shows all points where $y = 2/3$, which is the balancing point for x. When the population proportion of hard-liner women is below this line (that is, when $y < 2/3$), the proportion of hard-liner men, x, increases (motion horizontal and rightward) and decreases for population proportions above it, when $y > 2/3$ (motion horizontal and leftward).

When we combine the motions of x and y, we can follow their dynamic paths to determine the location of the population equilibrium. From a starting point in the bottom-left quadrant of Figure 13.11, for example, the dynamics

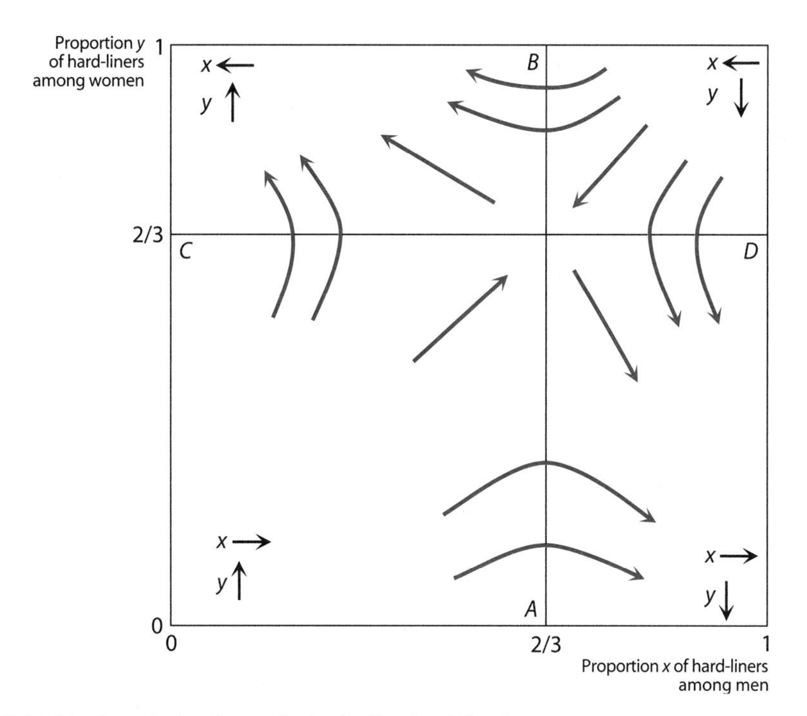

FIGURE 13.11 Population Dynamics in the Battle of the Sexes

entail both y and x increasing. This joint movement (to the northeast) continues until either $x = 2/3$ and y begins to decrease (motion now to the southeast) or $y = 2/3$ and x begins to decrease (motion now to the northwest). Similar processes in each quadrant yield the curved dynamic paths shown in the diagram. The vast majority of these paths lead to either the southeast or northwest corners of the diagram; that is, they converge either to (1, 0) or (0, 1). Thus in most cases evolutionary dynamics will lead to a configuration in which one species is entirely hard-line and the other is entirely compromising. Which species will be which type depends on the initial conditions. Note that the population dynamics starting from a situation with a small value of x and a larger value of y are more likely to cross the *CD* line first and head for (0, 1)—all hard-line women, $y = 1$—than to hit the *AB* line first and head for (1, 0); similar results follow for a starting position with a small y but a larger x. The species that starts out with more hard-liners will have the advantage of ending up all hard-line and getting the payoff of 2.

If the initial proportions are balanced just right, the dynamics may lead to the polymorphic point (2/3, 2/3). But unlike the polymorphic outcome in chicken, the polymorphism in the battle of the sexes is unstable. Most chance departures will set in motion a cumulative process that leads to one of the two extreme equilibria; those are the two ESSs for this game. This is a general property—such multispecies games can have only ESSs that are monomorphic for each species.

6 THE HAWK-DOVE GAME

The **hawk-dove game** was the first example biologists studied in their development of the theory of evolutionary games. It has instructive parallels with our analyses so far of the prisoners' dilemma and chicken, so we describe it here to reinforce and improve your understanding of the concepts.

The game is played not by birds of these two species, but by two animals of the same species, and Hawk and Dove are merely the names for their strategies. The context is competition for a resource. The Hawk strategy is aggressive and fights to try to get the whole resource of value V. The Dove strategy is to offer to share but to avoid a fight. When two Hawk types meet each other, they fight. Each animal is equally likely (probability 1/2) to win and get V or to lose, be injured, and get $-C$. Thus the expected payoff for each is $(V - C)/2$. When two Dove types meet, they share without a fight, so each gets $V/2$. When a Hawk type meets a Dove type, the latter retreats and gets a 0, whereas the former gets V. Figure 13.12 shows the payoff table.

		B	
		Hawk	Dove
A	Hawk	$(V-C)/2, (V-C)/2$	$V, 0$
	Dove	$0, V$	$V/2, V/2$

FIGURE 13.12 Payoff Table for the Hawk-Dove Game

The analysis of the game is similar to that for the prisoners' dilemma and chicken games, except that the numerical payoffs have been replaced by algebraic symbols. We will compare the equilibria of this game when the players rationally choose to play Hawk or Dove and then compare the outcomes when players are acting mechanically and success is being rewarded with faster reproduction.

A. Rational Strategic Choice and Equilibrium

1. If $V > C$, then the game is a prisoners' dilemma in which the Hawk strategy corresponds to "defect" and Dove corresponds to "cooperate." Hawk is the dominant strategy for each, but (Dove, Dove) is the jointly better outcome.

2. If $V < C$, then it's a game of chicken. Now $(V - C)/2 < 0$ and so Hawk is no longer a dominant strategy. Rather, there are two pure-strategy Nash equilibria: (Hawk, Dove) and (Dove, Hawk). There is also a mixed-strategy equilibrium, where B's probability p of choosing Hawk is such as to keep A indifferent:

$$p(V - C)/2 + (1 - p)V = p \times 0 + (1 - p)V/2$$
$$p = V/C.$$

B. Evolutionary Stability for $V > C$

We start with an initial population predominantly of Hawks and test whether it can be invaded by mutant Doves. Following the convention used in analyzing such games, we could write the population proportion of the mutant phenotype as m, for mutant, but for clarity in our case we will use d for mutant Dove. The population proportion of Hawks is then $(1 - d)$. Then, in a match against a randomly drawn opponent, a Hawk will meet a Dove a proportion d of the time and get V on each of those occasions and will meet another Hawk a proportion $(1 - d)$ of the time and get $(V - C)/2$ on each of those occasions. Therefore the fitness of a Hawk is $[dV + (1 - d)(V - C)/2]$. Similarly, the fitness of one of the mutant doves is $[d(V/2) + (1 - d) \times 0]$. Because $V > C$, it follows that $(V - C)/2 > 0$. Also, $V > 0$ implies that $V > V/2$. Then, for any value of d between 0 and 1, we have

$$dV + (1 - d)(V - C)/2 > d(V/2) + (1 - d) \times 0,$$

and so the Hawk type is fitter. The Dove mutants cannot successfully invade. The Hawk strategy is evolutionary stable, and the population is monomorphic (all Hawk).

The same holds true for any population proportion of Doves for all values of d. Therefore, from any initial mix, the proportion of Hawks will grow and they will predominate. In addition, if the population is initially all Doves, mutant Hawks can invade and take over. Thus the dynamics confirm that the Hawk strategy is the only ESS. This algebraic analysis affirms and generalizes our earlier finding for the numerical example of the prisoners' dilemma of restaurant pricing (Figure 13.1).

C. Evolutionary Stability for $V < C$

If the initial population is again predominantly Hawks, with a small proportion d of Dove mutants, then each has the same fitness function derived in Section 6.B. When $V < C$, however, $(V - C)/2 < 0$. We still have $V > 0$, and so $V > V/2$. But because d is very small, the comparison of the terms with $(1 - d)$ is much more important than that of the terms with d, so

$$d(V/2) + (1 - d) \times 0 > dV + (1 - d)(V - C)/2.$$

Thus the Dove mutants are fitter than the predominant Hawks and can invade successfully.

But if the initial population is almost all Doves, then we must consider whether a small proportion h of Hawk mutants can invade. (Note that, because the mutant is now a Hawk, we have used h for the proportion of the mutant invaders.) The Hawk mutants have a fitness of $[h(V - C)/2 + (1 - h)V]$ compared with $[h \times 0 + (1 - h)(V/2)]$ for the Doves. Again $V < C$ implies that $(V - C)/2 < 0$, and $V > 0$ implies that $V > V/2$. But, when h is small, we get

$$h(V - C)/2 + (1 - h)V > h \times 0 + (1 - h)(V/2).$$

This inequality shows that Hawks are fitter and will successfully invade a Dove population. Thus mutants of each type can invade populations of the other type. The population cannot be monomorphic, and neither pure phenotype can be an ESS. The algebra again confirms our earlier finding for the numerical example of chicken (Figures 13.6 and 13.7).

What happens in the population then when $V < C$? There are two possibilities. In one, every player follows a pure strategy, but the population has a stable mix of players following different strategies. This is the polymorphic equilibrium developed for chicken in Section 13.3. The other possibility is that every player uses a mixed strategy. We begin with the polymorphic case.

D. $V < C$: Stable Polymorphic Population

When the population proportion of Hawks is h, the fitness of a Hawk is $h(V - C)/2 + (1 - h)V$, and the fitness of a Dove is $h \times 0 + (1 - h)(V/2)$. The Hawk type is fitter if

$$h(V - C)/2 + (1 - h)V > (1 - h)(V/2),$$

which simplifies to:

$$\begin{aligned} h(V - C)/2 + (1 - h)(V/2) &> 0 \\ V - hC &> 0 \\ h &< V/C. \end{aligned}$$

The Dove type is then fitter when $h > V/C$, or when $(1 - h) < 1 - V/C = (C - V)/C$. Thus each type is fitter when it is rarer. Therefore we have a stable polymorphic equilibrium at the balancing point, where the proportion of Hawks in the population is $h = V/C$. This is exactly the probability with which each individual member plays the Hawk strategy in the mixed-strategy Nash equilibrium of the game under the assumption of rational behavior, as calculated in Section 6.A. Again, we have an evolutionary "justification" for the mixed-strategy outcome in chicken.

We leave it to you to draw a graph similar to that in Figure 13.7 for this case. Doing so will require you to determine the dynamics by which the population proportions of each type converge to the stable equilibrium mix.

E. $V < C$: Each Player Mixes Strategies

Recall the equilibrium mixed strategy of the rational-play game calculated earlier in Section 6.A in which $p = V/C$ was the probability of choosing to be a Hawk, while $(1 - p)$ was the probability of choosing to be a Dove. Is there a parallel in the evolutionary version, with a phenotype playing a mixed strategy? Let us examine this possibility. We still have H types who play the pure Hawk strategy and D types who play the pure Dove strategy. But now a third phenotype called M can exist; such a type plays a mixed strategy in which it is a Hawk with probability $p = V/C$ and a Dove with probability $1 - p = 1 - V/C = (C - V)/C$.

When an H or a D meets an M, their expected payoffs depend on p, the probability that M is playing H, and on $(1 - p)$, the probability that M is playing D. Then each player gets p times her payoff against an H, plus $(1 - p)$ times her payoff against a D. So when an H type meets an M type, she gets the expected payoff

$$\begin{aligned} p\,\frac{V - C}{2} + (1 - p)V &= \frac{V}{C}\,\frac{V - C}{2} - \frac{C - V}{C}\,V \\ &= -\frac{1}{2}\,\frac{V}{C}(C - V) + \frac{V}{C}(C - V) \\ &= V\,\frac{(C - V)}{2C}. \end{aligned}$$

And when a D type meets an M type, she gets

$$p \times 0 + (1 - p)\,\frac{V}{2} = \frac{C - V}{V}\,\frac{V}{2} = \frac{V(C - V)}{V}.$$

The two fitnesses are equal. This should not be a surprise; the proportions of the mixed strategy are determined to achieve exactly this equality. Then an M type meeting another M type also gets the same expected payoff. For brevity of future reference, we call this common payoff K, where $K = V(C - V)/2C$.

But these equalities create a problem when we test M for evolutionary stability. Suppose the population consists entirely of M types and that a few mutants of the H type, constituting a very small proportion h of the total population, invade. Then the typical mutant gets the expected payoff $h(V - C)/2 + (1 - h)K$. To calculate the expected payoff of an M type, note that she faces another M type a fraction $(1 - h)$ of the time and gets K in each instance. She then faces an H type for a fraction h of the interactions; in these interactions she plays H a fraction p of the time and gets $(V - C)/2$, and she plays D a fraction $(1 - p)$ of the time and gets 0. Thus the M type's total expected payoff (fitness) is

$$hp(V - C)/2 + (1 - h)K.$$

Because h is very small, the fitnesses of the M types and the mutant H types are almost equal. The point is that when there are very few mutants, both the H type and the M type meet only M types most of the time, and in this interaction the two have equal fitness as we just saw.

Evolutionary stability hinges on whether the original population M type is fitter than the mutant H when each is matched against one of the few mutants. Algebraically, M is fitter than H against other mutant H types when $pV(C - V)/2C = pK > (V - C)/2$. In our example here, this condition holds because $V < C$ (so $(V - C)$ is negative) and because K is positive. Intuitively, this condition tells us that an H-type mutant will always do badly against another H-type mutant because of the high cost of fighting, but the M type fights only part of the time and therefore suffers this cost only a fraction p of the time. Overall, the M type does better when matched against the mutants.

Similarly, the success of a Dove invasion against the M population depends on the comparison between a mutant Dove's fitness and the fitness of an M type. As before, the mutant faces another D a fraction d of the time and faces an M a fraction $(1 - d)$ of the time. An M type also faces another M type a fraction $(1 - d)$ of the time; but a fraction d of the time, the M faces a D and plays H a fraction p of these times, thereby gaining pV, and plays D a fraction $(1 - p)$ of these times, thereby gaining $(1 - p)V/2$. The Dove's fitness is then $[dV/2 + (1 - d)K]$, while the fitness of the M type is $d \times [pV + (1 - p)V/2] + (1 - d)K$. The final term in each fitness expression is the same, so a Dove invasion is successful only if $V/2$ is greater than $pV + (1 - p)V/2$. This condition does not hold; the latter expression includes a weighted average of V and $V/2$ that must exceed $V/2$ whenever $V > 0$. Thus the Dove invasion cannot succeed either.

This analysis tells us that M is an ESS. Thus if $V < C$, the population can exhibit either of two evolutionary stable outcomes. One entails a mixture of types (a stable polymorphism), and the other entails a single type that mixes its strategies in the same proportions that define the polymorphism.

7 THREE PHENOTYPES IN THE POPULATION

If there are only two possible phenotypes (strategies), we can carry out static checks for ESS by comparing the type being considered with just one type of mutant. We can show the dynamics of the population in an evolutionary game with graphs similar to those in Figures 13.4, 13.7, and 13.9. Now we illustrate how the ideas and methods can be used if there are three (or more) possible phenotypes and what new considerations arise.

A. Testing for ESS

Let us reexamine the thrice-repeated prisoners' dilemma of Section 13.2.A.II and Figure 13.3 by introducing a third possible phenotype. This strategy, labeled N, never defects. Figure 13.13 shows the fitness table with the three strategies—A, T, and N.

To test whether any of these strategies is an ESS, we consider whether a population of all one type can be invaded by mutants of one of the other types. An all-A population, for example, cannot be invaded by mutant N or T types; so A is an ESS. An all-N population can be invaded by type-A mutants, however; N lets itself get fooled thrice (shame on it). So N cannot be an ESS.

What about T? An all-T population cannot be invaded by A. But when faced with type-N mutants, the T types find themselves equally matched; notice that

		COLUMN		
		A	T	N
ROW	A	864, 864	936, 792	1080, 648
	T	792, 936	972, 972	972, 972
	N	648, 1080	972, 972	972, 972

FIGURE 13.13 Thrice-Repeated Prisoners' Dilemma with Three Types ($100s)

the four cells showing T and N competing only with each other show identical payoffs for both phenotypes. In this situation the mutant N types would not proliferate, but they would not die out either. A small proportion of mutants could coexist with the (almost) all-T population. Thus T does not satisfy either of the criteria for being an ESS, but it does exhibit some resistance to invasion.

We recognize the resilience shown by the T type in our example by introducing the concept of a **neutral ESS.**[12] In contrast to the standard ESS, in which a member of the main population needs to be strictly fitter than a mutant in a population with a small proportion of mutants, neutral stability requires only that a member of the main population have at least as high a fitness as does a mutant. Then the mutant proportion does not increase but can stay at an initially small level. This is the case when our all-T population is invaded by a small number of mutant N types. In the game illustrated in Figure 13.13, then, we have one standard ESS, strategy A, and one neutral ESS, strategy T.

Let us consider further the situation when an all-T population is invaded by type-N mutants. If the proportion of mutants is sufficiently small, the two types can coexist happily. But if the mutant population is too large a proportion of the full population, then type-A mutants can invade; A types do well against N but poorly against T. To be specific, consider a population with proportions x of N and $(1 - x)$ of T. The fitness of each of these types is 972. The fitness of a type-A mutant in this population is $936(1 - x) + 1{,}080x = 144x + 936$. This exceeds 972 if $144x > 972 - 936 = 36$, or $x > 1/4$. Thus we can have T as a neutral ESS coexisting with some small proportion of N-type mutants, but only so long as the proportion of Ns is less than 25%.

B. Dynamics

To motivate our discussion of dynamics in games with three possible phenotypes, we turn to another well-known game, rock–paper–scissors (RPS). In rational game-theoretic play of this game, each player simultaneously chooses one of the three available actions, either rock (make a fist), paper (lay your hand flat), or scissors (make a scissorlike motion with two fingers). The rules of the game state that rock beats ("breaks") scissors, scissors beat ("cut") paper, and paper beats ("covers") rock; identical actions tie. If players choose different actions, the winner gets a payoff of 1 and the loser gets a payoff of -1; ties yield both players 0.

For an evolutionary example, we turn to the situation faced by the side-blotched lizards living along the California coast. That species supports three types of male mating behavior, each type associated with a particular throat color. Males with blue throats guard a small number of female mates and fend

[12] Weibull describes neutral stability as a weakening of the standard evolutionary stability criteria in his *Evolutionary Game Theory* (p. 46).

		COLUMN			
		Yellow-throated sneaker	Blue-throated guarder	Orange-throated aggressor	q-mix
ROW	Yellow-throated sneaker	0	−1	1	$-q_2 + (1 - q_1 - q_2)$
	Blue-throated guarder	1	0	−1	$q_1 - (1 - q_1 - q_2)$
	Orange-throated aggressor	−1	1	0	$-q_1 + q_2$

FIGURE 13.14 Payoffs in the Three-Type Evolutionary Game

off advances made by yellow-throated males who attempt to sneak in and mate with unguarded females. The yellow-throated sneaking strategy works well against males with orange throats, who maintain large harems and are often out aggressively pursuing additional mates; those mates tend to belong to the blue-throated males, which can be overpowered by the orange-throat's aggression.[13] Their interactions can be modeled by using the payoff structure of the RPS game shown in Figure 13.14. We include a column for a q-mix to allow us to consider the evolutionary equivalent of the game's mixed-strategy equilibrium, a mixture of types in the population.[14]

Suppose q_1 is the proportion of lizards in the population that are yellow throated, q_2 the proportion of blue throats, and the rest, $(1 - q_1 - q_2)$, the proportion of orange throats. The right-hand column of the table shows each Row player's payoffs when meeting this mixture of phenotypes; that is, just Row's fitness. Suppose, as has been shown to be true in the side-splotched lizard population, that the proportion of each type in the population grows when its fitness is positive and declines when it is negative.[15] Then

$$q_1 \text{ increases if and only if } \quad -q_2 + (1 - q_1 - q_2) > 0, \quad \text{or } q_1 + 2q_2 < 1.$$

The proportion of yellow-throated types in the population increases when q_2, the proportion of blue-throated types, is small or when $(1 - q_1 - q_2)$, the

[13]For more information about the side-blotched lizards, see Kelly Zamudio and Barry Sinervo, "Polygyny, Mate-Guarding, and Posthumous Fertilizations As Alternative Mating Strategies," *Proceedings of the National Academy of Sciences*, vol. 97, no. 26 (December 19, 2000), pp. 14427–14432.

[14]One exercise in Chapter 7 considers the rational game-theoretic equilibrium of a version of the RPS game. You should be able to verify relatively easily that the game has no equilibrium in pure strategies.

[15]A little more care is necessary to ensure that the three proportions sum to 1, but that can be done, and we hide the mathematics so as to convey the ideas in a simple way. In the exercises, we develop the dynamics more rigorously for readers with sufficient mathematical training.

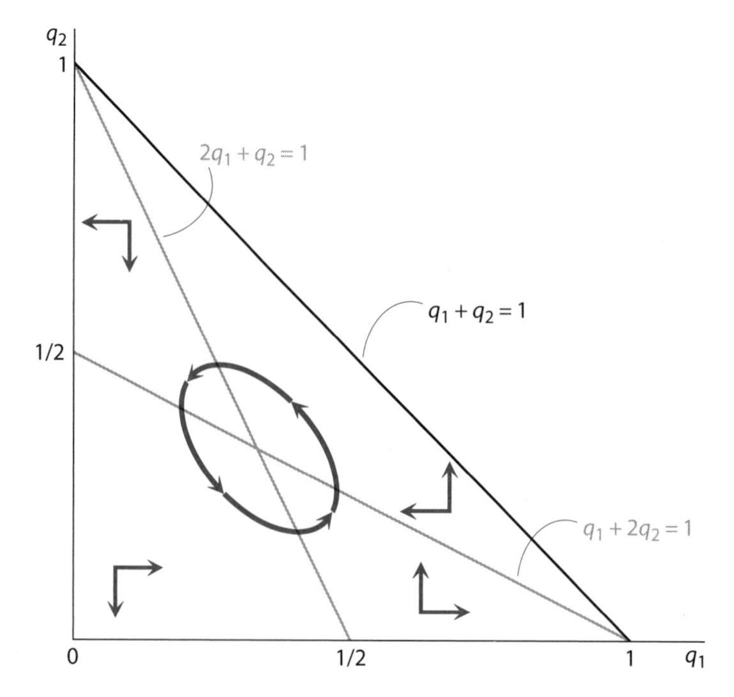

FIGURE 13.15 Population Dynamics in the Evolutionary RPS Game

proportion of orange-throated types, is large. This makes sense; yellow throats do poorly against blue throats but well against orange throats. Similarly, we see that

$$q_2 \text{ increases if and only if } \quad q_1 - (1 - q_1 - q_2) > 0, \quad \text{or } 2q_1 + q_2 > 1.$$

Blue-throated males do better when the proportion of yellow-throated competitors is large or the proportion of orange-throated types is small.

Figure 13.15 shows graphically the population dynamics and resulting equilibria for this game. The triangular area defined by the axes and the line $q_1 + q_2 = 1$ contains all the possible equilibrium combinations of q_1 and q_2. There are also two straight lines within this area. The first is $q_1 + 2q_2 = 1$ (the flatter one), which is the balancing line for q_1; for combinations of q_1 and q_2 below this line, q_1 (the proportion of yellow-throated players) increases; for combinations above this line, q_1 decreases. The second, steeper line is $2q_1 + q_2 = 1$, which is the balancing line for q_2. To the right of this line (when $2q_1 + q_2 > 1$), q_2 increases; to the left of the line (when $2q_1 + q_2 < 1$), q_2 decreases. Arrows on the diagram show directions of motion of these population proportions; red curves show typical dynamic paths. The general idea is the same as that of Figure 13.13.

On each of the two gray lines, one of q_1 and q_2 neither increases nor decreases. Therefore the intersection of the two lines represents the point where q_1, q_2, and therefore also $(1 - q_1 - q_2)$, are all constant; this point thus corresponds to a polymorphic equilibrium. It is easy to check that here $q_1 = q_2 = 1 - q_1 - q_2 = 1/3$. These proportions are the same as the probabilities in the rational mixed-strategy equilibrium of the RPS game.

Is this polymorphic outcome stable? In general, we cannot say. The dynamics indicate paths (shown in Figure 13.15 as a single ellipse) that wind around it. Whether these paths wind in a decreasing spiral toward the intersection (in which case we have stability) or in an expanding spiral away from the intersection (indicating instability) depends on the precise response of the population proportions to the fitnesses. It is even possible that the paths circle as drawn, neither approaching nor departing from the equilibrium.

Evidence suggests that the side-splotched lizard population is cycling around the evenly split polymorphic equilibrium point, with one type being slightly more common for a period of a few years but then being overtaken by its stronger competitor. Whether the cycle is approaching the stable equilibrium remains a topic for future study. At least one other example of an RPS-type interaction in an evolutionary game entails three strains of food-poisoning-related *E. coli* bacteria. Each strain displaces one of the others but is displaced by the third, as in the three-type game described earlier. Scientists studying the competition among the three strains have shown that a polymorphic equilibrium can persist if interactions between pairs stay localized, with small clumps of each strain shifting position continuously.[16]

8 SOME GENERAL THEORY

We now generalize the ideas illustrated in Section 6 to get a theoretical framework and set of tools that can then be applied further. This generalization unavoidably requires some slightly abstract notation and a bit of algebra. Therefore we cover only monomorphic equilibria in a single species. Readers who are adept at this level of mathematics can readily develop the polymorphism cases with two species by analogy. Readers who are not prepared for this material or interested in it can omit this section without loss of continuity.[17]

[16]The research on *E. coli* is reported in Martin Nowak and Karl Sigmund, "Biodiversity: Bacterial Game Dynamics," *Nature*, vol. 418 (July 11, 2002), p. 138. If the three strains were forcibly dispersed on a regular basis, a single strain could take over in a matter of days; the "winning" strain out-multiplied a second strain, which could quickly kill off the third.

[17]Conversely, readers who want more details can find them in Maynard Smith, *Evolution and the Theory of Games*, especially pp. 14–15. John Maynard Smith is a pioneer in the theory of evolutionary games.

We consider random matchings from a single species whose population has available strategies I, J, K. . . . Some of them may be pure strategies; some may be mixed. Each individual member is hardwired to play just one of these strategies. We let $E(\text{I, J})$ denote the payoff to an I player in a single encounter with a J player. The payoff of an I player meeting another of her own type is $E(\text{I, I})$ in the same notation. We write $W(\text{I})$ for the fitness of an I player. This is just her expected payoff in encounters with randomly picked opponents, when the probability of meeting a type is just the proportion of this type in the population.

Suppose the population is all I type. We consider whether this can be an evolutionary stable configuration. To do so, we imagine that the population is invaded by a few J-type mutants; so the proportion of mutants in the population is a very small number, m. Now the fitness of an I type is

$$W(\text{I}) = mE(\text{I, J}) + (1 - m)E(\text{I, I}),$$

and the fitness of a mutant is

$$W(\text{J}) = mE(\text{J, J}) + (1 - m)E(\text{J, I}).$$

Therefore the difference in fitness between the population's main type and its mutant type is

$$W(\text{I}) - W(\text{J}) = m[E(\text{I, J}) - E(\text{J, J})] + (1 - m)[E(\text{I, I}) - E(\text{J, I})].$$

Because m is very small, the main type's fitness will be higher than the mutant's if the second half of the preceding expression is positive; that is,

$$W(\text{I}) > W(\text{J}) \quad \text{if } E(\text{I, I}) > E(\text{J, I}).$$

Then the main type in the population cannot be invaded; it is fitter than the mutant type when each is matched against a member of the main type. This forms the **primary criterion** for evolutionary stability. Conversely, if $W(\text{I}) < W(\text{J})$, owing to $E(\text{I, I}) < E(\text{J, I})$, the J-type mutants will invade successfully, and an all-I population cannot be evolutionary stable.

However, it is possible that $E(\text{I, I}) = E(\text{J, I})$, as indeed happens if the population initially consists of a single phenotype that plays a strategy of mixing between the pure strategies I and J (a monomorphic equilibrium with a mixed strategy), as was the case in our final variant of the Hawk-Dove game (Section 6.E). Then the difference between $W(\text{I})$ and $W(\text{J})$ is governed by how each type fares against the mutants.[18] When $E(\text{I, I}) = E(\text{J, I})$, we get $W(\text{I}) > W(\text{J})$ if $E(\text{I, J}) > E(\text{J, J})$. This is the **secondary criterion** for the evolu-

[18]If the initial population is polymorphic and m is the proportion of J types, then m may not be "very small" any more. The size of m is no longer crucial, however, because the second term in $W(\text{I}) - W(\text{J})$ is now assumed to be zero.

tionary stability of I, to be invoked only if the primary one is inconclusive—that is, only if $E(\mathrm{I}, \mathrm{I}) = E(\mathrm{J}, \mathrm{I})$.

If the secondary condition is invoked—because $E(\mathrm{I}, \mathrm{I}) = E(\mathrm{J}, \mathrm{I})$—there is the additional possibility that it may also be inconclusive. That is, it may also be the case that $E(\mathrm{I}, \mathrm{J}) = E(\mathrm{J}, \mathrm{J})$. This is the case of *neutral stability* introduced in Section 7. If both the primary and secondary conditions for the evolutionary stability of I are inconclusive, then I is considered a neutral ESS.

Note that the primary criterion carries a punch. It says that if the strategy I is evolutionary stable, then for all other strategies J that a mutant might try, $E(\mathrm{I}, \mathrm{I}) \geq E(\mathrm{J}, \mathrm{I})$. This means that I is the best response to itself. In other words, if the members of this population suddenly started playing as rational calculators, everyone playing I would be a Nash equilibrium. *Evolutionary stability thus implies Nash equilibrium of the corresponding rationally played game!*[19]

This is a remarkable result. If you were dissatisfied with the rational behavior assumption underlying the theory of Nash equilibria given in earlier chapters and you came to the theory of evolutionary games looking for a better explanation, you would find that it yields the same results. The very appealing biological description—fixed nonmaximizing behavior, but selection in response to resulting fitness—does not yield any new outcomes. If anything, it provides a backdoor justification for Nash equilibrium. When a game has several Nash equilibria, the evolutionary dynamics may even provide a good argument for choosing among them.

However, your reinforced confidence in Nash equilibrium should be cautious. Our definition of evolutionary stability is static rather than dynamic. It only checks whether the configuration of the population (monomorphic, or polymorphic in just the right proportions) that we are testing for equilibrium cannot be successfully invaded by a small proportion of mutants. It does not test whether, starting from an arbitrary initial population mix, all the unwanted types will die out and the equilibrium configuration will be reached. And the test is carried out for those particular classes of mutants that are deemed logically possible; if the theorist has not specified this classification correctly and some type of mutant that she overlooked could actually arise, that mutant might invade successfully and destroy the supposed equilibrium. Our remark at the end of the twice-played prisoners' dilemma in Section 2.A warned of this possibility, and you will see in the exercises how it can arise. Finally, in Section 7 we saw how evolutionary dynamics can fail to converge at all.

[19]In fact, the primary criterion is slightly stricter than the standard definition of Nash equilibrium, which conforms more closely to that of neutral stability.

9 PLAYING THE FIELD

We have thus far looked at situations where each game is played between just two players who are randomly chosen from the population. There are other situations, however, when the whole population plays at once. In biology, a whole flock of animals with a mixture of genetically determined behaviors may compete for some resource or territory. In economics or business, many firms in an industry, each following the strategy dictated by its corporate culture, may compete all with all.

Such evolutionary games stand in the same relation to the rationally played collective-action games of Chapter 12 as do the pair-by-pair played evolutionary games of the preceding sections to the rationally played two-person games of Chapters 4 through 8. Just as we converted the expected payoff graphs of those chapters into the fitness diagrams in Figures 13.4, 13.7, and 13.9, we can convert the graphs for collective-action games (Figures 12.6 through 12.8) into fitness graphs for evolutionary games. For example, consider an animal species all of whose members come to a common feeding ground. There are two phenotypes: one fights for food aggressively, and the other hangs around and sneaks what it can. If the proportion of aggressive ones is small, they will do better; but, if there are too many of them, the sneakers will do better by ignoring the ongoing fights. This will be a collective chicken game whose fitness diagram will be exactly like Figure 12.7. Because no new principles or techniques are required, we leave it to you to pursue this idea further.

10 EVOLUTION OF COOPERATION AND ALTRUISM

Evolutionary game theory rests on two fundamental ideas: first, that individual organisms are engaged in games with others in their own species or with members of other species and, second, that the genotypes that lead to higher-payoff (fitter) strategies proliferate while the rest decline in their proportions of the population. These ideas suggest a vicious struggle for survival like that depicted by some interpreters of Darwin who understood "survival of the fittest" in a literal sense and who conjured up images of a "nature red in tooth and claw." In fact, nature shows many instances of cooperation (in which individual animals behave in a way that yields greater benefit to everyone in a group) and even altruism (in which individual animals incur significant costs in order to benefit others). Beehives and ant colonies are only the most obvious examples. Can such behavior be reconciled with the perspective of evolutionary games?

Biologists use a fourfold classification of the ways in which cooperation and altruism can emerge among selfish animals (or phenotypes or genes). Lee Dugatkin names the four categories (1) family dynamics, (2) reciprocal transactions, (3) selfish teamwork, and (4) group altruism.[20]

The behavior of ants and bees is probably the easiest to understand as an example of family dynamics. All the individual members of an ant colony or a beehive are closely related and have genes in common to a substantial extent. All worker ants in a colony are full sisters and therefore have half their genes in common; the survival and proliferation of one ant's genes is helped just as much by the survival of two of its sisters as by its own survival. All worker bees in a hive are half-sisters and therefore have a quarter of their genes in common. An individual ant or bee does not make a fine calculation of whether it is worthwhile to risk its own life for the sake of two or four sisters, but the underlying genes of those groups whose members exhibit such behavior (phenotype) will proliferate. The idea that evolution ultimately operates at the level of the gene has had enormous implications for biology, although it has been misapplied by many people, just as Darwin's original idea of natural selection was misapplied.[21] The interesting idea is that a "selfish gene" may prosper by behaving unselfishly in a larger organization of genes, such as a cell. Similarly, a cell and its genes may prosper by participating cooperatively and accepting their allotted tasks in a body.

Reciprocal altruism can arise among unrelated individual members of the same or different species. This behavior is essentially an example of the resolution of prisoners' dilemmas through repetition in which the players use strategies that are remarkably like tit-for-tat. For example, some small fish and shrimp thrive on parasites that collect in the mouths and gills of some large fish; the large fish let the small ones swim unharmed through their mouths for this "cleaning service." A more fascinating, although gruesome, example is that of vampire bats, who share blood with those who have been unsuccessful in their own hunting. In an experiment in which bats from different sites were brought together and selectively starved, "only bats that were on the verge of starving (i.e., would die within twenty-four hours without a meal) were given blood by any other bat in the experiment. But, more to the point, individuals were given a blood meal only from bats they already knew from their site. . . . Furthermore, vampires were much more likely to regurgitate blood to the specific individual(s)

[20]See his excellent exposition in *Cheating Monkeys and Citizen Bees: The Nature of Cooperation in Animals and Humans* (Cambridge: Harvard University Press, 1999).

[21]In this very brief account, we cannot begin to do justice to all the issues and the debates. An excellent popular account, and the source of many examples cited in this section, is Matt Ridley, *The Origins of Virtue* (New York: Penguin, 1996). We should also point out that we do not examine the connection between genotypes and phenotypes in any detail or the role of sex in evolution. Another book by Ridley, *The Red Queen* (New York: Penguin, 1995), gives a fascinating treatment of this subject.

from their site that had come to their aid when they needed a bit of blood."[22] Once again, it is not to be supposed that each animal consciously calculates whether it is in its individual interest to continue the cooperation or to defect. Instead, the behavior is instinctive.

Selfish teamwork arises when it is in the interests of each individual organism to choose cooperation when all others are doing so. In other words, this type of cooperative behavior applies to the selection of the good outcome in assurance games. Dugatkin argues that populations are more likely to engage in selfish teamwork in harsh environments than in mild ones. When conditions are bad, the shirking of any one animal in a group could bring disaster to the whole group, including the shirker. Then in such conditions each animal is crucial for survival, and none shirk so long as others are also pulling their weight. In milder environments, each may hope to become a free rider on the others' effort without thereby threatening the survival of the whole group, including itself.

The next step goes beyond biology and into sociology: a body (and its cells and ultimately its genes) may benefit by behaving cooperatively in a collection of bodies—namely, a society. This brings us to the idea of group altruism, which suggests that we should see some cooperation even among individual members of a group who are not close relatives. We do indeed find instances of it. Groups of predators such as wolves are a case in point, and groups of apes often behave like extended families. Even among species of prey, cooperation arises when individual fishes in a school take turns looking out for predators. And cooperation can also extend across species.

The general idea is that a group whose members behave cooperatively is more likely to succeed in its interactions with other groups than one whose members seek benefit of free-riding within the group. If, in a particular context of evolutionary dynamics, between-group selection is a stronger force than within-group selection, then we will see group altruism.[23]

An instinct is hardwired into an individual organism's brain by genetics, but reciprocity and cooperation can arise from more purposive thinking or experimentation within the group and can spread by socialization—through explicit instruction or observation of the behavior of elders—instead of genetics. The relative importance of the two channels—nature and nurture—will differ from one species to another and from one situation to another. One would expect socialization to be relatively more important among humans, but there are instances of its role among other animals. We cite a remarkable one. The expedition that Robert F. Scott led to the South Pole in 1911–1912 used teams of Siberian dogs. This group of dogs, brought together and trained for this specific purpose, developed

[22]Dugatkin, *Cheating Monkeys,* p. 99.

[23]Group altruism used to be thought impossible according to the strict theory of evolution that emphasizes selection at the level of the gene, but the concept is being revived in more sophisticated formulations. See Dugatkin, *Cheating Monkeys,* pp. 141–145 for a fuller discussion.

within a few months a remarkable system of cooperation and sustained it by using punishment schemes. "They combined readily and with immense effect against any companion who did not pull his weight, or against one who pulled too much . . . their methods of punishment always being the same and ending, if unchecked, in what they probably called justice, and we called murder."[24]

This is an encouraging account of how cooperative behavior can be compatible with evolutionary game theory and one that suggests that dilemmas of selfish actions can be overcome. Indeed, scientists investigating altruistic behavior have recently reported experimental support for the existence of such *altruistic punishment,* or *strong reciprocity* (as distinguished from reciprocal altruism), in humans. Their evidence suggests that people are willing to punish those who don't pull their own weight in a group setting, even when it is costly to do so and when there is no expectation of future gain. This tendency toward strong reciprocity may even help to explain the rise of human civilization if groups with this trait were better able to survive the traumas of war and other catastrophic events.[25] Despite these findings, strong reciprocity may not be widespread in the animal world. "Compared to nepotism, which accounts for the cooperation of ants and every creature that cares for its young, reciprocity has proved to be scarce. This, presumably, is due to the fact that reciprocity requires not only repetitive interactions, but also the ability to recognize other individuals and keep score."[26] In other words, precisely the conditions that our theoretical analysis in Section 2.D of Chapter 11 identified as being necessary for a successful resolution of the repeated prisoners' dilemma are seen to be relevant in the context of evolutionary games.

SUMMARY

The biological theory of evolution parallels the theory of games used by social scientists. Evolutionary games are played by behavioral *phenotypes* with genetically predetermined, rather than rationally chosen, strategies. In an evolutionary game, phenotypes with higher *fitness* survive repeated *interactions* with others to reproduce and to increase their representation in the population. A population containing one or more phenotypes in certain proportions is called *evolutionary stable* if it cannot be *invaded* successfully by other, *mutant* phenotypes or if it is the limiting outcome of the dynamics of proliferation of fitter phenotypes. If one phenotype maintains its dominance in the population

[24]Apsley Cherry-Garrard, *The Worst Journey in the World* (London: Constable, 1922; reprinted New York: Carroll and Graf, 1989), pp. 485–486.

[25]For the evidence on altruistic punishment, see Ernst Fehr and Simon Gachter, "Altruistic Punishment in Humans," *Nature,* vol. 415 (January 10, 2002), pp. 137–140.

[26]Ridley, *Origins of Virtue,* p. 83.

when faced with an invading mutant type, that phenotype is said to be an *evolutionary stable strategy,* and the population consisting of it alone is said to exhibit *monomorphism.* If two or more phenotypes coexist in an evolutionary stable population, it is said to exhibit *polymorphism.*

When the theory of evolutionary games is applied more generally to nonbiological games, the strategies followed by individual players are understood to be standard operating procedures or rules of thumb, instead of being genetically fixed. The process of reproduction stands for more general methods of transmission including socialization, education, and imitation; and *mutations* represent experimentation with new strategies.

Evolutionary games may have payoff structures similar to those analyzed in Chapters 4 and 7, including the prisoners' dilemma and chicken. In each case, the *evolutionary stable strategy* mirrors either the pure-strategy Nash equilibrium of a game with the same structure played by rational players or the proportions of the equilibrium mixture in such a game. In a prisoners' dilemma, "always defect" is evolutionary stable; in chicken, types are fitter when rare, and so there is a polymorphic equilibrium; in the assurance game, types are less fit when rare, and so the polymorphic configuration is unstable and the equilibria are at the extremes. When play is between two different types of members of each of two different species, a more complex but similarly structured analysis is used to determine equilibria.

The *hawk-dove game* is the classic biological example. Analysis of this game parallels that of the prisoners' dilemma and chicken versions of the evolutionary game; evolutionary stable strategies depend on the specifics of the payoff structure. The analysis can also be performed when more than two types interact or in very general terms. This theory shows that the requirements for evolutionary stability yield an equilibrium strategy that is equivalent to the Nash equilibrium obtained by rational players.

KEY TERMS

evolutionary stable (495)
evolutionary stable strategy (ESS) (498)
fitness (495)
genotype (495)
hawk-dove game (516)
interaction (496)
invasion by a mutant (495)
monomorphism (498)
mutation (495)
neutral ESS (522)
phenotype (495)
playing the field (496)
polymorphism (498)
primary criterion (526)
secondary criterion (526)
selection (495)

SOLVED EXERCISES

S1. Two travelers buy identical handcrafted souvenirs and pack them in their respective suitcases for their return flight. Unfortunately, the airline manages to lose both suitcases. Because the airline doesn't know the value of the lost souvenirs, it asks each traveler to report independently a value. The airline agrees to pay each traveler an amount equal to the minimum of the two reports. If one report is higher than the other, the airline takes a penalty of $20 away from the traveler with the higher report and gives $20 to the traveler with the lower report. If the reports are equal to one another, there is no reward or penalty. Neither traveler remembers exactly how much the souvenir cost, so that value is irrelevant; each traveler simply reports the value that her type determines she should report.

There are two types of travelers. The High type always reports $100, and the Low type always reports $50. Let h represent the proportion of High types in the population.

(a) Draw the payoff table for the game played between two travelers selected at random from the population.

(b) Graph the fitness of the High type, with h on the horizontal axis. On the same figure, graph the fitness of the Low type.

(c) Describe all of the equilibria of this game. For each equilibrium, state whether it is monomorphic or polymorphic and whether it is stable.

S2. In Section 7.A, we considered testing for ESSs (evolutionary stable strategies) in the thrice-repeated restaurant-pricing prisoners' dilemma.

(a) Explain completely (using Figure 13.13) why an all-type-A population cannot be invaded by either N- or T-type mutants.

(b) Explain why an all-N-type population can be invaded by type A mutants, and to what extent it can be invaded by type T mutants. Relate this explanation to the concept of neutral stability in the chapter.

(c) Finally, explain why an all-T-type population cannot be invaded by type A mutants but can be invaded by mutants that are type N.

S3. Consider a population in which there are two phenotypes: natural-born cooperators (who do not confess under questioning) and natural-born defectors (who confess readily). If two members of this population are drawn at random, their payoffs in a single play are the same as those of the husband-wife prisoners' dilemma game of Chapter 4, reproduced below. In repeated interactions there are two strategies available in the population, as there were in the restaurant-dilemma game of Section 13.2. The two strategies are A (always confess) and T (play tit-for-tat, starting with not confessing).

		COLUMN	
		Confess	Not
ROW	Confess	10 yr, 10 yr	1 yr, 25 yr
	Not	25 yr, 1 yr	3 yr, 3 yr

(a) Suppose that a pair of players plays this dilemma twice in succession. Draw the payoff table for the twice-repeated dilemma.

(b) Find all of the ESSs in this game.

(c) Now add a third possible strategy, N, which never confesses. Draw the payoff table for the twice-repeated dilemma with three possible strategies and find all of the ESSs of this new version of the game.

S4. In the assurance (meeting-place) game in this chapter, the payoffs were meant to describe the value of something material that the players gained in the various outcomes; they could be prizes given for a successful meeting, for example. Then other individual persons in the population might observe the expected payoffs (fitness) of the two types, see which was higher, and gradually imitate the fitter strategy. Thus the proportions of the two types in the population would change. But we can make a more biological interpretation. Suppose the column players are always female and the row players always male. When two of these players meet successfully, they pair off, and their children are of the same type as the parents. Therefore the types would proliferate or die off as a result of successful or unsuccessful meetings. The formal mathematics of this new version of the game makes it a "two-species game" (although the biology of it does not). Thus, the proportion of S-type females in the population—call this proportion x—need not equal the proportion of S-type males—call this proportion y.

(a) Examine the dynamics of x and y by using methods similar to those used in the chapter for the battle-of-the-sexes game.

(b) Find the stable outcome or outcomes of this dynamic process.

S5. Recall from Exercise S1 the travelers reporting the value of their lost souvenirs. Assume that a third traveler phenotype exists in the population. The third traveler type is a mixer; she plays a mixed strategy, sometimes reporting a value of $100 and sometimes reporting a value of $50.

(a) Use your knowledge of mixed strategies in rationally played games to posit a reasonable mixture for the mixer phenotype to use in this game.

(b) Draw the three-by-three payoff table for this game when the mixer type uses the mixed strategy that you found in part (a).

(c) Determine whether the mixer phenotype is an ESS of this game. (Hint: Test whether a mixer population can be invaded successfully by either the High type or the Low type.)

S6. Consider a simplified model in which everyone gets electricity either from solar power or from fossil fuels, which are both in relatively inelastic supply. (In the case of solar power, think of the required equipment as being in inelastic supply.) The upfront costs of using solar energy are high, so when the price of fossil fuels is low (that is, when few people are using fossil fuels and there is a high demand for solar equipment), the cost of solar can be prohibitive. On the other hand, when many individuals are using fossil fuels, the demand for them (and thus the price) is high, whereas the demand (and thus the price) for solar energy is relatively lower. Assume the payoff table for the two types of energy consumers to be as follows:

		COLUMN	
		Solar	Fossil fuels
ROW	Solar	2, 2	3, 4
	Fossil fuels	4, 3	2, 2

(a) Describe all possible ESS of this game in terms of s, the proportion of solar users, and explain why each is either stable or unstable.

(b) Suppose there are important economies of scale in producing solar equipment, such that the cost savings increase the payoffs in the (solar, solar) cell of the table to (y, y) where $y > 2$. How large would y need to be for the polymorphic equilibrium to have $s = 0.75$?

S7. There are two types of racers—tortoises and hares—who race against one another in randomly drawn pairs. In this world, hares beat tortoises every time without fail. If two hares race they tie, and they are completely exhausted by the race. When two tortoises race they also tie, but they enjoy a pleasant conversation along the way. The payoff table is as follows (where $c > 0$):

		COLUMN	
		Tortoise	Hare
ROW	Tortoise	c, c	−1, 1
	Hare	1, −1	0, 0

(a) Assume that the proportion of tortoises in the population, t, is 0.5. For what values of c will tortoises have greater fitness than hares?

(b) For what values of c will tortoises be fitter than hares if $t = 0.1$?

(c) If $c = 1$, will a single hare successfully invade a population of pure tortoises? Explain why or why not.

(d) In terms of t, how large must c be for tortoises to have greater fitness than hares?

(e) In terms of c, what is the level of t in a polymorphic equilibrium? For what values of c will such an equilibrium exist? Explain.

S8. Consider a population with two types, X and Y, with a payoff table as follows:

		COLUMN	
		X	*Y*
ROW	*X*	2, 2	5, 3
	Y	3, 5	1, 1

(a) Find the fitness for X as a function of x, the proportion of X in the population, and the fitness for Y as a function of x.

Assume that the population dynamics from generation to generation conform to the following model:

$$x_{t+1} = x_t \times F_{Xt} \,/\, [x_t \times F_{Xt} + (1 - x_t) \times F_{Yt}],$$

where x_t is the proportion of X in the population in period t, x_{t+1} is the proportion of X in the population in period $t + 1$, F_{Xt} is the fitness of X in period t, and F_{Yt} is the fitness of Y in period t.

(b) Assume that x_0, the proportion of X in the population in period 0, is 0.2. What are F_{X0} and F_{Y0}?

(c) Find x_1, using x_0, F_{X0}, F_{Y0}, and the model given above.

(d) What are F_{X1} and F_{Y1}?

(e) Find x_2 (rounded to five decimal places).

(f) What are F_{X2} and F_{Y2} (rounded to five decimal places)?

S9. Consider an evolutionary game between green types and purple types with a payoff table as follows:

		COLUMN	
		Green	Purple
ROW	Green	*a*, *a*	4, 3
	Purple	3, 4	2, 2

Let g be the proportion of greens in the population.

(a) In terms of g, what is the fitness of the purple type?

(b) In terms of g and a, what is the fitness of the green type?

(c) Graph the fitness of the purple types against the fraction g of green types in the population. On the same diagram, show three lines for the fitness of the green types when $a = 2$, 3, and 4. What can you conclude from this graph about the range of values of a that guarantees a stable polymorphic equilibrium?

(d) Assume that a is in the range found in part (c). In terms of a, what is the proportion of greens, g, in the stable polymorphic equilibrium?

S10. Prove the following statement: "If a strategy is strictly dominated in the payoff table of a game played by rational players, then in the evolutionary version of the same game it will die out, no matter what the initial population mix. If a strategy is weakly dominated, it may coexist with some other types but not in a mixture of all types."

UNSOLVED EXERCISES

U1. Consider a survival game in which a large population of animals meet and either fight over or share a food source. There are two phenotypes in the population: one always fights, and the other always shares. For the purposes of this question, assume that no other mutant types can arise in the population. Suppose that the value of the food source is 200 calories and that caloric intake determines each player's reproductive fitness. If two sharing types meet one another, they each get half the food, but if a sharer meets a fighter, the sharer concedes immediately, and the fighter gets all the food.

(a) Suppose that the cost of a fight is 50 calories (for each fighter) and that when two fighters meet, each is equally likely to win the fight and the food or to lose and get no food. Draw the payoff table for the game played between two random players from this population. Find all of the ESSs in the population. What type of game is being played in this case?

(b) Now suppose that the cost of a fight is 150 calories for each fighter. Draw the new payoff table and find all of the ESSs for the population in this case. What type of game is being played here?

(c) Using the notation of the Hawk-Dove game of Section 13.6, indicate the values of V and C in parts (a) and (b) and confirm that your answers to those parts match the analysis presented in the chapter.

U2. Suppose that a single play of a prisoners' dilemma has the following payoffs:

		PLAYER 2	
		Cooperate	Defect
PLAYER 1	Cooperate	3, 3	1, 4
	Defect	4, 1	2, 2

In a large population in which each member's behavior is genetically determined, each player will be either a defector (that is, always defects in any play of a prisoners' dilemma game) or a tit-for-tat player. (In multiple rounds of a prisoners' dilemma, she cooperates on the first play, and on any subsequent play she does whatever her opponent did on the preceding play.) Pairs of randomly chosen players from this population will play "sets" of n single plays of this dilemma (where $n \geq 2$). The payoff to each player in one whole set (of n plays) is the sum of her payoffs in the n plays.

Let the population proportion of defectors be p and the proportion of tit-for-tat players be $(1 - p)$. Each member of the population plays sets of dilemmas repeatedly, matched against a new, randomly chosen opponent for each new set. A tit-for-tat player always begins each new set by cooperating on its first play.

(a) Show in a two-by-two table the payoffs to a player of each type when, in one set of plays, each player meets an opponent of each of the two types.

(b) Find the fitness (average payoff in one set against a randomly chosen opponent) for a defector.

(c) Find the fitness for a tit-for-tat player.

(d) Use the answers to parts (b) and (c) to show that, when $p > (n-2)/(n-1)$, the defector type has greater fitness and that, when $p < (n-2)/(n-1)$, the tit-for-tat type has greater fitness.

(e) If evolution leads to a gradual increase in the proportion of the fitter type in the population, what are the possible eventual equilibrium outcomes of this process for the population described in this exercise? (That is, what are the possible equilibria, and which are evolutionary stable?) Use a diagram with the fitness graphs to illustrate your answer.

(f) In what sense does more repetition (larger values of n) facilitate the evolution of cooperation?

U3. Suppose that in the twice-repeated prisoners' dilemma of Exercise S3, a fourth possible type (type S) also can exist in the population. This type does not confess on the first play and confesses on the second play of each episode of two successive plays against the same opponent.

(a) Draw the four-by-four fitness table for the game.

(b) Can the newly conceived type S be an ESS of this game?

(c) In the three-types game of Exercise S3, A and T were both ESS, but T was only neutrally stable because a small proportion of N mutants could coexist. Show that in the four-types game here, T cannot be ESS.

U4. Following the pattern of Exercise S4, analyze an evolutionary version of the tennis point game (Figure 4.15). Regard servers and receivers as separate species, and construct a figure like Figure 13.11. What can you say about the ESS and its dynamics?

U5. Recall from Exercise U1 the population of animals fighting over a food source worth 200 calories. Assume that, as in part (b) of that exercise, the cost of a fight is 150 calories per fighter. Assume also that a third phenotype exists in the population. That phenotype is a mixer; it plays a mixed strategy, sometimes fighting and sometimes sharing.

(a) Use your knowledge of mixed strategies in rationally played games to posit a reasonable mixture for the mixer phenotype to use in this game.

(b) Draw the three-by-three payoff table for this game when the mixer type uses the mixed strategy that you found in part (a).

(c) Determine whether the mixer phenotype is an ESS of this game. (Hint: Test whether a mixer population can be invaded successfully by either the fighting type or the sharing type.)

U6. Consider an evolutionary version of the game between Baker and Cutler, from Exercise U1 of Chapter 11. This time Baker and Cutler are not two individuals but two separate species. Each time a Baker meets a Cutler, they play the following game. The Baker chooses the total prize to be either \$10 or \$100. The Cutler chooses how to divide the prize chosen by the Baker: the Cutler can choose either a 50:50 split or a 90:10 split in the Cutler's own favor. The Cutler moves first, and the Baker moves second.

There are two types of Cutlers in the population: type F chooses a fair (50:50) split, whereas type G chooses a greedy (90:10) split. There are also two types of Bakers: type S simply chooses the large prize (\$100) no matter what the Cutler has done, whereas type T chooses the large prize (\$100) if the Cutler chooses a 50:50 split, but the small prize (\$10) if the Cutler chooses a 90:10 split.

Let f be the proportion of type F in the Cutler population, so that $(1 - f)$ represents the proportion of type G. Let s be the proportion of type S in the Baker population, so that $(1 - s)$ represents the proportion of type T.

(a) Find the fitness of the Cutler types F and G in terms of s.

(b) Find the fitness of the Baker types S and T in terms of f.

(c) For what value of s are types F and G equally fit?

(d) For what value of f are types S and T equally fit?

(e) Use the answers above to sketch a graph displaying the population dynamics. Assign f as the horizontal axis and s as the vertical axis.

(f) Describe all of the equilibria of this evolutionary game, and indicate which ones are stable.

U7. Recall Exercise S7. Hares, it turns out, are very impolite winners. Whenever hares race tortoises they mercilessly mock their slow-footed (and easily defeated) rivals. The poor tortoises leave the race not only in defeat, but with their tender feelings crushed by the oblivious hares. The payoff table is thus:

		COLUMN	
		Tortoise	Hare
ROW	Tortoise	c, c	−2, 1
	Hare	1, −2	0, 0

(a) For what values of c are tortoises fitter than hares if t, the proportion of tortoises in the population, is 0.5? How does this compare with the answer in Exercise S7, part (a)?

(b) For what values of c are tortoises fitter than hares if $t = 0.1$? How does this compare with the answer in Exercise S7, part (b)?

(c) If $c = 1$, will a single hare successfully invade a population of pure tortoises? Explain why or why not.

(d) In terms of t, how large must c be for tortoises to be fitter than hares?

(e) In terms of c, what is the level of t in a polymorphic equilibrium? For what values of c will such an equilibrium exist? Explain.

(f) Will the polymorphic equilibria found to exist in part (e) be stable? Why or why not?

U8. **(Use of spreadsheet software recommended)** This problem explores more thoroughly the generation-by-generation population dynamics seen in Exercise S8. Since the math can quickly become very complicated and tedious, it is much easier to do this analysis with the aid of a spreadsheet.

Again, consider a population with two types, X and Y, with a payoff table as follows:

		COLUMN	
		X	Y
ROW	X	2, 2	5, 3
	Y	3, 5	1, 1

Recall that the population dynamics from generation to generation are given by:

$$x_{t+1} = x_t \times F_{Xt} / [x_t \times F_{Xt} + (1 - x_t) \times F_{Yt}],$$

where x_t is the proportion of X in the population in period t, x_{t+1} is the proportion of X in the population in period $t + 1$, F_{Xt} is the fitness of X in period t, and F_{Yt} is the fitness of Y in period t.

Use a spreadsheet to extend these calculations to many generations. [Hint: Assign three horizontally adjacent cells to hold the values of x_t, F_{Xt}, and F_{Yt}, and have each successive row represent a different period ($t = 0$, 1, 2, 3, . . .). Use spreadsheet formulas to relate F_{Xt} and F_{Yt} to x_t and x_{t+1} to x_t, F_{Xt}, and F_{Yt} according to the population model given above.]

(a) If there are initially equal proportions of X and Y in the population in period 1 (that is, if $x_0 = 0.5$), what is the proportion of X in the next generation, x_1? What are F_{X1} and F_{Y1}?

(b) Use a spreadsheet to extend these calculations to the next generation, and the next, and so on. To four decimal places, what is the value of x_{20}? What are F_{X20} and F_{Y20}?

(c) What is x^*, the equilibrium level of x? How many generations does it take for the population to be within 1% of x^*?

(d) Answer the questions in part (b), but with a starting value of $x_0 = 0.1$.

(e) Repeat part (b), but with $x_0 = 1$.

(f) Repeat part (b), but with $x_0 = 0.99$.

(g) Are monomorphic equilibria possible in this model? If so, are they stable? Explain.

U9. Consider an evolutionary game between green types and purple types, with a payoff table as follows:

		COLUMN	
		Green	Purple
ROW	Green	*a*, *a*	*b*, *c*
	Purple	*c*, *b*	*d*, *d*

In terms of the parameters *a*, *b*, *c*, and *d*, find the conditions that will guarantee a stable polymorphic equilibrium.

U10. (Optional, for mathematically trained students) In the three-type evolutionary game of Section 7.B and Figure 13.14, let $q_3 = 1 - q_1 - q_2$ denote the proportion of the orange-throated aggressor types. Then the dynamics of the population proportions of each type of lizard can be stated as

$$q_1 \text{ increases if and only if } -q_2 + q_3 > 0$$

and

$$q_2 \text{ increases if and only if } q_1 - q_3 > 0.$$

We did not state this explicitly in the chapter, but a similar rule for q_3 is

$$q_3 \text{ increases if and only if } -q_1 + q_2 > 0.$$

(a) Consider the dynamics more explicitly. Let the speed of change in a variable x in time t be denoted by the derivative dx/dt. Then suppose

$$dq_1/dt = -q_2 + q_3,\ dq_2/dt = q_1 - q_3, \text{ and } dq_3/dt = -q_1 + q_2.$$

Verify that these derivatives conform to the preceding statements regarding the population dynamics.

(b) Define $X = (q_1)^2 + (q_2)^2 + (q_3)^2$. Using the chain rule of differentiation, show that $dX/dt = 0$, that is, show that X remains constant over time.

(c) From the definitions of the entities, we know that $q_1 + q_2 + q_3 = 1$. Combining this fact with the result from part (b), show that over time, in three-dimensional space, the point (q_1, q_2, q_3) moves along a circle.

(d) What does the answer to part (c) indicate regarding the stability of the evolutionary dynamics in the colored-throated lizard population?

Mechanism Design

JAMES MIRRLEES WON THE NOBEL PRIZE in Economics in 1996 for his pioneering work on optimal nonlinear income taxation and related policy issues. Many noneconomists, and some economists too, found his work difficult to understand. But *The Economist* magazine gave a brilliant characterization of the broad importance and relevance of the work. It said that Mirrlees showed us "how to deal with someone who knows more than you do."[1]

We have already seen some of the ways in which such asymmetric information affects the analysis of games, in Chapter 9. But the underlying problem for Mirrlees differed slightly from the situations we considered earlier. In his work, one player (the government) needed to devise a set of rules so that the other players' (the taxpayers') incentives were aligned with the first player's goals. Models with this general framework, in which a less-informed player works to create motives for the more-informed player to take actions beneficial to the less informed, now abound and are relevant to a wide range of social and economic interactions. Generally, the less-informed player is called the *principal* while the more-informed is called the *agent;* hence these models are termed *principal-agent* models. And the process that the principal uses to devise the correct set of incentives for the agent is known as **mechanism design.**

In Mirrlees's model, the government seeks a balance between efficiency and equity. It wants the more productive members of society to contribute effort to increase total output; it can then redistribute the proceeds to benefit the poorer

[1]"Economics Focus: Secrets and the Prize," *The Economist,* October 12, 1996.

members. If the government knew the exact productive potential of every person and could observe the quantity and quality of his effort, it could simply order everyone to contribute according to their ability, and it could distribute the fruits of their work according to people's needs. But such detailed information is costly or even impossible to obtain, and such redistribution schemes can be equally difficult to enforce. Each person has a good idea of his abilities and needs and chooses his own effort level, but stands to benefit by concealing this information from the government. Pretending to have less ability and more needs will enable him to get away with paying less tax or getting larger checks from the government; the incentive to provide effort is reduced if the government takes part of the yield. The government must calculate its tax policy, or design its fiscal mechanism, taking into account these problems of information and incentives. Mirrlees's contribution was to solve this complex mechanism design problem within the principal-agent framework.

The economist William Vickrey shared the 1996 Nobel Prize in Economics with Mirrlees for his own work in mechanism design in the presence of asymmetric information. Vickrey is best known for designing an auction mechanism to elicit truthful bidding, a topic we will study in greater detail in Chapter 17. But his work extended to other mechanisms, such as congestion pricing on highways, and he and Mirrlees laid the groundwork for extensive research in the subfield.

Indeed, in the last thirty years, the general theory of mechanism design has made great advances. The 2007 Nobel Prize in Economics was awarded to Leonid Hurwicz, Roger Myerson, and Eric Maskin for their contributions to it. Their work, and that of many others, has taken the theory and applied it to numerous specific contexts, including the design of compensation schemes, insurance policies, and of course tax schedules and auctions. In this chapter we will develop a few prominent applications, using our usual method of numerical examples followed by exercises.

1 PRICE DISCRIMINATION

A firm generally sells to diverse customers with different levels of willingness to pay for its product. Ideally, the firm would like to extract from each customer the maximum that he would be willing to pay. If the firm could do so, charging each customer an individualized price based on willingness to pay, economists would say that it was practicing perfect (or first-degree) **price discrimination.**

Such perfect price discrimination may not be possible for many reasons. The most general underlying reason is that even a customer who is willing to pay a lot prefers to pay less. Therefore the customer will prefer a lower price, and this firm may have to compete with other firms or resellers who undercut its high price. But even if there are no close competitors, the firm usually does not know how much each individual customer is willing to pay, and the customers will try to get away with pretending to be unwilling to pay a high price so as to secure a lower price. Sometimes even if the firm could detect the willingness to pay, it may be illegal to practice blatant first-degree price discrimination based on the identity of the buyer. In such situations, the firm must devise a product line and prices so that the customers' choices of what they buy (and therefore what they pay) go some way toward the firm's goal of increasing its profit by way of price discrimination.

In our terminology of asymmetric information games developed in Chapter 9, the process by which the firm identifies customer willingness to pay from purchase decisions involves *screening* to achieve *separation of types* (by self-selection). The firm does not know each customer's *type* (willingness to pay), so it tries to acquire this information from their actions. An example that should be familiar to most readers is that of airlines. These firms try to separate business flyers, who are willing to pay more for their tickets, from tourists, who are not willing to pay that much, by offering low prices in return for various restrictions on fares that the business flyers are not willing to accept such restrictions include advance-purchase and minimum-stay requirements.[2] We develop this particular example in more detail to make the ideas more precise and quantifiable.

We consider the pricing decisions of a firm called Pie-In-The-Sky (PITS), an airline running a service from Podunk to South Succotash. It carries some business passengers and some tourists; the former type are willing to pay a higher price than the latter for any particular ticketed seat. To serve the tourists profitably without having to offer the same low price to the business passengers, PITS has to develop a way of creating different versions of the same flight; it then needs to price these options in such a way that each type will choose a different version. As mentioned above, the airline could distinguish between the two types of passengers by offering restricted and unrestricted fares. The practice of offering first-class and economy-class tickets is another way to distinguish between the two groups; we will use that practice as our example.

Suppose that 30% of PITS's customers are businesspeople and 70% are tourists. The table in Figure 14.1 shows the (maximum) willingness to pay for each type of customer for each class of service, along with the costs of providing the two types of service and the potential profits available under each option.

[2]This type of pricing policy, offering different prices to different groups of customers on the basis of some distinguishable characteristic, is generally known as *third-degree* price discrimination. It is thus differentiated from the *first-degree* discrimination described earlier. There is also *second-degree* price discrimination, which occurs when firms charge different unit prices for customers purchasing different quantities of a product.

Type of service	PITS's cost	Reservation price		PITS's potential profit	
		Tourist	Business	Tourist	Business
Economy	100	140	225	40	125
First	150	175	300	25	150

FIGURE 14.1 Airline Price Discrimination

We begin by setting up a ticket-pricing scheme that is ideal from PITS's point of view. Suppose it knows the type of each individual customer; salespeople determine customer type, for example, by observing their style of dress when they come to make their reservations. Also suppose that there are no legal prohibitions on differential pricing and no possibility that lower-priced tickets can be resold to other passengers. (Actual airlines prevent such resale by requiring positive ID for each ticketed passenger.) Then PITS could practice perfect (first-degree) price discrimination.

How much would PITS charge to each type of customer? It could sell a first-class ticket to each business person at \$300 for a profit of \$300 − \$150 = \$150 per ticket, or sell him an economy ticket at \$225, for a profit of \$225 − \$100 = \$125 per ticket. The former is better for PITS, so it would want to sell \$300 first-class tickets to these business customers. It could sell a first class ticket to each tourist at \$175, for a profit of \$175 − \$150 = \$25, or an economy ticket at \$140, for a profit of \$140 − \$100 = \$40. Here the latter is better for PITS, so it would want to sell \$140 economy-class tickets to the tourists. Ideally, PITS would like to sell only first-class tickets to business travelers and only economy-class tickets to tourists, in each case at a price equal to the relevant group's maximum willingness to pay. PITS' total profit per 100 customers from this strategy would be

$$(140 - 100) \times 70 + (300 - 150) \times 30 = 40 \times 70 + 150 \times 30 = 2{,}800 + 4{,}500 = 7{,}300.$$

Thus, PITS's best possible outcome earns it a profit of \$7,300 for every 100 customers it serves.

Now turn to the more realistic scenario in which PITS cannot identify the type of each customer or is not allowed to use the information for purposes of overt discrimination. How can it use the different ticket versions to screen its customers?

The first thing PITS should realize is that the pricing scheme devised above will not be the most profitable in the absence of identifying information about each customer. Most important, it cannot charge the business travelers their full \$300 willingness to pay for first-class seats while charging only \$140 for an economy-class seat. Then the business people could buy economy-class seats, for which they are willing to pay \$225, and get an extra benefit, or "consumer surplus" in the jargon of economics, of \$225 − \$140 = \$85. They might use this surplus, for example, for better food or accommodation on their travels. Paying the maximum \$300 that they are willing to pay for a first-class seat would leave them no consumer surplus. Therefore they would switch to economy-class in this situation, and screening would fail. PITS' profit per 100 customers would drop to $(140 - 100) \times 100 = \$4{,}000$.

The maximum that PITS will be able to charge for first-class tickets must give business travelers at least as much extra benefit as the \$85 they can get if they buy an economy-class ticket. Thus, the price of first-class tickets can be at most \$300 − \$85 = \$215. (Perhaps it should be \$214 to give business travelers a definite positive reason to choose first class, but we will ignore the trivial difference.) PITS can still charge \$140 for an economy-class ticket to extract as much profit as possible from the tourists, so its total profit in this case (from every 100 customers) would be

$$(140 - 100) \times 70 + (215 - 150) \times 30 = 40 \times 70 + 65 \times 30 = 2{,}800 + 1{,}950 = 4{,}750.$$

This profit is more than the \$4,000 that PITS would get if it tried unsuccessfully to implement its perfect discrimination scheme despite its limited information, but less than the \$7,300 it would get if it had full information and successfully practiced perfect price discrimination.

By pricing first-class seats at \$215 and economy-class seats at \$140, PITS can successfully screen and separate the two types of travelers on the basis of their self-selection of the two types of services. But PITS must sacrifice some profit to achieve this indirect discrimination. PITS loses this profit because it must charge the business travelers less than their full willingness to pay. As a result, its profit per 100 passengers drops from the \$7,300 it could achieve if it had full and complete information, to the \$4,750 it achieves from the indirect discrimination based on self-selection. The difference, \$2,550, is precisely 85 times 30, where 85 is the drop in the first-class fare below the business travelers' full willingness to pay for this service, and 30 is the number of these business travelers per 100 passengers served.

Our analysis shows that, in order to achieve separation with its ticket-pricing mechanism, PITS has to keep the first-class fare sufficiently low to give the business travelers enough incentive to choose this service. Those travelers have the option of choosing economy class if it provides more benefit (or surplus) to them; PITS has to ensure that they do not "defect" to making the choice that PITS intends for the tourists. Such a requirement, or constraint, on the screener's strategy arises in all problems of mechanism design and is called an *incentive-compatibility constraint.*

The only way PITS could charge business travelers more than \$215 without inducing their defection would be to increase the economy-class fare. For example, if the first-class fare is \$240 and the economy-class fare is \$165, then business travelers get equal consumer surplus from each class; their surplus is \$300 − \$240 from first class and \$225 − \$165 from economy class, or \$60 from each. At those higher prices, they are still (only just) willing to buy first-class tickets, and PITS could enjoy higher profits from each first-class ticket sale.

But at \$140 the economy-class fare is already at the limit of the tourists' willingness to pay. If PITS raised that fare to \$165, say, it would lose these customers altogether. In order to keep these customers willing to buy, PITS's pricing mechanism must meet an additional requirement, namely the tourists' *participation constraint.*

PITS's pricing strategy is thus squeezed between the participation constraint of the tourists and the incentive-compatibility constraint of the business people. If it charges X for economy and Y for first class, it must keep $X < 140$ to ensure that the tourists still buy tickets, and it must keep $225 - X < 300 - Y$, or $Y < X + 75$, to ensure that the business travelers choose first-class and not economy. Subject to these constraints, PITS wants to charge prices that are as high as possible. Therefore its profit-maximizing screening strategy is to make X as close to 140 and Y as close to 215 as possible. Ignoring the small differences that are needed to preserve the $<$ signs, let us call the prices 140 and 215. Then charging \$215 for first-class seats and \$140 for economy-class seats is the solution to PITS's mechanism-design problem.

This pricing strategy being optimal for PITS depends on the specific numbers in our example. If the proportion of business travelers were much higher, say 50%, PITS would have to revise its optimal ticket prices. With 50% of its customers being business people, the sacrifice of \$85 on each business traveler may be too high to justify keeping the few tourists. PITS may do better not to serve the tourists at all, that is, to violate their participation constraint and to raise the price of first-class service. Indeed, the strategy of discrimination by screening with these percentages of travelers yields PITS a profit, per 100 customers, of

$$(140 - 100) \times 50 + (215 - 150) \times 50 = 40 \times 50 + 65 \times 50 = 2{,}000 + 3{,}250 = 5{,}250.$$

The strategy of serving only business travelers in \$300 first-class seats would yield a profit (per 100 customers) of

$$(300 - 150) \times 50 = 150 \times 50 = 7{,}500,$$

which is higher than with the screening prices. Thus, if there are only relatively few customers with low willingness to pay, the seller might find it better not to serve them at all than to offer sufficiently low prices to the mass of high-paying customers to prevent their switching to the low-priced version.

Precisely what proportion of business travelers constitutes the borderline between the two cases? We leave this as an exercise for you at the end of the chapter. And we will just point out that an airlines' decision to offer low tourist fares may be a profit-maximizing response to the existence of asymmetric information, rather than an indication of some soft spot for vacationers!

2 SOME TERMINOLOGY

We have now seen one example of mechanism design in action. There are many others, of course, and we will see additional ones in later sections. We pause briefly here, however, to set out the specifics of the terminology used in most models of this type.

Mechanism-design problems are broadly of two kinds. In the first, which is similar to the price-discrimination example above, one player is better informed (in the example, the customer knows his own willingness to pay), and his information affects the payoff of the other player (in the example, the airline's pricing and therefore its profits). The less-informed player designs a scheme in which the better-informed player must make some choice that will reveal the information, albeit at some cost to the first (in the example, the airline's inability to charge the business fliers their full willingness to pay).

In the second kind of mechanism-design problem, one player takes some action that is not observable to others. For example, an employer cannot observe the quality, or sometimes even the quantity, of the effort an employee exerts, and an insurance company cannot observe all the actions that an insured driver or homeowner takes to reduce the risk of an accident or robbery. In the language of Chapter 9, this problem is one of *moral hazard.* The less-informed player designs a scheme—for example, profit sharing for the employee or deductibles and copayments for insurance—that aligns the other player's incentives to some extent with those of the mechanism designer.

In each case, the less-informed player designs the mechanism; he is called the **principal** in the strategic game. The more-informed player is then called the **agent**; this is most accurate in the case of the employee and less so in the cases of the customer or the insured, but the jargon has become established and we will adopt it. The game is then called a **principal-agent,** or **agency,** problem.

The principal in each case designs the mechanism to maximize his own payoff, subject to two types of constraints. First, the principal knows that the agent will utilize the mechanism to maximize his own (the agent's) payoff. In other words, the principal's mechanism has to be consistent with the agent's incentives. As we saw in Chapter 9, Section 4.B, this is called the *incentive-compatibility constraint.* Second, given that the agent responds to the mechanism in his own best interests, the agency relationship has to give the agent at least as much expected utility as he would get elsewhere, for example by working for someone else, or by driving instead of flying. In Chapter 9, we termed this the *participation constraint.* We saw specific examples of both constraints in the airline price-discrimination story in the previous section; we will meet many other examples and applications later in this chapter.

3 COST-PLUS AND FIXED-PRICE CONTRACTS

When writing procurement contracts for the acquisition of certain services, perhaps highway or office-space construction, governments and firms face mechanism-design problems of the kind we have been describing in this chapter. Two common methods for writing such contracts are "cost-plus" and "fixed-price." In a cost-plus contract, the supplier of the services is paid a sum equal to his cost, plus an allowance for normal profit. In a fixed-price contract, a specific price for the services is agreed on in advance; the supplier keeps any extra profit if his actual cost turns out to be less than anticipated, and he bears the loss if his actual costs are higher.

Each type of contract has its own good and bad points. The cost-plus contract appears not to give the supplier excessive profit; this characteristic is especially important for public-sector procurement contracts, where the citizens are the ones who ultimately pay for the procured services. But the supplier typically has better information about his cost than does the buyer of his services; therefore the supplier can be tempted to overstate the cost or to pad the costs in order to extract some benefit from the wasteful excess. The fixed-price contract, on the other hand, gives the supplier every incentive to keep the cost at a minimum and thus to achieve an efficient use of resources. But with this kind of public-sector contract, society has to pay the set price and give away any excess profit (to the supplier). The optimal procurement mechanism should balance these two considerations.

A. Highway Construction: Full Information

We will consider the example of a state government designing a procurement mechanism for a road-construction project. Specifically, suppose that a major highway is to be built by the state's road contractor and that the government has to decide how many lanes it should have.[3] More lanes yield more social benefit in the form of faster travel and fewer accidents (at least up to a point, beyond

[3]Generally numerous contractors could be competing for the highway-construction contract. For this example, we restrict ourselves to the case in which there is only one contractor.

which the harm to the countryside will be too great). To be specific, we suppose that the social value V (measured in billions of dollars) from having N lanes on the highway is given by the formula:

$$V = 15N - N^2/2.$$

The cost of construction per lane, including an allowance for normal profit, could be either \$3 billion or \$5 billion per lane, depending on the types of soil and minerals located in the construction zone. For now, we will assume that the government can identify the construction cost as well as the contractor. So it chooses N and writes a contract to maximize the benefit to the state (V) net of the fee paid to the contractor (call it F); that is, the government's objective is to maximize net benefit, G, where $G = V - F$.[4]

Suppose first that the government knows the actual cost is 3 (billion dollars per lane of highway). At this cost level, the government has to pay $3N$ to the contractor for an N-lane highway. The government then chooses N to maximize G, as above, where the appropriate formula in this situation is:

$$G = V - F = 15N - N^2/2 - 3N = 12N - N^2/2.$$

Recall that in the appendix to Chapter 5 we gave a formula for finding the correct value to maximize this type of function. Specifically, the solution to the problem of choosing X to maximize

$$Y = A + BX - CX^2$$

is $X = B/(2C)$. Here Y is V, X is N, and $A = 0$, $B = 12$ and $C = 1/2$. Applying our solution formula yields the government's optimal choice of $N = 12/(2 \times 1/2) = 12$. The best highway to choose therefore has 12 lanes, and the cost of that 12-lane highway is \$36 billion. So the government offers the contract: "Build a 12-lane highway and we will pay you \$36 billion."[5] This price includes normal profit, so the contractor is happy to take the contract.

Similarly, if the cost is \$5 billion per lane, the optimal N will be 10. The government will offer a \$50 billion contract for the 10-lane highway. And the contractor will accept the contract.

[4]In reality the cost per lane would not have only two discrete values, but could take any value along a continuous range of possibilities. The probabilities of each value would then correspondingly form a density function on this range. Our methods will not always yield an integer solution, N, for each possible cost along this range. But we leave these matters to more advanced treatments and confine ourselves to this simple illustrative example.

[5]In reality there will be many clauses specifying quality, timing, inspections, etc. We leave out these details to keep the exposition of the basic idea of mechanism design simple.

B. Highway Construction: Asymmetric Information

Now suppose that the contractor knows how to assess the relevant terrain to determine the actual per lane building cost, but the government does not. The government can only estimate what the cost will be. We will assume that it thinks that there is a two-thirds probability of the cost being 3 (billion dollars per lane) and a one-third probability of the cost being 5.

What if the government tries to go ahead with the ideal optimum and offers a pair of contracts: "12-lane highway for \$36 billion" and "10-lane highway for \$50 billion"? If the cost is really only \$3 billion per lane, the contractor will get more profit by taking the latter contract even though that one was designed for the situation in which the cost is \$5 billion per lane. The true cost of the 10-lane highway would be only \$30 billion, and the contractor would earn \$20 billion in excess profit.[6]

This outcome is not very satisfying. The contracts offered do not give the contractor sufficient incentive to choose between them on the basis of cost; he will always take the \$50 billion contract. There must be a better way for the government to design its procurement contract system.

So now we allow the government the freedom to design a more general mechanism to separate the types of projects. Suppose it offers a pair of contracts: "Contract L: Build N_L lanes and get paid R_L dollars" and "Contract H: Build N_H lanes and get paid R_H dollars." If contracts L and H are designed correctly, when cost is low (\$3 billion per lane) the contractor will pick contract L (L stands for "low"), and when cost is high (\$5 billion per lane) he will pick contract H (H stands for "high"). The numbers that the symbols N_L, R_L, N_H, and R_H represent must satisfy certain conditions for this screening mechanism to work.

First, under each contract, the contractor facing the relevant cost (low for contract L and high for contract H) must receive enough to cover his cost (inclusive of normal profit). Otherwise he will not agree to the terms; he will not participate in the contract. Thus the contract must satisfy two *participation constraints*: $3\,N_L \leq R_L$ for the contractor when the cost is 3, and $5\,N_H \leq R_H$ for the contractor when the cost is 5.

Next, the government needs the two contracts to be such that a contractor who knows his cost is low would not benefit by taking contract H and vice versa. That is, the contracts must also satisfy two incentive-compatibility

[6]If multiple contractors are competing for the job, the ones not selected may spill the beans about the true cost here. But for large highway projects (as for many other large government projects, such as defense contracts), there are often only a few potential contractors, and they do better by colluding among themselves and not revealing the private information. For simplicity, we keep the analysis confined to the case where there is just one contractor.

constraints. For example, if the true cost is low, contract L will yield excess profit $R_L - 3\ N_L$, whereas contract H will yield $R_H - 3\ N_H$. (Note that in the latter expression, the number of lanes and the payment are as specified in the H contract, but the contractor's cost is still only 3, not 5.) To be incentive compatible for the low-cost case, the contracts must keep the latter expression no larger than the former. Thus, we need $R_L - 3N_L \geq R_H - 3N_H$. Similarly, if the true cost is high, the contractor's excess profit from the L contract must be no larger than his excess profit from the H contract, so to be incentive compatible, we need $R_H - 5N_H \geq R_L - 5N_L$.

The government wants to maximize the net expected social value of the payment, and uses the probabilities of the two types as weights to calculate the expectation. Therefore the government's objective here is to maximize

$$G = (2/3)[15N_L - (N_L)^2/2 - R_L] + (1/3)[15N_H - (N_H)^2/2 - R_H].$$

The problem looks formidable, with four choice variables and four inequality constraints. But it simplifies greatly, because two of the constraints are redundant, and the other two must hold as exact equalities, allowing us to solve and substitute for two of the variables.

Note that if the participation constraint when cost is high, $5N_H \leq R_H$, and the incentive compatibility constraint when cost is low, $R_L - 3N_L \geq R_H - 3N_H$, both hold, then we can get the following string of inequalities (where we have used the fact that N_H will be positive):

$$R_L - 3N_L \geq R_H - 3N_H \geq 5N_H - 3N_H \geq 5N_H \geq 0.$$

The first and last expressions in the inequality string tell us that $R_L - 3N_L \geq 0$. Therefore we need not consider the participation constraint when cost is low, $3N_L \leq R_L$, separately; it is automatically satisfied when the two other constraints are satisfied.

It is also intuitive that the high-cost firm will not want to pretend to be low cost; it would get compensated for the smaller cost while incurring the larger cost. However, this intuition needs to be verified by the rigorous logic of the analysis. Therefore we proceed as follows. We will begin by ignoring the second incentive compatibility constraint, $R_H - 5N_H \geq R_L - 5N_L$, and we will solve the problem with just the remaining two constraints. Then we will return and verify that the solution to the two-constraint problem satisfies the ignored third constraint anyway. So our solution must also be the solution to the three-constraint problem. (If something better was available, it would also work better for the less-constrained problem.)

Thus we have two constraints to consider: $5N_H \leq R_H$ and $R_L - 3N_L \geq R_H - 3N_H$. Write these as $R_H \geq 5N_H$ and $R_L \geq R_H + 3(N_L - N_H)$. Then observe that R_L and R_H each enter negatively in the government's objective; it wants to make them as

	N_L	R_L	N_H	R_L
Perfect Information	12	36	10	50
Asymmetric Information	12	48	6	30

FIGURE 14.2 Highway-Building Contract Values

small as is compatible with the constraints. This result is achieved by satisfying each constraint with equality. So we set $R_H = 5N_H$ and $R_L = R_H + 3(N_L - N_H) = 3N_L + 2N_H$. These expressions for the contract payments can now be substituted into the objective function, G. This substitution yields:

$$G = (2/3)[15N_L - (N_L)^2/2 - 3N_L - 2N_H] + (1/3)[15N_H - (N_H)^2/2 - 5N_H]$$
$$= 8N_L - (N_L)^2/3 + 2N_H - (N_H)^2/6.$$

The objective function now splits cleanly into two parts; one (the first two terms) involves only N_L, and the other (the second two terms) involves only N_H. We can apply our maximization formula separately to each part. In the N_L part, the $A = 0$, $B = 8$ and $C = 1/3$, so the optimal $N_L = 8/(2 \times 1/3) = 24/2 = 12$. In the N_H part, the $A = 0$ again, $B = 2$ and $C = 1/6$, so the optimal $N_H = 2/(2 \times 1/6) = 12/2 = 6$.

Now we can use the optimal values for N_L and N_H to derive the optimal payment (R) values, using the formulas for R_L and R_H that we derived just above. Substituting $N_L = 12$ and $N_H = 6$ into those formulas gives us $R_H = 5 \times 6 = 30$ and $R_L = 3 \times 12 + 2 \times 6 = 48$. We thus have optimal values for all of the unknowns in the government's objective function. But remember that we ignored one of the incentive-compatibility constraints, so we need to go back to that now.

We must ensure that the ignored third constraint, $R_H - 5\,N_H \geq R_L - 5\,N_L$, holds with our calculated values for the Rs and the Ns. In fact, it does. The left-hand side of the expression equals $30 - 5 \times 6 = 0$. And the right-hand side equals $48 - 5 \times 12 = -12$, so the constraint is indeed satisfied.

Our solution indicates that the government should offer the following two contracts: "Contract L: Build 12 lanes and get paid 48 (billion dollars)" and "Contract H: Build 6 lanes and get paid 30 (billion dollars)." How can we interpret this solution so as best to understand the intuition for it? The intuition is most easily seen when we compare the solution here with the ideal one we found in Section 3.A under full information about costs. Figure 14.2 shows the comparisons in the optimal N and R values.

The optimal mechanism under asymmetric information differs in two important respects from the one we found when information was perfect. First, although the contract intended to be chosen if the contractor's cost is low has the same number of lanes (12) as in the full-information case, its payment to the contractor is larger in the asymmetric case (48 instead of 36). Second, the high-cost asymmetric-information contract has a smaller number of lanes (six instead of 10) but pays just the full cost for that number ($30 = 6 \times 5$). Both of these differences separate the types.

Under asymmetric information, the contractor may be tempted to pretend that the cost is high when it is in fact low. The optimal payment mechanism then incorporates both a carrot for truthfully admitting to low cost and a stick for trying to pretend to be high cost. The carrot is the excess profit, $48 - 36 = 12$, that comes from the admission, made implicitly by the choice of contract L. The stick is the reduction in excess profit from contract H, achieved by reducing the number of lanes that will be constructed in that case. The ideal high-cost mechanism would have the highway be 10 lanes and would pay $50 billion; the contractor whose true cost is low would make excess profit of $50 - 3 \times 10 = \$20$ billion. In the information-constrained optimal contract, only six lanes are constructed, and the contractor is paid $30 billion. If the true cost is low, he makes an excess profit of $30 - 3 \times 6 = \$12$ billion. His benefit from the pretense (implicitly made by the choice of contract H even though his true cost is low) is reduced. In fact it is reduced exactly to the amount that he is guaranteed by the carrot part of the mechanism, thereby exactly offsetting his temptation to pretend high cost.

4 EVIDENCE CONCERNING INFORMATION REVELATION MECHANISMS

The mechanisms considered so far have the common feature that the agent has some private information, which we called the player's type in Chapter 9. Further, the principal requires the agent to take some action that is designed to reveal this information. In the terminology of Chapter 9, these mechanisms are examples of screening for the separation of types by self-selection.

We see such mechanisms everywhere. Those for price discrimination are the most ubiquitous. All firms have customers who are diverse in their willingness to pay for the firms' products. As long as a customer is willing to pay more than the firm's incremental cost of supplying the product to him, the firm can turn a profit by dealing with this customer. But this customer's willingness to pay may be relatively low in comparison to that of other potential buyers. If a firm must charge the same price to all of its customers, including those who would have been willing to pay more than this one, charging this customer's willingness to pay means the firm has to sacrifice some profit from its higher-willingness customers. Ideally, the firm would like to discriminate by giving a price break to the less-willing customers without giving the same break to the more-willing ones.

The ability of a firm to practice price discrimination may be limited for reasons other than those of information. It may be illegal to price discriminate. Competition from other firms may limit this firm's ability to charge high prices to some of its customers. And if the product can be bought by one customer and resold to others, such competition from other buyers may be just as effective a constraint on discriminatory pricing as competition from other firms. But here we focus on the information reasons for price discrimination, keeping the other reasons in the background of the discussion.

Your local coffee shop probably has a "frequent-drinker card"; for every ten cups you buy, you get one free. Why is it in the firm's interest to do this? Frequent drinkers are more likely to be locals, who have the time and incentive to search out the best deals in the neighborhood. To attract those customers away from other competing coffee shops, this one must offer a sufficiently attractive price. In contrast, infrequent customers are more likely to be strangers in the town, or in a hurry, and have less time and incentive to search for the best deals; when they need a cup of coffee and see a coffee shop, they are willing to pay whatever the price is (within reason). So posting a higher price and giving out frequent-drinker cards enables this coffee shop to give a price break to the price-sensitive regular customers without giving the same price break to the occasional buyers. If you don't have the card, you are revealing yourself as the latter type, willing to pay a higher price.

In a similar manner, many restaurants offer fixed-price three-course menus or blue-plate specials, as well as regular à la carte offerings. This strategy enables them to separate diverse customer types with different tastes for soups, salads, main courses, desserts, and so on.

Book publishers start selling new books in a hardback version and issue a paperback version a year or more later. The price difference between the two versions is generally far greater than the difference in the costs of production of the two kinds of books. The idea behind the pricing scheme is to separate two types of customers, those who need or want to read the book immediately and are willing to pay more for the privilege, and those who are willing to wait until they can get a better price.

We invite you to look for other examples of such screening mechanisms for price discrimination in your own purchases. They appear in myriad ways. You can also read good accounts of such practices. One good source is Tim Harford's *Undercover Economist.*[7]

There is a lot of research literature on procurement mechanisms of the kind we sketched in Section 3.[8] These models pertain to situations where the buyer confronts just one potential seller whose cost is private information. This type

[7]Tim Harford, *The Undercover Economist: Exposing Why the Rich Are Rich, the Poor Are Poor—and Why You Can Never Buy a Decent Used Car!* (New York: Oxford University Press, 2005). The first two chapters give examples of pricing mechanisms.

[8]Jean-Jacques Laffont and Jean Tirole, *A Theory of Incentives in Procurement and Regulation* (Cambridge, Mass.: MIT Press, 1993) is the classic of this literature.

of interaction accurately describes how contracts for major defense weapons systems or very specialized equipment are designed; there is usually only one reliable supplier of such products or services. However, in reality buyers often have the choice of several suppliers, and mechanisms that set the suppliers in competition with each other are beneficial to the buyer. Many such mechanisms take the form of auctions. For example, construction contracts are often awarded by inviting bids and choosing the bidder that offers to do the job for the lowest price (after adjusting for the promised quality of the work and speed of completion, or other relevant known attributes of the bid). We will give some examples and discussion of such mechanisms in the chapter on auctions.

5 INCENTIVES FOR EFFORT: THE SIMPLEST CASE

We now turn from the first type of mechanism design problem, those in which the principal's goal is to achieve information revelation, to the second type, in which there is moral hazard. The principal's goal in such situations is to write a contract that will induce the best effort level from the agent, even though that effort level is unobservable by the principal.

A. Managerial Supervision

Suppose you are the owner of a company that is undertaking a new project. You have to hire a manager to supervise it. The success of the project is uncertain, but good supervision can increase the probability of success. Managers are only human, though; they will try to get away with as little effort as they can! If their effort is observable, you can write a contract that compensates the manager for his trouble sufficiently to bring forth good supervisory effort.[9] But if you cannot observe the effort, you have to try to give him incentives based on success of the project, for example a bonus. Unless good effort absolutely guarantees success, however, such bonuses make the manager's income uncertain. And the manager is likely to be averse to risk, so you have to compensate him for facing such risk. You have to design your compensation policy to maximize your own expected profit, recognizing that the manager's choice of effort depends on the nature and amount of the compensation. This is a mechanism-design

[9]Most important, if a dispute arises, you or the manager must be able to prove to a third party, such as an arbitrator or a court, whether the manager made the stipulated effort or shirked. This condition, often called *verifiability,* is more stringent than mere observability by the parties to the contract (you and the manager). We intend such public observability or verifiability when we use the more common term *observability.*

problem whose solution is intended to cope with the moral-hazard problem of the manager's shirking.

Let us consider a numerical example. Suppose that if the project succeeds, it will earn the company a profit of $1 million over material and wage costs. If it fails, the profit will be zero. With good supervision, the probability of success is one-half, but if supervision is poor, the probability of success is only one-quarter.

As mentioned above, the manager is risk averse. We saw in the Appendix to Chapter 7 how risk aversion can be captured by a concave utility function. So let us take a simple case, where the manager's utility u from income y (measured in millions of dollars) is the square-root function: $u = \sqrt{y}$. Suppose also that the manager gets disutility 0.1 from the extra effort that is needed for good supervision. Finally, suppose that if the manager does not work for you, he can get another job that does not require any extra effort and that pays $90,000 dollars, or 0.09 million, yielding utility $\sqrt{0.09} = 0.3$. Thus, if you want to hire the manager without requiring good supervision, you have to pay at least $90,000. If you want good supervision, you have to guarantee the manager at least as much utility as he could get from taking the other job; you must pay the y that ensures $\sqrt{y} - 0.1$ is at least 0.3, or $\sqrt{y} \geq 0.4$, or $y \geq 0.16$, or $160,000.

If effort is observable, you can write one of two contracts: (1) I pay you $90,000, and I don't care if you shirk; or (2) I pay you $160,000 and you have to make a good supervisory effort. This second contract can be enforced by a court, so if the manager accepts it he will in fact make good effort. Your expected profit from each contract depends on the probability that the project succeeds with the specified level of effort. So expected profit from the first is $(1/4) \times 1 - 0.09 = 0.160$ or $160,000, and that from the second is $(1/2) \times 1 - 0.16 = 0.340 =$ $340,000. Therefore you are better off paying the manager to provide good effort. In an ideal world of full information, you will use the second contract.

Now consider the more realistic scenario in which the manager's effort is not observable. This situation presents no extra problems if you would like the manager to exert low effort, and the first contract above applies. But if you would like good supervisory effort, you must use an incentive mechanism based on the only observable, namely success or failure of the project. So suppose you offer a contract that pays the manager x if the project fails and y if it succeeds. (Note that x may be zero, but if that is optimal, it should emerge from the solution. In fact it will not be zero, because of the manager's risk aversion.)

To induce the manager to choose high effort, you must ensure that his expected utility from doing so is higher than his expected utility from shirking. With high effort, he can guarantee a one-half chance that the project succeeds, and he therefore faces a one-half chance that it fails. With ordinary effort, he can guarantee only a one-quarter chance of success (a three-quarter chance of failure). So your contract must ensure the following:

$$(1/2)\sqrt{y} + (1/2)\sqrt{x} - 0.1 > (1/4)\sqrt{y} + (3/4)\sqrt{x}, \text{ or}$$
$$(1/4)[\sqrt{y} - \sqrt{x}] \geq 0.1, \text{ or } \sqrt{y} - \sqrt{x} \geq 0.4.$$

This expression is the *incentive-compatibility constraint* in this problem.

Next, you have to ensure that the manager gets enough expected utility to be willing to work for you *in the way you want* (exerting high supervisory effort) rather than taking his other possible offer. So his expected utility from accepting your job and exerting high effort must exceed his utility from the alternate job; your contract must then satisfy the following:

$$(1/2)\sqrt{y} + (1/2)\sqrt{x} - 0.1 \geq 0.3, \text{ or } \sqrt{y} + \sqrt{x} \geq 0.8.$$

This expression is the *participation constraint* for your contract intended to elicit high supervisory effort.

Subject to these constraints, you want to maximize your expected profit, Π. You calculate that expected profit under the assumption that by meeting the constraints above you are eliciting high supervisory effort. Thus, you assume that your project succeeds with probability one-half and your expected profit expression is:

$$\Pi = (1/2)(1 - y) + (1/2)(0 - x) = [1 - y - x]/2.$$

The mathematics in this problem becomes much easier if we work with the square roots of x and y instead of x and y themselves (that is, we work with the utilities of income instead of the incomes). Write these utilities as $X = \sqrt{x}$ and $Y = \sqrt{y}$, so $x = X^2$ and $y = Y^2$. Then you want to maximize

$$\Pi = [1 - Y^2 - X^2]/2.$$

subject to the participation constraint

$$Y + X \geq 0.8$$

and the incentive-compatibility constraint

$$Y - X \geq 0.4 .$$

Both X and Y enter negatively into the expression for your expected profit, so you want to make both as small as is compatible with the constraints. The participation constraint eventually holds with equality when both X and Y are made small. What about the incentive-compatibility constraint? If it does not also eventually hold with equality, then it does not constrain the choices and can be ignored. Let us suppose that is the case. Then we can substitute $X = 0.8 - Y$ from the participation constraint into your profit expression and write

$$\begin{aligned}\Pi &= [1 - Y^2 - X^2]/2 = [1 - Y^2 - (0.8 - Y)^2]/2 \\ &= [1 - Y^2 - 0.64 + 1.6Y - Y^2]/2 \\ &= [0.36 + 1.6Y - 2Y^2]\,2 = 0.18 + 0.8Y - Y^2.\end{aligned}$$

To maximize this profit expression, we again use the formula from the Appendix to Chapter 5; we have $B = 0.8$ and $C = 1$. This yields the optimal $Y = 0.8/(2 \times 1) = 0.4$. Then $X = 0.8 - 0.4 = 0.4$ also.

This solution implies that if the incentive compatibility constraint is ignored, the optimal mechanism requires equal payment to the manager whether the project succeeds or fails. This payment is just enough to give the manager a utility of $0.4 = 0.3 + 0.1$ (his utility from easy work elsewhere plus compensation for the disutility of the extra effort for high supervision) to meet the participation constraint. This result is intuitive and in keeping with our discussion of optimal risk bearing in Chapter 9, Section 1. The manager is risk averse and you are risk neutral (concerned with expected profit alone), so it is efficient for you to bear all of the risk and to keep the manager's income nonrandom.[10]

But if the manager gets the same income whether the project succeeds or fails, he has no incentive to make the unobservable effort. So the ignored incentive-compatibility constraint is not going to be fulfilled automatically, and we must make sure that X and Y do satisfy it. We therefore need both of the constraints to hold with equality: $Y + X = 0.8$ and $Y - X = 0.4$. Adding the two constraints together, we get $2Y = 1.2$ or $Y = 0.6$; this result immediately yields $X = 0.2$. Translating from utilities into dollar amounts, we have $x = X^2 = 0.04$ and $y = Y^2 = 0.36$. Thus the manager should be paid \$40,000 if the project fails and \$360,000 if it succeeds. The payment for failure is less than the \$90,000 he would be paid for the low-effort contract 1 in the full-information case, and the payment for success is more than the \$160,000 for the high-effort contract 2 in the full-information case. Thus the manager faces a combination of a stick (low pay if the project fails) and a carrot (high pay if it succeeds), just as does the contractor in the highway construction example of Section 3.

With this scheme, you (the owner) make an expected profit of:

$$\Pi = [1 - 0.36 - 0.04]/2 = 0.30,$$

or \$300,000. This amount is less than the \$340,000 you would make in the full-information ideal, when you could write an enforceable contract stipulating high effort. The \$40,000 difference is an unavoidable cost of the information asymmetry.

The manager's compensation scheme can be described as a base salary of \$40,000 and a success bonus of \$320,000, or equivalently, a \$40,000 salary and a 32% share in the operating profit of \$1 million. It would not be desirable for you to rely on profit sharing alone, offering the manager no base salary. Why not? If the salary component were zero, then in the event of the project's success you

[10]The case in which the owner is also risk averse can be treated by similar methods.

would have to pay the manager an amount y defined by $(1/2)\sqrt{y} - 0.1 = 0.3$, or $y = 0.64$, or \$640,000 to ensure his participation. Your expected profit would be

$$\Pi = [1 - 0.64 - 0]/2 = 0.180, \text{ or } \$180{,}000.$$

Thus, your profit in this case would be \$120,000 lower than when you offered a \$40,000 base salary with bonus (and a full \$160,000 below what you could earn in the full-information case). The reduction in profit is due to the fact that the manager is risk averse. A pure-bonus scheme makes his income very risky, so to ensure his participation you have to make the bonus so large that it cuts into your profit. The optimal asymmetric information payment scheme balances the stick and the carrot optimally to provide enough incentive for the manager to make high supervisory effort, but without imposing too much risk or his income.

B. Insurance Provision

Moral hazard can arise in other relationships beyond those in the labor market described above. Insurance markets in particular are subject to problems of moral hazard. And insurance companies must determine whether and how to offer appropriate insurance contracts that encourage their clients to take appropriate actions to reduce their likelihood of needing to file a claim with the company. For example, insurers would like those to whom they sell health insurance to continue regular wellness visits to their physicians and those to whom they sell car insurance to continue to practice defensive-driving techniques.[11] Because the insurance company cannot usually observe the clients' actions, however, creating the appropriate insurance policy will require an understanding of the theory of mechanism design in the face of asymmetric information.

Here we return to our example of a farmer facing the risk of crop failure due to some bad-weather outcome, such as a drought. We met this farmer originally in Chapter 9, Section 1. There we supposed that the farmer's income would be \$160,000 if the weather proved favorable and \$40,000 if not. When the two possibilities are equally likely, probability 0.5 each, the farmer's expected income is $0.5 \times \$160{,}000 + 0.5 \times \$40{,}000 = \$100{,}000$. The farmer faces considerable risk around this average value, however, and if he is risk averse he will care about the expected utility of the outcomes rather than just about his expected income.

Suppose then that the farmer is indeed risk averse. His utility function is $u = \sqrt{I}$, where I represents his income. The farmer therefore gets utility of $400 = \sqrt{160{,}000}$ if the weather is good (wet) and utility of $200 = \sqrt{40{,}000}$ if the weather is bad (dry). His expected utility is then $0.5 \times 400 + 0.5 \times 200 = 300$.

[11]Indeed, insurance companies regard the policyholders' failure to take such risk-reducing precautions as immoral behavior; this is the origin of the term *moral hazard*.

What would happen if this farmer could avoid the risk associated with a year of drought? Specifically, what would his situation be if he could ensure that his income was always \$100,000 (the expected value here) rather than \$160,000 half of the time and \$40,000 the other half of the time? Ignoring for a moment how he could make this happen, we note that the farmer gets utility of about $316 \approx \sqrt{100{,}000}$ every year under this outcome. The farmer therefore would enjoy higher expected utility ($316 > 300$) if he could find a way to smooth his income (and his utility) across the good and bad weather years.

One possible way for the farmer to achieve income smoothing is by way of insurance. A risk-neutral insurance company could offer the farmer a contract where the farmer pays the company \$60,000 in good-weather years and the insurer pays the farmer \$60,000 in bad-weather years. Because the probability of each outcome is 50%, the company's expected profit from this contract is exactly zero, making it just willing to offer the contract to the farmer. The farmer is strictly better off accepting the contract, however; his expected utility rises. So an insurance contract that is full (completely covers the cost of a bad outcome) and fair (priced just to offset the cost of the farmer's claims) would be acceptable to both parties.

So far this example has no information problem. But the farmer could take various actions to reduce the probability of the low income level associated with drought. He may be able to construct some water-catchment basins, for example, that would allow him to water his crops in all but the driest of years. However, the construction and maintenance of the basins will be at some cost to the farmer. If the basins are of good quality and well maintained, they will help protect the farmer from the risks associated with a drought. If the basins are shoddy ones that leak and are not well cared for, they will fail to do their job and so do not reduce the risk of crop failure from drought. If the farmer is well insured and the quality of the basins and their level of maintenance is not observable by simple inspection, he may be tempted to shirk the task to save himself the costs; this potential for shirking is the source of moral hazard in our example.

Suppose that the farmer's disutility of making the extra effort to construct and maintain high-quality water basins is 25,[12] and that with them in existence the farmer reduces the probability of the bad outcome to 25%. Then the farmer's expected income with the basins is $0.75 \times \$160{,}000 + 0.25 \times \$40{,}000 = \$130{,}000$ and his expected utility (in the absence of insurance) is $0.75 \times \sqrt{160{,}000} + 0.25 \times \sqrt{40{,}000} - 25 = 0.75 \times 400 + 0.25 \times 200 - 25 = 350 - 25 = 325$. The farmer's expected utility is higher with the basins than without them ($325 > 300$), so if no

[12] Formally, the farmer's utility function is now $u = \sqrt{I} - E$, where I is again income, and E the disutility of effort, 25 if the basins are of good quality and 0 if they are shoddy.

insurance is available, the farmer will definitely want to make the risk-reducing effort to construct the water basins.

The farmer could still benefit from insurance in this case. An income-smoothing policy that guaranteed him \$130,000 every year would ensure an expected utility of $360(\approx \sqrt{130{,}000}) - 25 = 335$, even when he builds and maintains high-quality basins. This utility is higher than the 325 he receives when he builds the basins but has no insurance, so the farmer would definitely prefer the insurance.

Suppose that a full and fair insurance contract could be written that stipulated that the farmer exert the effort necessary to reduce the probability of a bad outcome to 25%. Suppose further that the insurance company could verify the farmer's effort by sending an insurance agent to the farm to check on the water basins. Then the contract that guaranteed the farmer \$130,000 income each year would entail the farmer's paying the insurance company \$30,000 in a good-weather year and the insurer paying the farmer \$90,000 in a bad-weather year. As before, the insurance company reaps an expected profit of exactly zero with this contract ($0.75 \times 30{,}000 - 0.25 \times 90{,}000 = 0$) but the farmer's expected utility increases (to 335), so both parties will agree to the contract.

If the insurer cannot verify the farmer's effort, then the situation changes. The farmer could cheat and accept the "pay \$30,000 in a good year, get \$90,000 in a bad year" insurance contract but not make the stipulated effort (build shoddy basins and provide no maintenance). Then the probability of having a bad year reverts to 50%, but the farmer's income is \$130,00 every year. His expected utility from accepting such a contract but making no effort is $360(\approx \sqrt{130{,}000})$, which is better than all of the other possibilities we have so far considered. Of course, the insurance company does badly in this case. Its expected profit is $0.5 \times \$30{,}000 - 0.5 \times \$90{,}000 = -\$30{,}000$. The insurer cannot survive this contract, given the moral-hazard problem, and so will not offer it to the farmer.

Does this mean that the farmer cannot get insurance at all when he has the option to build and maintain water basins, but his insurer can't verify their quality and maintenance? No. But it does mean that he cannot get full insurance. There is still the option of a *partial* insurance contract in which the insurance company takes on a part, but not all, of the risk associated with a bad outcome.

Recall that when the farmer can build and maintain high-quality basins, full insurance entails his paying the insurer \$30,000 in a good year and receiving \$90,000 in a bad year. This contract gave the farmer no incentive actually to build or maintain the basins and left the insurer with a negative expected profit. To design the optimal insurance scheme here, the insurance company needs to determine the right X to require as payment from the farmer in a good year (leaving the farmer $\$160{,}000 - X$) and the right Y to pay out to the farmer in a bad year (boosting the farmer's income to $40,000 + Y$). Then the optimal mechanism must maximize the insurance company's expected profit, given X, Y, and the probabilities of the different outcomes while ensuring both that the farmer retains the incentive to build the catch basins and that he is willing to accept the insurance contract.

Because calculating the optimal values for X and Y here is quite complex, we will instead consider a specific pair of numbers that offers the farmer some insurance, gives him enough incentive to make the effort to reduce his risk, and breaks even for the company. Suppose the insurance company offers a contract that goes one-third of the way toward full insurance. Such a contract would stipulate a payment from the farmer of $10,000 in a good year (leaving him with $150,000) and a payment to the farmer of $30,000 in a bad year (bringing him to $70,000). If the farmer does build and maintain the high-quality basins, then this contract leaves the insurance company with an expected profit of $0.75 \times \$10{,}000 - 0.25 \times \$30{,}000 = \$7{,}500 - \$7{,}500 = 0$, so the company is just willing to offer insurance at this level.

But will the farmer make the stipulated effort? In other words, is the contract incentive compatible? It is if the farmer's expected utility with the insurance and the effort exceed his expected utility of accepting the insurance but not putting out the effort. That is, the contract must satisfy the following inequality:[13]

$$0.75 \times \sqrt{150{,}000} + 0.25 \times \sqrt{70{,}000} - 25 > 0.50 \times \sqrt{150{,}000} + 0.50 \times \sqrt{70{,}000}.$$

Calculating out the values of the two expressions yields (approximately) $331 > 326$, which is true. So the partial insurance contract is incentive compatible; it will induce the appropriate bad-outcome-reducing effort on the part of the farmer.

And does the contract satisfy the participation constraint as well? Yes. It must provide the farmer with an expected utility at least as large as he could achieve in the absence of insurance. That level, which we calculated earlier, is 325; here he gets 331. The farmer is better off with this partial insurance contract than with no insurance at all, and both parties will agree to this contract.

Evidence supporting this theory of insurance and moral hazard can be found in any of your insurance contracts. Most policies come with various requirements of deductibles and copayments that leave some of the policy holder's risk uninsured in order to reduce moral hazard.

[13]Compare the first term on the left-hand side with the first term on the right-hand side of the incentive-compatibility constraint. Exerting the effort to construct high-quality basins raises the coefficient multiplying the high utility, $\sqrt{150{,}000}$, from 0.50 to 0.75. Similarly, comparing the second terms on either side, you will see that failure to make the effort raises the coefficient multiplying the low utility, $\sqrt{70{,}000}$, from 0.25 to 0.50. These differences are analogous to the carrot and stick aspects of the incentive scheme in Section 5.A above.

6 INCENTIVES FOR EFFORT: EVIDENCE AND EXTENSIONS

The theme of the managerial-effort-incentive scheme of Section 5.A was the tradeoff between giving the manager a more powerful incentive to provide the optimal effort level and requiring him to bear more of the risk in the firm's profit. This tradeoff is an important consideration in practice, but it must be considered in combination with other features of the relationship between the firm and its employee. Most of these other features have to do with multiple dimensions of the activities that go on within the firm. The quality and quantity of effort are not just a matter of good or bad, and outcomes are not just a matter of success or failure; each can range over many possibilities, and such entities as hours and profits can vary continuously. The firm has many employees, and the overall outcome for the firm depends on some combination of their actions. Most firms have multiple outputs, and each employee performs multiple tasks. And the firm and its employees interact over a long period of time, not just for one project or over a short duration. All of these features correspondingly require more complex incentive schemes. In this section we outline a few of these and refer you to a rich body of literature for further details.[14] The mathematics of these schemes gets correspondingly complex, so we will merely give you the intuition behind them and leave formal rigorous analyses to more advanced courses.

A. Nonlinear Incentive Schemes

Can the optimal managerial effort scheme always be characterized by a base salary with a profit-share component? No. If there are three possible outcomes—failure, modest success, and huge success—then the percentage bonus for going from failure to modest success may not equal that for going from modest to huge success. So the optimal scheme may be nonlinear.

Suppose we alter the managerial supervision example in Section 5.A to allow for three possible outcomes: profit over material and wage cost of 0, $500,000, or $1 million. Suppose also that good supervisory effort yields probabilities of success

[14]Canice Prendergast, "The Provision of Incentives in Firms," *Journal of Economic Literature,* vol. 37, no. 1 (March 1999), pp. 7–63, is an excellent survey of the theory and practice of incentive mechanisms. Prendergast gives references to the original research literature from which many findings and anecdotes are mentioned in this section, so we will not repeat the specific citations. James N. Baron and David M. Kreps, *Strategic Human Resources: Frameworks for General Managers* (New York: Wiley, 1999) is a wider-ranging book on personnel management, combining perspectives from economics, sociology, and social psychology. Chapters 8, 11, and 16 and Appendixes C and D are closest to the concerns of this chapter and this book.

of one-sixth, one-third and one-half for the three possible outcomes in the same order. Poor supervision reverses the probabilities of success to one-half, one-third and one-sixth, respectively. Then a somewhat harder calculation along the same lines as above, which we relegate to an optional exercise, shows that the optimal payments are $30,625 for failure, $160,000 for the modest success, and $225,625 for the top outcome. If we interpret this payment scheme as a $30,625 base salary with a bonus for success, then the bonus is $129,375 for achieving the $500,000 profit and $195,000 for achieving the $1 million profit. The bonus represents a 26% share of profits for the first level of success but only a 13% share for the second level.

Special forms of nonlinear schemes are often used in practice. The most common of such schemes incorporates a stipulated, fixed bonus that is paid if a certain performance standard or quota is achieved. When might such a scheme be desirable?

A quota-bonus scheme constitutes a powerful incentive if it can be set at such a level that an increase in the worker's effort substantially increases the probability of meeting the quota. To illustrate such a case, consider a firm that wants each salesman to produce $1 million in sales, and it is willing to pay up to $100,000 for this level of performance. If it pays a flat 10% commission, the salesman's incremental effort in pushing sales from $900,000 to $1 million will bring him $10,000. But if the firm offers a wage of $60,000 and a bonus of $40,000 for meeting the quota of $1 million, then this last bit of effort pushes the salesman up to his quota and earns him an extra $40,000. Thus the quota gives the salesman a much stronger incentive to make the incremental effort.

But the quota-bonus scheme is not without its drawbacks. The level at which the quota is set must be judged quite precisely. Suppose the firm misjudges and sets the quota at $1.2 million, and the salesman knows that the probability of reaching that level of sales, even with superhuman effort, is quite small. The salesman may then give up, make very little effort, and settle for earning just the base salary. The salesman's resulting sales may fall far short of even $1 million. Conversely, the pure quota-bonus scheme gives him no incentive to go beyond the $1 million level. Finally, the quota must be applied over a specific period, usually the calendar year. This requirement produces even more perverse incentives. A salesman who has bad luck in the first few months of a year will realize that he has no chance of making his quota that year, so he will take things easy for the rest of the year. If on the other hand he has very good luck and meets the quota by July, again he has no incentive to exert himself for the rest of the year. And he may be able to manipulate the scheme by conspiring with his customers to shift sales from one year to another to improve his chances of making the quota in both years. A linear scheme like the one with profit sharing described above is less open to such manipulation.

Therefore firms usually combine a quota scheme with a more graduated piecewise linear-payment scheme. For example, the salesman may get a base salary, a low rate of commission for sales between $500,000 and $1 million, a higher rate of commission for sales between $1 million and $2 million, and so on.

Managers of mutual funds, for example, are rewarded for good performance over a calendar year. These rewards come from their firm in the form of bonuses but also from the public when they invest more in those specific funds. If these reward schemes are nonlinear, the managers respond by changing the risk profile of their funds' portfolios. We saw in the Appendix to Chapter 7 that a person with a concave utility function is risk averse and one with a convex utility function is a risk lover. In the same way that a risk-loving individual prefers risky situation to safe ones, a manager facing a convex reward scheme will take excessive risk with his fund's portfolio.

B. Incentives in Teams

Rarely do the employees of a firm act as individuals on separate tasks. Salesmen working in distinct assigned regions come closest to being so separate, although even in that case the performance of an individual salesman is affected by the support of others in the office. Usually people work in teams, and the outcome for the team and for each member depends on the efforts of all. A firm's profit as a whole, for example, depends on the performance of all of its workers and managers. This interaction creates special problems for the design of incentives.

When one worker's earnings depend on the profit of the firm as whole, each worker will see only a weak link between his effort and the aggregate profit, and each will have only a small fractional share in that aggregate profit. This share is a very weak incentive for other worker's to exert effort. Even in a smaller team, each member will be tempted to shirk and become a free rider on the efforts of the others. This outcome mirrors the prisoners' dilemma of collective action we saw in the street-garden example of Chapters 3 and 4, and throughout Chapter 12. If the team is small and stable over a sufficiently long time, we can expect its members to resolve the dilemma by devising internal and perhaps nonmonetary schemes of rewards and punishments like the ones we saw in Chapter 12, Section 4.

In another context, the existence of many workers on a team can sharpen incentives. Suppose a firm has many workers performing similar tasks, perhaps selling different components from the firm's product line. If there is a common (positively correlated) random component to each worker's sales, perhaps based on the strength of the underlying economy, then the sales of one worker relative to those of another worker are a good indicator of their relative effort levels. For example, the efforts of workers 1 and 2, denoted by x_1 and x_2, might be related to their sales, y_1 and y_2, according to the formulas: $y_1 = x_1 + r$ and $y_2 = x_2 + r$, where r represents the common random error in sales (or the common "luck factor," to use the terminology of Chapter 9, Section 1.C). In this case, it follows that $y_2 - y_1 = x_2 - x_1$ with no randomness; that is, the difference in observed sales will exactly equal the difference in exerted effort across workers 1 and 2.

The firm employing these workers can then reward them according to their relative outcomes. This payment scheme entails no risk for the workers. The tradeoff we considered in Section 5, between providing optimal effort and sharing in the profits of the firm, vanishes. Now, if the first worker has a poor sales record and tries to blame it on bad luck, the firm can respond, "Then how come this other worker achieved so much more? Luck was common to the two of you, so you must have made less effort." Of course if the two workers can collude they can defeat the firm's purpose, but otherwise the firm can implement a powerful incentive scheme by setting workers in competition with each other. An extreme example of such a scheme is a tournament in which the best performer gets a prize.

Tournaments also help mitigate another potential moral-hazard problem. In reality, the criteria of success are themselves not easily or publicly observable. Then the owner of the firm may be tempted to claim that no one has performed well enough and that no one should be paid a bonus. A tournament with a prize that must be awarded to someone, or a given aggregate bonus pool that must be distributed among the workers, eliminates this moral hazard on the part of the principal.

C. Multiple Tasks and Outcomes

Employees usually perform several tasks for their employers. These various tasks lead to several measurable outcomes of employee effort. Incentives for providing effort to the different tasks then interact. And this interaction makes mechanism design more complex for the firm.

The outcome of each of an agent's tasks depends partly on the agent's effort and partly on chance. That is why an outcome-based incentive scheme generally inflicts some risk on the agent's payoff. If the chance element is small, then the risk to the agent is small and the incentive to exert effort can be made more powerful. Of course, the outcomes of different tasks are likely to be affected by chance to different extents. So if the principal considers the tasks one at a time, he will use powerful incentives for effort on the tasks that have smaller elements of chance, and weaker incentives for effort on the tasks where outcomes are more uncertain indicators of the agent's effort. But the powerful incentive on one task will divert the agent's effort away from the other task, further weakening the performance on that task. To avoid this substitution of effort toward the task with the powerful incentive, the principal has to weaken the incentive on that task, too.

An example of this can be found in our own lives. Professors are supposed to do research as well as teaching. There are many accurate indicators of good research: publications in and appointments to editorial positions for prestigious journals, elections to scientific academies, and so on. By contrast, good teaching can only be observed less accurately and with long lags. Students often need years of experience to recognize the value of what they learned in college; in the short term, they may be more impressed by showmanship than by scholarship. If these two tasks required of faculty members were considered in isolation, university administrators would attach powerful incentives to research and weaker incentives to teaching. But if they did so, professors would divert their efforts away from teaching and toward research (even more than they already do in some institutions). Therefore the imprecise observation of teaching outcomes forces deans and presidents to offer only weak incentives for research as well.

The most cited example of a situation with multiple tasks and outcomes occurs in school teaching. Some outcomes of teaching, such as test scores, are precisely observable, whereas other valuable aspects of education, such as ability to work in teams or speak in public, are less accurately measurable. If teachers are rewarded on the basis of their students' test scores, they will "teach to the test," and the other dimensions of their students' education will get ignored. Such "gaming" of an incentive scheme also extends to sports. If a baseball hitter is rewarded for hitting home runs, he will neglect other aspects of batting (taking pitches, sacrifice bunts, etc.) that can sometimes be better for his team's chances of winning a game. Similarly, salesmen may sacrifice long-term customer relationships in favor of driving home a sale to meet a short-term sales goal.

If this problem of dysfunctional effects of some incentives on other tasks is too severe, other systems of rewarding tasks may be needed. A more holistic but subjective measure of performance, for example the boss's overall evaluation, may be used. This alternative is not without its own problems; workers may then divert their effort into activities that find favor with the boss!

D. Incentives over Time

Many employment relationships last for a long time, and that opens up opportunities for the firm to devise incentive schemes where performance at one time is rewarded at a later time. Firms regularly use promotions, seniority-based salaries, and other forms of deferred compensation. In effect, workers are underpaid relative to their performance in the earlier stages of their careers with the firm and overpaid in later years. The prospect of future rewards motivates younger workers to exert good effort and also induces them to stay with the firm, thus reducing job turnover. Of course the firm may be tempted to renege on its implicit promise of overpayment in later years; therefore such schemes must be credible if they are to be effective. They are more likely to be used effectively in firms that have a long record of stability and a reputation for treating their senior workers well.

A different way that the prospect of future compensation can keep workers motivated is through the use of an "efficiency wage." The firm pays a worker more than the going wage, and the excess is a surplus, or economic rent, for the worker. So long as the worker makes good effort, he will go on earning this surplus. But if he shirks, he may be detected, at which point he will be fired and will have to go back to the general labor market, where he can earn only the going wage.

The firm faces a problem in mechanism design when it tries to determine the appropriate efficiency wage level. Suppose the going wage is w_0, and the firm's efficiency wage is $w > w_0$. Let the monetary equivalent of the worker's subjective cost of making good effort be *e*. Each pay period the worker has the choice of whether to make this effort. If the worker shirks, he saves *e*. But with probability *p*, the shirking will be detected. If it is discovered that he has been shirking, the worker will lose the surplus $(w - w_0)$, starting in the next pay period and continuing indefinitely. Let *r* be the rate of interest from one period to the next. Then if the worker shirks today, the expected discounted present value of the worker's loss in the next pay period is $p\,(w - w_0)/(1 + \mathrm{r})$. And the worker loses $w - w_0$ with probability p in all future pay periods. A calculation similar to the ones we performed for repeated games in Chapter 11 and its Appendix shows that the total expected discounted present value of the future loss to the worker is

$$p\left[\frac{w-w_0}{1+r}+\frac{w-w_0}{(1+r)^2}+\ldots\right]=p(w-w)_0\,\frac{1/(1+r)}{1-1/(1+r)}=\frac{p(w-w_0)}{r}.$$

To deter shirking, the firm needs to make sure that this expected loss is at least as high as the worker's immediate gain from shirking, *e*. Therefore the firm must pay an efficiency wage that satisfies:

$$\frac{p(w-w_0)}{r}\geq e \text{ or } w-w_0\geq {}^{er}\!/_{p} \text{ or } w\geq w_0+{}^{er}\!/_{p}.$$

The smallest efficiency wage is the one that makes this expression hold with equality. And the more accurately the firm can detect shirking, that is the higher is *p*, the smaller its excess over the going wage needs to be.

A repeated relationship may also enable the firm to design a sharper incentive scheme in another way. In any one period, as we explained above, the worker's observed outcome is a combination of the worker's effort and an element of chance. But if the outcome is poor year after year, the worker cannot credibly blame bad luck year after year. Therefore the average outcome over a long period can, by the law of large numbers, be used as a more accurate measure of the worker's average effort, and the worker can be rewarded or punished accordingly.

SUMMARY

The study of *mechanism design* can be summed up as learning "how to deal with someone who knows more than you do." Such situations occur in numerous contexts, usually in interactions involving a more-informed player, called the *agent,* and a less-informed player, called the *principal,* who wants to design a mechanism to give the agent the correct incentives to help the principal attain his goal.

Mechanism design problems are of two types. The first type involves information revelation, in which the principal creates a scheme to screen information from the agent. The second type involves moral hazard, in which the principal creates a scheme to elicit the optimal level of an observable action by the agent. In all cases, the principal attempts to maximize its own objective function subject to the incentive compatibility and participation constraints of the agent.

Firms use information-revelation schemes in creating pricing structures that separate customers by their willingness to pay for the firm's product. Procurement contracts are also often designed to separate projects, or contractors, according to various levels of cost. Evidence of both price discrimination and screening with procurement contracts can be seen in actual markets.

When facing moral hazard, employers must devise contracts that encourage their employees to provide optimal effort. Similarly, insurance companies must write policies that give their clients the right incentives to protect against the insured bad outcome's occurring. In some simple situations, optimal contracts will be linear schemes, but in the presence of more complex relationships, nonlinear schemes may be more beneficial. Incentive systems designed for workers in teams, or when relationships continue over time, are correspondingly more complex than those written for simpler situations.

KEY TERMS

agent (549)
mechanism design (543)
price discrimination (544)
principal (549)
principal-agent (agency) problem (549)

SOLVED EXERCISES

S1. Firms that provide insurance to clients to protect them from the costs associated with theft or accident must necessarily be interested in the behavior of their policyholders. Sketch some ideas for the creation of an incentive scheme that such a firm might use to deter and detect fraud or lack of care on the part of its policyholders.

S2. Some firms sell goods and services either singly or in bundles in order to increase their own profit by separating consumers with different preferences.

(a) List three examples of quantity discounts offered by firms.

(b) How do quantity discounts allow firms to screen consumers by their preferences?

S3. Omniscient Wireless Limited (OWL) is planning to roll out a new nationwide, broadband, wireless telephone service next month. The firm has conducted market research indicating that its 10 million potential consumers are in two segments, which they call the Light segment and the Regular segment. Light users have less demand for wireless-phone service and in particular they seem unlikely to have any value for more than 300 minutes of calls per month. Regular users have more demand for mobile-phone service generally and have high value for more than 300 minutes per month. OWL analysts have determined that the best plans to offer to consumers entail 300 minutes per month and 600 minutes per month, respectively. They estimate that 50% of users are Light and 50% are Regular, and that each type has the following willingness to pay for each type of service:

	300 minutes	600 minutes
Light user (50%)	$20	$30
Regular user (50%)	$25	$70

OWL's cost per additional minute of wireless service is negligible, so the subscription cost to the company is $10 per user, no matter which plan the user chooses.

Each potential customer calculates the net payoff (benefit *minus* price) that she would get from each of the usage plans and buys the plan that would give the higher net payoff, so long as this payoff is not negative. If both plans give equal, nonnegative net payoffs for a buyer, she goes for 600 minutes; if both plans have negative net payoffs for a buyer, she does not purchase. OWL wants to maximize its expected profit per potential customer.

(a) Suppose the firm were to offer only the 300-minute plan, but not the 600-minute plan. What would be the optimal price to charge, and what would be the average profit per potential customer?

(b) Suppose instead that the firm were to offer only the 600-minute plan. What would be the optimal price, and what would be the average profit per potential customer?

(c) Suppose the firm wanted to offer both plans. Suppose further that it wanted the Light users to purchase the 300-minute plan and the Regular users to purchase the 600-minute plan. Write down the incentive-compatibility constraint for the Light user.

(d) Similarly, write down the incentive-compatibility constraint for the Regular user.

(e) Use the results from parts (c) and (d) to calculate the optimal pair of prices to charge for the 300-minute and 600-minute services, so that each user type will purchase its intended service plan. What would be the average profit per potential customer?

(f) Consider the outcomes described in parts (a), (b), and (e). For each of the three situations, describe whether it is a separating outcome, a pooling outcome, or a semiseparating outcome.

S4. Mictel corporation has a world monopoly on the production of personal computers. It can make two kinds of computers: low end and high end. One-fifth of the potential buyers are casual users, and the rest are intensive users.

The costs of production of the two kinds of machines, as well as the benefits gained from the two by the two types of prospective buyers, are given in the following table. All figures are in thousands of dollars.

Each type of buyer calculates the net payoff (benefit *minus* price) that

		COST	BENEFIT FOR USER TYPE	
			Casual	Intensive
PC TYPE	Low-end	1	4	5
	High-end	3	5	8

he would get from each type of machine and buys the type that would give the higher net payoff, provided that this payoff is nonnegative. If both types give equal, nonnegative net payoffs for a buyer, he goes for the high end; if both types have negative net payoff for a buyer, he does not purchase.

Mictel wants to maximize its expected profit.

(a) If Mictel were omniscient, then, when a prospective customer came along, knowing his type, the company could offer to sell him just one type of machine at a stated price, on a take-it-or-leave-it basis. What machine would Mictel offer, and at what price, to what buyer?

In fact, Mictel does not know the type of any particular buyer. It just makes its catalog available for all buyers to choose from.

(b) First, suppose the company produces just the low-end machines and sells them for price *x*. What value of *x* will maximize its profit? Why?

(c) Next, suppose Mictel produces just the high-end machines and sells them for price *y*. What value of *y* will maximize its profit? Why?

(d) Finally, suppose the company produces both types of machines, selling the low-end ones for price *x* and the high-end ones for price *y*. What incentive-compatibility constraints on *x* and *y* must the company satisfy if it wants the casual users to buy the low-end machines and the intensive users to buy the high-end machines?

(e) What participation constraints must *x* and *y* satisfy for the casual users to be willing to buy the low-end machines and for the intensive users to be willing to buy the high-end machines?

(f) Given the constraints in parts **(d)** and **(e)**, what values of *x* and *y* will maximize the expected profit when the company sells both types of machines? What is the company's expected profit from this policy?

(g) Putting it all together, decide what production and pricing policy the company should pursue.

S5. Redo Exercise **S4**, assuming that one-half of Mictel's customers are casual users.

S6. Using the insights gained in Exercises **S4** and **S5**, solve Exercise **S4** for the general case in which the proportion of casual users is *c* and the proportion of intensive users is (1 – *c*). The answers to some parts will depend on the value of *c*. In these instances, list all relevant cases and how they depend on *c*.

S7. Sticky Shoe, the discount movie theater, sells popcorn and soda at its concession counter. Cameron, Jessie, and Sean are regular patrons of Sticky Shoe, and the valuations of each for popcorn and soda are as follows

	Popcorn	Soda
Cameron	\$3.50	\$3.00
Jessica	\$4.00	\$2.50
Sean	\$1.50	\$3.50

There are 2,997 other residents of Harkinsville who see movies at Sticky Shoe. One-third of them have valuations identical to Cameron, one-third to Jessica, and one-third to Sean. If a customer is indifferent between buying and not, she buys. It costs Sticky Shoe essentially nothing to produce each additional order of popcorn or soda.

(a) If Sticky Shoe sets separate prices for popcorn and soda, what price should it set for each concession to maximize its profit? How much profit does Sticky Shoe make selling concessions separately?

(b) What does each type of customer (Cameron, Jessica, Sean) buy when Sticky Shoe sets separate profit-maximizing prices for popcorn and soda?

(c) Instead of selling the concessions separately, Sticky Shoe decides always to sell the popcorn and soda together in a combo, charging a single price for both. What single combo price would maximize its profit? How much profit does Sticky Shoe make selling only combos?

(d) What does each type of customer buy when Sticky Shoe sets a single profit-maximizing price for a popcorn and soda combo? How does this compare with the answer in part **(b)?**

(e) Which pricing scheme does each customer type prefer? Why?

(f) If Sticky Shoe sold the concessions both as a combo and separately, which products (popcorn, soda, or the combo) does it want to sell to each customer type? How can Sticky Shoe make sure that each customer type purchases exactly the product that it intends for him or her to purchase?

(g) What prices—for the popcorn, soda, and combo—would Sticky Shoe set to maximize its profit? How much profit does Sticky Shoe make selling the concessions at these three prices?

(h) How do the answers to parts **(a)**, **(c)**, and **(g)** differ? Explain why.

S8. Section 5.A of this chapter discusses the principal-agent problem in the context of a company deciding whether and how to induce a manager to put in high effort to increase the chances that the project succeeds. The value of a successful project is $1 million; the probability of success given high effort is 0.5; the probability of success given low effort is 0.25. The manager's utility is the square root of compensation (measured in millions of dollars), and his disutility from exerting high effort is 0.1. However, the reservation wage of the manager is now $160,000.

(a) What contract does the company offer if it wants only low effort from the manager?

(b) What is the expected profit to the company when it induces low managerial effort?

(c) What contract pair (y, x)—where y is the salary given for a successful project and x is the salary given for a failed project—should the company offer the manager to induce high effort?

(d) What is the company's expected profit when it induces high effort?

(e) Which level of effort does the company want to induce from its manager? Why?

S9. A company has purchased fire insurance for its main factory. The probability of a fire in the factory without a fire-prevention program is 0.01. The probability of a fire in a factory with a fire-protection program is 0.001. If a fire occurred, the value of the loss would be $300,000. A fire-prevention program would cost $80 to run, but the insurance company cannot costlessly observe whether or not the prevention program has been implemented.

(a) Why does moral hazard arise in this situation? What is its source?

(b) Can the insurance company eliminate the moral hazard problem? If so, how? If not, explain why not.

S10. Mozart moved from Salzburg to Vienna in 1781, hoping for a position at the Habsburg court. Instead of applying for a position, he waited for the emperor to call him, because "if one makes any move oneself, one receives less pay." Discuss this situation using the theory of games with asymmetric information, including theories of signaling and screening.

S11. **(Optional, requires calculus)** You are Oceania's Minister for Peace, and it is your job to purchase war materials for your country. The net benefit, measured in Oceanic dollars, from quantity Q of these materials is $2Q^{1/2} - M$, where M is the amount of money paid for the materials.

There is just one supplier—Baron Myerson's Armaments (BMA). You do not know BMA's cost of production. Everyone knows that BMA's cost per unit of output is constant, and that it is equal to 0.10 with probability $p = 0.4$ and equal to 0.16 with probability $1 - p$. Call BMA "low cost" if its unit cost is 0.10 and "high cost" if it is 0.16. Only BMA knows its true cost type with certainty.

In the past, your ministry has used two kinds of purchase contracts: cost plus and fixed price. But cost-plus contracts create an incentive for BMA to overstate its costs, and fixed-price contracts may compensate the firm more than is necessary. You decide to offer a menu of two possibilities:

Contract 1: Supply us quantity Q_1, and we will pay you money M_1.
Contract 2: Supply us quantity Q_2, and we will pay you money M_2.

The idea is to set Q_1, M_1, Q_2, and M_2 such that a low-cost BMA will find contract 1 more profitable, and a high-cost BMA will find contract 2 more profitable. If another contract is exactly as profitable, a low-cost BMA will choose contract 1, and a high-cost BMA will choose contract 2. Further, regardless of its cost, BMA will need to receive at least zero economic profit in any contract it accepts.

(a) Write expressions for the profit of a low-cost BMA and a high-cost BMA when it supplies quantity Q and is paid M.

(b) Write the incentive-compatibility constraints to induce a low-cost BMA to select contract 1 and a high-cost BMA to select contract 2.

(c) Give the participation constraints for each type of BMA.

(d) Assuming that each of the BMA types chooses the contract designed for it, write the expression for Oceania's expected net benefit.

Now your problem is to choose Q_1, M_1, Q_2, and M_2 to maximize the expected net benefit found in part (d) subject to the incentive-compatibility (IC) and participation constraints (PC).

(e) Assume that $Q_1 > Q_2$, and further assume that constraints IC_1 and PC_2 bind—that is, they will hold with equalities instead of weak inequalities. Use these constraints to derive lower bounds on your feasible choices of M_1 and M_2 in terms of Q_1 and Q_2.

(f) Show that when IC_1 and PC_2 bind, IC_2 and PC_1 are automatically satisfied.

(g) Substitute out for M_1 and M_2, using the expressions found in part (e) to express your objective function in terms of Q_1, and Q_2.

(h) Write the first-order conditions for the maximization, and solve them for Q_1, and Q_2.

(i) Solve for M_1 and M_2.

(j) What is Oceania's expected net benefit from offering this menu of contracts?

(k) What general principles of screening are illustrated in the menu of contracts you found?

S12. **(Optional)** Revisit Oceania's problem in Exercise **S11** to see how the optimal menu found in that problem compares with some alternative contracts.

(a) If you decided to offer a single fixed-price contract that was intended to attract only the low-cost BMA, what would it be? That is, what single (Q, M) pair would be optimal if you knew BMA was low cost?

(b) Would a high-cost BMA want to accept the contract offered in part (a)? Why or why not?

(c) Given the probability that BMA is low cost, what would the expected net benefit to Oceania be from offering the contract in part (a)? How does this compare with the expected net benefit from offering a menu of contracts, as found in part (j) of Exercise **S11**?

(d) What single fixed-price contract would you offer to a high-cost BMA?

(e) Would a low-cost BMA want to accept the contract found in part (d)? What would its profit be if it did?

(f) Given your answer in part (e), what would be the expected net benefit to Oceania from offering the contract in part (d)? How does this compare with the expected net benefit from offering a menu of contracts, found in part (j) of Exercise **S11**?

(g) Consider the case in which an industrial spy within BMA has promised to divulge the true per-unit cost, so that Oceania could offer the optimal single, fixed-price contract geared toward BMA's true type. What would Oceania's expected net benefit be if it knew that it was going to learn BMA's true type? How does this compare with parts (c) and (f) of this exercise, and with part (j) of Exercise **S11**?

UNSOLVED EXERCISES

U1. What problems of moral hazard and/or adverse selection arise in your dealings with each of the following? In each case, outline some appropriate incentive schemes and/or signaling and screening strategies to cope with these problems. No mathematical analysis is expected, but you should state clearly the economic reasoning of why and how your suggested methods work.

(a) Your financial adviser tells you what stocks to buy or sell.

(b) You consult a realtor when you are selling your house.

(c) You visit your doctor, whether for routine check-ups or treatments.

U2. MicroStuff is a software company that sells two popular applications, WordStuff and ExcelStuff. It doesn't cost anything for MicroStuff to make each additional copy of its applications. MicroStuff has three types of potential customers, represented by Ingrid, Javiera, and Kathy. There are 100 million potential customers of each type, whose valuations for each application are as follows:

	WordStuff	ExcelStuff
Ingrid	100	20
Javiera	30	100
Kathy	80	0

(a) If MicroStuff sets separate prices for WordStuff and ExcelStuff, what price should it set for each application to maximize its profit? How much profit does MicroStuff earn with these prices?

(b) What does each type of customer (Ingrid, Javiera, Kathy) buy when MicroStuff sets profit-maximizing, separate prices for WordStuff and ExcelStuff?

(c) Instead of selling the applications separately, MicroStuff decides always to sell WordStuff and ExcelStuff together in a bundle, charging a single price for both. What single price for the bundle would maximize its profit? How much profit does MicroStuff make selling its software only in bundles?

(d) What does each type of customer buy when MicroStuff sets a single, profit-maximizing price for a bundle of WordStuff and ExcelStuff? How does this compare with the answer in part (b)?

(e) Which pricing scheme does each customer type prefer? Why?

(f) If MicroStuff sold the applications both as a bundle and separately, which products (WordStuff, ExcelStuff, or the bundle) would it want to sell to each customer type? How can MicroStuff make sure that each customer type purchases exactly the product that it intends for them to purchase?

(g) What prices—for WordStuff, ExcelStuff, and the bundle—would MicroStuff set to maximize its profit? How much profit does MicroStuff make selling the products with these three prices?

(h) How do the answers to parts (a), (c), and (g) differ? Explain why.

U3. Consider a managerial effort example similar to the one in Section 5. The value of a successful project is $420,000; the probabilities of success are 1/2 with good supervision and 1/4 without. The manager is risk-neutral, not risk-averse as in the text, so his expected utility equals his expected income minus his disutility of effort. He can get other jobs paying $90,000, and his disutility for exerting the extra effort for good supervision on your project is $100,000.

(a) Show that inducing high effort would require the firm to offer a compensation scheme with a negative base salary; that is, if the project fails, the manager pays the firm an amount stipulated in the scheme.

(b) How might a negative base salary be implemented in reality?

(c) Show that if a negative base salary is not feasible, then the firm does better to settle for the low-pay, low-effort situation.

U4. Cheapskates is a very minor-league professional hockey team. Its facilities are large enough to accommodate all of the 1,000 fans who might want to watch its home games. It can provide two types of seats—ordinary and luxury. There are also two sorts of fans: 60% of the fans are blue-collar fans, and the rest are white-collar fans. The costs of providing each type of seat and the fans' willingness to pay for each type of seat are given in the following table (measured in dollars).

		COST	Willingness to Pay	
			Blue-Collar	White-Collar
SEAT TYPE	Ordinary	4	12	14
	Luxury	8	15	22

Each fan will buy at most one seat, depending on the consumer surplus he would get (maximum willingness to pay minus the actual price paid) from the two kinds. If the surplus for both is negative, then he won't buy any. If at least one kind gives him nonnegative surplus, then he will buy the kind that gives him the larger surplus. If the two kinds give him equal, nonnegative surplus, then the blue-collar fan will buy the ordinary kind of seat, and the white-collar fan will buy the luxury kind.

The team owners provide and price their seating to maximize profit, measured in thousands of dollars per game. They set prices for each kind of seat, sell as many tickets as are demanded at these prices, and then provide the numbers and types of seats of each kind for which the tickets have sold.

(a) First, suppose the team owners can identify the type of each individual fan who arrives at the ticket window (presumably by the color of his collar) and can offer him just one type of seat at a stated price, on a take-it-or-leave-it basis. What is the owners' maximum profit, π^*, under this system?

(b) Now, suppose that the owners cannot identify any individual fan, but they still know the proportion of blue-collar fans. Let the price of an ordinary seat be X and the price of a luxury seat be Y. What are the incentive-compatibility constraints that will ensure that the blue-collar fans buy the ordinary seats and the white-collar fans buy the luxury seats? Graph these constraints on an X-Y coordinate plane.

(c) What are the participation constraints for the fans' decisions on whether to buy tickets at all? Add these constraints to the graph in part (b).

(d) Given the constraints in parts (b) and (c), what prices X and Y maximize the owners' profit, π_2, under this price system? What is π_2?

(e) The owners are considering whether to set prices so that only the white-collar fans will buy tickets. What is their profit, π_w, if they decide to cater to only the white-collar fans?

(f) Comparing π_2 and π_w, determine the pricing policy that the owners will set. How does their profit achieved from this policy compare with the case of full information, where they earn π^*?

(g) What is the "cost of coping with the information asymmetry" in part (f)? Who bears this cost? Why?

U5. Redo Exercise **U4** above, assuming that 10% of the fans are blue-collar.

U6. Using the insights you gained in Exercises **U4** and **U5**, solve Exercise **U4** for the general case where a fraction *B* of the fans are blue collar and fraction (1 − *B*) are white collar. The answers to some parts will depend on the value of *B*. In these instances, list all relevant cases and how they depend on *B*.

U7. In many situations, agents exert effort in order to get promoted to a better-paid position, where the reward for that position is fixed and where agents compete among themselves for those positions. Tournament theory considers a group of agents competing for a fixed set of prizes. In this case, all that matters for winning is one's positions relative to others, rather than one's absolute level of performance.

(a) Discuss the reasons that a firm might wish to employ the tournament scheme described above. Consider the effects on the incentives of both the firm and the workers.

(b) Discuss the reasons that a firm might *not* wish to employ the tournament scheme described above.

(c) State one specific prediction of tournament theory and provide an example of empirical evidence in support of that prediction.

U8. Repeat Exercise **S8** with the following adjustments: Due to the departure of some of their brightest engineers, the probability of success given a high managerial effort is only 0.4, and the probability of success given a low managerial effort is reduced to 0.24.

U9. **(Optional)** A teacher wants to find out how confident the students are about their own abilities. He proposes the following scheme: "After you answer this question, state your estimate of the probability that you are right. I will then check your answer to the question. Suppose you have given the probability estimate *x*. If your answer is actually correct, your grade will be $\log(x)$. If incorrect, it will be $\log(1 - x)$." Show that this scheme will elicit the students' own truthful estimates—that is, if the truth is *p*, show that a student's stated estimate $x = p$.

U10. **(Optional)** Redo Exercise **S11**, but assume that the probability that BMA is low cost is 0.6.

U11. **(Optional)** Repeat Exercise **S11**, but assume that a low-cost BMA has a per-unit cost of 0.2, and a high-cost BMA has a per-unit cost of 0.38. Let the probability that BMA is low cost be 0.4.

U12. **(Optional)** Revisit the situation in which Oceania is procuring arms from BMA. (See Exercise **S11**.) Now consider the case in which BMA has three possible cost types: c_1, c_2, and c_3, where $c_3 > c_2 > c_1$. BMA has cost c_1 with probability p_1, cost c_2 with probability p_2, and cost c_3 with probability p_3, where $p_1 + p_2 + p_3 = 1$. In what follows, we will say that BMA is of type i if its cost is c_i, for $i = 1, 2, 3$.

You offer a menu of three possibilities: "Supply us quantity Q_i, and we will pay you M_i," for $i = 1$, 2, and 3. Assume that more than one contract is equally profitable, so that a BMA of type i will choose contract i. To meet the participation constraint, contract i should give BMA of type i nonnegative profit.

(a) Write an expression for the profit of type-i BMA when it supplies quantity Q and is paid M.

(b) Give the participation constraints for each BMA type.

(c) Write the six incentive-compatibility constraints. That is, for each type i give separate expressions that state that the profit that BMA receives under contract i is greater than or equal to the profit it would receive under the other two contracts.

(d) Write down the expression for Oceania's expected net benefit, B. This is the objective function (what you want to maximize).

Now your problem is to choose the three Q_i and the three M_i to maximize expected net benefit, subject to the incentive-compatibility (IC) and participation constraints (PC).

(e) Begin with just three constraints: the IC constraint for type 2 to prefer contract 2 over contract 3, the IC constraint for type 1 to prefer contract 1 over contract 2, and the participation constraint for type 3. Assume that $Q_1 > Q_2 > Q_3$. Use these constraints to derive lower bounds on your feasible choices of M_1, M_2, M_3 in terms of c_1, c_2, and c_3 and Q_1, Q_2, and Q_3. (Note that two or more of the *c*s and *Q*s may appear in the expression for the lower bound for each of the *M*s.)

(f) Prove that these three constraints—the two ICs and one PC in part (e)—will be binding at the optimum.

(g) Now prove that when the three constraints in part (e) are binding, the other six constraints (the remaining four ICs and two PCs) are automatically satisfied.

(h) Substitute out for the M_i to express your objective function in terms of the three Q_i only.

(i) Write the first-order conditions for the maximization, and solve for each of the Q_i. That is, take the three partial derivatives $\partial Q_i / \partial B$, set them equal to zero, and solve for Q_i.

(j) Show that the assumption made above, $Q_1 > Q_2 > Q_3$, will be true at the optimum if:

$$\frac{c_3 - c_2}{c_2 - c_1} > \frac{p_1 p_3}{p_2}.$$

PART FOUR

Applications to Specific Strategic Situations

Brinkmanship

The Cuban Missile Crisis

In Chapter 1, we explained that our basic approach was neither pure nor pure case study, but a combination in which theoretical ideas were developed by using features of particular cases or examples. Thus we ignored those aspects of each case that were incidental to the concept being developed. However, after you have learned the theoretical ideas, a richer mode of analysis becomes available to you in which factual details of a particular case are more closely integrated with game-theoretic analysis to achieve a fuller understanding of what has happened and why. Such *theory-based case studies* have begun to appear in diverse fields—business, political science, and economic history.[1]

Here we offer an example from political and military history—namely, nuclear brinkmanship in the Cuban missile crisis of 1962. Our choice is motivated by the sheer drama of the episode, the wealth of factual information that has become available, and the applicability of an important concept from game theory. You may think that the risk of nuclear war died with the dissolution of the Soviet Union and that therefore our case is a historical curiosity. But nuclear arms races continue in many parts of the world, and such rivals as India and Pakistan or Iran and Israel may find use for the lessons

[1]Two excellent examples of theory-based studies are Pankaj Ghemawat, *Games Businesses Play: Cases and Models* (Cambridge: MIT Press, 1997), and Robert H. Bates, Avner Greif, Margaret Levi, Jean-Laurent Rosenthal, and Barry Weingast, *Analytic Narratives* (Princeton: Princeton University Press, 1998).

taken from the Cuban crisis. More important for many of you, brinkmanship must be practiced in many everyday situations—for example, strikes and marital disputes. Although the stakes in such games are lower than those in a nuclear confrontation between superpowers, the same principles of strategy apply.

In Chapter 10, we introduced the concept of brinkmanship as a strategic move; here is a quick reminder. A *threat* is a response rule, and the threatened action inflicts a cost on both the player making the threat and the player whose action the threat is intended to influence. However, if the threat succeeds in its purpose, this action is not actually carried out. Therefore there is no apparent upper limit to the cost of the threatened action. But the risk of *errors*—that is, the risk that the threat may fail to achieve its purpose or that the threatened action may occur by accident—forces the strategist to use the minimal threat that achieves its purpose. If a smaller threat is not naturally available, a large threat can be scaled down by making its fulfillment probabilistic. You do something in advance that creates a probability, but not certainty, that the mutually harmful outcome will happen if the opponent defies you. If the need actually arose, you would not take that bad action if you had the full freedom to choose. Therefore you must arrange in advance to let things get out of your control to some extent. *Brinkmanship* is the creation and deployment of such a probabilistic threat; it consists of a deliberate loss of control.

The word *brinkmanship* is often used in connection with nuclear weapons. The Cuban missile crisis of 1962, when the world came as close to an unaccidental nuclear war as it ever has, is often offered as the classic example of brinkmanship. We will use the crisis as an extended case study to explicate the concept. In the process, we will find that many popular interpretations and analyses of the crisis are simplistic. A deeper analysis reveals brinkmanship to be a subtle and dangerous strategy. It also shows that many conflicts in business and personal interactions—such as strikes and breakups of relationships—are examples of brinkmanship gone wrong. Therefore a clear understanding of the strategy, as well as its limitations and risks, is very important to all game players, which includes just about everyone.

1 A BRIEF NARRATIVE OF EVENTS

We begin with a brief story of the unfolding of the crisis. Our account draws on several books, including some that were written with the benefit of documents and statements released since the collapse of the Soviet Union.[2] We cannot hope to do justice to the detail, let alone the drama, of the events. President Kennedy said at the time of the crisis: "This is the week when I earn my salary." Much more than a president's salary stood in the balance. We urge you to read the books that tell the story in vivid detail and to talk to any relatives who lived through it to get their firsthand memories.[3]

In late summer and early fall of 1962, the Soviet Union (USSR) started to place medium- and intermediate-range ballistic missiles (MRBMs and IRBMs) in Cuba. The MRBMs had a range of 1,100 miles and could hit Washington, DC; the IRBMs, with a range of 2,200 miles, could hit most of the major U.S. cities and military installations. The missile sites were guarded by the latest Soviet SA-2-type surface-to-air missiles (SAMs), which could shoot down U.S. high-altitude U-2 reconnaissance planes. There were also IL-28 bombers and tactical nuclear weapons called Luna by the Soviets and FROG (free rocket over ground) by the United States, which could be used against invading troops.

This was the first time that the Soviets had ever attempted to place their missiles and nuclear weapons outside Soviet territory. Had they been successful, it would have increased their offensive capability against the United States manyfold. It is now believed that the Soviets had fewer than 20, and perhaps as few as "two or three," operational intercontinental ballistic missiles (ICBMs) in their own country capable of reaching the United States (*War*, 464, 509–510). Their initial placement in Cuba had about 40 MRBMs and IRBMs, which was a substantial increase. But the United States would still have retained vast superiority in the nuclear balance between the superpowers. Also, as the Soviets built up their submarine fleet, the relative importance of land-based missiles near

[2]Our sources include Robert Smith Thompson, *The Missiles of October* (New York: Simon & Schuster, 1992); James G. Blight and David A. Welch, *On the Brink: Americans and Soviets Reexamine the Cuban Missile Crisis* (New York: Hill and Wang, 1989); Richard Reeves, *President Kennedy: Profile of Power* (New York: Simon & Schuster, 1993); Donald Kagan, *On the Origins of War and the Preservation of Peace* (New York: Doubleday, 1995); Aleksandr Fursenko and Timothy Naftali, *One Hell of a Gamble: The Secret History of the Cuban Missile Crisis* (New York: Norton, 1997); and last, latest, and most direct, *The Kennedy Tapes: Inside the White House During the Cuban Missile Crisis*, ed. Ernest R. May and Philip D. Zelikow (Cambridge: Harvard University Press, 1997). Graham T. Allison's *Essence of Decision: Explaining the Cuban Missile Crisis* (Boston: Little Brown, 1971) remains important not only for its narrative, but also for its analysis and interpretation. Our view differs from his in some important respects, but we remain in debt to his insights. We follow and extend the ideas in Avinash Dixit and Barry Nalebuff, *Thinking Strategically* (New York: Norton, 1991), chap. 8.

When we cite these sources to document particular points, we do so in parentheses in the text, in each case using a key word from the title of the book followed by the appropriate page number or page range. The key words have been underlined in the sources given here.

[3]For those of you with no access to firsthand information or those who seek a beginner's introduction to both the details and the drama of the missile crisis, we recommend the film *Thirteen Days* (2000, New Line Cinema).

the United States would have decreased. But the missiles had more than mere direct military value to the Soviets. Successful placement of missiles so close to the United States would have been an immense boost to Soviet prestige throughout the world, especially in Asia and Africa, where the superpowers were competing for political and military influence. Finally, the Soviets had come to think of Cuba as a "poster child" for socialism. The opportunity to deter a feared U.S. invasion of Cuba and to counter Chinese influence in Cuba weighed importantly in the calculations of the Soviet leader and Premier, Nikita Khrushchev. (See *Gamble,* 182–183, for an analysis of Soviet motives.)

U.S. surveillance of Cuba and of shipping lanes during the late summer and early fall of 1962 had indicated some suspicious activity. When questioned about it by U.S. diplomats, the Soviets denied any intentions to place missiles in Cuba. Later, faced with irrefutable evidence, they said that their intention was defensive, to deter the United States from invading Cuba. It is hard to believe this, although we know that an offensive weapon *can* serve as a defensive deterrent threat.

An American U-2 "spy plane" took photographs over western Cuba on Sunday and Monday, October 14 and 15. When developed and interpreted, they showed unmistakable signs of construction on MRBM launching sites. (Evidence of IRBMs was found later, on October 17.) These photographs were shown to President Kennedy the following day (October 16). He immediately convened an ad hoc group of advisers, which later came to be called the Executive Committee of the National Security Council (ExComm), to discuss the alternatives. At the first meeting (on the morning of October 16), he decided to keep the matter totally secret until he was ready to act, mainly because if the Soviets knew that the Americans knew, they might speed up the installation and deployment of the missiles before the Americans were ready to act, but also because spreading the news without announcing a clear response would create panic in the United States.

Members of ExComm who figured most prominently in the discussions were the Secretary of Defense, Robert McNamara; the National Security Adviser, McGeorge Bundy; the Chairman of the Joint Chiefs of Staff, General Maxwell Taylor; the Secretary of State, Dean Rusk, and Undersecretary George Ball; the Attorney General Robert Kennedy (who was also the President's brother); the Secretary of the Treasury, Douglas Dillon (also the only Republican in the Cabinet); and Llewellyn Thompson, who had recently returned from being U.S. Ambassador in Moscow. During the two weeks that followed, they would be joined by or would consult with several others, including the U.S. Ambassador to the United Nations, Adlai Stevenson; the former Secretary of State and a senior statesman of U.S. foreign policy, Dean Acheson; and the Chief of the U.S. Air Force, General Curtis LeMay.

In the rest of that week (October 16 through 21), the ExComm met numerous times. To preserve secrecy, the President continued his normal schedule, including travel to speak for Democratic candidates in the midterm congressional elections that were to be held in November 1962. He kept in constant touch with ExComm. He dodged press questions abut Cuba and persuaded one or two trusted media owners or editors to preserve the facade of business as usual. ExComm's own attempts to preserve secrecy in Washington sometimes verged on the comic, as when almost a dozen of them had to pile into one limo, because the sight of several government cars going from the White House to the State Department in a convoy could cause speculation in the media.

Different members of ExComm had widely differing assessments of the situation and supported different actions. The military Chiefs of Staff thought that the missile placement changed the balance of military power substantially; Defense Secretary McNamara thought it changed "not at all" but regarded the problem as politically important nonetheless (*Tapes,* 89). President Kennedy pointed out that the first placement, if ignored by the United States, could grow into something much bigger and that the Soviets could use the threat of missiles so close to the United States to try to force the withdrawal of the U.S., British, and French presence in West Berlin. Kennedy was also aware that it was a part of the *geopolitical* struggle between the United States and the Soviet Union (*Tapes,* 92).

It now appears that he was very much on the mark in this assessment. The Soviets planned to expand their presence in Cuba into a major military base (*Tapes,* 677). They expected to complete the missile placement by mid-November. Khrushchev had planned to sign a treaty with Castro in late November, then travel to New York to address the United Nations and issue an ultimatum for a settlement of the Berlin issue (*Tapes,* 679; *Gamble,* 182), using the missiles in Cuba as a threat for this purpose. Khrushchev thought Kennedy would accept the missile placement as a *fait accompli.* Khrushchev appears to have made these plans on his own. Some of his top advisers privately thought them too adventurous, but the top governmental decision-making body of the Soviet Union, the Presidium, supported him, although its response was largely a rubber stamp (*Gamble,* 180). Castro was at first reluctant to accept the missiles, fearing that they would trigger a U.S. invasion (*Tapes,* 676–678), but in the end he, too, accepted them. The prospect gave him great confidence and lent some swagger to his statements about the United States (*Gamble,* 186–187, 229–230).

In all ExComm meetings up to and including the one on the morning of Thursday, October 18, everyone appears to have assumed that the U.S. response would be purely military. The only options that they discussed seriously during this time were (1) an air strike directed exclusively at the missile sites and (probably) the SAM sites nearby, (2) a wider air strike including Soviet and Cuban aircraft parked at airfields, and (3) a full-scale invasion of Cuba. If anything, attitudes hardened when the evidence of the presence of the longer-range

IRBMs arrived. In fact, at the Thursday meeting, Kennedy discussed a timetable for air strikes to commence that weekend (*Tapes*, 148).

McNamara had first mentioned a blockade toward the end of the meeting on Tuesday, October 16, and developed the idea (in a form uncannily close to the course of action actually taken) in a small group after the formal meeting had ended (*Tapes*, 86, 113). Ball argued that an air strike without warning would be a "Pearl Harbor" and that the United States should not do it (*Tapes*, 115); he was most importantly supported by Robert Kennedy (*Tapes*, 149). The civilian members of ExComm further shifted toward the blockade option when they found that what the military Joint Chiefs of Staff wanted was a massive air strike; the military regarded a limited strike aimed at only the missile sites so dangerous and ineffective that "they would prefer taking no military action than to take that limited strike" (*Tapes*, 97).

Between October 18 and Saturday, October 20, the majority opinion within ExComm gradually coalesced around the idea of starting with a blockade, simultaneously issuing an ultimatum with a short deadline (from 48 to 72 hours was mentioned), and proceeding to military action if necessary after this deadline expired. International law required a declaration of war to set up a blockade, but this problem was ingeniously resolved by proposing to call it a "naval quarantine" of Cuba (*Tapes*, 190–196).

Some people held the same positions throughout these discussions (from October 16 through 21)—for example, the military Chiefs of Staff constantly favored a major air strike—but others shifted their views, at times dramatically. Bundy initially favored doing nothing (*Tapes*, 172) and then switched toward a preemptive surprise air attack (*Tapes*, 189). President Kennedy's own positions also shifted away from an air strike toward a blockade. He wanted the U.S. response to be firm. Although his reasons undoubtedly were mainly military and geopolitical, as a good domestic politician he was also fully aware that a weak response would hurt the Democratic party in the imminent congressional elections. On the other hand, the responsibility of starting an action that might lead to nuclear war weighed very heavily on him. He was impressed by the CIA's assessment that some of the missiles were already operational, which increased the risk that any air strike or invasion could lead to the Soviets' firing these missiles and to large U.S. civilian casualties (*Gamble*, 235). In the second week of the crisis (October 22 through 28), his decisions seemed constantly to favor the lowest-key options discussed by ExComm.

By the end of the first week's discussions, the choice lay between a blockade and an air strike, two position papers were prepared, and in a straw vote on October 20 the blockade won 11 to 6 (*War*, 516). Kennedy made the decision to start by imposing a blockade and announced it in a television address to the nation on Monday, October 22. He demanded a halt to the shipment of Soviet missiles to Cuba and a prompt withdrawal of those already there.

Kennedy's speech brought the whole drama and tension into the public arena. The United Nations held several dramatic but unproductive debates. Other world leaders and the usual busybodies of international affairs offered advice and mediation.

Between October 23 and October 25, the Soviets at first tried bluster and denial; Khrushchev called the blockade "banditry, a folly of international imperialism" and said that his ships would ignore it. The Soviets, in the United Nations and elsewhere, claimed that their intentions were purely defensive and issued statements of defiance. In secret, they explored ways to end the crisis. This exploration included some direct messages from Khrushchev to Kennedy. It also included some very indirect and lower-level approaches by the Soviets. In fact, as early as Monday, October 22—before Kennedy's TV address—the Soviet Presidium had decided not to let this crisis lead to war. By Thursday, October 25, they had decided that they were willing to withdraw from Cuba in exchange for a promise by the United States not to invade Cuba, but they had also agreed to "look around" for better deals (*Gamble,* 241, 259). The United States did not know any of the Soviet thinking about this.

In public as well as in private communications, the USSR broached the possibility of a deal concerning the withdrawal of U.S. missiles from Turkey and of Soviet ones from Cuba. This possibility had already been discussed by ExComm. The missiles in Turkey were obsolete; so the United States wanted to remove them anyway and replace them with a Polaris submarine stationed in the Mediterranean Sea. But it was thought that the Turks would regard the presence of U.S. missiles as a matter of prestige and so it might be difficult to persuade them to accept the change. (The Turks might also correctly regard missiles, fixed on Turkish soil, as a firmer signal of the U.S. commitment to Turkey's defense than an offshore submarine, which could move away on short notice; see *Tapes,* 568.)

The blockade went into effect on Wednesday, October 24. Despite their public bluster, the Soviets were cautious in testing it. Apparently, they were surprised that the United States had discovered the missiles in Cuba before the whole installation program was completed; Soviet personnel in Cuba had observed the U-2 overflights but had not reported them to Moscow (*Tapes,* 681). The Soviet Presidium ordered the ships carrying the most sensitive materials (actually the IRBM missiles) to stop or turn around. But it also ordered General Issa Pliyev, the commander of the Soviet troops in Cuba, to get his troops combat-ready and to use all means except nuclear weapons to meet any attack (*Tapes,* 682). In fact, the Presidium twice prepared (then canceled without sending) orders authorizing him to use tactical nuclear weapons in the event of a U.S. invasion (*Gamble,* 242–243, 272, 276). The U.S. side saw only that several Soviet ships (which were actually carrying oil and other nonmilitary cargo) continued to sail toward the blockade zone. The U.S. Navy showed some moderation in its enforcement of the blockade. A tanker was allowed to pass without being boarded; another tramp steamer carrying industrial cargo was boarded but allowed to proceed after only a cursory inspection. But tension was mounting, and neither side's actions were as cautious as the top-level politicians on both sides would have liked.

On the morning of Friday, October 26, Khrushchev sent Kennedy a conciliatory private letter offering to withdraw the missiles in exchange for a U.S. promise not to invade Cuba. But later that day he toughened his stance. It seems that he was emboldened by two items of evidence. First, the U.S. Navy was not being excessively aggressive in enforcing the blockade. It had let through some obviously civilian freighters; they boarded only one ship, the *Marucla,* and let it pass after a cursory inspection. Second, some dovish statements had appeared in U.S. newspapers. Most notable among them was an article by the influential and well-connected syndicated columnist Walter Lippman, who suggested the swap whereby the United States would withdraw its missiles in Turkey in exchange for the USSR's withdrawing its missiles in Cuba (*Gamble,* 275). Khrushchev sent another letter to Kennedy on Saturday, October 26, offering this swap, and this time he made the letter public. The new letter was presumably a part of the Presidium's strategy of "looking around" for the best deal. Members of ExComm concluded that the first letter was Khrushchev's own thoughts but that the second was written under pressure from hard-liners in the Presidium—or was even evidence that Khrushchev was no longer in control (*Tapes,* 498, 512–513). In fact, both of Khrushchev's letters were discussed and approved by the Presidium (*Gamble,* 263, 275).

ExComm continued to meet, and opinions within it hardened. One reason was the growing feeling that the blockade by itself would not work. Kennedy's television speech had imposed no firm deadline, and, as we know, in the absence of a deadline a compellent threat is vulnerable to the opponent's procrastination. Kennedy had seen this quite clearly and as early as Monday, October 22: in the morning ExComm meeting preceding his speech, he commented, "I don't think we're gonna be better off if they're just sitting there" (*Tapes,* 216). But a hard, short deadline was presumably thought to be too rigid. By Thursday, others in ExComm were realizing the problem; for example, Bundy said, "A plateau here is the most dangerous thing" (*Tapes,* 423). The hardening of the Soviet position, as shown by the public "Saturday letter" that followed the conciliatory private "Friday letter," was another concern. More ominously, that Friday, U.S. surveillance had discovered that there were tactical nuclear weapons (FROGs) in Cuba (*Tapes,* 475). This discovery showed the Soviet presence there to be vastly greater than thought before, but it also made invasion more dangerous to U.S. troops. Also on Saturday, a U.S. U-2 plane was shot down over Cuba. (It now appears that this was done by the local commander, who interpreted his orders more broadly than Moscow had intended [*War,* 537; *Tapes,* 682].)

In addition, Cuban antiaircraft defenses fired at low-level U.S. reconnaissance planes. The grim mood in ExComm throughout that Saturday was well encapsulated by Dillon: "We haven't got but one more day" (*Tapes,* 534).

On Saturday, plans leading to escalation were being put in place. An air strike was planned for the following Monday, or Tuesday at the latest, and Air Force reserves were called up (*Tapes,* 612–613). Invasion was seen as the inevitable culmination of events (*Tapes,* 537–538). A tough private letter to Khrushchev from President Kennedy was drafted and was handed over by Robert Kennedy to the Soviet Ambassador in Washington, Anatoly Dobrynin. In it, Kennedy made the following offer: (1) The Soviet Union withdraws its missiles and IL-28 bombers from Cuba with adequate verification (and ships no new ones). (2) The United States promises not to invade Cuba. (3) The U.S. missiles in Turkey will be removed after a few months, but this offer is void if the Soviets mention it in public or link it to the Cuban deal. An answer was required within 12 to 24 hours; otherwise "there would be drastic consequences" (*Tapes,* 605–607).

On the morning of Sunday, October 28, just as prayers and sermons for peace were being offered in many churches in the United States, Soviet radio broadcast the text of a letter that Khrushchev was sending to Kennedy, in which he announced that construction of the missile sites was being halted immediately and that the missiles already installed would be dismantled and shipped back to the Soviet Union. Kennedy immediately sent a reply welcoming this decision, which was broadcast to Moscow by the Voice of America radio. It now appears that Khrushchev's decision to back down was made before he received Kennedy's letter through Dobrynin but that the letter only reinforced it (*Tapes,* 689).

That did not quite end the crisis. The U.S. Joint Chiefs of Staff remained skeptical of the Soviets and wanted to go ahead with their air strike (*Tapes,* 635). In fact, the construction activity at the Cuban missile sites continued for a few days. Verification by the United Nations proved problematic. The Soviets tried to make the Turkey part of the deal semipublic. They also tried to keep the IL-28 bombers in Cuba out of the withdrawal. Not until November 20 was the deal finally clinched and the withdrawal begun (*Tapes,* 663–665; *Gamble,* 298–310).

2 A SIMPLE GAME-THEORETIC EXPLANATION

At first sight, the game-theoretic aspect of the crisis looks very simple. The United States wanted the Soviet Union to withdraw its missiles from Cuba; thus the U.S. objective was to achieve compellence. For this purpose, the United States deployed a threat: Soviet failure to comply would eventually lead to a nuclear war between the superpowers. The blockade was a starting point of this inevitable process and an action that demonstrated the credibility of U.S. resolve. In other words, Kennedy took Khrushchev to the brink of disaster. This was sufficiently frightening to Khrushchev that he complied. The prospect of nuclear annihilation was equally frightening to Kennedy, but that is in the nature of a threat. All that is needed is that the threat be sufficiently costly to the other side to induce it to act in accordance with our wishes; then we don't have to carry out the bad action anyway.

A somewhat more formal statement of this argument proceeds by drawing a game tree like that shown in Figure 15.1. The Soviets have installed the missiles, and now the United States has the first move. It chooses between doing nothing and issuing a threat. If the United States does nothing, this is a major military and political achievement for the Soviets; so we score the payoffs as −2 for the United States and 2 for the Soviets. If the United States issues its threat, the Soviets get to move, and they can either withdraw or defy. Withdrawal is a humiliation (a substantial minus) for the Soviets and a reaffirmation of U.S. military superiority (a small plus); so we score it 1 for the United States and −4 for the Soviets. If the Soviets defy the U.S. threat, there will be a nuclear war. This outcome is terrible for both, but particularly bad for the United States, which as a democracy cares more for its citizens; so we score this −10 for the United States and −8 for the Soviets. This quantification is very rough guesswork, but the conclusions do not depend on the precise numbers that we have chosen. If you disagree with our choice, you can substitute other numbers you think to be a more accurate representation; as long as the *relative* ranking of the outcomes is the same, you will get the same subgame-perfect equilibrium.

Now we can easily find the subgame-perfect equilibrium. If faced with the U.S. threat, the Soviets get −4 from withdrawal and −8 by defiance; so they prefer to withdraw. Looking ahead to this outcome, the United States reckons on getting 1 if it issues the threat and −2 if it does not; therefore it is optimal for the

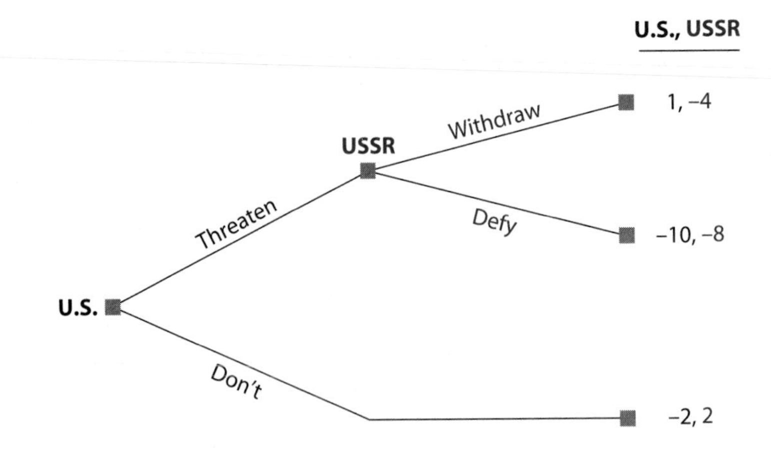

FIGURE 15.1 The Simple-Threat Model of the Crisis

United States to make the threat. The outcome gives payoffs of 1 to the United States and −4 to the Soviets.

But a moment's further thought shows this interpretation to be unsatisfactory. One might start by asking why the Soviets would deploy the missiles in Cuba at all, when they could look ahead to this unfolding of the subsequent game in which they would come out the losers. But more important, several facts about the situation and several events in the course of its unfolding do not fit into this picture of a simple threat.

Before explaining the shortcomings of this analysis and developing a better explanation, however, we digress to an interesting episode in the crisis that sheds light on the requirements of successful compellence. As pointed out in Chapter 10, a compellent threat must have a deadline; otherwise the opponent can nullify it by procrastination. The discussion of the crisis at the U.N. Security Council on Tuesday, October 23, featured a confrontation between U.S. Ambassador Adlai Stevenson and Soviet Ambassador Valerian Zorin. Stevenson asked Zorin point-blank whether the USSR had placed and was placing nuclear missiles in Cuba. "Yes or no—don't wait for the translation—yes or no?" he insisted. Zorin replied: "I am not in an American courtroom. . . . You will have your answer in due course," to which Stevenson retorted, "I am prepared to wait for my answer until hell freezes over." This was dramatic debating; Kennedy, watching the session on live television, remarked, "Terrific. I never knew Adlai had it in him" (*Profile,* 406). But it was terrible strategy. Nothing would have suited the Soviets better than to keep the Americans "waiting for their answer" while they went on completing the missile sites. "Until hell freezes over" is an unsuitable deadline for compellence.

3 ACCOUNTING FOR ADDITIONAL COMPLEXITIES

Let us return to developing a more satisfactory game-theoretic argument. As we pointed out before, the idea that a threat has only a lower limit on its size—namely, that it be large enough to frighten the opponent—is correct only if the threatener can be absolutely sure that everything will go as planned. But almost all games have some element of uncertainty. You cannot know your opponent's value system for sure, and you cannot be completely sure that the players' intended actions will be accurately implemented. Therefore a threat carries a twofold risk. Your opponent may defy it, requiring you to carry out the costly threatened action; or your opponent may comply, but the threatened action may occur by mistake anyway. When such risks exist, the cost of threatened action to oneself becomes an important consideration.

The Cuban missile crisis was replete with such uncertainties. Neither side could be sure of the other's payoffs—that is, of how seriously the other regarded the relative costs of war and of losing prestige in the world. Also, the choices of "blockade" and "air strike" were much more complex than the simple phrases suggest, and there were many weak links and random effects between an order in Washington or Moscow and its implementation in the Atlantic Ocean or in Cuba.

Graham Allison's excellent book *Essence of Decision* brings out all of these complexities and uncertainties. They led him to conclude that the Cuban missile crisis cannot be explained in game-theoretic terms. He considers two alternatives: one explanation based on the fact that bureaucracies have their set rules and procedures; another based on the internal politics of U.S. and Soviet governance and military apparatuses. He concludes that the political explanation is best.

We broadly agree but interpret the Cuban missile crisis differently. It is not the case that game theory is inadequate for understanding and explaining the crisis; rather, the crisis was *not a two-person game*—United States versus USSR, or Kennedy versus Khrushchev. Each of these two "sides" was itself a complex coalition of players with differing objectives, information, actions, and means of communication. The players within each side were engaged in other games, and some members were also directly interacting with their counterparts on the other side. In other words, the crisis can be seen as a complex many-person game with alignments into two broad coalitions. Kennedy and Khrushchev can be regarded as the top-level players in this game, but each was subject to constraints of having to deal with others in his own coalition with divergent views and information, and neither had full control over the actions of these others. We argue that this more subtle game-theoretic perspective is not only a good way to look at the crisis, but also essential in understanding how to practice brinkmanship. We begin with some items of evidence that Allison emphasizes, as well as others that emerge from other writings.

First, there are several indications of divisions of opinion on each side. On the U.S. side, as already noted, there were wide differences within ExComm. In addition, Kennedy found it necessary to consult others such as former President Eisenhower and leading members of Congress. Some of them had very different views; for example, Senator William Fulbright said in a private meeting that the blockade "seems to me the worst alternative" (*Tapes,* 271). The media and the political opposition would not give the President unquestioning support for too long either. Kennedy could not have continued on a moderate course if the opinion among his advisers and the public became decisively hawkish.

Individual people also *shifted* positions in the course of the two weeks. For example, McNamara was at first quite dovish, arguing that the missiles in Cuba were not a significant increase in the Soviet threat (*Tapes,* 89) and favoring blockade and negotiations (*Tapes,* 191), but ended up more hawkish, claiming that Khrushchev's conciliatory letter of Friday, October 26, was "full of holes" (*Tapes,* 495, 585) and urging an invasion (*Tapes,* 537). Most important, the U.S. military chiefs always advocated a far more aggressive response. Even after the crisis was over and everyone thought the United States had won a major round in the cold war, Air Force General Curtis LeMay remained dissatisfied and wanted action: "We lost! We ought to just go in there today and knock 'em off," he said (*Essence,* 206; *Profile,* 425).

Even though Khrushchev was the dictator of the Soviet Union, he was not in full control of the situation. Differences of opinion on the Soviet side are less well documented, but, for what it is worth, later memoirists have claimed that Khrushchev made the decision to install the missiles in Cuba almost unilaterally, and, when he informed them, they thought it a reckless gamble (*Tapes,* 674; *Gamble,* 180). There were limits to how far he could count on the Presidium to rubber-stamp his decisions. Indeed, two years later, the disastrous Cuban adventure was one of the main charges leveled against Khrushchev when the Presidium dismissed him (*Gamble,* 353–355). It has also been claimed that Khrushchev wanted to defy the U.S. blockade, and only the insistence of First Deputy Premier Anastas Mikoyan led to the cautious response (*War,* 521). Finally, on Saturday, October 27, Castro ordered his antiaircraft forces to fire on all U.S. planes overflying Cuba and refused the Soviet ambassador's request to rescind the order (*War,* 544).

Various parties on the U.S. side had very different information and a very different understanding of the situation, and at times this led to actions that were inconsistent with the intentions of the leadership or even against their explicit orders. The concept of an "air strike" to destroy the missiles is a good example. The nonmilitary people in ExComm thought this would be very narrowly targeted and would not cause significant Cuban or Soviet casualties, but the Air Force intended a much broader attack. Luckily, this difference came out in the open early, leading ExComm to decide against an air strike and the President to turn down an appeal by the Air Force (*Essence,* 123, 209). As for the blockade, the U.S. Navy had set procedures for this action. The political leadership wanted a different and softer process: form the ring closer to Cuba to give the Soviets more time to reconsider, allow the obviously nonmilitary cargo ships to pass unchallenged, and cripple but not sink the ships that defy challenge. Despite McNamara's explicit instructions, however, the Navy mostly followed its standard procedures (*Essence,* 130–132). The U.S. Air Force created even greater dangers. A U-2 plane drifted "accidentally" into Soviet air space and almost caused a serious setback. General Curtis LeMay, acting without the President's knowledge or authorization, ordered the Strategic Air Command's nuclear bombers to fly past their "turnaround" points and some distance toward Soviet air space to positions where they would be detected by Soviet radar. Fortunately, the Soviets responded calmly; Khrushchev merely protested to Kennedy.[4]

There was similar lack of information and communication, as well as weakness of the chain of command and control, on the Soviet side. For example, the construction of the missiles was left to the standard bureaucratic procedures. The Soviets, used to construction of ICBM sites in their own country where they did not face significant risk of air attack, laid out the sites in Cuba in a similar way, where they would have been much more vulnerable. At the height of the crisis, when the Soviet SA-2 troops saw an overflying U.S. U-2 plane on Friday, October 26, Pliyev was temporarily away from his desk and his deputy gave the order to shoot it down; this incident created far more risk than Moscow would have wished (*Gamble,* 277–288). And at numerous other points—for example, when the U.S. Navy was trying to get the freighter *Marucla* to stop and be boarded—the people involved might have set off an incident with alarming consequences by taking some action in fear of the immediate situation. Even more dramatically, it was revealed that a Soviet submarine crew, warned to surface when approaching the quarantine line on October 27, did consider firing a nuclear-tipped torpedo that it carried on board (unknown to the U.S. Navy). The firing-authorization rule required the approval of three officers, only two of whom agreed; the third officer himself may have prevented all-out nuclear war.[5]

All these factors made the outcome of any decision by the top-level commander on each side somewhat *unpredictable.* This gave rise to a substantial risk of the "threat going wrong." In fact, Kennedy thought that the chances of the blockade leading to war were "between one out of three and even" (*Essence,* 1).

As we pointed out, such uncertainty can make a simple threat too large to be acceptable to the threatener. We will take one particular form of the uncertainty—namely, U.S. lack of knowledge of the Soviets' true motives—and analyze its effect formally, but similar conclusions hold for all other forms of uncertainty.

Reconsider the game shown in Figure 15.1. Suppose the Soviet payoffs from withdrawal and defiance are the opposite of what they were before: -8 for withdrawal and -4 for defiance. In this alternative scenario, the Soviets are hard-liners. They prefer nuclear annihilation to the prospect of a humiliating withdrawal and

[4]Richard Rhodes, *Dark Sun: The Making of the Hydrogen Bomb* (New York: Simon & Schuster, 1995), pp. 573–575. LeMay, renowned for his extreme views and his constant chewing of large unlit cigars, is supposed to be the original inspiration, in the 1963 movie *Dr. Strangelove,* for General Jack D. Ripper, who orders his bomber wing to launch an unprovoked attack on the Soviet Union.

[5]This story became public in a conference held in Havana, Cuba, in October 2002, to mark the 40th anniversary of the missile crisis. See Kevin Sullivan, "40 Years After Missile Crisis, Players Swap Stories in Cuba," *Washington Post,* October 13, 2002, p. A28. Vadim Orlov, who was a member of the Soviet submarine crew, identified the officer who refused to fire the torpedo as Vasili Arkhipov, who died in 1999.

the prospect of living in a world dominated by the capitalist United States; their slogan is "Better dead than red-white-and-blue." We show the game tree for this case in Figure 15.2. Now, if the United States makes the threat, the Soviets defy it. So the United States stands to get -10 from the threat but only -2 if it makes no threat and accepts the presence of the missiles in Cuba. It takes the lesser of the two evils. In the subgame-perfect equilibrium of this version of the game, the Soviets "win" and the U.S. threat does not work.

In reality, when the United States makes it move, it does not know whether the Soviets are hard-liners, as in Figure 15.2, or softer, as in Figure 15.1. The United States can try to estimate the probabilities of the two scenarios, for example, by studying past Soviet actions and reactions in different situations. We can regard Kennedy's statement that the probability of the blockade leading to war was between one-third and one-half this estimate of the probability that the Soviets are hard-line. Because the estimate is imprecise over a range, we work with a general symbol, p, for the probability, and examine the consequences of different values of p.

The tree for this more complex game is shown in Figure 15.3. The game starts with an outside force (here labeled "Nature") determining the Soviets' type. Along the upper branch of Nature's choice, the Soviets are hard-line. This leads to the upper node, where the United States makes its decision whether to issue its threat, and the rest of the tree is exactly like the game in Figure 15.2. Along the lower branch of Nature's choice, the Soviets are soft. This leads to the lower node, where the United States makes its decision whether to issue its threat, and the rest of the tree is exactly like the game in Figure 15.1. But the United States does not know from which node it is making its choice. Therefore the two U.S. nodes are enclosed in an "information set." Its significance is that the United States cannot take different actions at the nodes within the set, such as issuing the threat only if the Soviets are soft. It must take the same action at

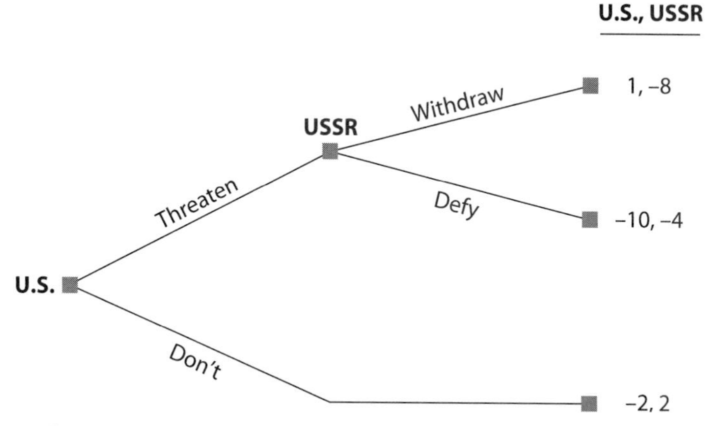

FIGURE 15.2 The Game with Hard-Line Soviets

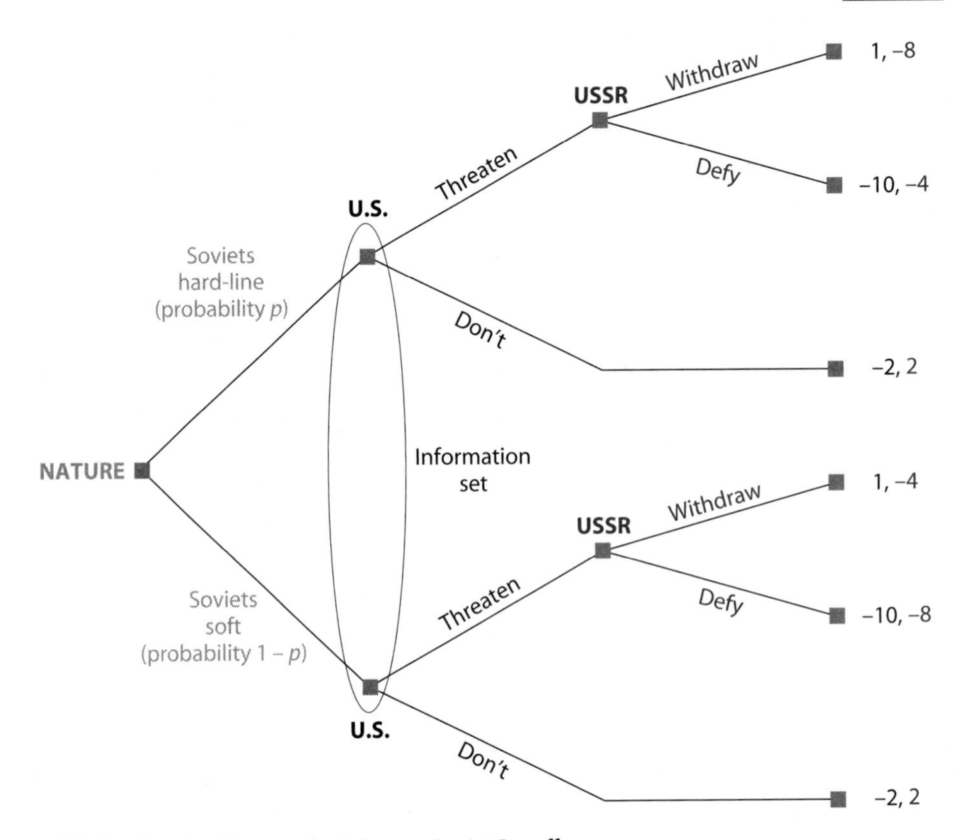

FIGURE 15.3 The Threat with Unknown Soviet Payoffs

both nodes, either threatening at both nodes or not threatening at both. It must make this decision in the light of the probabilities that the game might in truth be "located" at the one node or the other—that is, by calculating the *expected* payoffs of the two actions.

The Soviets themselves know what type they are. So we can do some rollback near the end of the game. Along the upper path, the hard-line Soviets will defy a U.S. threat, and along the lower path, the soft Soviets will withdraw in the face of the threat. Therefore the United States can look ahead and calculate that a threat will get a -10 if the game is actually moving along the upper path (a probability of p), and a 1 if it is moving along the lower path (a probability of $1 - p$). The expected U.S. payoff from making the threat is therefore $-10p + (1 - p) = 1 - 11p$.

If the United States does not make the threat, it gets a -2 along either path; so its expected payoff is also -2. Comparing the expected payoffs of the two actions, we see that the United States should make the threat if $1 - 11p > -2$, or $11p < 3$, or $p < 3/11 = 0.27$.

If the threat were sure to work, the United States would not care how bad its payoff could be if the Soviets defied it, whether -10 or even far more negative. But the risk that the Soviets might be hard-liners and thus defy a threat makes the -10 relevant in the U.S. calculations. Only if the probability, p, of the Soviets' being hard-line is small enough will the United States find it acceptable to make the threat. Thus the upper limit of 3/11 on p is also the upper limit of this U.S. tolerance, given the specific numbers that we have chosen. If we choose different numbers, we will get a different upper limit; for example, if we rate a nuclear war as -100 for the United States, then the upper limit on p will be only 3/101. But the idea of a large threat being "too large to make" if the probability of its going wrong is above a critical limit holds in general.

In this instance, Kennedy's estimate was that p lay somewhere in the range from 1/3 to 1/2. The lower end of this range, 0.33, is unfortunately just above our upper limit 0.27 for the risk that the United States is willing to tolerate. Therefore the simple bald threat "if you defy us, there will be nuclear war" is too large, too risky, and too costly for the United States to make.

4 A PROBABILISTIC THREAT

If an outright threat of war is too large to be tolerable and if you cannot find another, naturally smaller threat, then you can reduce the threat by creating merely a probability rather than a certainty that the dire consequences for the other side will occur if it does not comply. However, this does not mean that you decide after the fact whether to take the drastic action. If you had that freedom, you would choose to avoid the terrible consequences, and your opponents would know or assume this, so the threat would not be credible in the first place. You must relinquish some freedom of action and make a credible commitment. In this case, you must commit to a probabilistic device.

When making a *simple threat,* one player says to the other player: "If you don't comply, something will *surely* happen that will be very bad for you. By the way, it will also be bad for me, but my threat is credible because of my reputation [or through delegation or other reasons]." With a **probabilistic threat,** one player says to the other, "if you don't comply, there is a *risk* that something very bad for you will happen. By the way, it will also be very bad for me, but later I will be powerless to reduce that risk."

Metaphorically, a probabilistic threat of war is a kind of Russian roulette (an appropriate name in this context). You load a bullet into one chamber of a revolver and spin the barrel. The bullet acts as a "detonator" of the mutually costly war. When you pull the trigger, you do not know whether the chamber in the firing path is loaded. If it is, you may wish you had not pulled the trigger, but by then it will be too late. Before the fact, you would not pull the trigger if you knew that the bullet was in that chamber (that is, if the certainty of the dire action was too costly), but you are willing to pull the trigger knowing that there is only a 1 in 6 chance—in which the threat has been reduced by a factor of 6, to a point where it is now tolerable.

Brinkmanship is the creation and control of a suitable risk of this kind. It requires two apparently inconsistent things. On the one hand, you must let matters get enough out of your control that you will not have full freedom after the fact to refrain from taking the dire action, and so your threat will remain credible. On the other hand, you must retain sufficient control to keep the risk of the action from becoming too large and your threat too costly. Such "controlled lack of control" looks difficult to achieve, and it is. We will consider in Section 5 how the trick can be performed. Just one hint: All the complex differences of judgment, the dispersal of information, and the difficulties of enforcing orders, which made a simple threat too risky, are exactly the forces that make it possible to create a risk of war and therefore make brinkmanship credible. The real difficulty is not how to lose control, but how to do so in a controlled way.

We first focus on the mechanics of brinkmanship. For this purpose, we slightly alter the game of Figure 15.3 to get Figure 15.4. Here, we introduce a different kind of U.S. threat. It consists of choosing and fixing a probability, q, such that if the Soviets defy the United States, war will occur with that probability. With the remaining probability, $(1 - q)$, the United States will give up and agree to accept the Soviet missiles in Cuba. Remember that if the game gets to the point where the Soviets defy the United States, the latter does not have a choice in the matter. The Russian-roulette revolver has been set for the probability, q, and chance determines whether the firing pin hits a loaded chamber (that is, whether nuclear war actually happens).

Thus nobody knows the precise outcome and payoffs that will result if the Soviets defy this brinkmanship threat, but they know the probability, q, and can calculate expected values. For the United States, the outcome is -10 with the probability q and -2 with the probability $(1 - q)$, so the expected value is

$$-10q - 2(1 - q) = -2 - 8q.$$

For the Soviets, the expected payoff depends on whether they are hard-line or soft (and only they know their own type). If hard-line, they get a -4 from war, which happens with probability q, and a 2 if the United States gives up, which happens with the probability $(1 - q)$. The Soviets' expected payoff is $-4q + 2(1 - q) = 2 - 6q$. If they were to withdraw, they would get a -8, which is clearly worse no matter what value q takes between 0 and 1. Thus the hard-line Soviets will defy the brinkmanship threat.

The calculation is different if the Soviets are soft. Reasoning as before, we see that they get the expected payoff $-8q + 2(1 - q) = 2 - 10q$ from defiance,

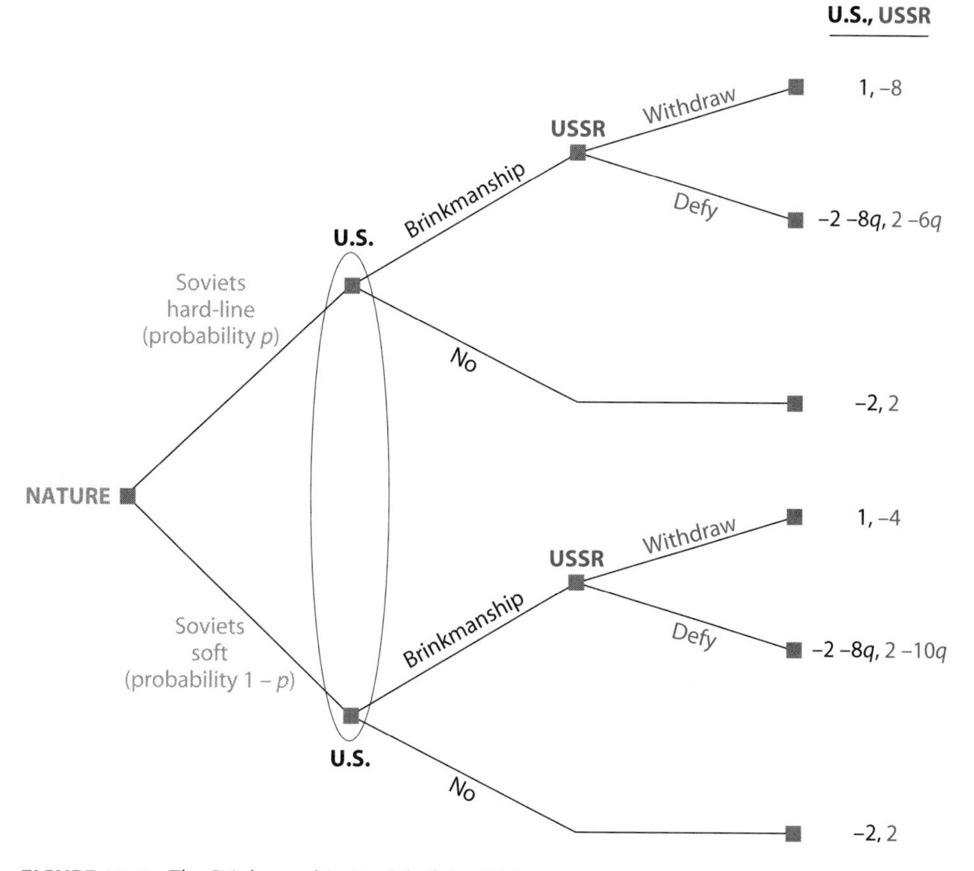

FIGURE 15.4 The Brinkmanship Model of the Crisis

and the sure payoff -4 if they withdraw. For them, withdrawal is better if $-4 > 2 - 10q$, or $10q > 6$, or $q > 0.6$. Thus U.S. brinkmanship must contain at least a 60% probability of war; otherwise it will not deter the Soviets, even if they are the soft type. We call this lower bound on the probability q the **effectiveness condition.**

Observe how the expected payoffs for U.S. brinkmanship and Soviet defiance shown in Figure 15.4 relate to the simple-threat model of Figure 15.3; the latter can now be thought of as a special case of the general brinkmanship-threat model of Figure 15.4, corresponding to the extreme value $q = 1$.

We can solve the game shown in Figure 15.4 in the usual way. We have already seen that along the upper path the Soviets, being hard-line, will defy the United States and that along the lower path the soft Soviets will comply with U.S. demands if the effectiveness condition is satisfied. If this condition is not satisfied, then both types of Soviets will defy the United States; so the latter would do better never to make this threat at all. So let us proceed by assuming that the soft Soviets will comply; we look at the U.S. choices. Basically, how risky can the U.S. threat be and still remain tolerable to the United States?

If the United States makes the threat, it runs the risk, p, that it will encounter the hard-line Soviets, who will defy the threat. Then the expected U.S. payoff will be $(-2 - 8q)$, as calculated before. The probability is $(1 - p)$ that the United States will encounter the soft-type Soviets. We are assuming that they comply; then the United States gets a 1. Therefore the expected payoff to the United States from the probabilistic threat, assuming that it is effective against the soft-type Soviets, is

$$(-2 - 8q) \times p + 1 \times (1 - p) = -8pq - 3p + 1.$$

If the United States refrains from making a threat, it gets a -2. Therefore the condition for the United States to make the threat is

$$-8pq - 3p + 1 > -2$$

$$\text{or} \quad q < \frac{3}{8}\frac{1-p}{p} = 0.375(1-p)/p.$$

That is, the probability of war must be small enough to satisfy this expression or the United States will not make the threat at all. We call this upper bound on q the **acceptability condition.** Note that p enters the formula for the maximum value of q that will be acceptable to the United States; the larger the chance that the Soviets will not give in, the smaller the risk of mutual disaster that the United States finds acceptable.

If the probabilistic threat is to work, it should satisfy both the effectiveness condition and the acceptability condition. We can determine the appropriate level of the probability of war by using Figure 15.5. The horizontal axis is the probability, p, that the Soviets are hard-line, and the vertical axis is the probability, q, that war will occur if they defy the U.S. threat. The horizontal line $q = 0.6$ gives the lower limit of the effectiveness condition; the threat should be such that its associated (p, q) combination is above this line if it is to work even against the soft-type Soviets. The curve $q = 0.375(1 - p)/p$ gives the upper limit of the acceptability condition; the threat should be such that (p, q) is below this curve if it is to be tolerable to the United States even with the assumption that it works against the soft-type Soviets. Therefore an effective and acceptable threat should fall somewhere between these two lines, above and to the left of their point of intersection, at $p = 0.38$ and $q = 0.6$ (shown as a gray "wedge" in Figure 15.5).

The curve reaches $q = 1$ when $p = 0.27$. For values of p less than this value, the dire threat (certainty of war) is acceptable to the United States and is effective against the soft-type Soviets. This just confirms our analysis in Section 3.

For values of p in the range from 0.27 to 0.38, the dire threat with $q = 1$ puts (p, q) to the right of the acceptability condition and is too large to be tolerable to the United States. But a scaled-down threat can be found. For this range of values of p, some values of q are low enough to be acceptable to the United States and yet high enough to compel the soft-type Soviets. Brinkmanship (using a

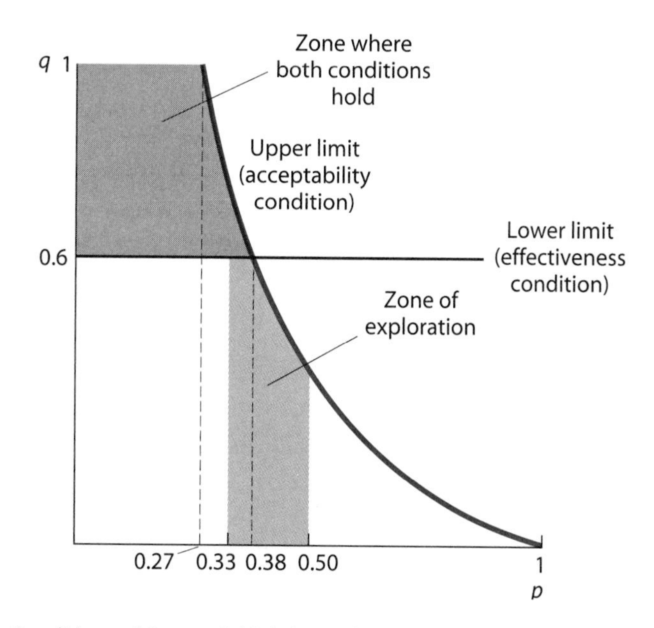

FIGURE 15.5 Conditions of Successful Brinkmanship

probabilistic threat) can do the job in this situation, whereas a simple dire threat would be too risky.

If p exceeds 0.38, then no value of q satisfies both conditions. If the probability that the Soviets will never give in is greater than 0.38, then any threat large enough to work against the soft-type Soviets ($q \geq 0.6$) creates a risk of war too large to be acceptable to the United States. If $p \geq 0.38$, therefore, the United States cannot help itself by using the brinkmanship strategy.

5 PRACTICING BRINKMANSHIP

If Kennedy has a very good estimate of the probability, p, of the Soviets being hard-liners, and is very confident about his ability to control the risk, q, that the blockade will lead to nuclear war, then he can calculate and implement his best strategy. As we saw in Section 3, if $p < 0.27$, the dire threat of a certainty of war is acceptable to Kennedy. (Even then he will prefer to use the smallest effective threat—namely, $q = 0.6$.) If p is between 0.27 and 0.38, then he has to use brinkmanship. Such a threat has to have the risk of disaster $0.6 < q < 0.375 (1 - p)/p$, and again Kennedy prefers the smallest of this range—namely, $q = 0.6$. If $p > 0.38$, then he should give in.

In practice Kennedy does not know p precisely; he only estimates that it lies within the range from 1/3 to 1/2. Similarly, he cannot be confident about the exact location of the critical value of q in the acceptability condition. That depends on the numbers used for the Soviet payoffs in various outcomes—for example, −8 (for war) versus −4 (for compliance)—and Kennedy can only estimate these values. Finally, he may not even be able to control the risk created by his brinkmanship action very precisely. All these ambiguities make it necessary to proceed cautiously.

Suppose Kennedy thinks that $p = 0.35$ and issues a threat backed by an action that carries the risk $q = 0.65$. The risk is greater than what is needed to be effective—namely, 0.6. The limit of acceptability is $0.375 \times (1 - 0.35)/0.35 = 0.7$, and the risk $q = 0.65$ is less than this limit. Thus, according to Kennedy's calculations, the risk satisfies both of the conditions—effectiveness and acceptability. However, suppose Kennedy is mistaken. For example, if he has not realized that LeMay might actually defy orders and take an excessively aggressive action, then q may in reality be higher than Kennedy thinks it is; for example, q may equal 0.8, which Kennedy would regard as too risky. Or suppose p is actually 0.4; then Kennedy would regard even $q = 0.65$ as too risky. Or Kennedy's experts may have misestimated the values of the Soviet payoffs. If they rate the humiliation of withdrawal as −5 instead of −4, then the threshold of the effectiveness condition will actually be $q = 0.7$, and Kennedy's threat with $q = 0.65$ will go wrong.

All that Kennedy knows is that the general shape of the effectiveness and acceptability conditions is like that shown in Figure 15.5. He does not know p for sure. Therefore he does not know exactly what value of q to choose to fulfill both the effectiveness and the acceptability conditions; indeed, he does not even know if such a range exists for the unknown true value of p: it might be greater than or less than the borderline value of 0.38 that divides the two cases. And he is not able to fix q very precisely; therefore, even if he knew p, he would not be able to act confident of his willingness to tolerate the resulting risk.

With such hazy information, imprecise control, and large risks, what is Kennedy to do? He has to *explore* the boundaries of the Soviets' risk tolerance as well as his own. It would not do to start the exploration with a value of q that might turn out to be too high. Instead, Kennedy must explore the boundaries "from below"; he must start with something quite safe and gradually increase the level of risk to see "who blinks first." That is exactly how brinkmanship is practiced in reality.

We explain this with the aid of Figure 15.5. Observe the green-shaded area. Its left and right boundaries, $p = 1/3$ and $p = 1/2$, correspond to the limits of Kennedy's estimated range of p. The lower boundary is the horizontal axis ($q = 0$). The upper boundary is composed of two segments. For $p < 0.38$, this segment corresponds to the effectiveness condition; for $p > 0.38$, it corresponds to the acceptability condition. Remember that Kennedy does not know the precise positions of these boundaries but must grope toward them from below. Therefore the red-shaded region is where he must start the process.

Suppose Kennedy starts with a very safe action—say, q equaling approximately 0.01 (1%). In our context of the Cuban missile crisis, we can think of this as his television speech, which announced that a quarantine would soon go into effect. At this juncture, the point with coordinates (p, q) lies somewhere near the bottom edge of the shaded region. Kennedy does not know exactly where, because he does not know p for sure. But the overwhelming likelihood is that at this point the threat is quite safe but also ineffective. Therefore Kennedy escalates it a little bit. That is, he moves the point (p, q) in a vertically upward direction from wherever it was initially. This could be the actual start of the quarantine. If that proves to be still safe but ineffective, he jacks up the risk one more notch. This could be the leaking of information about bombing plans.

As he proceeds in this way, eventually his exploration will encounter one of the boundaries of the red-shaded area in Figure 15.5, and which boundary this is depends on the value of p. One of two things comes to pass. Either the threat becomes serious enough to deter the Soviets; this happens if the true value of p is less than its true critical value, here 0.38. On the diagram, we see this as a movement out of the red-shaded area and into the area in which the threat is both acceptable *and* effective. Then the Soviets concede and Kennedy has won. Or the threat becomes too risky for the United States; this happens if $p > 0.38$. Kennedy's exploration in this case pushes him above the acceptability condition. Then Kennedy decides to concede, and Khrushchev has won. Again we point out that because Kennedy is not sure of the true value of p, he does not know in advance which of these two outcomes will prevail. As he gradually escalates the risk, he may get some clues from Soviet behavior that enable him to make his estimate of p somewhat more precise. Eventually he will reach sufficient precision to know which part of the boundary he is headed toward and therefore whether the Soviets will concede or the United States must be the player to do so.

Actually, there are two possible outcomes only so long as the ever-present and steadily increasing mutual risk of disaster does not come to pass while Kennedy is groping through the range of ever more risky military options. Therefore there is a third possibility—namely, that the explosion occurs before either side recognizes that it has reached its limit of tolerance of risk and climbs down. This continuing and rising risk of a very bad outcome is what makes brinkmanship such a delicate and dangerous strategy.

Thus brinkmanship in practice is the **gradual escalation of the risk of mutual harm.** It can be visualized vividly as **chicken in real time.** In our analysis of chicken in Chapter 4, we gave each player a simple binary choice: either go straight or swerve. In reality, the choice is usually one of timing. The two cars are rushing toward each other, and either player can choose to swerve at any time. When the cars are very far apart, swerving ensures safety. As they get closer together, they face an ever-increasing risk that they will collide anyway, and even swerving will not avoid a collision. As the two players continue to drive toward one another, each is exploring the limit of the other's willingness to take this risk and is perhaps at the same time exploring his own limit. The one who hits that limit first swerves. But there is always the risk that they have left it long enough and are close enough that, even after choosing Swerve, they can no longer avoid the collision.

Now we see why, in the Cuban missile crisis, the very features that make it inaccurate to regard it as a two-person game make it easier to practice such brinkmanship. The blockade was a relatively small action, unlikely to start a nuclear war at once. But once Kennedy set the blockade in motion, its operation, escalation, and other features were not totally under his control. So Kennedy was not saying to Khrushchev, "If you defy me (cross a sharp brink), I will coolly and deliberately launch a nuclear war that will destroy both out peoples." Rather, he was saying, "The wheels of the blockade have started to turn and are gathering their own momentum. The more or longer you defy me, the more likely it is that some operating procedure will slip up, the political pressure on me will rise to a point where I must give in, or some hawk will run amok. If this risk comes to pass, I will be unable to prevent nuclear war, no matter how much I may regret it at that point. Only you can now defuse the tension by complying with my demand to withdraw the missiles."

We believe that this perspective gives a much better and deeper understanding of the crisis than can most analyses based on simple threats. It tells us why the *risk* of war played such an important role in all discussions. It even makes Allison's compelling arguments about bureaucratic procedures and internal divisions on both sides an integral part of the picture: these features allow the top-level players on both sides credibly to lose some control—that is, to practice brinkmanship.

One important condition remains to be discussed. In Chapter 10, we saw that every threat has an associated implicit promise—namely, that the bad consequence will not take place if your opponent complies with your wishes. The same is required for brinkmanship. If, as you are increasing the level of risk, your opponent does comply, you must be able to "go into reverse"—begin reducing the risk immediately and quite quickly remove it from the picture. Otherwise the opponent would not gain anything by compliance. This may have been a problem in the Cuban missile crisis. If the Soviets feared that Kennedy could not control hawks such as LeMay ("We ought to just go in there today and knock 'em off"), they would gain nothing by giving in.

To reemphasize and sum up, brinkmanship is the strategy of exposing your rival and yourself to a gradually increasing risk of mutual harm. The actual occurrence of the harmful outcome is not totally within the threatener's control.

Viewed in this way, brinkmanship is everywhere. In most confrontations—for example, between a company and a labor union, a husband and a wife, a parent and a child, and the President and Congress—one player cannot be sure of the other party's objectives and capabilities. Therefore most threats carry a risk of error, and every threat must contain an element of brinkmanship. We hope that we have given you some understanding of this strategy and that we have impressed on you the risks that it carries. You will have to face up to brinkmanship or to conduct it yourself on many occasions in your personal and professional lives. Please do so carefully, with a clear understanding of its potentialities and risks.

To help you do so, we now recapitulate the important lessons learned from the handling of the Cuban missile crisis, reinterpreted as a labor union leadership contemplating a strike in pursuit of its wage demand, unsure whether this will result in the whole firm's shutting down:

1. Start small and safe. Your first step should not be an immediate walkout; it should be to schedule a membership meeting at a date a few days or weeks hence, while negotiations continue.
2. Raise the risks gradually. Your public and private statements, as well as the stirring up of the sentiments of the membership, should induce management to believe that acceptance of its current low-wage offer is becoming less and less likely. If possible, stage small incidents—for example, a few one-day strikes or local walkouts.
3. As this process continues, read and interpret signals in management's actions to figure out whether the firm has enough profit potential to afford the union's high-wage demand.
4. Retain enough control over the situation; that is, retain the power to induce your membership to ratify the agreement that you will reach with management; otherwise management will think that the risk will not deescalate even if it concedes to your demands.

SUMMARY

In some game situations, the risk of error in the presence of a threat may call for the use of as small a threat as possible. When a large threat cannot be reduced in other ways, it can be scaled down by making its fulfillment probabilistic. Strategic use of *probabilistic threat,* in which you expose your rival and yourself to an increasing risk of harm, is called brinkmanship.

Brinkmanship requires a player to relinquish control over the outcome of the game without completely losing control. You must create a threat with a risk level that is both large enough to be effective in compelling or deterring your rival and small enough to be acceptable to you. To do so, you must determine the levels of risk tolerance of both players through a *gradual escalation of the risk of mutual harm.*

The Cuban missile crisis of 1962 serves as a case study in the use of brinkmanship on the part of President Kennedy. Analyzing the crisis as an example of a simple threat, with the U.S. blockade of Cuba establishing credibility, is inadequate. A better analysis accounts for the many complexities and uncertainties inherent in the situation and the likelihood that a simple threat was too risky. Because the actual crisis included numerous political and military players, Kennedy was able to achieve "controlled loss of control" by ordering the blockade and gradually letting incidents and tension escalate, until Khrushchev yielded in the face of the rising risk of nuclear war.

KEY TERMS

acceptability condition (604)
chicken in real time (607)
effectiveness condition (603)
gradual escalation of the risk of mutual harm (607)
probabilistic threat (601)

SOLVED EXERCISES

S1. Consider a game between a union and the company that employs the union membership. The union can threaten to strike (or not) to get the company to meet its wage and benefits demands. When faced with a threatened strike, the company can choose to concede to the demands of the union or to defy its threat of a strike. The union, however, does not know the company's profit position when it decides whether to make its threat; it does not know whether the company is sufficiently profitable to meet its demands—and the company's assertions in this matter cannot be believed. Nature determines whether the company is profitable; the probability that the firm is unprofitable is p.

The payoff structure is as follows: (i) When the union makes no threat, the union gets a payoff of 0 (regardless of the profitability of the company). The company gets a payoff of 100 if it is profitable but a payoff of 10 if it is unprofitable. A passive union leaves more profit for the company if there is any profit to be made. (ii) When the union threatens to strike and the company concedes, the union gets 50 (regardless of the profitability of the company) and the company gets 50 if it is profitable but -40 if it is not. (iii) When the union threatens to strike and the company defies the union's threat, the union must strike and gets -100 (regardless of the profitability of the company). The company gets -100 if it is profitable and -10 if it is not. Defiance is very costly for a profitable company but not so costly for an unprofitable one.

(a) What happens when the union uses the pure threat to strike unless the company concedes to the union's demands?

(b) Suppose that the union sets up a situation in which there is some risk, with probability $q < 1$, that it will strike after the company defies its threat. This risk may arise from the union leadership's imperfect ability to keep the membership in line. Draw a game tree similar to Figure 15.4. for this game.

(c) What happens when the union uses brinkmanship, threatening to strike with some probability q unless the company accedes to its demands?

(d) Derive the effectiveness and acceptability conditions for this game, and determine the values for p and q for which the union can use a pure threat, brinkmanship, or no threat at all.

S2. Scenes from many movies illustrate the concept of brinkmanship. Analyze the following descriptions from this perspective. What are the risks the two sides face? How do those risks increase during the course of the execution of the brinkmanship threat?

(a) In the 1980 film *The Gods Must Be Crazy*, the only survivor of a rebel team that tried to assassinate the president of an African country has been captured and is being interrogated. He stands blindfolded with his back to the open door of a helicopter. Above the noise of the helicopter rotors, an officer asks him, "Who is your leader? Where is your hideout?" The man does not answer, and the officer pushes him out of the door. In the next scene, we see that although its engine is running, the helicopter is actually on the ground, and the man has fallen 6 feet on his back. The officer appears at the door and says, laughing, "Next time it will be a little higher."

(b) In the 1998 film *A Simple Plan*, two brothers remove some of a \$4.4 million ransom payment that they find in a crashed airplane. After many intriguing twists of fate, the remaining looter, Hank, finds himself in conference with an FBI agent. The agent, who suspects but cannot prove that Hank has some of the missing money, fills Hank in on the story of the money's origins and tells him that the FBI possesses the serial numbers of about 1 of every 10 of the bills in that original ransom payment. The agent's final words to Hank are, "Now it's simply a matter of waiting for the numbers to turn up. You can't go around passing \$100 bills without eventually sticking in someone's memory."

S3. In this exercise, we provide a couple examples of the successful use of brinkmanship, where "success" is indicative of the two sides' reaching a mutually acceptable deal. For each example, (i) identify the interests of the parties; (ii) describe the nature of the uncertainty inherent in the situation; (iii) give the strategies the parties used to escalate the risk of disaster; (iv) discuss whether the strategies were good ones; and (v) **(Optional)** if you can, set up a small mathematical model of the kind presented in this chapter. In each case, we provide a few readings to get you started; you should locate more by using the resources of your library and resources on the World Wide Web such as Lexis-Nexis.

(a) The Uruguay Round of international trade negotiations that started in 1986 and led to the formation of the World Trade Organization in 1994. *Reading*: John H. Jackson, *The World Trading System,* 2nd ed. (Cambridge, MA: MIT Press, 1997), pp. 44–49 and chaps. 12 and 13.

(b) The Camp David accords between Israel and Egypt in 1978. *Reading*: William B. Quandt, *Camp David: Peacemaking and Politics* (Washington, DC: Brookings Institution, 1986).

S4. The following examples illustrate the unsuccessful use of brinkmanship, where brinkmanship is considered "unsuccessful" when the mutually bad outcome (disaster) occurs. Answer the questions outlined in Exercise S3 for the following situations:

(a) The confrontation between the regime and the student prodemocracy demonstrators in Beijing in June 1989. *Readings*: Donald Morrison, ed., *Massacre in Beijing: China's Struggle for Democracy* (New York: Time Magazine Publications, 1989); Suzanne Ogden, Kathleen Hartford, L. Sullivan, and D. Zweig, eds., *China's Search for Democracy: The Student and Mass Movement of 1989* (Armonk, NY: M.E. Sharpe, 1992).

(b) The Caterpillar strike, from 1991 to 1998. *Readings*: "The Caterpillar Strike: Not Over Till It's Over," *Economist,* February 28, 1998; "Caterpillar's Comeback," *Economist,* June 20, 1998; Aaron Bernstein, "Why Workers Still Hold a Weak Hand," *BusinessWeek,* March 2, 1998.

S5. Answer the questions listed in Exercise S3 for these potential cases for brinkmanship in the future:

(a) A Taiwanese declaration of independence from the People's Republic of China. *Reading:* Ian Williams, "Taiwan's Independence," *Foreign Policy in Focus,* December 20, 2006. Available at www.fpif.org/fpiftxt/3815.

(b) The militarization of space, for example, the positioning of weapons in space or the shooting down of satellites. *Reading*: "Disharmony in the Spheres," *Economist,* January 17, 2008. Available at www.economist.com/displaystory.cfm?story_id=10533205.

UNSOLVED EXERCISES

U1. In the chapter, we argue that the payoff to the United States is −10 when (either type) Soviets defy the U.S. threat; these payoffs are illustrated in Figure 14.3. Suppose now that this payoff is in fact −12 rather than −10.

(a) Incorporate this change in payoff into a game tree similar to the one in Figure 15.4.

(b) Using the payoffs from your game tree in part (a), find the effectiveness condition for this version of the U.S.–USSR brinkmanship game.

(c) Using the payoffs from part (a), find the acceptability condition for this game.

(d) Draw a diagram similar to that in Figure 15.5, illustrating the effectiveness and acceptability conditions found in parts (b) and (c).

(e) For what values of p, the probability that the Soviets are hard-line, is the pure threat ($q = 1$) acceptable? For what values of p is the pure threat unacceptable but brinkmanship still possible?

(f) If Kennedy was correct in believing that p lay between 1/3 and 1/2, does your analysis of this version of the game suggest than an effective *and* acceptable probabilistic threat existed? Use this example to explain how a game theorist's assumptions about player payoffs can have a major effect on the predictions that arise from the theoretical model.

U2. Answer the questions from Exercise S2 for the following movies:

(a) In the 1941 movie classic *The Maltese Falcon,* the hero, Sam Spade (Humphrey Bogart), is the only person who knows the location of the immensely valuable gem-studded falcon figure, and the villain, Caspar Gutman (Sydney Greenstreet), is threatening to torture him for that information. Spade points out that torture is useless unless the threat of death lies behind it, and Gutman cannot afford to kill Spade, because then the information dies with him. Therefore he may as well not bother with the threat of torture. Gutman replies, "That is an attitude, sir, that calls for the most delicate judgment on both sides, because, as you know, sir, men are likely to forget in the heat of action where their best interests lie and let their emotions carry them away."

(b) The 1925 Soviet classic *The Battleship Potemkin* (set in the summer of 1905) closes with a squadron of ships from the tsar's Black Sea fleet chasing the mutinous and rebellious crew of the *Potemkin.* The tension mounts as the ships draw ever closer. Men on each side race to their battle stations, load and aim the huge guns, and wait nervously for the order to fire on their countrymen. Neither side wants to attack the other, but neither wants to back down or to die without defending itself. The tsar's ships have orders to take the *Potemkin* by any means necessary, and the crew knows it will be tried for treason if it surrenders.

U3. Answer the questions in Exercise **S3** for these examples of successful brinkmanship:

(a) The negotiations between the South African apartheid regime and the African National Congress to establish a new constitution with majority rule, 1989 to 1994. *Reading*: Allister Sparks, *Tomorrow Is Another Country* (New York: Hill and Wang, 1995).

(b) Peace in Northern Ireland: disarmament of the IRA in July 2005, the St. Andrews Agreement of October 2006, the elections of March 2007, and the power-sharing government of Ian Paisley and Martin McGuinness. *Reading*: "The Thorny Path to Peace and Power Sharing," CBC News, March 26, 2007. Available at www.cbc.ca/news/background/northern-ireland/timeline.html.

U4. Answer the questions in Exercise **S3** for these examples of unsuccessful brinkmanship:

(a) The U.S. budget confrontation between President Clinton and the Republican-controlled Congress in 1995. *Readings*: Sheldon Wolin, "Democracy and Counterrevolution," *Nation*, April 22, 1996; David Bowermaster, "Meet the Mavericks," *U.S. News and World Report*, December 25, 1995–January 1, 1996; "A Flight that Never Seems to End," *Economist*, December 16, 1995.

(b) The television writers' strike of 2007–2008. *Readings*: "Writers Guild of America," online archive of the *New York Times* on the Writers Guild and the strike. Available at http://topics.nytimes.com/top/reference/timestopics/organizations/w/writers_guild_of_america/index.html; "Writers Strike: A Punch from the Picket Line." Available at http://writers-strike.blogspot.com.

U5. Answer the questions in Exercise **S3** for these potential cases of future brinkmanship:

(a) The stationing of an American antiballistic missile launch site in Poland with an accompanying radar site in the Czech Republic, ostensibly intended to intercept missiles from Iran but angering Russia. *Reading*: "Q&A: US Missile Defence," *BBC News*, August 20, 2008. Available at http://news.bbc.co.uk/2/hi/europe/6720153.stm.

(b) Deterring Iran from obtaining nuclear weapons. *Readings*: James Fallows, "The Nuclear Power Beside Iraq," *Atlantic*, May 2006. Available at www.theatlantic.com/doc/200605/fallows-iran); James Fallows, "Will Iran Be Next?" *Atlantic*, December 2004. Available at www.theatlantic.com/doc/200412/fallows.

Strategy and Voting

VOTING IN ELECTIONS may seem outside the purview of game theory at first glance, but no one who remembers the 2000 U.S. presidential election should be surprised to discover that there are many strategic aspects to elections and voting. The presence in that election of the third-party candidate Ralph Nader had a number of strategic implications for the election outcome. A group called "Citizens for Strategic Voting" took out paid advertisements in major newspapers to explain voting strategy to the general public. And analysts have suggested that Nader's candidacy may have been the element that swung the final tally in George W. Bush's favor.

It turns out that there is a lot more to participating in an election than deciding what you like and voting accordingly. Because many different types of voting procedures can be used to determine a winner in an election, there can be many different election outcomes. One particular procedure might lead to an outcome that you prefer, in which case you might apply your strategic abilities to see that this procedure is used. Or you might find that the procedure used for a specific election is open to manipulation; voters might be able to alter the outcome of an election by misrepresenting their preferences. If other voters behave strategically in this way, then it may be in your best interest to do so as well. An understanding of the strategic aspects of voting can help you determine your best course of action in such situations.

In the following sections, we look first at the various available voting rules and procedures and then at some nonintuitive or paradoxical results that can arise in certain situations. We consider some of the criteria used to judge the performance of the various voting methods. We also address strategic behavior of voters and the scope for outcome manipulation. Finally, we present two different versions of a well-known result known as the *median voter theorem*—as a two-person zero-sum game with discrete strategies and with continuous ones.

1 VOTING RULES AND PROCEDURES

Numerous election procedures are available to help choose from a slate of alternatives (that is, candidates or issues). With as few as three available alternatives, election design becomes interestingly complex. We describe in this section a variety of procedures from three broad classes of voting, or vote-aggregation, methods. The number of possible voting procedures is enormous, and the simple taxonomy that we provide here can be broadened extensively by allowing elections based on a combination of procedures; a considerable literature in both economics and political science deals with just this topic. We have not attempted to provide an exhaustive survey but rather to give a flavor of that literature. If you are interested, we suggest you consult the broader literature for more details on the subject.[1]

A. Binary Methods

Vote aggregation methods can be classified according to the number of options or candidates considered by the voters at any given time. **Binary methods** require voters to choose between only two alternatives at a time. In elections in which there are exactly two candidates, votes can be aggregated by using the well-known principle of **majority rule,** which simply requires that the alternative with a majority of votes wins. When dealing with a slate of more than two alternatives, **pairwise voting**—a method consisting of a repetition of binary votes—can be used. Pairwise procedures are **multistage;** they entail voting on pairs of alternatives in a series of majority votes to determine which is most preferred.

One pairwise procedure, in which each alternative is put up against each of the others in a round-robin of majority votes, is called the **Condorcet method,** after the 18th-century French theorist Jean Antoine Nicholas Caritat, marquis de Condorcet. He suggested that the candidate who defeats each of the others

[1]The classic textbook on this subject, which was instrumental in making game theory popular in political science, is William Riker, *Liberalism Against Populism* (San Francisco: W. H. Freeman, 1982). A general survey is the symposium on "Economics of Voting," *Journal of Economic Perspectives,* vol. 9, no. 1 (Winter 1995). An important early research contribution is Michael Dummett, *Voting Procedures* (Oxford: Clarendon Press, 1984). Donald Saari, *Chaotic Elections* (Providence, RI: American Mathematical Society, 2000), develops some new ideas that we use later in this chapter.

in such a series of one-on-one contests should win the entire election; such a candidate, or alternative, is now termed a **Condorcet winner.** Other pairwise procedures produce "scores" such as the **Copeland index,** which measures an alternative's win–loss record in a round-robin of contests. The first round of the World Cup soccer tournament uses a type of Copeland index to determine which teams from each group move on to the second round of play.[2]

Another well-known pairwise procedure, used when there are three possible alternatives, is the **amendment procedure,** required by the parliamentary rules of the U.S. Congress when legislation is brought to a vote. When a bill is brought before Congress, any amended version of the bill must first win a vote against the original version of the bill. The winner of that vote is then paired against the status quo and members vote on whether to adopt the version of the bill that won the first round; majority rule can then be used to determine the winner. The amendment procedure can be used to consider any three alternatives by pairing two in a first-round election and then putting the third up against the winner in a second-round vote.

B. Plurative Methods

Plurative methods allow voters to consider three or more alternatives simultaneously. One group of plurative voting methods applies information on the positions of alternatives on a voter's ballot to assign points used when tallying ballots; these voting methods are known as **positional methods.** The familiar **plurality rule** is a special-case positional method in which each voter casts a single vote for her most-preferred alternative. That alternative is assigned a single point when votes are tallied; the alternative with the most votes (or points) wins. Note that a plurality winner need *not* gain a majority, or 51%, of the vote. Thus, for instance, in the 1994 Maine gubernatorial election, the independent candidate Angus King captured the governorship with only 36% of the vote; his Democratic, Republican, and Green Party opponents gained 34%, 23%, and 6% of the vote, respectively. Another special-case positional method, the **antiplurality method,** asks voters to vote against one of the available alternatives or, equivalently, to vote for all but one. For counting purposes, the alternative voted against is allocated -1 point, or else all alternatives except that one receive 1 point while the alternative voted against receives 0.

One of the best-known positional methods is the **Borda count,** named after Jean-Charles de Borda, a fellow countryman and contemporary of Condorcet. Borda described the new procedure as an improvement on plurality rule.

[2]Note that such indices, or scores, must have precise mechanisms in place to deal with ties; World Cup soccer uses a system that undervalues a tie to encourage more aggressive play. See Barry Nalebuff and Jonathan Levin, "An Introduction to Vote Counting Schemes," *Journal of Economic Perspectives,* vol. 9, no. 1 (Winter 1995), pp. 3–26.

The Borda count requires voters to rank-order all of the possible alternatives in an election and to indicate their rankings on their ballot cards. Points are assigned to each alternative on the basis of its position on each voter's ballot. In a three-person election, the candidate at the top of a ballot gets 3 points, the next candidate 2 points, and the bottom candidate 1 point. After the ballots are collected, each candidate's points are summed, and the one with the most points wins the election. A Borda count procedure is used in a number of sports-related elections, including professional baseball's Cy Young Award and college football's championship elections.

Many other positional methods can be devised simply by altering the rule used for the allocation of points to alternatives based on their positions on a voter's ballot. One system might allocate points in such a way as to give the top-ranked alternative relatively more than the others—for example, 5 points for the most-preferred alternative in a three-way election but only 2 and 1 for the second- and third-ranked options. In elections with larger numbers of candidates—say, eight—the top two choices on a voter's ballot might receive preferred treatment, gaining 10 and 9 points, respectively, while the others receive 6 or fewer.

An alternative to the positional plurative methods is the relatively recently invented **approval voting** method, which allows voters to cast a single vote for each alternative of which they "approve."[3] Unlike positional methods, approval voting does not distinguish between alternatives on the basis of their positions on the ballot. Rather, all approval votes are treated equally, and the alternative that receives the most approvals wins. In elections in which more than one winner can be selected (in electing a school board, for instance), a threshold level of approvals is set in advance and alternatives with more than the required minimum approvals are elected. Proponents of this method argue that it favors relatively moderate alternatives over those at either end of the spectrum; opponents claim that unwary voters could elect an unwanted novice candidate by indicating too many "encouragement" approvals on their ballots. Despite these disagreements, several professional societies have adopted approval voting to elect their officers, and some states have used or are considering using this method for public elections.

C. Mixed Methods

Some multistage voting procedures combine plurative and binary voting in **mixed methods.** The **majority runoff** procedure, for instance, is a two-stage method used to decrease a large group of possibilities to a binary decision. In a

[3]Unlike many of the other methods that have histories going back several centuries, the approval voting method was designed and named by then–graduate student Robert Weber in 1971; Weber is now a professor of managerial economics and decision sciences at Northwestern University, specializing in game theory.

first-stage election, voters indicate their most-preferred alternative, and these votes are tallied. If one candidate receives a majority of votes in the first stage, she wins. However, if there is no majority choice, a second-stage election pits the two most-preferred alternatives against each other. Majority rule chooses the winner in the second stage. French presidential elections use the majority runoff procedure, which can yield unexpected results if three or four strong candidates split the vote in the first round. In the spring of 2002, for example, the far-right candidate Le Pen came in second ahead of France's socialist Prime Minister Jospin in the first round of the presidential election. This result aroused surprise and consternation among French citizens, 30% of whom hadn't even bothered to vote in the election and some of whom had taken the first round as an opportunity to express their preference for various candidates of the far and fringe left. Le Pen's advance to the runoff election led to considerable political upheaval, although he lost in the end to the incumbent president, Chirac.

Another mixed procedure consists of voting in successive **rounds.** Voters consider a number of alternatives in each round of voting, with the worst-performing alternative eliminated after each stage. Voters then consider the remaining alternatives in a next round. The elimination continues until only two alternatives remain; at that stage, the method becomes binary and a final majority runoff determines a winner. A procedure with rounds is used to choose sites for the Olympic Games.

One could eliminate the need for successive rounds of voting by having voters indicate their preference orderings on the first ballot. Then a **single transferable vote** method can be used to tally votes in later rounds. With a single transferable vote, each voter indicates her preference by ordering all candidates on a single initial ballot. If no alternative receives a majority of all first-place votes, the bottom-ranked alternative is eliminated and all first-place votes for that candidate are "transferred" to the candidate listed second on those ballots; similar reallocation occurs in later rounds as additional alternatives are eliminated until a majority winner emerges. The city of San Francisco voted in March of 2002 to use this voting method, more commonly called **instant runoff,** in all future municipal elections.

The single transferable vote is sometimes combined with **proportional representation** in an election. Proportional representation implies that a state electorate consisting of 55% Republicans, 25% Democrats, and 20% Independents, for example, would yield a body of representatives mirroring the party affiliations of that electorate. In other words, 55% of the U.S. Representatives from such a state would be Republican, and so on; this result contrasts starkly with the plurality rule method, which would elect *all* Republicans (assuming that the voter mix in each district exactly mirrors the overall voter mix in the state). Candidates who attain a certain quota of votes are elected, and others who fall below a certain quota are eliminated, depending on the exact specifications of the voting procedure. Votes for those candidates who are eliminated are again transferred by using the voters' preference orderings. This procedure continues until an appropriate number of candidates from each party is elected.

Clearly, there is room for considerable strategic thinking in the choice of a vote aggregation method, and strategy is also important even after the rule has been chosen. We examine some of the issues related to rule making and agenda setting in Section 2. Furthermore, strategic behavior on the part of voters, often called **strategic voting** or **strategic misrepresentation of preferences,** can also alter election outcomes under any set of rules, as we will see later in this chapter.

2 VOTING PARADOXES

Even when people vote according to their true preferences, specific conditions on voter preferences and voting procedures can give rise to curious outcomes. In addition, election outcomes can depend critically on the type of procedure used to aggregate votes. This section describes some of the most famous of the curious outcomes—the so-called voting paradoxes—as well as some examples of how election results can change under different vote-aggregation methods with no change in voter preferences and no strategic voting.

A. The Condorcet Paradox

The **Condorcet paradox** is one of the most famous and important of the voting paradoxes.[4] As mentioned earlier, the Condorcet method calls for the winner to be the candidate who gains a majority of votes in each round of a round-robin of pairwise comparisons. The paradox arises when no Condorcet winner emerges from this process.

To illustrate the paradox, we construct an example in which three people vote on three alternative outcomes by using the Condorcet method. Consider three city councillors (Left, Center, and Right) who are asked to rank their preferences for three alternative welfare policies, one that extends the welfare benefits currently available (call this one Generous, or G), another that decreases available benefits (Decreased, or D), and yet another that maintains the status quo (Average, or A). They are then asked to vote on each pair of policies to establish a council ranking, or a **social ranking.** This ranking is meant to describe how the council as a whole judges the merits of the possible welfare systems.

[4]It is so famous that economists have been known to refer to it as *the* voting paradox. Political scientists appear to know better, in that they are far more likely to use its formal name. As we will see, there are any number of possible voting paradoxes, not just the one named for Condorcet.

LEFT	CENTER	RIGHT
G > A > D	A > D > G	D > G > A

FIGURE 16.1 Councillor Preferences over Welfare Policies

Suppose Councillor Left prefers to keep benefits as high as possible, whereas Councillor Center is most willing to maintain the status quo but concerned about the state of the city budget and so least willing to extend welfare benefits. Finally, Councillor Right most prefers reducing benefits but prefers an increase in benefits to the status quo; she expects that extending benefits will soon cause a serious budget crisis and turn public opinion so much against benefits that a more permanent state of low benefits will result, whereas the status quo could go on indefinitely. We illustrate these preference orderings in Figure 16.1 where the "curly" greater-than symbol, $\succ$, is used to indicate that one alternative is preferred to another. (Technically, $\succ$ is referred to as a *binary ordering relation.*)

With these preferences, if Generous is paired against Average, Generous wins. In the next pairing, of Average against Decreased, Average wins. And in the final pairing of Generous against Decreased, the vote is again 2 to 1, this time in favor of Decreased. Therefore, if the council votes on alternative pairs of policies, a majority prefer Generous over Average, Average over Decreased, *and* Decreased over Generous. No one policy has a majority over both of the others. The group's preferences are cyclical: $G \succ A \succ D \succ G$.

This cycle of preferences is an example of an **intransitive ordering** of preferences. The concept of rationality is usually taken to mean that individual preference orderings are **transitive** (the opposite of intransitive). If someone is given choices A, B, and C and you know that she prefers A to B and B to C, then transitivity implies that she also prefers A to C. (The terminology comes from the transitivity of numbers in mathematics; for instance, if $3 > 2$ and $2 > 1$, then we know that $3 > 1$.) A transitive preference ordering will not cycle as does the social ordering derived in our city council example; hence, we say that such an ordering is intransitive.

Notice that all of the *councillors* have transitive preferences over the three welfare policy alternatives but the *council* does not. This is the Condorcet paradox: even if all individual preference orderings are transitive, there is no guarantee that the social-preference ordering induced by Condorcet's voting procedure also will be transitive. The result has far-reaching implications for public servants, as well as for the general public. It calls into question the basic notion of the "public interest," because such interests may not be easily defined or may not even exist. Our city council does not have any well-defined set of group preferences over the welfare policies. The lesson is that societies, institutions, or other large groups of people should not always be analyzed as if they acted like individuals.

The Condorcet paradox can even arise more generally. There is no guarantee that the social ordering induced by *any* formal group-voting process will be transitive just because individual preferences are. However, some estimates have shown that the paradox is most likely to arise when large groups of people are considering large numbers of alternatives. Smaller groups considering smaller numbers of alternatives are more likely to have similar preferences over those alternatives; in such situations, the paradox is much less likely to appear.[5] In fact, the paradox arose in our example because the council completely disagreed not only about which alternative was best but also about which was worst. The smaller the group, the less likely such outcomes are to occur.

B. The Agenda Paradox

The second paradox that we consider also entails a binary voting procedure, but this example considers the ordering of alternatives in that procedure. In a parliamentary setting with a committee chair who determines the specific order of voting for a three-alternative election, substantial power over the final outcome lies with the chair. In fact, the chair can take advantage of the intransitive social-preference ordering that arises from some sets of individual preferences and, by selecting an appropriate agenda, manipulate the outcome of the election in any manner she desires.

Consider again the city councillors Left, Center, and Right, who must decide among Generous, Average, and Decreased welfare policies. The councillors' preferences over the alternatives were shown in Figure 16.1. Let us now suppose that one of the councillors has been appointed chair of the council by the mayor, and the chair is given the right to decide which two welfare policies get voted on first and which goes up against the winner of that initial vote. With the given set of councillor preferences and common knowledge of the preference orderings, the chair can get any outcome that she wants. If Left were chosen chair, for example, she could orchestrate a win for Generous by setting Average against Decreased in the first round, with the winner to go up against Generous in round two. The result that any final ordering can be obtained by choosing an appropriate procedure is known as the **agenda paradox.**

With the set of preferences illustrated in Figure 16.1, then, we get not only the intransitive social preference ordering of the Condorcet paradox but also the result that the outcome of a binary procedure could be any of the possible alternatives. The only determinant of the outcome in such a case is the ordering of the agenda. This result implies that setting the agenda is the real game here, and because the chair sets the agenda, the appointment or election of the chair

[5]See Peter Ordeshook, *Game Theory and Political Theory* (Cambridge, UK: Cambridge University Press, 1986), p. 58.

is the true outlet for strategic behavior. Here, as in many other strategic situations, what appears to be the game (in this case, choice of a welfare policy) is not the true game at all; rather, those participating in the game engage in strategic play at an earlier point (deciding the identity of the chair) and vote according to set preferences in the eventual election.

However, the preceding demonstration of the agenda setter's power assumes that in the first round, voters choose between the two alternatives (Average and Decreased) on the basis only of their preferences between these two alternatives, with no regard for the eventual outcome of the procedure. Such behavior is called **sincere voting;** actually, myopic or nonstrategic voting would be a better name. If Center is a strategic game player, she should realize that if she votes for Decreased in the first round (even though she prefers Average between the pair presented at that stage), then Decreased will win the first round and will also win against Generous in the second round with support from Right. Center prefers Decreased over Generous as the eventual outcome. Therefore she should do this rollback analysis and vote strategically in the first round. But should she, if everyone else is also voting strategically? We examine the game of strategic voting and find its equilibrium in Section 4.

C. The Reversal Paradox

Positional voting methods also can lead to paradoxical results. The Borda count, for example, can yield the **reversal paradox** when the slate of candidates open to voters changes. This paradox arises in an election with at least four alternatives when one of them is removed from consideration after votes have been submitted, making recalculation necessary.

Suppose there are four candidates for a (hypothetical) special commemorative Cy Young Award to be given to a retired major-league baseball pitcher. The candidates are Steve Carlton (SC), Sandy Koufax (SK), Robin Roberts (RR), and Tom Seaver (TS). Seven prominent sportswriters are asked to rank these pitchers on their ballot cards. The top-ranked candidate on each card will get 4 points; decreasing numbers of points will be allotted to candidates ranked second, third, and fourth.

Across the seven voting sportswriters, there are three different preference orderings over the candidate pitchers; these preference orderings, with the number of writers having each ordering, are shown in Figure 16.2. When the votes are tallied, Seaver gets $(2 \times 3) + (3 \times 2) + (2 \times 4) = 20$ points; Koufax gets $(2 \times 4) + (3 \times 3) + (2 \times 1) = 19$ points; Carlton gets $(2 \times 1) + (3 \times 4) + (2 \times 2) = 18$ points; and Roberts gets $(2 \times 2) + (3 \times 1) + (2 \times 3) = 13$ points. Seaver wins the election, followed by Koufax, Carlton, and Roberts in last place.

Now suppose it is discovered that Roberts is not really eligible for the commemorative award, because he never actually won a Cy Young Award, having

ORDERING 1 (2 voters)	ORDERING 2 (3 voters)	ORDERING 3 (2 voters)
Koufax > Seaver > Roberts > Carlton	Carlton > Koufax > Seaver > Roberts	Seaver > Roberts > Carlton > Koufax

FIGURE 16.2 Sportswriter Preferences over Pitchers

reached the pinnacle of his career in the years just before the institution of the award in 1956. This discovery requires points to be recalculated, ignoring Roberts on the ballots. The top spot on each card now gets 3 points, while the second and third spots receive 2 and 1, respectively. Ballots from sportswriters with preference ordering 1, for example, now give Koufax and Seaver 3 and 2 points, respectively, rather than 4 and 3 from the first calculation; those ballots also give Carlton a single point for last place.

Adding votes with the revised point system shows that Carlton receives 15 points, Koufax receives 14 points, and Seaver receives 13 points. Winner has turned loser as the new results reverse the standings found in the first election. No change in preference orderings accompanies this result. The only difference in the two elections is the number of candidates being considered.

D. Change the Voting Method, Change the Outcome

As should be clear from the preceding discussion, election outcomes are likely to differ under different sets of voting rules. As an example, consider 100 voters who can be broken down into three groups on the basis of their preferences over three candidates (A, B, and C). Preferences of the three groups are shown in Figure 16.3. With the preferences as shown, and depending on the vote-aggregation method used, any of these three candidates could win the election.

With simple plurality rule, candidate A wins with 40% of the vote, even though 60% of the voters rank her lowest of the three. Supporters of candidate A would obviously prefer this type of election. If they had the power to choose the voting method, then plurality rule, a seemingly "fair" procedure, would win the election for A in spite of the majority's strong dislike for that candidate.

The Borda count, however, would produce a different outcome. In a Borda system with 3 points going to the most-preferred candidate, 2 points to the

GROUP 1 (40 voters)	GROUP 2 (25 voters)	GROUP 3 (35 voters)
A > B > C	B > C > A	C > B > A

FIGURE 16.3 Group Preferences over Candidates

middle candidate, and 1 to the least-preferred candidate, A gets 40 first-place votes and 60 third-place votes, for a total of 40(3) + 60(1) = 180 points. Candidate B gets 25 first-place votes and 75 second-place votes, for a total of 25(3) + 75(2) = 225 points; and C gets 35 first-place votes, 25 second-place votes, and 40 third-place votes, for a total of 35(3) + 25(2) + 40(1) = 195 points. In this procedure, B wins, with C in second place and A last. Candidate B would also win with the antiplurality vote, in which electors cast votes for all but their least-preferred candidate.

And what about candidate C? She can win the election if a majority or an instant-runoff system is used. In either method, A and C, with 40 and 35 votes in the first round, survive to face each other in the runoff. The majority-runoff system would call voters back to the polls to consider A and C; the instant runoff system would eliminate B and reallocate B's votes (from group 2 voters) to the next preferred alternative, candidate C. Then, because A is the least-preferred alternative for 60 of the 100 voters, candidate C would win the runoff election 60 to 40.

Another example of how different procedures can lead to different outcomes can be seen in the voting for the site of the Summer Olympics. The actual voting procedure for Olympics site selection is a mixed method consisting of multiple rounds of plurality rule with elimination and then a final majority-rule vote. Each country representative participating in the election casts one vote, and the candidate city with the lowest vote total is eliminated after each round. Majority rule is used to choose between the final two candidate cities. In voting for the sites of both the 1996 and 2000 games, one site led through most of the early rounds and would have won a single plurality vote; for the 1996 games, Athens (Greece) led early, and for the 2000 site, Beijing led early. In both cases, these plurality winners gained little additional support as other cities were eliminated and neither front-runner was eventually chosen in either election. Each actually retained a plurality through the next-to-last round of voting, but after the final elimination round, Athens saw the needed extra votes go to Atlanta, and Beijing watched them go to Sydney. It is worth noting that the Olympics site-selection procedure has since been changed!

3 GENERAL IDEAS PERTAINING TO PARADOXES

The discussion of the various voting paradoxes in Section 2 suggests that voting methods can suffer from a number of faults that lead to unusual, unexpected, or even unfair outcomes. In addition, this suggestion leads us to ask: Is there one voting system that satisfies certain regularity conditions, including transitivity, and that is the most "fair"—that is, most accurately captures the preferences of the electorate? Kenneth Arrow's **impossibility theorem** tells us that the answer to this question is no.[6]

The technical content of Arrow's theorem makes it beyond our scope to prove completely. But the sense of the theorem is straightforward. Arrow argued that no preference-aggregation method could satisfy all six of the critical principles that he identified:

1. The social or group ranking must rank all alternatives (be complete).
2. It must be transitive.
3. It should satisfy a condition known as *positive responsiveness,* or the Pareto property. Given two alternatives, A and B, if the electorate unanimously prefers A to B, then the aggregate ranking should place A above B.
4. The ranking must not be imposed by external considerations (such as customs) independent of the preferences of individual members of the society.
5. It must not be dictatorial—no single voter should determine the group ranking.
6. And it should be independent of irrelevant alternatives; that is, no change in the set of candidates (addition to or subtraction from) should change the rankings of the unaffected candidates.

Often the theorem is abbreviated by imposing the first four conditions and focusing on the difficulty of simultaneously obtaining the last two; the simplified form states that we cannot have independence of irrelevant alternatives (IIA) without dictatorship.[7]

You should be able to see immediately that some of the voting methods considered earlier do not satisfy all of Arrow's principles. The requirement of IIA, for example, is violated by the single transferable-vote procedure as well as by the Borda count. Borda's procedure is, however, nondictatorial and consistent, and it satisfies the Pareto property. All of the other systems that we have considered satisfy IIA but break down on one of the other principles.

Arrow's theorem has provoked extensive research into the robustness of his conclusion to changes in the underlying assumptions. Economists, political scientists, and mathematicians have searched for a way to reduce the number of criteria or relax Arrow's principles minimally to find a procedure that satisfies the criteria without sacrificing the core principles; their efforts have been largely unsuccessful. Most economic and political theorists now accept the idea that some form of compromise is necessary when choosing a vote- or preference-aggregation method. Here are a few prominent examples, each

[6]A full description of this theorem, often called "Arrow's General Possibility Theorem," can be found in Kenneth Arrow, *Social Choice and Individual Values,* 2nd ed. (New York: Wiley, 1963).

[7]See Walter Nicholson's treatment of Arrow's impossibility theorem in his *Microeconomic Theory,* 7th ed. (New York: Dryden Press, 1998), pp. 764–766, for more detail at a level appropriate for intermediate-level economics students.

representing the approach of a particular field—political science, economics, and mathematics.

A. Black's Condition

As the discussion in Section 2.A showed, the pairwise voting procedure does not satisfy Arrow's condition on transitivity of the social ranking, even when all individual rankings are transitive. One way to surmount this obstacle to meeting Arrow's conditions, as well as a way to prevent the Condorcet paradox, is to place restrictions on the preference orderings held by individual voters. Such a restriction, known as the requirement of **single-peaked preferences,** was put forth by the political scientist Duncan Black in the late 1940s.[8] Black's seminal paper on group decision making actually predates Arrow's impossibility theorem and was formulated with the Condorcet paradox in mind, but voting theorists have since shown its relevance to Arrow's work; in fact, the requirement of single-peaked preferences is sometimes referred to as **Black's condition.**

For a preference ordering to be single peaked, it must be the case that the alternatives being considered can be ordered along some specific dimension (for example, the expenditure level associated with each policy). To illustrate this requirement, we draw a graph in Figure 16.4 with the specified dimension on the horizontal axis and a voter's preference ranking (or payoff) on the vertical axis. For the single-peaked requirement to hold, each voter must have a single ideal or most-preferred alternative, and alternatives "farther away" from the most-preferred point must provide steadily lower payoffs. The two voters in Figure 16.4, Mr. Left and Ms. Right, have different ideal points along the policy dimension, but for each, the payoff falls steadily as the policy moves away from his or her ideal point.

Black shows that if preferences of each voter are single peaked, then the pairwise (majority) voting procedure must produce a transitive social ordering. The Condorcet paradox is prevented, and pairwise voting satisfies Arrow's transitivity condition.

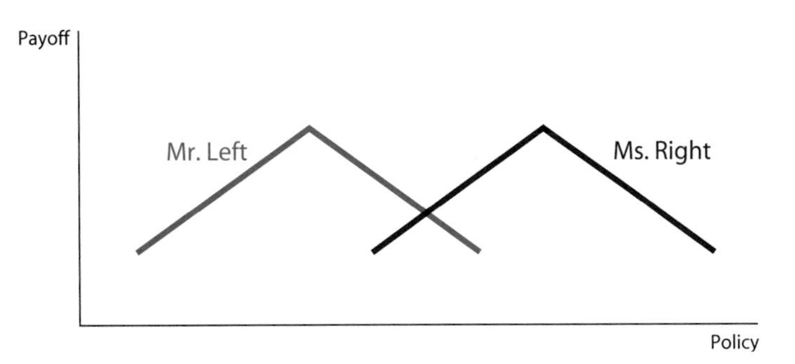

FIGURE 16.4 Single-Peaked Preferences

[8]Duncan Black, "On the Rationale of Group Decision-Making," *Journal of Political Economy,* vol. 56, no. 1 (Feb. 1948), pp. 23–34.

B. Robustness

An alternative, more recent method of compromise with Arrow comes from the economic theorists Partha Dasgupta and Eric Maskin.[9] They suggest a new criterion called **robustness** by which to judge voting methods. Robustness is measured by considering how often a voting procedure that is nondictatorial and that satisfies IIA as well as the Pareto property also satisfies the requirement of transitivity of its social ranking: For how many sets of voter-preference orderings does such a procedure satisfy transitivity?

With the use of the robustness criterion, simple majority rule can be shown to be *maximally robust*—that is, it is nondictatorial, satisfies IIA and Pareto, and provides transitive social rankings for the largest possible set of voter-preference orderings. Behind majority rule on the robustness scale lie other voting procedures, including the Borda count and plurality rule. The robustness criterion is appealing in its ability to establish one of the most commonly used voting procedures—the one most often associated with the democratic process—as a candidate for the best aggregation procedure.

C. Intensity Ranking

Another class of attempts to escape from Arrow's negative result focuses on the difficulty of satisfying Arrow's IIA requirement. A recent theory of this kind comes from the mathematician Donald Saari.[10] He suggests that a vote-aggregation method might use more information about voters' preferences than is contained in their mere ordering of any pair of alternatives, X and Y; rather, it could take into account each individual voter's *intensity* of preferences between that pair of alternatives. This intensity can be measured by counting the number of other alternatives, Z, W, V, . . . that a voter places between X and Y. Saari therefore replaces the IIA condition, number 6 of Arrow's principles, with a different one, which he labels IBI (intensity of binary independence) and which we will number 6′:

6′. Society's relative ranking of any two alternatives should be determined only by (1) each voter's relative ranking of the pair and (2) the intensity of this ranking.

This condition is weaker than IIA because it is like applying IIA only to such additions or deletions of "irrelevant" alternatives that do not change the intensity

[9]See Partha Dasgupta and Eric Maskin, "On the Robustness of Majority Rule," *Journal of the European Economic Association*, vol. 6 (2008), pp. 949–973.

[10]For more precise information about Saari's work on Arrow's theorem, see D. Saari, "Mathematical Structure of Voting Paradoxes I: Pairwise Vote," *Economic Theory*, vol. 15 (2000), pp. 1–53. Additional information on this result and on the robustness of the Borda count can be found in D. Saari, *Chaotic Elections* (Providence, RI: American Mathematical Society, 2000).

of people's preferences between the "relevant" ones. With this revision, the Borda count satisfies the modified Arrow theorem. It is the only one of the positional voting methods that does so.

Saari also hails the Borda count as the only procedure that appropriately observes ties within collections of ballots, a criterion that he argues is essential for a good aggregation system to satisfy. Ties can occur two ways: through **Condorcet terms** or through **reversal terms** within voter-preference orderings. In a three-candidate election among alternatives A, B, and C, the Condorcet terms are the preference orderings A > B > C, B > C > A, and C > A > B. A set of three ballots with these preferences appearing on one ballot apiece should logically offset each other, or constitute a tie. Reversal terms are preference orderings that contain a reversal in the location of a *pair* of alternatives. In the same election, two ballots with preference orderings of A > B > C and B > A > C should logically lead to a tie in a pairwise contest between A and B. Only the Borda procedure treats such collections of ballots—those with Condorcet terms or reversal terms—as tied. Although the Borda count can lead to the reversal paradox, as shown in the preceding section, it retains many proponents. The *only* time that the Borda procedure produces paradoxical results is when alternatives are dropped from consideration after ballots have been collected. Because such results can be prevented by using only ballots for the full set of candidates, the Borda procedure has gained favor in some circles as one of the best vote-aggregation methods.

Other researchers have made different suggestions regarding criteria that a good aggregation system should satisfy. Some of them include the *Condorcet criterion* (that a Condorcet winner should be selected by a voting system, if such a winner exists), the *consistency criterion* (that an election including all voters should elect the same alternative as would two elections held for an arbitrary division of the entire set of voters), and lack of manipulability (a voting system should not encourage manipulability—strategic voting—on the part of voters). We cannot consider each of these suggestions at length, but we do address strategic manipulation by voters in the next two sections.

4 STRATEGIC MANIPULATION OF VOTES

Several of the voting systems that we have considered yield considerable scope for strategic misrepresentation of preferences by voters. In Section 2.B, we showed how the power of an agenda-setting Left chair can be countered by a Center councillor voting in the first round against her true preference, so as to knock out her least-preferred alternative and send a more preferred one into the second round. More generally, voters can choose to vote for candidates, issues, or policies that are not actually their most-preferred outcomes among the alternatives presented in an early round if such behavior can alter the final election results in their favor. In this section, we consider a number of ways in which strategic voting behavior can affect elections.

A. Plurality Rule

Plurality-rule elections, often perceived as the fairest by many voters, still provide opportunities for strategic behavior. In Presidential elections, for instance, there are generally two major candidates in contention. When such a race is relatively close, there is potential for a third candidate to enter the race and divert votes away from the leading candidate; if the entry of this third player truly threatens the chances of the leader winning the election, the late entrant is called a **spoiler.**

Spoilers are generally believed to have little chance to win the whole election, but their role in changing the election outcome is undisputed. In elections with a spoiler candidate, those who prefer the spoiler to the leading major candidate but least prefer the trailing major candidate may do best to strategically misrepresent their preferences to prevent the election of their least-favorite candidate. That is, you should vote for the leader in such a case even though you would prefer the spoiler because the spoiler is unlikely to garner a plurality; voting for the leader then prevents the trailing candidate, your least favorite, from winning.[11]

Evidence shows that there has been a great deal of strategic voting in recent U.S. Presidential elections—the 1992 and 2000 races in particular. In 1992, Ross Perot ran a strong third-party campaign for much of the primary season, only to drop out of the race at midsummer and reappear in the fall. His reappearance at such a late date cast him as a spoiler in the then close race between the incumbent, President George Bush, and his Democratic challenger, Bill Clinton. Perot's role in dividing the conservative vote and clinching Clinton's victory that year has been debated, but it is certainly possible that Clinton would not have won without Perot on the ballot. A more interesting theory suggests significant misrepresentation of preferences at the polls. A *Newsweek* survey claimed that if more voters had believed Perot was capable of winning the election, he might have done so; a plurality of 40% of voters surveyed said they would have voted for Perot (instead of Bush or Clinton) if they had thought he could have won.[12]

Ralph Nader played a similar role in the 2000 Presidential election between the Democratic Vice-president, Al Gore, and the Republican challenger, George W. Bush. Nader was more concerned about garnering 5% of the popular

[11]Note that an approval-voting method would not suffer from this same problem.

[12]"Ross Reruns," *Newsweek*, Special Election Recap Issue, November 18, 1996, p. 104.

vote so that his Green Party could qualify for federal matching election funds than he was about winning the presidency, but his candidacy made an already close election even tighter. The 2000 election will be remembered more for the extraordinarily close outcome in Florida and the battle for that state's electoral votes. But Nader was pulling needed votes from Gore's camp in the lead-up to the election, causing enough concern that public discussions (on radio and television and in major metropolitan newspapers) arose regarding the possibilities for strategic voting. Specifically, several groups (as well as a number of Web sites) advocated "vote swapping" schemes designed to gain Nader his needed votes without costing Gore the electoral votes of any of his key states. Nader voters in key Gore states (such as Pennsylvania, Michigan, and Maine) were asked to "swap" their votes with Gore supporters in a state destined to go to George W. Bush (such as Texas or Wyoming); a Michigan Nader supporter could vote for Gore while her Nader vote was cast in Texas. Evidence on the efficacy of these strategies is mixed. We do know that Nader failed to win his 5% of the popular vote but that Gore carried all of Pennsylvania, Michigan, and Maine.

In elections for legislatures, where many candidates are chosen, the performance of third parties is very different under a system of proportional representation of the whole population in the whole legislature from that under a system of plurality in separate constituencies. Britain has the constituency and plurality system. In the past 50 years, the Labor and Conservative parties have shared power. The Liberal Party, despite sizable third-place support in the electorate, has suffered from strategic voting and therefore has had disproportionately few seats in Parliament. Italy has had the nationwide list and proportional representation system; there is no need to vote strategically in such a system, and even small parties can have significant presence in the legislature. Often no party has a clear majority of seats, and small parties can affect policy through bargaining for alliances.

A party cannot flourish if it is largely ineffective in influencing a country's political choices. Therefore we tend to see just two major parties in countries with the plurality system and several parties in those with the proportional representation system. Political scientists call this observation *Duverger's law.*

In the legislature, the constituency system tends to produce only two major parties—often one of them with a clear majority of seats and therefore more decisive government. But it runs the risk that the minority's interests will be overlooked—that is, of producing a "tyranny of the majority." A proportional representation system gives more of a voice to minority views. But it can produce inconclusive bargaining for power and legislative gridlock. Interestingly, each country seems to believe that its system performs worse and considers switching to the other; in Britain, there are strong voices calling for proportional representation, and Italy has been seriously considering a constituency system.

B. Pairwise Voting

When you know that you are bound by a pairwise method such as the amendment procedure, you can use your prediction of the second-round outcome to determine your optimal voting strategy in the first round. It may be in your interest to appear committed to a particular candidate or policy in the first round, even if it is not your most-preferred alternative, so that your least-favorite alternative cannot win the entire election in the second round.

We return here to our example of the city council with an agenda-setting chair; again, all three preference rankings are assumed to be known to the entire council. Suppose Councillor Left, who most prefers the Generous welfare package, is appointed chair and sets the Average and Decreased policies against each other in a first vote, with the winner facing off against the Generous policy in the second round. If the three councillors vote strictly according to their preferences, shown in Figure 16.1, Average will beat Decreased in the first vote and Generous will then beat Average in the second vote; the chair's preferred outcome will be chosen. The city councillors are likely to be well-trained strategists, however, who can look ahead to the final round of voting and use rollback to determine which way to vote in the opening round.

In the scenario just described, Councillor Center's least-preferred policy will be chosen in the election. Therefore, rollback analysis says that she should vote strategically in the first round to alter the election's outcome. If Center votes for her most-preferred policy in the first round, she will vote for the Average policy, which will then beat Decreased in that round and lose to Generous in round two. However, she could instead vote strategically for the Decreased policy in the first round, which would lift Decreased over Average on the first vote. Then, when Decreased is set up against Generous in the second round, Generous will lose to Decreased. Councillor Center's misrepresentation of her preference ordering with respect to Average and Decreased helps her to change the winner of the election from Generous to Decreased. Although Decreased is not her most-preferred outcome, it is better than Generous from her perspective.

This strategy works well for Center if she can be sure that no other strategic votes will be cast in the election. Thus we need to analyze both rounds of voting fully to verify the Nash equilibrium strategies for the three councillors. We do so by using rollback on the two simultaneous-vote rounds of the election, starting with the two possible second-round contests, A versus G or D versus G. In the following analysis, we use the abbreviated names of the policies, G, A, and D.

Figure 16.5 illustrates the outcomes that arise in each of the possible second-round elections. The two tables in Figure 16.5a show the winning policy (not payoffs to the players) when A has won the first round and is pitted against G; the tables in Figure 16.5b show the winning policy when L has won the first

(a) A versus G election

Right votes:

A

		CENTER	
		A	G
LEFT	A	A	A
	G	A	G

G

		CENTER	
		A	G
LEFT	A	A	G
	G	G	G

(b) D versus G election

Right votes:

D

		CENTER	
		D	G
LEFT	D	D	D
	G	D	G

G

		CENTER	
		D	G
LEFT	D	D	G
	G	G	G

FIGURE 16.5 Election Outcomes in Two Possible Second-Round Votes

round. In both cases, Councillor Left chooses the row of the final outcome, Center chooses the column, and Right chooses the actual table (left or right).

You should be able to establish that each councillor has a dominant strategy in each second-round election. In the A-versus-G election, Left's dominant strategy is to vote for G, Center's dominant strategy is to vote for A, and Right's dominant strategy is to vote for G; G will win this election. If the councillors consider D versus G, Left's dominant strategy is still to vote for G, and Right and Center both have a dominant strategy to vote for D; in this vote, D wins. A quick check shows that all of the councillors vote according to their true preferences in this round. Thus these dominant strategies are all the same: "Vote for the alternative that I prefer." Because there is no future to consider in the second-round vote, the councillors simply vote for whichever policy ranks higher in their preference ordering.[13]

We can now use the results from our analysis of Figure 16.5 to consider optimal voting strategies in the first round of the election, in which voters choose

[13]It is a general result in the voting literature that voters faced with pairs of alternatives will always vote truthfully at the last round of voting.

between policies A and D. Because we know how the councillors will vote in the next round given the winner here, we can show the outcome of the entire election in the tables in Figure 16.6.

As an example of how we arrived at these outcomes, consider the G in the upper-left-hand cell of the right-hand table in Figure 16.6. The outcome in that cell is obtained when Left and Center both vote for A in the first round while Right votes for D. Thus A and G are paired in the second round, and as we saw in Figure 16.5, G wins. The other outcomes are derived in similar fashion.

Given the outcomes in Figure 16.6, Councillor Left (who is the chair and has set the agenda) has a dominant strategy to vote for A in this round. Similarly, Councillor Right has a dominant strategy to vote for D. Neither of these councillors misrepresent their preferences or vote strategically in either round. Councillor Center, however, has a dominant strategy to vote for D here even though she strictly prefers A to D. As the preceding discussion suggested, she has a strong incentive to misrepresent her preferences in the first round of voting; and she is the only one who votes strategically. Center's behavior changes the winner of the election from G (the winner without strategic voting) to D.

Remember that the chair, Councillor Left, set the agenda in the hope of having her most-preferred alternative chosen. Instead, her *least*-preferred alternative has prevailed. It appears that the power to set the agenda may not be so beneficial after all. But Councillor Left should anticipate the strategic behavior. Then she can choose the agenda so as to take advantage of her understanding of games of strategy. In fact, if she sets D against G in the first round and then the winner against A, the Nash equilibrium outcome is G, the chair's most-preferred outcome. With that agenda, Right misrepresents her preferences in the first round to vote for G over D to prevent A, her least-preferred outcome, from winning. You should verify that this is Councillor Left's best agenda-setting strategy. In the full voting game where setting the agenda is considered an initial, prevoting round, we should expect to see the Generous welfare policy adopted when Councillor Left is chair.

We can also see an interesting pattern emerge when we look more closely at voting behavior in the strategic version of the election. There are pairs of

Right votes:

A

		CENTER	
		A	D
LEFT	A	G	G
	D	G	D

D

		CENTER	
		A	D
LEFT	A	G	D
	D	D	D

FIGURE 16.6 Election Outcomes Based on First-Round Votes

councillors who vote "together" (the same as one another) in both rounds. Under the original agenda, Right and Center vote together in both rounds, and in the suggested alternative (D versus G in the first round), Right and Left vote together in both rounds. In other words, a sort of long-lasting coalition has formed between two councillors in each case.

Strategic voting of this type appears to have taken place in Congress on more than one occasion. One example was a federal school-construction-funding bill considered in 1956.[14] Before being brought to a vote against the status quo of no funding, the bill was amended in the House of Representatives to require that aid be offered only to states with no racially segregated schools. Under the parliamentary voting rules of Congress, a vote on whether to accept the so-called Powell Amendment was taken first, with the winning version of the bill considered afterward. Political scientists who have studied the history of this bill argue that opponents of school funding strategically misrepresented their preferences regarding the amendment to defeat the original bill. A key group of Representatives voted for the amendment but then joined opponents of racial integration in voting against the full bill in the final vote; the bill was defeated. Voting records of this group indicate that many of them had voted against racial integration matters in other circumstances, implying that their vote for integration in this case was merely an instance of strategic voting and not an indication of their true feelings regarding school integration.

C. Strategic Voting with Incomplete Information

The preceding analysis showed that sometimes committee members have incentives to vote strategically to prevent their least-preferred alternative from winning an election. Our example assumed that the council members knew the possible preference orderings and how many other councillors had those preferences. Now suppose information is incomplete; each council member knows the possible preference orderings, her own actual ordering, and the probabilities that each of the others have a particular ordering, but not the actual distribution of the different preference orderings among the other councillors. In this situation, each councillor's strategy needs to be conditioned on her beliefs about that distribution and on her beliefs about how truthful other voters will be.[15]

For an example, suppose that we still have a three-member council considering the three alternative welfare policies described earlier according to the

[14]A more complete analysis of the case can be found in Riker, *Liberalism Against Populism*, pp. 152–157.

[15]This result can be found in P. Ordeshook and T. Palfrey, "Agendas, Strategic Voting, and Signaling with Incomplete Information," *American Journal of Political Science*, vol. 32, no. 2 (May 1988), pp. 441–466. The structure of the example to follow is based on Ordeshook and Palfrey's analysis.

(original) agenda set by Councillor Left; that is, the Council considers policies A and D in the first round with the winner facing G in the second round. We assume that there are still three different possible preference orderings, as illustrated in Figure 16.1, and that the councillors know that these orderings are the only possibilities. The difference is that no one knows for sure exactly how many councillors have each set of preferences. Rather, each councillor knows her own type, and she knows that there is some positive probability of observing each type of voter (Left, Center, or Right), with the probabilities p_L, p_C, and p_R summing to 1.

We saw earlier that all three councillors vote truthfully in the last round of balloting. We also saw that Left- and Right-type councillors vote truthfully in the first round as well. This result remains true in the incomplete information case. Right-type voters prefer to see D win the first-round election; given this preference, Right always does at least as well by voting for D over A (if both other councillors have voted the same way) and sometimes does better by voting this way (if the other two votes split between D and A). Similarly, Left-type voters prefer to see A survive to vie against G in round two; these voters always do at least as well as otherwise—and sometimes do better—by voting for A over D.

At issue then is only the behavior of the Center-type voters. Because they do not know the types of the other councillors and because they have an incentive to vote strategically for some preference distributions—specifically the case in which it is known for certain that there is one voter of each type—their behavior will depend on the probabilities that the various voter types may occur within the council. We consider here one of two polar cases in which a Center-type voter believes that other Center types will vote truthfully, and we look for a symmetric, pure-strategy Nash equilibrium. The case in which she believes that other Center types will vote strategically is taken up in the exercises.

To make outcome comparisons possible, we will specify payoffs for the Center-type voter associated with the possible winning policies. Center-type preferences are A > D > G. Suppose that, if A wins, Center types receive a payoff of 1 and, if G wins, Center types receive a payoff of 0. If D wins, Center types receive some intermediate-level payoff, call it u, where $0 < u < 1$.

Now suppose our Center-type councillor must decide how to vote in the first round (A versus D) in an election in which she believes that both other voters vote truthfully, regardless of their type. If both voters choose either A or D, then Center's vote is immaterial to the final outcome; she is indifferent between A and D. If the other two voters split their votes, however, then Center can influence the election outcome. Her problem is that she needs to decide whether to vote truthfully herself.

If the other two voters split between A and D and if both are voting truthfully, then the vote for D must have come from a Right-type voter. But the vote for A could have come from *either* a Left type *or* a (truthful) Center type. If the A vote came from a Left-type voter, then Center knows that there is one voter of each type. If she votes truthfully for A in this situation, A will win the first round but lose to G in the end; Center's payoff will be 0. If Center votes strategically for D, D beats A and G, and Center's payoff is u. On the other hand, if the A vote came from a Center-type voter, then Center knows there are two Center types and a Right type but no Left type on the Council. In this case, a truthful vote for A helps A win the first round, and then A also beats G by a vote of 2 to 1 in round two; Center gets her highest payoff of 1. If Center were to vote strategically for D, D would win both rounds again and Center would get u.

To determine Center's optimal strategy, we need to compare her expected payoff from truthful voting with her expected payoff from strategic voting. With a truthful vote for A, Center's payoff depends on how likely it is that the other A vote comes from a Left type or a Center type. Those probabilities are straightforward to calculate. The probability that the other A vote comes from a Left type is just the probability of a Left type being one of the remaining voters, or $p_L/(p_L + p_C)$; similarly, the probability that the A vote comes from a Center type is $p_C/(p_L + p_C)$. Then Center's payoffs from truthful voting are 0 with probability $p_L/(p_L + p_C)$ and 1 with probability $p_C/(p_L + p_C)$, so the expected payoff is $p_C/(p_L + p_C)$. With a strategic vote for D, D wins regardless of the identity of the third voter—D wins with certainty—and so Center's expected payoff is just u. Center's final decision is to vote truthfully as long as $p_C/(p_L + p_C) > u$.

Note that Center's decision-making condition is an intuitively reasonable one. If the probability of there being more Center-type voters is large or relatively larger than the probability of having a Left-type voter, then the Center types vote truthfully. Voting strategically is useful to Center only when she is the only voter of her type on the Council.

We add two additional comments on the existence of imperfect information and its implications for strategic behavior. First, if the number of councillors, n, is larger than three but odd, then the expected payoff to a Center type from voting strategically remains equal to u, and the expected payoff from voting truthfully is $[p_C/(p_L + p_C]^{(n-1)/2}$.[16] Thus, a Center type should vote truthfully only when $[p_C/(p_L + p_C)]^{(n-1)/2} > u$. Because $p_C/(p_L + p_C) < 1$ and $u > 0$, this inequality will *never* hold for large enough values of n. This result tells us that a symmetric truthful-voting equilibrium can never persist in a large enough council! Second, imperfect information about the preferences of other voters yields additional

[16]A Center type can affect the election outcome only if all other votes are split evenly between A and D. Thus, there must be exactly $(n - 1)/2$ Right-type voters choosing D in the first round and $(n - 1)/2$ other voters choosing A. If those A voters are Left types, then A won't win the second-round election, and Center will get 0 payoff. For Center to get a payoff of 1, it must be true that all of the other A voters are Center types. The probability of this occurring is $[p_C/(p_L + p_C)]^{(n-1)/2}$; then Center's expected payoff from voting truthfully is as stated. See Ordeshook and Palfrey, p. 455.

scope for strategic behavior. With agendas that include more than two rounds, voters can use their early-round votes to signal their types. The extra rounds give other voters the opportunity to update their prior beliefs about the probabilities p_C, p_L, and p_R and a chance to act on that information. With only two rounds of pairwise votes, there is no time to use any information gained during round one, because truthful voting is a dominant strategy for all voters in the final round.

5 GENERAL IDEAS PERTAINING TO MANIPULATION

The extent to which a voting procedure is susceptible to strategic misrepresentation of preferences, or strategic manipulability by voters of the types illustrated in Section 4, is another topic that has generated considerable interest among voting theorists. Arrow does not require nonmanipulability in his theorem, but the literature has considered how such a requirement would relate to Arrow's conditions. Similarly, theorists have considered the scope for manipulability in various procedures, producing rankings of voting methods.

The economist William Vickrey, perhaps better known for his work on auctions (see Chapter 17), did some of the earliest work considering strategic behavior of voters. He pointed out that procedures satisfying Arrow's IIA assumption were most immune to strategic manipulation. He also set out several conditions under which strategic behavior is more likely to be attempted and be successful. In particular, he noted that situations with smaller numbers of informed voters and smaller sets of available alternatives may be most susceptible to manipulation, given a voting method that is itself manipulable. This result means, however, that weakening the IIA assumption to help voting procedures satisfy Arrow's conditions makes way for more manipulable procedures. In particular, Saari's intensity ranking version of IIA (called IBI), mentioned in Section 3.C, may allow more procedures to satisfy a modified version of Arrow's theorem but may simultaneously allow more manipulable procedures to do so.

Like Arrow's general result on the impossibility of preference aggregation, the general result on manipulability is a negative one. Specifically, the **Gibbard-Satterthwaite theorem** shows that if there are three or more alternatives to consider, the only voting procedure that prevents strategic voting is dictatorship: one voter is assigned the role of dictator, and her preferences determine the election outcome.[17] Combining the Gibbard-Satterthwaite outcome

[17]For the theoretical details on this result, see A. Gibbard, "Manipulation of Voting Schemes: A General Result," *Econometrica*, vol. 41, no. 4 (July 1973), pp. 587–601, and M. A. Satterthwaite, "Strategy-Proofness and Arrow's Conditions," *Journal of Economic Theory*, vol. 10 (1975), pp. 187–217. The theorem carries both their names because each proved the result independent of the other.

with Vickrey's discussion of IIA may help the reader understand why Arrow's theorem is often reduced to a consideration of which procedures can simultaneously satisfy nondictatorship and IIA.

Finally, some theorists have argued that voting systems should be evaluated not on their ability to satisfy Arrow's conditions but on their tendency to encourage manipulation. The relative manipulability of a voting system can be determined by the amount of information about the preferences of other voters that is required by voters to manipulate an election successfully. Some research based on this criterion suggests that of the procedures so far discussed, plurality rule is the most manipulable (that is, requires the least information). In decreasing order of manipulability are approval voting, the Borda count, the amendment procedure, majority rule, and the Hare procedure (single transferable vote).[18]

It is important to note that the classification of procedures by level of manipulability depends only on the amount of information necessary to manipulate a voting system and is not based on the ease of putting such information to good use or whether manipulation is most easily achieved by individual voters or groups. In practice, the manipulation of plurality rule by *individual* voters is quite difficult.

6 THE MEDIAN VOTER THEOREM

All of the preceding sections have focused on the behavior, strategic and otherwise, of voters in multiple alternative elections. However, strategic analysis can also be applied to *candidate* behavior in such elections. Given a particular distribution of voters and voter preferences, candidates will, for instance, need to determine optimal strategies in building their political platforms. When there are just two candidates in an election, when voters are distributed in a "reasonable" way along the political spectrum, and when each voter has "reasonably" consistent (meaning singled-peaked) preferences, the **median voter theorem** tells us that both candidates will position themselves on the political spectrum at the same place as the median voter. The **median voter** is the "middle" voter in that distribution—more precisely, the one at the 50th percentile.

The full game here has two stages. In the first stage, candidates choose their locations on the political spectrum. In the second stage, voters elect one of the candidates. The general second-stage game is open to all of the varieties of strategic misrepresentation of preferences discussed earlier. Hence we have

[18]H. Nurmi's classification can be found in his *Comparing Voting Systems* (Norwell, MA: D. Reidel, 1987).

reduced the choice of candidates to two for our analysis to prevent such behavior from arising in equilibrium. With only two candidates, second-stage votes will directly correspond to voter preferences, and the first-stage location decision of the candidates remains the only truly interesting part of the larger game. It is in that stage that the median voter theorem defines Nash equilibrium behavior.

A. Discrete Political Spectrum

Let us first consider a population of 90 million voters, each of whom has a preferred position on a five-point political spectrum: Far Left (FL), Left (L), Center (C), Right (R), and Far Right (FR). We suppose that these voters are spread symmetrically around the center of the political spectrum. The **discrete distribution** of their locations is shown by a **histogram,** or bar chart, in Figure 16.7. The height of each bar indicates the number of voters located at that position. In this example, we have supposed that, of the 90 million voters, 40 million are Left, 20 million are Far Right, and 10 million each are Far Left, Center, and Right.

Voters will vote for the candidate who publicly identifies herself as being closer to their own position on the spectrum in an election. If both candidates are politically equidistant from a group of like-minded voters, each voter flips a coin to decide which candidate to choose; this process gives each candidate one-half of the voters in that group.

Now suppose there is an upcoming Presidential election between a former First Lady (Claudia) and a former First Lady hopeful (Dolores), each now

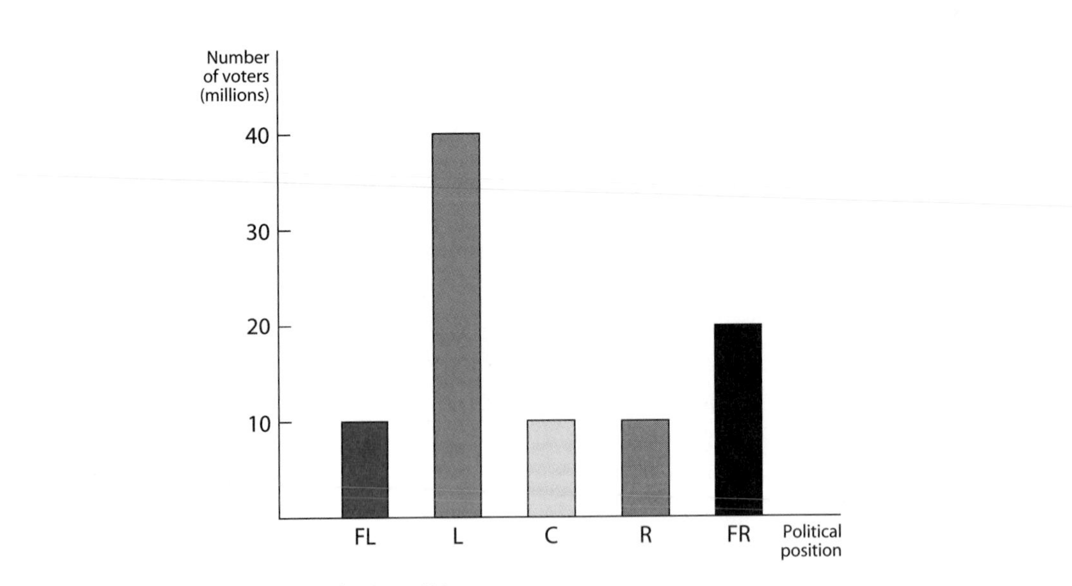

FIGURE 16.7 Discrete Distribution of Voters

running for office on her own.[19] Under the configuration of voters illustrated in Figure 16.7, we can construct a payoff table for the two candidates showing the number of votes that each can expect to receive under all of the different combinations of political platform choices. This five-by-five table is shown in Figure 16.8, with totals denoted in millions of votes. The candidates will choose their optimal location strategies to maximize the number of votes that they receive (and thus increase the chances of winning).[20]

Here is how the votes are allocated. When both candidates choose the *same* position (the five cells along the top-left to bottom-right diagonal of the table), each candidate gets exactly one-half of the votes; because all voters are equidistant from each candidate, all of them flip coins to decide their choices, and each candidate garners 45 million votes. When the two candidates choose *different* positions, the more-left candidate gets all the votes at or to the left of her position while the more-right candidate gets all the votes at or to the right of her position. In addition, each candidate gets the votes in central positions closer to her than to her rival, and the two of them split the votes from any voters in a central position equidistant between them. Thus, if Claudia locates herself at L while Dolores locates herself at FR, Claudia gets the 40 million votes at L, the 10 million at FL, *and* the 10 million at C (because C is closer to L than to FR). Dolores gets the 20 million votes at FR and the 10 million at R (because R is closer to FR than to L). The payoff is (60, 30). Similar calculations determine the outcomes in the rest of the table.

		DOLORES				
		FL	L	C	R	FR
CLAUDIA	FL	45, 45	10, 80	30, 60	50, 40	55, 35
	L	80, 10	45, 45	50, 40	55, 35	60, 30
	C	60, 30	40, 50	45, 45	60, 30	65, 25
	R	40, 50	35, 55	30, 60	45, 45	70, 20
	FR	35, 55	30, 60	25, 65	20, 70	45, 45

FIGURE 16.8 Payoff Table for Candidates' Positioning Game

[19]Any resemblance between our hypothetical candidates and actual past or possible future candidates in the United States is not meant to imply an analysis or prediction of their performances relative to the Nash equilibrium. Nor is our distribution of voters meant to typify U.S. voter preferences.

[20]To keep the analysis simple, we ignore the complications created by the electoral college and suppose that only the popular vote matters.

The table in Figure 16.8 is large, but the game can be solved very quickly. We begin with the now familiar search for dominant, or dominated, strategies for the two players. Immediately we see that for Claudia, FL is dominated by L and FR is dominated by R. For Dolores, too, her FL is dominated by L and FR by R. With these extreme strategies eliminated, for each candidate her R is dominated by C. With the two R strategies gone, C is dominated by L for each candidate. The only remaining cell in the table is (L, L); this is the Nash equilibrium.

We now note three important characteristics of the equilibrium in the candidate-location game. First, both candidates locate at the *same* position in equilibrium. This illustrates the **principle of minimum differentiation,** a general result in all two-player games of locational competition, whether it be political platform choice by presidential candidates, hotdog-cart location choices by street vendors, or product feature choices by electronics manufacturing firms.[21] When the persons who vote for or buy from you can be arranged on a well-defined spectrum of preferences, you do best by looking as much like your rival as possible. This explains a diverse collection of behaviors on the part of political candidates and businesses. It may help you understand, for example, why there is never just one gas station at a heavily traveled intersection or why all brands of four-door sedans (or minivans or sport utility vehicles) seem to look the same even though every brand claims to be coming out continually with a "new" look.

Second and perhaps most crucial, both candidates locate at the position of the median voter in the population. In our example, with a total of 90 million voters, the median voter is number 45 million from each end. The numbers within one location can be assigned arbitrarily, but the location of the median voter is clear; here, the median voter is located at the L position on the political spectrum. So that is where both candidates locate themselves, which is the result predicted by the median voter theorem.

Third, observe that the location of the median voter need not coincide with the geometric center of the spectrum. The two will coincide if the distribution of voters is symmetric, but the median voter can be to the left of the geometric center if the distribution is skewed to the left (as is true in Figure 16.7) and to the right if the distribution is skewed to the right. This helps explain why state political candidates in Massachusetts, for example, *all* tend to be more liberal than candidates for similar positions in Texas or South Carolina.

The median voter theorem can be expressed in different ways. One version states simply that the position of the median voter is the equilibrium-location position of the candidates in a two-candidate election. Another version says that the position that the median voter most prefers will be the Condorcet winner; this position will defeat every other position in a pairwise contest. For

[21]Economists learn this result within the context of Hotelling's model of spatial location. See Harold Hotelling, "Stability in Competition," *Economic Journal*, vol. 39, no. 1 (March 1929), pp. 41–57.

example, if M is this median position and L is any position to the left of M, then M will get all the votes of people who most prefer a position at or to the right of M, plus some to the left of M but closer to M than to L. Thus M will get more than 50% of the votes. The two versions amount to the same thing because, in a two-candidate election, both seeking to win a majority will adopt the Condorcet-winner position. These interpretations are identical. In addition, to guarantee that the result holds for a particular population of voters, the theorem (in either form) requires that each voter's preferences be "reasonable," as suggested earlier. *Reasonable* here means "single peaked," as in Black's condition described in Section 3.A and Figure 16.4. Each voter has a unique, most-preferred position on the political spectrum, and her utility (or payoff) decreases away from that position in either direction.[22] In actual U.S. Presidential elections, the theorem is borne out by the tendency for the main candidates to make very similar promises to the electorate.

B. Continuous Political Spectrum

The median voter theorem can also be proved for a continuous distribution of political positions. Rather than having five, three, or any finite number of positions from which to choose, a **continuous distribution** assumes there are effectively an infinite number of political positions. These political positions are then associated with locations along the real number line between 0 and 1.[23] Voters are still distributed along the political spectrum as before, but because the distribution is now continuous rather than discrete, we use a voter **distribution function** rather than a histogram to illustrate voter locations. Two common functions—the

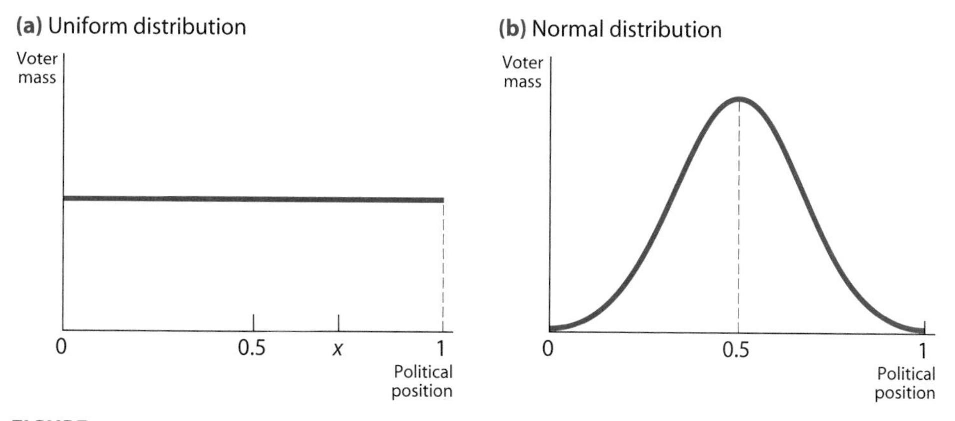

FIGURE 16.9 Continuous Voter Distributions

[22]However, the distribution of voters' ideal points along the political spectrum does not have to be single peaked, as indeed the histogram in Figure 16.7 is not—there are two peaks at L and FR.

[23]This construction is the same one used in Chapters 12 and 13 for analyzing large populations of individual members.

uniform distribution and the (symmetric) **normal distribution**—are illustrated in Figure 16.9.[24] The area under each curve represents the total number of votes available; at any given point along the interval from 0 to 1, such as x in Figure 16.9a, the number of votes up to that point is determined by finding the area under the distribution function from 0 to x. It should be clear that the median voter in each of these distributions is located at the center of the spectrum, at position 0.5.

It is not feasible to construct a payoff table for our two candidates in the continuous-spectrum case; such tables must necessarily be finitely dimensioned and thus cannot accommodate an infinite number of possible strategies for players. We can, however, solve the game by applying the same strategic logic that we used for the discrete (finite) case discussed in Section 5.A.

Consider the options of Claudia and Dolores as they contemplate the possible political positions open to them. Each knows that she must find her Nash equilibrium strategy—her best response to the equilibrium strategy of her rival. We can define a set of strategies that are best responses quite easily in this game, even though the complete set of possible strategies is impossible to delineate.

Suppose Dolores locates at a random position on the political spectrum, such as x in Figure 16.9a. Claudia can then calculate how the votes will be split for all possible positions that she might choose. If she chooses a position to the left of x, she gets all the votes to her left and half of the votes lying between her position and Dolores's. If she locates to the right of x, she gets all the votes to her right and half of the votes lying between her position and x. Finally, if she, too, locates at x, she and Dolores split the votes 50–50. These three possibilities effectively summarize all of Claudia's location choices, given that Dolores has chosen to locate at x.

But which of the response strategies just outlined is Claudia's "best" response? The answer depends on the location of x relative to the median voter. If x is to the right of the median, then Claudia knows that her best response will be to maximize the number of votes that she gains, which she can do by locating an infinitely small bit to the left of x.[25] In that case, she effectively gets all the votes from 0 to x, and Dolores gets those from x to 1. When x is to the right of the median, as in Figure 16.9a, then the number of voters represented by the area under the distribution curve from 0 to x is by definition larger than the number of voters from x to 1, so Claudia would win the election. Similarly,

[24] We do not delve deeply into the mechanics underlying distribution theory or the integral calculus required to calculate the exact proportion of the voting population lying to the left or right of any particular position on the continuous political spectrum. Here we present only enough information to convince you that the median voter theorem continues to hold in the continuous case.

[25] Such a location, infinitesimally removed from x to the left, is feasible in the continuous case. In our discrete example, candidates had to locate at exactly the same position.

if x is to the left of the median, Claudia's best response will be to locate an infinitely small bit to the right of x and thus gain all the votes from x to 1. When x is exactly at the median, Claudia does best by also choosing to locate at x. The best-response strategies for Dolores are constructed exactly the same way and, given the location of her rival, are exactly the same as those described for Claudia. Graphically, these best-response curves lie just above and below the 45° line up to the position of the median voter, at which point they lie exactly on the 45° line. (Claudia's best response to Dolores's location at that of the median voter is to locate in the same place; the same is true in reverse for Dolores.) Beyond the position of the median voter, the best-response curves switch sides of the 45° line.

We now have complete descriptions of the best-response strategies for both candidates. The Nash equilibrium occurs at the intersection of the best-response curves; this intersection lies at the position of the median voter. You can think this through intuitively by picking any starting location for one of the candidates and applying the best-response strategies over and over until each candidate is located at a position that represents her best response to the position chosen by her rival. If Dolores were contemplating locating at x in Figure 16.9a, Claudia would want to locate just to the left of x, but then Dolores would want to locate just to the left of that, and so on. Only when the two candidates locate exactly at the median of the distribution (whether the distribution is uniform or normal or some other kind) do they find that their decisions are best responses to each other. Again we see that the Nash equilibrium is for both candidates to locate at the position of the median voter.

More complex mathematics is needed to prove the continuous version of the median voter theorem to the satisfaction of a true mathematician. For our purposes, however, the discussion given here should convince you of the validity of the theorem in both its discrete and continuous forms. The most important limitation of the median voter theorem is that it applies when there is just one issue, or on a one-dimensional spectrum of political differences. If there are two or more dimensions—for example, if being conservative versus liberal on social issues does not coincide with being conservative versus liberal on economic issues—then the population is spread out in a two-dimensional "issue space" and the median voter theorem no longer holds. The preferences of every individual voter can be single peaked, in the sense that the individual voter has a most-preferred point and her payoff value drops away from this point in all directions, like the height going away from the peak of a hill. But we cannot identify a median voter in two dimensions, such that exactly the same number of voters have their most-preferred point to the one side of the median voter position as to the other side. In two dimensions, there is no unique sense of side, and the numbers of voters to the two sides can vary, depending on just how we define "side."

SUMMARY

Elections can be held with the use of a variety of different voting procedures that alter the order in which issues are considered or the manner in which votes are tallied. Voting procedures are classified as *binary, plurative,* or *mixed methods.* Binary methods include *majority rule,* as well as *pairwise* procedures such as the *Condorcet method* and the *amendment procedure. Positional methods* such as *plurality rule* and the *Borda count,* as well as *approval voting,* are plurative methods. And *majority runoffs, instant runoffs,* and *proportional representation* are mixed methods.

Voting paradoxes (such as the *Condorcet,* the *agenda,* and *the reversal paradox*) show how counterintuitive results can arise owing to difficulties associated with aggregating preferences or to small changes in the list of issues being considered. Another paradoxical result is that outcomes in any given election under a given set of voter preferences can change, depending on the voting procedure used. Certain principles for evaluating voting methods can be described, although Arrow's *impossibility theorem* shows that no one system satisfies all of the criteria at the same time. Researchers in a broad range of fields have considered alternatives to the principles that Arrow identified.

Voters have scope for strategic behavior in the game that chooses the voting procedure or in an election itself through the *misrepresentation of their own preferences.* Voters may strategically misrepresent preferences to achieve their most-preferred or to avoid their least-preferred outcome. In the presence of imperfect information, voters may decide whether to vote strategically on the basis of their beliefs about others' behavior and their knowledge of the distribution of preferences.

Candidates also may behave strategically in building a political platform. A general result known as the *median voter theorem* shows that in elections with only two candidates, both locate at the preference position of the *median voter.* This result holds when voters are distributed along the preference spectrum either *discretely* or *continuously.*

KEY TERMS

agenda paradox (622)
amendment procedure (617)
antiplurality method (617)
approval voting (618)
binary method (616)
Black's condition (627)
Borda count (617)
Condorcet method (616)
Condorcet paradox (620)
Condorcet terms (629)
Condorcet winner (617)
continuous distribution (643)

Copeland index (617)
discrete distribution (640)
distribution function (643)
Gibbard-Satterthwaite theorem (638)
histogram (640)
impossibility theorem (626)
instant runoff (619)
intransitive ordering (621)
majority rule (616)
majority runoff (618)
median voter (639)
median voter theorem (639)
mixed method (618)
multistage procedure (616)
normal distribution (644)
pairwise voting (616)
plurality rule (617)
plurative method (617)
positional method (617)
principle of minimum differentiation (642)
proportional representation (619)
reversal paradox (623)
reversal terms (629)
robustness (628)
rounds (619)
sincere voting (623)
single-peaked preferences (627)
single transferable vote (619)
social ranking (620)
spoiler (630)
strategic misrepresentation of preferences (620)
strategic voting (620)
transitive ordering (621)
uniform distribution (644)

SOLVED EXERCISES

S1. Consider a vote being taken by three roommates, A, B, and C, who share a triple dorm room. They are trying to decide which of three elective courses to take together this term. (Each roommate has a different major and is taking required courses in her major for the rest of her courses.) Their choices are Philosophy, Geology, and Sociology, and their preferences for the three courses are as shown here:

A	B	C
Philosophy	Sociology	Geology
Geology	Philosophy	Sociology
Sociology	Geology	Philosophy

The roommates have decided to have a two-round vote and will draw straws to determine who sets the agenda. Suppose A sets the agenda and wants the Philosophy course to be chosen. How should she set the agenda to achieve this outcome if she knows that everyone will vote truthfully in all rounds? What agenda should she use if she knows that they will all vote strategically?

S2. Suppose that voters 1 through 4 are being asked to consider three different candidates—A, B, and C—in a Borda-count election. Their preferences are:

1	2	3	4
A	A	B	C
B	B	C	B
C	C	A	A

Assume that voters will cast their votes truthfully (no strategic voting). Find a Borda weighting system—a number of points to be allotted to the first, second, and third preferences—in which candidate A wins.

S3. Consider a group of 50 residents attending a town meeting in Massachusetts. They must choose one of three proposals for dealing with town garbage. Proposal 1 asks the town to provide garbage collection as one of its services; Proposal 2 calls for the town to hire a private garbage collector to provide collection services; and Proposal 3 calls for residents to be responsible for their own garbage. There are three types of voters. The first type prefers Proposal 1 to Proposal 2 and Proposal 2 to Proposal 3; there are 20 of these voters. The second type prefers Proposal 2 to Proposal 3 and Proposal 3 to Proposal 1; there are 15 of these voters. The third type prefers Proposal 3 to Proposal 1 and Proposal 1 to Proposal 2; there are 15 of them.

(a) Under a plurality voting system, which proposal wins?

(b) Suppose voting proceeds with the use of a Borda count in which voters list the proposals, in order of preference, on their ballots. The proposal listed first (or at the top) on a ballot gets three points; the proposal listed second gets two points; and the proposal listed last gets one point. In this situation, with no strategic voting, how many points does each proposal gain? Which proposal wins?

(c) What strategy can the second and third types of voters use to alter the outcome of the Borda-count vote in part (b) to one that both types prefer? If they use this strategy, how many points does each proposal get, and which wins?

S4. During the Cuban missile crisis, serious differences of opinion arose within the ExComm group advising President John F. Kennedy, which we summarize here. There were three options: Soft (a blockade), Medium (a limited air strike), and Hard (a massive air strike or invasion). There were also three groups in ExComm. The civilian doves ranked the alternatives Soft best, Medium next, and Hard last. The civilian hawks preferred Medium best, Hard next, and Soft last. The military preferred Hard best, but they felt "so strongly about the dangers inherent in the limited strike that they would prefer taking no military action rather than to take that limited strike." [Ernest R. May and Philip D. Zelikow, eds., *The Kennedy Tapes: Inside the White House During the Cuban Missile Crisis* (Cambridge, MA: Harvard University Press, 1997), p.97.] In other words, they ranked Soft second and Medium last. Each group constituted about one-third of ExComm, and so any two of the groups would form a majority.

(a) If the matter were to be decided by a majority vote in ExComm and the members voted sincerely, which alternative, if any, would win?

(b) What outcome would arise if members voted strategically? What outcome would arise if one group had agenda-setting power? (Model your discussion in these two cases after the analysis found in Sections 2.B and 4.B.)

S5. In his book *A Mathematician Reads the Newspaper*, John Allen Paulos gives the following caricature based on the 1992 Democratic presidential primary caucuses. There are five candidates: Jerry Brown, Bill Clinton, Tom Harkin, Bob Kerrey, and Paul Tsongas. There are 55 voters, with different preference orderings concerning the candidates. There are six different orderings, which we label I through VI. The preference orderings (1 for best to 5 for worst), along with the numbers of voters with each ordering, are shown in the following table; the candidates are identified by the first letters of their last names.[26]

		GROUPS AND THEIR SIZES					
		I 18	II 12	III 10	IV 9	V 4	VI 2
RANKING	1	T	C	B	K	H	H
	2	K	H	C	B	C	B
	3	H	K	H	H	K	K
	4	B	B	K	C	B	C
	5	C	T	T	T	T	T

(a) First, suppose that all voters vote sincerely. Consider the outcomes of each of several different election rules. Show each of the following outcomes: (i) Under the plurality method (the one with the most first preferences), Tsongas wins. (ii) Under the runoff method (the top two first preferences go into a second round), Clinton wins. (iii) Under the elimination method (at each round, the one with the fewest first preferences in that round is eliminated, and the rest go into the next round), Brown wins. (iv) Under the Borda-count method (5 points for first preference, 4 for second, and

[26]John Allen Paulos, *A Mathematician Reads the Newspaper* (New York: Basic Books, 1995), pp. 104–106.

so on; the candidate with the most points wins), Kerrey wins. (v) Under the Condorcet method (pairwise comparisons), Harkin wins.

(b) Suppose that you are a Brown, Kerrey, or Harkin supporter. Under the plurality method, you would get your worst outcome. Can you benefit by voting strategically? If so, how?

(c) Are there opportunities for strategic voting under each of the other methods as well? If so, explain who benefits from voting strategically and how they can do so.

S6. As mentioned in the chapter, some localities (such as San Francisco) have replaced runoff elections and even primaries with instant runoff voting to save time and money. Most jurisdictions have implemented a two-stage system in which if a candidate fails to receive a majority of votes in the first round, a second runoff election is held weeks later between the two candidates who earned the most votes.

For instance, France employs a two-stage system for its presidential elections. No primaries are held. Instead, all candidates from all parties are on the ballot in the first round, which usually guarantees a second round, since it is very difficult for a single candidate to earn a majority of votes among such a large field. Although a runoff in the French presidential election is always expected, it doesn't mean that French elections are not without the occasional surprise. In 2002, the country was shocked when the right-wing candidate Jean-Marie Le Pen beat the socialist Lionel Jospin to take second place and thus advance to the runoff election against the first-round winner (and incumbent) Jacques Chirac. It had been widely assumed that Jospin would take second, setting up a runoff between himself and Chirac.

Instant runoff voting can be explained in five steps:

1. Voters rank all candidates according to their preferences.
2. The votes are counted.
3. If a candidate has earned a majority of the votes, that candidate is the winner. If not, go to step 4.
4. Eliminate candidate(s) with the fewest votes. (Eliminate more than one candidate at the same time only if they tie for the fewest votes.)
5. Redistribute votes from eliminated candidates to the next-ranked choices on those ballots. Once this is done, return to step 2.

(a) Instant runoff voting is slowly gaining traction. It is used in a half-dozen cities in the United States and is being considered by a dozen more (as of 2008). Given the potential savings in money and time, it might be surprising that the institution isn't more widely adopted. Why might some oppose instant runoff voting? (Hint: Which candidates, parties, and interests benefit from the two-stage system that is currently in place?)

(b) What other concerns or criticisms might be raised about instant runoff voting?

S7. An election has three candidates and takes place under the plurality rule. There are numerous voters, spread along an ideological spectrum from left to right. Represent this spread by a horizontal straight line whose extreme points are 0 (left) and 1 (right). Voters are uniformly distributed along this spectrum; so the number of voters in any segment of the line is proportional to the length of that segment. Thus, a third of the voters are in the segment from 0 to 1/3, a quarter in the segment from 1/2 to 3/4, and so on. Each voter votes for the candidate whose declared position is closest to the voter's own position. The candidates have no ideological attachment and take up any position along the line, each seeking only to maximize her share of votes.

(a) Suppose you are one of the three candidates. The leftmost of the other two is at point x, and the rightmost is at the point $(1 - y)$, where $x + y < 1$ (so the rightmost candidate is a distance y from 1). Show that your best response is to take up the following positions under the given conditions:
   (i) just slightly to the left of x if $x > y$ and $3x + y > 1$;
   (ii) just slightly to the right of $(1 - y)$ if $y > x$ and $x + 3y > 1$; and
   (iii) exactly halfway between the other candidates if $3x + y < 1$ and $x + 3y < 1$.

(b) In a graph with x and y along the axes, show the areas (the combination of x and y values) where each of the response rules ((i) to (iii) in part (a)) is best for you.

(c) From your analysis, what can you conclude about the Nash equilibrium of the game where the three candidates each choose positions?

UNSOLVED EXERCISES

U1. Repeat Exercise S1 for the situation in which B sets the agenda and wants to ensure that Sociology wins.

U2. Repeat Exercise S2 to find a Borda weighting system in which candidate B wins.

U3. Every year college football's Heisman Trophy is awarded by means of a Borda-count voting system. Each voter submits first-, second-, and third-place votes, worth three points, two points, and one point, respectively. Thus the Borda-count point scheme used may be called (3-2-1), where the first digit is the point value of a first-place vote, the second digit denotes the value of a second-place vote, and the third digit gives the point value of a third-place vote. In 2004, the vote totals for the top five under the Borda system were as follows:

Player	1st Place	2nd Place	3rd Place
Leinhart (USC)	267	211	102
Peterson (Oklahoma)	154	180	175
White (Oklahoma)	171	149	146
Smith (Utah)	98	112	117
Bush (USC)	118	80	83

(a) Compare the Borda point scores of Leinhart and Peterson. By what margin of Borda points did Leinhart win?

(b) It seems only fair that a point scheme should give a first-place vote at least as much weight as a second-place vote and a second-place vote at least as much weight as a third-place vote. That is, for a point scheme (x-y-z), we should have $x \geq y \geq z$. Given this "fairness" restriction, is there any point scheme under which Leinhart would have lost? If so, provide such a scheme. If not, explain why not.

(c) Even though White had more first-place votes than Peterson, Peterson had a higher Borda-count total. If first-place votes were weighted enough, White's edge in first-place votes could give him a higher Borda count. Assume that second-place votes are worth 2 points and third-place votes are worth 1 point, so that the point scheme is (x-2-1). What is the lowest integer value of x such that White gets a higher Borda count than Peterson?

(d) Suppose that the above vote data represent truthful voting. For simplicity, let's suppose that the election were a simple plurality vote instead of a Borda count. Note that Leinhart and Bush are both from USC, whereas Peterson and White are both from Oklahoma. Suppose that, due to Oklahoma loyalty, those voters who prefer White all have Peterson as their second choice. If these voters were to vote strategically in a plurality election, could they change the outcome of the election? Explain.

(e) Similarly, suppose that due to USC loyalty, those voters who prefer Bush all have Leinhart as their second choice. If all four voting groups (Leinhart, Peterson, White, Bush) were to vote strategically in a plurality election, who would be the winner of the Heisman Trophy?

(f) In 2004, there were 923 Heisman voters. Under the actual (3-2-1) system, what is the minimum integer number of first-place votes that it would have taken to guarantee victory (that is, without the help of any second- or third-place votes)? Note that a player's name may appear on a ballot only once.

U4. Olympic skaters complete two programs in their competition, one short and one long. In each program, the skaters are scored and then ranked by a panel of nine judges. The skaters' positions in the rankings are used to determine their final scores. A skater's ranking depends on the number of judges placing her first (or second or third); the skater judged to be best by the most judges is ranked number one, and so on. In the calculation of a skater's final score, the short program gets half the weight of the long program. That is, Final score = 0.5 (Rank in short program) + Rank in long program. The skater with the lowest final score wins the gold medal. In the event of a tie, the skater judged best in the long program by the most judges takes the gold. In the 2002 women's individual figure-skating competition in Salt Lake City, Michelle Kwan was in first place after the short program. She was followed by Irina Slutskaya, Sasha Cohen, and Sarah Hughes, who were in second, third, and fourth places, respectively. In the long program, the judges' cards for these four skaters were as follows:

		JUDGE NUMBER								
		1	2	3	4	5	6	7	8	9
KWAN	Points	11.3	11.5	11.7	11.5	11.4	11.5	11.4	11.5	11.4
	Rank	2	3	2	2	2	3	3	2	3
SLUTSKAYA	Points	11.3	11.7	11.8	11.6	11.4	11.7	11.5	11.4	11.5
	Rank	3	1	1	1	4	1	2	3	2
COHEN	Points	11.0	11.6	11.5	11.4	11.4	11.4	11.3	11.3	11.3
	Rank	4	2	4	3	3	4	4	4	4
HUGHES	Points	11.4	11.5	11.6	11.4	11.6	11.6	11.3	11.6	11.6
	Rank	1	4	3	4	1	2	1	1	1

(a) At the Olympics, Slutskaya skated last of the top skaters. Use the information from the judges' cards to determine the judges' long-program ranks for Kwan, Cohen, and Hughes *before* Slutskaya skated. Then, using the standings already given for the short program in conjunction with your calculated ranks for the long program, determine the final scores and standings among these three skaters *before* Slutskaya skated. (Note that Kwan's rank in the short program was 1, and so her partial score after the short program is 0.5.)

(b) Given your answer to part (a), what would have been the final outcome of the competition if the judges had ranked Slutskaya's long program above all three of the others?

(c) Use the judges' cards to determine the actual final scores for all four skaters *after* Slutskaya skated. Who won each medal?

(d) What important principle, of those identified by Arrow, does the Olympic figure-skating scoring system violate? Explain.

U5. The 2008 presidential nomination season saw 21 Republican primaries and caucuses on Super Tuesday—February 5, 2008. By that day—only a month after the Iowa caucus that began the process—more than half of the Republican contenders had dropped out the race, leaving only four: John McCain, Mitt Romney, Mike Huckabee, and Ron Paul. McCain, Romney, and Huckabee had each previously won at least one state. McCain had beaten Romney in Florida the week before the big day, and at that point it looked like only the two of them stood a realistic chance of wining the nomination. In this primary season, as is typical for the Republican Party, nearly every GOP contest (whether primary or caucus) was winner-take-all, so winning a given state would earn a candidate all of the delegates allotted to that state by the Republican National Committee.

The West Virginia caucus was the first contest to reach a conclusion on Super Tuesday, since the caucus took place in the afternoon, it was brief, and the state is in the eastern time zone. News of the result was available hours before the close of polls in many of the states voting that day.

The following problem is based on the results of that West Virginia caucus. As we might expect, the caucusers did not all share the same preferences over the candidates. Some favored McCain, whereas others liked Romney or Huckabee. The voters also certainly had varied preferences about whom they wanted to win if their favorite candidate did not. Simplifying substantially from reality (but based on the actual voting), assume that there were seven types of West Virginia caucus goers that day, whose prevalence and preferences were as follows:

	I (16%)	**II (28%)**	**III (13%)**	**IV (21%)**	**V (12%)**	**VI (6%)**	**VII (4%)**
1st	McCain	Romney	Romney	Huckabee	Huckabee	Paul	Paul
2nd	Romney	McCain	Huckabee	Romney	McCain	Romney	Huckabee
3rd	Huckabee	Huckabee	McCain	McCain	Romney	Huckabee	Romney
4th	Paul	Paul	Paul	Paul	Paul	McCain	McCain

At first, no one knew the distribution of preferences of those in attendance at the caucus, so everyone voted truthfully. Thus, Romney won a plurality of the votes in the first round with 41%.

After each round of this caucus, if no candidate wins a majority, the candidate with the least number of votes is dropped from consideration, and his or her supporters vote for one of the remaining candidates in the following rounds.

(a) What would the results of the second round have been under truthful (nonstrategic) voting for the remaining three candidates?

(b) If West Virginia had held pairwise votes among the four candidates, which one would have been the Condorcet winner with truthful voting?

(c) In reality, the results of the second round of the caucus were:

Huckabee: 52%
Romney: 47%
McCain: 1%

Given the preferences of the McCain voters, why might this have happened? (Hint: how would the outcome have been different if West Virginia had voted last on Super Tuesday?)

(d) After the fact Romney's campaign cried foul and accused the McCain and Huckabee supporters of making a backroom deal (see Susan Davis, "Romney Cries Foul in W. Va. Loss," *Wall Street Journal*, February 5, 2008. Available at http://blogs.wsj.com/washwire/2008/02/05/huckabee-wins-first-super-tuesday-contest/?mod=WSJBlog). Should Romney's campaign have suspected collusion between the McCain and Huckabee camps in this case? Explain why or why not.

U6. Return to the discussion of instant runoff voting in Exercise S6.

(a) Consider the following IRV ballots of five voters:

	Ana	**Bernard**	**Cindy**	**Desmond**	**Elizabeth**
1st	Jack	Jack	Kate	Locke	Locke
2nd	Kate	Kate	Locke	Kate	Jack
3rd	Locke	Locke	Jack	Jack	Kate

Which—if any—of the five voters have an incentive to vote strategically? If so, who and why? If not, explain why not.

(b) Consider the following table, which gives the IRV ballots of a small town of seven citizens voting on five policy proposals put forward by the mayor:

	Anderson	Brown	Clark	Davis	Evans	Foster	García
1st	V	V	W	W	X	Y	Z
2nd	W	X	V	X	Y	X	Y
3rd	X	W	Y	V	Z	Z	X
4th	Y	Y	X	Y	V	W	W
5th	Z	Z	Z	Z	W	V	V

Assuming that all candidates (or policies) that tie for the fewest votes are eliminated at the same time, under what conditions is an eventual majority winner guaranteed? Put another way, under what conditions might there not be an unambiguous majority winner? (Hint: How important is it for Evans, Foster, and García to fill out their ballots completely?) How will these conditions change if Harris moves into town and votes?

U7. Recall the three-member council considering three alternative welfare policies in Section 4.C. There, three councillors (Left, Center, and Right) considered policies A and D in a first-round vote, with the winner facing policy G in a second-round election. But no one knows for sure exactly how many councillors have each set of possible preferences. The possible preference orderings are shown in Figure 16.1. Each councillor knows her own type, and she knows the probabilities of observing each type of voter, p_L, p_C, and p_R (with $p_L + p_C + p_R = 1$). The behavior of the Center-type voters in the first-round election is the only unknown in this situation and will depend on the probabilities that the various preference types occur. Suppose here that a Center-type voter believes (in contrast with the case considered in the chapter) that other Center types will vote strategically; suppose further that the Center type's payoffs are as in Section 4.C: 1 if A wins, 0 if G wins, and $0 < u < 1$ if D wins.

(a) Under what configuration of the other two votes does the Center-type voter's first-round vote matter to the outcome of the election? Given her assumption about the behavior of other Center-type voters, how would she identify the source of the first-round votes?

(b) Following the analysis in Section 4.C, determine the expected payoff to the Center type when she votes truthfully. Compare this with her expected payoff when she votes strategically. What is the condition under which the Center type votes strategically?

Bidding Strategy and Auction Design

IN THIS CHAPTER, we consider a topic that is becoming increasingly relevant in many of our lives—bidding strategy and auction design. Although you might argue that you have never been, nor do you expect ever to be, at Christie's or Sotheby's or an antique auction in the wilds of Vermont or a livestock auction in Illinois, you probably participate in more auctions than you imagine. With the phenomenal recent growth in online auction sites, for example, millions of people now buy regularly at auctions. And other transactions in which you participate may also be classified as auctions.

Auctions entail the transfer of a particular object from a seller to a buyer (or bidder) for a certain price (or bid). Considered in this simple form, many market transactions resemble auctions. For example, stores such as Filene's Basement in Boston use a clever pricing strategy to keep customers coming back for more: they reduce the prices on items remaining on the racks successively each week until either the goods are purchased or the price gets so low that they donate the items to charity. Shoppers love it. Little do they realize that they are participating in what is known as a descending, or Dutch, auction—one of the types of auctions described in detail in this chapter.

Even if you do not personally participate in many auctions, your life is greatly influenced by them. Since 1994, the Federal Communications Commission (FCC) has auctioned off large parts of the electromagnetic broadcasting spectrum in almost 70 different auctions. The kind of television that you watch for the next few decades, along with radio broadcasting options and the kinds of cellular phones that you use, will be affected by this process and its outcomes.

These auctions have already raised approximately $78 billion in government revenues, as of November 2008, with the most recent auction in mid 2008 bringing in just over $21 million. There are five more auctions to go, one of which is scheduled for the fall of 2008. Because these revenues have made significant contributions to the federal budget, they have affected important macroeconomic magnitudes, such as interest rates. International variables also have been affected not only by the U.S. spectrum auctions, but also by similar auctions in at least six European countries as well as in Australia and New Zealand. Understanding how auctions work will help you understand these important events and their implications.

From a strategic perspective, auctions have several characteristics of interest. Most crucial is the existence of asymmetric information between seller and bidders, as well as among bidders. Thus, signaling and screening can be important components of strategy for both bidders and sellers. In addition, optimal strategies for both bidders and sellers will depend on their levels of aversion to risk. We will also see that under some specific circumstances, expected payoffs to the seller as well as to the winning bidder are the same across auction types.

This chapter explores the various types of auctions, as well as the strategies that you might want to employ as either the buyer or the seller in such situations. The formal theory of auctions relies on advanced calculus to derive its results, but we eschew most of this difficult mathematics in favor of more intuitive descriptions of optimal behavior and strategy choice.[1] As an example of how auction design and bidding theory influence modern commerce, we provide a detailed discussion of the use of auctions on the Internet.

1 TYPES OF AUCTIONS

Auctions differ in the methods used for the submission of bids and for the determination of the final price paid by the winner. These aspects of an auction, that are set in advance by the seller, are known as auction *rules.* In addition, auctions can be classified according to the type of object being auctioned and how it is valued; this determines the auction *environment.* Here we categorize the various auction rules and environments, describing their characteristics and mechanics.

A. Auction Rules

The seller generally determines the rules that will govern the auction. She has to do this with only limited knowledge of the bidders' willingness to pay. Therefore

[1]A reference list of sources for additional information on the theory and practice of auctions can be found in the final section of the chapter.

the seller is in much the same position as the firm that tried to practice price discrimination, or the government procurement officer that tried to find out the contractors' cost, in Chapter 14. In other words, in choosing the rules the seller is designing the mechanism of the auction. This mechanism-design approach can be developed into a theory of optimal auctions, and also tell us when two or more such mechanisms will be equivalent. We must leave the general theory for more advanced texts; here we will study and compare a few specific mechanisms that are most frequently and prominently used in reality.[2]

The four major categories of auction rules can be divided into two groups. The first group is known as **open outcry.** Under this type of auction rule, bidders call out or otherwise make their bids in public. All bidders are able to observe bids as they are made. This type perhaps best fits the popular vision of the way in which auctions work—an image that includes feverish bidders and an auctioneer. But open outcry auctions can be organized in two ways. Only one of them would ever demonstrate "feverish" bidding.

The **ascending,** or **English,** version of an open-outcry auction conforms best to this popular impression of auctions. Ascending auctions were and still are the norm at English auction houses such as Christie's and Sotheby's, from which they take their alternate name. The auction houses have a conventional auctioneer who starts at a low price and calls out successively higher prices for an item, waiting to receive a bid at each price before going on. When no further bids can be obtained, the item goes to the most recent, highest, bidder. Thus, any number of bidders can take part in English auctions, although only the top bidder gains the item up for sale. And the bidding process may not literally entail the actual outcry of bids, because the mere nod of a head or the flick of a wrist is common bidding behavior in such auctions. A large majority of the existing Internet auction sites now run what are essentially ascending auctions (in virtual, rather than real, time) for almost any item imaginable.

The other type of open outcry auction is the **Dutch,** or **descending,** auction. Dutch auctions, which get their name from the way in which tulips and other flowers are auctioned in the Netherlands, work in the opposite direction from that of English auctions. The auctioneer starts at an extremely high price and calls out successively lower prices until one of the assembled potential bidders accepts the price, makes a bid, and takes the item. Because of the desire or need for speed, Dutch flower auctions, as well as auctions for other agricultural or perishable goods (such as the daily auction at the Sydney Fish Market), use a

[2]Roger Myerson's paper "Optimal Auction Design," *Mathematics of Operations Research,* vol. 6, no. 1 (February 1981), pp. 58–73, was a pioneering contribution to the general theory of auctions, and an important part of the work that won him the Nobel Prize in Economics in 2007. Paul Klemperer, *Auctions: Theory and Practice* (Princeton, NJ: Princeton University Press, 2004) is an excellent modern treatment of the theory.

"clock" that ticks down (counterclockwise) to ever lower prices until one bidder "stops the clock" and collects her merchandise. In many cases, the auction clock displays considerable information about the lot of goods currently for sale in addition to the falling price of those goods. And unlike in the English auction, there is no feverish bidding in a Dutch auction, because only the one person who "stops the clock" takes any action.

The second group of auction rules requires sales to occur by **sealed bid.** In these auctions, bidding is done privately and bidders cannot observe any of the bids made by others; in many cases, only the winning bid is announced. Bidders in such auctions, as in Dutch auctions, have only one opportunity to bid. (Technically, you could submit multiple bids, but only the highest one would be relevant to the auction outcome.) Sealed-bid auctions have no need for an auctioneer. They require only an overseer who opens the bids and determines the winner.

Within sealed-bid auctions, there are two rules for determining the price paid by the high bidder. In a **first-price** sealed-bid auction, the highest bidder wins the item and pays a price equal to her bid. In a **second-price,** sealed-bid auction, the highest bidder wins the item but pays a price equal to the bid of the second-highest bidder. A second-price rule can be extremely useful for eliciting truthful bids, as we will see in Section 4. Such an auction is often termed a **Vickrey auction** after the Nobel-prize-winning economist who first noted this particular characteristic. We will also see that the sealed-bid auctions are each similar, in regard to bidding strategy and expected payoffs, to one of the open-outcry auctions; first-price sealed-bid auctions are similar to Dutch auctions, and second-price, sealed-bid auctions are similar to English auctions.

Other, less common, configurations also can be used to sell goods at auction. For example, you could set up an auction in which the highest bidder wins but the top two bidders pay their bids, or one in which the high bidder wins but all bidders pay their bids, a procedure discussed in Section 5. Although we do not attempt to consider all possible combinations in this chapter, we do analyze several of the most common auction schemes by using examples that bring out important strategic concepts.

B. Auction Environments

Finally, there are a number of ways in which bidders may value an item up for auction. The main distinction in such auction environments is based on the difference between *common-* and *private-value* objects. In a **common-value,** or **objective-value,** auction, the value of the object is the same for all the bidders, but each bidder generally knows only an imprecise estimate of it. Bidders may have some sense of the distribution of possible values, but each must form her own estimate before bidding. For example, an oil-drilling tract has a given amount of oil that should produce the same revenue for all companies, but each company has only its own expert's estimate of the amount of oil contained under the tract. Similarly, each bond trader has only an estimate of the future course of interest rates. In such auctions, signaling and screening can play an important role.

In a common-value auction, each bidder should be aware of the fact that other bidders possess some (however sketchy) information about an object's value, and she should attempt to infer the contents of that information from the actions of rival bidders. In addition, she should be aware of how her own actions might signal her private information to those rival bidders. When bidders' estimates of an object's value are influenced by their beliefs about other bidders' estimates, we have an environment in which bids are said to be *correlated* with each other. This situation has implications for both buyers and sellers as we will see later in this chapter.

In a **private-value,** or **subjective-value,** auction, bidders each determine their own individual value for an object. In this case, bidders place different values on an object. For example, a gown worn by Princess Diana or a necklace worn by Jacqueline Bouvier Kennedy Onassis may have sentimental value to some bidders. Bidders know their own private valuations in such auction environments but do not know each other's valuations of an object. Similarly, the seller does not know any of the bidders' valuations. Bidders and sellers may each be able to formulate rough estimates of others' valuations and, as above, can use signals and screens to attempt to improve their final outcomes. The information problem is relevant, then, not only to bidding strategies, but also to the seller's strategy in designing the form of auction to identify the highest valuation and to extract the best price.

Other, less common, configurations also can be used to sell goods at auction. For example, you could set up an auction in which the highest bidder wins but the top two bidders pay their bids, or one in which the high bidder wins but all bidders pay their bids, a procedure discussed in Section 5. We do not attempt to consider all possible combinations here. Rather, we analyze several of the most common auction schemes by using examples that bring out important strategic concepts.

2 THE WINNER'S CURSE

A standard but often ignored outcome arises in common-value auctions. Recall that such auctions entail the sale of an object whose value is fixed and identical for all bidders, although each bidder can only estimate it. The **winner's curse** is a warning to bidders that if they win the object in the auction, they are likely to have paid more than it is worth.

Suppose you are a corporate raider bidding for Targetco. Your experts have studied this company and produced estimates that, in the hands of the current management, it is worth somewhere between 0 and $10 billion, all values in this range being equally likely. The current management knows the precise figure, but of course it is not telling you. You believe that whatever Targetco is worth under existing management, it will be worth 50% more under your control. What should you bid?

You might be inclined to think that, on average, Targetco is worth $5 billion under existing management and thus $7.5 billion, on average, under yours. If so, then a bid somewhere between $5 billion and $7.5 billion should be profitable. But such a bidding strategy reckons without the response of the existing management to your bid. If Targetco is actually worth more than your bid, the current owners are not going to accept the bid. You are going to get the company only if its true worth is toward the lower end of the range.

Suppose you bid amount b. Your bid will be accepted and you will take over the management of Targetco if it is worth somewhere between 0 and b under the current management; on average, you can expect the company to be currently worth $b/2$ if your bid is accepted. In your hands, the average worth will be 50% more than the current worth, or $(1.5)(b/2) = 0.75b$. Because this value is always less than b, you would win the takeover battle only when it was not worth winning! Many raiders seem to have discovered this fact too late.

This result is not unlike that faced by the purchaser of a used car, which we discussed in Chapter 9. The theory of adverse selection in markets with asymmetric information is directly applicable to the common-value auction described here. Just as the average value of a used car will always fall below the price attached to the "good" cars, so will the average worth of Targetco in your hands always fall below your bid.

But corporate raiders, often engaged with target firms in one-on-one negotiations resembling auctions with only one bidder, are not the only ones affected by the winner's curse. Similar problems arise when you are competing with other bidders in a common-value auction and all of you have separate estimates for the object's value.

Consider a lease for the oil- or gas-drilling rights on a tract of land (or sea).[3] At the auction for this lease, you win only if your rivals make estimates of the value of the lease that are lower than your estimate. You should recognize this fact and try to learn from it.

Suppose the true value of the lease, unknown to any of the bidders, is $1 billion. (In this case, the seller probably does not know the true value of the tract either.) Suppose there are 10 oil companies in the bidding. Each company's experts

[3]For example, the United States auctions leases for offshore oil-drilling rights, including rights in the Gulf of Mexico and off the coast of Alaska. The state of Pennsylvania auctioned leases for natural gas–drilling rights on almost a quarter of a million acres of state forest land in 2002; this was also the first online, real-time, anonymous auction.

estimate the value of the tract with an error of $100 million, all numbers in this range being equally likely. If all 10 of the estimates could be pooled, their arithmetic average would be an unbiased and much more accurate indicator of the true value than any single estimate. But when each bidder sees only one estimate, the largest of these estimates is biased: on average, it will be $1.08 billion, right near the upper end of the range.[4] Thus, the winning company is likely to pay too much, unless it recognizes the problem and adjusts its bid downward to compensate for this bias. The exact calculation required to determine how far to shade down your bid without losing the auction is difficult, however, because you must also recognize that all the other bidders will be making the same adjustment.

We do not pursue the advanced mathematics required to create an optimal bidding strategy in the common-value auction. However, we can provide you with some general advice. If you are bidding on an item, the question "Would I be willing to purchase the lease for $1.08 billion, given what I know before submitting my bid?" is very different from the question "Would I still be willing to purchase the lease for $1.08 billion, given what I know before submitting my bid and given the knowledge that I will be able to purchase the lease only if no one else is willing to bid $1.08 billion for it?"[5] Even in a sealed-bid auction, it is the second question that reveals correct strategic thinking, because you win with any given bid only when all others bid less—only when all other bidders have a lower estimate of the value of the object than you do.

If you do not take the winner's curse into account in your bidding behavior, you should expect to lose substantial amounts, as indicated by the numerical calculations performed above for bidding on the hypothetical Targetco. How real is this danger in practice? Richard Thaler has marshaled a great deal of evidence to show that the danger is very real indeed.[6]

The simplest experiment to test the winner's curse is to auction a jar of pennies. The prize is objective, but each bidder forms a subjective estimate of how many pennies there are in the jar and therefore of the size of the prize; this experiment is a pure example of a common-value auction. Most teachers have conducted such experiments with students and found significant overbidding. In a similar but related experiment, M.B.A. students were asked to bid for a hypothetical company instead of a penny jar. The game was repeated, with feedback after each round on the true value of the company. Only 5 of 69 students learned to bid less over time; the average bid actually went up in the later rounds.

[4]The 10 estimates will, on average, range from $0.9 billion to $1.1 billion ($100 million on either side of $1 billion). The low and high estimates will, on average, be at the extremes of the distribution.

[5]See Steven Landsburg, *The Armchair Economist* (New York: Free Press, 1993), p. 175.

[6]Richard Thaler, "Anomalies: The Winner's Curse," *Journal of Economic Perspectives*, vol. 2, no. 1 (Winter 1988), pp. 191–201.

Observations of reality confirm these findings. There is evidence that winners of oil- and gas-drilling leases at auctions take substantial losses on their leases. Baseball players who as free agents went to new teams were found to be overpaid in comparison with those who re-signed with their old teams.

We repeat: the precise calculations that show how much you should shade down your bidding to take into account the winner's curse are beyond the scope of this text; the articles cited in Section 9 contain the necessary mathematical analysis. Here we merely wish to point out the problem and emphasize the need for caution. When your willingness to pay depends on your expected ability to make a profit from your purchase or on the expected resale value of the item, be wary.

This analysis shows the importance of the prescriptive role of game theory. From observational and experimental evidence, we know that many people fall prey to the winner's curse. By doing so, they lose a lot of money. Learning the basics of game theory would help them anticipate the winner's curse and prevent attendant losses.

3 BIDDING STRATEGIES

We turn now to private-value auctions and a discussion of optimal bidding strategies. Suppose you are interested in purchasing a particular lot of Chateau Margaux 1952 Bordeaux wine. Consider some of the different possible auction procedures that could be used to sell the wine.

Suppose first that you are participating in a standard, English auction. Your optimal bidding strategy is straightforward, given that you know your valuation V. Start at any step of the bidding process. If the last bid made by a rival bidder—call it r—is at or above V, you are certainly not willing to bid higher; so you need not concern yourself with any further bids. Only if the last bid is still below V do you bid at all. In that case, you can add a penny (or the smallest increment allowed by the auction house) and bid r plus one cent. If the bidding ends there, you get the wine for r (or virtually r), and you make an effective profit of $V - r$. If the bidding continues, you repeat the process, substituting the value of the new last bid for r. In this type of auction, the high bidder gets the wine for (virtually) the valuation of the second-highest bidder. How close the final price is to the second-highest valuation will be determined by the minimum bid increment defined in the auction rules.

Now suppose the wine auction is first price, sealed bid, and you suspect that you are a very high value bidder. You need to decide whether to bid V or something other than V. Should you put in a bid equal to the full value V that you place on the object?

Remember that the high bidder in this auction will be required to pay her bid. In that case, you should not in fact bid V. Such a bid would be sure to give you zero profit, and you could do better by reducing your bid somewhat. If you bid a little less than V, you run the risk of losing the object should a rival bidder make a bid above yours but below V. But as long as you do not bid so low that this outcome is guaranteed, you have a positive probability of making a positive profit. Your optimal bidding strategy entails **shading** your bid. Calculus would be required to describe the actual strategy required here, but an intuitive understanding of the result is simple. An increase in shading (a lowering of your bid from V) provides both an advantage and a disadvantage to you; it increases your profit margin if you obtain the wine, but it also lowers your chances of being the high bidder and therefore of actually obtaining the wine. Your bid is optimal when the last bit of shading just balances these two effects.

What about a Dutch auction? Your bidding strategy in this case is similar to that for the first-price, sealed-bid auction. Consider your bidding possibilities. When the price called out by the auctioneer is above V, you choose not to bid. If no one has bid by the time the price gets down to V, you may choose to do so. But again, as in the sealed-bid case, you have two options. You can bid now and get zero profit or wait for the price to drop lower. Waiting a bit longer will increase the profit that you take from the sale, but it also increases your risk of losing the wine to a rival bidder. Thus shading is in your interest here as well, and the precise amount of shading depends on the same cost-benefit analysis described in the preceding paragraph.

Finally, there is the second-price, sealed-bid auction. In that auction, the cost-benefit analysis regarding shading is different from that in the preceding three types of auctions. This result is due to the fact that the advantage gained from shading, the increase in your profit margin, is zero in this auction. You do not improve your profit by shading your bid, because your profit is determined by the second-highest bid, not your own. Because second-price, sealed-bid auctions have this interesting property, all of the next section deals with the analysis of bidding strategies in such auctions.

4 VICKREY'S TRUTH SERUM

We have just seen that bidding strategies in private-value English auctions differ from those for first-price, sealed-bid auctions. The high bidder in the English auction can get the object for essentially the valuation of the second-highest bidder. In the sealed-bid auction, the high bidder pays her bid, regardless of the distance between it and the next highest bid. Strategic bidders in a sealed-bid auction recognize this fact and attempt to retain some profit (surplus) for themselves by shading their bids. All else being equal, sellers would prefer bids that were not shaded downward. They are thus faced with a problem in mechanism design in which they want to induce information revelation; the sellers want to induce bidders to reveal their true valuations with their bids.

William Vickrey showed that rather than using a first-price auction when selling a subjectively (private) valued object with sealed bids, truthful revelation of valuations from bidders would arise if the seller used a modified version of the sealed-bid scheme; his suggestion was to modify the sealed-bid auction so that it more closely resembles its open-outcry counterpart.[7] That is, the highest bidder should get the object for a price equal to the second-highest bid—a second-price, sealed-bid auction. Vickrey showed that, with these rules, every bidder has a dominant bidding strategy to bid her true valuation. Thus we facetiously dub it **Vickrey's truth serum.**

However, we saw in Chapter 14 that there is usually a cost to using a mechanism that extracts information. Auctions are no exception. Buyers reveal the truth about their valuations in an auction using Vickrey's scheme only because it gives them some profit from doing so. The second-price, sealed-bid auction mechanism reduces the profit for the seller, just as the shading of bids does in a first-price auction, and just as the information-revelation mechanisms we studied in Chapter 14 did for the principals in those cases. The relative merit of the two procedures from the seller's point of view therefore depends on which one entails a greater reduction in her profit. We consider this matter later in Section 6; but first we explain how Vickrey's scheme works.

Suppose you are an antique-china collector, and you have discovered that a local estate auction will be selling off a 19th-century Meissen "Blue Onion" tea set in a sealed-bid, second-price auction. As someone experienced with vintage china but lacking this set for your collection, you value it at $3,000, but you do not know the valuations of the other bidders. If they are inexperienced, they may not realize the considerable value of the set. If they have sentimental attachments to Meissen or the "Blue Onion" pattern, they may value it more highly than the value that you have calculated.

The rules of the auction allow you to bid any real-dollar value for the tea set. We will call your bid b and consider all of its possible values. Because you are not constrained to a small, specific set of bids, we cannot draw a finite payoff matrix for this bidding game, but we can logically deduce the optimal bid.

The success of your bid will obviously depend on the bids submitted by others interested in the tea set, primarily because you need to consider whether your bid will win. The outcome thus depends on all rival bids, but only the

[7]Vickrey was one of the most original minds in economics in the past four decades. In 1996, he won the Nobel Prize for his work on mechanism design in auctions and truth-revealing procedures. Sadly, he died just 3 days after the prize was announced.

largest bid among them will affect your outcome. We call this largest bid r and disregard all bids below r.

What is your optimal value of b? We will look at bids both above and below $3,000 to determine whether any option other than exactly $3,000 can yield you a better outcome than bidding your true valuation.

We start with $b > 3{,}000$. There are three cases to consider. First, if your rival bids less than $3,000 (so $r < 3{,}000$), then you get the tea set at the price r. Your profit, which depends only on what you pay relative to your true valuation, is $(3{,}000 - r)$, which is what it would have been had you simply bid $3,000. Second, if your rival's bid falls between your actual bid and your true valuation (so $3{,}000 < r < b$), then you are forced to take the tea set for more than it is worth to you. Here you would have done better to bid $3,000; you would not have gotten the tea set, but you would not have given up the $(r - 3{,}000)$ in lost profit either. Third, your rival bids even more than you do (so $b < r$). You still do not get the tea set, but you would not have gotten it even had you bid your true valuation. Putting together the reasoning of the three cases, we see that bidding your true valuation is never worse, and sometimes better, than bidding something higher.

What about the possibility of shading your bid slightly and bidding $b < 3{,}000$? Again, there are three situations. First, if your rival's bid is lower than yours (so $r < b$), then you are the high bidder, and you get the tea set for r. Here you could have gotten the same result by bidding $3,000. Second, if your rival's bid falls between 3,000 and your actual bid (so $b < r < 3{,}000$), your rival gets the tea set. If you had bid $3,000 in this case, you would have gotten the tea set, paid r, and still made a profit of $(3{,}000 - r)$. Third, your rival's bid could have been higher than $3,000 (so $3{,}000 < r$). Again, you do not get the tea set but, if you had bid $3,000, you still would not have gotten it, so there would have been no harm in doing so. Again, we see that bidding your true valuation, then, is no worse, and sometimes better, than bidding something lower.

If truthful bidding is never worse and sometimes better than bidding either above or below your true valuation, then you do best to bid truthfully. That is, no matter what your rival bids, it is always in your best interest to be truthful. Put another way, bidding your true valuation is your dominant strategy whether you are allowed discrete or continuous bids.

Vickrey's remarkable result that truthful bidding is a dominant strategy in second-price, sealed-bid auctions has many other applications. For example, if each member of a group is asked what she would be willing to pay for a public project that will benefit the whole group, each has an incentive to understate her own contribution—to become a "free rider" on the contributions of the rest. We have already seen examples of such effects in the collective-action games of Chapter 12. A variant of the Vickrey scheme can elicit the truth in such games as well.

5 ALL-PAY AUCTIONS

We have considered most of the standard auction types discussed in Section 1 but none of the more creative configurations that might arise. Here we consider a common-value, sealed-bid, first-price auction in which every bidder, win or lose, pays to the auctioneer the amount of her bid. An auction where the losers also pay may seem strange. But in fact, many contests result in this type of outcome. In political contests, all candidates spend a lot of their own money and a lot of time and effort for fund raising and campaigning. The losers do not get any refunds on all their expenditures. Similarly, hundreds of competitors spend four years of their lives preparing for an event at the next Olympic games. Only one wins the gold medal and the attendant fame and endorsements; two others win the far less valuable silver and bronze medals; the efforts of the rest are wasted. The tournaments we discussed in Chapter 14, Secion 6.B, are similar. Once you start thinking along these lines, you will realize that such all-pay auctions are, if anything, more frequent in real life than situations resembling the standard formal auctions where only the winner pays.

How should you bid (that is, what should your strategy be for expenditure of time, effort, and money) in an **all-pay auction?** Once you decide to participate, your bid is wasted unless you win, so you have a strong incentive to bid very aggressively. In experiments, the sum of all the bids often exceeds the value of the prize by a large amount, and the auctioneer makes a handsome profit.[8] In that case, everyone's submitting extremely aggressive bids cannot be the equilibrium outcome; it seems wiser to stay out of such destructive competition altogether. But if everyone else did that, then one bidder could walk away with the prize for next to nothing; thus, not bidding cannot be an equilibrium strategy either. This analysis suggests that the equilibrium lies in mixed strategies.

Consider a specific auction with n bidders. To keep the notation simple, we choose units of measurement so that the common-value object (prize) is worth 1. Bidding more than 1 is sure to bring a loss, and so we restrict bids to those between 0 and 1. It is easier to let the bid be a continuous variable x, where x can take on any (real) value in the interval [0, 1]. Because the equilibrium will be in mixed strategies, each person's bid, x, will be a continuous random variable. Because you win the object only if all other bidders submit bids below yours, we can express your equilibrium mixed strategy as $P(x)$, the probability that your

[8]One of us (Dixit) has auctioned $10 bills to his Games of Strategy class and made a profit of as much as $60 from a 20-student section. At Princeton there is a tradition of giving the professor a polite round of applause at the end of a semester. Once Dixit offered $20 to the student who kept applauding continuously the longest. This is an open-outcry, all-pay auction with payments in kind (applause). Although most students dropped out between 5 and 20 minutes, three went on for 4 1/2 hours!

bid takes on a value less than x; for example, $P(1/2) = 0.25$ would mean that your equilibrium strategy entailed bids below 1/2 one-quarter of the time (and bids above 1/2 three-quarters of the time).[9]

As usual, we can find the mixed-strategy equilibrium by using an indifference condition. Each bidder must be indifferent about the choice of any particular value of x, given that the others are playing their equilibrium mixes. Suppose you, as one of the n bidders, bid x. You win if all of the remaining $(n - 1)$ are bidding less than x. The probability of anyone else bidding less than x is $P(x)$; the probability of two others bidding less than x is $P(x) \times P(x)$, or $[P(x)]^2$; the probability of all $(n - 1)$ of them bidding less than x is $P(x) \times P(x) \times P(x) \cdots$ multiplied $(n - 1)$ times, or $[P(x)]^{n-1}$. Thus with a probability of $[P(x)]^{n-1}$, you win 1. Remember that you pay x no matter what happens. Therefore, your net expected payoff for any bid of x is $[P(x)]^{n-1} - x$. But you could get 0 for sure by bidding 0. Thus, because you must be indifferent about the choice of any particular x, including 0, the condition that defines the equilibrium is $[P(x)]^{n-1} - x = 0$. In a full mixed-strategy equilibrium, this condition must be true for all x. Therefore the equilibrium mixed-strategy bid is $P(x) = x^{1/(n-1)}$.

A couple of sample calculations will illustrate what is implied here. First, consider the case in which $n = 2$; then $P(x) = x$ for all x. Therefore the probability of bidding a number between two given levels x_1 and x_2 is $P(x_2) - P(x_1) = x_2 - x_1$. Because the probability that the bid lies in any range is simply the length of that range, any one bid must be just as likely as any other bid. That is, your equilibrium mixed-strategy bid should be random and uniformly distributed over the whole range from 0 to 1.

Next let $n = 3$. Then $P(x) = \sqrt{x}$. For $x = 1/4$, $P(x) = 1/2$; so the probability of bidding 1/4 or less is 1/2. The bids are no longer uniformly distributed over the range from 0 to 1; they are more likely to be in the lower end of the range.

Further increases in n reinforce this tendency. For example, if $n = 10$, then $P(x) = x^{1/9}$, and $P(x)$ equals 1/2 when $x = (1/2)^9 = 1/512 = 0.00195$. In this situation, your bid is as likely to be smaller than 0.00195 as it is to be anywhere within the whole range from 0.00195 to 1. Thus your bids are likely to be very close to 0.

Your average bid should correspondingly be smaller the larger the number n. In fact, a more precise mathematical calculation shows that if everyone bids according to this strategy, the average or expected bid of any one player will be just $(1/n)$.[10] With n players bidding, on average, $1/n$ each, the total

[9] $P(x)$ is called the *cumulative probability distribution function* for the random variable x. The more familiar probability density function for x is its derivative, $P'(x) = p(x)$. Then $p(x)dx$ denotes the probability that the variable takes on a value in a small interval from x to $x + dx$.

[10] The expected bid of any one player is calculated as the expected value of x, by using the probability density function, $p(x)$. In this case, $p(x) = P'(x) = [1/(n - 1)]x^{(2-n)/(n-1)}$, and the expected value of x is the sum, or integral, of this from 0 to 1, namely $\int x\, p(x)\, dx = 1/n$.

expected bid is 1, and the auctioneer makes zero expected profit. This calculation provides more precise confirmation that the equilibrium strategy eliminates overbidding.

The idea that your bid should be much more likely to be close to 0 when the total number of bidders is large makes excellent intuitive sense, and the finding that equilibrium bidding eliminates overbidding lends further confidence to the theoretical analysis. Unfortunately, many people in actual all-pay auctions either do not know or forget this theory and bid to excess.

Interestingly, philanthropists have figured out how to take this tendency to overbid and harness it for social benefit. Building on the historical lessons learned from prizes offered in 1919 by a New York hotelier for the first transatlantic flight (won by Charles Lindbergh in 1927) and even earlier, in 1714, by the British government for a method to precisely measure longitude for sea navigation (eventually awarded to John Harrison in the 1770s), several U.S. and international foundations have begun offering incentive prizes for various socially worthwhile innovations. One foundation in particular, the X Prize Foundation, has as its sole purpose the provision of incentive prizes; its first prize was awarded in 2004 for the first private space flight. Its latest prize competition, begun in 2006, solicits a design for an environmentally sound, superefficient automobile. Some foundation experts estimate that as much as 40 times the amount of money that would otherwise be devoted to a particular innovation gets spent when incentive prizes are available. Thus the tendency to overbid in all-pay auctions can actually have a beneficial impact on society (if not on the individual pursuing the prize).[11]

6 HOW TO SELL AT AUCTION

Bidders are not the only auction participants who need to consider their optimal strategies carefully. An auction is really a sequential-play game in which the first move is the setting of the rules; bidding starts only in the second round of moves. It falls to the sellers, then, to determine the path that later bidding will follow by choosing a particular auction rule or mechanism.

As a seller interested in auctioning off your prized art collection or even your home, you must decide on the best auction mechanism or rule to use. To guarantee yourself the greatest profit from your sale, you must look ahead to the

[11]For more on incentive prizes, see "Incentivizing Prizes: How Foundations Can Utilize Prizes to Generate Solutions to the Most Intractable Social Problems" by Matthew Leerberg (Duke University Center for the Study of Philanthropy and Voluntarism Working Paper, spring 2006). Information on the X Prize Foundation is available at www.xprize.org.

predicted outcome of the different auction mechanisms before making a choice. One concern of many sellers is that an item will go to a bidder for a price lower than the value that the seller places on the object. To counter this concern, most sellers insist on setting a **reserve price** for auctioned objects; they reserve the right to withdraw the object from the sale if no bid higher than the reserve price is obtained.

Beyond setting a reserve price, however, what can sellers do to determine the type of auction mechanism that might net them the most profit possible? One possibility is to use Vickrey's suggested scheme, a second-price, sealed-bid auction. According to him, this kind of auction elicits truthful bidding from potential buyers. Does this effect make it a good auction type from the seller's perspective?

In a sense, the seller in such a second-price auction is giving the bidder a profit margin to counter the temptation to shade down the bid in the hope of a larger profit. But this outcome then reduces the seller's revenue, just as shading down in a first-price, sealed-bid auction would. Which type of auction mechanism is ultimately better for the seller actually turns out to depend on the bidders' attitudes toward risk and their beliefs about the value of the object for sale. The relative merits of different mechanisms in reality can also depend on other issues such as the possibility of collusion among bidders, and the choice can also involve political considerations when selling public property such as the airwave spectrum or drilling rights. Thus, the auction environment is critical to seller revenue.[12]

A. Risk-Neutral Bidders and Independent Estimates

The least complex configuration of bidder risk attitudes and beliefs occurs when there is risk neutrality (no risk aversion) and when bidder estimates about the value of the object for sale remain independent of each other. As we said in the Appendix to Chapter 7, risk-neutral people care only about the expected monetary value of their outcomes, regardless of the level of uncertainty associated with those outcomes. Independence in estimates means that a bidder is not influenced by the estimates of other bidders when determining how much an object is worth to her; the bidder has decided independently exactly how much the object is worth to her. In this case, there can be no winner's curse. If these conditions for bidders hold, sellers can expect the same average revenue (over a large number of trials) from any of the four primary types of auction: English, Dutch, and first- and second-price sealed-bid.

[12]Klemperer's book, especially chapters 3 and 4, has detailed discussions and warnings on all these issues.

This **revenue equivalence** result implies not that all of the auctions will yield the same revenue for every item sold, but that the auctions will yield the same selling price on average in the course of numerous auctions. We can see the equivalence quite easily between second-price, sealed-bid auctions and English auctions. We have already seen that, in the second-price auction, each bidder's dominant strategy is to bid her true valuation. The highest bidder gets the object for the second-highest bid, and the seller gets a price equal to the valuation of the second-highest bidder. Similarly, in an English auction, bidders drop out as the price increases beyond their valuations, until only the first- and second-highest-valuation bidders remain. When the price reaches the valuation of the second-highest bidder, that bidder also will drop out, and the remaining (highest-valuation) bidder will take the object for just a cent more than the second-highest bid. Again, the seller gets a price (essentially) equivalent to the valuation of the second-highest bidder.

More advanced mathematical techniques are needed to prove that revenue equivalence can be extended to Dutch and first-price, sealed-bid auctions as well, but the intuition should be clear. In all four types of auctions, in the absence of any risk aversion on the part of bidders, the highest-valuation bidder should win the auction and pay on average a price equal to the second-highest valuation. If the seller is likely to use a particular auction mechanism repeatedly, she need not be overly concerned about her choice of auction structure; all four would yield her the same expected price.

Experimental and field evidence has been collected to test the validity of the revenue-equivalent theorem in actual auctions. The results of laboratory experiments tend to show Dutch auction prices lower, on average, than first-price, sealed-bid auction prices for the same items being bid on by the same group of bidders, possibly owing to some positive utility associated with the suspense factor in Dutch auctions. These experiments also find evidence of overbidding (bidding above your known valuation) in second-price, sealed-bid auctions but not in English auctions. Such behavior suggests that bidders go higher when they have to specify a price, as they do in sealed-bid auctions; these auctions seem to draw more attention to the relationship between the bid price and the probability of ultimately winning the item. Field evidence from Internet auctions finds literally opposite results, with Dutch auction revenue as much as 30% higher, on average, than first-price, sealed-bid revenue. Additional bidder interest in the Dutch auctions or impatience in the course of a 5-day auction could explain the anomaly. The Internet-based field evidence did find near revenue equivalence for the other two auction types.

B. Risk-Averse Bidders

Here we continue to assume that bids and beliefs are uncorrelated but incorporate the possibility that auction outcomes could be affected by bidders' attitudes toward risk. In particular, suppose bidders are risk averse. They may be much more concerned, for example, about the losses caused by underbidding—losing the object—than by the costs associated with bidding at or close to their true valuations. Thus, risk-averse bidders generally want to win if possible without ever overbidding.

What does this preference structure do to the types of bids that they submit in first-price versus second-price (sealed-bid) auctions? Again, think of the first-price auction as being equivalent to the Dutch auction. Here risk aversion leads bidders to bid earlier rather than later. As the price drops to the bidder's valuation and beyond, there is greater and greater risk in waiting to bid. We expect risk-averse bidders to bid quickly, not to wait just a little bit longer in the hope of gaining those extra few pennies of profit. Applying this reasoning to the first-price, sealed-bid auction, we expect bidders to shade down their bids by less than they would if they were not risk averse: too much shading actually increases the risk of not gaining the object, which risk-averse bidders would want to avoid.

Compare this outcome with that of the second-price auction, where bidders pay a price equal to the second-highest bid. Bidders bid their true valuations in such an auction but pay a price less than that. If they shade their bids only slightly in the first-price auction, then those bids will tend to be close to the bidders' true valuations—and bidders pay their bids in such auctions. Thus bids will be shaded somewhat, but the price ultimately paid in the first-price auction will probably exceed what would be paid in the second-price auction. When bidders are risk averse, the seller then does better to choose a first-price auction rather than a second-price auction.

The seller does better with the first-price auction in the presence of risk aversion only in the sealed-bid case. If the auction is English, the bidders' attitudes toward risk are irrelevant to the outcome. Thus, risk aversion does not alter the outcome for the seller in these auctions.

C. Correlated Estimates

Now suppose that in determining their own valuations of an object, bidders are influenced by the estimates (or by their beliefs about the estimates) of other bidders. Such a situation is relevant for common-value auctions, such as those for oil or gas exploration considered in Section 2. Suppose your experts have not presented a glowing picture of the future profits to be gleaned from the lease on a specific tract of land. You are therefore pessimistic about its potential benefits, and you have constructed an estimate V of its value that you believe corresponds to your pessimism.

Under the circumstances, you may be concerned that your rival bidders also have received negative reports from their experts. When bidders believe their valuations are all likely to be similar, either all relatively low or all relatively high, for example, we say that those beliefs or estimates of value are positively correlated. Recall that we introduced the concept of correlation in Section 1 of Chapter 9, where we considered correlated risks and insurance. In the current context, the likelihood that your rivals' estimates also are unfavorable may magnify the effect of your pessimism on your own valuation. If you are participating in a first-price, sealed-bid auction, you may be tempted to shade down your bid even more than you would in the absence of correlated beliefs. If bidders are optimistic and valuations generally high, correlated estimates may lead to less shading than when estimates are independent.

However, the increase in the shading of bids that accompanies correlated low (or pessimistic) bids in a first-price auction should be a warning to sellers. With positively correlated bidder beliefs, the seller may want to avoid the first-price auction and take advantage of Vickrey's recommendation to use a second-price structure. We have just seen that this auction mechanism encourages truthful revelation, and when correlated estimates are possible, the seller does even better to avoid auctions in which there might be any additional shading of bids.

An English auction will have the same ultimate outcome as the second-price, sealed-bid auction, and a Dutch auction will have the same outcome as a first-price, sealed-bid auction. Thus a seller facing bidders with correlated estimates of an object's value also should prefer the English to the Dutch version of the open-outcry auction. If you are bidding on the oil-land lease in an English auction and the price is nearing your estimate of the lease's value but your rivals are still bidding feverishly, you can infer that their estimates are at least as high as yours—perhaps significantly higher. The information that you obtain from observing the bidding behavior of your rivals may convince you that your estimate is too low. You might even increase your own estimate of the land's value as a result of the bidding process. Your continuing to bid may provide an impetus for further bidding by other bidders, and the process may continue for a while. If so, the seller reaps the benefits. More generally, the seller can expect a higher selling price in an English auction than in a first-price, sealed-bid auction when bidder estimates are correlated. For the bidders, however, the effect of the open bidding is to disperse additional information and to reduce the effect of the winner's curse.

The discussion of correlated estimates assumes that a fairly large number of bidders take part in the auction. But an English auction can be beneficial to the seller if there are only two bidders, both of whom are particularly enthusiastic about the object for sale. They will bid against each other as long as possible, pushing the price up to the lower of the valuations, both of which were high from the start. The same auction can be disastrous for the seller, however, if one of the bidders has a very low valuation; the other is then quite likely to have a valuation considerably higher than the first. In this case, we say that bidder valuations are negatively correlated. We encourage any seller facing a small number of bidders with potentially very different valuations to choose a Dutch or first-price, sealed-bid structure. Either of them would reduce the possibility of the high-valuation bidder gaining the object for well under her true valuation; that is, either type would transfer the available profit from the buyer to the seller.

7 SOME ADDED TWISTS TO CONSIDER

A. Multiple Objects

When you think about an auction of a group of items, such as a bank's auctioning repossessed vehicles or estate sales auctioning the contents of a home, you probably envision the auctioneer bringing each item to the podium individually and selling it to the highest bidder. This process is appropriate when each bidder has independent valuations for each item. However, independent valuations may not always be an appropriate way to model bidder estimates. Then, if bidders value specific groups or whole packages of items higher than the sum of their values for the component items, the choice of auctioning the lots separately or together makes a big difference to bidding strategies as well as to outcomes.

Consider a real-estate developer named Red who is interested in buying a very large parcel of land on which to build a townhouse community for professionals. Two townships, Cottage and Mansion, are each auctioning a land parcel big enough to suit her needs. Both parcels are essentially square in shape and encompass 4 square acres. The mayor of Cottage has directed that the auctioneer sell the land as quarter-acre blocks, one at a time, starting at the perimeter of the land and working inward, selling the corner lots first and then the lots on the north, south, east, and west borders in that order. At the same time, the mayor of Mansion has directed that the auctioneer attempt to sell the land in her town first as a full 4-acre block, then as two individual 2-acre lots, and then as four 1-acre lots after that, if no bids exceed the set reserve prices.

Through extensive market analyses, Red has determined that the blocks of land in Cottage and Mansion would provide the same value to her. However, she has to obtain the full 4 acres of land in either town to have enough room for her planned development. The auctions are being held on the same day at the same time. Which should she attend?

It should be clear that her chances of acquiring a 4-acre block of land for a reasonable price—less than or equal to her valuation—are much better in Mansion than in Cottage. In the Mansion auction, she would simply wait to see how bidding proceeded, submitting a final high bid if the second-highest offer fell below her valuation of the property. In the Cottage auction, she would need to win each and every one of the 16 parcels up for sale. Under the circumstances, she should expect rival bidders interested in owning land in Cottage to become more intent on their goals—perhaps even joining forces—as the number of available parcels decreases in the course of the auction. Red would have to bid aggressively enough to win parcels in the early rounds while being conservative enough to ensure that she did not exceed her total valuation by the end of the auction. The difficulties in crafting a bidding strategy for such an auction are numerous, and the probability of being unable to obtain every parcel profitably is quite large—hence Red's preference for the Mansion auction.

Note that, from the seller's point of view, the Cottage auction is likely to bring in greater revenue than the Mansion auction if an adequate number of bidders are interested in small pieces of land. If the only bidders, are all developers like Red, however, they might be hesitant even to participate in the Cottage auction for fear of being beaten in just one round. In that case, the Mansion-type auction mechanism is better for the seller.

The township of Cottage could allay the fears of developers by revising the rules for its auction. In particular, it would not need to auction each parcel individually. Instead it could hold a single auction in which all parcels would be available simultaneously. Such an auction could be run so that each bidder could specify the number of parcels that she wanted and the price that she was willing to pay per parcel. The bidder with the highest total-value bid—determined by multiplying the number of parcels desired by the price for each—would win the desired number of parcels. If parcels remained after the high bidder took her land, additional parcels would be won in a similar way until all the land was sold. This mechanism gives bidders interested in larger parcels an opportunity to bid, potentially against each other, for blocks of the land. Thus Cottage might find this type of auction more lucrative in the end.

B. Defeating the System

We saw earlier which auction mechanism is best for the seller, given different assumptions about how bidders felt toward risk and whether their estimates were correlated. There is always an incentive for bidders, though, to come up with a bidding strategy that defeats the seller's efforts. The best-laid plans for a profitable auction can almost always be defeated by an appropriately clever bidder or, more often, group of bidders.

Even the Vickrey second-price, sealed-bid auction can be defeated if there are only a few bidders in the auction, all of whom can collude among themselves. By submitting one high bid and a lowball second-highest bid, collusive bidders can obtain an object for the second-bid price. This outcome results only if no other bidders submit intermediate bids or if the collusive bidders are able to prevent such an occurrence. The possibility of collusion highlights the need for the seller's reserve prices, although they only partly offset the problem in this case.

First-price, sealed-bid auctions are less vulnerable to bidder collusion for two reasons. The potential collusive group engages in a multiperson prisoners' dilemma game in which each bidder has a temptation to cheat. In such cheating, an individual bidder might submit her own high bid so as to win the object for herself, reneging on any obligation to share profits with group members. Collusion among bidders in this type of auction is also difficult to sustain because cheating (that is, making a different bid from that agreed to within the collusive group) is easy to do but difficult for other buyers to detect. Thus, the sealed-bid nature of the auction prevents detection of a cheater's behavior, and hence punishment, until the bids are opened and the auction results announced; at that point, it is simply too late. However, there may be more scope for sustaining collusion if a particular group of bidders participates in a number of similar auctions over time, so that they engage in the equivalent of a repeated game.

Other tricky bidding schemes can be created to meet the needs of specific individual bidders or groups of bidders in any particular type of auction. One very clever example of bid rigging arose in an early U.S. Federal Communications Commission auction of the U.S. airwave spectrum, specifically for personal cellular service (Auction 11, August 1996–January 1997). After watching prices soar in some of the earlier auctions, bidders were apparently eager to reduce the price of the winning bids. The solution, used by three firms (later sued by the Department of Justice), was to signal their intentions to go after licenses for certain geographic locations by using the FCC codes or telephone area codes for those areas as the last three digits of their bids. The FCC has claimed that this practice significantly reduced the final prices on these particular licenses. In addition, other signaling devices were apparently used in earlier broadband auctions. While some firms literally announced their intentions to win a particular license, others used a variety of strategic bidding techniques to signal their interest in specific licenses or to dissuade rivals from horning in on their territories. In the first broadband auction, for example, GTE and other firms apparently used the code-bidding technique of ending their bids with the numbers that spelled out their names on a touch-tone telephone keypad!

We note briefly here that fraudulent behavior is not merely the territory of bidders at auction. Sellers also can use underhanded practices to inflate the final bid price of pieces that they are attempting to auction. **Shilling,** for example, occurs when a seller is able to plant false bids at her own auction. Possible only in English auctions, shilling can be done with the use of an agent who works for the seller and who pretends to be a regular bidder. On Internet auction sites, shilling is actually easier, because a seller can register a second identity and log in and bid in her own auction; all Internet auctions have rules and oversight mechanisms designed to prevent such behavior. Sellers in second-price, sealed-bid auctions can also benefit if they inflate the level of the (not publicly known) second-highest bid.

C. Information Disclosure

Finally, we consider the possibility that the seller has some private information about an object that might affect the bidders' valuations of that object. Such a situation arises when the quality or durability of a particular object, such as an automobile, a house, or a piece of electronic equipment, is of great importance to the buyers. Then the seller's past experience with the object may be a good predictor of the future benefits that will accrue to the winning bidder.

As we saw in Chapter 9, the more informed player in an asymmetric information game must decide whether to reveal or conceal her private information. In the auction context, a seller must carefully consider any temptation to conceal information. If the bidders know that the seller has some private information, they are likely to interpret any failure to disclose that information as a signal that the information is unfavorable. Even if the seller's information is unfavorable, she may be better off revealing it; bidders' beliefs might be worse than the actual information. Thus honesty is often the best policy.

Honesty can also be in the seller's interest for another reason. When she has private information about a common-value object, she should disclose that information to sharpen the bidders' estimates of the value of the object. The more confident the bidders are that their valuations are correct, the more likely they are to bid up to those valuations. Thus, disclosure of private seller information in a common-value auction can help not only the seller by reducing the amount of shading done by bidders but also the bidders by reducing the effects of the winner's curse.

8 AUCTIONS ON THE INTERNET

Auctions have become a significant presence in the world of e-commerce. As many as 40 million people make purchases at online auction sites each month, with as much as $60 billion per year in sales being transacted. Auction sites themselves, which make money on listing fees, seller commissions, and advertising revenues, make $1 billion to $2 billion per year. With the Internet auction traffic and revenue increasing annually, this particular area of e-commerce is of considerable interest to many industry analysts as well as to consumers and to auction theorists.

Internet auction sites have been in existence slightly more than a decade. The eBay site began operation in September 1995, shortly after the advent of

Onsale.com in May of that year.[13] A large number of auction sites now exist, with between 100 and 150 different sites available; precise numbers change frequently as new sites are created, as mergers are consummated between existing sites, and as smaller, unprofitable sites shut down. These sites, both small and large, sell an enormous variety of items in many different ways.

The majority of items on the larger, most often used sites, such as eBay and uBid, are goods that are classified as "collectibles," which can mean anything from an antique postcard to a melamine bowl acquired by saving cereal-box tops. But there are also specialty auction sites that deal with items ranging from postage stamps, wine, and cigars to seized property from police raids, medical equipment, and large construction equipment (scissorlift, anyone?). Most of these items, regardless of the type of site, would be considered "used." Thus, consumers have access to what might be called the world's largest garage sale, all at their fingertips.

This information regarding items sold is consistent with one hypothesis in the literature that suggests that Internet auctions are most useful for selling goods available in limited quantity, for which there is unknown demand, and for which the seller cannot easily determine an appropriate price. The auction process can effectively find a "market price" for these goods. Sellers of such goods then have their best profit opportunity online, where a broad audience can supply formerly unknown demand parameters. And consumers can obtain desired but obscure items, presumably with profit margins of their own. Economists would call this matching process, in which each agent receives a positive profit from a trade, *efficient.* (Efficient mechanisms are discussed in greater detail in Chapter 19.)

In addition to selling many different categories of goods, Internet auctions employ a variety of auction rules. Many sites actually offer several auction types and allow a seller to choose her auction's rules when she lists an item for sale. The most commonly used rules are those for English and second-price, sealed-bid auctions; one or both of them are offered by the majority of auction sites.

The sites that offer true English auctions post the high bid as soon as it is received; at the end of the auction, the winner pays her bid. In addition, many sites appear to use the English auction format but allow what is known as **proxy bidding.** The proxy-bidding process actually makes the auction second-price, sealed-bid rather than English.

With proxy bidding, a bidder enters the maximum price that she is willing to pay (her **reservation price**) for an item. Rather than displaying this maximum

[13]Onsale merged with Egghead.com in 1999. Amazon bought the assets of the merged company late in 2001. The three auction sites originally available as Onsale, Egghead, and Amazon are now simply "Amazon.com Auctions."

price, the auction site displays only one bid increment above the most recent high bid. The proxy-bidding system then bids for the buyer, outbidding others by a single bid increment, until the buyer's maximum price is reached. This system allows the auction winner to pay just one bid increment over the second highest bid rather than paying her own bid. Most of the large auction sites, including eBay, uBid, Amazon, and Yahoo, offer proxy-bidding. In addition, some sites offer explicit second-price auctions in which actual high bids are posted, but the winner pays a price equal to the second-highest bid.

True Dutch auctions at online auction sites are quite rare. Some sites that were reported in the literature as having descending price auctions, some with clocks, are no longer in existence. Only a few retail sites now offer the equivalent of Dutch auctions. Land's End, for example, posts some overstocked items each weekend in a special area of its Web site; it then reduces the prices on these items three times in the next week, removing unsold items at week's end. CNET Shopper posts recent price decreases for electronic goods but does not promise additional reductions within a specified time period. One suspects that the dearth of Dutch auctions can be attributed to the real-time nature of such auctions, which makes them difficult to reproduce online.

However, several sites offer auctions *called* Dutch auctions. These auctions, along with a companion type known as **Yankee auctions,** actually offer multiple (identical) units in a single auction. Similar to the auction described in Section 7.A for the available land parcels in the township of Cottage, these auctions offer bidders the option to bid on one or more of the units. Online sites use the terminology "Yankee auction" to refer to the system that we described for the Cottage auction; bidders with the highest total-value bid(s) win the items, and each bidder pays her bid price per unit. The "Dutch auction" label is reserved for auctions in which bids are ranked by total value, but at the end of the auction, all bidders pay the lowest winning bid price for their units.

The Internet has also made it possible to create and apply auction rules that would previously have been impractical. The newest such auction is one in which the *lowest unmatched bid* is the one that wins the item; the object trades at the winning bid price. How can a seller afford such an auction? She can do so simply by auctioning a fairly valuable item, such as a piece of real estate or a sizeable quantity of gold bullion, and charging a small fee for each bid. In such auctions, available as of the summer of 2007 at www.humraz.com, bidding continues until a specified number of bids is obtained, at which time the lowest unmatched bid is awarded the object. Information about the current status of ongoing auctions, including the range in which the current lowest unmatched can be found, is continually updated. And the required number of bids is determined using a profit margin (based on the assessed value of the object in most cases) and the per-bid fee. The Humraz site has successfully auctioned several lots of gold bullion with winning bids below £1 for gold valued at almost £2000.

The success of this type of online auction remains to be seen, but initial interest in the site has been significant.

Although the Internet does allow for creativity, most Internet auctions tend to be quite similar in their rules and outcomes to traditional live auctions. Strategic issues such as those considered in Sections 6 and 7 are relevant to both. There are some benefits to online auctions as well as costs. Online auctions are good for buyers because they are easy to "attend" and they provide search engines that make it simple to identify items of interest. Similarly, sellers are able to reach a wide audience and often have the convenience of choosing the rules of their own auctions. On the other hand, online auction sales can suffer from the fact that buyers cannot inspect goods before they must bid and because buyer and seller must each trust the other to pay or deliver as promised.

The most interesting difference between live and online auctions, though, is the way in which the auctions must end. Live (English and Dutch) auctions cease when no additional bids can be obtained. Online auctions need to have specific auction-ending rules. In virtually all cases, online auctions end after a specified period of time, often 7 days. Different sites, though, use different rules about how hard a line they take on the ending time. Some sites, such as Amazon, allow an extension of the auction if a new bid is received within 10 minutes of the posted auction end time. Others, such as eBay, end their auctions at the posted ending time, regardless of how many bids are received in the closing minutes.

Considerable interest has been generated in the auction literature by the differences in bidder behavior under the two ending-time rules. Evidence has been gathered by Alvin Roth and Axel Ockenfels that shows that eBay's hard end time makes it profitable for bidders to bid late. This behavior, referred to as *sniping*, is found in both private-value and common-value auctions.

Strategically, late bidders on eBay gain by avoiding bidding wars with others who do not use the proxy-bidding system and update their own bids throughout the auction. In addition, they gain by protecting any private information that they hold regarding the common valuation of a good. (Even on eBay, frequent users can identify bidders who seem to know a good item when they see it, particularly in the antiques category. Bidders with this information may want to hide it from others to keep the final sale price down.) Thus, buyers of goods that require special valuation skills, such as antiques, may prefer eBay to Amazon. By the same token, sellers of such goods may get better prices at Amazon—but only if the expert buyers are willing to bid there. The evidence over time suggests that repeat bidders on eBay tend to bid nearer the end in later auctions; bidders on Amazon learn that they can bid earlier. This finding is evidence of movement toward equilibrium bidding behavior in the same way that bidder strategy in the different auctions conforms to our expectations about payoff-maximizing behavior.

9 ADDITIONAL READING ON THE THEORY AND PRACTICE OF AUCTIONS

Much of the literature on the theory of auctions is quite mathematically complex. Some general insights into auction behavior and outcomes can be found in Paul Milgrom, "Auctions and Bidding: A Primer"; Orley Ashenfelter, "How Auctions Work for Wine and Art"; and John G. Riley, "Expected Revenues from Open and Sealed Bid Auctions," all in the *Journal of Economic Perspectives,* vol. 3, no. 3 (Summer 1989), pp. 3–50. These papers should be readable by those of you with a reasonably strong background in calculus.

More complex information on the subject also is available. R. Preston McAfee and John McMillan have an overview paper, "Auctions and Bidding," in the *Journal of Economic Literature,* vol. 25 (June 1987), pp. 699–738. A more recent review of the literature can be found in Paul Klemperer, "Auction Theory: A Guide to the Literature," in the *Journal of Economic Surveys,* vol. 13, no. 3 (July 1999), pp. 227–286. Both of these pieces contain some of the high-level mathematics associated with auction theory but also give comprehensive references to the rest of the literature. Klemperer's book *Auctions: Theory and Practice* (Princeton, NJ: Princeton University Press, 2004) has a more recent and somewhat less mathematically complex survey in Chapter 1.

Vickrey's original article containing the details on truthful bidding in second-price auctions is "Counterspeculation, Auctions, and Competitive Sealed Tenders," *Journal of Finance,* vol. 16, no. 1 (March 1961), pp. 8–37. This paper was one of the first to note the existence of revenue equivalence. A more recent study gathering a number of the results on revenue outcomes for various auction types is J. G. Riley and W. F. Samuelson, "Optimal Auctions," *American Economic Review,* vol. 71, no. 3 (June 1981), pp. 381–392. A very readable history of the "Vickrey" second-price auction is David Lucking-Reiley, "Vickrey Auctions in Practice: From Nineteenth-Century Philately to Twenty-First-Century E-Commerce," *Journal of Economic Perspectives,* vol. 14, no. 3 (Summer 2000), pp. 183–192.

Some of the experimental evidence on auction behavior is reviewed in John H. Kagel, "Auctions: A Survey of Experimental Research" in John Kagel and Alvin Roth, eds., *The Handbook of Experimental Economics* (Princeton: Princeton University Press, 1995), pp. 501–535. More recent evidence on revenue equivalence is provided in David Lucking-Reiley, "Using Field Experiments to Test Equivalence Between Auction Formats: Magic on the Internet," *American Economic Review,* vol. 89, no. 5 (December 1999), pp. 1063–1080. Other evidence on behavior in online auctions is presented in Alvin Roth and Axel Ockenfels, "Last-Minute

Bidding and the Rules for Ending Second-Price Auctions: Evidence from eBay and Amazon Auctions on the Internet," *American Economic Review,* vol. 92, no. 4 (September 2002), pp. 1093–1103.

For information specific to bid rigging in the Federal Communications Commission's airwave spectrum auctions and auction design matters, see Peter Cramton and Jesse Schwartz, "Collusive Bidding: Lessons from the FCC Spectrum Auctions," *Journal of Regulatory Economics,* vol. 17 (May 2000), pp. 229–252, and Paul Klemperer, "What Really Matters in Auction Design," *Journal of Economic Perspectives,* vol. 16, no. 1 (Winter 2002), pp. 169–189. A survey of Internet auctions can be found in David Lucking-Reiley, "Auctions on the Internet: What's Being Auctioned, and How?" *Journal of Industrial Economics,* vol. 48, no. 3 (September 2000), pp. 227–252. Klemperer's book also has many details of successes and errors in spectrum auctions; see especially his Chapters 5–7.

SUMMARY

In addition to the standard *first-price, open-outcry, ascending,* or *English,* auction, there are also *Dutch,* or *descending,* auctions as well as *first-price* and *second-price, sealed-bid* auctions. Objects for bid may have a single *common value* or many *private values* specific to each bidder. With common-value auctions, bidders often win only when they have overbid, falling prey to the *winner's curse.* In private-value auctions, optimal bidding strategies, including decisions about when to *shade* down bids from your true valuation, depend on the auction type used. In the familiar first-price auction, there is a strategic incentive to underbid.

Vickrey showed that sellers can elicit true valuations from bidders by using a second-price, sealed-bid auction. Generally, sellers will choose the mechanism that guarantees them the most profit; this choice will depend on bidder risk attitudes and bidder beliefs about an object's value. If bidders are risk neutral and have independent valuation estimates, all auction types will yield the same outcome. Decisions regarding how to auction a large number of objects, individually or as a group, and whether to disclose information are nontrivial. Sellers must also be wary of bidder collusion or fraud.

The Internet is currently one of the largest venues for items sold at auction. Online auction sites sell many different types of goods and use a variety of different types of auctions. The main strategic difference between live and online auctions arises owing to the hard ending times imposed at some sites, particularly eBay. Late bidding and sniping have been shown to be rational responses to this end-of-auction rule.

KEY TERMS

all-pay auction (668)
ascending auction (659)
common value (660)
descending auction (659)
Dutch auction (659)
English auction (659)
first-price auction (660)
objective value (660)
open outcry (659)
private value (661)
proxy bidding (679)
reservation price (679)
reserve price (671)
revenue equivalence (672)
sealed bid (660)
second-price auction (660)
shading (665)
shilling (677)
subjective value (661)
Vickrey auction (660)
Vickrey's truth serum (666)
winner's curse (661)
Yankee auction (680)

SOLVED EXERCISES

S1. A house painter has a regular contract to work for a builder. On these jobs, her cost estimates are generally right: sometimes a little high, sometimes a little low, but correct on average. When her regular work is slack, she bids competitively for other jobs. "Those are different," she says. "They almost always end up costing more than I estimate." If we assume that her estimating skills do not differ between the two types of jobs, what can explain the difference?

S2. Consider an auction where n identical objects are offered, and there are $(n + 1)$ bidders. The actual value of an object is the same for all bidders and equal for all objects, but each bidder gets only an independent estimate, subject to error, of this common value. The bidders submit sealed bids. The top n bidders get one object each, and each pays what she has bid. What considerations will affect your bidding strategy? How?

S3. You are in the market for a used car and see an ad for the model that you like. The owner has not set a price but invites potential buyers to make offers. Your prepurchase inspection gives you only a very rough idea of the value of the car; you think it is equally likely to be anywhere in the range of $1,000 to $5,000 (so your calculation of the average of this value is $3,000). The current owner knows the exact value and will accept your offer if it exceeds that value. If your offer is accepted and you get the car, then you will find out the truth. But you have some special repair skills and know that when you own the car, you will be able to work on it and increase its value by a third (33.3 . . . %) of whatever it is worth.

(a) What is your expected profit if you offer \$3,000? Should you make such an offer?

(b) What is the highest offer that you can make without losing money on the deal?

S4. In this problem, we consider a special case of the first-price sealed-bid auction and show what the equilibrium amount of bid shading should be. Consider a first-price sealed-bid auction with n risk-neutral bidders. Each bidder has a private value independently drawn from a uniform distribution on [0,1]. That is, for each bidder, all values between 0 and 1 are equally likely. The complete strategy of each bidder is a "bid function" that will tell us, for any value v, what amount $b(v)$ that bidder will choose to bid. Deriving the equilibrium bid functions requires solving a differential equation, but instead of asking you to derive the equilibrium using a differential equation, this problem proposes a candidate equilibrium and asks you to confirm that it is indeed a Nash equilibrium.

It is proposed that the equilibrium-bid function for $n = 2$ is $b(v) = v/2$ for each of the two bidders. That is, if we have two bidders, each should bid half her value, which represents considerable shading.

(a) Suppose you're bidding against just one opponent whose value is uniformly distributed on [0, 1], and who always bids half her value. What is the probability that you will win if you bid $b = 0.1$? If you bid $b = 0.4$? If you bid $b = 0.6$?

(b) Put together the answers to part (a). What is the correct mathematical expression for Pr(win), the probability that you win, as a function of your bid b?

(c) Find an expression for the expected profit you make when your value is v and your bid is b, given that your opponent is bidding half her value. Remember that there are two cases: either you win the auction, or you lose the auction. You need to average the profit between these two cases.

(d) What is the value of b that maximizes your expected profit? This should be a function of your value v.

(e) Use your results to argue that it is a Nash equilibrium for both bidders to follow the same bid function $b(v) = v/2$.

S5. **(Optional)** This question looks at the equilibrium bidding strategies of all-pay auctions, in which bidders have private values for the good, as opposed to the discussion in Section 17.5, where the all-pay auction is for a good with a publicly known value. For the all-pay auction with private values distributed uniformly between 0 and 1, the Nash equilibrium bid function is $b(v) = [(n - 1)/n]v^n$.

(a) Plot graphs of $b(v)$ for the case $n = 2$ and for the case $n = 3$.

(b) Are the bids increasing in the number of bidders, or decreasing in the number of bidders? Your answer might depend on n and v. That is, bids are sometimes increasing in n, and sometimes decreasing in n.

(c) Prove that the function given above is really the Nash-equilibrium bid function. Use a similar approach to that of Exercise **S4**. Remember that in an all-pay auction, you pay your bid even when you lose, so your payoff is $v - b$ when you win, and $-b$ when you lose.

UNSOLVED EXERCISES

U1. "In the presence of very risk-averse bidders, a person selling her house in an auction will have a high expected profit by using a first-price, sealed-bid auction." True or false? Explain your answer.

U2. Suppose that three risk-neutral bidders are interested in purchasing a Princess Beanie Baby. The bidders (numbered 1 through 3) have valuations of \$12, \$14, and \$16, respectively. The bidders will compete in auctions as described in parts **(a)** through **(d)**; in each case, bids can be made in \$1 increments at any value from \$5 to \$25.

(a) Which bidder wins an open-outcry English auction? What are the final price paid and the profit to the winning bidder?

(b) Which bidder wins a second-price sealed-bid auction? What are the final price paid and the profit to the winning bidder? Contrast your answer here with that for part **(a)**. What is the cause of the difference in profits in these two cases?

(c) In a sealed-bid first-price auction, all the bidders will bid a positive amount (at least \$1) less than their true valuations. What is the likely outcome in this auction? Contrast your answer with those for parts **(a)** and **(b)**. Does the seller of the Princess Beanie Baby have any clear reason to choose one of these auction mechanisms over the other?

(d) Risk-averse bidders would reduce the shading of their bids in part **(c)**; assume, for the purposes of this question, that they do not shade at all. If that were true, what would be the winning price (and profit for the bidder) in part **(c)**? Does the seller care about which type of auction she chooses? Why?

U3. You are a turnaround artist, specializing in identifying underperforming companies, buying them, improving their performance and stock price, and then selling them. You have found such a prospect, Sicco. This company's marketing department is mediocre; you believe that if you take over the company, you will increase its value by 75% of whatever it was before. But its accounting department is very good; it can conceal assets, liabilities, and transactions to a point where the company's true value is hard for outsiders to identify. (But insiders know the truth perfectly.) You think that the company's value in the hands of its current management is somewhere between $10 million and $110 million, uniformly distributed over this range. The current management will sell the company to you if, and only if, your bid exceeds the true value known to them.

(a) If you bid $110 million for the company, your bid will surely succeed. Is your expected profit positive?

(b) If you bid $50 million for the company, what is the probability that your bid succeeds? What is your expected profit if you do succeed in buying the company? Therefore, at the point in time when you make your bid of $50 million, what is your expected profit? (Warning: In calculating this expectation, don't forget the probability of your getting the company.)

(c) What should you bid if you want to maximize your expected profit? (Hint: Assume it is X million dollars. Carry out the same analysis as in part (b) above, and find an algebraic expression for your expected profit as seen from the point in time when you are making your bid. Then choose X to maximize this expression.)

U4. The idea of the winner's curse can be expressed slightly differently from its usage in the chapter: "The only time your bid matters is when you win, which happens when your estimate is higher than the estimates of all the other bidders. Therefore you should focus on this case. That is, you should always act as if all the others have received estimates lower than yours, and use this 'information' to revise your own estimate." Here we ask you to apply this idea to a very different situation.

A jury consists of 12 people who hear and see evidence presented at a trial and collectively reach a verdict of guilt or innocence. Simplifying the process somewhat, assume that the jurors hold a single simultaneous vote to determine the verdict. Each juror is asked to vote Guilty or Not guilty. The accused is convicted if all 12 vote Guilty and is acquitted if one or more vote Not guilty; this is known as the unanimity rule. Each juror's objective is to arrive at a verdict that is the most accurate verdict in light of the evidence, but each juror interprets the evidence in accord with her own thinking and experience. Thus, she arrives at an estimate of the guilt or the innocence of the accused that is individual and private.

(a) If jurors vote truthfully—that is, in accordance with their individual private estimates of the guilt of the accused—will the verdict be Not guilty more often under a unanimity rule or under a majority rule, where the accused is convicted if seven jurors vote Guilty? Explain. What might we call the "juror's curse" in this situation?

(b) Now consider the case in which each juror votes strategically, taking into account the potential problems of the juror's curse and using all the devices of information inference that we have studied. Are individual jurors more likely to vote Guilty under a unanimity rule when voting truthfully or strategically? Explain.

(c) Do you think strategic voting to account for the juror's curse would produce too many Guilty verdicts? Why or why not?

U5. **(Optional)** This exercise is a continuation of Exercise S4; it looks at the general case where n is any positive integer. It is proposed that the equilibrium-bid function with n bidders is $b(v) = v(n - 1)/n$. For $n = 2$, we have the case explored in Exercise S4: each of the bidders bids half of her value. If there are nine bidders ($n = 9$), then each should bid 9/10 of her value, and so on.

(a) Now there are $n - 1$ other bidders bidding against you, each using the bid function $b(v) = v(n - 1)/n$. For the moment, let's focus on just one of your rival bidders. What is the probability that she will submit a bid less than 0.1? Less than 0.4? Less than 0.6?

(b) Using the above results, find an expression for the probability that the other bidder has a bid less than your bid amount b.

(c) Recall that there are $n - 1$ other bidders, all using the same bid function. What is the probability that your bid b is larger than *all* of the other bids? That is, find an expression for Pr(win), the probability that you win, as a function of your bid b.

(d) Use this result to find an expression for your expected profit when your value is v and your bid is b.

(e) What is the value of b that maximizes your expected profit?

Use your results to argue that it is a Nash equilibrium for all n bidders to follow the same bid function $b(v) = v(n - 1)/n$.

U6. **(Optional)** In Exercises S4 and U5, we showed that the symmetric equilibrium where everyone uses the bid function $b(v) = v(n - 1)/n$ is indeed a Nash equilibrium when everyone has their values drawn from a uniform distribution on [0,1]. Now we will show how to derive a first-price sealed-bid equilibrium-bid function by solving a differential equation.

Let's consider the general case in which the distribution of bidder values has a CDF (cumulative distribution function) of $F(v)$. That is, $F(v)$ is the probability that a given bidder's value is less than or equal to the number v. For now, this will be a general, unspecified function. Later, we'll substitute in a specific function that corresponds to the case of a uniform distribution in which all values between 0 and 1 are equally likely. Let's assume that some strictly increasing bid function $b(v)$ exists, such that if all n bidders use this function, we will have a Nash equilibrium. We can use the definition of Nash equilibrium to figure out an equation that must be true for $b(v)$.

(a) Let's start by finding the probability that I have the highest of the n values. If my value is v, then what is the probability that all $n - 1$ other bidders have values less than mine? I'm asking you to use reasoning similar to that in part (c) of Exercise U5, but your probability expression should now contain the general function $F(v)$.

(b) We have supposed $b(v)$ to be strictly increasing in v, so it must have an inverse function. Let $V(b)$ be this inverse function. This is the function that, for a given bid amount b, will tell you the value of the person who submitted the bid, assuming that person followed the equilibrium-bid function $b(v)$. (For example, if $b(v) = 0.5v$, then $V(b) = 2b$.) Now, suppose that I know everyone else is using the bid function $b(v)$. Suppose I have a value of v, and I choose to bid some number x, which may or not be equal to $b(v)$. Then, if someone else assumes I am following the bid function $b(v)$, it will look as though I have a value of $V(x)$.

Since $b(v)$ is strictly increasing, the highest bid among my rivals will be submitted by the bidder with the highest value. So to win the auction, I need to bid an amount x such that my apparent value $V(x)$ is higher than all of my rivals' values. Thus, if we substitute $v = V(x)$ into our answer for part (a), we should have an expression for the probability that I win the auction with bid amount x. Use this to write an expression for my expected profit if I bid amount x. This should be similar to your answer for part (a) of Exercise U5 above, but will use the general functions F and V.

(c) Now I want to choose my bid amount to maximize my expected profit. Write the appropriate first-order condition. Since we are using general functions F and V, your first-order condition should include their derivatives F' and V'. You will also need to remember the chain rule from calculus class: the derivative of $f(g(x))$ with respect to x is equal to $f'(g(x))$ times $g'(x)$.

(d) Now let's impose the Nash-equilibrium condition. If the bid function $b(v)$ is a symmetric Nash equilibrium, then I should want to follow it when everyone else does. That means that in the first-order condition in part (c) for the optimal value of my bid x, if my value is v, I should want to bid $x = b(v)$. Similarly, my apparent value $V(x)$ will equal my actual value v. Substitute $x = b(v)$ and $V(x) = v$ into the above equation. Show how to rearrange your result into the following differential equation:

$$b'(v) + (n-1)[F'(v)/F(v)]b(v) = (n-1)[F'(v)/F(v)]v$$

(e) The above equation is called a "differential equation" because it tells us a rule that must be satisfied by the unknown function $b(v)$ for all possible values of v. Remember that $F(v)$ is a known function that describes the distribution of bidders' values; we just haven't specified what $F(v)$ is yet.

On the other hand, $b(v)$ is an unknown function. The above equation gives us the relationship that must always hold true for the relationship between the function $b(v)$ and its derivative $b'(v)$, for all values of v.

Let's make an analogy to ordinary algebraic equations. When we look at an algebraic equation, we want to find the unknown value x that satisfies the equation. Here, we have something similar but more complicated: we have an equation that must be satisfied by an unknown function, and we want to find the entire shape of this function. There is no general recipe for solving a differential equation; different techniques work well for different types of equations. This equation turns out to be one that can be solved by multiplying by an "integrating factor" and then integrating both sides of the equation.

So to solve this equation, multiply both sides by the factor $F(v)^{n-1}$. Then integrate both sides of the equation. You should get an equation showing that $b(v)$ equals some expression involving the function $F(v)$ as well as an integral sign, plus some arbitrary constant C of integration. There shouldn't be any b on the right side of the equation.

(f) We haven't yet figured out the value of the arbitrary constant of integration. Restating this in math jargon, we haven't figured out the appropriate "boundary condition" for this differential equation. So our solution is not yet complete, because there are an infinite number of different values of C that we could use.

It turns out it helps to specify the integral on the right side as a definite integral, from the lowest possible value of 0 up to some arbitrary value v. Since the variable v currently appears both inside and outside the integral sign, we need to be careful. Inside the integral, you should replace v with some other variable u, to keep track of the fact that u is just a variable of integration, rather than being the same as the v outside the integral sign. Then the integral sign should have a lower limit 0 and an upper limit v, so that we are integrating over u from 0 to v.

Now what is the appropriate "boundary condition"? To figure this out, consider what a bidder with a value of 0 should do. If he bids more than 0, he risks winning the auction and getting a negative payoff, so he wouldn't want to do that in equilibrium. But he can't bid less than zero, by the rules of the auction. So a bidder with $v = 0$ must submit a bid of 0. That is, our boundary condition is $b(0) = 0$.

Substitute this boundary condition into our solution for $b(v)$, and use this to find the appropriate value of the constant C. You should get a result saying that $b(v)$ equals v, minus some bid-shading amount. The bid-shading amount is the ratio of two expressions. In the numerator is the integral of $F(u)^{n-1}$, integrated from 0 to v. In the denominator is the integral $F(v)^{n-1}$.

Note that this result gives us a general solution for a first-price sealed-bid auction. You tell me some probability distribution $F(v)$ of values, and I'll tell you what the equilibrium-bid function must be. $F(v)$ can have any shape at all, so long as it is a valid CDF. It must start at $F(0) = 1$, end at $F(\text{infinity}) = 1$, and be weakly increasing for all v in between 0 and infinity.

(g) Now we have a general result, but this general function $F(v)$ is rather abstract and hard to think about. So let's go back to the specific case of a uniform distribution of values from 0 to 1. Then $F(v) = v$. Substitute this value of F into your solution, and show that you get the same bid function $b(v)$ as in Exercise **U5.** You have now derived the bid function from scratch!

18 Bargaining

PEOPLE ENGAGE IN BARGAINING throughout their lives. Children start by negotiating to share toys and to play games with other children. Couples bargain about matters of housing, child rearing, and the adjustments that each must make for the other's career. Buyers and sellers bargain over price, workers and bosses over wages. Countries bargain over policies of mutual trade liberalization; superpowers negotiate mutual arms reduction. And the two original authors of this book had to bargain with one another—generally very amicably—about what to include or exclude, how to structure the exposition, and so forth. To get a good result from such bargaining, the participants must devise good strategies. In this chapter, we raise and explicate some of these basic ideas and strategies.

All bargaining situations have two things in common. First, the total payoff that the parties to the negotiation are capable of creating and enjoying as a result of reaching an agreement should be greater than the sum of the individual payoffs that they could achieve separately—the whole must be greater than the sum of the parts. Without the possibility of this excess value, or "surplus," the negotiation would be pointless. If two children considering whether to play together cannot see a net gain from having access to a larger total stock of toys or from one another's company in play, then it is better for each to "take his toys and play by himself." The world is full of uncertainty, and the expected benefits may not materialize. But when engaged in bargaining, the parties must at least perceive some gain therefrom: when he agreed to sell his soul to the Devil, Faust thought the benefits of knowledge and power that he gained were worth the price that he would eventually have to pay.

The second important general point about bargaining follows from the first: it is not a zero-sum game. When a surplus exists, the negotiation is about how to divide it up. Each bargainer tries to get more for himself and leave less for the others. This may appear to be zero sum, but behind it lies the danger that if the agreement is not reached, no one will get any surplus at all. This mutually harmful alternative, as well as *both* parties' desire to avoid it, is what creates the potential for the threats—explicit and implicit—that make bargaining such a strategic matter.

Before the advent of game theory, one-on-one bargaining was generally thought to be a difficult and even indeterminate problem. Observation of widely different outcomes in otherwise similar-looking situations lent support to this view. Management–union bargaining over wages yields different outcomes in different contexts; different couples make different choices. Theorists were not able to achieve any systematic understanding of why one party gets more than another and attributed this result to vague and inexplicable differences in "bargaining power."

Even the simple theory of Nash equilibrium does not take us any further. Suppose two people are to split \$1. Let us construct a game in which each is asked to announce what he would want. The moves are simultaneous. If their announcements x and y add up to 1 or less, each gets what he announced. If they add up to more than 1, neither gets anything. Then *any* pair (x, y) adding to 1 constitutes a Nash equilibrium in this game: *given* the announcement of the other, each player cannot do better than to stick to his own announcement.[1]

Further advances in game theory have brought progress along two quite different lines, each using a distinct mode of game-theoretic reasoning. In Chapter 2, we distinguished between cooperative-game theory, in which the players decide and implement their actions jointly, and noncooperative-game theory, in which the players decide and take their actions separately. Each of the two lines of advance in bargaining theory uses one of these two approaches. One approach views bargaining as a *cooperative* game, in which the parties find and implement a solution jointly, perhaps, using a neutral third party such as an arbitrator for enforcement. The other approach views bargaining as a *noncooperative* game, in which the parties choose strategies separately and we look for an equilibrium. However, unlike our earlier simple

[1]As we saw in Chapter 5 (Section 3.B), this type of game can be used as an example to bolster the critique that the Nash-equilibrium concept is too imprecise. In the bargaining context, we might say that the multiplicity of equilibria is just a formal way of showing the indeterminacy that previous analysts had claimed.

game of simultaneous announcements, whose equilibrium was indeterminate, here we impose more structure and specify a sequential-move game of offers and counteroffers, which leads to a determinate equilibrium. As in Chapter 2, we emphasize that the labels "cooperative" and "noncooperative" refer to joint versus separate actions, not to nice versus nasty behavior or to compromise versus breakdown. The equilibria of noncooperative bargaining games can entail a lot of compromise.

1 NASH'S COOPERATIVE SOLUTION

In this section we present Nash's cooperative-game approach to bargaining. First we present the idea in a simple numerical example; then we develop the more general algebra.[2]

A. Numerical Example

Imagine two Silicon Valley entrepreneurs, Andy and Bill. Andy produces a microchip set that he can sell to any computer manufacturer for $900. Bill has a software package that can retail for $100. The two meet and realize that their products are ideally suited to each other and that, with a bit of trivial tinkering, they can produce a combined system of hardware and software worth $3,000 in each computer. Thus together they can produce an extra value of $2,000 per unit, and they expect to sell millions of these units each year. The only obstacle that remains on this path to fortune is to agree to a division of the spoils. Of the $3,000 revenue from each unit, how much should go to Andy and how much to Bill?

Bill's starting position is that without his software, Andy's chip set is just so much metal and sand, so Andy should get only the $900 and Bill himself should get $2,100. Andy counters that without his hardware, Bill's programs are just symbols on paper or magnetic signals on a diskette, so Bill should get only $100, and $2,900 should go to him, Andy.

Watching them argue, you might suggest they "split the difference." But that is not an unambiguous recipe for agreement. Bill might offer to split the profit on each unit equally with Andy. Under this scheme, each will get a profit of $1,000, meaning that $1,100 of the revenue goes to Bill and $1,900 to Andy. Andy's response might be that they should have an equal percentage of profit on their contribution to the joint enterprise. Thus Andy should get $2,700 and Bill $300.

The final agreement depends on their stubbornness or patience if they negotiate directly with one another. If they try to have the dispute arbitrated by a

[2]John F. Nash, Jr., "The Bargaining Problem," *Econometrica,* vol. 18, no. 2 (1950), pp. 155–162.

third party, the arbitrator's decision depends on his sense of the relative value of hardware and software and on the rhetorical skills of the two principals as they present their arguments before the arbitrator. For the sake of definiteness, suppose the arbitrator decides that the division of the profit should be 4 : 1 in favor of Andy; that is, Andy should get four-fifths of the surplus while Bill gets one-fifth, or Andy should get four times as much as Bill. What is the actual division of revenue under this scheme? Suppose Andy gets a total of x and Bill gets a total of y; thus Andy's profit is $(x - 900)$ and Bill's is $(y - 100)$. The arbitrator's decision implies that Andy's profit should be four times as large as Bill's; so $x - 900 = 4(y - 100)$, or $x = 4y + 500$. The total revenue available to both is \$3,000; so it must also be true that $x + y = 3{,}000$, or $x = 3{,}000 - y$. Then $x = 4y + 500 = 3{,}000 - y$, or $5y = 2{,}500$, or $y = 500$, and thus $x = 2{,}500$. This division mechanism leaves Andy with a profit of $2{,}500 - 900 = \$1{,}600$ and Bill with $500 - 100 = \$400$, which is the 4 : 1 split in favor of Andy that the arbitrator wants.

We now develop this simple data into a general algebraic formula that you will find useful in many practical applications. Then we go on to examine more specifics of what determines the ratio in which the profits in a bargaining game get split.

B. General Theory

Suppose two bargainers, A and B, seek to split a total value v, which they can achieve if and only if they agree on a specific division. If no agreement is reached, A will get a and B will get b, each by acting alone or in some other way acting outside of this relationship. Call these their *backstop* payoffs or, in the jargon of the Harvard Negotiation Project, their **BATNAs (best alternative to a negotiated agreement).**[3] Often a and b are both zero, but, more generally, we only need to assume that $a + b < v$, so that there is a positive **surplus** $(v - a - b)$ from agreement; if this were not the case, the whole bargaining would be moot because each side would just take up its outside opportunity and get its BATNA.

Consider the following rule: each player is to be given his BATNA plus a share of the surplus, a fraction h of the surplus for A and a fraction k for B, such that $h + k = 1$. Writing x for the amount that A finally ends up with, and similarly y for B, we translate these statements as

$$x = a + h(v - a - b) = a(1 - h) + h(v - b)$$
$$x - a = h(v - a - b)$$

and

$$y = b + k(v - a - b) = b(1 - k) + k(v - a)$$
$$y - b = k(v - a - b)$$

[3] See Roger Fisher and William Ury, *Getting to Yes*, 2nd ed. (New York: Houghton Mifflin, 1991).

We call these expressions the Nash formulas. Another way of looking at them is to say that the surplus $(v - a - b)$ gets divided between the two bargainers in the proportions of $h\!:k$, or

$$\frac{y - b}{x - a} = \frac{k}{h}$$

or, in slope-intercept form,

$$y = b + \frac{k}{h}(x - a) = \left(b - \frac{ak}{h}\right) + \frac{k}{h}x$$

To use up the whole surplus, x and y must also satisfy $x + y = v$. The Nash formulas for x and y are actually the solutions to these last two simultaneous equations.

A geometric representation of the **Nash cooperative solution** is shown in Figure 18.1. The backstop, or BATNA, is the point P, with coordinates (a, b). All points (x, y) that divide the gains in proportions $h\!:k$ between the two players lie along the straight line passing through P and having slope k/h; this slope is just the line $y = b + (k/h)(x - a)$ that we derived earlier. All points (x, y) that use up the whole surplus lie along the straight line joining $(v,0)$ and $(0,v)$; this line is the second equation that we derived—namely, $x + y = v$. The Nash solution is at the intersection of the lines, at the point Q. The coordinates of this point are the parties' payoffs after the agreement.

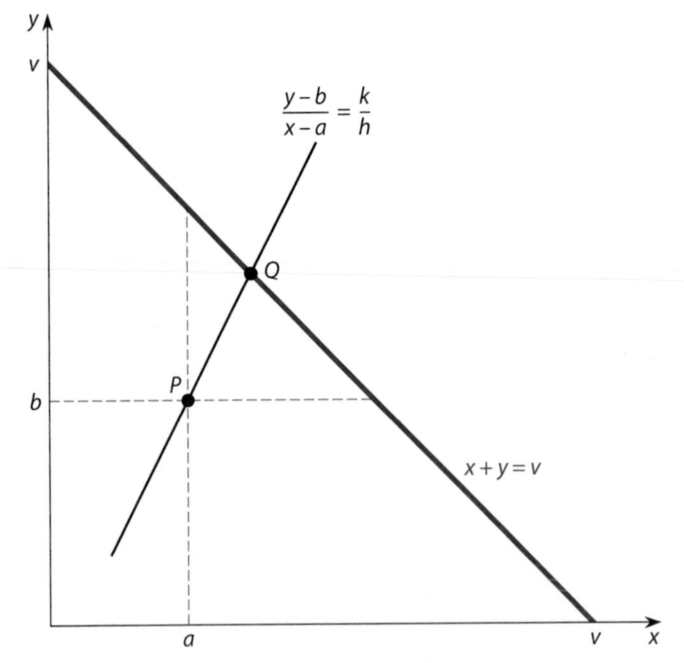

FIGURE 18.1 The Nash Bargaining Solution in the Simplest Case

The Nash formula says nothing about how or why such a solution might come about. And this vagueness is its merit—it can be used to encapsulate the results of many different theories taking many different perspectives.

At the simplest, you might think of the Nash formula as a shorthand description of the outcome of a bargaining process that we have not specified in detail. Then h and k can stand for the two parties' relative bargaining strengths. This shorthand description is a cop-out; a more complete theory should explain where these bargaining strengths come from and why one party might have more than the other. We do so in a particular context later in the chapter. In the meantime, by summarizing any and all of the sources of bargaining strength in these numbers h and k, the formula has given us a good tool.

Nash's own approach was quite different—and indeed different from the whole approach to game theory that we have taken thus far in this book. Therefore it deserves more careful explanation. In all the games that we have studied so far, the players chose and played their strategies separately from each other. We have looked for equilibria in which each player's strategy was in his own best interest, given the strategies of the others. Some such outcomes were very bad for some or even all of the players, the prisoners' dilemma being the most prominent example. In such situations, there was scope for the players to get together and agree that all would follow some particular strategy. But in our framework, there was no way in which they could be sure that the agreement would hold. After reaching an agreement, the players would disperse, and, when it was each player's turn to act, he would actually take the action that served his own best interest. The agreement on joint action would unravel in the face of such separate temptations. True, in considering repeated games in Chapter 11, we found that the implicit threat of the collapse of an ongoing relationship might sustain an agreement, and, in Chapter 9, we did allow for communication by signals. But individual action was of the essence, and any mutual benefit could be achieved only if it did not fall prey to the selfishness of separate individual actions. In Chapter 2, we called this approach to game theory *noncooperative*, emphasizing that the term signified how actions are taken, not whether outcomes are jointly good. The important point, again, is that any joint good has to be an equilibrium outcome of separate action in such games.

What if joint action *is* possible? For example, the players might take all their actions immediately after the agreement is reached, in one another's presence. Or they might delegate the implementation of their joint agreement to a neutral third party, or to an arbitrator. In other words, the game might be *cooperative* (again in the sense of joint action). Nash modeled bargaining as a cooperative game.

The thinking of a collective group that is going to implement a joint agreement by joint action can be quite different from that of a set of individual people who know that they are *interacting* strategically but are *acting* noncooperatively. Whereas the latter set will think in terms of an equilibrium and then delight or grieve, depending on whether they like the results, the former can think first of what is a good outcome and then see how to implement it. In other words, the theory defines the outcome of a cooperative game in terms of some general principles or properties that seem reasonable to the theorist.

Nash formulated a set of such principles for bargaining and proved that they implied a unique outcome. His principles are roughly as follows: (1) the outcome should be invariant if the scale in which the payoffs are measured changes linearly; (2) the outcome should be **efficient;** and (3) if the set of possibilities is reduced by removing some that are irrelevant in the sense that they would not be chosen anyway, then the outcome should not be affected.

The first of these principles conforms to the theory of expected utility, which we discussed briefly in the Appendix to Chapter 7. We saw there that a nonlinear rescaling of payoffs represents a change in a player's attitude toward risk and a real change in behavior; a concave rescaling implies risk aversion, and a convex rescaling implies risk preference. A linear rescaling, being the intermediate case between these two, represents no change in the attitude toward risk. Therefore it should have no effect on expected payoff calculations and no effect on outcomes.

The second principle simply means that no available mutual gain should go unexploited. In our simple example of A and B splitting a total value of v, it would mean that x and y has to sum to the full amount of v available, and not to any smaller amount; in other words, the solution has to lie on the $x + y = v$ line in Figure 18.1. More generally, the complete set of logically conceivable agreements to a bargaining game, when plotted on a graph as in Figure 18.1, will be bounded above and to the right by the subset of agreements that leave no mutual gain unexploited. This subset need not lie along a straight line such as $x + y = v$ (or $y = v - x$); it could lie along any curve of the form $y = f(x)$.

In Figure 18.2, all of the points on and below (that is, "south" and to the "west" of) the thick red curve labeled $y = f(x)$ constitute the complete set of conceivable outcomes. The curve itself consists of the efficient outcomes; there are no conceivable outcomes that include more of both x and y than the outcomes on $y = f(x)$; so there are no unexploited mutual gains left. Therefore we call the curve $y = f(x)$ the **efficient frontier** of the bargaining problem.

We can illustrate a curved efficient frontier using the example of efficient risk allocation from Chapter 9, Section1.A. Two farmers, each with a square root utility function, face the risk that equally likely good or bad weather would make their incomes either \$160,000 or \$40,000, yielding each an expected utility of

$$1/2 \times \sqrt{160{,}000} + 1/2 \times \sqrt{40{,}000} = 1/2 \times 400 + 1/2 \times 200 = 300.$$

But their risks are perfectly negatively correlated. One gets good weather only when the other gets bad, so their combined income is \$200,000 no matter which of them gets the good weather. If they negotiate so that the first of them

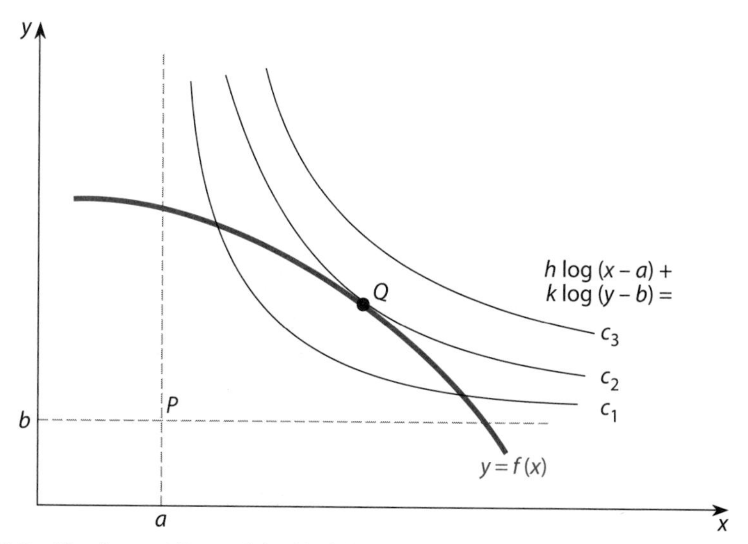

FIGURE 18.2 The General Form of the Nash Bargaining Solution

gets z of the combined income and the other gets the remaining $(200{,}000 - z)$, their respective utilities x and y will be

$$x = \sqrt{z} \quad \text{and} \quad y = \sqrt{200{,}000 - z}.$$

Therefore, we can describe the set of possible risk-sharing outcomes by the equation

$$x^2 + y^2 = z + (200{,}000 - z) = 200{,}000.$$

This equation defines a quarter-circle in the positive quadrant and represents the efficient frontier of the farmers' bargaining problem. The BATNA of each farmer is the expected utility 300 he would get if the two are not able to come to any risk-sharing agreement. Substituting this value into the equation above yields $300^2 + 300^2 = 90{,}000 + 90{,}000 = 180{,}000 < 200{,}000$. So the farmers' BATNA point lies inside the quarter-circle efficient frontier.

The third principle also seems appealing. If an outcome that a bargainer wouldn't have chosen anyway drops out of the picture, what should it matter? This assumption is closely connected to the "independence of irrelevant alternatives" assumption of Arrow's impossibility theorem, which we met in Chapter 16, Section 3, but we must leave the development of this connection to more advanced treatments of the subject.

Nash proved that the cooperative outcome that satisfied all three of these assumptions could be characterized by the mathematical maximization problem: choose x and y to

$$\text{maximize } (x - a)^h (y - b)^k \quad \text{subject to } y = f(x).$$

Here x and y are the outcomes, a and b the backstops, and h and k two positive numbers summing to 1, which are like the bargaining strengths of the Nash formula. The values for h and k cannot be determined by Nash's three assumptions alone; thus they leave a degree of freedom in the theory and in the outcome. Nash actually imposed a fourth assumption on the problem—that of symmetry between the two players; this additional assumption led to the outcome $h = k = 1/2$ and fixed a unique solution. We have given the more general formulation that subsequently became common in game theory and economics.

Figure 18.2 also gives a geometric representation of the objective of the maximization. The black curves labeled c_1, c_2, and c_3 are the level curves, or contours, of the function being maximized; along each such curve, $(x - a)^h(y - b)^k$ is constant and equals c_1, c_2, or c_3 (with $c_1 < c_2 < c_3$) as indicated. The whole space could be filled with such curves, each with its own value of the constant, and curves farther to the "northeast" would have higher values of the constant.

It is immediately apparent that the highest possible value of the function is at that point of tangency, Q, between the efficient frontier and one of the level curves.[4] The location of Q is defined by the property that the contour passing through Q is tangent to the efficient frontier. This tangency is the usual way to illustrate the Nash cooperative solution geometrically.[5]

In our example of Figure 18.1, we can also derive the Nash solution mathematically; to do so requires calculus, but the ends here are more important—at least to the study of games of strategy—than the means. For the solution, it helps to write $X = x - a$ and $Y = y - b$. Thus X is the amount of the surplus that goes to A, and Y is the amount of the surplus that goes to B. The efficiency of the outcome guarantees that $X + Y = x + y - a - b = v - a - b$, which is just the total surplus and which we will write as S. Then $Y = S - X$, and

$$(x - a)^h(y - b)^k = X^h Y^k = X^h(S - X)^k.$$

In the Nash solution, X takes on the value that maximizes this function. Elementary calculus tells us that the way to find X is to take the derivative of this expression with respect to X and set it equal to zero. Using the rules of calculus for taking the derivatives of powers of X and of the product of two functions of X, we get

$$hX^{h-1}(S - X)^k - X^h k(S - X)^{k-1} = 0.$$

[4]One and only one of the (convex) level curves can be tangential to the (concave) efficient frontier; in Figure 18.2, this level curve is labeled c_2. All lower-level curves (such as c_1) cut the frontier in two points; all higher-level curves (such as c_3) do not meet the frontier at all.

[5]If you have taken an elementary microeconomics course, you will have encountered the concept of social optimality, illustrated graphically by the tangent point between the production possibility frontier of an economy and a social indifference curve. Our Figure 18.2 is similar in spirit; the efficient frontier in bargaining is like the production possibility frontier, and the level curves of the objective in cooperative bargaining are like social indifference curves.

When we cancel the common factor $X^{h-1}(S-X)^{k-1}$, this equation becomes

$$h(S-X) - kX = 0$$
$$hY - kX = 0$$
$$kX = hY$$
$$\frac{X}{h} = \frac{Y}{k}.$$

Finally, expressing the equation in terms of the original variables x and y, we have $(x - a)/h = (y - b)/k$, which is just the Nash formula. The punch line: Nash's three conditions lead to the formula we originally stated as a simple way of splitting the bargaining surplus.

The three principles, or desired properties, that determine the Nash cooperative-bargaining solution are simple and even appealing. But in the absence of a good mechanism to make sure that the parties take the actions stipulated by the agreement, these principles may come to nothing. A player who can do better by strategizing on his own than by using the Nash solution may simply reject the principles. If an arbitrator can enforce a solution, the player may simply refuse to go to arbitration. Therefore Nash's cooperative solution will seem more compelling if it can be given an alternative interpretation—namely, as the Nash equilibrium of a noncooperative game played by the bargainers. This can indeed be done, and we will develop an important special case of it in Section 5.

2 VARIABLE-THREAT BARGAINING

In this section, we embed the Nash cooperative solution within a specific game—namely, as the second stage of a sequential-play game. We assumed in Section 1 that the players' backstops (BATNAs) a and b were fixed. But suppose there is a first stage to the bargaining game in which the players can make strategic moves to manipulate their BATNAs within certain limits. After they have done so, the Nash cooperative outcome starting from those BATNAs will emerge in a second stage of the game. This type of game is called **variable-threat bargaining.** What kind of manipulation of the BATNAs is in a player's interest in this type of game?

We show the possible outcomes from a process of manipulating BATNAs in Figure 18.3. The originally given backstops (a and b) are the coordinates for the game's backstop point P; the Nash solution to a bargaining game with these backstops is at the outcome Q. If player A can increase his BATNA to move the game's backstop point to P_1, then the Nash solution starting there leads to the outcome Q', which is better for player A (and worse for B). Thus a strategic move that improves one's own BATNA is desirable. For example, if you have a good job

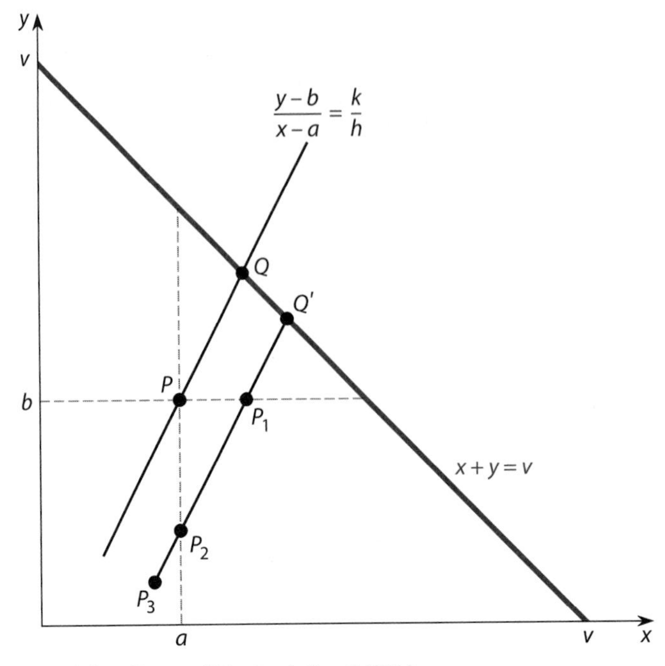

FIGURE 18.3 Bargaining Game of Manipulating BATNAs

offer in your pocket—a higher BATNA—when you go for an interview at another company, you are likely to get a better offer from that employer than you would if you did not have the first alternative.

The result that improving your own BATNA can improve your ultimate outcome is quite obvious, but the next step in the analysis is less so. It turns out that if player A can make a strategic move that *reduces* player B's BATNA and moves the game's backstop point to P_2, the Nash solution starting there leads to the *same* outcome Q' that was achieved after A increased his own BATNA to get to the backstop point P_1. Therefore this alternative kind of manipulation is equally in player A's interest. As an example of decreasing your opponent's BATNA, think of a situation in which you are already working and want to get a raise. Your chances are better if you can make yourself indispensable to your employer so that without you his business has much worse prospects; his low outcome in the absence of an agreement—not offering you a raise and your leaving the firm—may make him more likely to accede to your wishes.

Finally and even more dramatically, if player A can make a strategic move that lowers *both players' BATNAs* so that the game's backstop point moves to P_3, that again has the same result as each of the preceding manipulations. This particular move is like using a threat that says, "This will hurt you more than it hurts me."

In general, the key for player A is to shift the game's BATNA point to somewhere below the line PQ. The farther southeast the BATNA point is moved, the better it is for player A in the eventual outcome. As is usual with threats, the idea is not actually to suffer the low payoff but merely to use its prospect as a lever to get a better outcome.

The possibility of manipulating BATNAs in this way depends on the context. We offer just one illustration. In 1980 there was a baseball players' strike. It took a very complicated form. The players struck in spring training, then resumed working (playing, really) when the regular season began in April, and went on strike again starting on Memorial Day. A strike is costly to both sides, employers and employees, but the costs differ. During spring training the players do not have salaries, but the owners make some money from vacationing spectators. At the start of the regular season, in April and May, the players get salaries but the weather is cold and the season is not yet exciting; therefore the crowds are small, and so the cost of a strike to the owners is low. The crowds start to build up from Memorial Day onward, which raises the cost of a strike to the owners, but the salaries that the players stand to lose stay the same. So we see that the two-piece strike was very cleverly designed to lower the BATNA of the owners *relative to* that of the players as much as possible.[6]

One puzzle remains: Why did the strike occur at all? According to the theory, everyone should have seen what was coming; a settlement more favorable to the players should have been reached so that the strike would have been unnecessary. A strike that actually happens is a threat that has "gone wrong." Some kind of uncertainty—asymmetric information or brinkmanship—must be responsible.

3 ALTERNATING-OFFERS MODEL I: TOTAL VALUE DECAYS

Here we move back to the more realistic noncooperative-game theory and think about the process of individual strategizing that may produce an equilibrium in a bargaining game. Our standard picture of this process is one of **alternating offers.** One player—say, A—makes an offer. The other—say, B—either accepts it or makes a counteroffer. If he does the latter, then A can either accept it or come back with another offer of his own. And so on. Thus we have a sequential-move game and look for its rollback equilibrium.

To find a rollback equilibrium, we must start at the end and work backward. But where is the end point? Why should the process of offers and counteroffers ever terminate? Perhaps more drastic, why would it ever start? Why would

[6]See Larry DeBrock and Alvin Roth, "Strike Two: Labor-Management Negotiations in Major League Baseball," *Bell Journal of Economics*, vol. 12, no. 2 (Autumn 1981), pp. 413–425.

the two bargainers not stick to their original positions and refuse to budge? It is costly to both if they fail to agree at all, but the benefit of an agreement is likely to be smaller to the one who makes the first or the larger concession. The reason that anyone concedes must be that continuing to stand firm would cause an even greater loss of benefit. This loss takes one of two broad forms. The available pie, or surplus, may **decay** (shrink) with each offer, a possibility that we consider in this section. The alternative possibility is that time has value and **impatience** is important, and so a delayed agreement is worth less; we examine this possibility in Section 5.

Consider the following story of bargaining over a shrinking pie. A fan arrives at a professional football (or basketball) game without a ticket. He is willing to pay as much as $25 to watch each quarter of the game. He finds a scalper who states a price. If the fan is not willing to pay this price, he goes to a nearby bar to watch the first quarter on the big-screen TV. At the end of the quarter, he comes out, finds the scalper still there, and makes a counteroffer for the ticket. If the scalper does not agree, the fan goes back to the bar. He comes out again at the end of the second quarter, when the scalper makes him yet another offer. If that offer is not acceptable to the fan, he goes back into the bar, emerging at the end of the third quarter to make yet another counteroffer. The value of watching the rest of the game is declining as the quarters go by.[7]

Rollback analysis enables us to predict the outcome of this alternating-offers bargaining process. At the end of the third quarter, the fan knows that, if he does not buy the ticket then, the scalper will be left with a small piece of paper of no value. So the fan will be able to make a very small offer that, for the scalper, will still be better than nothing. Thus, on his last offer, the fan can get the ticket almost for free. Backing up one period, we see that, at the end of the second quarter, the scalper has the initiative in making the offer. But he must look ahead and recognize that he cannot hope to extract the whole of the remaining two quarters' value from the fan. If the scalper asks for more than $25—the value of the *third* quarter to the fan—the fan will turn down the offer because he knows that he can get the fourth quarter later for almost nothing, so the scalper can ask for $25 at most. Now consider the situation at the end of the first quarter. The fan knows that if he does not buy the ticket now, the scalper can expect to get only $25 later, and so $25 is all that the fan needs to offer now to secure the ticket. Finally, before the game even begins, the scalper can look ahead and ask for $50; this $50 includes the $25 value of the *first* quarter to the fan plus the $25 for which the fan can get the remaining three quarters' worth. Thus the two will strike an immediate agreement, and the ticket will

[7]Just to keep the argument simple, we imagine this process as one-on-one bargaining. Actually, there may be several fans and several scalpers, turning the situation into a *market*. We analyze interactions in markets in detail in Chapter 19.

change hands for $50, but the price is determined by the full forward-looking rollback reasoning.[8]

This story can be easily turned into a more general argument for two bargainers, A and B. Suppose A makes the first offer to split the total surplus, which we call v (in some currency—say, dollars). If B refuses the offer, the total available drops by x_1 to $(v - x_1)$; B offers a split of this amount. If A refuses B's offer, the total drops by a further amount x_2 to $(v - x_1 - x_2)$; A offers a split of this amount. This offer and counteroffer process continues until finally—say, after 10 rounds—$v - x_1 - x_2 - \cdots - x_{10} = 0$, so the game ends. As usual with sequential-play games, we begin our analysis at the end.

If the game has gone to the point where only x_{10} is left, B can make a final offer whereby he gets to keep "almost all" of the surplus, leaving a measly cent or so to A. Left with the choice of that or absolutely nothing, A should accept the offer. To avoid the finicky complexity of keeping track of tiny cents, let us call this outcome "x_{10} to B, 0 to A." We will do the same in the other (earlier) rounds.

Knowing what is going to happen in round 10, we turn to round 9. Here A is to make the offer, and $(x_9 + x_{10})$ is left. A knows that he must offer at least x_{10} to B or else B will refuse the offer and take the game to round 10, where he can get that much. Bargainer A does not want to offer any more to B. So, on round 9, A will offer a split where he keeps x_9 and leaves x_{10} to B.

Then on the round before, when $x_8 + x_9 + x_{10}$ is left, B will offer a split where he gives x_9 to A and keeps $(x_8 + x_{10})$. Working backward, on the very first round, A will offer a split where he keeps $(x_1 + x_3 + x_5 + x_7 + x_9)$ and gives $(x_2 + x_4 + x_6 + x_8 + x_{10})$ to B. This offer will be accepted.

You can remember these formulas by means of a simple trick. *Hypothesize* a sequence in which all offers are refused. (This sequence is *not* what actually happens.) Then add up the amounts that would be destroyed by the refusals of one player. This total is what the other player gets in the actual equilibrium. For example, when B refused A's first offer, the total available surplus dropped by x_1, and x_1 became part of what went to A in the equilibrium of the game.

If each player has a positive BATNA, the analysis must be modified somewhat to take them into account. At the last round, B must offer A at least the BATNA a. If x_{10} is greater than a, B is left with $(x_{10} - a)$; if not, the game must terminate before this round is reached. Now at round 9, A must offer B the larger of the two amounts—the $(x_{10} - a)$ that B can get in round 10 or the BATNA b that B can get outside this agreement. The analysis can proceed all the way back to round 1 in this way; we leave it to you to complete the rollback reasoning for this case.

[8]To keep the analysis simple, we omitted the possibility that the game might get exciting, and so the value of the ticket might actually increase as the quarters go by. The uncertainty makes the problem much more complex but also more interesting. The ability to deal with such possibilities should inspire you to go beyond this book or course to study more advanced game theory.

We have found the rollback equilibrium of the alternating-offers bargaining game, and, in the process of deriving the outcome, we have also described the full strategies—complete contingent plans of action—behind the equilibrium—namely, what each player *would* do if the game reached some later stage. In fact, actual agreement is immediate on the very first offer. The later stages are not reached; they are off-equilibrium nodes and paths. But as usual with rollback reasoning, the foresight about what rational players would do at those nodes if they were reached is what informs the initial action.

The other important point to note is that *gradual decay* (several potential rounds of offers) leads to a more even or fairer split of the total than does *sudden decay* (only one round of bargaining permitted). In the latter, no agreement would result if B turned down A's very first offer; then, in a rollback equilibrium, A would get to keep (almost) the whole surplus, giving B an "ultimatum" to accept a measly cent or else get nothing at all. The subsequent rounds give B the credible ability to refuse a very uneven first offer.

4 EXPERIMENTAL EVIDENCE

The theory of this particular type of bargaining process is fairly simple, and many people have staged laboratory or classroom experiments that create such conditions of decaying totals, to observe what the experimental subjects actually do. We mentioned some of them briefly in Chapter 3 when considering the validity of rollback reasoning; now we examine them in more detail in the context of bargaining.[9]

The simplest bargaining experiment is the **ultimatum game,** in which there is only one round: player A makes an offer and, if B does not accept it, the bargaining ends and both get nothing. The general structure of these games is as follows. A pool of players is brought together, either in the same room or at computer terminals in a network. They are paired; one person in the pair is designated to be the *proposer* (the one who makes the offer or is the seller who posts a price), and the other to be the *chooser* (the one who accepts or refuses the offer or is the customer who decides whether to buy at that price). The pair is given a fixed surplus, usually $1 or some other sum of money, to split.

Rollback reasoning suggests that A should offer B the minimal unit—say, 1 cent out of a dollar—and that B should accept such an offer. Actual results are dramatically different. In the case in which the subjects are together in a

[9]For more details, see Douglas D. Davis and Charles A. Holt, *Experimental Economics* (Princeton, NJ: Princeton University Press, 1993), pp. 263–269, and *The Handbook of Experimental Economics*, ed. John H. Kagel and Alvin E. Roth (Princeton, NJ: Princeton University Press, 1995), pp. 255–274.

room and the assignment of the role of proposer is made randomly, the most common offer is a 50:50 split. Very few offers worse than 75:25 are made (with the proposer to keep 75% and the chooser to get 25%), and if made, they are often rejected.

This finding can be interpreted in one of two ways. Either the players cannot or do not perform the calculation required for rollback or the payoffs of the players include something other than what they get out of this round of bargaining. Surely the calculation in the ultimatum game is simple enough that anyone should be able to do it, and the subjects in most of these experiments are college students. A more likely explanation is the one that we put forth in Chapters 3 (Section 6) and 5 (Section 3)—that the theory, which assumed payoffs to consist only of the sum earned in this one round of bargaining, is too simplistic.

Participants can have payoffs that include other things. They may have self-esteem or pride that prevents them from accepting a very unequal split. Even if the proposer A does not include this consideration in his own payoff, if he thinks that B might, then it is a good strategy for A to offer enough to make it likely that B will accept. Proposer A balances his higher payoff with a smaller offer to B against the risk of getting nothing if B rejects an offer deemed too unequal.

A second possibility is that, when the participants in the experiment are gathered in a room, the anonymity of pairing cannot be guaranteed. If the participants come from a group such as college students or villagers who have ongoing relationships outside this game, they may value those relationships. Then the proposers fear that, if they offer too unequal a split in this game, those relationships may suffer. Therefore they would be more generous in their offers than the simplistic theory suggests. If this is the explanation, then ensuring greater anonymity should enable the proposers to make more unequal offers, and experiments do find this to be the case.

Finally, people may have a sense of fairness drilled into them during their nurture and education. This sense of fairness may have evolutionary value for society as a whole and may therefore have become a social norm. Whatever its origin, it may lead the proposers to be relatively generous in their offers, quite irrespective of the fear of rejection. One of us (Skeath) has conducted classroom experiments of several different ultimatum games. Students who had partners previously known to them with whom to bargain were noticeably "fairer" in their split of the pie. In addition, several students cited specific cultural backgrounds as explanations for behavior that was inconsistent with theoretical predictions.

Experimenters have tried variants of the basic game to differentiate between these explanations. The point about ongoing relationships can be handled by stricter procedures that visibly guarantee anonymity. Doing so by itself has some effect on the outcomes but still does not produce offers as extreme as those predicted by the purely selfish rollback argument of the theory. The remaining explanations—namely, "fear of rejection" and the "ingrained sense of fairness"—remain to be sorted out.

The fear of rejection can be removed by considering a variant called the *dictator game.* Again, the participants are matched in pairs. One person—say, A—is designated to determine the split, and the other—say, B—is simply a passive recipient of what A decides. Now the split becomes decidedly more uneven, but even here a majority of As choose to keep no more than 70%. This result suggests a role for an ingrained sense of fairness.

But such a sense has its limits. In some experiments, a sense of fairness was created simply when the experimenter randomly assigned roles of proposer and chooser. In one variant, the participants were given a simple quiz, and those who performed best were made proposers. This created a sense that the role of proposer had been earned, and the outcomes did show more unequal splits. When the dictator game was played with earned rights and with stricter anonymity conditions, most As kept everything, but some (about 5%) still offered a 50:50 split.

One of us (Dixit) carried out a classroom experiment in which students in groups of 20 were gathered together in a computer cluster. They were matched randomly and anonymously in pairs, and each pair tried to agree on how to split 100 points. Roles of proposer and chooser were not assigned; either could make an offer or accept the other's offer. Offers could be made and changed at any time. The pairs could exchange messages instantly with their matched opponent on their computer screens. The bargaining round ended at a random time between 3 and 5 minutes; if agreement was not reached in time by a pair, both got zero. There were 10 such rounds with different random opponents each time. Thus the game itself offered no scope for cooperation through repetition. In a classroom context, the students had ongoing relationships outside the game, but they did not generally know or guess with whom they were playing in any round, even though no great attempt was made to enforce anonymity. Each student's score for the whole game was the sum of his point score for the 10 rounds. The stakes were quite high, because the score accounted for 5% of the course grade!

The highest total of points achieved was 515. Those who quickly agreed on 50:50 splits did the best, and those who tried to hold out for very uneven scores or who refused to split a difference of 10 points or so between the offers and ran the risk of time running out on them did poorly.[10] It seems that moderation and fairness do get rewarded, even as measured in terms of one's own payoff.

[10]Those who were best at the mathematical aspects of game theory, such as problem sets, did a little worse than the average, probably because they tried too hard to eke out an extra advantage and met resistance. And women did slightly better than men.

5 ALTERNATING-OFFERS MODEL II: IMPATIENCE

Now we consider a different kind of cost of delay in reaching an agreement. Suppose the actual monetary value of the total available for splitting does not decay, but players have a "time value of money" and therefore prefer early agreement to later agreement. They make offers alternately as described in Section 3, but their time preferences are such that money now is better than money later. For concreteness, we will say that both bargainers believe that having only 95 cents right now is as good as having $1 one round later.

A player who prefers having something right away to having the same thing later is impatient; he attaches less importance to the future relative to the present. We came across this idea in Chapter 11 (Section 2) and saw two reasons for it. First, player A may be able to invest his money—say, $1—now and get his principal back along with interest and capital gains at a rate of return r, for a total of $(1 + r)$ in the next period (tomorrow, next week, next year, or whatever is the length of the period). Second, there may be some risk that the game will end between now and the next offer (as in the sudden end at a time between 3 and 5 minutes in the classroom game described earlier). If p is the probability that the game continues, then the chance of getting a dollar next period has an expected value of only p now.

Suppose we consider a bargaining process between two players with zero BATNAs. Let us start the process with one of the two bargainers—say, A—making an offer to split $1. If the other player, B, rejects A's offer, then B will have an opportunity to make his own offer one round later. The two bargainers are in identical situations when each makes his offer, because the amount to be split is always $1. Thus in equilibrium the amount that goes to the person currently in charge of making the offer (call it x) is the same, regardless of whether that person is A or B. We can use rollback reasoning to find an equation that can be solved for x.

Suppose A starts the alternating offer process. He knows that B can get x in the next round when it is B's turn to make the offer. Therefore, A must give B at least an amount that is equivalent, in B's eyes, to getting x in the next round; A must give B at least $0.95x$ now. (Remember that, for B, 95 cents received now is equivalent to $1 received in the next round; so $0.95x$ now is as good as x in the next round.) Player A will not give B any more than is required to induce B's acceptance. Thus A offers B exactly $0.95x$ and is left with $(1 - 0.95x)$. But the amount that A gets when making the offer is just what we called x. Therefore $x = 1 - 0.95x$, or $(1 + 0.95)x = 1$, or $x = 1/1.95 = 0.512$.

Two things about this calculation should be noted. First, even though the process allows for an unlimited sequence of alternating offers and counteroffers, in the equilibrium the very first offer A makes gets accepted and the bargaining ends. Because time has value, this outcome is efficient. The cost of delay governs how much A must offer B to induce acceptance; it thus affects A's rollback reasoning. Second, the player who makes the first offer gets more than half of the pie—namely, 0.512 rather than 0.488. Thus each player gets more when he makes the first offer than when the other player makes the first offer. But this advantage is far smaller than that in an ultimatum game with no future rounds of counteroffers.

Now suppose the two players are not equally patient (or impatient, as the case may be). Player B still regards $1 in the next round as being equivalent to 95 cents now, but A regards it as being equivalent to only 90 cents now. Thus A is willing to accept a smaller amount to get it sooner; in other words, A is more impatient. This inequality in rates of impatience can translate into unequal equilibrium payoffs from the bargaining process. To find the equilibrium for this example, we write x for the amount that A gets when he starts the process and y for what B gets when he starts the process.

Player A knows that he must give B at least $0.95y$; otherwise B will reject the offer in favor of the y that he knows he can get when it becomes his turn to make the offer. Thus the amount that A gets, x, must be $1 - 0.95y$; $x = 1 - 0.95y$. Similarly, when B starts the process, he knows that he must offer A at least $0.90x$, and then $y = 1 - 0.90x$. These two equations can be solved for x and y:

$$\begin{aligned} x &= 1 - 0.95(1 - 0.9x) \\ [1 - 0.95(0.9)]x &= 1 - 0.95 \\ 0.145x &= 0.05 \\ x &= 0.345 \end{aligned} \qquad \text{and} \qquad \begin{aligned} y &= 1 - 0.9(1 - 0.95y) \\ [1 - 0.9(0.95)]y &= 1 - 0.9 \\ 0.145y &= 0.10 \\ y &= 0.690 \end{aligned}$$

Note that x and y do not add up to 1, because each of these amounts is the payoff to a given player when he makes the first offer. Thus, when A makes the first offer, A gets 0.345 and B gets 0.655; when B makes the first offer, B gets 0.69 and A gets 0.31. Once again, each player does better when he makes the first offer than when the other player makes the first offer, and once again the difference is small.

The outcome of this case with unequal rates of impatience differs from that of the preceding case with equal rates of impatience in a major way. With unequal rates of impatience, the more impatient player, A, gets a lot less than B even when he is able to make the first offer. We expect that the person who is willing to accept less to get it sooner ends up getting less, but the difference is very dramatic. The proportion of A's shares and B's shares is almost 1:2.

As usual, we can now build on these examples to develop the more general algebra. Suppose A regards $1 immediately as being equivalent to $$(1 + r)$ one offer later or, equivalently, A regards $$1/(1 + r)$ immediately as being equivalent to $1 one offer later. For brevity, we substitute a for $1/(1 + r)$ in the calculations that follow. Likewise, suppose player B regards \$1 today as being equivalent to \$$(1 + s)$ one offer later; we use b for $1/(1 + s)$. If r is high (or equivalently, if a is low), then player A is very impatient. Similarly, B is impatient if s is high (or if b is low).

Here we look at bargaining that takes place in alternating rounds, with a total of \$1 to be divided between two players, both of whom have zero BATNAs. (You can do the even more general case easily once you understand this one.) What is the rollback equilibrium?

We can find the payoffs in such an equilibrium by extending the simple argument used earlier. Suppose A's payoff in the rollback equilibrium is x when he makes the first offer; B's payoff in the rollback equilibrium is y when he makes the first offer. We look for a pair of equations linking the values x and y and then solve these equations to determine the equilibrium payoffs.[11]

When A is making the offer, he knows that he must give B an amount that B regards as being equivalent to y one period later. This amount is $by = y/(1 + s)$. Then, after making the offer to B, A can keep only what is left: $x = 1 - by$.

Similarly, when B is making the offer, he must give A the equivalent of x one period later—namely, ax. Therefore $y = 1 - ax$. Solving these two equations is now a simple matter. We have $x = 1 - b(1 - ax)$, or $(1 - ab)x = 1 - b$. Expressed in terms of r and s, this equation becomes

$$x = \frac{1-b}{1-ab} = \frac{s+rs}{r+s+rs}.$$

Similarly, $y = 1 - a(1 - by)$, or $(1 - ab)y = 1 - a$. This equation becomes

$$y = \frac{1-a}{1-ab} = \frac{r+rs}{r+s+rs}.$$

Although this quick solution might seem a sleight of hand, it follows the same steps used earlier, and we soon give a different reasoning yielding exactly the same answer. First, let us examine some features of the answer.

First note that, as in our simple unequal-impatience example, the two magnitudes x and y add up to more than 1:

$$x + y = \frac{r+s+2rs}{r+s+rs} > 1$$

Remember that x is what A gets when he has the right to make the first offer, and y is what B gets when he has the right to make the first offer. When A makes the

[11] We are taking a shortcut; we have simply assumed that such an equilibrium exists and that the payoffs are uniquely determined. More rigorous theory proves these conditions. For a step in this direction, see John Sutton, "Non-Cooperative Bargaining: An Introduction," *Review of Economic Studies*, vol. 53, no. 5 (October 1986), pp. 709–724. The fully rigorous (and quite difficult) theory is given in Ariel Rubinstein, "Perfect Equilibrium in a Bargaining Model," *Econometrica*, vol. 50, no. 1 (January 1982), pp. 97–109.

first offer, B gets $(1 - x)$, which is less than y; this just shows A's advantage from being the first proposer. Similarly, when B makes the first offer, B gets y and A gets $(1 - y)$, which is less than x.

However, usually r and s are small numbers. When offers can be made at short intervals such as a week or a day or an hour, the interest that your money can earn from one offer to the next or the probability that the game ends precisely within the next interval is quite small. For example, if r is 1% (0.01) and s is 2% (0.02), then the formulas yield $x = 0.668$ and $y = 0.337$; so the advantage of making the first offer is only 0.005. (A gets 0.668 when making the first offer, but $1 - 0.337 = 0.663$ when B makes the first offer; the difference is only 0.005.) More formally, when r and s are each small compared with 1, then their product rs is very small indeed; thus we can ignore rs to write an approximate solution for the split that does not depend on which player makes the first offer:

$$x = \frac{s}{r+s} \quad \text{and} \quad y = \frac{r}{r+s}.$$

Now $x + y$ is approximately equal to 1.

Most important, x and y in the approximate solution are the shares of the surplus that go to the two players, and $y/x = r/s$; that is, the shares of the players are inversely proportional to their rates of impatience as measured by r and s. If B is twice as impatient as A, then A gets twice as much as B; so the shares are 1/3 and 2/3, or 0.333 and 0.667, respectively. Thus we see that patience is an important advantage in bargaining. Our formal analysis supports the intuition that, if you are very impatient, the other player can offer you a quick but poor deal, knowing that you will accept it.

This effect of impatience hurts the United States in numerous negotiations that our government agencies and diplomats conduct with other countries. The American political process puts a great premium on speed. The media, interest groups, and rival politicians all demand results and are quick to criticize the administration or the diplomats for any delay. Under this pressure to deliver, the negotiators are always tempted to come back with results of any kind. Such results are often poor from the long-run U.S. perspective; the other countries' concessions often have loopholes, and their promises are less than credible. The U.S. administration hails the deals as great victories, but they usually unravel after a few years. The financial crisis of 2008 offers another and dramatic example. When the housing boom collapsed, some major financial institutions that held mortgage-backed assets faced bankruptcy. That led them to curtail credit, which in turn threatened to throw the U.S. economy into a severe recession. The crisis exploded in September, in the midst of a presidential election campaign. The Treasury, the Federal Reserve, and political leaders in Congress all wanted to act fast. This impatience led them to offer far more generous terms of rescue to many financial insitutions, when a slower process would have yielded an outcome that cost the taxpayers much less and offered them much better prospects of sharing in future gains on the assets being rescued.

Individuals who suffer losses are in a much weaker position when they negotiate with insurance companies on coverage. The companies often make lowball offers of settlement to people who have suffered a major loss; knowing that their clients urgently want to make a fresh start and are therefore very impatient.

As a conceptual matter, the formula $y/x = r/s$ ties our noncooperative game approach to bargaining to the cooperative approach of the Nash solution discussed in Section 1. The formula for shares of the available surplus that we derived there becomes, with zero BATNAs, $y/x = k/h$. In the cooperative approach, the shares of the two players stood in the same proportions as their bargaining strengths, but these strengths were assumed to be given somehow from the outside. Now we have an explanation for the bargaining strengths in terms of some more basic characteristics for the players—h and k are inversely proportional to the players' rates of impatience r and s. In other words, Nash's cooperative solution can also be given an alternative and perhaps more satisfactory interpretation as the rollback equilibrium of a noncooperative game of offers and counteroffers, if we interpret the abstract bargaining-strength parameters in the cooperative solution correctly in terms of the players' characteristics, such as impatience.

Finally, note that agreement is once again immediate—the very first offer is accepted. As usual, the whole rollback analysis disciplines by making the first proposer recognize that the other would credibly reject a less adequate offer.

To conclude this section, we offer an alternative derivation of the same (precise) formula for the equilibrium offers that we derived earlier. Suppose this time that there are 100 rounds; A is the first proposer and B the last. Start the backward induction in the 100th round; B will keep the whole dollar. Therefore in the 99th round, A will have to offer B the equivalent of $1 in the 100th round—namely, b, and A will keep $(1 - b)$. Then proceed backward:

In round 98, B offers $a(1 - b)$ to A and keeps $1 - a(1 - b) = 1 - a + ab$.
In round 97, A offers $b(1 - a + ab)$ to B and keeps $1 - b(1 - a + ab) = 1 - b + ab - ab^2$.
In round 96, B offers $a(1 - b + ab - ab^2)$ to A and keeps $1 - a + ab - a^2b + a^2b^2$.
In round 95, A offers $b(1 - a + ab - a^2b + a^2b^2)$ to B and keeps $1 - b + ab - ab^2 + a^2b^2 - a^2b^3$.

Proceeding in this way and following the established pattern, we see that, in round 1, A gets to keep

$$1 - b + ab - ab^2 + a^2b^2 - a^2b^3 + \cdots + a^{49}b^{49} - a^{49}b^{50} = (1 - b)[1 + ab + (ab)^2 + \cdots + (ab)^{49}]$$

The consequence of allowing more and more rounds is now clear. We just get more and more of these terms, growing geometrically by the factor *ab* for every two offers. To find A's payoff when he is the first proposer in an infinitely long sequence of offers and counteroffers, we have to find the limit of the infinite geometric sum. In the Appendix to Chapter 11 we saw how to sum such series. Using the formula obtained there, we get the answer

$$(1 - b)[1 + ab + (ab) + (ab)^2 + \cdots + (ab)^{49} + \cdots] = \frac{1 - b}{1 - ab}$$

This is exactly the solution for x that we obtained before. By a similar argument, you can find B's payoff when he is the proposer and, in doing so, improve your understanding and technical skills at the same time.

6 MANIPULATING INFORMATION IN BARGAINING

We have seen that the outcomes of a bargain depend crucially on various characteristics of the parties to the bargain, most important their BATNAs and their impatience. We have proceeded thus far by assuming that the players knew each other's characteristics as well as their own. In fact, we have assumed that each player knew that the other knew, and so on; that is, the characteristics were common knowledge. In reality, we often engage in bargaining without knowing the other side's BATNA or degree of impatience; sometimes we do not even know our own BATNA very precisely.

As we saw in Chapter 9, a game with such uncertainty or informational asymmetry has associated with it an important game of signaling and screening of strategies for manipulating information. Bargaining is replete with such strategies. A player with a good BATNA or a high degree of patience wants to signal this fact to the other. However, because someone without these good attributes will want to imitate them, the other party will be skeptical and will examine the signals critically for their credibility. And each side will also try screening, by using strategies that induce the other to take actions that will reveal its characteristics truthfully.

In this section, we look at some such strategies used by buyers and sellers in the housing market. Most Americans are active in the housing market several times in their lives, and many people are professional estate agents or brokers who have even more extensive experience in the matter. Moreover, housing is one of the few markets in the United States where haggling or bargaining over price is accepted and even expected. Therefore considerable experience of strategies is available. We draw on this experience for many of our examples and interpret it in the light of our game-theoretic ideas and insights.[12]

When you contemplate buying a house in a new neighborhood, you are unlikely to know the general range of prices for the particular type of house in which you are interested. Your first step should be to find this range so that you can then determine your BATNA. And that does not mean looking at newspaper ads or realtors' listings, which indicate only asking prices. Local newspapers and some Internet sites list recent actual transactions and the actual prices; you should check them against the asking prices of the same houses to get an idea of the state of the market and the range of bargaining that might be possible.

Next comes finding out (screening) the other side's BATNA and level of impatience. If you are a buyer, you can find out why the house is being sold and how long it has been on the market. If it is empty, why? And how long has it been that way? If the owners are getting divorced or have moved elsewhere and are financing another house on an expensive bridge loan, it is likely that they have a low BATNA or are rather impatient.

You should also find out other relevant things about the other side's preferences, even though these preferences may seem irrational to you. For example, some people consider an offer too far below the asking price an insult and will not sell at any price to someone who makes such an offer. Norms of this kind vary across regions and times. It pays to find out what the common practices are.

Most important, the *acceptance* of an offer more accurately reveals a player's true willingness to pay than anything else and therefore is open to exploitation by the other player. A brilliant game-theorist friend of ours tried just such a ploy. He was bargaining for a floor lamp. Starting with the seller's asking price of $100, the negotiation proceeded to a point where our friend made an offer to buy the lamp for $60. The seller said yes, at which point our friend thought: "This guy is willing to sell it for $60, so his true rock-bottom price must be even lower. Let me try to find out whether it is." So our friend said, "How about $55?" The seller got very upset, refused to sell for any price, and asked our friend to leave the store and never come back.

The seller's behavior conformed to the norm that it is utmost bad faith in bargaining to renege on an offer once it is accepted. It makes good sense as a norm in the whole context of all bargaining games that take place in society. If an offer on the table cannot be accepted in good faith by the other player without fear of the kind of exploitation attempted by our friend, then each bargainer will wait to get the other to accept an offer, thereby revealing the limit of his true rock-bottom acceptance level, and the whole process of bargains will grind to

[12]We have taken the insights of practitioners from Andrée Brooks, "Honing Haggling Skills," *New York Times*, December 5, 1993.

a halt. Therefore such behavior has to be disallowed. Making it a social norm to which people adhere instinctively, as the seller in the example did, is a good way for society to achieve this aim.

The offer may explicitly say that it is open only for a specified and limited time; this stipulation can be part of the offer itself. Job offers usually specify a deadline for acceptance; stores have sales for limited periods. But in that case the offer is truly a *package* of price and time, and reneging on either dimension provokes a similar instinctive anger. For example, customers get quite angry if they arrive at a store in the sale period and find an advertised item unavailable. The store must offer a rain check, which allows the customer to buy the item at its sale price when next available at the regular price; even this offer causes some inconvenience to the customer and risks some loss of goodwill. The store can specify "limited quantities, no rain checks" very clearly in its advertising of the sale; even then, many customers get upset if they find that the store has run out of the item.

Next on our list of strategies to use in one-on-one bargaining, as in the housing market, comes signaling your own high BATNA or patience. The best way to signal patience is to *be* patient. Do not come back with counteroffers too quickly. "Let the sellers think they've lost you." This signal is credible because someone not genuinely patient would find it too costly to mimic the leisurely approach. Similarly, you can signal a high BATNA by starting to walk away, a tactic that is common in negotiations at bazaars in other countries and some flea markets and tag sales in the United States.

Even if your BATNA is low, you may commit yourself to not accepting an offer below a certain level. This constraint acts just like a high BATNA, because the other side cannot hope to get you to accept anything less. In the housing context, you can claim your inability to concede any further by inventing (or creating) a tightwad parent who is providing the down payment or a spouse who does not really like the house and will not let you offer any more. Sellers can try similar tactics. A parallel in wage negotiations is the *mandate*. A meeting is convened of all the workers who pass a resolution—the mandate—authorizing the union leaders to represent them at the negotiation but with the constraint that the negotiators must not accept an offer below a certain level specified in the resolution. Then, at the meeting with the management, the union leaders can say that their hands are tied; there is no time to go back to the membership to get their approval for any lower offer.

Most of these strategies entail some risk. While you are signaling patience by waiting, the seller of the house may find another willing buyer. As employer and union wait for one another to concede, tensions may mount so high that a strike that is costly to both sides nevertheless cannot be prevented. In other words, many strategies of information manipulation are instances of brinkmanship. We saw in Chapter 14 how such games can have an outcome that is bad for both parties. The same is true in bargaining. *Threats* of breakdown of negotiations or of strikes are strategic moves intended to achieve quicker agreement or a better deal for the player making the move; however, an *actual* breakdown or strike is an instance of the threat "gone wrong." The player making the threat—initiating the brinkmanship—must assess the risk and the potential rewards when deciding whether and how far to proceed down this path.

7 BARGAINING WITH MANY PARTIES AND ISSUES

Our discussion thus far has been confined to the classic situation where two parties are bargaining about the split of a given total surplus. But many real-life negotiations include several parties or several issues simultaneously. Although the games get more complicated, often the enlargement of the group or the set of issues actually makes it easier to arrive at a mutually satisfactory agreement. In this section, we take a brief look at such matters.[13]

A. Multi-Issue Bargaining

In a sense, we have already considered multi-issue bargaining. The negotiation over price between a seller and a buyer always comprises *two* things: (1) the object offered for sale or considered for purchase and (2) money. The potential for mutual benefit arises when the buyer values the object more than the seller does—that is, when the buyer is willing to give up more money in return for getting the object than the seller is willing to accept in return for giving up the object. Both players can be better off as a result of their bargaining agreement.

The same principle applies more generally. International trade is the classic example. Consider two hypothetical countries, Freedonia and Ilyria. If Freedonia can convert 1 loaf of bread into 2 bottles of wine (by using less of its resources such as labor and land in the production of bread and using them to produce more wine instead) and Ilyria can convert 1 bottle of wine into 1 loaf of bread (by switching its resources in the opposite direction), then between them they can create more goods "out of nothing." For example, suppose that Freedonia can produce 200 more bottles of wine if it produces 100 fewer loaves of bread and that Ilyria can produce 150 more loaves of bread if it produces 150 fewer bottles of wine. These switches in resource utilization create an extra 50 loaves of bread and 50 bottles of wine relative to what the two countries produced originally. This extra bread and wine is the "surplus" that they can create if they can agree

[13]For a more thorough treatment, see Howard Raiffa, *The Art and Science of Negotiation* (Cambridge, MA: Harvard University Press, 1982), parts III and IV.

on how to divide it between them. For example, suppose Freedonia gives 175 bottles of wine to Ilyria and gets 125 loaves of bread. Then each country will have 25 more loaves of bread and 25 more bottles of wine than it did before. But there is a whole range of possible exchanges corresponding to different divisions of the gain. At one extreme, Freedonia may give up all the 200 extra bottles of wine that it has produced in exchange for 101 loaves of bread from Ilyria, in which case Ilyria gets almost all the gain from trade. At the other extreme, Freedonia may give up only 151 bottles of wine in exchange for 150 loaves of bread from Ilyria, and so Freedonia gets almost all the gain from trade.[14] Between these limits lies the frontier where the two can bargain over the division of the gains from trade.

The general principle should now be clear. When two or more issues are on the bargaining table at the same time and the two parties are willing to trade more of one against less of the other at different rates, then a mutually beneficial deal exists. The mutual benefit can be realized by trading at a rate somewhere between the two parties' different rates of willingness to trade. The division of gains depends on the choice of the rate of trade. The closer it is to one side's willingness ratio, the less that side gains from the deal.

Now you can also see how the possibilities for mutually beneficial deals can be expanded by bringing more issues to the table at the same time. With more issues, you are more likely to find divergences in the ratios of valuation between the two parties and are thereby more likely to locate possibilities for mutual gain. In regard to a house, for example, many of the fittings or furnishings may be of little use to the seller in the new house to which he is moving, but they may be of sufficiently good fit and taste that the buyer values having them. Then if the seller cannot be induced to lower the price, he may be amenable to including these items in the original price to close the deal.

However, the expansion of issues is not an unmixed blessing. If you value something greatly, you may fear putting it on the bargaining table; you may worry that the other side will extract big concessions from you, knowing that you want to protect that one item of great value. At the worst, a new issue on the table may make it possible for one side to deploy threats that lower the other side's BATNA. For example, a country engaged in diplomatic negotiations may be vulnerable to an economic embargo; then it would much prefer to keep the political and economic issues distinct.

[14]Economics uses the concept *ratio of exchange,* or price, which here is expressed as the number of bottles of wine that trade for each loaf of bread. The crucial point is that the possibility of gain for both countries exists with any ratio that lies between the 2:1 at which Freedonia can just convert bread into wine and the 1:1 at which Ilyria can do so. At a ratio close to 2:1, Freedonia gives up almost all of its 200 extra bottles of wine and gets little more than the 100 loaves of bread that it sacrificed to produce the extra wine; thus Ilyria has almost all of the gain. Conversely, at a ratio close to 1:1, Freedonia has almost all of the gain. The issue in the bargaining is the division of gain and therefore the ratio or the price at which the two should trade.

B. Multiparty Bargaining

Having many parties simultaneously engaged in bargaining also may facilitate agreement, because instead of having to look for pairwise deals, the parties can seek a circle of concessions. International trade is again the prime example. Suppose the United States can produce wheat very efficiently but is less productive in cars, Japan is very good at producing cars but has no oil, and Saudi Arabia has a lot of oil but cannot grow wheat. In pairs, they can achieve little, but the three together have the potential for a mutually beneficial deal.

As with multiple issues, expanding the bargaining to multiple parties is not simple. In our example, the deal would be that the United States would send an agreed amount of wheat to Saudi Arabia, which would send its agreed amount of oil to Japan, which would in turn ship its agreed number of cars to the United States. But suppose that Japan reneges on its part of the deal. Saudi Arabia cannot retaliate against the United States, because, in this scenario, it is not offering anything to the United States that it can potentially withhold. Saudi Arabia can only break its deal to send oil to Japan, an important party. Thus enforcement of multilateral agreements may be problematic. The General Agreement on Tariffs and Trade (GATT) between 1946 and 1994, as well as the World Trade Organization (WTO) since then, have indeed found it difficult to enforce their agreements and to levy punishments on countries that violate the rules.

SUMMARY

Bargaining negotiations attempt to divide the *surplus* (excess value) that is available to the parties if an agreement can be reached. Bargaining can be analyzed as a *cooperative* game in which parties find and implement a solution jointly or as a (structured) *noncooperative* game in which parties choose strategies separately and attempt to reach an equilibrium.

Nash's cooperative solution is based on three principles of the outcomes' invariance to linear changes in the payoff scale, *efficiency,* and invariance to removal of irrelevant outcomes. The solution is a rule that states the proportions of division of surplus, beyond the backstop payoff levels (also called *BATNAs* or *best alternatives to a negotiated agreement*) available to each party, based on relative bargaining strengths. Strategic manipulation of the backstops can be used to increase a party's payoff.

In a noncooperative setting of *alternating offer and counteroffer,* rollback reasoning is used to find an equilibrium; this reasoning generally includes a first-round offer that is immediately accepted. If the surplus value *decays* with refusals, the sum of the (hypothetical) amounts destroyed owing to the refusals of a single player is the payoff to the other player in equilibrium. If delay in agreement is costly owing to *impatience,* the equilibrium offer shares the surplus roughly in inverse proportion to the parties' rates of *impatience.* Experimental evidence indicates that players often offer more than is necessary to reach an agreement in such games; this behavior is thought to be related to player anonymity as well as beliefs about fairness.

The presence of information asymmetries in bargaining games makes signaling and screening important. Some parties will wish to signal their high BATNA levels or extreme patience; others will want to screen to obtain truthful revelation of such characteristics. When more issues are on the table or more parties are participating, agreements may be easier to reach, but bargaining may be riskier or the agreements more difficult to enforce.

KEY TERMS

alternating offers (703)
best alternative to a negotiated agreement (BATNA) (695)
decay (704)
efficient frontier (698)
efficient outcome (698)
impatience (704)
Nash cooperative solution (696)
surplus (695)
ultimatum game (706)
variable-threat bargaining (701)

SOLVED EXERCISES

S1. Consider the bargaining situation between Compaq Computer Corporation and the California businessman who owned the Internet address www.altavista.com.[15] Compaq, which had recently taken over Digital Equipment Corporation, wanted to use this man's Web address for Digital's Internet search engine, which at that time had the address www.altavista.digital.com. Compaq and the businessman apparently negotiated long and hard during the summer of 1998 over a selling price for the latter's address.

Although the businessman was the "smaller" player in this game, the final agreement appeared to entail a \$3.35 *million* price tag for the Web address in question. Compaq confirmed the purchase in August and began using the address in September but refused to divulge any of the financial details of the settlement. Given this information, comment on the likely values of the BATNAs for these two players, their bargaining strengths or levels of impatience, and whether a cooperative outcome appears to have been attained in this game.

[15]Details regarding this bargaining game were reported in "A Web Site by Any Other Name Would Probably Be Cheaper," *Boston Globe,* July 29, 1998, and in Hiawatha Bray's "Compaq Acknowledges Purchase of Web Site," *Boston Globe,* August 12, 1998.

S2. Ali and Baba are bargaining to split a total that starts out at $100. Ali makes the first offer, stating how the $100 will be divided between them. If Baba accepts this offer, the game is over. If Baba rejects it, a dollar is withdrawn from the total, so it is now only $99. Then Baba gets the second turn to make an offer of a division. The turns alternate in this way, a dollar being removed from the total after each rejection. Ali's BATNA is $2.25 and Baba's BATNA is $3.50. What is the rollback-equilibrium outcome of the game?

S3. Two hypothetical countries, Euphoria and Militia, are holding negotiations to settle a dispute. They meet once a month, starting in January, and take turns making offers. Suppose the total at stake is 100 points. The government of Euphoria is facing reelection in November. Unless the government produces an agreement at the October meeting, it will lose the election, which it regards as being just as bad as getting zero points from an agreement. The government of Militia does not really care about reaching an agreement; it is just as happy to prolong the negotiations or even to fight, because it would be settling for anything significantly less than 100.

(a) What will be the outcome of the negotiations? What difference will the identity of the first mover make?

(b) In light of your answer to part (a), discuss why actual negotiations often continue right down to the deadline.

UNSOLVED EXERCISES

U1. Recall the variant of the pizza pricing game in Exercise U2, part (b) in Chapter 11, in which one store (Donna's Deep Dish) was much larger than the other (Pierce's Pizza Pies). The payoff table for that version of the game is:

		PIERCE'S PIZZA PIES	
		High	Low
DONNA'S DEEP DISH	High	156, 60	132, 70
	Low	150, 36	130, 50

The noncooperative dominant-strategy equilibrium is (High, Low), yielding profits of 132 to Donna's and 70 to Pierce's, for a total of 202. If the two could achieve (High, High), their total profit would be 156 + 60 = 216, but Pierce's would not agree to this pricing.

Suppose the two stores can reach an enforceable agreement whereby both charge High and Donna's pays Pierce's a sum of money. The alternative to this agreement is simply the noncooperative dominant-strategy equilibrium.

They bargain over this agreement, and Donna's has 2.5 times as much bargaining power as Pierce's. In the resulting agreement, what sum will Donna's pay to Pierce's?

U2. Consider two players who bargain over a surplus initially equal to a whole-number amount V, using alternating offers. That is, Player 1 makes an offer in round 1; if Player 2 rejects this offer, she makes an offer in round 2; if Player 1 rejects this offer, she makes an offer in round 3; and so on. Suppose that the available surplus decays by a constant value of $c = 1$ each period. For example, if the players reach agreement in round 2, they divide a surplus of $V - 1$; if they reach agreement in round 5, they divide a surplus of $V - 4$. This means that the game will be over after V rounds, because at that point there will be nothing left to bargain over. (For comparison, remember the football-ticket example, in which the value of the ticket to the fan started at \$100 and declined by \$25 per quarter over the four quarters of the game.) In this problem, we will first solve for the rollback equilibrium to this game, and then solve for the equilibrium to a generalized version of this game in which the two players can have BATNAs.

(a) Let's start with a simple version. What is the rollback equilibrium when $V = 4$? In which period will they reach agreement? What payoff x will Player 1 receive, and what payoff y will Player 2 receive?

(b) What is the rollback equilibrium when $V = 5$?

(c) What is the rollback equilibrium when $V = 10$?

(d) What is the rollback equilibrium when $V = 11$?

(e) Now we're ready to generalize. What is the rollback equilibrium for any whole-number value of V? (Hint: You may want to consider even values of V separately from odd values.)

Now consider BATNAs. Suppose that if no agreement is reached by the end of round V, Player A gets a payoff of a and Player B gets a payoff of b. Assume that a and b are whole numbers satisfying the inequality $a + b < V$, so that the players can get higher payoffs from reaching agreement than they can by not reaching agreement.

(f) Suppose that $V = 4$. What is the rollback equilibrium for any possible values of a and b? (Hint: You may need to write down more than one formula, just as you did in part (e). If you get stuck, try assuming specific values for a and b, and then change those values to see what happens. In order to roll back, you'll need to figure out the turn at which the value of V has declined to the point where a negotiated agreement would no longer be profitable for the two bargainers.)

(g) Suppose that $V = 5$. What is the rollback equilibrium for any possible values of a and b?

(h) For any whole-number values of a, b, and V, what is the rollback equilibrium?

(i) Relax the assumption that a, b, and V are whole numbers: let them be any nonnegative numbers such that $a + b < V$. Also relax the assumption that the value of V decays by exactly 1 each period: let the value decay each period by some constant amount $c > 0$. What is the rollback equilibrium to this general problem?

U3. Let x be the amount that player A asks for, and let y be the amount that B asks for, when making the first offer in an alternating-offers bargaining game with impatience. Their rates of impatience are r and s, respectively.

(a) If we use the approximate formulas $x = s/(r + s)$ for x and $y = r/(r + s)$ for y, and if B is twice as impatient as A, then A gets two-thirds of the surplus and B gets one-third. Verify that this result is correct.

(b) Let $r = 0.01$ and $s = 0.02$, and compare the x and y values found by using the approximation method with the more exact solutions for x and y found by using the formulas $x = (s + rs)/(r + s + rs)$ and $y = (r + rs)/(r + s + rs)$ derived in the chapter.

Markets and Competition

An economy is a system engaged in production and consumption. Its primary resources, such as labor, land, and raw materials, are used in the production of goods and services, including some that are then used in the production of other goods and services. The end points of this chain are the goods and services that go to the ultimate consumers. The system has evolved differently in different places and times, but always and everywhere it must create methods of linking its component parts—that is to say, a set of institutions and arrangements that enable various suppliers, producers, and buyers to deal with each other.

In Chapters 17 and 18, we examined two such institutions: auctions and bargaining. In auctions, one seller generally deals with several actual or potential buyers, although the reverse arrangement—in which one buyer deals with several actual or potential sellers—also exists; an example is a construction or supply project being offered for tender. Bargaining generally confronts one buyer with one seller.

That leaves perhaps the most ubiquitous economic institution of all—namely, the *market,* where several buyers and several sellers can deal simultaneously. A market can conjure up the image of a bazaar in a foreign country, where several sellers have their stalls, several buyers come, and much bargaining is going on at the same time between pairs of buyers and sellers. But most markets operate differently. A town typically has two or more supermarkets, and the households in the town decide to shop at one (or perhaps buy some items at one and some at another) by comparing availability, price, quality, and so on.

Supermarkets are generally not located next to each other; nevertheless, they know that they have to compete with each other for the townspeople's patronage, and this competition affects their decisions on what to stock, what prices to charge, and so forth. There is no overt bargaining between a store and a customer; prices are posted, and one can either pay that price or go elsewhere. But the existence of another store limits the price that each store can charge.

In this chapter, we will briefly examine markets from a strategic viewpoint. How should, and how do, a group of sellers and buyers act in this environment? What is the equilibrium of their interaction? Are buyers and sellers efficiently matched to one another in a market equilibrium?

We have already seen a few examples of strategies in markets. Remember the pricing game of the two restaurants introduced in Chapter 4? Because the strategic interaction pits two sellers against each other, we call it a *duopoly* (from the Greek *duo,* meaning "two," and *polein,* "to sell"). We found that the equilibrium was a prisoners' dilemma, in which both players charge prices below the level that would maximize their joint profit. In Chapter 11, we saw some ways, most prominently through the use of a repeated relationship, whereby the two restaurants can resolve the prisoners' dilemma and sustain tacit collusion with high prices; they thus become a *cartel,* or in effect a *monopoly.*

But even when the restaurants do not collude, their prices are still higher than their costs because each restaurant has a somewhat captive clientele whom it can exploit and profit from. When there are more than two sellers in a market—and as the number of sellers increases—we would expect the prisoners' dilemma to worsen, or the competition among the sellers to intensify, to the point where prices are as low as they can be. If there are hundreds or thousands of sellers, each might be too small to have a truly strategic role, and the game-theoretic approach would have to be replaced with one or more suitable for dealing with many individually small buyers and sellers. In fact, that is what the traditional economic "apparatus" of *supply and demand* does—it assumes that each such small buyer or seller is unable to affect the market price. Each party simply decides how much to buy or sell at the prevailing price. Then the price adjusts, through some invisible and impersonal market, to the level that equates supply and demand and thus "clears" the market. Remember the distinction that we drew in Chapter 2 between an individual decision-making situation and a game of strategic interaction. The traditional economic approach regards the market as a collection of individual decision makers who interact not strategically but through the medium of the price set by the market.

We can now go beyond this cursory treatment to see precisely how an increase in the number of sellers and buyers intensifies the competition among them and leads to an outcome that looks like an impersonal market in which the price adjusts to equate demand and supply. That is to say, we can derive our supply-and-demand analysis from the more basic strategic considerations of bargaining and competition. This approach will deepen your understanding of the market mechanism even if you have already studied it in elementary economics courses and textbooks. If you have not, this chapter will allow you to leapfrog ahead of others who have studied economics but not game theory.

In addition to the discussion of the market mechanism, we will show when and how competition yields an outcome that is socially desirable; we also consider what is meant by the phrase "socially desirable." Finally, in Sections 4 and 5, we examine some alternative mechanisms that attempt to achieve efficient or fair allocation where markets cannot or do not function well.

1 A SIMPLE TRADING GAME

We begin with a situation of pure bargaining between two people, and then we introduce additional bargainers and competition. Imagine two people, a seller S and a buyer B, negotiating over the price of a house. Seller S may be the current owner, who gets a payoff with a monetary equivalent of \$100,000 from living in the house. Or S may be a builder whose cost of constructing the house is \$100,000. In either interpretation, this amount represents the seller's BATNA, or in the language of auctions, his reserve price. From living in the same house, B can get a payoff of \$300,000, which is the maximum that he would be willing to pay. Right now, however, B does not own a house; so his BATNA is 0. Throughout this section, we assume that the sellers' reserve prices and the buyers' willingness to pay are common knowledge for all the participants; so the negotiation games are not plagued by the additional difficulty of information manipulation. We consider the latter in connection with the market mechanism later in the chapter.

If the house is sold for a price p, the seller gets p but gives up the house; so his surplus, measured in thousands of dollars, is $(p - 100)$. The buyer gets to live in the house, which gives him a benefit of 300, for which he pays p; so his surplus is $(300 - p)$. The two surpluses add up to

$$(300 - p) + (p - 100) = 300 - 100 = 200$$

regardless of the specific value of p. Thus 200 is the total surplus that the two can realize by striking a deal—the extra value that exists when S and B reach an agreement with one another.

The function of the price in this agreement is to divide up the available surplus between the two. For S to agree to the deal, he must get a nonnegative surplus from it; therefore p must be at least 100. Similarly, for B to agree to the deal, p cannot exceed 300. The full range of the negotiation for p lies between these two limits, but we cannot predict p precisely without specifying more details of the bargaining process. In Chapter 18, we saw two ways to model this process. The Nash cooperative model determined an outcome, given the relative bargaining strengths of S and B; the noncooperative game of alternating offers and counteroffers fixed an outcome, given the relative patience of the two.

Here we develop a different graphical illustration of the BATNAs, the surplus, and the range of negotiation from the approach used in Chapter 18. This device will help us to relate ideas based on bargaining to those of trading in markets. Figure 19.1 shows quantity on the horizontal axis and money measured in thousands of dollars on the vertical axis. The curve *S*, which represents the seller's behavior, consists of three vertical and horizontal red lines that together are shaped like a rising step.[1] The curve *B*, which represents buyer's behavior, consists of three vertical and horizontal gray lines, shaped like a falling step.

Here's how to construct the seller's curve. First, locate the point on the vertical axis corresponding to the seller's reserve price, here 100. Then draw a line from 0 to 100 on the vertical axis, indicating that no house is available for prices below 100. From there, draw a horizontal line (shown in green in Figure 19.1) of length 1; this line shows that the quantity that the seller wants to sell is exactly one house. Finally, from the right-hand end of this line, draw a line vertically upward to the region of very high money amounts. This line emphasizes that in this example, there is no other seller, and so no quantity greater than 1 is available for sale, no matter how much money is offered for it.

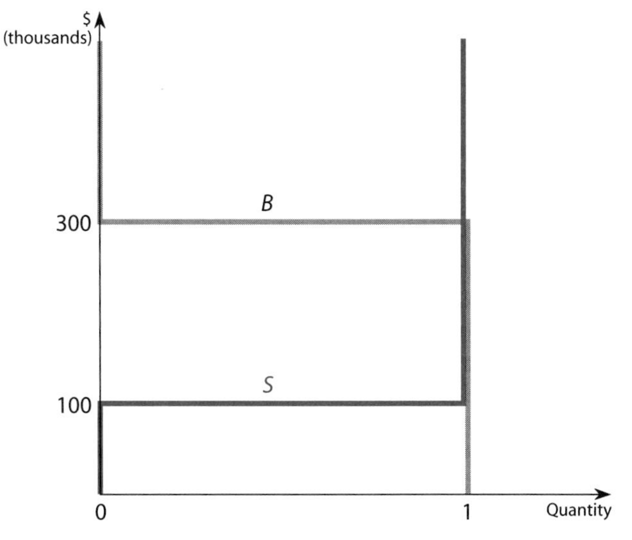

FIGURE 19.1 One Buyer, One Seller

[1]The word *curve* is more commonly used for a smoothly curving shape; therefore you may think it strange to have a "curve" consisting of three (or later, more) straight-line segments joined end to end at right angles. But we use it, for want of a better word.

In a similar manner, the buyer's bargaining curve *B* is constructed. As before, first locate the point on the vertical axis representing the buyer's willingness to pay, here 300. From there, draw a vertical line going upward along the axis, indicating that, for prices above 300, the buyer is not willing to purchase any house at all. Next, draw a horizontal line (shown in gray in Figure 19.1) of length 1, representing the quantity (one house) that the buyer wants to buy. Finally, from the right end of this horizontal line, draw a line going vertically downward to the horizontal axis, suggesting that, in this example, there is no other buyer, and so no quantity greater than 1 finds any customers, no matter how low the asking price.

Note that the two step-shaped curves *S* and *B* overlap along their vertical segments, at quantity 1 and at money amounts ranging between 100 and 300. The overlap indicates the range of negotiation for bargaining between buyer and seller. The horizontal coordinate of the overlapping segment (here 1) is the amount traded; the vertical segment of overlap (here from 100 to 300) represents the range of prices at which the final bargaining agreement can take place.

Those of you who have had an introductory course in economics will recognize the *S* as the seller's **supply curve** and *B* as the buyer's **demand curve.** For those unfamiliar with these concepts, we can explain them here briefly. Suppose the trades are initiated and negotiated not by the buyers and sellers themselves, as in the two-person game depicted in Figure 19.1, but in an organized market. A neutral "market maker," whose job it is to bring buyers and sellers together by determining a price at which they can trade, finds out how much will be on offer for sale and how much will be demanded at any price. The market maker then adjusts the price so that the market clears—that is, the quantity demanded equals the quantity supplied. This market maker is sometimes a real person—for example, the manager of the trading desk for a particular security in financial markets; at other times he is a hypothetical construct in economic theorizing. The same conclusions follow in either case.

Let us see how we can apply the market concept to the outcome in our bargaining example. First, consider the supply side. Curve *S* in Figure 19.1 tells us that for any price below 100, nothing is offered for sale and that for any price above 100, exactly one house is offered. This curve also shows that when the price is exactly 100, the seller would be just as happy selling the house as he would be remaining in it as the original owner (or not building it at all if he is the builder), so the quantity offered for sale could be 0 or 1. Thus this curve indicates the quantity offered for sale at each possible price. In our example, the fractional quantities between 0 and 1 do not have any significance, but for other objects of exchange such as wheat or oil, it is natural to think of the supply curve as showing continuously varying quantities as well.

Next consider the demand side. The graph of the buyer's willingness to pay, labeled *B* in Figure 19.1, indicates that, for any price above 300, there is no willing buyer, but for any price below 300, there is a buyer willing to buy exactly one house. At a price of exactly 300, the buyer is just as happy buying as not, so he may purchase either 0 or 1 item.

The demand curve and the supply curve coincide at quantity 1 over the range of prices from 100 to 300. When the market maker chooses any price in this range, exactly one buyer and exactly one seller will appear. The market maker can then bring them together and consummate the trade; the market will have cleared. We explained this outcome earlier as arising from a bargaining agreement, but economists would call it a **market equilibrium.** Thus the market is an institution or arrangement that brings about the same outcome as would direct bargaining between the buyer and the seller. Moreover, the range of prices that clears the market is the same as the range of negotiation in bargaining.

The institution of a market, where buyers and sellers respond passively to a price set by a market maker, seems strange when there is just one buyer and one seller. Indeed, the very essence of markets is competition, which demands the presence of several sellers and several buyers. Therefore we now gradually extend the scope of our analysis by introducing more of each type of agent. At each step we examine the relationship between direct negotiation and market equilibrium.

Let us consider a situation in which there is just the one original buyer, *B*, who is still willing to pay a maximum of $300,000 for a house. But now we introduce a second seller, S_2, who has a house for sale identical with that offered by the first seller, whom we now call S_1. Seller S_2 has a BATNA of $150,000. It may be higher than S_1's because S_2 is an owner who places a higher subjective value on living in his house than S_1 does. Or S_2 is a builder with a higher cost of construction than S_1's cost.

The existence of S_2 means that S_1 cannot hope to strike a deal with B for any price higher than 150. If S_1 and B try to make a deal at a price of even 152, for example, S_2 could undercut this price with an offer of 151 to the buyer. S_2 would still get a positive surplus instead of being left out of the bargain and getting 0. Similarly, S_2 cannot hope to charge more than 150, because S_1 could undercut that price even more eagerly. Thus the presence of the competing seller narrows the range of bargaining in this game from (100, 300) to (100, 150). The equilibrium trade is still between the original two players, S_1 and B, because it will take place at a price somewhere *between* 100 and 150, which cuts S_2 out of the trade. The presence of the second seller thus drives down the price that the original seller can get. Where the price settles between 100 and 150 depends on the relative bargaining powers of B and S_1. Even though S_2 drops out of the picture at any price below 150, if B has a lot of bargaining power for other reasons not mentioned here (for example, relative patience), then he may refuse an offer from S_1 that is close to 150 and hold out for a figure much closer to 100. If S_1 has relatively little bargaining power, he may have to accept such a counteroffer.

We show the demand and supply curves for this case in Figure 19.2a. The demand curve is the same as before, but the supply curve has two steps. For any price below 100, neither seller is willing to sell, and so the quantity is zero; this is the vertical line along the axis from 0 to 100. For any price from 100 to 150, S_1 is willing to sell, but S_2 is not. Thus the quantity supplied to the market is 1, and the vertical line at this quantity extends from 100 to 150. For any price above 150, both sellers are willing to sell and the quantity is 2; the vertical line at this quantity, from the price of 150 upward, is the last segment of the supply curve. Again we draw horizontal segments from quantity 0 to quantity 1 at price 100 and from quantity 1 to quantity 2 at price 150 to show the supply curve as an unbroken entity.

The overlap of the demand and supply curves is now a vertical segment representing a quantity of 1 and prices ranging from 100 to 150. This is exactly the range of negotiation that we found earlier when considering bargaining instead of a market. Thus the two approaches predict the same range of outcomes: one house changes hands, the seller with the lower reserve price makes the sale, and the final price is somewhere between \$100,000 and \$150,000.

What if both sellers had the same reserve price of \$100,000? Then neither would be able to get any price above 100. For example, if S_1 asked for \$101,000, then S_2, facing the prospect of making no sale at all, could counter by asking only \$100,500. The range of negotiation would be eliminated entirely. One seller would succeed in selling his house for \$100,000. The other would be left out, but at this price he is just as happy not selling the house as he is selling it.

We show this case in the context of market supply and demand in Figure 19.2b. Neither seller offers his house for sale for any price below 100, and both

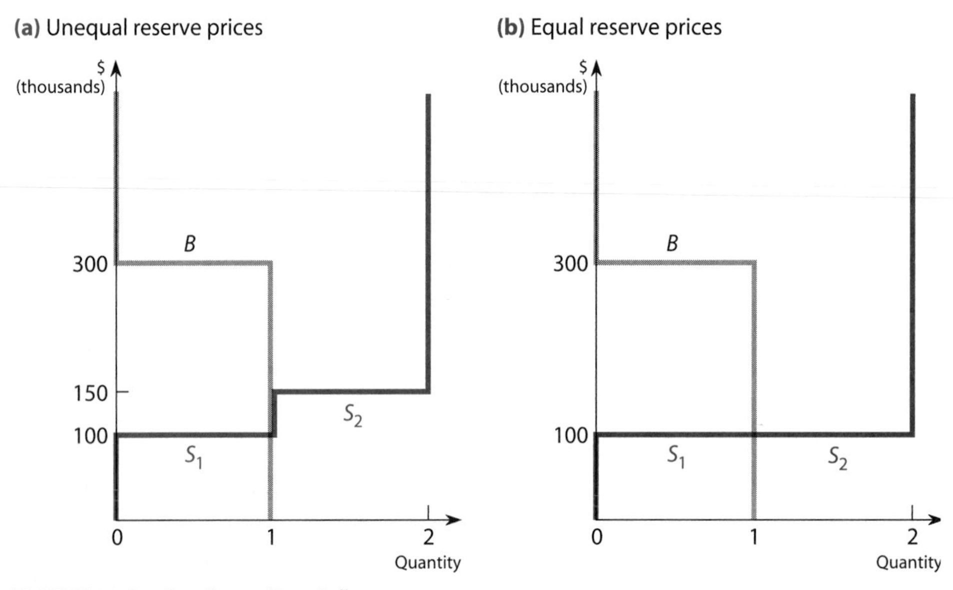

FIGURE 19.2 One Buyer, Two Sellers

sellers offer houses as soon as the price rises above 100. Therefore the supply curve has a horizontal segment of length 2 at height 100; at its right-hand end starts the terminal vertical upward line of indefinite length. The demand curve has not changed, and so now the two curves overlap only at the point (1,100). The range of prices is now reduced to one value, and this price represents the market equilibrium. One house changes hands, one of the two sellers (it does not matter which) makes the sale, and the price is exactly 100. The competition between the two sellers with identical reserve prices prevents either from getting more than the reserve price; the buyer reaps the benefit of the competition between the sellers in the form of a large surplus—namely, $300 - 100 = 200$.

The situation in the market could be reversed so that there is just one seller, S, but a second buyer, B_2, who is willing to pay $200,000 for a house. In this case, the first buyer, now labelled B_1, cannot hope to acquire the house for any less than 200. You should be able to figure out why: B_2 will bid the price up to at least 200; so if B_1 is to get the house, he will have to offer more than that amount. The range of possible prices on which S and B_1 could agree now extends from 200 to 300. This range is again narrower than the original 100 to 300 that we had with only one buyer and one seller. We leave you to do the explicit analysis by using the negotiation approach and the market supply-and-demand curves.

Proceeding with our pattern of introducing more buyers and sellers into our bargaining or market, we next consider a situation with two of each type of trader. Suppose the sellers, S_1 and S_2, have reserve prices of 100 and 150, respectively, and the buyers, B_1 and B_2, have a maximum willingness to pay of 300 and 200, respectively. We should expect the range of negotiation in the bargaining game to extend from 150 to 200 and can prove it as follows.

Each buyer and each seller is looking for the best possible deal for himself. They can make and receive tentative proposals while continuing to look around for better ones, until no one can find anything better, at which point the whole set of deals is finalized. Consider one such tentative proposal, in which one house—say, S_1's—would sell at a price $p_1 < 150$ and the other, S_2's, at $p_2 > 150$. The buyer who would pay the higher price—say, B_1—can then approach S_1 and offer him $(p_1 + p_2)/2$. This offer is greater than p_1 and so is better for S_1 than the original tentative proposal. It is also less than p_2 and so is better for B_1 as well. This offer would be worse for the other two traders, but because B_1 and S_1 are free to choose their partners and trades, B_2 and S_2 can do nothing about it. Thus any price lower than 150 cannot prevail. Similarly, you can check that any price higher than 200 cannot prevail either.

All of this is a fairly straightforward continuation of our reasoning in the earlier cases with fewer traders. But with two of each type of trader, an interesting new feature emerges: the two houses (assumed to be identical) must both sell for the same price. If they did not, the buyer who was scheduled to pay more in the original proposal could strike a mutually better deal with the seller who was getting less.

Note the essential difference between competition and bargaining here. You *bargain* with someone on the "other side" of the deal; you *compete* with someone on the "same side" of the deal for an opportunity to strike a deal with someone on the other side. Buyers *bargain* with sellers; buyers *compete* with other buyers, and sellers *compete* with other sellers.

We draw the supply and demand curves for the two-seller, two-buyer case in Figure 19.3a. Now each curve has three vertical segments and two horizontal lines joining the vertical segments at the reserve prices of the two traders on that side of the market. The curves overlap at the quantity 2 and along the range of prices from 150 to 200. This overlap is where the market equilibrium must lie. The two houses will change hands at the same price; it does not matter whether S_1 sells to B_1 and S_2 to B_2, or S_1 sells to B_2 and S_2 to B_1.

We move finally to the case of three buyers and three sellers. Suppose the first two sellers, S_1 and S_2, have the same respective reserve prices of 100 and 150 as before, but the third, S_3, has a higher reserve price of 220. Similarly, the first two buyers, B_1 and B_2, have the same willingness to pay as before, 300 and 200, respectively, but the third buyer's willingness to pay is lower, only 140. The new buyer and seller do not alter the negotiation or its range at all; only two houses get traded, between sellers S_1 and S_2 and buyers B_1 and B_2, for a price somewhere between 150 and 200 as before. B_3 and S_3 do not get to make a trade.

To see why B_3 and S_3 are left out of the trading, suppose S_3 does come to an agreement to sell his house. Only B_1 could be the buyer because neither B_2 nor B_3 is willing to pay S_3's reserve price. Suppose the agreed-on price is 221; so B_1

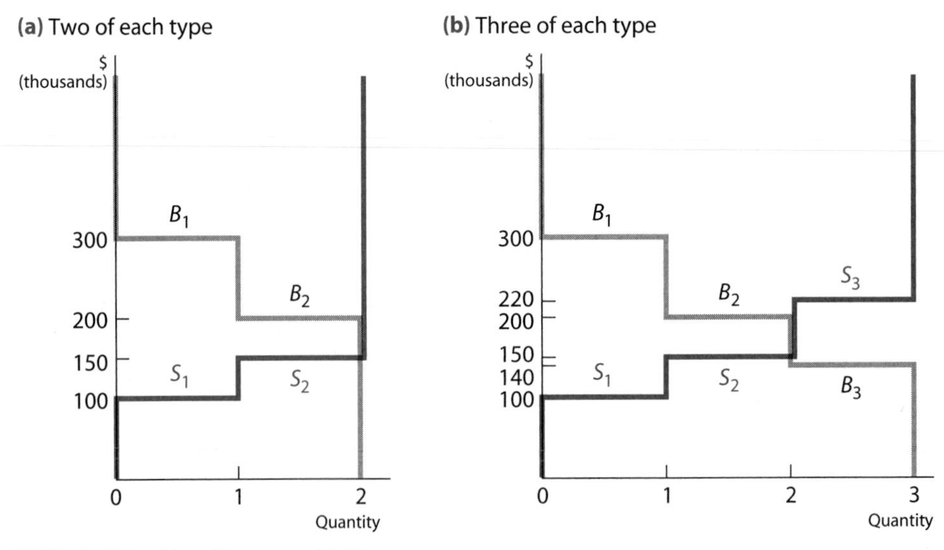

FIGURE 19.3 More Buyers and Sellers

gets a surplus of 79 and S_3 gets only 1. Then consider how the others must pair off. S_1, with his reserve price of 100, must sell to B_3, with his willingness to pay of 140; and S_2, with his reserve price of 150, must sell to B_2, with his willingness to pay of 200. The surplus from the S_1 to B_3 sale is 40, and the surplus from the S_2 to B_2 sale is 50. Therefore S_1 can get at most almost 40 in surplus from his sale, and S_2 can get at most 50 from his; either seller could benefit by approaching B_1 and striking a mutually beneficial alternative deal. For example, S_2 could approach B_1 and say, "With your willingness to pay of 300 and my reserve price of 150, we have a total available surplus of 150 between us. You are getting 79 in your deal with S_3; I am getting less than 50 in my deal with B_2. Let us get together; you could have 90 and I could have 60, and we both would be better off." B_1 would accept this deal, and his tentative arrangement with S_3 would fall through.

The numbers derived here are specific to our example, but the argument itself is perfectly general. For any number of buyers and sellers, first rank sellers by increasing reserve prices and buyers by decreasing willingness to pay. Then pair buyers and sellers of the same rank. When you find the pairing number n such that the nth buyer's willingness to pay is less than the nth seller's reserve price, you know that all buyers and sellers ranked n or more will not get to make a trade.

We can see this argument even more clearly by using market supply-and-demand-curve analysis as in Figure 19.3b. Construction of the supply-and-demand curves follows the same steps as before, so we do not describe it in detail here. The two curves overlap at the quantity 2, and the prices in the overlap range from 150 to 200. Therefore, in the market equilibrium, two houses change hands at a price somewhere between these two limits. Given the range of possible prices, we know that B_3's willingness to pay, 140, will be less than the market price, and S_3's reserve price, 220, will be higher than the market price. Therefore, just as we could say in the general case that the nth buyer and seller do not get to trade, we can assert that these two agents will not be active in the market equilibrium. Once again, it does not matter whether S_1 sells to B_1 and S_2 to B_2, or S_1 sells to B_2 and S_2 to B_1.

Suppose for the sake of definiteness that the market price in this case is 175. Then the surpluses for the traders are 75 for S_1, 25 for S_2, 125 for B_1, 25 for B_2, and 0 for S_3 and B_3; the total surplus is 250. You can check that the total surplus will be the same no matter where the price ends up within the range from 150 to 200.

But couldn't we pair differently and do better—especially if all three houses get sold? The answer is no. The negotiation process or the market mechanism produces the largest possible surplus. As before, to come up with an alternate solution we have to pair B_1 with S_3, B_3 with S_1, and B_2 with S_2. Then the B_1 and S_3 pairing generates a surplus of $300 - 200 = 80$ and the B_3 and S_1 pairing generates $140 - 100 = 40$, whereas the remaining pairing between B_2 and S_2 generates $200 - 150 = 50$. The total surplus in this alternative scenario comes to only 170, which is 80 less than the surplus of the 250 associated with the negotiation outcome or the market equilibrium. The problem is that B_3 is willing to pay only 140, a price below the equilibrium range, but S_3 has a reserve price of 220, above the equilibrium price range; between them, they bring to the game a negative surplus of $140 - 220 = -80$. If they get an opportunity to trade, there is no escaping the fact that the total surplus will go down by 80.

Although the outcome of the negotiation process or the market equilibrium yields the highest total surplus, it does not necessarily guarantee the most equal distribution of surplus: B_1, B_2, S_1, and S_2 share the whole 250, but B_3 and S_3 get nothing. In our alternate outcome, everyone got something, but the pairings could not survive the process of renegotiation. If you are sufficiently concerned about equity, you might prefer the alternative outcome despite the level of surplus generated. You could then restrict the process of renegotiation to bring about your preferred outcome. Better still, you could let the negotiation or market mechanisms go their way and generate the maximum surplus; then you could take some of that surplus away from those who enjoy it and redistribute it to the others by using a tax-and-transfer policy. The subject of public economics considers such issues in depth.

Let us pause here to sum up what we have accomplished in this section. We have developed a method of thinking about how negotiations between buyers and sellers proceed to secure mutually advantageous trades. We also linked the outcomes of this bargaining process with those of the market equilibrium based on supply-and-demand analysis. Although our analysis was of simple numerical examples with small numbers of buyers and sellers, the ideas have broad-ranging implications, which we develop in a more general way in the sections that follow. In the process, we also take the negotiation idea beyond that of the trade context. As we saw in Section 7 of Chapter 18, with many agents and many goods and services available to trade, the best deals may be multilateral. A good general theory must allow not just pairs, but also triplets or more general groupings of the participants. Such groups—sometimes called *coalitions*—can get together to work out tentative deals as the individual people and groups continue the search for better alternatives. The process of deal making stops only when no person or coalition can negotiate anything better for itself. Then the tentative deals become final and are consummated. We take up the subject of creating mutually advantageous deals in the next section.

2 THE CORE

We now introduce some notation and a general theory for analyzing a process of negotiation and renegotiation for mutually advantageous deals. Then we apply it in a simple context. We will find that, as the number of participants becomes large, the complex process of coalition formation, re-formation, and deal making produces an outcome very close to what a traditional market would produce. Thus we will see the market is an institution that can simplify the complex process of deal making among large numbers of traders.

Consider a general game with n players, labeled simply 1, 2, . . . n, such that the set of all players is $N = \{1, 2, 3, \ldots, n\}$. Any subset of players in N is a **coalition** C. When it is necessary to specify the members of a coalition in full, we list them in braces; $C = \{1, 2, 7\}$ might be one such coalition. We allow the N to be a subset of itself and call it the **grand coalition** G; thus $G = N$. A simple formula in combinatorial mathematics gives the total number of subsets of N that include one or two or three or . . . n people out of the n members of N. This number of possible subsets is $(2^n - 1)$; there are thus $(2^n - 1)$ possible coalitions altogether.

For the simple two-person trading game discussed in Section 1, $n = 2$, and so we can imagine three coalitions: two single-member ones of one buyer and one seller, respectively, and the grand coalition of the two together. When there are two sellers and one buyer, we have seven $(2^3 - 1)$ coalitions: three trivial ones consisting of one player each; three pairs, $\{S_1, B\}$, $\{S_2, B\}$, and $\{S_1, S_2\}$; and the grand coalition, $\{S_1, S_2, B\}$.

A coalition can strike deals among its own members to exploit all available mutual advantage. For any coalition C, we let the function $v(C)$ be the total surplus that its members can achieve on their own, regardless of what the players not in the coalition (that is, in the set $N - C$) do. This surplus amount is called the **security level** of the coalition. When we say "regardless of what the players not in the coalition do," we visualize the worst-case scenario—namely, what would happen to members of C if the others (those not in C) took the action that was the worst from C's perspective. This idea is the analogue of an individual player's minimax, which we studied in Chapters 7 and 8.

The calculation of $v(C)$ comes from the specific details of a game. As an example, consider our simple game of house trading from Section 1. No individual player can achieve any surplus on his own without some cooperation from a player on the other side, and so the security levels of all single-member coalitions are zero; $v(\{S_1\}) = v(\{S_2\}) = v(\{B\}) = 0$. The coalition $\{S_1, B\}$ can realize a total surplus of $300 - 100 = 200$, and S_2 on his own cannot do anything to reduce the surplus that the other two can enjoy (arson being outside the rules of the game); so $v(\{S_1, B\}) = 200$. Similarly, $v(\{S_2, B\}) = 300 - 150 = 150$. The two sellers on their own cannot achieve any extra payoff without including the buyer, so $v(\{S_1, S_2\}) = 0$.

In a particular game, then, for any coalition C, we can find its security level, $v(C)$, which is also called the **characteristic function** of the game. We assume that the characteristic function is common knowledge and that there is no asymmetry of information.

Next, we consider possible outcomes of the game. An **allocation** is a list of surplus amounts, $(x_1, x_2, \ldots x_n)$, for the players. An allocation is **feasible** if it is logically conceivable within the rules of the game—that is, if it arises from any deal that could in principle be made. Any feasible allocation may be proposed as a tentative deal, but the deal can be upset if some coalition can break away and do better for itself. We say that a feasible allocation is **blocked** by a coalition C if

$$v(C) > \sum_{i \text{ in } C} x_i.$$

This inequality says that the security level of the coalition—the total surplus that it can minimally guarantee for its members—exceeds the sum of the surpluses that its members get in the proposed tentative allocation. If this is true, then the coalition can form and will strike a bargain among its members that leaves all of them better off than in the proposed allocation. Therefore the implicit threat by the coalition to break away (back out of the deal if the surplus gained is not high enough) is credible; that is how the coalition can upset (block) the proposed allocation.

The set of allocations that cannot be blocked by any coalition contains all of the possible stable deals, or the bargaining range of the game. This set cannot be reduced further by any groups searching for better deals; we call this set of allocations the **core** of the game.

A. Numerical Example

We have already the blocking in the simple two-person trading game in Section 1, although we did not use this terminology. However, we can now see that what we called the "range of negotiation" corresponds exactly to the core. So we can translate, or restate, our results as follows. With one buyer and one seller, the range of negotiation is from 100 to 300. This outcome corresponds to a core consisting of all allocations ranging from the one (with price 100) where the buyer gets surplus 200 and the seller gets 0 to the one (with price 300) where the buyer gets 0 and the seller gets 200.

Now we use our new analytical tools to consider the case in which one buyer, willing to pay as much as 300, meets two sellers having equal reserve prices of 100. What is the core of the game among these three participants?

Consider an allocation of surpluses—say, x_1 and x_2—to the two sellers and y to the buyer. For this allocation to be in the core, no coalition should be able to achieve more for its members than what they receive here. We know that each single-member coalition can achieve only zero; therefore a core allocation must satisfy the conditions

$$x_1 \geq 0, \quad x_2 \geq 0, \quad \text{and} \quad y \geq 0.$$

Next, the coalitions of one buyer and one seller can each achieve a total of 300 − 100 = 200, but the coalition of the two sellers can achieve only zero. Therefore, for the proposed allocation to survive being blocked by a two-person coalition, it must satisfy

$$x_1 + y \geq 200, \quad x_2 + y \geq 200, \quad \text{and} \quad x_1 + x_2 \geq 0.$$

Finally, all three participants together can achieve a surplus of 200; therefore, if the proposed allocation is not to be blocked by the grand coalition, we must have

$$x_1 + x_2 + y \geq 200.$$

But because the total surplus available is only 200, $x_1 + x_2 + y$ cannot be greater than 200. Therefore we get the equation $x_1 + x_2 + y = 200$.

Given this last equation, we can now argue that x_1 must be zero. If x_1 were strictly positive and we combined the equation $x_1 > 0$ with one of the two-person inequalities, $x_2 + y \geq 200$, then we would get $x_1 + x_2 + y > 200$, which is impossible. Similarly, we find $x_2 = 0$ also. Then it follows that $y = 200$. In other words, the sellers get no surplus—it all goes to the buyer. To achieve such an outcome, one of the houses must sell for its reserve price $p = 100$; it does not matter which.

We conducted the analysis in Section 2.A by using our intuition about competition. When both sellers have the same reserve price, the competition between them becomes very fierce. If one has negotiated a deal to sell at a price of 102, the other is ready to undercut and offer the buyer a deal at 101 rather than go without a trade. In this process, the sellers "beat each other down" to 100. In this section, we have developed some general theory about the core and applied it in a mechanical and mathematical way to get a result that leads back to the same intuition. We hope this two-way traffic between mathematical theory and intuitive application reinforces both in your mind. The purely intuitive thinking becomes harder as the problem grows more complex, with many objects, many buyers, and many sellers. The mathematics provides a general algorithm that we can apply to all of these problems; you will have ample opportunities to use it. But having solved the more complex problems by using the math, you should also pause and relate the results to intuition, as we have done here.

B. Some Properties of the Core

EFFICIENCY We saw earlier that an allocation is in the core if it cannot be blocked by *any* coalition—including the grand coalition. Therefore, if a core allocation gives n players surpluses of $(x_1, x_2, \ldots x_n)$, the sum of these surpluses must be at least as high as the security level that the grand coalition can achieve:

$$x_1 + x_2 + \cdots + x_n \geq v(\{1, 2, \ldots, n\})$$

But the grand coalition can achieve anything that is feasible. Therefore for *any* feasible allocation of surpluses—say, $(z_1, z_2, \ldots z_n)$—it must be true that

$$v(\{1, 2, \ldots, n\}) \geq z_1 + z_2 + \cdots + z_n.$$

These two inequalities together imply that it is impossible to find another feasible allocation that will give everyone a higher payoff than a core allocation; we *cannot* have

$$z_1 > x_1, \quad z_2 > x_2, \ldots, \quad \text{and} \quad z_n > x_n$$

simultaneously. For this reason, the core allocation is said to be an **efficient allocation.** (In the jargon of economists, it is called *Pareto efficient,* after Wilfredo Pareto, who first introduced this concept of efficiency.)

Efficiency sounds good, but it is only one of a number of good properties that we might want a core allocation to have. Efficiency simply says that no other allocation will improve everyone's payoff simultaneously; the core allocation could still leave some people with very low payoffs and some with very high payoffs. That is, efficiency says nothing about the equity or fairness or distributive justice of a core allocation. In our trading game, a seller got zero surplus if there was another seller with the same reserve price; otherwise he had hope of some positive surplus. This is a matter of luck, not fairness. And we saw that some sellers with high reserve prices and some buyers with a low willingness to pay could get cut out of the deals altogether because the others could get greater total surplus; this outcome also sacrifices equity for the sake of efficiency.

A core allocation will always yield the maximum total surplus, but it does not regulate the distribution of that surplus. In our two-buyer, two-seller example, the buyer with the higher willingness to pay (B_1) gets more surplus because competition from the other buyer (B_2) means that the two pay the same price. Similarly, the lower-reserve-price seller (S_1) gets the most surplus—or, equivalently, the lower-cost builder (S_1) makes more profit.

In our simple trading game, the core was generally a range rather than a single point. Similarly, in other games there may be many allocations in the core. All will be efficient but will entail different distributions of the total payoff among the players, some fairer than others. The concept of efficiency says nothing about how we might choose one particular core allocation when there are many. Some other theory or mode of analysis is needed for that. Later in the chapter, we take a brief look at other mechanisms that do pay attention to the distribution of payoffs among the players.

EXISTENCE Is there a guarantee that all games will have a core? Alas, no. In games where the players have negative spillover effects (externalities) on each other, it may be possible for every tentatively proposed allocation to be upset by a coalition of players who gang together to harm the others. Then the core may be empty.

An extreme example is the *garbage game*. There are n players, each with one bag of garbage. Each can dump it in his own yard or in a neighbor's yard; he can even divide it up and dump some in one yard and some in others. The payoff from having b bags of garbage in your yard is $-b$, where b does not have to be an integer. Then a coalition of m people has the security level $-(n - m)$ as long as $m < n$, because they get rid of their m bags in nonmembers' yards. But in the worst-case scenario that underlies the calculation of the security level, all $(n - m)$ nonmembers dump their bags in members' yards. Therefore the characteristic function of the game is given by $v(m) = -(n - m) = m - n$ so long as $m < n$. But the grand-coalition members have no nonmembers on whom to dump their garbage, so $v(n) = -n$.

Now it is easy to see that any proposed allocation can be blocked. The most severe test is for a coalition of $(n - 1)$ people, which has a very high security level of -1 because they are dumping all their bags on the one unfortunate person excluded from the coalition and suffer only his one bag among the $(n - 1)$ of them. Suppose person n is currently the excluded one. Then he can get together with any $(n - 2)$ members of the coalition—say, the last $(n - 2)$—and offer them an even better deal. The new coalition of $(n - 1)$—namely, 2 through n—should dump its bags on 1. Person n will accept the whole of 1's bag. Thus persons 2 through $(n - 1)$ will be even better off than in the original coalition of the first $(n - 1)$ because they have 0 surplus to share among them rather than -1. And person n will be dramatically better off with 1 bag on his lawn than in the original situation in which he was the excluded person and suffered $(n - 1)$ bags. Thus the original allocation can be blocked. The same is true of every allocation; therefore the core is nonexistent, or empty.

The problem arises because of the nature of "property rights," which are implicit in the game. People have the right to dump their garbage on others, just as in some places and at some times firms have a right to pollute the air or the water or people have the right to smoke. In such a regime of "polluter's rights," negative externalities go unchecked and an unstable process of coalition formation and re-formation can result, leading to an empty core. In an alternative system where people have the right to remain free from the pollution caused by others, this problem does not arise. In our game, if the property right is changed so that no one can dump garbage on anyone else without that person's consent, then the core is not empty; the allocation in which every person keeps his own bag in his own yard and gets a payoff of -1 cannot be blocked.

C. Discussion

COOPERATIVE OR NONCOOPERATIVE GAME? The concept of the core was developed for an abstract formulation of a game, starting with the characteristic function. No details are provided about just *how* a coalition C can achieve total surplus equal to its security level $v(C)$. Those details are specific to each application. In the same way, when we find a core allocation, nothing is said about how the individual actions that give rise to that allocation are implemented. Implicitly, we assume that some external referee sees to it that all players carry out their assigned roles and actions. Therefore the core must be understood as a concept within the realm of cooperative game theory, where actions are jointly chosen and implemented.

But the freedom of individual players to join coalitions or to break up existing coalitions and form new ones introduces a strong noncooperative element into the picture. That is why the core is often a good way to think about competition. However, for competition to take place freely, it is important that all buyers and sellers have equal and complete freedom to form and re-form coalitions. What if only the sellers are active players in the game, and the buyers must passively accept their stated prices and cannot search for better deals? Then the core of the sellers' game that maximizes their total profit entails forming the grand coalition of sellers, which is a cartel or a monopoly. This problem is often observed in practice. There are numerous buyers in the market at any given time, but because each individual buyer is small and in the market infrequently, they find it difficult to get together or to negotiate with sellers for a better deal. Sellers are fewer in number and have repeated relationships among themselves; they are more likely to form coalitions. That is why antitrust laws try to prevent cartels and to preserve competition, whereas similar laws to prevent combination on the buyers' side are almost nonexistent.[2]

ADDITIVE AND NONADDITIVE PAYOFFS We defined the security level of a coalition as the minimum sum total of payoffs that could be guaranteed for its members. When we add individual payoffs in this way, we are making two assumptions: (1) that the payoffs are measured in some comparable units, money being the most frequently used yardstick of this kind; and (2) that the coalition somehow solves its own internal bargaining problem. Even when it has a high total payoff, a coalition can break up if some members get a lot while others get very little—unless the former can compensate the latter in a way that keeps everyone better off by being within the coalition than outside it. Such compensation must take the form of transfers of the unit in which payoffs are measured, again typically money.

If money (or something else that is desired by all and measured in comparable units for all) is not transferable, or if people's payoffs are nonlinear rescalings of money amounts—for example, because of risk aversion—then a coalition's

[2]In labor markets in some European countries such as Sweden, the coalition of all buyers (owners or management) deals with the coalition of all sellers (unions), but that is not the way that the concept of core assumes the process will work. The core would allow coalitions of smaller subsets of companies and workers to form and re-form, constantly seeking a better deal for themselves. In practice there is pure bargaining between two fixed coalitions.

security level cannot be described by its total payoff alone. We then have to keep track of the separate payoffs of all individual members. In other words, the characteristic function $v(C)$ is no longer a number but a set of vectors listing all members' payoff combinations that are possible for this coalition regardless of what others do. The theory of the core for this situation, technically called the case of *nontransferable utility,* can be developed, and you will find it in more advanced treatments of cooperative game theory.[3]

3 THE MARKET MECHANISM

We have already looked at supply and demand curves for markets with small numbers of traders. Generalizing them for more realistic markets is straightforward, as we see in Figure 19.4. There can be hundreds or thousands of potential buyers and sellers of each commodity; on each side of the market, they span a range of reserve prices and willingness to pay. For each price p, measured on the vertical axis, the supply curve shows the quantity of all sellers whose reserve price is below p—those who are willing to sell at this price. When we increase p enough that it exceeds a new seller's reserve price, that seller's quantity comes onto the market, and so the quantity supplied increases; we show this quantity as a horizontal segment of the curve at the relevant value of p. The rising-staircase

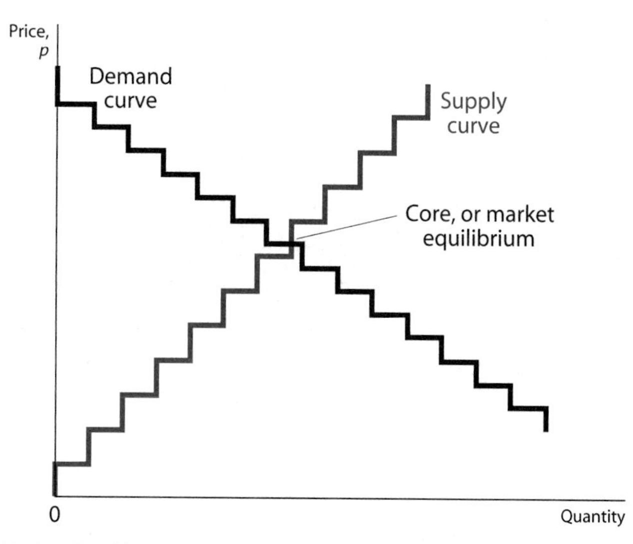

FIGURE 19.4 Market Equilibrium with Many Buyers and Sellers

[3]R. Duncan Luce and Howard Raiffa, *Games and Decisions* (New York: Wiley, 1957), especially chaps. 8 and 11, remains a classic treatment of cooperative games with transferable utility. An excellent modern general treatment that also allows nontransferable utility is Roger B. Myerson, *Game Theory* (Cambridge: Harvard University Press, 1991), chap. 9.

form that results from this procedure is the supply curve. Similarly, the descending-staircase graph on the buyers' side shows, for each price, the quantity that the buyers are willing to purchase (demand); therefore it is called the demand curve.

If all the sellers have access to the same inputs to production and the same technology of production, they may have the same costs and therefore the same reserve price. In that case, the supply curve may be horizontal rather than stepped. However, among the buyers there are generally idiosyncratic differences of taste, and so the demand curve is always a step-shaped curve.

Market equilibrium occurs where the two curves meet, which can happen in three possible ways. One is a vertical overlap at one quantity. This overlap indicates that exactly that quantity is the amount traded, and the price (common to all the trades) can be anything in the range of overlap. The second possibility is that the two curves overlap on a horizontal segment. Then the price is uniquely determined, but the quantity can vary over the range shown by the overlap. A horizontal overlap further means that the unique price is simultaneously the reserve price of one or more sellers and the willingness to pay of one or more buyers; at this price, the two types are equally happy making or not making a trade, and so it does not matter where along the overlap the quantity traded "settles." The third possibility is that a horizontal segment of one curve cuts a vertical segment of the other curve. Then there is only one point common to both curves, and the price and the quantity are both uniquely determined.

In each case, the price (even if not uniquely determined) is one at which, if our neutral market maker were to call it out, the quantity demanded would equal the quantity supplied, and the market would clear. Any such price at which the market clears is called a market equilibrium price. It is the market maker's purpose to find such a price, and he does so by a process of trial and error. He calls out a price and asks for offers to buy and sell. All of these offers are tentative at first. If the quantity demanded at the current price is greater than the quantity supplied, the market maker tries out a slightly higher price; if demand is less than supply, he tries a slightly lower price. The market maker adjusts the trial price in this way until the market clears. Only then are the tentative offers to buy and sell made actual, and they are put into effect at that market equilibrium price.

When thousands of buyers are ranked according to their willingness to pay and hundreds or thousands of sellers according to their reserve prices, the differences between two successive trades on each side are quite small. Therefore each step of the demand curve falls only a little and that of the supply curve rises only a little. With these types of supply and demand curves, even when there is a vertical overlap, the range of indeterminacy of the price in the market equilibrium is small, and the market-clearing price will be almost uniquely determined.

We showed graphically in simple examples that the core of our trading game, where each seller has one unit to offer and each buyer wants to buy just one unit, coincides with the market equilibrium. Similar results are true for any number of buyers and sellers, each of whom may want to buy or sell multiple units at different prices, but the algebra is more complex than is worthwhile, and so we will omit it. If some sellers or some buyers want to trade more than one unit each, then the link between the core and the market equilibrium is slightly weaker. A market equilibrium always lies in the core, but the core may have other allocations that are not market equilibria. However, in large markets with numerous buyers and sellers, there are relatively few such extra allocations, and the core and the market outcomes virtually coincide. Because this result comes from rather advanced microeconomic theory, we must leave it as an assertion, hoping that the simple examples worked out here give you confidence in its truth.[4]

A. Properties of the Market Mechanism

EFFICIENCY The correspondence between the outcomes of the core and of the market equilibrium has an immediate and important implication. We saw in Section 2 that a core allocation is efficient; it maximizes the total possible surplus. Therefore the market equilibrium also must be efficient, a fact that accounts in part for the conceptual appeal of the market.

Like the core, the market does not guarantee a fair or just distribution of income or wealth. A seller's profit depends on how far the market price happens to be above his cost, and a buyer's surplus depends on how far the market price happens to be below his willingness to pay.

INFORMATION AND MANIPULABILITY The market maker typically does not know the reserve price of any seller or the willingness to pay of any buyer. He sets a price and asks traders to disclose some of this information through their offers to sell and buy. Do they have the incentive to reply truthfully? In other words, is the market mechanism manipulable? This question is quite central to the strategic perspective of this book and relates to the discussion of information in Chapter 9.

First consider the case in which each seller and each buyer wants to trade just one unit. For simplicity, consider the two-buyer, two-seller example from Section 1, in which seller S_1 has a reserve price of 100. He could pretend to have a higher reserve price (by not offering to sell his house until the market maker calls out a price appropriately higher than 100) or a lower one. That is, he could

[4]For a discussion of the general theory about the relationship between the core and the market equilibrium, see Andreu Mas-Colell, Michael D. Whinston, and Jerry R. Green, *Microeconomic Theory* (New York: Oxford University Press, 1995), pp. 652–660.

pretend that his supply curve is different from what it is. Would that be in his interest?

Just as the market maker does not know anyone's reserve price or willingness to pay, S_1 does not know this information about anyone else. So he does not know what the market-clearing price will be. Suppose he pretends that his reserve price is 120. It might be that the market-clearing price exceeds 120, in which case he would get to make the sale, as he would have anyway, and would get the market price as he would have had he shown his reserve price to be 100; his exaggeration would make no difference. But the market-clearing price might be somewhere between 100 and 120, in which case the market maker would not take up S_1's offer to sell, and S_1 would lose a deal that would have been profitable. In this instance, exaggeration can only hurt him. Similarly, he might understate his reserve price and offer to sell when the market maker calls out a price of 80. If the market-clearing price ended up below 80, the understatement would make no difference. But if it ended up somewhere between 80 and 100, then the market maker would hold S_1 to his offer, and the sale would net a loss to S_1. Thus understatement, like exaggeration, can only hurt him. In other words, truthful revelation is the dominant strategy for S_1. The same is true for all buyers and sellers.

It is no coincidence that this assertion is similar to the argument in Chapter 17 that truthful bidding is the dominant strategy in a private-value, second-price auction. A crucial thing about the second-price auction is that what you pay when you win depends not on what you bid, but on what someone else bid. Over- or underbidding can affect your chances of winning only in an adverse way; you might fail to win when it would have been good and might win when it would be bad. Similarly, in the market, if you get to make a sale or a purchase, the price depends not on your reserve price or your willingness to pay, but on that of someone else. Misstating your values can affect your chances of making a trade only in an adverse way.

This argument does not work if some person has several units to trade and is a large trader who can affect the price. For example, a seller might have a reserve price of 100 and 15 units for sale. At a price of 100, there might be 15 willing buyers, but the seller might pretend to have only 10 units. Then the market maker would believe there was some scarcity, and he would set a higher market-clearing price at which the number of willing buyers matched the number of units the seller chose to put up for sale. If the buyers' willingness to pay rises sufficiently rapidly as the quantity goes down (that is, if the demand curve is steep), then the seller in question might make more profit by selling 10 units at the higher price than he would by selling 15 units at a lower price. Then the misstatement would work to his advantage, and the market mechanism would be manipulable. In the jargon of economics, this is an instance of the seller exercising some **monopoly power.** By offering less than he actually has (or can produce) at a particular price, the seller is driving up the market equilibrium price. Similarly, a multiunit buyer can understate his demand to drive the price down and thereby get greater surplus on the units that he does buy; he is exercising **monopsony power.**

However, if the market is so large that any one seller's attempt to exercise monopoly power can alter the price only very slightly, then such actions will not be profitable to the seller. When each seller and each buyer is small in relation to the market, each has negligible monopoly and monopsony power. In the jargon of economics, competition becomes *perfect;* in our jargon of game theory, the market mechanism becomes **nonmanipulable,** or **incentive compatible.**

B. Experimental Evidence

We have told a "story" of the process by which a market equilibrium is reached. However, as we noted earlier, the market maker who played the central role in that story—calling out prices, collecting offers to buy and sell, and adjusting the price until the market cleared—does not actually exist in most markets. Instead, the buyers and sellers have to search for trading partners and attempt to strike deals on their own. The process is somewhat like the formation and re-formation of coalitions in the theory of the core but even more haphazard and unorganized. Can such a process lead to a market-clearing outcome?

Through observation of actual markets, we can readily find many instances where markets seem to function well and many where they fail. But situations of the former kind differ in many ways from those of the latter, and many of these differences are not easily observable. Therefore the evidence does not help us understand whether and how markets can solve the problems of coordinating the actions of many buyers and sellers to arrive at an equilibrium price. Laboratory experiments, on the other hand, allow us to observe outcomes under controlled conditions, varying just those conditions whose effects we want to study. Therefore they are a useful approach and a valuable item of evidence for judging the efficacy of the market mechanism.

In the past half-century, many such experiments have been carried out. Their objectives and methods vary, but in some form all bring together a group of subjects who are asked to trade a tangible or intangible "object." Each is informed of his own valuation and given real incentives to act on this valuation. For example, would-be sellers who actually consummate a trade are given real monetary rewards equal to their profit or surplus, which is equivalent to the price that they succeed in obtaining *minus* the cost that they are assigned.

The results are generally encouraging for the theory. In experiments with only two or three traders on one side of the market, attempts to exercise monopoly or monopsony power are observed. But otherwise, even with relatively small numbers of traders (say, five or six on each side), trading prices generally converge quite quickly to the market equilibrium that the experimenter has calculated, knowing the valuations that he has assigned to the traders. If anything, the experimental results are "too good." Even when the traders know nothing about the valuations of other traders, have no understanding of the theory of market equilibrium, and have no prior experience in trading, they arrive at the equilibrium configuration relatively quickly. In recent work, the focus has shifted away from asking *whether* the market mechanism works to asking *why* it works so much better than we have any reason to expect.[5]

There is one significant exception—namely, markets for assets. The decision to buy or sell a long-lived asset must take into account not merely the relation between the price and your own valuation, but also the expected movement of the price in the future. Even if you don't expect your dot.com stock ever to pay any dividend, you might buy and hold it for a while if you believe that you can then sell it to someone else for a sufficiently higher price, thereby making a capital gain. You might come to this belief because you have observed rising prices for a while. But prices might be rising because others are buying dot.com stock, too, similarly expecting to resell it with a capital gain. If many people simultaneously try to sell the stocks to obtain those expected gains, the price will fall, the gains will fail to be realized, and the expectations will collapse. Then we will say that the dot.com market has experienced a *speculative bubble* that has burst. Such bubbles are observed in experimental markets, as they are in reality.[6]

4 THE SHAPLEY VALUE

The core has an important desirable property of efficiency, but it also has some undesirable ones. Most basically, some games have no core; many others have very large cores consisting of whole ranges of outcomes, and so the concept does not determine an outcome uniquely. Other concepts that do better in these respects have been constructed. The best known is the Shapley value, named after Lloyd Shapley of UCLA.

Like Nash's bargaining solution, discussed in Chapter 18, the Shapley value is a cooperative concept. It is similarly grounded in a set of axioms or assumptions that, it is argued, a solution should satisfy. And it is the unique solution that conforms to all these axioms. The desirability of the properties is a matter of

[5]Douglas D. Davis and Charles A. Holt, *Experimental Economics* (Princeton, NJ: Princeton University Press, 1993), chap. 3.

[6]Shyam Sunder, "Experimental Asset Markets: A Survey," *Handbook of Experimental Economics*, ed. John H. Kagel and Alvin E. Roth (Princeton, NJ: Princeton University Press, 1995), pp. 468–474.

judgment; other game theorists have specified other combinations of properties, each of which leads to its own unique solution. But the solutions for different combinations of properties are different, and not all the properties can be equally desirable. Perhaps more important, each player will judge the properties not by their innate attractiveness, but by what the outcome gives him. If an outcome is not satisfactory for some players and they believe they can do better by insisting on a different game, the proposed solution may be a nonstarter.

Therefore any cooperative solution should not be accepted as the likely actual outcome of the game solely on the basis of the properties that seem desirable to some game theorists or in the abstract. Rather, it should be judged by its actual performance for prediction or analysis. From this practical point of view, the Shapley value turns out to do rather well. We do not go into the axiomatic basis of the concept here. We simply state its formula, briefly interpret and motivate it, and demonstrate it in action.

Suppose a game has n players and its characteristic function is $v(C)$, which is the minimum total payoff that coalition C can guarantee to its members. The **Shapley value** is an allocation of payoffs u_i to each player i as defined by

$$u_i = \sum_C \frac{(n - k)!(k - 1)!}{n!}[v(C) - v(C - \{i\})], \tag{19.1}$$

where $n!$ denotes the product $1 \times 2 \times \cdots \times n$; where the sum is taken over all the coalitions C that have i as a member, and where, in each term of the sum, k is the size of the coalition C.

The idea is that each player should be given a payoff equal to the average of the "contribution" that he makes to each coalition to which he could belong, where all coalitions are regarded as equally likely in a suitable sense. First consider the size k of the coalition—from 1 to n. Because all sizes are equally likely, a particular size coalition occurs with probability $1/n$. Then the $(k - 1)$ partners of i in a coalition of size k can be chosen from among the remaining $(n - 1)$ players in any of

$$\frac{(n - 1)!}{[(n - 1) - (k - 1)]!(k - 1)!} = \frac{(n - 1)!}{[(n - k)!(k - 1)]!}$$

ways. The reciprocal of this expression is the probability of any one such choice. Combining that reciprocal with $(1/n)$ gives us the probability of a particular coalition C of size k containing member i, and that is what appears as the built-up fraction in each term of the sum on the right-hand side of the formula for u_i. What multiplies the fraction is the difference between the security level of the coalition C—namely, $v(C)$—and the security level that the remaining people would have if i were removed from the coalition—namely, $v(C - \{i\})$. This term measures the contribution that i makes to C.

The idea that each player's payoff should be commensurate with his contribution has considerable appeal. Most important, if a player must make some effort to generate this contribution, then such a payment scheme gives him the correct incentive to make that effort. In the jargon of economics, each person's incremental contribution to the economy is called his *marginal product;* therefore the concept of tying payoffs to contributions is called the *marginal productivity theory of distribution.* A market mechanism would automatically reward each participant for his contribution. The Shapley value can be understood as a way of implementing this principle—of achieving a marketlike outcome—at least approximately when an actual market cannot be arranged, for example, within units of a single larger organization.

The formula embodies much more than the general principle of marginal productivity payment. Averaging over all coalitions to which a player could belong and regarding all coalitions as equally likely are very specific procedures that may or may not have any counterpart in reality. Nevertheless, the Shapley value often produces outcomes that have some realistic features. We now examine two specific examples of the Shapley value in action, one from politics and one from economics.

A. Power in Legislatures and Committees

Suppose a 100-member legislature consists of four parties, Red, Blue, Green, and Brown. The Reds have 43 seats, the Blues 33, the Greens 16, and the Browns 8. Each party is a coherent block that votes together, and so each can be regarded as one player. A majority is needed to pass any legislation; therefore no party can get legislation through without the help of another block, and so no party can govern on its own. In this situation the power of a party will depend on how crucial that party is to the formation of a majority coalition. The Shapley value provides a measure of just that power. Therefore in this context it is called the **power index.** The index was used in this way by Shapley and Martin Shubik. A similar index, but with a different assumption about coalition formation, was devised by John Banzhaf.[7]

Give the value 1 to any coalition (including the grand coalition) that has a majority and 0 to any coalition without a majority. Then, in the formula for the Shapley value of party i (Eq. 19.1), the contribution that it makes to a coalition C—namely, $v(C) - v(C - \{i\})$—is 1 if the coalition C has a majority but $C - \{i\}$ if it does not. When party i's contribution to C is 1, we say that party i is a *pivotal*

[7]See Lloyd S. Shapley and Martin Shubik, "A Method for Evaluating the Distribution of Power in a Committee System," *American Political Science Review,* vol. 48, no. 3 (September 1954), pp. 787–792, and John Banzhaf, "Mathematical Analysis of Voting Power and Effective Representation," *George Washington Law Review,* vol. 36, no. 4 (1968), pp. 808–823.

member of the majority coalition C. In all other cases—namely, if coalition C does not have a majority at all or if C would have a majority even without i—the contribution of i to C is zero.

We can now list all the majority coalitions and which party is pivotal in each. None of the four possible (albeit trivial) one-party "coalitions" has a majority. Three of the six possible two-party coalitions have a majority—namely, {Red, Blue}, {Red, Green}, and {Red, Brown}. In each case both members are pivotal because the loss of either would mean the loss of the majority. Of the four possible three-party coalitions, in three of them—namely, {Red, Blue, Green}, {Red, Blue, Brown}, and {Red, Green, Brown}—only Red is pivotal. In the fourth three-party coalition—namely, {Blue, Green, Brown}—all three parties are pivotal. In the grand coalition, no party is pivotal.

In the Shapley value formula, the term corresponding to each two-party coalition takes the value $(4 - 2)!(2 - 1)!/4! = 2! \times 1!/4! = 2/24 = 1/12$, and each three-party coalition also gets $(4 - 3)!(3 - 1)! = 1! \times 2!/4! = 1/12$.

With this information we can calculate the Shapley value of each party. We have

$$u_{\text{Red}} = \frac{1}{12} \times 3 + \frac{1}{12} \times 3 = \frac{1}{2}$$

because the Red party is pivotal in three two-party coalitions and three three-party ones and each such coalition gets a weight of 1/12. Similarly

$$u_{\text{Blue}} = \frac{1}{12} \times 1 + \frac{1}{12} \times 1 = \frac{1}{6}$$

and $u_{\text{Green}} = u_{\text{Brown}} = 1/6$ likewise, because Blue, Green, and Brown are each pivotal in exactly one two-party and one three-party coalition.

Even though the Blues have almost twice as many members as the Greens, who in turn have twice as many as the Browns, the three parties have equal power. The reason is that they are all equally crucial in the formation of a majority coalition, either one at a time with Red, or the three of them joined together against Red.

We do observe in reality that small parties in multiparty legislatures enjoy far more power than their number proportions in the legislature might lead us to believe, which the Shapley value index shows in a dramatic way. It does make some unappealing assumptions; for example, it takes all coalitions to be equally likely and all contributions to have the same value 1 if they create a majority and the same value 0 if they do not. Therefore we should interpret the power index not as a literal or precise quantitative measure but as a rough indication of political power levels. More realistic analysis can sometimes show that small parties have even greater power than the Shapley measure might suggest. For example, the two largest parties often have ideological antagonisms that rule out a big coalition between them. If coalitions that include both Red and Blue together are not possible, then Green and Brown will have even greater power because of their ability to construct a majority—one of them together with Red or both of them together with Blue.

B. Allocation of Joint Costs

Probably the most important practical use of the Shapley value is for allocating the costs of a joint project among its several constituents. Suppose a town is contemplating the construction of a multipurpose auditorium building. The possible uses are as a lecture hall, a theater, a concert hall, and an opera house. A single-purpose lecture hall would cost \$1 million. A theater would need more sophisticated stage and backstage facilities and would cost \$4 million. A concert hall would need better acoustics than the lecture hall but less sophisticated staging than the theater, and so the cost of a pure concert hall would be \$3 million. An opera house would need both staging and acoustics, and the cost of the two together, whether for a theater–concert combination or for opera, would be \$6 million. If the opera house is built, it can also be used for any of the three other purposes, and a theater or a concert hall can also be used for lectures.

The town council would like to recoup the construction costs by charging the users. The tickets for each type of activity will therefore include an amount that corresponds to an appropriate share of the building cost attributable to that activity. How are these shares to be apportioned? This is a mirror image of the problem of distributing payoffs according to contributions—we want to charge each use for the extra cost that it entails. And the Shapley value offers an answer to the problem. For each activity—say, the theater—we list all the combinations in which it could take place and what it would cost to construct the facility for that combination, as well as for that combination minus the theater. This analysis tells us the contribution of the theater to the cost of building for that combination. Assuming all combinations for this activity to be equally likely and averaging, we get the overall cost share of the theater.

First, we calculate the probabilities of each combination of uses. The theater could be on its own with a probability of 1/4. It could be in a combination of two activities with a probability of 1/4. There are three ways to combine the theater with another activity, so the probability of each such combination is 1/12. Similarly, the probability of each three-activity combination including the theater is also 1/12. Finally, the probability of the theater's being in the grand coalition of all four activities is 1/4.

Next, we calculate the contributions that the theater makes to the cost of each combination. Here we just give an example. The theater–concert combination costs \$6 million, and the concert hall alone would cost \$3 million; therefore $6 - 3 = 3$ is the contribution of the theater to this combination. The complete list of combinations and contributions is shown in Figure 19.5. Each row focuses on one activity. The left-most column simply labels the activity: L for lecture, T for theater, C for concert, and O for opera. The successive columns to the right show one-, two-, three-, and four-activity combinations. The heading of each column shows the size and the probability of each combination of that size. The actual cells for each row in these columns list the combinations of that size that include the activity of that row (when there is only one activity we omit the listing), followed by the contribution of the particular activity to that combination. The right-most column gives the average of all these contributions with the use of the appropriate probabilities as weights—that is, the Shapley value.

It is interesting to note that, though the cost of building an opera facility is *6* times that of building a lecture hall (6 versus 1 in the bottom and top cells of the "Size 1" column), the cost share of opera in the combined facility is *13* times that of lectures (2.75 versus 0.25 in the "Shapley Value" column). This is because providing for lectures adds nothing to the cost of building for any other use, and so the contribution of lectures to all other combinations is zero. On the other hand, providing for opera raises the cost of *all* combinations except those that also include both theater and concert, of which there is only one.

Note also that the cost shares add up to 6, the total cost of building the combined facility—another useful property of the Shapley value. If you have sufficient mathematical facility, you can use Eq. (19.1) to show that the payoffs u_i of all the players sum to the characteristic function v of the grand coalition.

To sum up, the Shapley value allocates the costs of a joint project by calculating what the presence of each participant adds to the total cost. Each participant

ACTIVITY	COMBINATIONS				Shapley Value
	Size 1 Prob. = 1/4	Size 2 Prob. = 1/12	Size 3 Prob. = 1/12	Size 4 Prob. = 1/4	
L	1	LT: 4 − 4 = 0 LC: 3 − 3 = 0 LO: 6 − 6 = 0	LTC: 6 − 6 = 0 LTO: 6 − 6 = 0 LCO: 6 − 6 = 0	6 − 6 = 0	0.25
T	4	LT: 4 − 1 = 3 TC: 6 − 3 = 3 TO: 6 − 6 = 0	TLC: 6 − 3 = 3 TLO: 6 − 6 = 0 TCO: 6 − 6 = 0	6 − 6 = 0	1.75
C	3	CL: 3 − 1 = 2 CT: 6 − 4 = 2 CO: 6 − 6 = 0	CLT: 6 − 4 = 2 CLO: 6 − 6 = 0 CTO: 6 − 6 = 0	6 − 6 = 0	1.25
O	6	OL: 6 − 1 = 5 OT: 6 − 4 = 2 OC: 6 − 3 = 3	OLT: 6 − 4 = 2 OLC: 6 − 3 = 3 OTC: 6 − 6 = 0	6 − 6 = 0	2.75

FIGURE 19.5 Cost Allocation for Auditorium Complex

is then charged this addition to the cost (or in the jargon of economics his *marginal cost*).

The principle can be applied in reverse. Suppose some firms could create a larger total profit by cooperating than they could by acting separately. They could bargain over the division of the surplus (as we saw in Chapter 18), but the Shapley value provides a useful alternative in which each firm receives its marginal contribution to the joint enterprise. Adam Brandenburger and Barry Nalebuff, in their book *Co-opetition*, develop this idea in detail.[8]

5 FAIR DIVISION MECHANISMS

Neither the core nor the market mechanism nor the Shapley value guarantee any fairness of payoffs. Game theorists have examined other mechanisms that focus on fairness and have developed a rich line of theory and applications. Here we give you just a brief taste.[9]

The oldest and best-known **fair division mechanism** is "one divides, the other chooses." If A divides the cake into two pieces or the collection of objects available into two piles and B chooses one of the pieces or piles, the outcome will be fair. The idea is that, if A were to produce an unequal division, then B would choose the larger. Knowing this, A will divide equally. Such behavior then constitutes a rollback equilibrium of this sequential-move game.

A related mechanism has a referee who slowly moves a knife across the face of the cake—say, from left to right. At any point, either player can say "stop." The referee cuts the cake at that point and gives the piece to the left of the knife to the player who spoke.

We do not require the cake to be homogeneous—one part may have more raisins and another more walnuts—or all objects in the pile to be identical. The participants can differ in their preferences for different parts of the cake or different objects, and therefore they may have different perceptions about the relative values of portions and about who got a better deal. Then we can distinguish two concepts of fairness. First, the division should be *proportional,* meaning that each of n people believes that his share is at least $1/n$ of the total. Second, the division should be **envy free,** which means that no participant believes someone else got a better deal. With two people, the two concepts are

[8]Adam Brandenburger and Barry Nalebuff, *Co-opetition* (New York: Doubleday, 1996).

[9]For a thorough treatment, see Steven J. Brams and Alan D. Taylor, *Fair Division: From Cake-Cutting to Dispute Resolution* (New York: Cambridge University Press, 1996).

equivalent, but with more, envy-freeness is a stronger requirement, because each may think he got $1/n$ while also thinking that someone else got more than he did.

With more than two people, simple extensions of the "one cuts, the other chooses" procedure ensure proportionality. Suppose there are three people, A, B and C. Let any one person—say, A—divide the total into three parts. Ask B and C which portions they regard as acceptable—that is, of size at least 1/3. If both reject only one piece, then give that to A, leaving each of B and C a piece that he regards as acceptable. If they reject two pieces, give one of the rejected pieces to A and reassemble the remaining two pieces. Both B and C must regard this total as being of size at least 2/3 in size. Then use the "one cuts, the other chooses" procedure to divide this piece between them. Each gets what he regards as at least 1/2 of this piece, which is at least 1/3 of the initial whole. And A will make the initial cuts to create three pieces that are equal in his own judgment, to ensure himself of getting 1/3. With more people, the same idea can be extended by using mathematical induction. Similar, and quite complex, generalizations exist for the moving-knife procedure.

Ensuring envy-freeness with more than two people is harder. Here is how it can be done with three. Let A cut the cake into three pieces. If B regards one of the pieces as being larger than the other two, he "trims" the piece just enough so that there is at least a two-way tie for largest. The trimming is set aside and C gets to choose from the three pieces. If C chooses the trimmed piece, then B can choose either of the other two, and A gets the third. If C chooses an untrimmed piece, then B must take the piece that he trimmed, and A gets the third.

Leaving the trimming out of consideration for the moment, this division is envy free. Because C gets to choose any of the three pieces, he does not envy anyone. B created at least a two-way tie for what he believed to be the largest and gets one of the two (the trimmed piece if C does not choose it) or the opportunity to choose one of the two, and so B does not envy anyone. Finally, A will create an initial cut that he regards as equal, and so he will not envy anyone.

Then whoever of B or C did *not* get the trimmed piece divides the trimming into three portions. Whoever of B or C *did* get the trimmed piece takes the first pick, A gets the second pick, and the divider of the trimming gets the residual piece of it. A similar argument shows that the division of the whole cake is envy free. For examples of even more people, and for many other fascinating details and applications, read Brams and Taylor's *Fair Division* (cited earlier).

We do not get fairness without paying a price in the form of the loss of some other desirable property of the mechanism. In realistic environments with heterogeneous cakes or other objects and with diverse preferences, these fair division mechanisms generally are not efficient; they are also manipulable.

SUMMARY

We consider a market where each buyer and each seller attempts to make a deal with someone on the other side, in competition with others on his own side looking for similar deals. This is an example of a general game where *coalitions* of players can form to make tentative agreements for joint action. An outcome can be prevented or blocked if a coalition can form and guarantee its members better payoffs. The *core* consists of outcomes that cannot be blocked by any coalition.

In the trading game with identical objects for exchange, the core can be interpreted as a *market equilibrium* of *supply and demand.* When there are several buyers and sellers with slight differences in the prices at which they are willing to trade, the core shrinks and a competitive market equilibrium is determinate. This outcome is *efficient.* With few sellers or buyers, preference manipulation can occur and amounts to the exercise of *monopoly* or *monopsony power.* Laboratory experiments find that with as few as five or six participants, trading converges quickly to the market equilibrium.

The *Shapley value* is a mechanism for assigning payoffs to players based on an average of their contributions to possible coalitions. It is useful for allocating costs of a joint project and is a source of insight into the relative power of parties in a legislature.

Neither the core nor the Shapley value guarantees fair outcomes. Other *fair-division mechanisms* exist; they are often variants or generalizations of the "one cuts, the other chooses" idea. But they are generally inefficient and manipulable.

KEY TERMS

allocation (735)
blocking (736)
characteristic function (735)
coalition (735)
core (736)
demand curve (728)
efficient allocation (738)
envy-free allocation (752)
fair division mechanism (752)
feasible allocation (735)
grand coalition (735)
incentive-compatible market mechanism (745)
market equilibrium (729)
monopoly power (744)
monopsony power (745)
nonmanipulable market mechanism (745)
power index (748)
security level (735)
Shapley value (747)
supply curve (728)

SOLVED EXERCISES

S1. Consider the two buyers, B_1 and B_2, and one seller, S, in Section 1. The seller's reserve price is 100; B_1 has a willingness to pay of 300 and B_2 a willingness to pay of 200. (All numbers are dollar amounts in the thousands.)

(a) Use the negotiation approach to describe the equilibrium range of negotiation and the trade that takes place in this situation.

(b) Construct and illustrate the market supply and demand curves for this case. Show on your diagram the equilibrium range of prices and the equilibrium quantity traded.

S2. Suppose that there are four sellers and three buyers. Each seller has one unit to sell and a reserve price of 100. Each buyer wishes to buy one unit. One buyer is willing to pay as much as 400; each of the other two, as much as 300. Find the market equilibrium by drawing a figure. Find the core by setting up and solving all the no-blocking inequalities.

UNSOLVED EXERCISES

U1. For two house sellers with unequal reserve prices, 100 for S_1 and 150 for S_2, and one buyer B with a willingness to pay of 300, show that in a core allocation, the respective surpluses x_1, x_2, and y must satisfy $x_1 \leq 50$, $x_2 = 0$, and $y \geq 150$. Verify that this outcome means that S1 sells his house for a price between 100 and 150.

U2. In Exercise S2, suppose the four sellers get together and decide that only two of them will go to the market and that all four will share any surplus that they get there. Can they benefit by doing so? Can they benefit even more if one or three of them go to the market? What is the intuition for these results?

U3. An airport runway is going to be used by four types of planes: small corporate jets and commercial jets of the narrow-body, wide-body, and jumbo varieties. A corporate jet needs a runway 2,000 feet long, a narrow-body 6,000 feet, a wide-body 8,000 feet, and a jumbo 10,000 feet. The cost of constructing a runway is proportional to its length. Considering each *type* of aircraft as a player, use the Shapley value to allocate the costs of constructing the 10,000-foot runway among the four types. Is this a reasonable way to allocate the costs? Why or why not?

Glossary

Here we define the key terms that appear in the text. We aim for verbal definitions that are logically precise but not mathematical or detailed like those found in more advanced textbooks.

acceptability condition An upper bound on the probability of fulfillment in a brinkmanship threat, expressed as a function of the probability of error, showing the upper limit of risk that the player making the threat is willing to tolerate.

action node A node at which one player chooses an action from two or more that are available.

addition rule If the occurrence of X requires the occurrence of *any one* of several disjoint $Y, Z, \ldots$, then the probability of X is the sum of the separate probabilities of $Y, Z, \ldots$

adverse selection A form of information asymmetry where a player's type (available strategies, payoffs . . .) is his private information, not directly known to others.

agenda paradox A voting situation where the order in which alternatives are paired when voting in multiple rounds determines the final outcome.

agent The agent is the more-informed player in a principal-agent game of asymmetric information. The principal (less-informed) player in such games attempts to design a mechanism that aligns the agent's incentives with his own.

allocation A list of payoffs (sometimes also details of the underlying outcomes such as the quantities of commodities being consumed), one for each player in a game.

all-pay auction An auction in which each person who submits a bid must pay her highest bid amount at the end of the auction, even if she does not win the auction.

alternating offers A sequential move procedure of bargaining in which, if the offer made by one player is refused by the other, then the refuser gets the next turn to make an offer, and so on.

amendment procedure A procedure in which any amended version of a proposal must win a vote against the original version before the winning version is put to a vote against the status quo.

antiplurality method A positional voting method in which the electorate is asked to vote against one item on the slate (or to vote for all but one).

approval voting A voting method in which voters cast votes for all alternatives of which they approve.

ascending auction An open outcry auction in which the auctioneer accepts ascending bids during the course of the auction; the highest bid wins. Also called **English auction.**

assurance game A game where each player has two strategies, say cooperate and not, such that the best response of each is to cooperate if the other cooperates, not if not, and the outcome from (cooperate, cooperate) is better for both than the outcome of (not, not).

asymmetric information Information is said to be asymmetric in a game if some aspects of it—rules about what actions are permitted and the order of moves if any, payoffs as functions of the players strategies, outcomes of random choices by "nature," and of previous actions by the actual players in the game—are known to some of the players but are not common knowledge among all players.

babbling equilibrium In a game where communication among players (which does not affect their payoffs directly) is followed by their choices of actual strategies, a babbling equilibrium is one where the strategies are chosen ignoring the communication, and the communication at the first stage can be arbitrary.

backward induction Same as **rollback.**

battle of the sexes A game where each player has two strategies, say Hard and Soft, such that [1] (Hard, Soft) and (Soft, Hard) are both Nash equilibria, [2] of the two Nash equilibria, each player prefers the outcome where he is Hard and the other is Soft, and [3] both prefer the Nash equilibria to the other two possibilities, (Hard, Hard) and (Soft, Soft).

Bayes' theorem An algebraic formula for estimating the probabilities of some underlying event, by using knowledge of some consequences of it that are observed.

Bayesian Nash equilibrium A Nash equilibrium in an asymmetric information game where players use Bayes' theorem and draw correct inferences from their observations of other players' actions.

belief The notion held by one player about the strategy choices of the other players and used when choosing his own optimal strategy.

best alternative to a negotiated agreement (BATNA) In a bargaining game, this is the payoff a player would get from his other opportunities if the bargaining in question failed to reach an agreement.

best response The strategy that is optimal for one player, given the strategies actually played by the other players, or the belief of this player about the other players' strategy choices.

best-response analysis Finding the Nash equilibria of a game by calculating the best response functions or curves of each player, and solving them simultaneously for the strategies of all the players.

best-response curve A graph showing the best strategy of one player as a function of the strategies of the other player(s) over the entire range of those strategies.

best-response rule A function expressing the strategy that is optimal for one player, for each of the strategy combinations actually played by the other players, or the belief of this player about the other players' strategy choices.

binary method A class of voting methods in which voters choose between only two alternatives at a time.

Black's condition Same as the condition of **single-peaked preferences.**

blocking A coalition of players blocks an allocation if it can, using the strategies feasible for its members, ensuring a better outcome for all of its members, regardless of the strategy choices of nonmembers (players outside the coalition).

Borda count A positional voting method in which the electorate indicates its order of preference over a slate of alternatives. The winning alternative is determined by allocating points based on an alternative's position on each ballot.

branch Each branch emerging from a node in a game tree represents one action that can be taken at that node.

brinkmanship A threat that creates a risk but not certainty of a mutually bad outcome if the other player defies your specified wish as to how he should act, and then gradually increases this risk until one player gives in or the bad outcome happens.

cell-by-cell inspection Finding the Nash equilibria of a game by examining each cell in turn to see if any one player can do better by moving to another cell along his dimension of choice (row or column). Also called **enumeration.**

characteristic function A function that shows, for each coalition, the aggregate payoff its members can ensure for themselves, regardless of the strategy choices of nonmembers.

cheap talk equilibrium In a game where communication among players (which does not affect their payoffs directly) is followed by their choices of actual strategies, a cheap-talk equilibrium is one where the strategies are chosen optimally given the players' interpretation of the communication, and the communication at the first stage is optimally chosen by calculating the actions that will ensue.

chicken A game where each player has two strategies, say Macho and Wimp, such that [1] both (Macho, Wimp) and (Wimp, Macho) are Nash equilibria, [2] of the two, each prefers the outcome where he plays Macho and the other plays Wimp, and [3] the outcome (Macho, Macho) is the worst for both.

chicken in real time A game of chicken in which the choice to swerve is not once and for all, but where a decision must be made at any time, and as time goes on while neither driver has swerved, the risk of a crash increases gradually.

coalition In a game, a subset of the players that coordinates the strategy choices of the members of the subset.

coercion In this context, forcing a player to accept a lower payoff in an asymmetric equilibrium in a collective action game, while other favored players are enjoying higher payoffs. Also called **oppression** in this context.

collective action problem A problem of achieving an outcome that is best for society as a whole, when the interests of some or all individuals will lead them to a different outcome as the equilibrium of a noncooperative game.

combination rule A formula for calculating the probability of occurrence of *X*, which requires any one of several disjoint *Y*, *Z* etc., each of which in turn requires two or more other events to occur. A combination of the addition and multiplication rules defined in the glossary.

commitment An action taken at a pregame stage, stating what action you would take unconditionally in the game to follow.

common value An auction is called a common-value auction when the object up for sale has the same value to all bidders, but each bidder knows only an imprecise estimate of that value. Also called **objective value.**

compellence An attempt to induce the other player(s) to act to change the status quo in a specified manner.

complementary slackness A property of a mixed strategy equilibrium, saying that for each player, against the equilibrium mixture of the other players, all the strategies that are used in this player's mixture yield him equal payoff, and all the strategies that would yield him lower payoff are not used.

compound interest When an investment goes on for more than one period, compound interest entails calculating interest in any one period on the whole accumulation up to that point, including not only the principal initially invested but also the interest earned in all previous periods, which itself involves compounding over the period previous to that.

conditional probability The probability of a particular event X occurring, given that another event Y has already occurred, is called the conditional probability of X given (or conditioned on) Y.

Condorcet method A voting method in which the winning alterna-tive must beat each of the other alternatives in a round-robin of pairwise contests.

Condorcet paradox Even if all individual voter preference orderings are transitive, there is no guarantee that the social preference ordering generated by Condorcet's voting method will also be transitive.

Condorcet terms A set of ballots that would generate the Condorcet paradox and that should together logically produce a tied vote among three possible alternatives. In a three-candidate election among A, B, and C, the Condorcet terms are three ballots that show A preferred to B preferred to C; B preferred to C preferred to A; C preferred to A preferred to B.

Condorcet winner The alternative that wins an election run using the **Condorcet method.**

constant-sum game A game in which the sum of all players' payoffs is a constant, the same for all their strategy combinations. Thus there is a strict conflict of interests among the players—a higher payoff to one must mean a lower payoff to the collectivity of all the other players. If the payoff scales can be adjusted to make this constant equal to zero, then we have a **zero-sum game.**

contingent strategy In repeated play, a plan of action that depends on other players' actions in previous plays. (This is implicit in the definition of a strategy; the adjective "contingent" merely reminds and emphasizes.)

continuation The continuation of a strategy from a (noninitial) node is the remaining part of the plan of action of that strategy, applicable to the subgame that starts at this node.

continuous distribution A probability distribution in which the random variables may take on a continuous range of values.

continuous strategy A choice over a continuous range of real numbers available to a player.

contract In this context, a way of achieving credibility for one's strategic move by entering into a legal obligation to perform the committed, threatened, or promised action in the specified contingency.

convention A mode of behavior that finds automatic acceptance as a focal point, because it is in each individual's interest to follow it when others are expected to follow it too (so the game is of the Assurance type). Also called **custom.**

convergence of expectations A situation where the players in a noncooperative game can develop a common understanding of the strategies they expect will be chosen.

cooperative game A game in which the players decide and implement their strategy choices jointly, or where joint-action agreements are directly and collectively enforced.

coordination game A game with multiple Nash equilibria, where the players are unanimous about the relative merits of the equilibria, and prefer any equilibrium to any of the nonequilibrium possibilities. Their actions must somehow be coordinated to achieve the preferred equilibrium as the outcome.

Copeland index An index measuring an alternative's record in a round-robin of contests where different numbers of points are allocated for wins, ties, and losses.

core An allocation that cannot be blocked by any coalition.

credible A strategy is credible if its continuation at all nodes, on or off the equilibrium path, is optimal for the subgame that starts at that node.

custom Same as **convention.**

decay Shrinkage over time of the total surplus available to be split between the bargainers, if they fail to reach an agreement for some length of time during the process of their bargaining.

decision An action situation in a passive environment where a person can choose without concern for the reactions or responses of others.

decision node A decision node in a decision or game tree represents a point in a game where an action is taken.

decision tree Representation of a sequential decision problem facing one person, shown using nodes, branches, terminal nodes, and their associated payoffs.

demand curve A graph with quantity on the horizontal axis and price on the vertical axis, showing for each price the quantity that one buyer in a market (or the aggregate of all buyers depending on the context) will choose to buy.

descending auction An open outcry auction in which the auctioneer announces possible prices in descending order. The first person to accept the announced price makes her bid and wins the auction. Also called **Dutch auction.**

deterrence An attempt to induce the other player(s) to act to maintain the status quo.

diffusion of responsibility A situation where action by one or a few members of a large group would suffice to bring about an outcome that all regard as desirable, but each thinks it is someone else's responsibility to take this action.

discount factor In a repeated game, the fraction by which the next period's payoffs are multiplied to make them comparable with this period's payoffs.

discrete distribution A probability distribution in which the random variables may take on only a discrete set of values such as integers.

disjoint Events are said to be disjoint if two or more of them cannot occur simultaneously.

distribution function A function that indicates the probability that a variable takes on a value less than or equal to some number.

dominance solvable A game where iterated elimination of dominated strategies leaves a unique outcome, or just one strategy for each player.

dominant strategy A strategy X is dominant for a player if, for each permissible strategy configuration of the other players, X gives him a higher payoff than any of his other strategies. (That is, his best response function is constant and equal to X.)

dominated strategy A strategy X is dominated for a player if there is another strategy Y such that, for each permissible strategy configuration of the other players, Y gives him a higher payoff than X.

doomsday device An automaton that will under specified circumstances generate an outcome which is very bad for all players. Used for giving credibility to a severe threat.

Dutch auction Same as a **descending auction.**

effectiveness condition A lower bound on the probability of fulfillment in a brinkmanship threat, expressed as a function of the probability of error, showing the lower limit of risk that will induce the threatened player to comply with the wishes of the threatener.

effective rate of return Rate of return corrected for the probability of noncontinuation of an investment to the next period.

efficient allocation An allocation is called efficient if there is no other allocation that is feasible within the rules of the game, and if it yields a higher payoff to at least one player without giving a lower payoff to any player.

efficient frontier This is the north-east boundary of the set of feasible payoffs of the players, such that in a bargaining game it is not possible to increase the payoff of one person without lowering that of another.

efficient outcome An outcome of a bargaining game is called efficient if there is no feasible alternative that would leave one bargainer with a higher payoff without reducing the payoff of the other.

English auction Same as an **ascending auction.**

enumeration Same as **cell-by-cell inspection.**

envy-free allocation An allocation is called envy-free if no player would wish to have the outcome (typically quantities of commodities available for consumption) that someone else is getting.

equilibrium A configuration of strategies where each player's strategy is his best response to the strategies of all the other players.

equilibrium path of play The **path of play** actually followed when players choose their rollback equilibrium strategies in a sequential game.

evolutionary game A situation where the strategy of each player in a population is fixed genetically, and strategies that yield higher payoffs in random matches with others from the same population reproduce faster than those with lower payoffs.

evolutionary stable A population is evolutionary stable if it cannot be successfully invaded by a new mutant phenotype.

evolutionary stable strategy A phenotype or strategy which can persist in a population, in the sense that all the members of a population or species are of that type; the population is evolutionary stable (static criterion). Or, starting from an arbitrary distribution of phenotypes in the population, the process of selection will converge to this strategy (dynamic criterion).

expected payoff The probability-weighted average (statistical mean or expectation) of the payoffs of one player in a game, corresponding to all possible realizations of a random choice of nature or mixed strategies of the players.

expected utility The probability-weighted average (statistical mean or expectation) of the utility of a person, corresponding to all possible realizations of a random choice of nature or mixed strategies of the players in a game.

expected value The probability-weighted average of the outcomes of a random variable, that is, its statistical mean or expectation.

extensive form Representation of a game by a game tree.

external effect When one person's action alters the payoff of another person or persons. The effect or spillover is **positive** if one's action raises others' payoffs (for example, network effects), and **negative** if it lowers others' payoffs (for example, pollution or congestion). Also called **externality** or **spillover.**

external uncertainty A player's uncertainty about external circumstances such as the weather or product quality.

externality Same as **external effect.**

fair division mechanism A procedure for dividing some total quantity between two or more people according to some accepted notion of "fairness" such as equality or being envy-free.

feasible allocation An allocation that is permissible within the rules of the game: being within the constraints of resource availability and production technology, and perhaps also constraints imposed by incomplete information.

first-mover advantage This exists in a game if, considering a hypothetical choice between moving first and moving second, a player would choose the former.

first-price auction An auction in which the highest bidder wins and pays the amount of her bid.

fitness The expected payoff of a phenotype in its games against randomly chosen opponents from the population.

focal point A configuration of strategies for the players in a game, which emerges as the outcome because of the convergence of the players' expectations on it.

free rider A player in a collective action game who intends to benefit from the positive externality generated by others' efforts without contributing any effort of his own.

game An action situation where there are two or more mutually aware players, and the outcome for each depends on the actions of all.

game matrix A spreadsheetlike table whose dimension equals the number of players in the game; the strategies available to each player are arrayed along one of the dimensions (row, column, page, . . .); and each cell shows the payoffs of all the players in a specified order, corresponding to the configuration of strategies which yield that cell. Also called **game table** or **payoff table.**

game table Same as **game matrix.**

game tree Representation of a game in the form of nodes, branches, and terminal nodes and their associated payoffs.

genotype A gene or a complex of genes, which give rise to a phenotype and which can breed true from one generation to another. (In social or economic games, the process of breeding can be interpreted in the more general sense of teaching or learning.)

Gibbard-Satterthwaite theorem With three or more alternatives to consider, the only voting method that prevents strategic voting is dictatorship; one person is identified as the dictator and her preferences determine the outcome.

gradual escalation of the risk of mutual harm A situation where the probability of having to carry out the threatened action in a probabilistic threat increases over time, the longer the opponent refuses to comply with what the threat is trying to achieve.

grand coalition The set of all players in the game, acting as a coalition.

grim strategy A strategy of noncooperation forever in the future, if the opponent is found to have cheated even once. Used as a threat of punishment in an attempt to sustain cooperation.

hawk-dove game An evolutionary game where members of the same species or population can breed to follow one of two strategies, Hawk and Dove, and depending on the payoffs, the game between a pair of randomly chosen members can be either a prisoners' dilemma or chicken.

histogram A bar chart; data is illustrated by way of bars of a given height (or length).

impatience Preference for receiving payoffs earlier rather than later. Quantitatively measured by the discount factor.

imperfect information A game is said to have perfect information if each player, at each point where it is his turn to act, knows the full history of the game up to that point, including the results of any random actions taken by nature or previous actions of other players in the game, including pure actions as well as the actual outcomes of any mixed strategies they may play. Otherwise, the game is said to have imperfect information.

impossibility theorem A theorem that indicates that no preference aggregation method can satisfy the six critical principles identified by Kenneth Arrow.

incentive compatibility conditions (constraints) Constraints on an incentive scheme or screening device that makes it optimal for the agent (more-informed player) of each type to reveal his true type through his actions.

incentive-compatible market mechanism Same as **nonmanipulable market mechanism.**

incentive scheme This is a schedule of payments offered by a "principal" to an "agent," as a function of an observable outcome, that induces the agent in his own interests to choose the underlying unobservable action at a level the principal finds optimal (recognizing the cost of the incentive scheme).

incomplete information A game is said to have incomplete information if rules about what actions are permitted and the order of moves if any, payoffs as functions of the players' strategies, outcomes of random choices by "nature" and of previous actions by the actual players in the game are not common knowledge among the players. This is essentially a more formal term for **asymmetric information.**

independent events Events Y and Z are independent if the actual occurrence of one does not change the probability of the other occurring. That is, the conditional probability of Y occurring given that Z has occurred is the same as the ordinary or unconditional probability of Y.

infinite horizon A repeated decision or game situation that has no definite end at a fixed finite time.

information set A set of nodes among which a player is unable to distinguish when taking an action. Thus his strategies are restricted by the condition that he should choose the same action at all points of an information set. For this, it is essential that all the nodes in an information set have the same player designated to act, with the same number and similarly labeled branches emanating from each of these nodes.

initial node The starting point of a sequential-move game. (Also called the **root** of the tree.)

instant runoff Same as **single transferable vote.**

intermediate valuation function A rule assigning payoffs to nonterminal nodes in a game. In many complex games, this must be based on knowledge or experience of playing similar games, instead of explicit rollback analysis.

internalize the externality To offer an individual a reward for the external benefits he conveys on the rest of society, or to inflict a penalty for the external costs he imposes on the rest, so as to bring his private incentives in line with social optimality.

intransitive ordering A preference ordering that cycles and is not **transitive.** For example, a preference ordering over three alternatives A, B, and C is intransitive if A is preferred to B and B is preferred to C but it it not true that A is preferred to C.

invasion by a mutant The appearance of a small proportion of mutants in the population.

irreversible action An action that cannot be undone by a later action. Together with observability, this is an important condition for a game to be sequential-move.

iterated elimination of dominated strategies Considering the players in turns and repeating the process in rotation, eliminating all strategies that are dominated for one at a time, and continuing doing so until no such further elimination is possible.

leadership In a prisoners' dilemma with asymmetric players, this is a situation where a large player chooses to cooperate even though he knows that the smaller players will cheat.

locked in A situation where the players persist in a Nash equilibrium that is worse for everyone than another Nash equilibrium.

majority rule A voting method in which the winning alternative is the one that garners a majority (more than 50%) of the votes.

majority runoff A two-stage voting method in which a second round of voting ensues if no alternative receives a majority in the first round. The top two vote-getters are paired in the second round of voting to determine a winner.

marginal private gain The change in an individual's own payoff as a result of a small change in a continuous strategy variable that is at his disposal.

marginal social gain The change in the aggregate social payoff as a result of a small change in a continuous strategy variable chosen by one player.

market equilibrium An outcome in a market where the price is such that the aggregate quantity demanded equals the aggregate quantity supplied. Graphically, this is represented by a point where the demand and supply curves intersect.

maximin In a zero-sum game, for the player whose strategies are arrayed along the rows, this is the maximum over the rows of the minimum of his payoffs across the columns in each row.

mechanism design Mechanism design is the process by which a principal in a principal-agent problem devises the rules of their game to provide optimal (from the principal's perspective) incentives for the agent.

median voter The voter in the middle—at the 50th percentile—of a distribution.

median voter theorem If the political spectrum is one-dimensional and every voter has single-peaked preferences, then [1] the policy most preferred by the median voter will be the Condorcet winner, and [2] power-seeking politicians in a two-candidate election will choose platforms that converge to the position most preferred by the median voter. (This is also known as the **principle of minimum differentiation.**)

minimax In a zero-sum game, for the player whose strategies are arrayed along the columns, this is the minimum over the columns of the maximum of the other player's payoffs across the rows in each column.

minimax method Searching for a Nash equilibrium in a zero-sum game by finding the maximin and the minimax and seeing if they are equal and attained for the same row-column pair.

mixed method A multistage voting method that uses plurative and binary votes in different rounds.

mixed strategy A mixed strategy for a player consists of a random choice, to be made with specified probabilities, from his originally specified pure strategies.

modified addition rule If the occurrence of X requires the occurrence of one or both of Y and Z, then the probability of X is the sum of the separate probabilities of Y and Z minus the probability that *both* Y and Z occur.

modified multiplication rule If the occurrence of X requires the occurrence of both Y and Z, then the probability of X equals the product of two things: [1] the probability that Y alone occurs, and [2] the probability that Z occurs given that Y has already occurred, or the *conditional probability* of Z, conditioned on Y having already occurred.

monomorphism All members of a given species or population exhibit the same behavior pattern.

monopoly power A large seller's ability to raise the price by reducing the quantity offered for sale.

monopsony power A large buyer's ability to lower the price by reducing the quantity purchased.

moral hazard A situation of information asymmetry where one player's actions are not directly observable to others.

move An action at one node of a game tree.

multiplication rule If the occurrence of X requires the simultaneous occurrence of *all* the several independent $Y, Z, \ldots$, then the probability of X is the *product* of the separate probabilities of $Y, Z, \ldots$

multistage procedure A voting procedure in which there are multiple rounds of voting. Also called **rounds.**

mutation Emergence of a new genotype.

Nash cooperative solution This outcome splits the bargainers' surpluses in proportion to their bargaining powers.

Nash equilibrium A configuration of strategies (one for each player) such that each player's strategy is best for him, given those of the other players. (Can be in pure or mixed strategies.)

negatively correlated Two random variables are said to be negatively correlated if, as a matter of probabilistic average, when one is above its expected value, the other is below its expected value.

neutral ESS An evolutionary stable strategy (ESS) that persists in a population but that can coexist with a small number of mutants having the same fitness as the predominant type.

never a best response A strategy is never a best response for a player if, for each list of strategies that the other players choose (or for each list of strategies that this player believes the others are choosing), some other strategy is this player's best response. (The other strategy can be different for different lists of strategies of the other players.)

node This is a point from which branches emerge, or where a branch terminates, in a decision or game tree.

noncooperative game A game where each player chooses and implements his action individually, without any joint-action agreements directly enforced by other players.

nonexcludable benefits Benefits that are available to each individual, regardless of whether he has paid the costs that are necessary to secure the benefits.

nonmanipulable market mechanism A market mechanism is called nonmanipulable if no buyer has monopsony power and no seller has monopoly power, so no player can change the outcome by misrepresenting his true preferences in his buying or selling decision. Also called **incentive-compatible market mechanism.**

nonrival benefits Benefits whose enjoyment by one person does not detract anything from another person's enjoyment of the same benefits.

norm A pattern of behavior that is established in society by a process of education or culture, to the point that a person who behaves differently experiences a negative psychic payoff.

normal distribution A commonly used statistical distribution for which the **distribution function** looks like a bell-shaped curve.

normal form Representation of a game in a game matrix, showing the strategies (which may be numerous and complicated if the game has several moves) available to each player along a separate dimension (row, column, etc.) of the matrix and the outcomes and payoffs in the multidimensional cells. Also called **strategic form.**

objective value Same as **common value.**

observable action An action that other players know you have taken before they make their responding actions. Together with irreversibility, this is an important condition for a game to be sequential-move.

off-equilibrium path A path of play that does not result from the players' choices of strategies in a subgame perfect equilibrium.

off-equilibrium subgame A subgame starting at a node that does not lie on the equilibrium path of play.

open outcry An auction mechanism in which bids are made openly for all to hear or see.

opponent's indifference property An equilibrium mixed strategy of one player in a two-person game has to be such that the other player is indifferent among all the pure strategies that are actually used in his mixture.

oppression In this context, same as **coercion.**

option A right, but not an obligation, to take an action such as buying something, after some information pertinent to the action has been revealed.

pairwise voting A voting method in which only two alternatives are considered at the same time.

partially revealing equilibrium A perfect Bayesian equilibrium in a game of incomplete information, where the actions in the equilibrium convey some additional information about the players' types, but some ambiguity about these types remains. Also called **semi-separating equilibrium.**

participation condition (constraint) A constraint on an incentive scheme or a screening device that it should give the more-informed player an expected payoff at least as high as he can get outside this relationship.

path of play A route through the game tree (linking a succession of nodes and branches) that results from a configuration of strategies for the players that are within the rules of the game. (See also **equilibrium path of play.**)

payoff The objective, usually numerical, that a player in a game aims to maximize.

payoff table Same as **game matrix.**

penalty We reserve this term for one-time costs (such as fines) introduced into a game to induce the players to take actions that are in their joint interests.

perfect Bayesian equilibrium (PBE) An equilibrium where each player's strategy is optimal at all nodes given his beliefs, and beliefs at each node are updated using Bayes' rule in the light of the information available at that point including other players' past actions.

perfect information A game is said to have perfect information if players face neither strategic nor external uncertainty.

phenotype A specific behavior or strategy, determined by one or more genes. (In social or economic games, this can be interpreted more generally as a customary strategy or a rule of thumb.)

playing the field A many-player evolutionary game where all animals in the group are playing simultaneously, instead of being matched in pairs for two-player games.

pluralistic ignorance A situation of collective action where no individual knows for sure what action is needed, so everyone takes the cue from other people's actions or inaction, possibly resulting in persistence of wrong choices.

plurality rule A voting method in which two or more alternatives are considered simultaneously and the winning alternative is the one that garners the largest number of votes; the winner needs only gain more votes than each of the other alternatives and does not need 50% of the vote as would be true in **majority rule.**

plurative method Any voting method that allows voters to consider a slate of three or more alternatives simultaneously.

polymorphism An evolutionary stable equilibrium in which different behavior forms or phenotypes are exhibited by subsets of members of an otherwise identical population.

pooling equilibrium A perfect Bayesian equilibrium in a game of asymmetric information, where the actions in the equilibrium cannot be used to distinguish type.

pooling of types An outcome of a signaling or screening game in which different types follow the same strategy and get the same payoffs, so types cannot be distinguished by observing actions.

positional method A voting method that determines the identity of the winning alternative using information on the position of alternatives on a voter's ballot to assign points used when tallying ballots.

positive feedback When one person's action increases the payoff of another person or persons taking the same action, thus increasing their incentive to take that action too.

positively correlated Two random variables are said to be positively correlated if, as a matter of probabilistic average, when one is above its expected value, the other is also above its expected value, and vice versa.

power index A measure of an individual's ability to make a difference to the payoffs of coalitions which he could join, averaged across these coalitions.

present value The total payoff over time, calculated by summing the payoffs at different periods each multiplied by the appropriate discount factor to make them all comparable with the initial period's payoffs.

prevent exploitation A method of finding mixed strategy equilibria in zero-sum games. A player's mixture probabilities are found by equating his own expected payoffs from the mixture against all the pure strategies used in the opponent's mixture. This method is not valid for non-zero-sum games.

price discrimination Perfect, or first-degree, price discrimination occurs when a firm charges each customer an individualized price based on willingness to pay. In general, price discrimination refers to situations in which a firm charges different prices to different customers for the same product.

primary criterion Comparison of the fitness of a mutant with that of a member of the dominant population, when each plays against a member of the dominant population.

principal The principal is the less-informed player in a principal-agent game of asymmetric information. The principal in such games wants to design a mechanism that creates incentives for the more-informed player (agent) to take actions beneficial to the principal.

principal-agent (agency) problem A situation in which the less-informed player (principal) wants to design a mechanism that creates incentives for the more-informed player (agent) to take actions beneficial to himself (the principal).

principle of minimum differentiation Same as part [2] of the **median voter theorem.**

prisoners' dilemma A game where each player has two strategies, say Cooperate and Defect, such that [1] for each player, Defect dominates Cooperate, and [2] the outcome (Defect, Defect) is worse for both than the outcome (Cooperate, Cooperate).

private value A bidder's individual valuation of an object available at auction. Also called **subjective value.**

probabilistic threat A strategic move in the nature of a threat, but with the added qualification that if the event triggering the threat (the opponent's action in the case of deterrence, or inaction in the case of compellence) comes about, a chance mechanism is set in motion, and if its outcome so dictates, the threatened action is carried out. The nature of this mechanism and the probability with which it will call for the threatened action must both constitute prior commitments.

probability The probability of a random event is a quantitative measure of the likelihood of its occurrence. For events that can be observed in repeated trials, it is the long-run frequency with which it occurs. For unique events or other situations where uncertainty may be in the mind of a person, other measures are constructed, such as subjective probability.

promise An action by one player, say A, in a pre-game stage, establishing a response rule that, if the other player B chooses an action specified by A, then A will respond with a specified action that is costly to A and rewards B (gives him a higher payoff). (For this to be feasible, A must have the ability to move second in the actual game.)

proportional representation This voting system requires that the number of seats in a legislature be allocated in proportion to each party's share of the popular vote.

proxy-bidding A process by which a bidder submits her maximum bid (**reservation price**) for an item up for auction and a third party takes over bidding for her; the third party bids only the minimum increment above any existing bids and bids no higher than the bidder's specified maximum.

prune To use rollback analysis to identify and eliminate from a game tree those branches that will not be chosen when the game is rationally played.

punishment We reserve this term for costs that can be inflicted on a player in the context of a repeated relationship (often involving termination of the relationship) to induce him to take actions that are in the joint interests of all players.

pure coordination game A coordination game where the payoffs of each player are the same in all the Nash equilibria. Thus all players are indifferent among all the Nash equilibria, and coordination is needed only to ensure avoidance of a non-equilibrium outcome.

pure public good A good or facility that benefits all members of a group, when these benefits cannot be excluded from a member who has not contributed efforts or money to the provision of the good, and the enjoyment of the benefits by one person does not significantly detract from their simultaneous enjoyment by others.

pure strategy A rule or plan of action for a player that specifies without any ambiguity or randomness the action to take in each contingency or at each node where it is that player's turn to act.

rational behavior Perfectly calculating pursuit of a complete and internally consistent objective (payoff) function.

rational irrationality Adopting a strategy that is not optimal after the fact, but serves a rational strategic purpose of lending credibility to a threat or a promise.

rationalizability A solution concept for a game. A list of strategies, one for each player, is a rationalizable outcome of the game if each strategy in the list is rationalizable for the player choosing it.

rationalizable A strategy is called rationalizable for a player if it is his optimal choice given some belief about what (pure or mixed strategy) the other player(s) would choose, provided this belief is formed recognizing that the other players are making similar calculations and forming beliefs in the same way. (This concept is more general than that of the Nash equilibrium and yields outcomes that can be justified on the basis only of the players' common knowledge of rationality.)

refinement A restriction that narrows down possible outcomes when multiple Nash equilibria exist.

repeated play A situation where a one-time game is played repeatedly in successive periods. Thus the complete game is mixed, with a sequence of simultaneous-move games.

reputation Relying on the effect on payoffs in future or related games to make threats or promises credible, when they would not have been credible in a one-off or isolated game.

reservation price The maximum amount that a bidder is willing to pay for an item.

reserve price The minimum price set by the seller of an item up for auction; if no bids exceed the reserve, the item is not sold.

response rule A rule that specifies how you will act in response to various actions of other players.

revenue equivalence In the equilibrium of a private value auction where all bidders are risk-neutral and have independent valuations, all auction forms will yield the seller the same expected revenue.

reversal paradox This paradox arises in an election with at least four alternatives when one of these is removed from consideration after votes have been submitted and the removal changes the identity of the winning alternative.

reversal terms A set of ballots that would generate the **reversal paradox** and that should together logically produce a tied vote between a pair of alternatives. In a three-candidate election among A, B, and C, the reversal terms are two ballots that show a reversal in the location of a pair of alternatives. For example, one ballot with A preferred to B preferred to C and another with B preferred to A preferred to C should produce a tie between A and B.

risk-averse A decision-maker (or a player in a game) is called risk-averse if he prefers to replace a lottery of monetary amounts by the expected monetary value of the same lottery, but now received with certainty.
risk-neutral A decision-maker (or a player in a game) is called risk-neutral if he is indifferent between a lottery of monetary amounts and the expected monetary value of the same lottery, but now received with certainty.
robustness A measure of the number of sets of voter preference orderings that are nondictatorial, satisfy independence of irrelevant alternatives and the Pareto property, and also produce a transitive **social ranking.**
rollback Analyzing the choices that rational players will make at all nodes of a game, starting at the terminal nodes and working backward to the initial node.
rollback equilibrium The strategies (complete plans of action) for each player that remain after rollback analysis has been used to prune all the branches that can be pruned.
root Same as **initial node.**
rounds A voting situation in which votes take place in several stages. Also called **multistage.**
saddle point In this context, an equilibrium of a two-person zero-sum game where the payoff of one player is simultaneously maximized with respect to his own strategy choice and minimized with respect to the strategy choice of the other player.
salami tactics A method of defusing threats by taking a succession of actions, each sufficiently small to make it non-optimal for the other player to carry out his threat.
sanction Punishment approved by society and inflicted by others on a member who violates an accepted pattern of behavior.
screening Strategy of a less-informed player to elicit information credibly from a more-informed player.
screening devices Methods used for screening.
sealed bid An auction mechanism in which bids are submitted privately in advance of a specified deadline, often in sealed envelopes.
secondary criterion Comparison of the fitness of a mutant with that of a member of the dominant population, when each plays against a mutant.
second-mover advantage A game has this if, considering a hypothetical choice between moving first and moving second, a player would choose the latter.
second-price auction An auction in which the highest bidder wins the auction but pays a price equal to the value of the second-highest bid; also called a **Vickrey auction.**
security level The security level of a coalition is the total payoff or surplus its members can ensure for themselves, no matter what strategies other players who are not members of the coalition may choose.
selection The dynamic process by which the proportion of fitter phenotypes in a population increases from one generation to the next.
self-selection Where different types respond differently to a screening device, thereby revealing their type through their own action.
semiseparating equilibrium Same as **partially revealing equilibrium.**
separating equilibrium A perfect Bayesian equilibrium in a game of asymmetric information, where the actions in the equilibrium reveal player type.

separation of types An outcome of a signaling or screening game in which different types follow different strategies and get the different payoffs, so types can be identified by observing actions.

sequential moves The moves in a game are sequential if the rules of the game specify a strict order such that at each action node only one player takes an action, with knowledge of the actions taken (by others or himself) at previous nodes.

shading A strategy in which bidders bid slightly below their true valuation of an object.

Shapley value A solution concept for a cooperative game, where all coalitions to which an individual may belong are regarded equally likely, and each individual is awarded the average of his contributions to raising the aggregate payoffs of all these coalitions.

shilling A practice used by sellers at auction by which they plant false bids for an object they are selling.

signaling Strategy of a more-informed player to convey his "good" information credibly to a less-informed player.

signal jamming A situation in a signaling game where an informed player of a "bad" type mimics the strategy of a "good" type, thereby preventing separation or achieving pooling. This term is used particularly if the action in question is a mixed strategy.

signals Devices used for signaling.

simultaneous moves The moves in a game are simultaneous if each player must take his action without knowledge of the choices of others.

sincere voting Voting at each point for the alternative that you like best among the ones available at that point, regardless of the eventual outcome.

single-peaked preferences A preference ordering in which alternatives under consideration can be ordered along some specific dimension and each voter has a single ideal or most-preferred alternative with alternatives farther away from the most-preferred point providing steadily lower payoffs.

single transferable vote A voting method in which each voter indicates her preference ordering over all candidates on a single initial ballot. If no alternative receives a majority of all first-place votes, the bottom-ranked alternative is eliminated and all first-place votes for that candidate are "transferred" to the candidate listed second on those ballots; this process continues until a majority winner emerges. Also called **instant runoff.**

social optimum In a collective-action game where payoffs of different players can be meaningfully added together, the social optimum is achieved when the sum total of the players' payoffs is maximized.

social ranking The preference ordering of a group of voters that arises from aggregating the preferences of each member of the group.

spillover effect Same as **external effect.**

spoiler Refers to a third candidate who enters a two-candidate race and reduces the chances that the leading candidate actually wins the election.

strategic form Same as **normal form.**

strategic game See **game.**

strategic misrepresentation of preferences Refers to strategic behavior of voters when they use rollback to determine that they can achieve a better outcome for themselves by not voting strictly according to their preference orderings.

strategic moves Actions taken at a pregame stage that change the strategies or the payoffs of the subsequent game (thereby changing its outcome in favor of the player[s] making these moves).

strategic uncertainty A player's uncertainty about an opponent's moves made in the past or made at the same time as her own.

strategic voting Voting in conformity with your optimal rational strategy found by doing rollback analysis on the game tree of the voting procedure.

strategy A complete plan of action for a player in a game, specifying the action he would take at all nodes where it is his turn to act according to the rules of the game (whether these nodes are on or off the equilibrium path of play). If two or more nodes are grouped into one information set, then the specified action must be the same at all these nodes.

subgame A game comprising a portion or remnant of a larger game, starting at a noninitial node of the larger game.

subgame-perfect equilibrium A configuration of strategies (complete plans of action) such that their continuation in any subgame remains optimal (part of a rollback equilibrium), whether that subgame is on- or off-equilibrium. This ensures credibility of all the strategies.

subjective uncertainty A situation where one person is unsure in his mind about which of a set of events will occur, even though there may not be any chance mechanism such as a coin toss with objectively calculable probabilities that governs the outcome.

subjective value Same as **private value.**

successive elimination of dominated strategies Same as **iterated elimination of dominated strategies.**

supply curve A graph with quantity on the horizontal axis and price on the vertical axis, showing for each price the quantity that one seller in a market (or the aggregate of all sellers, depending on the context) will choose to buy.

surplus A player's surplus in a bargaining game is the excess of his payoff over his BATNA.

terminal node This represents an end point in a game tree, where the rules of the game allow no further moves, and payoffs for each player are realized.

threat An action by one player, say A, in a pre-game stage, establishing a response rule that, if the other player B chooses an action specified by A, then A will respond with a specified action that is damaging to B (gives him a lower payoff), and also costly to A to carry out after the fact. (For this to be possible, A must have the ability to be the second mover in the actual game.)

Tit-for-tat In a repeated prisoners' dilemma, this is the strategy of [1] cooperating on the first play and [2] thereafter doing each period what the other player did the previous period.

trading risk Inducing someone else to bear some of the risk to which one is exposed, in return for a suitable monetary or other compensation.

transitive ordering A preference ordering for which it is true that if option A is preferred to B and B is preferred to C, then A is also preferred to C.

trigger strategy In a repeated game, this strategy cooperates until and unless a rival chooses to defect, and then switches to noncooperation for a specified period.

type Players who possess different private information in a game of asymmetric information are said to be of different types.

ultimatum game A form of bargaining where one player makes an offer of a particular split of the total available surplus, and the other has only the all-or-nothing choice of accepting the offer or letting the game end without agreement, when both get zero surplus.

uniform distribution A common statistical distribution in which the **distribution function** is horizontal; data is distributed uniformly at each location along the range of possible values.

utility function In this context, a nonlinear scaling of monetary winnings or losses, such that its expected value (the expected utility) accurately captures a person's attitudes toward risk.

variable-threat bargaining A two-stage game where at the first stage you can take an action that will alter the BATNAs of both bargainers (within certain limits), and at the second stage bargaining results in the Nash solution on the basis of these BATNAs.

Vickrey auction Same as **sealed bid** auction.

Vickrey's truth serum Our name for the result that, in a second-price sealed bid auction, it is every bidder's dominant strategy to bid her true valuation.

winner's curse A situation in a common value auction, where although each person may make an unbiased estimate of the value, only the one with the highest estimate will bid high and win the object, and is therefore likely to have made an upward-biased (too high) estimate. A rational calculation of your bidding strategy will take this into account and lower your bid appropriately to counter this effect.

Yankee auction An auction in which multiple units of a particular item are available for sale; bidders can bid on one or more units at the same time.

zero-sum game A game where the sum of the payoffs of all players equals zero for every configuration of their strategy choices. (This is a special case of a **constant-sum game,** but in practice no different because adding a constant to all the payoff numbers of any one player makes no difference to his choices.)

Index